Remote Sensing Handbook, Volume II (Six Volume Set)

Volume II of the Six Volume *Remote Sensing Handbook*, Second Edition, is focused on digital image processing including image classification methods in land cover and land use. It discusses object-based segmentation and pixel-based image processing algorithms, change detection techniques, and image classification for a wide array of applications including land use/land cover, croplands, urban studies, processing hyperspectral remote sensing data, thermal imagery, light detection and ranging (LiDAR), geoprocessing workflows, frontiers of GIScience, and future pathways. This thoroughly revised and updated volume draws on the expertise of a diverse array of leading international authorities in remote sensing and provides an essential resource for researchers at all levels interested in using remote sensing. It integrates discussions of remote sensing principles, data, methods, development, applications, and scientific and social context.

FEATURES

- Provides the most up-to-date comprehensive coverage of digital image processing.
- Highlights object-based image analysis (OBIA) and pixel-based classification methods and techniques of digital image processing.
- Demonstrates practical examples of image processing for a myriad of applications such as land use/land cover, croplands, and urban.
- Establishes image processing using different types of remote sensing data that includes multispectral, radar, LiDAR, thermal, and hyperspectral.
- Highlights change detection, geoprocessing, and GIScience.

This volume is an excellent resource for the entire remote sensing and GIS community. Academics, researchers, undergraduate and graduate students, as well as practitioners, decision makers, and policymakers, will benefit from the expertise of the professionals featured in this book, and their extensive knowledge of new and emerging trends.

Remote Sensing Handbook, Volume II (Six Volume Set)

Image Processing, Change Detection, GIS, and Spatial Data Analysis

Second Edition

Edited by Prasad S. Thenkabail, PhD

CRC Press is an imprint of the
Taylor & Francis Group, an **informa** business

Second edition published 2025
by CRC Press
2385 NW Executive Center Drive, Suite 320, Boca Raton FL 33431

and by CRC Press
4 Park Square, Milton Park, Abingdon, Oxon, OX14 4RN

CRC Press is an imprint of Taylor & Francis Group, LLC

First edition published by CRC Press 2016

Library of Congress Cataloging-in-Publication Data
Names: Thenkabail, Prasad Srinivasa, 1958– editor.
Title: Remote sensing handbook / edited by Prasad S. Thenkabail ; foreword by Compton J. Tucker.
Description: Second edition. | Boca Raton, FL : CRC Press, 2025. | Includes bibliographical references and index. | Contents: v. 1. Remotely sensed data characterization, classification, and accuracies — v. 2. Image processing, change detection, GIS and spatial data analysis — v. 3. Agriculture, food security, rangelands, vegetation, phenology, and soils — v. 4. Forests, biodiversity, ecology, LULC, and carbon — v. 5. Water, hydrology, floods, snow and ice, wetlands, and water productivity — v. 6. Droughts, disasters, pollution, and urban mapping.
Identifiers: LCCN 2024029377 (print) | LCCN 2024029378 (ebook) | ISBN 9781032890951 (hbk ; v. 1) | ISBN 9781032890968 (pbk ; v. 1) | ISBN 9781032890975 (hbk ; v. 2) | ISBN 9781032890982 (pbk ; v. 2) | ISBN 9781032891019 (hbk ; v. 3) | ISBN 9781032891026 (pbk ; v. 3) | ISBN 9781032891033 (hbk ; v. 4) | ISBN 9781032891040 (pbk ; v. 4) | ISBN 9781032891453 (hbk ; v. 5) | ISBN 9781032891477 (pbk ; v. 5) | ISBN 9781032891484 (hbk ; v. 6) | ISBN 9781032891507 (pbk ; v. 6)
Subjects: LCSH: Remote sensing—Handbooks, manuals, etc.
Classification: LCC G70.4 .R4573 2025 (print) | LCC G70.4 (ebook) | DDC 621.36/780285—dc23/eng/20240722
LC record available at https://lccn.loc.gov/2024029377
LC ebook record available at https://lccn.loc.gov/2024029378

ISBN: 978-1-032-89097-5 (hbk)
ISBN: 978-1-032-89098-2 (pbk)
ISBN: 978-1-003-54115-8 (ebk)

DOI: 10.1201/9781003541158

Typeset in Times New Roman
by Apex CoVantage, LLC

Contents

PART II Change Detection

PART III Integrating Geographic Information Systems (GIS) and Remote Sensing in Spatial Modeling Framework for Decision Support

PART IV Summary and Synthesis of Volume II

Foreword

Satellite remote sensing has progressed tremendously since the first Landsat was launched on June 23, 1972. Since the 1970s, satellite remote sensing and associated airborne and in situ measurements have resulted in geophysical observations for understanding our planet through time. These observations have also led to improvements in numerical simulation models of the coupled atmosphere-land-ocean systems at increasing accuracies and predictive capabilities. This was made possible by data assimilation of satellite geophysical variables into simulation models, to update model variables with more current information. The same observations document the Earth's climate and have driven consensus that *Homo sapiens* are changing our climate through greenhouse gas emissions.

These accomplishments are the work of many scientists from a host of countries and a dedicated cadre of engineers who build and operate the instruments and satellites that collect geophysical observation data from satellites, all working toward the goal of improving our understanding of the Earth. This edition of *Remote Sensing Handbook* (Second Edition, Volumes I–VI) is a compendium of information for many research areas of the Earth system that have contributed to our substantial progress since the 1970s. The remote sensing community is now using multiple sources of satellite and in situ data to advance our studies of planet Earth. In the following paragraphs, I will illustrate how valuable and pivotal satellite remote sensing has been in climate system study since the 1970s. The chapters in *Remote Sensing Handbook* provide other specific studies on land, water, and other applications using Earth observation data of the past 60+ years.

The Landsat system of Earth-observing satellites led the way in pioneering sustained observations of our planet. From 1972 to the present, at least one and frequently two Landsat satellites have been in operation (Wulder et al. 2022; Irons et al. 2012). Starting with the launch of the first NOAA-NASA Polar Orbiting Environmental Satellites NOAA-6 in 1978, improved imaging of land, clouds, and oceans and atmospheric soundings of temperature were accomplished. The NOAA system of polar-orbiting meteorological satellites has continued uninterrupted since that time, providing vital observations for numerical weather prediction. These same satellites are also responsible for the remarkable records of sea surface temperature and land vegetation index from the Advanced Very High-Resolution Radiometers (AVHRR) that now span more than 46 years as of 2024, although no one anticipated valuable climate records from these instrument before the launch of NOAA-6 in 1978 (Cracknell 2001). AVHRR instruments are expected to remain in operation on the European MetOps satellites into 2026 and possibly beyond.

The successes of data from the AVHRR led to the MODerate resolution Imaging Spectrometer (MODIS) instruments on NASA's Earth Observing System of satellite platforms that improved substantially upon the AVHRR. The first of the EOS platforms, Terra, was launched in 2000, and the second of these platforms, Aqua, was launched in 2002. Both of these platforms are nearing their operational end-of-life and many of the climate data records from MODIS will be continued with the Visible Infrared Imaging Suite (VIIRS) instrument on the Joint Polar Satellite System (JPSS) meteorological satellites of NOAA. The first of these missions, the NPOES Preparation Project, was launched in 2012 with the first VIIRS instrument that is operating currently along with similar instruments on JPSS-1 (launched in 2017) and JPSS-2 (launched in 2022). However, unlike the morning/afternoon overpasses of MODIS, the VIIRS instruments are all in an afternoon overpass orbit. One of the strengths of the MODIS observations was morning and afternoon data from identical instruments.

Continuity of observations is crucial for advancing our understanding of the Earth's climate system. Many scientists feel the crucial climate observations provided by remote sensing satellites are among the most important satellite measurements because they contribute to documenting the current state of our climate and how it is evolving. These key satellite observations of our climate are second in importance only to the polar orbiting and geostationary satellites needed for numerical weather prediction that provide natural disaster alerts.

The current state of the art for remote sensing is to combine different satellite observations in a complementary fashion for what is being studied. Climate study is an example of using disparate observations from multiple satellites coupled with in situ data to determine if climate change is occurring and where it is occurring, and to identify the various component processes responsible.

1. **Planet warming quantified by satellite radar altimetry.** Remotely sensed climate observations provide the data to understand our planet and to identify what forces shape our climate. The primary sea level climate observations come from radar altimetry that started in late 1992 with TOPEX-Poseidon and has been continued by Jason-1, Jason-2, Jason-3, and Sentinel-6 to provide an uninterrupted record of global sea level. Changes in global sea level provide unequivocal evidence that our planet is warming, cooling, or staying at the same temperature. Radar altimetry from 1992 to date has shown global sea level increases of ~3.5 mm/y; hence our planet is warming (Figure 0.1). Sea level rise has two components, ocean thermal expansion and ice melt from the ice sheets of Greenland and Antarctica, and to a lesser extent for glacier concentrations in places like the Gulf of Alaska and Patagonia. The combination of GRACE and GRACE Follow-On gravity measurements quantifies the ice mass losses of Greenland and Antarctica to a high degree of accuracy. Combining the gravity data with the flotilla of almost 4,000 Argo floats provides the temperature data with the depth necessary to quantify ocean temperatures and isolate the thermal component of sea level rise.
2. **Our Sun is remarkedly stable in total solar irradiance.** Observations of total solar irradiance have been made from satellites since 1979 and show total solar irradiance has varied only ±1 part in 500 over the past 35 years, establishing that our Sun is not to blame for global warming (Figure 0.2).
3. **Determining ice sheet contributions to sea level rise.** Since 2002 gravity observations from the Gravity Recovery and Climate Experiment Satellite, or GRACE, mission and the GRACE Follow-On mission have been measured. GRACE data quantify ice mass changes from the Antarctic and Greenland ice sheets that constitute 98% of the ice mass on land

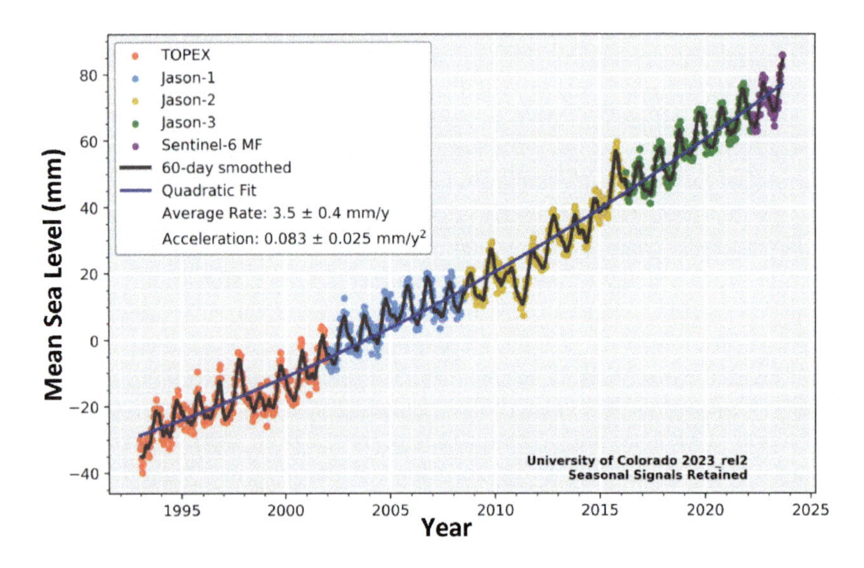

FIGURE 0.1 Seasonal sea level from five satellite radar altimeters from later 1992 to the present. Sea level is the unequivocal indicator of the Earth's climate—when sea level rises, the planet is warming; when sea level falls, the planet is cooling. (Nerem et al. 2018 updated to 2023; https://sealevel.colorado.edu/data/total-sea-level-change.)

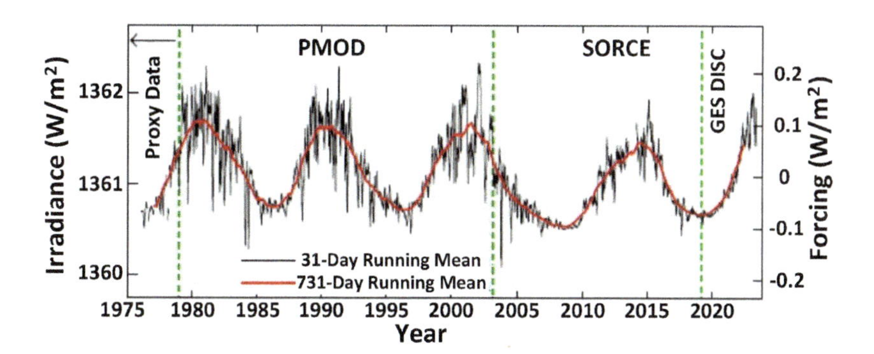

FIGURE 0.2 The Sun is not to blame for global warming, based on total solar irradiance observations from satellites. The few Watts/m^2 solar irradiance variations covary with the sunspot cycle. The luminosity of the Sun varies 0.2% over the course of the 11-year solar and sunspot cycle. The SORCE TSI dataset continues these important observations with improved accuracy on the order of ±0.035. (Kopp et al. 2024, and from https://lasp.colorado.edu/sorce/data/tsi-data/.)

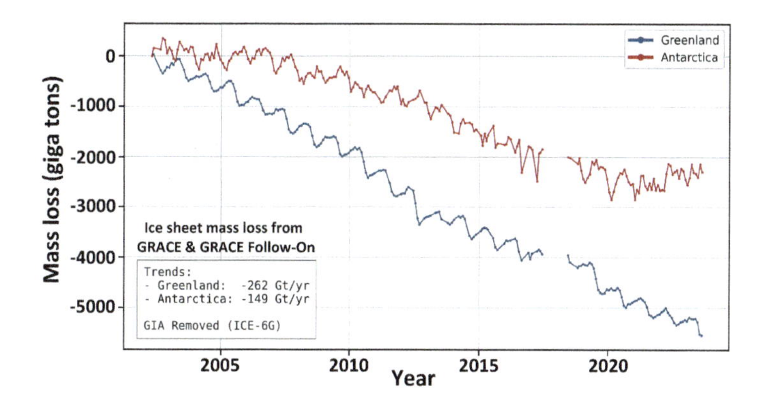

FIGURE 0.3 Sixty percent sea level rise is explained by mass balance of melting of ice measured by GRACE and GRACE Follow-On satellites. Ice mass variations are from 2003 to 2023 for the Antarctica and the Greenland ice sheets using gravity data (Croteau et al. 2021 updated to 2023). The Antarctic and Greenland Ice Sheets constitute 98% of the Earth's land ice.

(Luthcke et al. 2013). GRACE data are truly remarkable—their retrieval of variations in the Earth's gravity field is quantitatively and directly linked to mass variations. With GRACE data we are able for the first time to determine the mass balance with time of the Antarctic and Greenland ice sheets and concentrations of glaciers on land. GRACE data show sea level rise is 60% explained by ice sheet mass loss (Figure 0.3). GRACE data have many other uses, such as changes in groundwater storage. See http://www.csr.utexas.edu/grace/.

4. **Forty percent sea level rise explained by thermal expansion in the planet's oceans measured by in situ ~ 3,700 Argo drifting floats.** The other contributor to sea level rise is the thermal expansion or "steric" component of our planet's oceans. To document this necessitates using diving and drifting floats or buoys in the Argo network to record temperature with depth (Roemmich, The Argo Steering Team 2009 and Figure 0.4). Argo floats are deployed from ships; they then submerge and descend slowly to 1,000 m depth, recording temperature, pressure, and salinity as they descend. At 1,000 m depth, they drift for ten days continuing their measurements of temperature and salinity. After ten

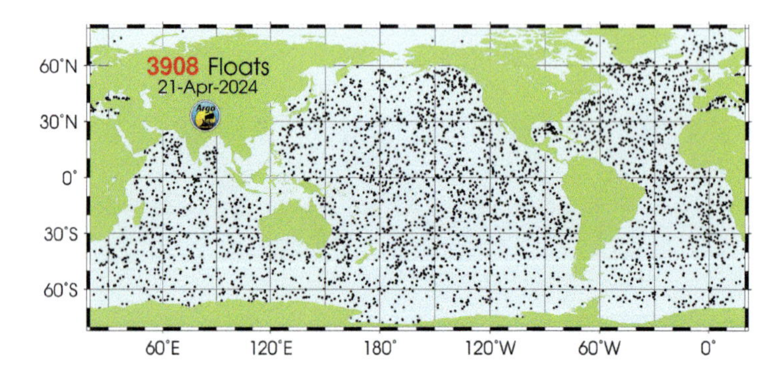

FIGURE 0.4 Forty percent sea level rise explained by thermal expansion in the planet's oceans measured in situ by ~3,908 drifting floats that were in operation on April 21, 2024. These floats provide the data needed to document thermal expansion of the oceans. (Roemmich & the Argo Float Team 2009, updated to 2024, and http://www.argo.ucsd.edu/.)

days, they slowly descend to 3,000 m and then ascend to the surface, all the time recording their measurements. At the surface, each float transmits all the data collected on the most recent excursion to a geostationary satellite and then descends again to repeat this process.

Argo temperature data show that 40% of sea level rise results from warming and thermal expansion of our oceans. Combining radar altimeter data, GRACE and GRACE Follow-On data, and Argo data provide confirmation of sea level rise and show what is responsible for it and in what proportions. With total solar irradiance being near constant, what is driving global warming can be determined. Analysis of surface in situ air temperature coupled with lower tropospheric air temperature and stratospheric temperature data from remote sensing infrared and microwave sounders show the surface and near-surface is warming while the stratosphere is cooling. This is an unequivocal confirmation that greenhouse gases are warming the planet.

Combining sea level radar altimetry, GRACE and GRACE Follow On gravity data to quantify ice sheet mass loses, and Argo floats to measure ocean temperatures with depth enables reconciliation of sea level increases with mass loss of ice sheets and ocean thermal expansion. The ice and steric expansion explains 95% of sea level rise (Figure 0.5).

5. **The global carbon cycle.** Many scientists are actively working to study the Earth's carbon cycle and there are several chapters in this *Remote Sensing Handbook* (Volumes I–VI) on various components under study.

Carbon cycles through reservoirs on the Earth's surface in plants and soils, exists in the atmosphere as gases such as carbon dioxide (CO_2), and exists in ocean water in phytoplankton and in marine sediments. CO_2 is released to the atmosphere from the combustion of fossil fuels, by land cover changes on the Earth's surface, by respiration of green plants, and by decomposition of carbon in dead vegetation and in soils, including carbon in permafrost.

Land gross primary production has been a MODIS product that is extended into the VIIRS era (Running et al. 2004; Román et al. 2024). MODIS data also provide burned area and CO_2 emissions from wildfire (Giglio et al. 2016). Oceanic gross primary production will be provided by the Plankton, Aerosol, Cloud, and ocean Ecosystem, or PACE, satellite that was launched in early 2024 (Gorman et al. 2019). This complements the GPP land portion of the carbon cycle and will enable global gross primary production to be determined by MODIS-VIIRS and PACE.

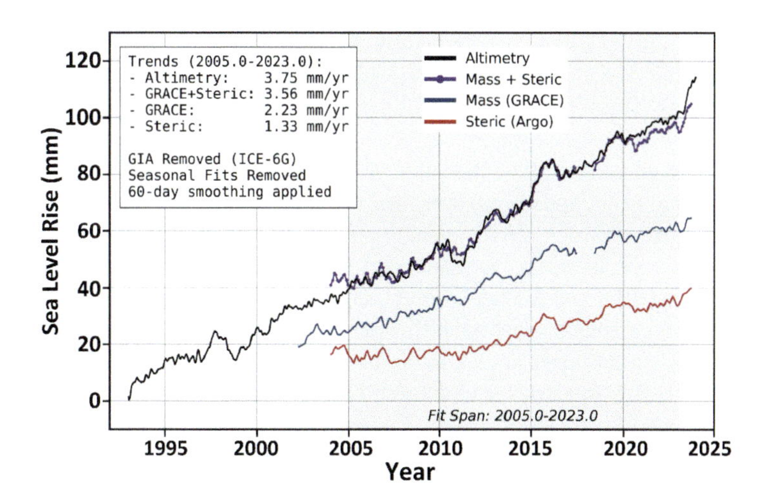

FIGURE 0.5 Sea level rise with the gravity tide mass loss and the Argo thermal expansion quantities added to the plot of global mean sea level. The GRACE and GRACE Follow On ice sheet gravity term and Argo thermal expansion term together explain 95% of sea level rise. (Croteau et al. 2021 updated to 2023.)

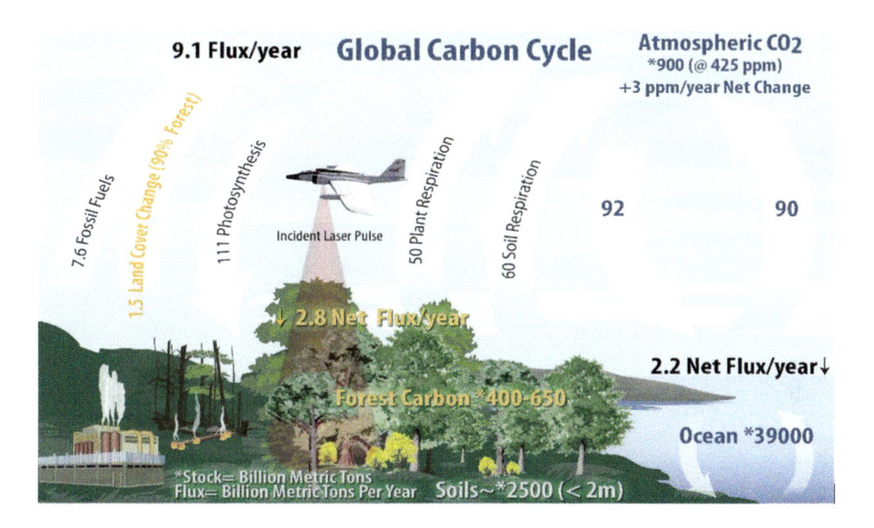

FIGURE 0.6 Global carbon cycle measurements from a multitude of satellite sensors. A representation of the global carbon cycle showing our best estimates of carbon fluxes and carbon reservoirs as of 2024. A series of satellite observations are needed simultaneously to understand the carbon cycle and its role in the Earth's climate system (Ciais et al. 2014 updated to 2023). The major unknowns in the global carbon cycle are fluxes between different reservoirs, oceanic gross primary production, carbon in soils, and the carbon in woody vegetation.

Furthermore, Harmonized Landsat-8, Landsat-9, and Sentinel-2 30 m data (HLS) provide multispectral time series data at 30 m with a revisit frequency of three days at the equator (Crawford et al. 2023; Masek et al. 2018). This will enable time series improvements in spatial detail to 30 m from the 250 m scale of MODIS. The revisit time of Sentinel-2 with 10 m data is five days at the equator, which is a major improvement from 30 m. Multispectral time series observations are the basis for providing gross primary production estimates on land that are also used for food security (Claverie et al. 2018).

Refinements in satellite multispectral spatial resolution to the 50 cm to 3-4 m scale provided by commercial satellite data have enabled tree carbon to be determined from large areas of trees outside of forests. NASA has started using commercial satellite data to complement MODIS, Landsat, and other observations. One of the uses for Planet 3–4 m and Maxar <1 m data has been for mapping trees outside of forests (Brandt et al. 2020; Reiner et al. 2023; Tucker et al. 2023). Tucker et al. (2023) mapped ten billion trees at the 50 cm scale over 10 million km^2 and converted them into carbon at the tree level with allometry. The value of Planet and Maxar (formerly Digital Globe) data allows carbon studies to be extended into areas with discrete trees and Huang et al. (2024) has successfully mapped one tree species across the entire Sahelian and Sudanian Zones of Africa.

The height of trees is an important measurement to determine their carbon content. For areas of contiguous tree crowns, GEDI and ICESat laser altimetry (Magruder et al. 2024) coupled with Landsat and Sentinel-2 observations enable improved estimates of carbon in these forests (Claverie et al. 2018).

The key to closing several uncertainties in the carbon cycle is to quantify fluxes among the various components. Passive CO_2 retrieval methods from the Greenhouse gases Observing SATellite (GOSAT) (Noël et al. 2021) and the Orbiting Carbon Observatory-2 (OCO-2) (Jacobs et al. 2024) are inadequate to provide this. Passive methods are not possible at night, in all seasons, and require specific Sun-target-sensor viewing perspectives and conditions. A recent development of the Aerosol and Carbon dioxide Detection Lidar (ACDL) instrument (Dai et al. 2024) by our Chinese colleagues offers a tenfold coverage improvement in CO_2 retrievals over those provided by OCO-2, and 20-fold coverage improvement over GOSAT. The reported uncertainty of ACDL is on the order of ±0.6 ppm.

Understanding the carbon cycle requires a "full court press" of satellite and in situ observations, because all of these observations must be made at the same time. Many of these measurements have been made over the past 30 to 40 years but new measurements are needed to quantify carbon storage in vegetation, to quantify CO_2 fluxes, to quantify land respiration, and to improve numerical carbon models. Similar work needs to be performed for the role of clouds and aerosols in climate and to improve our understanding of the global hydrological cycle.

The remote sensing community has made tremendous progress over the last five decades as captured in various chapters of *Remote Sensing Handbook* (Second Edition, Volumes I–VI). Handbook chapters provide comprehensive understanding of land and water studies through detailed methods, approaches, algorithms, synthesis, and key references. Every type of remote sensing data obtained from systems such as optical, radar, LiDAR, hyperspectral, and hyperspatial are presented and discussed in different Chapters. Chapters in this volume address remote sensing data characteristics, within and between sensor calibrations, classification methods, and accuracies by taking wide array of remote sensing data from numerous sensor platforms over last five decades. Volume I also brings in new remote sensing technologies such as radio occultation & reflectometry from the global navigation satellite system, or GPS, satellites, crowdsourcing, drones, cloud computing, artificial intelligence, machine learning, hyperspectral, radar, and remote sensing law. The chapters in *Remote Sensing Handbook* are written by leading remote sensing scientists of the world and ably edited by Dr. Prasad S. Thenkabail, senior scientist (ST) at U.S. Geological Survey (USGS) in Flagstaff, Arizona. The importance and the value of *Remote Sensing Handbook* is clearly demonstrated by the need for a second edition. The *Remote Sensing Handbook* (First Edition, Volumes I–III) was published in 2014, and now after ten years Remote Sensing Handbook (Second Edition, Volumes I–VI) with 91 chapters and nearly 3,500 pages will be published. It is certainly monumental work in remote sensing science, and for this I want to complement Dr. Prasad Thenkabail. Remote sensing is now important to a large number of scientific disciplines beyond our community, and I recommend *Remote Sensing Handbook* (Second Edition, six volumes) to not only remote sensers but also to the entire scientific community.

We can look forward in the coming decades to improving our quantitative understanding of the global carbon cycle, understanding the interaction of clouds and aerosols in our radiation budget, and understanding the global hydrological cycle.

by Compton J. Tucker
Satellite Remote Sensing Beyond 2025
NASA/Goddard Space Flight Center
Earth Science Division
Greenbelt, Maryland 20771 USA

REFERENCES

Brandt, M., Tucker, C.J., Kariryaa, A., et al. 2020. An unexpectedly large count of trees in the West African Sahara and Sahel. *Nature* 587: 78–82. http://doi.org/10.1038/s41586-020-2824-5

Ciais, P., et al. 2014. Current systematic carbon-cycle observations and the need for implementing a policy-relevant carbon observing system. *Biogeosciences* 11(13): 3547–3602.

Claverie, M., Ju, J, Masek, J.G., Dungan, J.L., Vermote, E.F., Roger, J.-C., Skakun, S.V., et al. 2018. The Harmonized Landsat and Sentinel-2 surface reflectance data set. *Remote Sensing of Environment* 219: 145–161. http://doi.org/10.1016/j.rse.2018.09.002

Cracknell, A. 2001. The exciting and totally unanticipated success of the AVHRR in applications for which it was never intended. *Advances in Space Research* 28: 233–240. http://doi.org/10.1016/S0273-1177(01)00349-0

Crawford, C.J., Roy, D.P., Arab, S., Barnes, C., Vermote, E., Hulley, G., et al. 2023. The 50-year Landsat collection 2 archive. *Science of Remote Sensing* 8(2023): 100103. ISSN 2666–0172. https://doi.org/10.1016/j.srs.2023.100103. https://www.sciencedirect.com/science/article/pii/S2666017223000287

Croteau, M.J., Sabaka, T.J., Loomis, B.D. 2021. GRACE fast mascons from spherical harmonics and a regularization design trade study. *Journal of Geophysical Research—Solid Earth* 126: e2021JB022113. https://doi.org/10.1029/2021JB022113.10.1029/2021JB022113

Dai, G., Wu, S., Long, W., Liu, J., Xie, Y., Sun, K., Meng, F., Song, X., Huang, Z., Chen, W. 2024. Aerosol and cloud data processing and optical property retrieval algorithms for the spaceborne ACDL/DQ-1. *Atmospheric Measurement Techniques* 17: 1879–1890, https://doi.org/10.5194/amt-17-1879-2024, 2024

Giglio, L., Schroeder, W., Justice, C.O. 2016. The collection 6 MODIS active fire detection algorithm and fire products. *Remote Sensing of Environment* 178: 31–41. http://doi.org/10.1016/j.rse.2016.02.054

Gorman, E.T., Kubalak, D.A., Patel, D., Dress, A., Mott, D.B., Meister, G., Werdell, P.J. 2019. The NASA Plankton, Aerosol, Cloud, ocean Ecosystem (PACE) mission: An emerging era of global, hyperspectral Earth system remote sensing. *Sensors, Systems, and Next Generation Satellites* XXIII: 11151. http://doi.org/10.1117/12.2537146

Huang, K. et al. 2024. Mapping every adult baobab (Adansonia digitata L.) across the Sahel to uncover the co-existence with rural livelihoods. *Nature Ecology and Evolution*. http://doi.org/10.21203/rs.3.rs-3243009/v1

Irons, J.R., Dwyer, J.L., Barsi, J.A. 2012. The next Landsat satellite: The Landsat data continuity mission. *Remote Sensing of Environment* 122: 11–21. http://doi.org/10.1016/j.rse.2011.08.026

Jacobs, N., et al. 2024. The importance of digital elevation model accuracy in X_{CO2} retrievals: Improving the Orbiting Carbon Observatory-2 Atmospheric Carbon Observations from Space version 11 retrieval product. *Atmospheric Measurement Techniques* 17(5): 1375–1401. http://doi.org/10.5194/amt-17-1375-2024

Kopp, G., Nèmec, N.E., Shapiro, A. 2024. Correlations between total and spectral solar irradiance variations. *Astrophysical Journal* 964(1). http://doi.org/10.3847/1538-4357/ad24e5

Luthcke, S.B., Sabaka, T.J., Loomis, B.D., Arendt, A.A., McCarthy, J.J., Camp, J. 2013. Antarctica, Greenland and Gulf of Alaska land-ice evolution from an iterated GRACE global mascon solution. *Journal of Glaciology* 59(216): 613–631. doi:10.3189/2013JoG12J147

Magruder, L.A., Farrell, S.L., Neuenschwander, A., Duncanson, L., Csatho, B., Kacimi, S., et al. 2024. Monitoring Earth's climate variables with satellite laser altimetry. *Nature Reviews of Earth and Environment* 5(2):120–136. http://doi.org/10.1038/s43017-023-00508-8

Masek, J., Ju, J., Roger, J.-C., Skakun, S., Claverie, M., Dungan, J. 2018. Harmonized Landsat/Sentinel-2 products for land monitoring. *IGARSS 2018—2018 IEEE International Geoscience and Remote Sensing Symposium*, Valencia, Spain, 2018, pp. 8163–8165. http://doi.org/10.1109/IGARSS.2018.8517760

Nerem, R.S., Beckley, B.D., Fasullo, J.T., Mitchum, G.T. 2018. Climate-change–driven accelerated sea-level rise detected in the altimeter era. *Proceeding of the National Academy of Sciences* 115(9): 2022–2025. http://doi.org/10.1073/pnas.1717312115

Noël, S., et al. 2021. XCO_2 retrieval for GOSAT and GOSAT-2 based on the FOCAL algorithm. *Atmospheric Measurement Techniques* 14(5):3837–3869. http://doi.org/10.5194/amt-14-3837-2021

Reiner, F., et al. 2023. More than one quarter of Africa's tree cover is found outside areas previously classified as forest. *Nature Communications*. http://doi.org/10.1038/s41467-023-37880-4

Roemmich, D., The Argo Steering Team. 2009. Argo—The challenge of continuing 10 years of progress. *Oceanography* 22(3): 46–55.

Román, M., et al. 2024. Continuity between NASA MODIS collection 6.1 and VIIRS collection 2 land products. *Remote Sensing of Environment* 302. http://doi.org/10.1016/j.rse.2023.113963

Running, S.W., Nemani, R.R., Heinsch, F.A., Zhao, M.S., Reeves, M., Hashimoto, H. 2004. A continuous satellite-derived measure of global terrestrial primary production. *Bioscience* 54(6)547–560. http://doi.org/10.1641/0006-3568

Tucker, C., Brandt, M., Hiernaux, P., Kariryaa, A., et al. 2023. Sub-continental-scale carbon stocks of individual trees in African drylands. *Nature* 615: 80–86. http://doi.org/10.1038/s41586-022-05653-6

Wulder, M.A., Roy, D.P., Radeloff, V.C., Loveland, T.R., Anderson, M.C., Johnson, D.M., et al. 2022. Fifty years of Landsat science and impacts. *Remote Sensing of Environment* 280(2022): 113195. ISSN 0034-4257. https://doi.org/10.1016/j.rse.2022.113195. https://www.sciencedirect.com/science/article/pii/S0034425722003054

Preface

The overarching goal of this six-volume, 91-chapter, about 3,500-page *Remote Sensing Handbook* (Second Edition, Volumes I–VI) was to capture and provide the most comprehensive state of the art of remote sensing science and technology development and advancement in the last 60+ years, by clearly demonstrating the (1) scientific advances, (2) methodological advances, and (3) societal benefits achieved during this period, as well as to provide a vision of what is to come in years ahead. The book volumes are, to date and to my best knowledge, the most comprehensive documentation of the scientific and methodological advances that have taken place in understanding remote sensing data, methods, and a wide array of applications. Written by 300+ leading global experts in the area, each chapter (1) focuses on specific topic (e.g., data, methods, and specific set of applications), (2) reviews existing state-of-the-art knowledge, (3) highlights the advances made, and (4) provides guidance for areas requiring future development. Chapters in the book cover a wide array of subject matter concerning remote sensing applications. *Remote Sensing Handbook* (Second Edition, Volumes I–VI) is planned as reference material for a broad spectrum of remote sensing scientists to understand the fundamentals as well as the latest advances, and a wide array of applications for land and water resource practitioners, natural and environmental practitioners, professors, students, and decision makers.

Special features of the six-volume *Remote Sensing Handbook* (Second Edition) include the following:

1. Participation of an outstanding group of remote sensing experts, an unparalleled team of writers for such a book project
2. Exhaustive coverage of a wide array of remote sensing science: data, methods, applications;
3. Each chapter led by a luminary and most chapters written by writing teams that further enriched the chapters
4. Broadening the scope of the book to make it ideal for expert practitioners as well as students
5. Global team of writers, global geographic coverage of study areas, and wide array of satellites and sensors
6. Plenty of color illustrations

Chapters in the book cover the following aspects of remote sensing:

State of the art on satellites, sensors, science, technology, and applications
Methods and techniques
Wide array of applications such as land and water applications, natural resources management, and environmental issues
Scientific achievements and advancements over last 60+ years
Societal benefits
Knowledge gaps
Future possibilities in the 21st century

Great advances have taken place over the last 60+ years in the study of planet Earth from remote sensing, especially using data gathered from the multitude of Earth Observation (EO) satellites launched by various governments as well as private entities. A large part of the initial remote sensing technology was developed and tested during the two world wars. In the 1950s, remote sensing slowly began its foray into civilian applications. But, during the years of cold war, remote sensing applications both in civilian and military increased swiftly. But, it was also an age when remote sensing was the domain of very few top experts, often having multiple skills in engineering, science, and computer technology. From the 1960s onward, there have been many governmental agencies that have initiated civilian remote sensing. The National Aeronautics and Space Administration (NASA) of the United States has been at the forefront of many of these efforts. Others who have provided leadership in civilian remote sensing include, but are not limited to, European Space Agency (ESA) of European Union; Indian Space Research Organization (ISRO) of India; the Centre national d'études spatiales (CNES) of France,; the Canadian Space Agency (CSA), Canada; the Japan Aerospace Exploration Agency (JAXA), Japan; the German Aerospace Center (DLR), Germany; the China National Space Administration (CNSA), China; the United Kingdom Space Agency (UKSA), UK; and Instituto Nacional de Pesquisas Espaciais (INPE), Brazil. Many private entities such as the Planet Labs PBC have launched and operate satellites. These government and private agencies and enterprises have launched, and continue to launch and operate, a wide array of satellites and sensors that capture data of the planet Earth in various regions of electromagnetic spectrum and in various spatial, radiometric, and temporal resolutions, routinely and repeatedly. However, the real thrust for remote sensing advancement came during the last decade of the 20th century and the beginning of the 21st century. These initiatives included the launch of series of new generation EO satellites to gather data more frequently and routinely, release of pathfinder datasets, web-enabling the data for free by many agencies (e.g., USGS release of the entire Landsat archives as well as real-time acquisitions of the world for free by making them web accessible), and providing processed data ready to users (e.g., the Harmonized Landsat and Sentinel-2 or HLS data, surface reflectance products of MODIS). Other efforts like Google Earth made remote sensing more popular and brought in a new platform for easy visualization and navigation of remote sensing data. Advances in computer hardware and software made it possible to handle big data. Crowdsourcing, web access, cloud computing such as in Google Earth Engine (GEE) platform, machine learning, deep learning, coding, artificial intelligence, mobile apps, and mobile platforms (e.g., drones) added new dimension to how remote sensing data is used. Integration with global positioning systems (GPS) and global navigation satellite systems (GNSS), and inclusion of digital secondary data (e.g., digital elevation, precipitation, temperature) in analysis has made remote sensing much more powerful. Collectively, these initiatives provided new vision in making remote sensing data more popular, more widely understood, and increasingly used for diverse applications, hitherto considered difficult. Freely available archival data when combined with more recent acquisitions has also enabled quantitative studies of change over space and time. ***Remote Sensing Handbook*** **(Volumes I–VI) is targeted to capture these vast advances in data, methods, and applications, so a remote sensing student, scientist, or a professional practitioner will have the most comprehensive, all-encompassing reference material in one place.**

Modern-day remote sensing technology, science, and applications are growing exponentially. This growth is as a result of combination of factors that include (1) advances and innovations in data capture, access, processing, computing, and delivery (e.g, big data analytics, harmonized and normalized data, inter-sensor relationships, web-enabling of data, cloud computing such as in Google Earth Engine (GEE), crowdsourcing, mobile apps, machine learning, deep learning, coding in Python and Java Script, and artificial intelligence); (2) an increasing number of satellites and sensors gathering data of the planet, repeatedly and routinely, in various portions of the electromagnetic spectrum as well as in array of spatial, radiometric, and temporal resolutions; (3) efforts at integrating data from multiple satellites and sensors (e.g., Sentinels with Landsat); (4) advances in data normalization, standardization, and harmonization (e.g., delivery of data in surface reflectance, inter-sensor calibration); (5) methods and techniques for handling very large data volumes (e.g., global mosaics); (6) quantum leap in computer hardware and software capabilities (e.g., ability to process several terabytes of data); (7) innovation in methods, approaches, and techniques leading to sophisticated algorithms (e.g., spectral matching techniques, neural network perceptron); and (8) development of new spectral indices to quantify and study specific land and water parameters (e.g., hyperspectral vegetation indices or HVIs). As a result of these all-round developments, remote sensing science is today very mature and is widely used in virtually every discipline of Earth sciences for quantifying, mapping, modeling, and monitoring our planet Earth. Such rapid advances are captured in a number of remote sensing and Earth science journals. However, students, scientists, and practitioners of remote sensing science and applications have significant difficulty in gathering a complete understanding of the various developments and advances that have taken place because of their vastness spread across the last 60+ years. Thereby, the chapters in *Remote Sensing Handbook* are designed to give a whole picture of scientific and technological advances of the last 60+ years.

Today, the science, art, and technology of remote sensing is truly ubiquitous and increasingly part of everyone's everyday life, often without the user even knowing it. Whether looking at your own home or farm (e.g., Figure 0.7), helping you navigate when you drive, visualizing a phenomenon occurring in a distant part of the world (e.g., Figure 0.7), monitoring events such as droughts and floods, reporting weather, detecting and monitoring troop movements or nuclear sites, studying deforestation, assessing biomass carbon, addressing disasters like earthquakes or tsunamis, and a host of other applications (e.g, precision farming, crop productivity, water productivity, deforestation, desertification, water resources management), remote sensing plays a key role. Already, many new innovations are taking place. Companies such as the Planet Labs PBC and Skybox are capturing very high spatial resolution imagery and even videos from space using a large number of microsatellite (CubeSat) constellations. Planet Labs also will soon launch hyperspectral satellites called Tanager. There are others (e.g., Pixxel, India) who have launched and continue to launch constellations of hyperspectral or other sensors. China is constantly putting a wide array of satellites into orbit. Just as the smartphone and social media connected the world, remote sensing is making the world our backyard (e.g., Figure 0.7). No place goes unobserved and no event gets reported without an image. True liberation for any technology and science comes when it is widely used by common people who often have no idea on how it all comes together, but understand the information provided intuitively. That is already happening (e.g., how we use smartphones is significantly driven by satellite data-driven maps and GPS-driven locations). These developments make it clear that not only do we need to understand the state of the art but also have a vision on where the future of remote sensing is headed. Thereby, in a nutshell, the goal of *Remote Sensing Handbook* (Volumes I–VI) is to cover the developments and advancement of six distinct eras (listed here) in terms of data characterization and processing as well as myriad land and water applications:

Pre-civilian remote sensing era of pre-1950s: World War I and II when remote sensing was a military tool

Technology demonstration era of 1950s and 1960s: Sputnik-I and NOAA AVHRR era of the 1950s and 1960s

Landsat era of 1970s: when first truly operational land remote sensing satellite (Earth Resources Technology Satellite or ERTS, later renamed Landsat) was launched and operated

Earth observation era of 1980s and 1990s: when a number of space agencies began launching and operating satellites (e.g., Landsat 4,5 by USA, SPOT-1,2 by France; IRS-1a, 1b by India)

Earth observation and new millennium era of 2000s: when data dissemination to user's became as important as launching, operating, and capturing data (e.g., MODIS terra/acqua, Landsat-8, Resourcesat)

Twenty-first century era starting 2010s: when new generation micro/Nano satellites or CubeSats (e.g., Planet Labs PBC, Skybox), hyperspectral satellite sensors (e.g., DESIS, PRISMA, EnMAP, upcoming NASA SBG) add to increasing constellation of multi-agency sensors (e.g., Sentinels, Landsat-8, 9, upcoming Landsat-Next)

FIGURE 0.7 Google Earth can be used to seamlessly navigate and precisely locate any place on Earth, often with very high spatial resolution data (VHRI; sub-meter to 5 m) from satellites such as IKONOS, Quickbird, and Geoeye (Note: this image is from one of the VHRI). Here, the editor in chief (EiC) of this *Remote Sensing Handbook* (Volumes I–VI) (Thenkabail) located his village home and surroundings, which has land cover such as secondary rainforests, lowland paddy farms, areca nut plantations, coconut plantations, minor roads, walking routes, open grazing lands, and minor streams (typically, first and second order) (Note: land cover based on ground knowledge of the EiC). The first primary school attended by the EiC is located precisely. Precise coordinates (13 45 39.22 Northern latitude, 75 06 56.03 Eastern longitude) of Thenkabail's village house on the planet are located, and the date of image acquisition (March 1, 2014) is noted. Google Earth Images are used for visualization as well as for numerous science applications such as accuracy assessment, reconnaissance, determining land cover, establishing land use, and for various ground surveys.

Motivation to take up editing the six Volume *Remote Sensing Handbook* (second edition) wasn't easy. It is a daunting work and requires an extraordinary commitment over two to three years. After repeated requests from Ms. Irma Shagla-Britton, manager and leader for Remote Sensing and GIS books of Taylor and Francis/CRC Press, and considerable thought, I finally agreed to take the challenge in 2022. Having earlier edited the three-volume *Remote Sensing Handbook*, published in 2014, I was pleased that the books were of considerable demand for a second edition. This was enough motivation. Further, I wanted to do something significant at this stage of my career that will make a considerable contribution to the global remote sensing community. When I edited the first edition during the 2012–2014 period, I was still recovering post colon cancer surgery and chemotherapy. But this second edition is a celebration of my complete recovery from the dreaded disease. I have not only fully recovered but never felt so completely full of health and vigor. This, naturally, gave me the sufficient energy and enthusiasm required to back my motivation to edit this monumental six-volume *Remote Sensing Handbook*. At least for me, this is the *magnum opus* that I feel proud to have accomplished and feel confident of the immense value for students, scientists, and professional practitioners of remote sensing who are interested in a standard reference on the subject. They will find these six volumes of *Remote Sensing Handbook*: "Complete and comprehensive coverage of the state-of-the-art remote sensing, capturing the advances that have taken place over last 60+ years, which will set the stage for a vision for the future."

Above all, I am indebted to some 300+ authors and co-authors of the chapters who have spent so much of their creative energy to work on the chapters, deliver them on time, and patiently address all edits and comments. These are amongst the very best remote sensing scientists from around the world. Extremely busy people, making time for the book project and making outstanding contributions. I went back to everyone who contributed to *Remote Sensing Handbook* (First Edition, three volumes) published in 2014 and requested them to revise their chapters. Most of the lead authors of the chapters agreed to revise, which was reassuring. However, some were not available, due to retirement or for other reasons. In such cases, I adopted two strategies: (1) invite a few new chapter authors to make up for this gap, and (2) update the chapters myself in other cases. I am convinced this strategy worked very well to ensure capturing the latest information and to maintain the integrity of every chapter. What was also important was to ensure that the latest advances in remote sensing science were adequately covered. The authors of the chapters amazed me by their commitment and attention to detail. First, the quality of each of the chapters was of the highest standards. Second, with very few exceptions, chapters were delivered on time. Third, edited chapters were revised thoroughly and returned on time. Fourth, all my requests on various formatting and quality enhancements were addressed. My heartfelt gratitude to these great authors for their dedication to quality science. It has been my great honor and privilege to work with these dedicated legends. Indeed, I call them my "heroes" in the true sense. These are highly accomplished, renowned, pioneering scientists of the highest merit in remote sensing science, and I am ever grateful to have their time, effort, enthusiasm, and outstanding intellectual contributions. I am indebted to their kindness and generosity. In the end, we had 300+ authors writing 91 chapters.

Overall, ***Remote Sensing Handbook* (Volumes I–VI)** took about two years, from the time book chapters and authors were identified to final publication of the book. The six volumes of *Remote Sensing Handbook* were designed in such a way that a reader can have all six volumes as standard reference or have individual volumes to study specific subject areas. The six volumes are:

Remote Sensing Handbook, Second Edition, Vol. I
Volume I*: Sensors, Data Normalization, Harmonization, Cloud Computing, and Accuracies—9781032890951*

Remote Sensing Handbook, Second Edition, Vol. II
Volume II*: Image Processing, Change Detection, GIS, and Spatial Data Analysis—9781032890975*

Remote Sensing Handbook, Second Edition, Vol. III
Volume III*: Agriculture, Food Security, Rangelands, Vegetation, Phenology, and Soils—9781032891019*

Remote Sensing Handbook, Second Edition, Vol. IV
Volume IV*: Forests, Biodiversity, Ecology, LULC, and Carbon—9781032891033*

Remote Sensing Handbook, Second Edition, Vol. V
Volume V*: Water, Hydrology, Floods, Snow and Ice, Wetlands, and Water Productivity—9781032891453*

Remote Sensing Handbook, Second Edition, Vol. VI
Volume VI*: Droughts, Disasters, Pollution, and Urban Mapping—9781032891484*

There are 18, 17, 17, 12, 13, and 14 chapters, respectively, in the six volumes.

A wide array of topics covered in the six volumes.

The topics covered in the **chapters of Volume I** include: (1) satellites and sensors, (2) global navigation satellite systems (GNSS), (3) remote sensing fundamentals, (4) data normalization, harmonization, and standardization, (5) vegetation indices and their within and across sensor calibration, (6) crowdsourcing, (7) cloud computing, (8) Google Earth Engine supported remote sensing, (9) accuracy assessments, and (10) remote sensing law.

The topics covered in the **chapters of Volume II** include: (1) digital image processing fundamentals and advances; (2) digital image classifications for applications such as urban, land use, and land cover; (3) hyperspectral image processing methods and approaches; (4) thermal infrared image processing principles and practices; (5) image segmentation; (6) object-oriented image analysis (OBIA), including geospatial data integration techniques in OBIA; (7) image segmentation in specific applications like land use and land cover; (8) LiDAR digital image processing; (9) change detection; and (10) integrating geographic information systems (GIS) with remote sensing in geoprocessing workflows, democratization of GIS data and tools, fronters of GIScience, and GIS and remote sensing policies.

The topics covered in the **chapters of Volume III** include: (1) vegetation and biomass, (2) agricultural croplands, (3) rangelands, (4) phenology and food security, and (5) soils.

The topics covered in the **chapters of Volume IV** include: (1) forests, (2) biodiversity, (3) ecology, (4) land use and land cover, and (5) carbon. Under each of the preceeding broad topics, there are one or more chapters.

The **chapters of Volume V** focus on hydrology, water resources, ice, wetlands, and crop water productivity. The chapters are broadly classified into (1) geomorphology, (2) hydrology and water resources, (3) floods, (4) wetlands, (5) crop water use and crop water productivity, and (6) snow and ice.

The **chapters of Volume VI** focus on water resources, disasters, and urban remote sensing. The chapters are broadly classified into (1) droughts and drylands, (2) disasters, (3) volcanoes, (4) fires, and (5) nightlights.

There are many ways to use the *Remote Sensing Handbook* (Second Edition, six volumes). A lot of thought went in organizing the volumes and chapters. So, you will see a "flow" from chapter to chapter and volume to volume. As you read through the chapters, you will see how they are interconnected and how reading all of them provides you with greater in-depth understanding. You will also realize, as someone deeply interested in one of the topics, that you will have greater interest in one volume. Having all sxi volumes as reference material is ideal for any remote sensing expert, practitioner, or student. However, you can also refer to individual volumes based on your interest. We have also made great attempts to ensure chapters are self-contained. That way, you can focus on a chapter and read it through, without having to be overly dependent on other chapters. Taking this perspective, a small amount of material (~5 to 10%) may be repeated across chapters. This is done deliberately. For example, when you are reading a chapter on LiDAR or Radar, you don't want to go all the way back to another chapter to understand characteristics of these data. Similarly, certain indices (e.g., vegetation condition index (VCI) or temperature condition index (TCI)) that are defined in one chapter (e.g., on drought) may be repeated in another chapter (also on drought). Such minor overlaps help the reader avoid going back to another chapter to understand a phenomenon or an index or a characteristic of a sensor. However, if you want a lot of details of these sensors or indices or phenomenon, then you will have to read the appropriate chapter where there is in-depth coverage of the topic.

Each volume has a summary chapter (the last chapter of each volume). The summary chapter can be read two ways: (1) either as last chapter to recapture the main points of each of the chapters, or (2) as an initial overview to get the first feeling for what is in the volume, before diving into to read each chapter in detail. I suggest the readers do it both ways: read it first before reading chapters in detail to gather an idea on what to expect in each chapter and then read it at the end to recapture what is being read in each of the chapters.

It has been a great honor as well as humbling experience to edit the *Remote Sensing Handbook* (Volumes I–VI). I truly enjoyed the effort, albeit felt overwhelmed at times with never-ending work. What an honor to work with luminaries in your field of expertise. I learned a lot from them and am very grateful for their support, encouragement, and deep insights. Also, it has been a pleasure working with outstanding professionals of Taylor and Francis Inc./CRC Press. There is no joy greater than being immersed in pursuit of excellence, knowledge gain, and knowledge capture. At the same time, I am happy it is over. If there will be a third edition in a decade or so from now, it will be taken up by someone else (individually or as a team) and certainly not me!

I expect the book to be a standard reference of immense value to any student, scientist, professional, and practical practitioners of remote sensing. Any book that has the privilege of 300+ of the best brains of truly outstanding and dedicated remote sensing scientists ought to be a *magnum opus* deserving to be standard reference on the subject.

Dr. Prasad S. Thenkabail, PhD
Editor in Chief (EiC)
Remote Sensing Handbook (Second Edition, Volumes I–VI)

Volume I: Sensors, Data Normalization, Harmonization, Cloud Computing, and Accuracies
Volume II: Image Processing, Change Detection, GIS, and Spatial Data Analysis
Volume III: Agriculture, Food Security, Rangelands, Vegetation, Phenology, and Soils
Volume IV: Forests, Biodiversity, Ecology, LULC and Carbon
Volume V: Water, Hydrology, Floods, Snow and Ice, Wetlands, and Water Productivity
Volume VI: Droughts, Disasters, Pollution, and Urban Mapping

About the Editor

Dr. Prasad S. Thenkabail, PhD, is a senior scientist with the U.S. Geological Survey (USGS), specializing in remote sensing science for agriculture, water, and food security. He is a world-recognized expert in remote sensing science with multiple major contributions in the field sustained for 40+ years. Dr. Thenkabail has conducted pioneering research in hyperspectral remote sensing of vegetation, global croplands mapping for water and food security, and crop water productivity. His work on hyperspectral remote sensing of agriculture and vegetation are widely cited. His papers on hyperspectral remote sensing are first of its kind and, collectively, they have (1) determined optimal hyperspectral narrowbands (OHNBs) in study of agricultural crops; (2) established hyperspectral vegetation indices (HVIs) to model and map crop biophysical and biochemical quantities; (3) created framework and sample data for the global hyperspectral imaging spectral libraries of crops (GHISA); (4) developed methods and techniques of overcoming Hughes' phenomenon; (5) demonstrated the strengths of hyperspectral narrowband (HNB) data in advancing classification accuracies relative to multispectral broadband (MBB) data; (6) showed advances one can make in modeling crop biophysical and biochemical quantities using HNB and HVI data relative to MBB data; and (7) created a body of work in understanding, processing, and utilizing HNB and HVI data in agricultural cropland studies. This body of work has become a widely referred reference worldwide. In studies of global croplands for food and water security, he has led the release of the world's first 30-m Landsat Satellite-derived global cropland extent product at 30 m (GCEP30; https://www.usgs.gov/apps/croplands/app/map) (Thenkabail et al., 2021; https://lpdaac.usgs.gov/news/release-of-gfsad-30-meter-cropland-extent-products/) and Landsat-derived global rainfed and irrigated area product at 30 m (LGRIP30; https://lpdaac.usgs.gov/products/lgrip30v001/) (Teluguntla et al., 2023). Earlier, he led producing the world's first global irrigated area map (https://lpdaac.usgs.gov/products/gfsad1kcdv001/ and https://lpdaac.usgs.gov/products/gfsad1kcdv001/) using multi-sensor satellite data at nominal 1 km. The global cropland datasets using satellite remote sensing demonstrates a "paradigm shift" in global cropland mapping using remote sensing through big data analytics, machine learning, and petabyte-scale cloud computing on the Google Earth Engine (GEE). The LGRIP30 and GCEP30 products are released through NASA's LP DAAC and published in USGS professional paper 1868 (Thenkabail et al., 2021). He has been principal investigator of many projects over the years, including the NASA-funded global food security support analysis data in the 30-m (GFSAD) project (www.usgs.gov/wgsc/gfsad30).

His career scientific achievements can be gauged by successfully making the list of the world's top 1% of scientists as per the Stanford study ranking world's scientists from across 22 scientific fields and 176 subfields based on deep analysis evaluating about ten million scientists based on SCOPUS data from Elsevier from 1996 to 2023 (Ioannidis and John, 2023; Ioannidis et al., 2020). Dr. Thenkabail was recognized as Fellow of the American Society of Photogrammetry and Remote Sensing (ASPRS) in 2023. Dr. Thenkabail has published more than 150 peer-reviewed scientific papers and edited 15 books. His scientific papers have won several awards over the years, demonstrating world-class highest quality research. These include: 2023 Talbert Abrams Grand Award, the highest scientific paper award of the ASPRS (with Itiya Aneece); 2015 ASPRS ERDAS award for best scientific paper in remote sensing (with Michael Marshall); 2008 John I. Davidson ASPRS President's Award for Practical Papers (with Pardha Teluguntla); and 1994 Autometric Award for the outstanding paper in remote sensing (with Dr. Andy Ward).

Dr. Thenkabail' s contributions to series of leading edited books places him as a world leader in remote sensing science. There are three seminal book-sets with a total of 13 volumes that he edited that have demonstrated his major contributions as an internationally acclaimed remote sensing scientist. These are (1) *Remote Sensing Handbook* (Second Edition, six-volume book-set, 2024) with 91 chapters and nearly 3,000 pages and for which he is the sole editor; (2) *Remote Sensing Handbook* (First Edition, three-volume book-set, 2015) with 82 chapters and 2,304 pages and for which he is the sole editor; and (3) *Hyperspectral Remote Sensing of Vegetation* (four-volume book-set, 2018) with 50 chapters and 1,632 pages that he edited as the chief-editor (co-editors: Prof. John Lyon and Prof. Alfredo Huete).

Dr. Thenkabail is at the center of rendering scientific service to the world's remote sensing community over long periods of service. This includes serving as editor in chief (2011–present) of *Remote Sensing Open Access Journal*; associate editor (2017–present) of *Photogrammetric Engineering and Remote Sensing* (PE&RS); Editorial Advisory Board (2016–present) of the International Society of Photogrammetry and Remote Sensing (ISPRS); and Editorial Board Member of Remote Sensing of Environment (2007–2016).

The USGS and NASA selected him as one of the three international members on the Landsat Science Team (2006–2011). He is an Advisory Board member of the online library collection to support the United Nations' Sustainable Development Goals (UN SDGs), and currently scientist for the NASA and ISRO (Indian Space Research Organization) Professional Engineer and Scientist Exchange Program (PESEP) program for 2022–2024. He was the chair, International Society of Photogrammetry and Remote Sensing (ISPRS) Working Group WG VIII/7 (land cover and its dynamics) from 2013–2016; played a vital role for USGS as global coordinator, Agricultural Societal Beneficial Area (SBA), Committee for Earth Observation (CEOS) (2010–2013) during which he co-wrote the global food security case study for the CEOS for the *Earth Observation Handbook* (EOS), Special Edition for the UN Conference on Sustainable Development, presented in Rio de Janeiro, Brazil; was the co-lead (2007–2011) of IEEE "Water for the World" initiative, a nonprofit effort funded by IEEE which worked in coordination with the Group on Earth Observations (GEO) in its GEO Water and GEO Agriculture initiatives.

Dr. Thenkabail worked as a postdoctoral researcher and research faculty at the Center for Earth Observation (YCEO), Yale University (1997–2003), and led remote sensing programs in three international organizations including the following:

- International Water Management Institute (IWMI), 2003–2008
- International Center for Integrated Mountain Development (ICIMOD), 1995–1997
- International Institute of Tropical Agriculture (IITA), 1992–1995

He began his scientific career as a scientist (1986–1988) working for the National Remote Sensing Agency (NRSA) (now renamed National Remote Sensing Center, or NRSC), Indian Space Research Organization (ISRO), Department of State, Government of India.

Dr. Thenkabail's work experience spans over 25 countries including East Asia (China), South-East Asia (Cambodia, Indonesia, Myanmar, Thailand, Vietnam), Middle East (Israel, Syria), North America (United States, Canada), South America (Brazil), Central Asia (Uzbekistan), South Asia (Bangladesh, India, Nepal, and Sri Lanka), West Africa (Republic of Benin, Burkina Faso, Cameroon, Central African Republic, Cote d'Ivoire, Gambia, Ghana, Mali, Nigeria, Senegal, and Togo), and Southern Africa (Mozambique, South Africa). Dr. Thenkabail is regularly invited as keynote speaker or invited speaker at major international conferences and at other important national and international forums every year.

Dr. Thenkabail obtained his PhD in agricultural engineering from The Ohio State University, USA, in 1992. He has a master's degree in hydraulics and water resources engineering, and a bachelor's degree in civil engineering (both from India). He has 168 publications including 15 books; 175+ peer-reviewed journal articles, book chapters, and professional papers/monographs; and 15+ significant major global and regional data releases.

REFERENCES

SCIENTIFIC PAPERS

https://scholar.google.com/citations?user=9IO5Y7YAAAAJ&hl=en

USGS Professional Paper, Data and Product Gateways, Interactive Viewers

Ioannidis, J.P.A., Boyack, K.W., and Baas, J. (2020) Updated science-wide author databases of standardized citation indicators. *PLoS Biol 18*(10), e3000918. https://doi.org/10.1371/journal.pbio.3000918

Teluguntla, P., Thenkabail, P., Oliphant, A., Gumma, M., Aneece, I., Foley, D., and McCormick, R. (2023). Landsat-Derived Global Rainfed and Irrigated-Cropland Product @ 30-m (LGRIP30) of the World (GFSADLGRIP30WORLD). The Land Processes Distributed Active Archive Center (LP DAAC) of NASA and USGS. p. 103. https://lpdaac.usgs.gov/news/release-of-lgrip30-data-product/ (download data, documents).

Thenkabail, P.S., Teluguntla, P.G., Xiong, J., Oliphant, A., Congalton, R.G., Ozdogan, M., Gumma, M.K., Tilton, J.C., Giri, C., Milesi, C., Phalke, A., Massey, R., Yadav, K., Sankey, T., Zhong, Y., Aneece, I., and Foley, D. (2021). Global Cropland-Extent Product at 30-m Resolution (GCEP30) Derived from Landsat Satellite Time-Series Data for the Year 2015 Using Multiple Machine-Learning Algorithms on Google Earth Engine Cloud: U.S. *Geological Survey Professional Paper 1868*, 63 pages, https://doi.org/10.3133/pp1868 (research paper). https://lpdaac.usgs.gov/news/release-of-gfsad-30-meter-cropland-extent-products/ (download data, documents). https://www.usgs.gov/apps/croplands/app/map (view data interactively).

Books

Remote Sensing Handbook (Second Edition, Six Volumes, 2024)

Thenkabail, Prasad. 2024. Remote Sensing Handbook (Second Edition, Six Volume Book-set), *Volume I: Sensors, Data Normalization, Harmonization, Cloud Computing, and Accuracies.* Taylor and Francis Inc./CRC Press, Boca Raton, London, New York. 978-1-032-89095-1 — CAT# T132478. Print ISBN: 9781032890951. eBook ISBN: 9781003541141. Pp. 581.

Thenkabail, Prasad. 2024. Remote Sensing Handbook (Second Edition, Six Volume Book-set), *Volume II: Image Processing, Change Detection, GIS, and Spatial Data Analysis.* Taylor and Francis Inc./CRC Press, Boca Raton, London, New York. 978-1-032-89097-5 — CAT# T133208. Print ISBN: 9781032890975. eBook ISBN: 9781003541158. Pp. 464.

Thenkabail, Prasad. 2024. Remote Sensing Handbook (Second Edition, Six Volume Book-set), *Volume III: Agriculture, Food Security, Rangelands, Vegetation, Phenology, and Soils.* Taylor and Francis Inc./CRC Press, Boca Raton, London, New York. 978-1-032-89101-9 — CAT# T133213. Print ISBN: 9781032891019; eBook ISBN: 9781003541165. Pp. 788.

Thenkabail, Prasad. 2024. Remote Sensing Handbook (Second Edition, Six Volume Book-set), *Volume IV: Forests, Biodiversity, Ecology, LULC, and Carbon.* Taylor and Francis Inc./CRC Press, Boca Raton, London, New York. 978-1-032-89103-3 — CAT# T133215. Print ISBN: 9781032891033. eBook ISBN: 9781003541172. Pp. 501.

Thenkabail, Prasad. 2024. Remote Sensing Handbook (Second Edition, Six Volume Book-set), *Volume V: Water, Hydrology, Floods, Snow and Ice, Wetlands, and Water Productivity.* Taylor and Francis Inc./CRC Press, Boca Raton, London, New York. 978-1-032-89145-3 — CAT# T133261. Print ISBN: 9781032891453. eBook ISBN: 9781003541400. Pp. 516.

Thenkabail, Prasad. Remote Sensing Handbook (Second Edition, Six Volume Book-set), *Volume VI: Droughts, Disasters, Pollution, and Urban Mapping.* Taylor and Francis Inc./CRC Press, Boca Raton, London, New York. 978-1-032-89148-4 — CAT# T133267. Print ISBN: 9781032891484; eBook ISBN: 9781003541417. Pp. 467.

Hyperspectral Remote Sensing of Vegetation (First Edition, Four Volumes, 2018)

Thenkabail, P.S., Lyon, G.J., and Huete, A. (Editors) (2018). *Hyperspectral Remote Sensing of Vegetation* (Second Edition, Four-Volume set).

Volume I: *Fundamentals, Sensor Systems, Spectral Libraries, and Data Mining for Vegetation.* CRC Press-Taylor and Francis Group, Boca Raton, London, New York. p. 449, Hardback ID: 9781138058545; eBook ID: 9781315164151.

Volume II: *Hyperspectral Indices and Image Classifications for Agriculture and Vegetation.* CRC Press-Taylor and Francis Group, Boca Raton, London, New York. p. 296. Hardback ID: 9781138066038; eBook ID: 9781315159331.

Volume III: *Biophysical and Biochemical Characterization and Plant Species Studies*. CRC Press-Taylor and Francis Group, Boca Raton, London, New York. p. 348. Hardback: 9781138364714; eBook ID: 9780429431180.

Volume IV: *Advanced Applications in Remote Sensing of Agricultural Crops and Natural Vegetation*. CRC Press-Taylor and Francis Group, Boca Raton, London, New York. p. 386. Hardback: 9781138364769; eBook ID: 9780429431166.

Remote Sensing Handbook (First Edition, Three Volumes, 2015)

Thenkabail, P.S. (Editor in Chief) (2015). *Remote Sensing Handbook.*

Volume I: *Remotely Sensed Data Characterization, Classification, and Accuracies*. Taylor and Francis Inc./CRC Press, Boca Raton, London, New York. ISBN 9781482217865—CAT# K22125. Print ISBN: 978-1-4822-1786-5; eBook ISBN: 978-1-4822-1787-2. p. 678.

Volume II: *Land Resources Monitoring, Modeling, and Mapping with Remote Sensing*. Taylor and Francis Inc./CRC Press, Boca Raton, London, New York. ISBN 9781482217957—CAT# K22130. p. 849.

Volume III: *Remote Sensing of Water Resources, Disasters, and Urban Studies*. Taylor and Francis Inc./CRC Press, Boca Raton, London, New York. ISBN 9781482217919—CAT# K22128. p. 673.

Hyperspectral Remote Sensing of Vegetation (First Edition, Single Volume, 2013)

Thenkabail, P.S., Lyon, G.J., and Huete, A. (Editors) (2012). *Hyperspectral Remote Sensing of Vegetation*. CRC Press/Taylor and Francis Group, Boca Raton, London, New York. p. 781 (80+ pages in color). http://www.crcpress.com/product/isbn/9781439845370.

Remote Sensing of Global Croplands for Food Security (First Edition, Single Volume, 2009)

Thenkabail, P., Lyon, G.J., Turral, H., and Biradar, C.M. (Editors) (2009). *Remote Sensing of Global Croplands for Food Security*. CRC Press/Taylor and Francis Group, Boca Raton, London, New York. p. 556 (48 pages in color). Published in June 2009.

IMAGES ABOVE: Snap-shots of the Editor-in-Chief's work and life.

Contributors

Selim Aksoy
Department of Computer Engineering
Bilkent University
Bilkent, Ankara, Turkey

Tareefa S. Alsumaiti
United Arab Emirates University

Aryabrata Basu
Department of Computer Science and Emerging Analytics Center
University of Arkansas at Little Rock
Little Rock, Arkansas

Thomas Blaschke
Interfaculty Department of Geoinformatics Z_GIS
University of Salzburg
Austria

Jeffrey C. Brunskill
Associate Professor
Department of Environmental, Geographical, and Geological Sciences
Bloomsburg University of Pennsylvania
Bloomsburg, Pennsylvania

Dongmei Chen
Laboratory of Geographic Information and Spatial Analysis
Department of Geography and Planning
Queen's University
Kingston, ON, Canada

Shih-Hong Chio
Associate Professor
Department of Land Economics
National Chengchi University
Taipei, Taiwan

Tzu-Yi Chuang
Department of Civil Engineering
National Taiwan University
Taipei, Taiwan

Jackson D. Cothren
Professor
University of Arkansas
Fayetteville, Arkansas

Jason Davis
Assistant Professor
University of Arkansas Division of Agriculture
Batesville, Arkansas

Stefan Dech
German Aerospace Center (DLR), Earth Observation Center (EOC), German Remote Sensing Data Center (DFD)
Oberpfaffenhofen, Wessling, Germany
and
Department for Geography and Geology
University of Wuerzburg
Wuerzburg, Germany

Qian Du
Mississippi State University
Mississippi

W. Fredrick Limp
University Professor
Emeritus University of Arkansas
Fayetteville, Arkansas

Michael Hagenlocher
Researcher
Department of Geoinformatics—Z_GIS
Paris-Lodron University Salzburg
Salzburg, Austria

Mohammad D. Hossain
Laboratory of Geographic Information and Spatial Analysis
Department of Geography and Planning
Queen's University
Kingston, ON, Canada

Pai-Hui Hsu
Assistant Professor
National Taiwan University
Taipei City, Taiwan

Jen-Jer Jaw
Department of Civil Engineering
National Taiwan University
Taipei, Taiwan

Maggi Kelly
Berkeley University of California
Berkeley, California

Stefan Kienberger
Senior Researcher, Working Group Leader 'Integrated Spatial Indicators'
Department of Geoinformatics—Z_GIS
Paris-Lodron University Salzburg
Salzburg, Austria

Barry J. Kronenfeld
Assistant Professor
Department of Geology and Geography
Eastern Illinois University
Charleston, IL, USA

Claudia Kuenzer
German Aerospace Center (DLR), Earth Observation Center (EOC), German Remote Sensing Data Center (DFD)
Oberpfaffenhofen, Wessling, Germany
and
Department for Geography and Geology
University of Wuerzburg
Wuerzburg, Germany

Stefan Lang
Division Head, Research Coordinator
Department of Geoinformatics—Z_GIS
Paris-Lodron University Salzburg
Salzburg, Austria

David P. Lanter
Assistant Professor
Temple University
Philadelphia, PA

Guiying Li
Key Laboratory for Humid Subtropical Eco-Geographical Processes of the Ministry of Education
Fujian Normal University
Fuzhou, China
and
Institute of Geography
Fujian Normal University
Fuzhou, China

Jun Li
School of Geography and Planning
Sun Yat-Sen University
Guangzhou, China

Wei Li
Beijing Institute of Technology
China

Shih-Yuan Lin
Assistant Professor
Department of Land Economics
National Chengchi University
Taipei, Taiwan

Rachel F. Linck
Masters Candidate Geography
University of Arkansas
Fayetteville, Arkansas

Dengsheng Lu
Key Laboratory for Humid Subtropical Eco-Geographical Processes of the Ministry of Education
Fujian Normal University
Fuzhou, China
and
Institute of Geography
Fujian Normal University
Fuzhou, China

Paul Merani
Department of Earth and Atmospheric Sciences
University of Nebraska–Lincoln

Helena Merschdorf
Interfaculty Department of Geoinformatics Z_GIS
University of Salzburg
Austria

Victor Mesev
Department of Geography
Florida State University
Tallahassee, Florida

Soe W. Myint
School of Geographical Sciences and Urban Planning
Arizona State University
Tempe, Arizona

Sunil Narumalani
Professor, Department of Earth and Atmospheric Sciences
University of Nebraska–Lincoln

Mutlu Ozdogan
Nelson Institute for Environmental Studies and Department of Forest and Wildlife Ecology University of Wisconsin
Madison, Wisconsin

Lena Pernkopf
Researcher
Department of Geoinformatics—Z_GIS
Paris-Lodron University Salzburg
Salzburg, Austria

Antonio Plaza
Hyperspectral Computing Laboratory
University of Extremadura
Cáceres, Spain

Dale Quattrochi
Earth Science Office
NASA Marshall Space Flight Center
Huntsville, AL, USA

Philipp Reiners
German Aerospace Center (DLR), Earth Observation Center (EOC), German Remote Sensing Data Center (DFD)
Oberpfaffenhofen, Wessling, Germany

Chiranjibi Shah
Mississippi State University
Mississippi

Xuan Shi
Department of Geosciences
University of Arkansas
Fayetteville, AR 72701

Gaurav Sinha
Assistant Professor
Department of Geography
Ohio University
Athens, Ohio

Hongjun Su
Hohai University
China

Yuliya Tarabalka
INRIA-SAM, AYIN Team
INRIA Sophia-Antipolis Méditerranée
Sophia-Antipolis Cedex, France

Tee-Ann Teo
Associate Professor
National Chiao Tung University
Hsinchu City, Taiwan

Prasad S. Thenkabail
U. S. Geological Survey (USGS)
Flagstaff, Arizona

Dirk Tiede
Department of Geoinformatics (Z_GIS)
Salzburg University
Salzburg, Austria

James C. Tilton
Goddard Space Flight Center, National Aeronautics and Space Administration (NASA)
Greenbelt, Maryland

Fuan Tsai
Associate Professor
Center for Space and Remote Sensing Research
National Central University
Zhongli, Taoyuan, Taiwan

Yi-Hsing Tseng
Professor
National Cheng Kung University
Tainan City, Taiwan

Jason A. Tullis
Professor
University of Arkansas
Fayetteville, Arkansas

Cheng-Kai Wang
PhD Candidate
National Cheng Kung University
Tainan City, Taiwan

Chi-Kuei Wang
Associate Professor National Cheng Kung University Tainan City, Taiwan

Miao Wang
Assistant Researcher
National Cheng Kung University
Tainan City, Taiwan

Elizabeth A. Wentz
School of Geographical Sciences and Urban Planning
Arizona State University
Tempe, Arizona

Ming-Der Yang
National Chung Hsing University
Taichung City, Taiwan

Sean G. Young
Assistant Professor
University of Texas Southwestern Medical Center
Dallas, TX, USA

May Yuan
University of Texas at Dallas
Richardson, TX, USA

Jianzhong Zhang
Freelance Remote Sensing Consultant
Harbin Engineering University
Heilongjiang, China

Ming Zhang
Key Laboratory for Humid Subtropical Eco-Geographical Processes of the Ministry of Education
Fujian Normal University
Fuzhou, China
and
Institute of Geography
Fujian Normal University
Fuzhou, China

Mingxing Zhou
Key Laboratory for Humid Subtropical Eco-Geographical Processes of the Ministry of Education
Fujian Normal University
Fuzhou, China
and
Institute of Geography
Fujian Normal University
Fuzhou, China

Acknowledgments

Remote Sensing Handbook (Second Edition, Volumes I–VI) brought together a galaxy of highly accomplished, renowned remote sensing scientists, professionals, and legends from around the world. The lead authors were chosen by me after careful review of their accomplishments and sustained publication record over the years. The chapters in the second edition were written/revised over a period of two years. All chapters were edited and revised.

Gathering such a galaxy of authors was the biggest challenge. These are all extremely busy people, and committing to a book project that requires substantial work is never easy. However, almost all those whom I requested agreed to write a chapter specific to their area of specialization, and only a few I had to convince to make time. The quality of the chapters should convince readers why these authors are such highly rated professionals and why they are so successful and accomplished in their fields of expertise. They not only wrote very high-quality chapters but also delivered them on time, addressed any editorial comments in a timely manner without complaints, and were extremely humble and helpful. Their commitment for quality science is what makes them special. I am truly honored to have worked with such great professionals.

I would like to mention the names of everyone who contributed and made ***Remote Sensing Handbook* (Second Edition, Volumes I–VI) possible**. In the end, we had 91 chapters, a little over 3,000 pages, and a little more than 400 authors. My gratitude goes to each one of them. These are well-known **"who is who"** in remote sensing science in the world. List of all authors are provided here. The names of the authors are organized chronologically for each volume and the chapters. Each lead author of the chapter is in bold type. **The names of the 400+ authors who contributed to six volumes are as follows:**

Volume I: Sensors, Data Normalization, Harmonization, Cloud Computing, and Accuracies—18 chapters written by 53 authors (editor in chief: Prasad S. Thenkabail):

Drs. Sudhanshu S. Panda, Mahesh Rao, Prasad S. Thenkabail, Debasmita Misra, and James P. Fitzgerald; **Mohinder S. Grewal**; **Kegen Yu**, Chris Rizos, and Andrew Dempster; **D. Myszor**, O. Antemijczuk, M. Grygierek, M. Wierzchanowski, K.A. Cyran; **Natascha Oppelt** and Arnab Muhuri; **Philippe M. Teillet**; **Philippe M. Teillet** and Gyanesh Chander; **Rudiger Gens** and Jordi Cristóbal Rosselló; **Aolin Jia** and Dongdong Wang; **Tomoaki Miura**, Kenta Obata, Hiroki Yoshioka, and Alfredo Huete; **Michael D. Steven**, Timothy J. Malthus, and Frédéric Baret; **Fabio Dell'Acqua** and Silvio Dell'Acqua; **Ramanathan Sugumaran**, James W. Hegeman, Vivek B. Sardeshmukh and Marc P. Armstrong; **Lizhe Wang**, Jining Yan, Yan Ma, Xiaohui Huang, Jiabao Li, Sheng Wang, Haixu He, Ao Long, and Xiaohan Zhang; **John E. Bailey** and Josh Williams; **Russell G. Congalton**; **P.J. Blount**; **Prasad S. Thenkabail**.

Volume II: Image Processing, Change Detection, GIS, and Spatial Data Analysis—17 chapters written by 64 authors (editor in chief: Prasad S. Thenkabail):

Sunil Narumalani and Paul Merani; **Mutlu Ozdogan**; **Soe W. Myint**, Victor Mesev, Dale Quattrochi, and Elizabeth A. Wentz; **Jun Li**, Paolo Gamba, and Antonio Plaza; **Qian Du**, Chiranjibi Shah, Hongjun Su, and Wei Li; **Claudia Kuenzer**, Philipp Reiners, Jianzhong Zhang, Stefan Dech; **Mohammad D. Hossain** and Dongmei Chen; **Thomas Blaschke**, Maggi Kelly, Helena Merschdorf; **Stefan Lang** and Dirk Tiede; **James C. Tilton**, Selim Aksoy, and Yuliya Tarabalka; **Shih-Hong Chio**, Tzu-Yi Chuang, Pai-Hui Hsu, Jen-Jer Jaw, Shih-Yuan Lin, Yu-Ching Lin, Tee-Ann Teo, Fuan Tsai, Yi-Hsing Tseng, Cheng-Kai Wang, Chi-Kuei Wang, Miao Wang, and Ming-Der Yang; **Guiying Li**, Mingxing Zhou, Ming Zhang, Dengsheng Lu; **Jason A. Tullis**, David P. Lanter, Aryabrata Basu, Jackson D. Cothren, Xuan Shi, W. Fredrick Limp, Rachel F. Linck, Sean G.

Young, Jason Davis, and Tareefa S. Alsumaiti; **Gaurav Sinha**, Barry J. Kronenfeld, and Jeffrey C. Brunskill; **May Yuan**; **Stefan Lang**, Stefan Kienberger, Michael Hagenlocher, and Lena Pernkopf; **Prasad S. Thenkabail.**

Volume III: Agriculture, Food Security, Rangelands, Vegetation, Phenology, and Soils: 17 chapters written by 110 authors (editor in chief: Prasad S. Thenkabail):

Alfredo Huete, Guillermo Ponce-Campos, Yongguang Zhang, Natalia Restrepo-Coupe, Xuanlong Ma; **Juan Quiros-Vargas,** Bastian Siegmann, Juliane Bendig, Laura Verena Junker-Frohn, Christoph Jedmowski, David Herrera, Uwe Rascher; **Frédéric Baret; Lea Hallik**, Egidijus Šarauskis, Ruchita Ingle, Indrė Bručienė, Vilma Naujokienė, Kristina Lekavičienė; **Clement Atzberger** and Markus Immitzer; **Agnès Bégué**, Damien Arvor, Camille Lelong, Elodie Vintrou, and Margareth Simoes; **Pardhasaradhi Teluguntla,** Prasad S. Thenkabail, Jun Xiong, Murali Krishna Gumma, Chandra Giri, Cristina Milesi, Mutlu Ozdogan, Russell G. Congalton, James C. Tilton, Temuulen Tsagaan Sankey, Richard Massey, Aparna Phalke, and Kamini Yadav; **Yuxin Miao**, David J. Mulla, Yanbo Huang; **Baojuan Zheng**, James B. Campbell, Guy Serbin, Craig S.T. Daughtry, Heather McNairn, and Anna Pacheco; **Prasad S. Thenkabail**, Itiya Aneece, Pardhasaradhi Teluguntla, Richa Upadhyay, Asfa Siddiqui, Justin George Kalambukattu, Suresh Kumar, Murali Krishna Gumma, Venkateswarlu Dheeravath; **Matthew C. Reeves,** Robert Washington-Allen, Jay Angerer, Raymond Hunt, Wasantha Kulawardhana, Lalit Kumar, Tatiana Loboda, Thomas Loveland, Graciela Metternicht, Douglas Ramsey, Joanne V. Hall, Trenton Benedict, Pedro Millikan, Angus Retallack, Arjan J.H. Meddens, William K. Smith, Wen Zhang; **E. Raymond Hunt Jr**, Cuizhen Wang, D. Terrance Booth, Samuel E. Cox, **Lalit Kumar**, and Matthew C. Reeves; Lalit Kumar, Priyakant Sinha, Jesslyn F Brown, R. Douglas Ramsey, Matthew Rigge, Carson A Stam, Alexander J. Hernandez, E. Raymond Hunt, Jr. and Matt Reeves; **Molly E. Brown**, Kirsten de Beurs, Kathryn Grace; **José A. M. Demattê**, Cristine L. S. Morgan, Sabine Chabrillat, Rodnei Rizzo, Marston H. D. Franceschini, Fabrício da S. Terra, Gustavo M. Vasques, Johanna Wetterlind, Henrique Bellinaso, Letícia G. Vogel; **E. Ben-Dor**, J. A. M. Demattê; **Prasad S. Thenkabail**.

Volume IV: Forests, Biodiversity, Ecology, LULC and Carbon—12 chapters written by 71 authors (editor in chief: Prasad S. Thenkabail):

E. H. Helmer, Nicholas R. Goodwin, Valéry Gond, Carlos M. Souza Jr., and Gregory P. Asner; **Juha Hyyppä**, Xiaowei Yu, Mika Karjalainen, Xinlian Liang, Anttoni Jaakkola, Mike Wulder, Markus Hollaus, Joanne C. White, Mikko Vastaranta, Jiri Pyörälä, Tuomas Yrttimaa, Ninni Saarinen, Josef Taher, Juho-Pekka Virtanen, Leena Matikainen, Yunsheng Wang, Eetu Puttonen, Mariana Campos, Matti Hyyppä, Kirsi Karila, Harri Kaartinen, Matti Vaaja, Ville Kankare, Antero Kukko, Markus Holopainen, Hannu Hyyppä, Masato Katoh, Eric Hyyppä; **Gregory P. Asner**, Susan L. Ustin, Philip A. Townsend, and Roberta E. Martin; **Sylvie Durrieu**, Cédric Véga, Marc Bouvier, Frédéric Gosselin, Jean-Pierre Renaud, Laurent Saint-André; **Thomas W. Gillespie**, Morgan Rogers, Chelsea Robinson, Duccio Rocchini; **Stefan Lang**, Christina Corbane, Palma Blonda, Kyle Pipkins, Michael Förster; **Conghe Song**, Jing Ming Chen, Taehee Hwang, Alemu Gonsamo, Holly Croft, Quanfa Zhang, Matthew Dannenberg, Yulong Zhang, Christopher Hakkenberg, Juxiang Li; **John Rogan** and Nathan Mietkiewicz; **Zhixin Qi**, Anthony Gar-On Yeh, Xia Li, Qianwen Lv; **R.A. Houghton; Wenge Ni-Meister; Prasad S. Thenkabail**.

Volume V: Water, Hydrology, Floods, Snow and Ice, Wetlands, and Water Productivity—13 chapters written by 60 authors (editor in chief: Prasad S. Thenkabail):

James B. Campbell and Lynn M. Resler; **Sadiq I. Khan**, Ni-Bin Chang, Yang Hong, Xianwu Xue, Yu Zhang; **Santhosh Kumar Seelan; Allan S. Arnesen**, Frederico T. Genofre, Marcelo P. Curtarelli, and Matheus Z. Francisco; **Allan S. Arnesen**, Frederico T. Genofre, Marcelo P. Curtarelli, and Matheus Z. Francisco; **Sandro Martinis**, Claudia Kuenzer, and André Twele; **Le Wang**, Jing

Miao, Ying Lu; **Chandra Giri; D. R. Mishra**, X. Yan, S. Ghosh, C. Hladik, J. L. O'Connell, H. J. Cho; **Murali Krishna Gumma**, Prasad S. Thenkabail, Pranay Panjala, Pardhasaradhi Teluguntla, Birhanu Zemadim Birhanu, Mangi Lal Jat; **Trent W. Biggs**, Pamela Nagler, Anderson Ruhoff, Triantafyllia Petsini, Michael Marshall, George P. Petropoulos, Camila Abe, Edward P. Glenn; **Antônio Teixeira**, Janice Leivas; Celina Takemura, Edson Patto, Edlene Garçon, Inajá Sousa, André Quintão, Prasad Thenkabail, and Ana Azevedo; **Hongjie Xie,** Tiangang Liang, Xianwei Wang, Guoqing Zhang, Xiaodong Huang, and Xiongxin Xiao; **Prasad. S. Thenkabail**.

Volume VI: Droughts, Disasters, Pollution, and Urban Mapping—14 Chapters written by 53 authors (editor in chief: Prasad S. Thenkabail):

Felix Kogan and Wei Guo; **F. Rembold**, M. Meroni, O. Rojas, C. Atzberger, F. Ham and E. Fillol; **Brian D. Wardlow**, Martha A. Anderson, Tsegaye Tadesse, Mark S. Svoboda, Brian Fuchs, Chris R. Hain, Wade T. Crow, and Matt Rodell; **Jinyoung Rhee**, Jungho Im, and Seonyoung Park; **Marion Stellmes**, Ruth Sonnenschein, Achim Röder, Thomas Udelhoven, Gabriel del Barrio, and Joachim Hill; **Norman Kerle; Stefan LANG**, Petra FÜREDER, Olaf KRANZ, Brittany CARD, Shadrock ROBERTS, Andreas PAPP; **Robert Wright; Krishna Prasad Vadrevu** and Kristofer Lasko; **Anupma Prakash**, Claudia Kuenzer, Santosh K. Panda, Anushree Badola, Christine F. Waigl; **Hasi Bagana**, Chaomin Chena, and Yoshiki Yamagata; **Yoshiki Yamagata**, Daisuke Murakami, Hajime Seya, and Takahiro Yoshida; **Qingling Zhang**, Noam Levin, Christos Chalkias, Husi Letu and Di Liu; **Prasad S. Thenkabail**.

The authors not only delivered excellent chapters, but they also provided valuable insights and inputs for me in many ways throughout the book project.

I was delighted when **Dr. Compton J. Tucker**, senior Earth scientist, Earth Sciences Division, Science and Exploration Directorate, NASA Goddard Space Flight Center (GSFC) agreed to write the foreword for the book. For anyone practicing remote sensing, Dr. Tucker needs no introduction. He has been a "godfather" of remote sensing and has inspired a generation of remote sensing scientists. I have been a student of his without ever really being one. I mean, I have not been his student in a classroom, but have followed his legendary work throughout my career. I remember reading his highly cited paper (now with citations nearing 7,700!):

- Tucker, C.J. (1979) "Red and Photographic Infrared Linear Combinations for Monitoring Vegetation," *Remote Sensing of Environment*, **8(2)**, 127–150.

I first read this paper in 1986 when I had just joined the National Remote Sensing Agency (NRSA; now NRSC), Indian Space Research Organization (ISRO). Ever since, Dr. Tucker's pioneering works have been a guiding light for me. After getting his PhD from the Colorado State University in 1975, Dr. Tucker joined NASA GSFC as a postdoctoral fellow in 1975 and became a full-time NASA employee in 1977. Ever since, he has conducted several path-finding research studies. He has used NOAA AVHRR, MODIS, SPOT Vegetation, and Landsat satellite data for studying deforestation, habitat fragmentation, desert boundary determination, ecologically coupled diseases, terrestrial primary production, glacier extent, and how climate affects global vegetation. He has authored or co-authored more than 280 journal articles that have been **cited more than 93,000 times**, is an adjunct professor at the University of Maryland, is a consulting scholar at the University of Pennsylvania's Museum of Archaeology and Anthropology, and has appeared in more than 20 radio and TV programs. He is a Fellow of the American Geophysical Union and has been awarded several medals and honors, including NASA's Exceptional Scientific Achievement Medal, the Pecora Award from the U.S. Geological Survey, the National Air and Space Museum Trophy, the Henry Shaw Medal from the Missouri Botanical Garden, the Galathea Medal from the Royal Danish Geographical Society, and the Vega Medal from the Swedish Society of Anthropology and Geography. He was the NASA representative to the U.S. Global Change Research Program from 2006 to 2009. He is instrumental in releasing the AVHRR 33-year (1982–2014) **Global Inventory Monitoring and Modeling Studies** (GIMMS) data. **I strongly recommend that everyone read his excellent foreword before reading the book.** In the foreword, Dr. Tucker demonstrates the importance of data from Earth Observation (EO) sensors from orbiting satellites to maintaining a reliable and consistent climate record. Dr. Tucker further highlights the importance of continued measurements of these variables of our planet in the new millennium through new, improved, and innovative EO sensors from Sun-synchronous and/or geostationary satellites.

I want to acknowledge with thanks for the encouragement and support received by my U.S. Geological Survey (USGS) colleagues. I would like to mention the late Mr. Edwin Pfeifer, Dr. Susan Benjamin (my director at the Western Geographic Science Center), Dr. Dennis Dye, Mr. Larry Gaffney, Mr. David F. Penisten, Ms. Emily A. Yamamoto, Mr. Dario D. Garcia, Mr. Miguel Velasco, Dr. Chandra Giri, Dr. Terrance Slonecker, Dr. Jonathan Smith, Timothy Newman, and Zhouting Wu. Of couse, my dear colleagues at USGS, Dr. Pardhasaradhi Teluguntla, Dr. Itiya Aneece, Mr. Adam Oliphant, and Mr. Daniel Foley, have helped me in numerous ways. I am ever grateful for their support and significant contributions to my growth and this body of work. Throughout my career, there have been many postdoctoral level scientists who have worked with me closely and contributed in my scientific growth in different ways. They include Dr. Murali Krishna Gumma, head of Remote Sensing at the International Crops Research Institute for the Semi-Arid Tropics; Dr. Jun Xiong, Geo ML ≠ ML with GeoData, Climate Corp., Dr. Michael Marshall, associate professor, University of Twente, Netherlands; Dr. Isabella Mariotto, former USGS postdoctoral researcher; Dr. Chandrashekar Biradar, country director, India for World Agroforestry; and numerous others. I am thankful for their contributions. I know I am missing many names: too numerous to mention them all, but my gratitude for them is the same as the names I have mentioned here.

There is a very special person I am very thankful for: the late Dr. Thomas Loveland. I first met Dr. Loveland at USGS, Sioux Falls, for an interview to work for him as a scientist in the late 1990s when I was still at Yale University. But even though I was selected, I was not able to join him as I was not a U.S. citizen at that time and working for USGS required that. He has been my mentor and pillar of strength over two decades, particularly during my Landsat Science Team days (2006–2011) and later once I joined USGS in 2008. I have watched him conduct Landsat Science Team meetings with great professionalism, insights, and creativity. I remember him telling my PhD advisor on me being hired at USGS: "We don't make mistakes!" During my USGS days, he was someone I could ask for guidance and seek advice and he would always be there to respond with kindness and understanding. And, above all, share his helpful insights. It is too sad that we lost him too early. I pray for his soul. Thank you, Tom, for your kindness and generosity.

Over the years, there are numerous people who have come into my professional life who have helped me grow. It is a tribute to their guidance, insights, and blessings that I am here today. In this regard, I need to mention a few names as gratitude: (1) Prof. G. Ranganna, my master's thesis advisor in India at the National Institute of Technology (NIT), Surathkal, Karnataka, India. Prof. Ranganna is 92 years old (2024) and I met him few months back and to this day he has remained my guiding light on how to conduct oneself with fairness and dignity in professional and personal conduct. Prof. Ranganna's trait of selflessly caring for his students throughout his life is something that influenced me to follow. (2) Prof. E.J. James, former director of the Center for Water Resources Development and Management (CWRDM), Calicut, Kerala, India. Prof. James was my master's thesis advisor in India, whose dynamic personality in professional and personal matters had an influence on me. Dr. James's always went out of his way to help his students in spite of his busy schedule. (3) The late Dr. Andrew Ward, my PhD advisor at The Ohio State University, Columbus, Ohio. He funded my PhD studies in the U.S. through grants. Through him I learned how to write scientific papers and how to become a thorough professional. He was a tough task master, your worst critic (to help you grow), but also a perfectionist who helped you grow as a peerless professional, and above all a very kind human being at the core. He would write you long memos on the flaws in your research, but then help you out of it by making you work double the time! To make you work harder, he would tell you, “You won’t get my sympathy.” Then when you accomplished the task, he would tell you, “You have paid back for your scholarship many times over!!” (4) Dr. John G. Lyon, also my PhD advisor at The Ohio State University, Columbus, Ohio. He was a peerless motivator, encouraged you to believe in yourself. (5) Dr. Thiruvengadachari, scientist at the National Remote Sensing Agency (NRSA), which is now the National Remote Sensing Center (NRSC), India. He was my first boss at the Indian Space Research Organization (ISRO) and through him I learned the initial steps in remote sensing science. I was just 25 years of age then and had joined NRSA after my master of engineering (hydraulics and water resources) and bachelor of engineering (civil engineering) degrees. The first day in office Dr. Thiruvengadachari asked me how much remote sensing did I know. I told him “zero” and instantly thought, he will ask me to leave the room. But his response was “very good!” and he gave me a manual on remote sensing from Laboratory for Applications of Remote Sensing (LARS), Purdue University, to study. Those were the days where there was no formal training in remote sensing in universities. So, my remote sensing lessons began working practically on projects and one of our first projects was “drought monitoring for India using NOAA AVHRR data.” This was an intense period of learning the fundamentals of remote sensing science for me by practicing on a daily basis. Data came in 9 mm tapes, data was read on massive computing systems, image processing was done mostly working on night shifts by booking time on centralized computing, fieldwork was conducted using false color composite (FCC) outputs and topographic maps (there was no global positioning systems or GPS), geographic information system (GIS) was in its infancy, and a lot of calculations were done using calculators, as we had just started working in IBM 286 computers with floppy disks. So, when I decided to resign my NRSA job and go to the United States to do my PhD, Dr. Thiruvengadachari told me, “Prasad, I am losing my right hand, but you can’t miss this opportunity.” Those initial wonderful days of learning from Dr. Thiruvengadachari will remain etched in my memory. I am also thankful to my very good old friend Shri C.J. Jagadeesha, who was my colleague at NRSA/NRSC, ISRO. He was a friend who encouraged me to grow as a remote sensing scientist through our endless rambling discussions over tea in Iranian restaurants outside NRSA those days and elsewhere.

I am ever grateful to my former professors at The Ohio State University, Columbus, Ohio, USA: the late Prof. Carolyn Merry, Dr. Duane Marble, and Dr. Michael Demers. They taught, and/or encouraged, and/or inspired, and/or gave me opportunities at the right time. The opportunity to work for six years at the Center for Earth Observation of the Yale University (YCEO) was incredibly important. I am thankful to Prof. Ronald G. Smith, director of YCEO for the opportunity, guidance, and kindness. At YCEO I learned and advanced myself as a remote sensing scientist. The opportunities I got working for the International Institute of Tropical Agriculture (IITA) based in Nigeria and the International Water Management Institute (IWMI) based in Sri Lanka, where I worked on remote sensing science pertaining to a number of applications such as agriculture, water, wetlands, food security, sustainability, climate, natural resources management, environmental issues, droughts, and biodiversity water, were extremely important in my growth as a remote sensing scientist—especially from the point of view of understanding the real issues on the ground in real-life situations. Finding solutions and applying one’s theoretical understanding to practical problems and seeing them work has its own nirvana.

As it is clear from the preceding, it is of great importantance to have guiding pillars of light at crucial stages of your education. That is where you become what you become in the end, grow, and make your own contributions. I am so blessed to have had these wonderful guiding lights come into my professional life at right time of my career (which also influenced me positively in my personal life). From that firm foundation, I could build on from what I learned, and through the confidence of knowledge and accomplishments pursue my passion for science and do several significant pioneering research projects throughout my career.

I mention all of the preceding as a gratitude for my ability today to edit such a monumental *Remote Sensing Handbook* (Second Edition, Volumes I–VI).

I am very thankful to Ms. Irma Shagla-Britton, manager and leader for Remote Sensing and GIS books at Taylor and Francis/CRC Press. Without her consistent encouragement to take on this responsibility of editing *Remote Sensing Handbook*, especially in trusting me to accomplish this momentous work over so many other renowned experts, I would never have gotten to work on this in the first place. Thank you, Irma. Sometimes you need to ask several times, before one can say yes to something!

I am very grateful to my wife (Sharmila Prasad), daughter (Spandana Thenkabail), and son-in-law (Tejas Mayekar) for their usual unconditional understanding, love, and support. My wife and daughter have always been pillars of my life, now joined by my equally loving son-in-law. I learned the values of hard work and dedication from my revered parents. This work wouldn't come through without their life of sacrifices to educate their children and their silent blessings. My father's vision in putting emphasis on education and sending me to the best of places to study in spite of our family's very modest income and my mother's endless hard work are my guiding light and inspiration. Of couse, there are many, many others to be thankful for, but too many to mention here. Finally, it must be noted that work of this magnitude, editing monumental Remote *Sensing Handbook* (Second Edition, Volumes I–VI) continuing from the three-volume first edition, requires blessings of almighty. I firmly believe nothing happens without the powers of the universe blessing you and providing needed energy, strength, health, and intelligence. To that infinite power my humble submission of everlasting gratefulness.

It has been my deep honor and great privilege to have edited the *Remote Sensing Handbook* (Second Edition, Volumes I–VI) after having edited the three-volume first edition that was published in 2014. Now, after ten years, we will have a six-volume second edition in the year 2024. A huge thanks to all the authors, publisher, family, friends, and everyone who made this huge task possible.

Dr. Prasad S. Thenkabail, PhD
Editor in Chief
Remote Sensing Handbook (Second Edition, Volumes I–VI)

Volume I: Sensors, Data Normalization, Harmonization, Cloud Computing, and Accuracies
Volume II: Image Processing, Change Detection, GIS, and Spatial Data Analysis
Volume III: Agriculture, Food Security, Rangelands, Vegetation, Phenology, and Soils
Volume IV: Forests, Biodiversity, Ecology, LULC and Carbon
Volume V: Water, Hydrology, Floods, Snow and Ice, Wetlands, and Water Productivity
Volume VI: Droughts, Disasters, Pollution, and Urban Mapping

Part I

Image Processing Methods and Approaches

1 Digital Image Processing
A Review of the Fundamental Methods and Techniques

Sunil Narumalani and Paul Merani

LIST OF ACRONYMS

AGB	Aboveground green biomass
ANN	Artificial neural networks
AOI	Area-of-interest
BV	Brightness value
CALMIT	Center for Advanced Land Management Information Technologies
DEM	Digital elevation model
EM	Electromagnetic
EPA	Environmental Protection Agency
EVI	Enhanced Vegetation Index
FT	Fourier transform
GIS	Geographic information systems
GCP	Ground control point
GPS	Global positioning system
ISODATA	Iterative self-organizing data analysis
LAI	Leaf area index
NDVI	Normalized Difference Vegetation Index
NIR	Near-infrared
OLI	Operational Land Imager
OIF	Optimum Index Factor
PCA	Principal components analysis
SR	Simple ratio
TBM	Three-band model
TM	Thematic mapper
VBV	Vegetation biophysical variable
VNIR	Visible through near-infrared
WDRVI	Wide Dynamic Range Vegetation Index

1.1 INTRODUCTION

The modern era of commercial digital remote sensing began with the launch of Landsat 1 in 1972 (Schowengerdt, 1997). For the first time ever, a satellite system was able to provide high spectral quality (4-bands), medium (80-m) spatial resolution data that could be analyzed by public experts for numerous terrestrial applications. Initially, much of the "digital processing" of these data could only be done by large, expensive computer systems that lay in the purview of government agencies,

DOI: 10.1201/9781003541158-2

and only large-format hard-copy images were made available for visual or analog analysis. With the technological evolution and computer revolution in the early 1970s, computing power began to slowly trickle down to the mainstream such as educational institutions, and research centers across the United States and Europe. As computing capabilities and their usage increased, their application to remotely sensed data also increased, and for almost two decades, from the mid-1970s to the early 1990s there continued to be sustained development and improvement of algorithms for extracting image data.

Remotely sensed data acquired in or converted to digital format are most often subject to analyses and information extraction using image-processing techniques developed over the past 50 years. The process encompasses a myriad of algorithms that an analyst can apply to get meaningful information from an image. This chapter describes some of the basic image processing methods applied to remotely sensed data.

Remote sensing instruments (or sensors) detect and record electromagnetic (EM) energy. There are two types of senor systems, *active sensors*, which emit energy directed towards a target of interest and then record the return; and *passive sensors* that record energy *either* emitted by or reflected from an object. Data may be acquired in single (panchromatic) or multiple (multi/hyper-spectral) bands of the EM spectrum. In either case, the data appear in a matrix of *x-columns* by *y-rows*, with each square of the matrix referenced as a "pixel" and have a grayscale value that indicates the EM energy recorded for that pixel (commonly referred to as the "brightness value" (BV)). The range of BVs recorded in each band across the image is dependent on the radiometric resolution of the sensor system (e.g., values ranging from 0–255 would be recorded for a sensor system with an 8-bit radiometric resolution). Jensen (2005) notes that most remote sensing studies are based on developing deterministic relationships between the EM signals recorded in various bands of the spectrum and the chemical or biophysical properties of the features being investigated.

1.2 IMAGE QUALITY ASSESSMENT: BASIC STATISTICS AND HISTOGRAM ANALYSIS

Many remote sensing data are of high-quality. However, on occasions, errors (or noise) are introduced into the data by numerous factors such as the environment (e.g., atmospheric scattering), random or systematic malfunction of the sensor system (e.g., an uncalibrated detector creates striping), or improper processing of the raw data prior to actual data analysis (e.g., inaccurate analog-to-digital conversion). Therefore, one of the initial tasks of an image analyst should be to assess its quality and statistical characteristics (Burger and Burge, 2022; Richards, 2022; Elachi and van Zyl, 2021; Amani et al., 2020; Chuvieco, 2020; Sabins and Ellis, 2020; Liang and Wang, 2019; Nixon and Aguado, 2019; Shao et al., 2019; Sun et al., 2019; Li et al., 2018; Tan and Jiang, 2018; Zhong et al., 2018; Dong and Chen, 2017; Girard, 2017; Shih, 2017; Woodhouse, 2017; Lillesand et al., 2015; Campbell and Wynne, 2011;

Jensen, 2005). This is normally accomplished by:

- Examining the frequency of occurrence of individual BVs in the image displayed in a histogram,
- Sample visual analysis of individual pixel BVs at specific locations or within a geographic area,
- Computing univariate descriptive statistics to determine if there are unusual anomalies in the image data, and
- Computing multivariate statistics to determine the amount of between-band correlation (e.g., to identify redundancy).

Some of the basic statistical information that an analyst may find useful is found in measures of central tendency such as the mean, standard deviation and variance. These statistics provide information about the range of BVs in each band, the relationship of the BVs in each band, representation of spectral characteristics of features being examined, and an indicator of values that can be used for image enhancement (e.g., using a histogram stretch using minimum-maximum values in a given band). However, for an in-depth analysis other statistical measures such as *variance* and *correlation* may be required to provide an insight into the data quality and redundancy.

Remote-sensing-derived spectral measurements for each pixel often change together in some predictable fashion because objects or features exhibit spectral behavioral patterns across the bands. For example clear, deep water would have low, steadily declining BVs across the blue, green, and red portions of the spectrum, until it reaches near zero in the near-infrared (NIR). If there is no relationship between the BVs in one band and that of another for a given pixel, it may imply that the values are mutually independent (e.g., reflective spectra versus temperature as observed in thermal infrared) or there may be an anomalous observation for a given feature (e.g., sedimentation present in deep water). In most cases spectral measurements of individual pixels may not be independent and a measure of their mutual interaction is reflected in the *covariance* or the joint variation of two variables about their common mean.

To estimate the degree of interrelation between variables in a manner not influenced by measurement units, the *correlation coefficient*, is commonly used. The correlation between two bands of remotely sensed data, $r_{k,l}$, is the ratio of their covariance ($\text{cov}_{k,l}$) to the product of their standard deviations ($s_k s_l$); thus:

$$r_{k,l} = \frac{cov_{k,l}}{s_k s_l} \tag{2.1}$$

If we square the correlation coefficient ($r_{k,l}$), we obtain the *sample coefficient of determination* (r^2), which expresses the proportion of the total variation in the values of "band l" that can be accounted for or explained by a linear relationship with the values of the random variable "band k." Thus, a correlation coefficient (r_{kl}) of 0.70 results in an R^2 value of 0.49, meaning that 49% of the total variation of the values of "band l" in the sample is accounted for by a linear relationship with values of "band k." In order to optimize usage of multiple bands, scientists prefer higher variance and lower correlation. Furthermore, the correlation and covariance information can be used for analysis by advanced image processing functions such as Principal Components Analysis (PCA) and image classification.

1.2.1 Histogram

The histogram is most fundamental and useful graphical representation of information content in an image. It tabulates frequencies of occurrences of each BV, displays them graphically and provides information on "contrast" within each band. Peaks and valleys often correspond to the dominant land cover types in an image and the information can be converted into "percent" representation to highlight information content by masking out BVs with high frequency (or vice versa) (Figure 1.1). Basically, histograms may provide the user with information on the quality of the image (e.g., high-contrast, low-contrast, bi-modal, multi-modal, etc.) and are often used to enhance imagery—e.g., brightening up darker areas in an image or conversely darkening up extremely bright areas of an image (Figure 1.2).

1.3 IMAGE ENHANCEMENT

To improve the appearance of an image for visual analysis or at times even for subsequent computer analysis, an analyst may prefer to apply select algorithms. A basic suite of algorithms that aid in enhancing an image include image reduction and magnification, spatial and spectral profiles, image

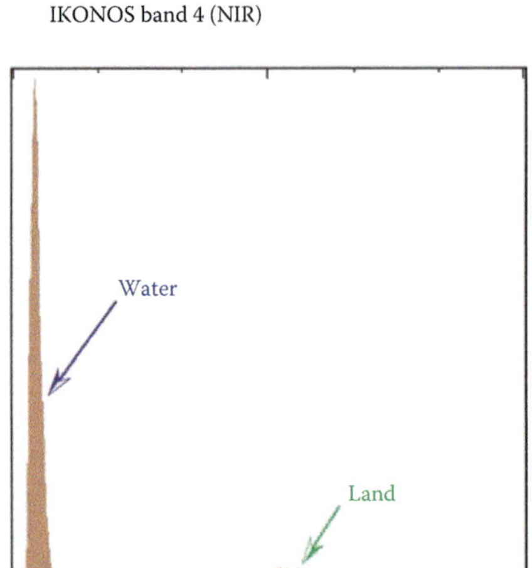

FIGURE 1.1 Histogram of the near-infrared band (Band 4) of the IKONOS sensor system. The near-infrared band is useful for land/water delination and this evident in histogram where clear, deep water pixels have very low reflectivity while the land pixels record higher BVs in the band. (Image courtesy of the Center for Advanced Land Management Information Technologies (CALMIT), University of Nebraska, Lincoln.)

contrasting, density slicing and composite generation (Table 1.1). Higher order image enhancement techniques include band ratioing, spatial filtering, edge enhancement and spectral image transformation.

Reduction and magnification operations are used to adjust the image scale visually in order to provide either a regional perspective (i.e., display of an entire scene at a small scale) or a zoom-in of an area-of-interest (AOI) for closer examination (Figure 1.3). They allow an analyst to derive image coordinates (*x*, *y* locations), per-pixel spectral data of features across, and gain an understanding of the spatial distribution of objects across the landscape.

An analyst may further their understanding of the image landscape by deriving *spatial* and *spectral* profiles along user-specified transects (Burger and Burge, 2022; Richards, 2022; Elachi and van Zyl, 2021; Amani et al., 2020; Chuvieco, 2020; Sabins and Ellis, 2020; Liang and Wang, 2019; Nixon and Aguado, 2019; Shao et al., 2019; Sun et al., 2019; Li et al., 2018; Tan and Jiang, 2018; Zhong et al., 2018; Dong and Chen, 2017; Girard, 2017; Shih, 2017; Woodhouse, 2017; Lillesand et al., 2015; Campbell and Wynne, 2011; Jensen, 2005). The pixels that lie along that transect can be measured and displayed to compare the spectral (BVs) or spatial differences (coordinate space). Multiple transects may be used to determine spatial patterns or trends. Transects can also be used assist in density slicing an image or a portion of it (Figure 1.4).

Density Slicing is a pseudocolor enhancement technique normally applied to a single band monochrome. It is considered an effective way of highlighting different but apparently homogeneous areas within an image by "slicing" the range of grayscale values (e.g., 0–255) and assigning different colors to each of those slices (Figure 1.5). This technique is often used in conjunction with a vegetation index such as the Normalized Difference Vegetation Index (NDVI) to highlight variations in the density of biomass.

Another effective method or visual analysis is composite generation. This method utilizes the three planes of a computer's display device (Red (R), Green (G), Blue (B)) and allows the analyst to place different bands of a multispectral image into the various planes. For example to generate a true color composite from Landsat Operational Land Imager (OLI) image one would insert Bands 4, 3, 2 into the R, G, B planes respectively, thus generating a true color image (Figure 1.6a). Similarly, to display a false color composite of the same image, OLI Bands 5, 4, 3 would be placed in the R, G,

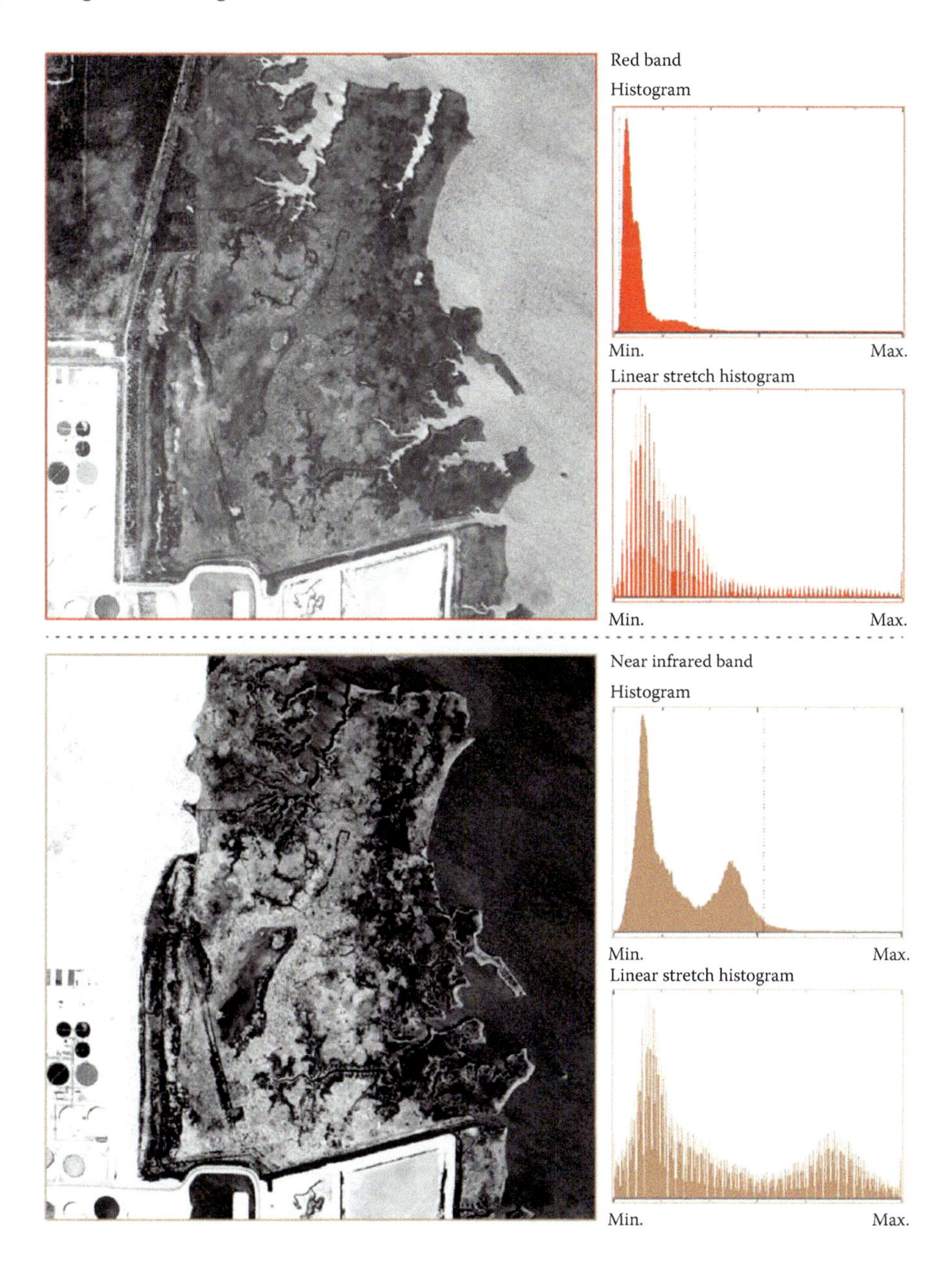

FIGURE 1.2 Example of histogram stretch performed on the red and near-infrared bands of a WorldView 2 image acquired for the Grand Bay, MS area. (Image courtesy of the Center for Advanced Land Management Information Technologies (CALMIT), University of Nebraska, Lincoln.)

B planes (Figure 1.6b). In Figure 1.6b, the near infrared band (OLI Band 5) is placed into the Red plane of the display and is often used for vegetation analysis because of high spectral reflectance of vegetation in the near and mid-infrared portion of the spectrum.

1.4 IMAGE PREPROCESSING

Fureder (2010) states that the operational use of remote sensing data is often limited due to sensor variation, atmospheric effects as well as topographically induced illumination effects. *Image*

TABLE 1.1

Image enhancement techniques, their effects and examples of application.

Image Enhancement Technique	Effect on Image	Example of Application
Image Magnification	Zooms into area of interest or closer observation of feature	
Image Reduction	Zooms out either partially or completely from an image to provide a large area perspective	Enable a geographic or spatial analysis of an entire landscape
Spatial Profiles	Draw a transect across an area of interest	Changes in features across the transect (e.g., forest to grasslands, to water)
Spectral Profiles	Draw a transect across an area of interest	Changes in spectral signature observed along the transect as variations in features occur
Density Slicing	Color coding a band based on BV ranges	Highlight specific features or a rapid visualization of potential land cover observed in the selected band
Image Composites	Representation of land cover features in various colors based on the band combinations used	Geologic highlighting a mineral may use a combination of certain bands based on the spectral properties of that mineral

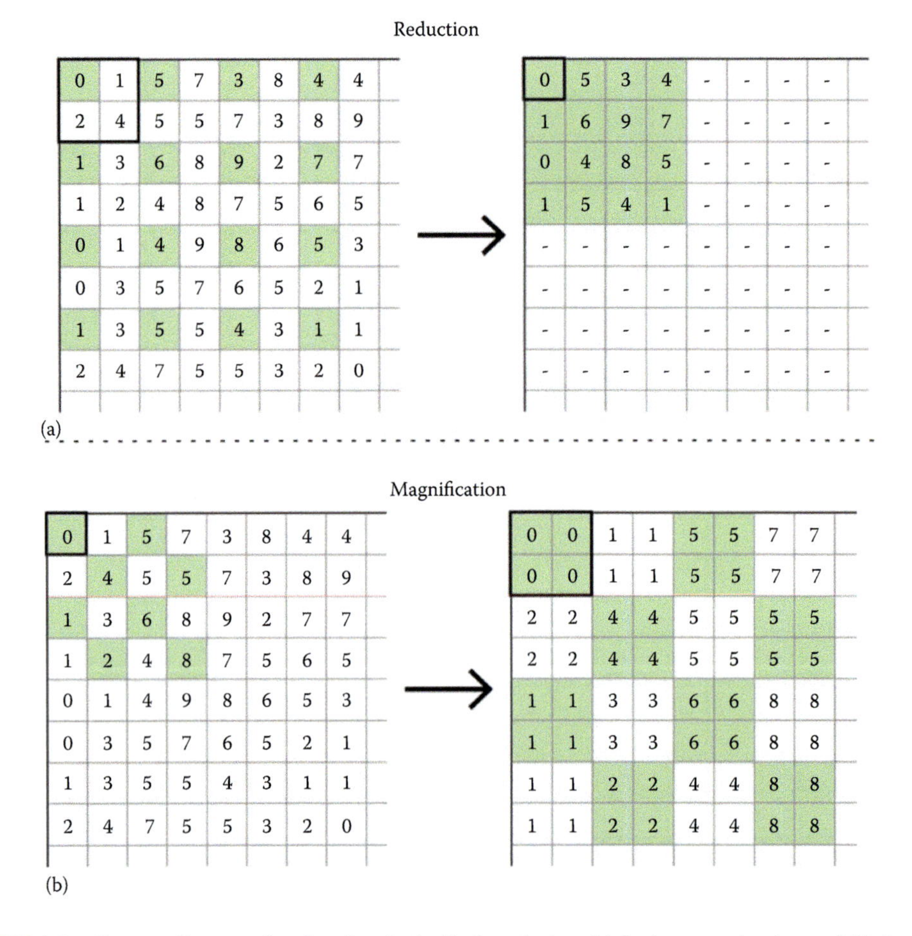

FIGURE 1.3 Concept diagram showing the pixels displayed when (a) 2× image reduction and (b) 2× image magnification is applied.

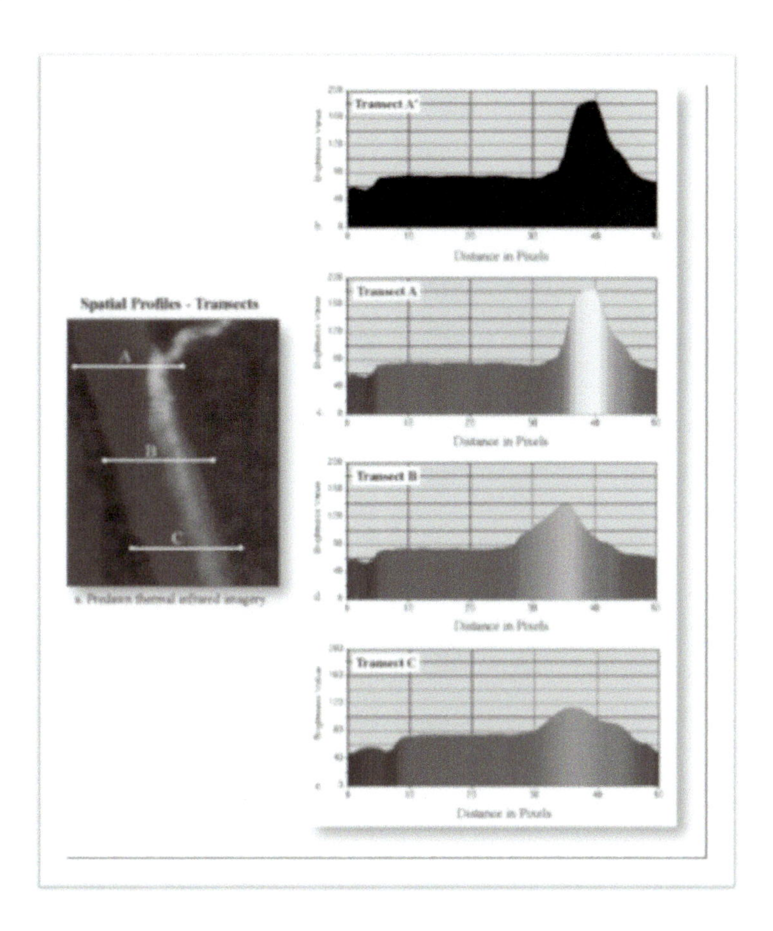

FIGURE 1.4 Example of spatial profiles acquired for the Savannah River over three transtects along the channel. (Source: Jensen, 2005.)

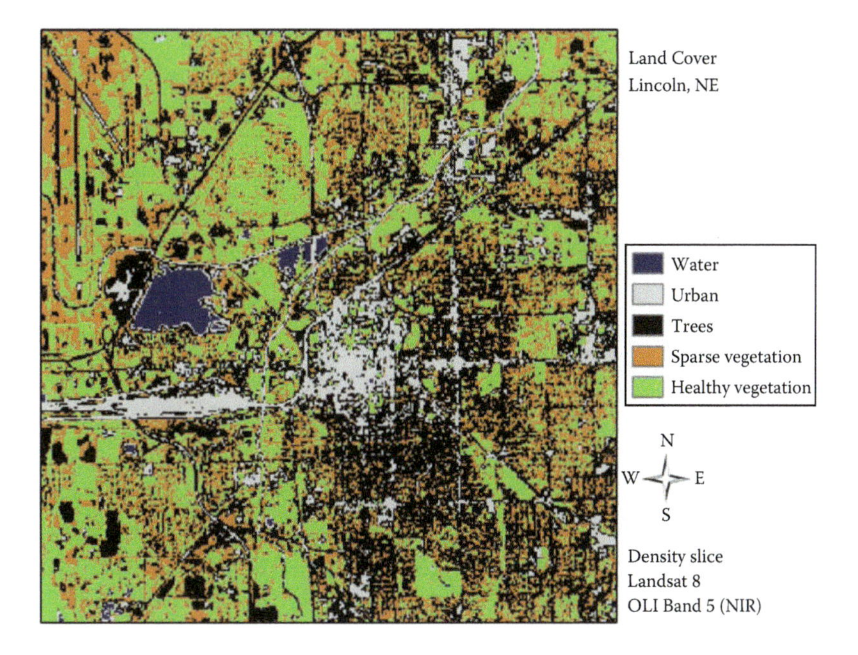

FIGURE 1.5 An example of a density sliced image acquired by the Landsat OLI system of Lincoln, NE.

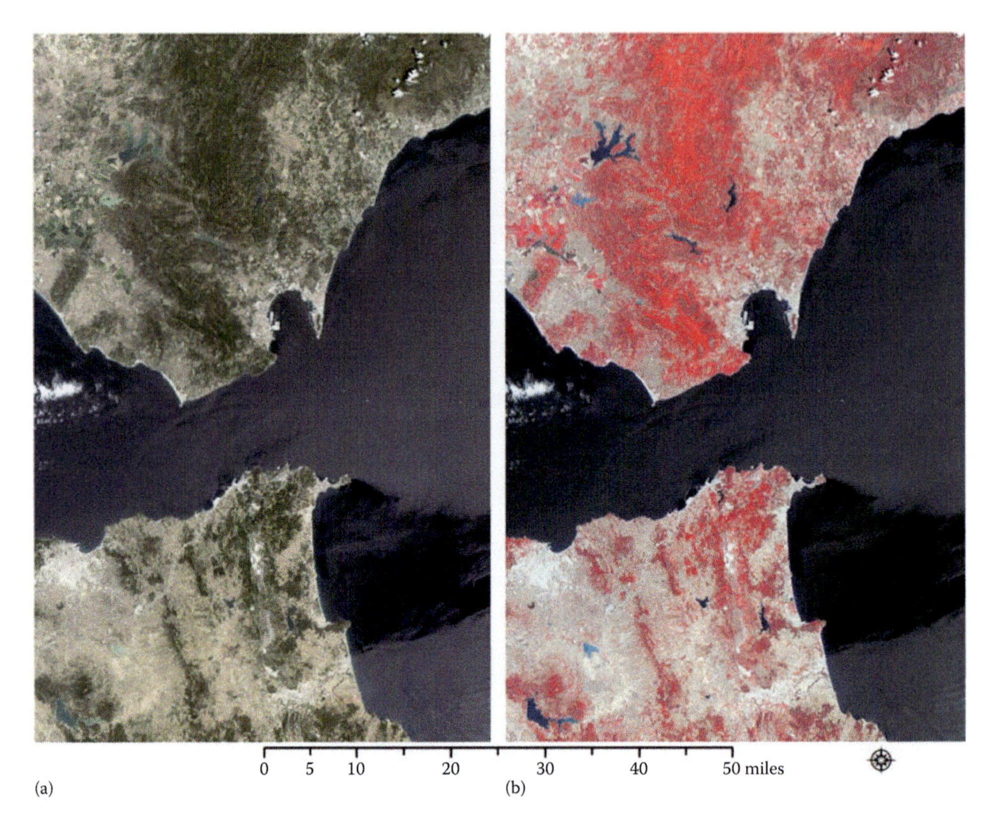

FIGURE 1.6 (a) True color image of the Gibraltor area in the Mediterranean Sea of the Landsat OLI sensor system. (b) Color infrared image of the same region. Note that healthy green vegetation appears bright red, urban areas shades of white and water (depending on the turbidity) from blue (turbid water) to dark gray (deep, clear water).

preprocessing is a preparatory phase that, in principle, improves image quality as the basis for later analyses that will extract information from the image (Burger and Burge, 2022; Richards, 2022; Elachi and van Zyl, 2021; Chuvieco, 2020; Sabins and Ellis, 2020; Liang and Wang, 2019; Tan and Jiang, 2018; Zhong et al., 2018; Dong and Chen, 2017; Girard, 2017; Shih, 2017; Woodhouse, 2017; Lillesand et al., 2015; Campbell and Wynne, 2011). Often known as image restoration, the process produces a corrected image that is as close as possible, both geometrically and radiometrically, to the radiant energy characteristics of the actual scene. This requires that internal and external errors be determined and corrected for. Internal errors are created by the sensor itself and are generally systematic and stationary (i.e., predictable and constant). External errors are due to perturbations and modulation of scene characteristics—that is, they are variable and are corrected by relating points on ground to sensor measurements.

1.4.1 Radiometric Correction

As radiation passes through the atmosphere it undergoes several different process including absorption, scattering, attenuation, and transmission (Lillesand et al., 2015; Campbell and Wynne, 2011; Jensen, 2005; Schowengerdt, 1997). Various methods of atmospheric correction can be applied ranging from detailed modeling of the conditions during data acquisition, to simple calculations based solely on the image data. Broadly, atmospheric correction can be divided into two types: (1) absolute and (2) relative. Absolute radiometric corrections turn the BVs into scaled surface reflectance values (Lillesand et al., 2015; Campbell and Wynne, 2011; Du et al., 2002) and attempt to model the atmosphere, as it would exist at the time of image acquisition. Several radiative transfer models have been developed as part of the absolute radiometric correction efforts including the Atmospheric CORrection Now by ImSpec (2002), ATmospheric REMoval program by the University of Colorado (Tanre et al., 1986), Fast Line-of-Sight Atmospheric Analysis of Spectral Hypercubes by Exelis (formerly Research Systems, 2003). It is important to note that application of these algorithms for a given scene and date requires knowledge of the spectral profile and atmospheric properties for that date and time. This information is extremely difficult to acquire, however, these models can provide a close approximation of the reflectance for the scene sans the atmosphere, versus an atmospherically uncorrected scene.

Relative atmospheric correction is often used if an analyst wants to normalize the BV amongst the various bands of a single scene or normalize multi-date imagery to a single/standard scene selected amongst the dataset (Burger and Burge, 2022; Tan and Jiang, 2018; Lillesand et al., 2015; Campbell and Wynne, 2011; Jensen, 2005). An example of the former method is to **examine the observed BVs**, in an area of shadow or for a very dark object (such as a large clear lake or an asphalt surface) and determine the minimum value. The correction is applied by subtracting the minimum observed value, determined for each specific band, from all pixel values in each respective band. Because scattering is wavelength dependent, the minimum values will vary from band to band. This method is based on the assumption that the reflectance from these features, if the atmosphere is clear, should be very small (if not zero.) If values are much greater than zero, then they are considered to have resulted from atmospheric scattering.

Multi-date image normalization techniques involve the selection of a base image and then transforming the spectral characteristics of all other images to have the same radiometric scale as the base image (Jensen, 2005). The method involves the selection of pseudo-invariant features (i.e., radiometric ground control points) from the base image, identifying the BVs of the same features across all the multi-date imagery, and normalizing them to the base image. The pseudo-invariant features need to meet specific spectral and spatial criteria (Liang and Wang, 2019; Nixon and Aguado, 2019; Tan and Jiang, 2018; Lillesand et al., 2015; Campbell and Wynne, 2011; Eckhardt et al., 1990; Hall et al., 1991; Jensen et al., 1995).

In addition to atmospheric effects, the landscape elements such as slope and aspect can cause radiometric distortion of the signal received by the sensor system (Jensen, 2005). Jensen (2005) describes four topographic correction methods including (1) Cosine Correction, (2) Minnaert Correction, (3) Statistical-Empirical Correction, and (4) C-Correction. Each correction method is based on illumination and requires a Digital Elevation Model (DEM) to determine how much illumination each pixel receives relative to its topography in the landscape. Much research continues on the removal of topographic effects from the scene. For example, Civco (1989) identifies several considerations including matching the DEM spatial resolution to that of the image, overcorrection of topographic effect because of the Lambertian surface assumption, ignoring that the diffuse component also illuminates the topography, strong anisotropy of apparent reflectance, consideration to wavelength and deeply shadowed areas (Kawata et al., 1988). Readers should refer to more detailed discussions of radiometric correction in Chapters 3 and 4 of this volume.

Noise in an image may be due to irregularities or errors that occur in the sensor response and/or data recording and transmission (van der Meer et al., 2009). Common forms of noise include systematic striping or banding and dropped lines. Early Landsat MSS data had substantial striping due to variations and drift in the response over time of the six MSS detectors. The "drift" was different for each of the six detectors, causing the same brightness to be represented differently by each detector (Fureder, 2010). The overall appearance was thus a "striped" effect. The corrective process made a relative correction among the six sensors to bring their apparent values in line with each other. Dropped lines occur when there are systems errors that result in missing or defective data along a scan line and is often "corrected" by replacing the line with the pixel values in the line above or below, or with the average of the two.

1.4.2 Geometric Correction

The stability of a remote sensing platform, the curvature of the earth, sensor orientation, topography, and other factors cause geometric distortion in an image. Consequently, geometric correction is applied to remove these distortions so that the image is planimetrically (x, y) correct and the displacement of objects as well as scale variations are minimized or removed entirely (Lillesand et al., 2015; Campbell and Wynne, 2011; Aronoff, 2005). *Geometrically corrected imagery* can be used to extract accurate distance, area, and direction (bearing) information and any information derived from such images can be related to other thematic information in a geographic information systems (GIS) or spatial decision support systems (SDSS) (Jensen, 2005).

Remotely sensed imagery collected from airborne or spaceborne sensors often contain internal and external geometric errors (Jensen, 2005). These can be *systematic* (predictable) or *nonsystematic* (random), and generally, systematic geometric error is generally easier to identify and correct than random geometric error. Some of these errors can be corrected by using ephemeris of the platform and known internal sensor distortion characteristics. Commercial satellite data (e.g., SPOT Image, Landsat, QuickBird, GeoEye, and others) already have much of the *systematic error* removed. Other errors can only be corrected by matching image coordinates of physical features recorded by the image to the geographic coordinates of the same features collected from a map or global positioning system (GPS).

Internal geometric errors are introduced by the remote sensing system itself or in combination with Earth rotation or curvature characteristics. These distortions are often systematic (predictable) and may be identified and corrected using pre-launch or in-flight platform ephemeris (i.e., information about the geometric characteristics of sensor and the Earth at data acquisition). Geometric distortions in imagery that can sometimes be corrected through analysis of sensor characteristics and ephemeris data include:

- Skew caused by Earth rotation effects,
- Scanning system–induced variation including ground resolution cell size, relief displacement, and tangential scale distortion.

External geometric errors are usually introduced by phenomena that vary in nature through space and time. The most important external variables that can cause geometric error in remote sensor data are random movements of the remote sensing platform at the time of data collection (i.e., altitude and attitude changes). Unless otherwise processed, however, *unsystematic random error* remains in the image, making it non-planimetric. To correct for these errors, two common geometric correction procedures are used to make the digital remote sensor data of value:

- *Image-to-map rectification*
- *Image-to-image registration*

Image-to-map rectification is the process by which the geometry of an image is made planimetric. Whenever accurate area, direction, and distance measurements are required, image-to-map geometric rectification should be performed, however, it may not remove all the distortion caused by topographic relief displacement in an image. The *image-to-map rectification* process normally involves selecting well-identified ground control points (GCPs) on an image and associating them with their planimetrically correct map counterparts (i.e., GCPs from a paper or digital map for which geographic coordinates can be derived—for example, meters northing and easting in a Universal Transverse Mercator map projection). Alternatives to obtaining accurate GCP map coordinate information for image-to-map rectification include:

- *Hard-copy planimetric maps* (e.g., USGS 7.5-minute 1:24,000-scale topographic maps) where GCP coordinates are extracted using simple ruler measurements or a coordinate digitizer;

- *Digital planimetric maps* (e.g., the USGS digital 7.5-minute topographic map series) where GCP coordinates are extracted directly from the digital map on the screen;
- *Digital orthophotoquads* that are already geometrically rectified (e.g., USGS digital orthophoto quarter quadrangles DOQQs)); and/or
- *Global positioning system (GPS) instruments* that may be taken into the field to obtain the coordinates of objects to within +20 cm if the GPS data are differentially corrected.

Once GCP map coordinates are obtained for several points on an image, spatial interpolation algorithms are applied to transform the image coordinates to the map coordinates, thus making the image planimetrically correct.

- Polynomial equations are used to convert the file coordinates into rectified map coordinates.
- Depending on the distortion of the imagery, complex (higher order) polynomial equations may be required to express the needed transformation.
- The degree of complexity of the polynomial is expressed as the order of the polynomial (i.e., the highest exponent used in the polynomial).

Intensity interpolation arises from the fact that there is no one-to-one relationship between the input image pixel location to the output pixel location. Therefore, a new BV has to be assigned to the rectified pixel. Three methods for such "resampling" (Figure 1.7):

- Nearest neighbor
- Bilinear interpolation
- Cubic convolution

The nearest neighbor algorithm assigns the BV of the closest input x, y to the output x, y. This method maintains the integrity of the data and is not computationally intensive. Unfortunately, the output image may not be aesthetically pleasing as it has a "block appearance." Bilinear interpolation derives the new BV of the pixel based on the weighted value of the four pixels nearest to those in the original image. This method produces a "smoother" image, but has a slight impact on the integrity of the data. Cubic convolution determines the output BV based on the weighted values of 16 input pixels surrounding the location of the original pixel. This method produces a "smoother" image but is computationally intensive and may have a considerable impact on the integrity of the data.

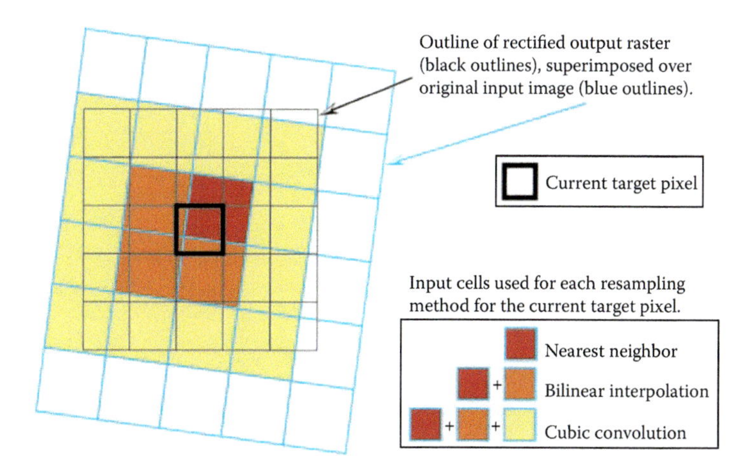

FIGURE 1.7 Schematic diagram showing the comparison between nearest neighbor, bilinear interpolation and cubic convolution algorithms for intensity interpolation. (Adapted from: http://marswiki.jrc.ec.europa.eu/wikicap/index.php/Resampling_techniques_in_image_processing.)

TABLE 1.2

Geometric correction methods and their comparative advantages and disadvantages.

Intensity Interpolation Technique	Pixels Used for Interpolation	Advantages	Disadvantages
Nearest Neighbor	1	• *Maintains data integrity* • *Not computationally intensive*	• *Blocky appearance*
Bilinear Interpolation	4	• *Minimal disruption to data integrity* • *Smooth appearance*	• *Contrast may be reduced* • *BVs are interpolated*
Cubic Convolution	16	• *Improved appearance*	• *BVs are highly manipulated*

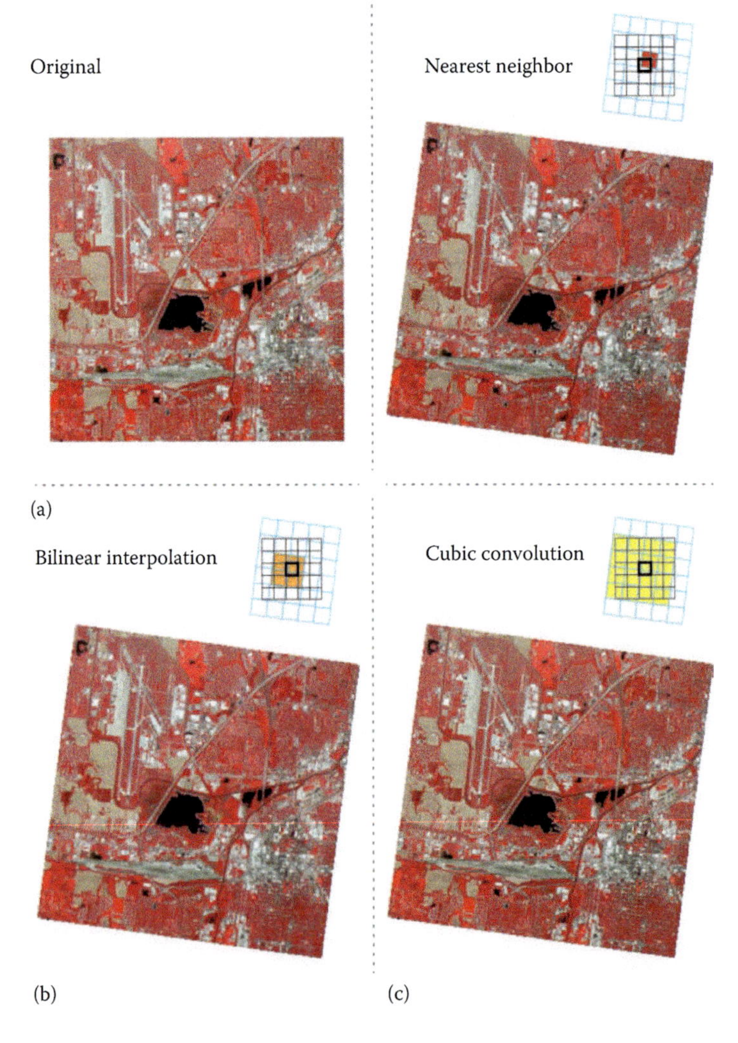

FIGURE 1.8 An example of three intensity interpolation techniques used for the geometric rectification of remotely sensed data. (a) The nearest-neighbor algorithm that uses the closest pixel location BV to assign to the corrected data; (b) the bilinear interpolation that uses a weighted average of the four spatially closest pixels to assign a BV to the corrected data; and (c) the cubic convolution algorithm that uses the weighted average of the 16 closest pixels to determine the BV of the rectified pixel.

Image-to-image registration is the translation and rotation alignment process by which two images of like geometry and of the same geographic area are positioned coincident with respect to one another so that corresponding elements of the same ground area appear in the same place on the registered images (Burger and Burge, 2022; Lillesand et al., 2015; Campbell and Wynne, 2011; Chen and Lee, 1992; Jensen, 2005). This type of geometric correction is used when it is *not* necessary to have each pixel assigned a unique *x*, *y* coordinate in a map projection. For example, we might want to make a cursory examination of two images obtained on different dates to see if any change has taken place. In such a case we need to only "register" the two (or more images) to a *base* image (selected among the available dataset) and perform a rapid visual analyses of the data.

1.5 PRINCIPAL COMPONENT ANALYSIS (PCA)

The majority of remote sensing data are acquired in many different bands leading to the generation of vast quantities of data. Because of the spectral proximity of some bands in a multispectral dataset and certainly in the case of hyperspectral imagery, there is often a high degree of correlation between bands, implying that there may be similar information content between them. For example, Landsat Thematic Mapper (TM) bands 2 and 3 (green and red respectively) typically have similar visual appearances because reflectances for the same cover types are almost equal. Image transformation techniques based on complex processing of the statistical characteristics of multi-band datasets can be used to reduce this redundancy and correlation between bands. One such transformation is called PCA whose objective is to reduce the dimensionality (i.e., the number of bands) in the data and compress as much of the information in the original bands into fewer bands. The "new" bands that result from this statistical procedure are called components. The process attempts to statistically maximize the amount of information (or variance) from the original data into the least number of *useful* new components.

PCA transforms the axes of the multispectral space such that it coincides with the directions of greatest correlation. Each of these new axes are orthogonal to one another, that is, they are at right angles and the component images are arranged such that the greatest amount of variance (or information) within the original dataset is contained within the first component and the amount of variance decreases with each component (Jensen, 2005; Figure 1.9). Transformation of original data on X_1 and X_2 axes onto PC_1 and PC_2 axes requires transformation coefficients that can be applied in linear fashion to original pixel values (Figure 1.9). These new axes are called first PC. The second

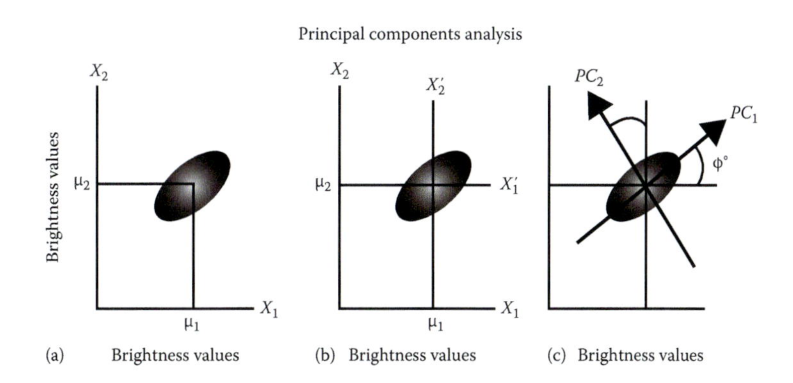

FIGURE 1.9 Concept diagram illustrating the PCA process: (a) the cluster of BVs from two bands of an image (b) a new coordinate system is defined by the X', and (c) the PCA transformation occurs by rotating to the new axis which is orthogonal to the original X' axis. The new axes are no longer the bands of the original image, but derivative components from those data. (Source: Jensen, 2005.)

PC is perpendicular (orthogonal) to PC1 (Burger and Burge, 2022; Lillesand et al., 2015; Campbell and Wynne, 2011; Gonzalez, 2014). Subsequent components contain decreasing amounts of the variance found in the dataset.

By computing the correlation between each band and each PC, it is possible to determine how each band "loads" or is associated with each PC. A linear combination of original BV and factor scores (eigenvectors) produces the new BV for each pixel of every PC. It is often the case that the majority of the information contained in a multispectral dataset can be represented by the first three or four PCA components. Higher-order components may be associated with noise in the original dataset.

1.6 SPATIAL FILTERING

For any given image, or part thereof, there are changes in BVs throughout the scene. The number of changes in BVs per unit distance for any particular part of the image is called spatial frequency—that is, the "roughness" of the tonal variations occurring in an image (Jensen, 2005). Figure 1.10a and b demonstrates the differences between low frequency (less "roughness") and high frequency (more "roughness") images. In a low frequency area, the changes in BVs are subtle over the given area, while the opposite is true in a high frequency image.

To extract quantitative information, "local" operations are performed (spatial filtering) and the BV of a given pixel is modified based on the values of neighboring pixels.

Depending on the features to be extracted, filters can be applied to an image. A filter (or a convolution mask/kernel) is a moving window function that defines a small sub-window with dimension of 3 × 3 or larger and usually with odd-numbered dimensions (e.g., 3 × 3, 5 × 5, 7 × 7, etc.) An example of a 3 × 3 window is shown in Figure 1.11, with pixels numbered from the top-left. In this example, pixel $c_{2,2}$ in the window is the center pixel, and odd-numbered window sizes ensure that there is always a center pixel in the sub-window.

Filtering involves computing a weighted average of the pixels in the moving window. The choice of weights determines how the filter affects the image. A window of weight values is called a convolution kernel. Multiplying each pixel in the moving window by its weight and summing all

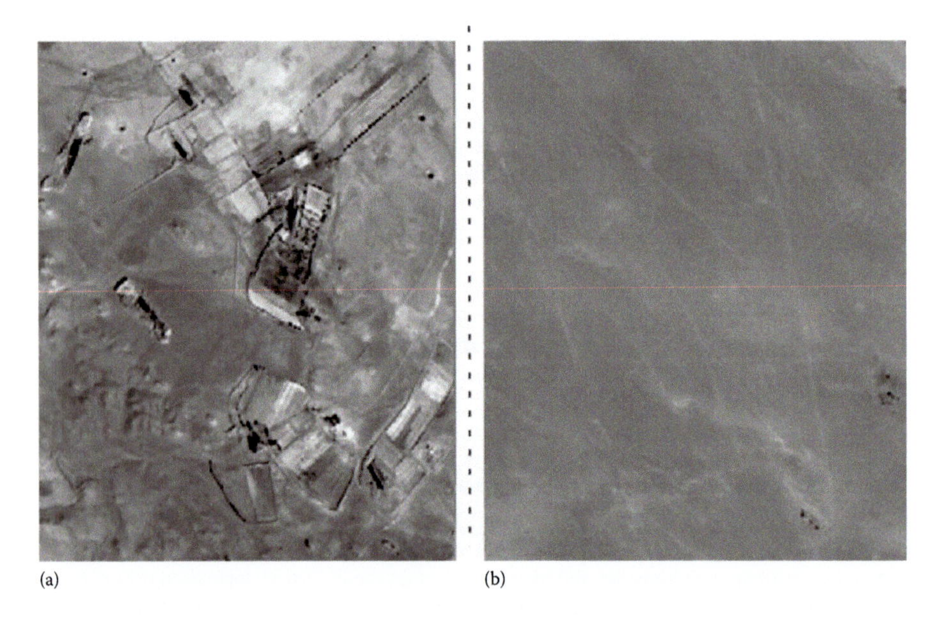

FIGURE 1.10 Examples of (a) high frequency image and (b) low frequency image.

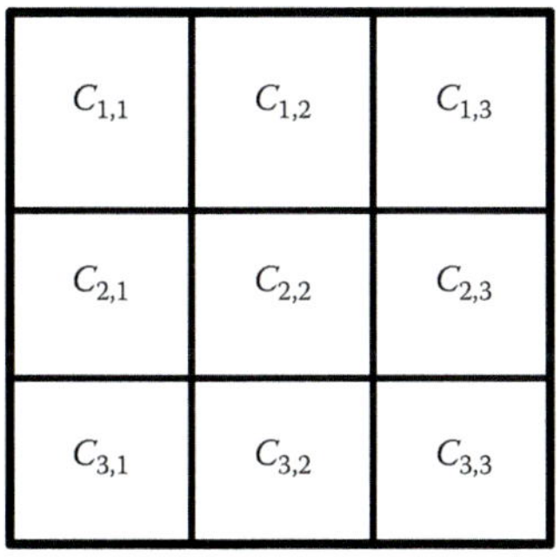

$C_{1,1}$	$C_{1,2}$	$C_{1,3}$
$C_{2,1}$	$C_{2,2}$	$C_{2,3}$
$C_{3,1}$	$C_{3,2}$	$C_{3,3}$

FIGURE 1.11 Schematic of 3 × 3 convolution kernel. Values are assigned to each cell depending on the type of information that an analyst may wish to extract.

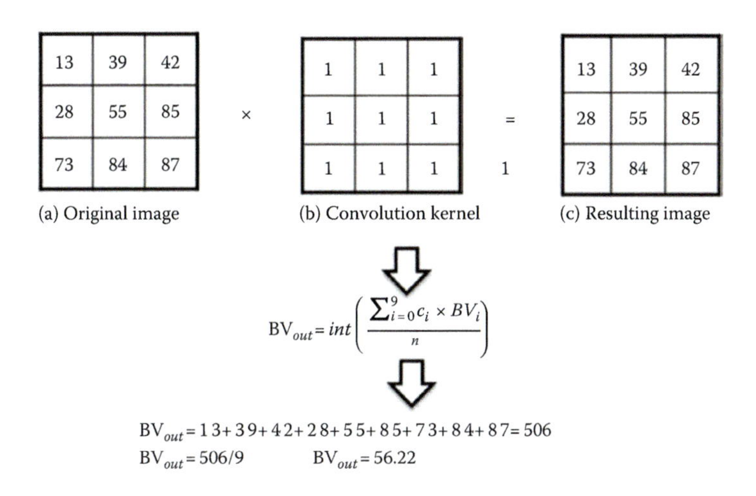

FIGURE 1.12 Schematic diagram showing the process and results of applying a low filter. The original center cell value was (a) 55. With the application of a convolution kernel (b) with equal weights for all nine cells, an output image (c) is created. Further, the formula described earlier sums the values of the nine cells and produced the average value of 56.22. Rounded off to the nearest whole number, the new value of the center cell will be *56*.

the products yields a new value for the center pixel (Figure 1.12). The values used in a convolution kernel defines whether the filter is *low pass* or *high pass.*

Low pass filters are designed to *emphasize low frequency features and de-emphasize the high frequency components of an image.* Thus, information changing very fast across a landscape (e.g., in an urban area) will be subdued while low frequency information (e.g., grassland, water) is preserved. Low pass filters are excellent for retaining low frequency information and are useful for removing "noise" (such as speckle) in an image (Jensen, 2005). Low pass filters make similar cover areas appear uniform and can be useful for boundary detection (Figure 1.13). Conversely, low pass filters do not preserve edges and larger window sizes lead to greater smoothing.

High pass filters emphasize the detailed high frequency components of an image and deemphasize the more general low frequency information. They enhance image details (infrequent information) and are useful where lower frequency information tends to "hide" parts of the scene of interest, for example, roads in an urban scene. When building a high pass filter, the center pixel of the kernel is given more weight. Consequently, if it is an edge then the pixel will be greatly enhanced because edges have higher pixel values.

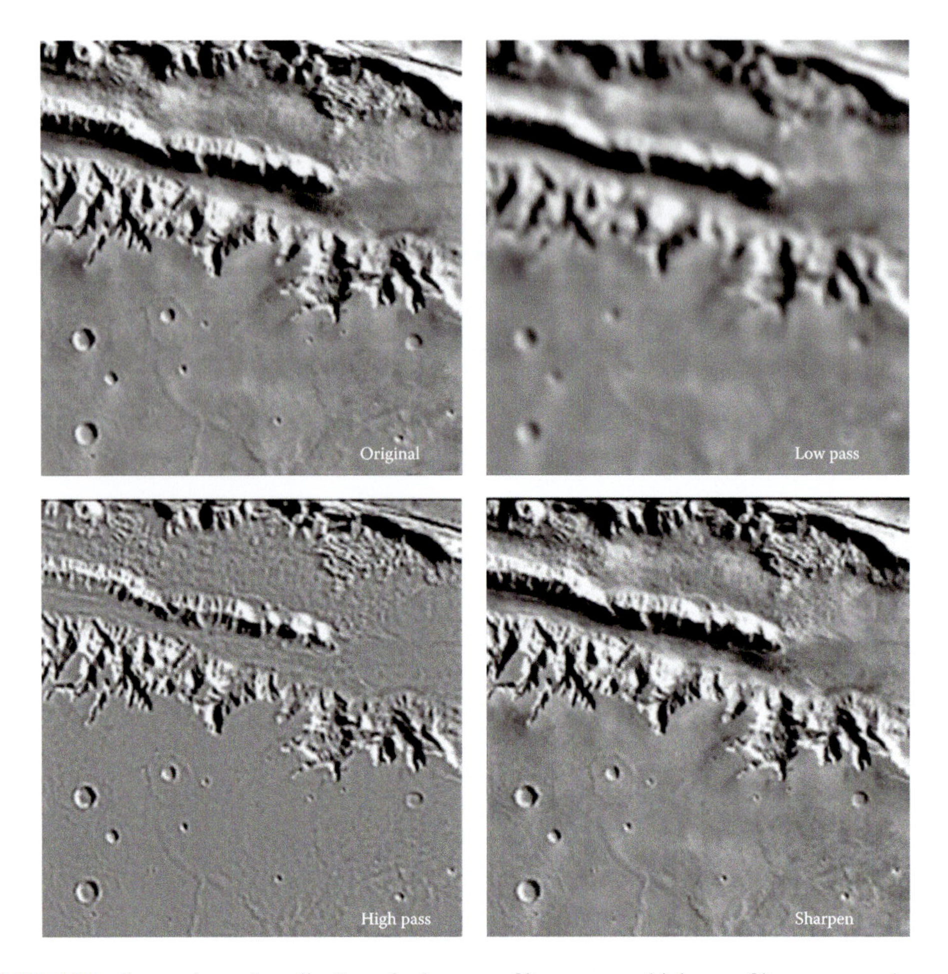

FIGURE 1.13 Comparison of application of a low pass filter versus a high pass filter versus a sharpening filter. (Source: USGS, 2007.)

Spatial filtering methods can also be used to remove noise in the data (e.g., striping or speckles). Jensen (2005) used the Fourier Transform (FT) on Landsat TM data to remove striping in a coastal area near Al Jubail, Saudi Arabia. The process involved the computation of a FT for the study area, modifying the resulting FT by selectively removing the points associated with the striping and then computing a reverse FT to derive a clean, destriped image. Destriping of an image can also be done by running a low pass and a high pass filter on an image and then adding the filter outputs. The resulting image will have its striping removed or minimized (USGS, 2007). Table 1.3 broadly describes the types of filters used on remotely sensed data and their utility.

1.7 BAND RATIOING AND VEGETATION INDICES

Another useful image processing technique exploits the relationships among the brightness values (BVs) of different bands, of image features. Mathematical expressions are applied to image bands in order to extract thematic information. These expressions may be a *simple ratio*, or complex equations, and are generally developed to target a specific feature of interest. Many such algorithms have been developed to highlight characteristics of land cover such as vegetation, soil, water, and urban areas, and the information extracted can be applied to a wide range of analyses (see Table 1.4 for examples of ratios and vegetation indices).

TABLE 1.3
Filtering techniques and their utility in analyzing remotely sensed data.

Filtering Technique	What it does	Filter Examples
High Frequency	– *Allows high frequency information to pass through* – *Suppresses low frequency information* – *Edges are sharp and small features stand out* – *Large features look suppressed*	– *Enhancing structural details* – *Bring out boundaries and edges*
Low Frequency	– *Allows low frequency information to pass through* – *Suppresses high frequency information* – *Edges get subdued* – *Larger features are enhanced* – *Smaller features begin to get smoothed*	– *Highlight larger features* – *Bring out information in larger features*
Edge Enhancement	– *Detect edges/boundaries between features*	– *Aid in automated feature extraction* – *Useful for geologic information, urban areas, boundaries, etc.*
Frequency Domain	– *Converts data from spatial to frequency domain*	– *Enhancement, compression* – *Noise removal, image restoration* – *Textural classification, quality assessment*

1.7.1 Band Ratio

The BVs of specific targets of interest vary from image to image depending on environmental factors including topography, slope of target surface, aspect ratio, solar angle, seasonal changes, atmospheric conditions, water content, substrate conditions, or shadowing. Such factors may significantly increase or decrease BV relative to what would be expected in laboratory conditions. This may make complex image analysis functions such as classification, feature discrimination, and change detection, difficult to perform. However, certain ratio transformations applied to two or more spectral bands can minimize such effects. In addition, these ratios may generate unique information, not otherwise attainable, through visual image analysis techniques. The mathematical expression of the band ratio is:

$$\boldsymbol{BR}, \boldsymbol{P_x} = \frac{\mathrm{BV}, P_x B_x}{\mathrm{BV}, P_x B_y}$$

where BR, P_x is the output value for a *pixel* (P_x) using the BV of two bands: *band x* (B_x) and *band y* (B_y). One obvious problem becomes clear that BR, $P_x = 0$ is a possible outcome. There are several methods to address this however, including assigning a value of 1 to any BV with a value of 0 or adding a small value to the denominator if it equals zero (such as 0.1).

While band ratios provide a new series of brightness values to evaluate, it is not always easy to determine which bands should be used to provide information on specific targets of interest. Sometimes, the decision process may be as simple as the analyst displaying the results of multiple ratios and choosing the resulting dataset that appears most visually appealing or informative. However, there are widely used techniques for determining optimum bands for ratioing, such as the Optimum Index Factor (OIF) and the Sheffield Index (Chavez et al., 1984; Sheffield, 1985).

TABLE 1.4
Examples of some of the common ratio and vegetation indices.

Index	Formula	Source
Simple Ratio	$SR = \frac{\rho_{red}}{\rho_{nir}}$	*Birth and McVey, 1968*
Normalized Difference Vegetation Index	$NDVI = \frac{(\rho_{nir} - \rho_{red})}{(\rho_{nir} + \rho_{red})}$	*Rouse et al., 1974*
Soil-adjusted Vegetation Index	$SAVI = \frac{(\rho_{nir} - \rho_{red})}{(\rho_{nir} + \rho_{red} + a) \times (1 + a)}$	*Huete, 1988*
Green Normalized Difference Vegetation Index	$GNDVI = \frac{(\rho_{nir} - \rho_{green})}{(\rho_{nir} + \rho_{green})}$	*Buschmann & Nagel, 1993*
Green Atmospherically Resistant Vegetation Index	$GARI = \frac{\rho_{nir} - [\rho_{green} - (\rho_{blue} - \rho_{red})]}{\rho_{nir} - [\rho_{green} + (\rho_{blue} - \rho_{red})]}$	*Gitelson et al., 1996*
Enhanced Vegetation Index	$EVI = 2.5 \times \frac{(\rho_{nir} - \rho_{red})}{(\rho_{nir} + 6(\rho_{red}) - 7.5(\rho_{blue}) + 1)}$	*Huete et al., 1996*
Visible Atmospherically Resistant Index	$VARI = \frac{(\rho_{green} - \rho_{red})}{(\rho_{green} + \rho_{red} - \rho_{blue})}$	*Gitelson et al., 2002*
Wide Dynamic Range Vegetation Index	$WDRVI = \frac{a \times (\rho_{nir} - \rho_{red})}{a \times (\rho_{nir} + \rho_{red})}$	*Gitelson, 2004*
Three-band Model	$TbM = [\rho(\lambda_1)^{-1} - \rho(\lambda_2)^{-1}] \times \rho(\lambda_3)$	*Gitelson et al., 2006*
Enhanced Vegetation Index 2	$EVI2 = 2.5 \times \frac{(\rho_{nir} - \rho_{red})}{(\rho_{nir} + 2.4(\rho_{red}) + 1)}$	*Jiang et al., 2008*

1.7.2 Vegetation Index (VI)

Vegetation is a critical component of the health and condition of the Earth's natural environment. The U.S. Environmental Protection Agency (EPA) cites vegetation characteristics and biophysical variables (such as biomass and percentage of cover) and key indicators of ecosystem health (EPA/600/s-05, 2010). Because of this, vegetation studies have been a popular subject of remote sensing research since the 1960s. Scientists have modeled biophysical characteristics of vegetation using digital imagery since the data became available and continue to do so today (Burger and Burge, 2022; Elachi and van Zyl, 2021; Sabins and Ellis, 2020; Liang and Wang, 2019; Tan and Jiang, 2018; Dong and Chen, 2017; Girard, 2017; Shih, 2017; Woodhouse, 2017; Lillesand et al., 2015; Campbell and Wynne, 2011; Hardisky et al., 1986; Gross and Klemas 1988; Zhang et al., 1997; Spanglet et al., 1998; Zhang et al., 2009; Mishra et al., 2012). Many of these studies involve the use of *vegetation indices*, complex mathematical equations applied to image bands to measure the relative "greenness" of image features, from which meaningful information may be extracted of the composition and characteristics of vegetation. Such *vegetation biophysical variables* (VBVs) may include, but are not limited to; aboveground green biomass (AGB), absorbed photosynthetic active radiation, concentration of chlorophyll (or other leaf pigment), leaf area index (LAI), and percent of substrate covered by vegetation (vegetation fraction). Often, these biophysical properties can be key indicators of vegetation health, ecosystem health, and other critical ecological factors.

In order to maximize the ability to extract meaningful information, a VI should have four characteristics (Burger and Burge, 2022; Elachi and van Zyl, 2021; Chuvieco, 2020; Sabins and Ellis, 2020; Liang and Wang, 2019; Nixon and Aguado, 2019; Lillesand et al., 2015; Campbell and Wynne, 2011; Running et al., 1994; Gitelson et al., 2006). *First*, the VI must be sensitive to VBVs of interest. It is helpful if the sensitivity demonstrates a predictable relationship between index and VBV. Preferably, a linear relationship which is applicable across a wide range of vegetation conditions, substrates, and species-types. *Second*, the impact of external variables such as atmospheric interaction, solar angle, and viewing angle must be minimized. This is necessary to compare multiple datasets with consistent spatial and temporal conditions. *Third*, the impact of internal variables such as canopy architecture, substrate, phonological changes, and nonphotosynthetic canopy components should be minimized. Such internal variables can contribute substantially to the recorded spectral response of vegetation in digital imagery and may course spatial resolution may contribute to poor results when comparing multiple image datasets. *Fourth*, accuracy assessment must be tied to a specific measureable VBV such as AGB, LAI, or vegetation fraction.

Since the 1960s several vegetation indices have been developed, though some of these provide redundant information content (Jensen, 2005; Flightriot.com, 2014). Subtle differences in algorithms are often adopted due to variability in sensor specifications and/or target characteristics. Vegetation however, is a spectrally unique surface feature due to the chlorophyll absorption and leaf reflectance characteristics in the visible through near-infrared (VNIR) regions of the spectrum. Many of the VIs developed early, target the inverse relationship between red and near-infrared reflectance. However, algorithms have been successfully developed using other characteristics of vegetation reflectance in a variety of environments and/or sensor specifications (Viña et al., 2011).

1.7.3 Simple Ratio (SR)

Birth and McVey (1968) described the *Simple Ratio (SR)*, one of the first documented VIs that provides a simple formula for measuring the ratio of red reflectance (ρ_{red} in % or dimensionless) to near-infrared reflectance (ρ_{nir}).

$$SR = \frac{\rho_{red}}{\rho_{nir}} \tag{2.2}$$

Green vegetation strongly reflects incident irradiation in the NIR region (40–60%) while absorbing up to 97% in the red region (Gitelson, 2004). As vegetation *greenness* declines, red reflectance increases and NIR reflectance decreases. By computing the ratio of red to NIR, this relationship can be quantified.

1.7.4 Normalized Difference Vegetation Index (NDVI)

One of the most noticeable problems with the simple ratio is that it is not normalized, making it difficult to compare results among different studies. Rouse et al. (1974) addressed this issue with the *Normalized Difference Vegetation Index (NDVI)*. NDVI is functionally equivalent to SR, and comparison plots reveal no scatter between SR and NDVI.

$$NDVI = \frac{(\rho_{nir} - \rho_{red})}{(\rho_{nir} + \rho_{red})} \tag{2.3}$$

NDVI is widely applied to spectral and image data for monitoring, analyzing, and mapping VBVs. There are several characteristics of NDVI that contribute to its utility and continuing popularity among vegetation experts:

- Seasonal and phonological changes in vegetation can be monitored.
- Normalized data makes comparisons more reliable.
- Ratioing reduces some cases of multiplicative noise cause by differences in solar angle, shadows, and topographic variations.

Conversely, a major disadvantage to NDVI is the nonlinear nature of the relationship between NDVI values and many VBVs. The index becomes saturated at high levels and as VBVs increase, NDVI shows little variation.

1.7.5 Enhanced Vegetation Index (EVI) 1 and 2

Several VIs are tailored to specific sensors and may be *tuned* to maximize the results of analysis at specific resolution characteristics. An example of this is the Enhanced Vegetation Index (EVI) which was developed by Huete et al. (2002) specifically for application to MODIS data. EVI is similar to NDVI, however, includes several coefficients in the equation to account for atmospheric scattering and reduce the saturation effects of NDVI at high values.

$$EVI = 2.5 \times \frac{(\rho_{nir} - \rho_{red})}{(\rho_{nir} + 6(\rho_{red}) - 7.5(\rho_{blue}) + 1)} \tag{2.4}$$

An updated version of the index, Enhanced Vegetation Index 2 (EVI2), was developed later by Jiang et al. (2008) for use with datasets that did not have sensitivity in the blue region of the spectrum.

$$EVI2 = 2.5 \times \frac{(\rho_{nir} - \rho_{red})}{(\rho_{nir} + 2.4(\rho_{red}) + 1)} \tag{2.5}$$

1.7.6 Wide Dynamic Range Vegetation Index (WDRVI)

Gitelson et al. (2003) proposed a simple adjustment to NDVI to compensate for the high-end saturation. The *Wide Dynamic Range Vegetation Index* (WDRVI) applies a weighted coefficient (*a*) to NDVI with a value of 0.1–0.2 to *linearize* the index relationship to VBVs.

$$WNDVI = \frac{a \times (\rho_{nir} - \rho_{red})}{a \times (\rho_{nir} + \rho_{red})} \tag{2.6}$$

1.7.7 Three-Band Model (TBM)

Most VIs focus on chlorophyll absorption characteristics, Gitelson et al. (2006) proposed an index that may be optimizable for other pigments, and potentially other features of interest. This *Three-Band Model (TBM)* requires the use of three spectral bands that must be identified as:

- Band 1 (λ_1): The band that is most sensitive to changes in VBV.
- Band 2 (λ_2): The band that is the most insensitive to changes in VBV.
- Band 3 (λ_3): The band that accounts for backscattering/noise among samples.

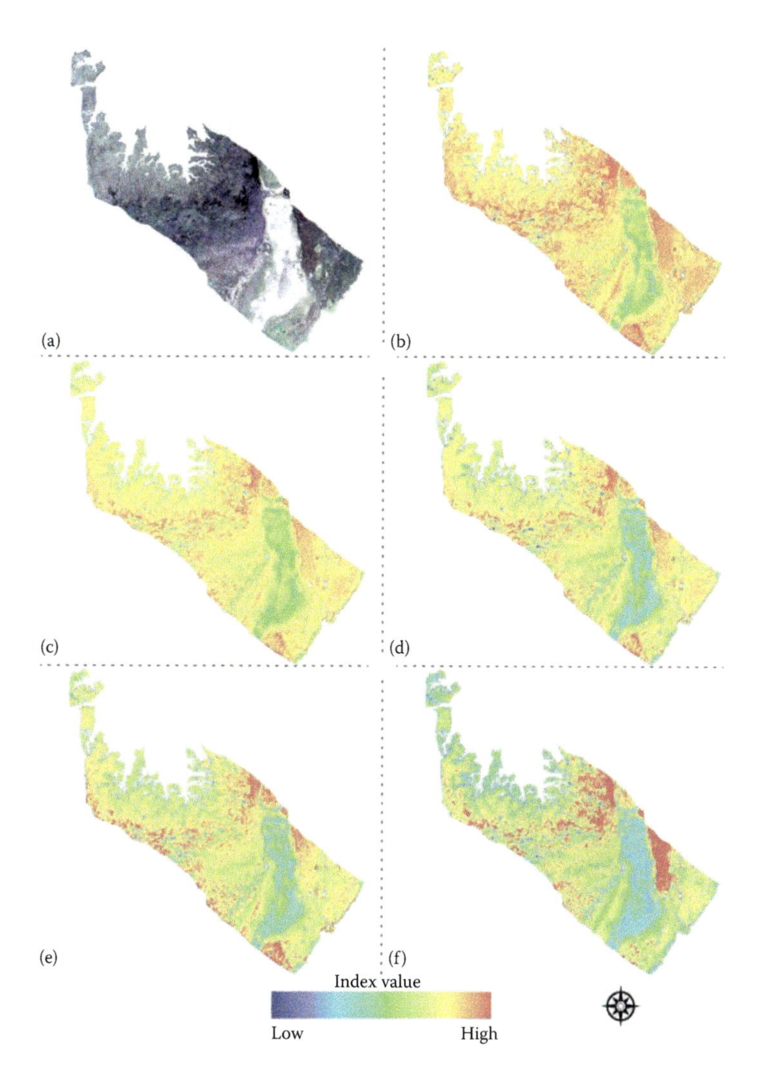

FIGURE 1.14 A suite of selected vegetation indices applied to an Airborne Imaging Spectroradiometer for Applications Eagle acquired over Grand Bay, Mississippi. (Image courtesy of the Center for Advanced Land Management Information Technologies (CALMIT), University of Nebraska, Lincoln.)

$$TbM = \left[\rho(\lambda_1)^{-1} - \rho(\lambda_2)^{-1}\right] \times \rho(\lambda_3) \tag{2.7}$$

TbM generates a linear relationship and, by accounting for backscatter/noise, normalizes the relationship between VBV and index value. Additionally, TbM has the potential to be applied to other features of interest, including nonvegetation surface-types.

1.8 IMAGE CLASSIFICATION

With the advent of digital image data and computer technology, algorithms were developed to enable the extraction of land-use/land-cover and biophysical information directly from the images (Jensen et al., 2009). Over the past five decades an numerous algorithms have been developed to aid analysts in their image interpretation processes. Broadly, these algorithms are based on parametric statistics (assuming normally distributed data), non-parametric statistics (does not require normally distributed data) and non-metric (capable of operating on real-valued data and nominal scaled data) (Burger and Burge, 2022; Richards, 2022; Elachi and van Zyl, 2021; Amani et al., 2020; Chuvieco, 2020; Sabins and Ellis, 2020; Liang and Wang, 2019; Nixon and Aguado, 2019; Shao et al., 2019; Sun et al., 2019; Li et al., 2018; Tan and Jiang, 2018; Zhong et al., 2018; Dong and Chen, 2017; Girard, 2017; Shih, 2017; Woodhouse, 2017; Lillesand et al., 2015; Campbell and Wynne, 2011; Jensen et al., 2009). Jensen (2005), and Lu and Weng (2007) provide an extensive taxonomic overview of several image classification methods, and this section will broadly review some of the more common methods used for thematic information extraction.

In general, image classification can be accomplished using either supervised or unsupervised techniques. Within each of these broad categories exist a host of different methods that can be applied toward specific imagery (e.g., high spatial resolution, hyperspectral, etc.) and/or have been

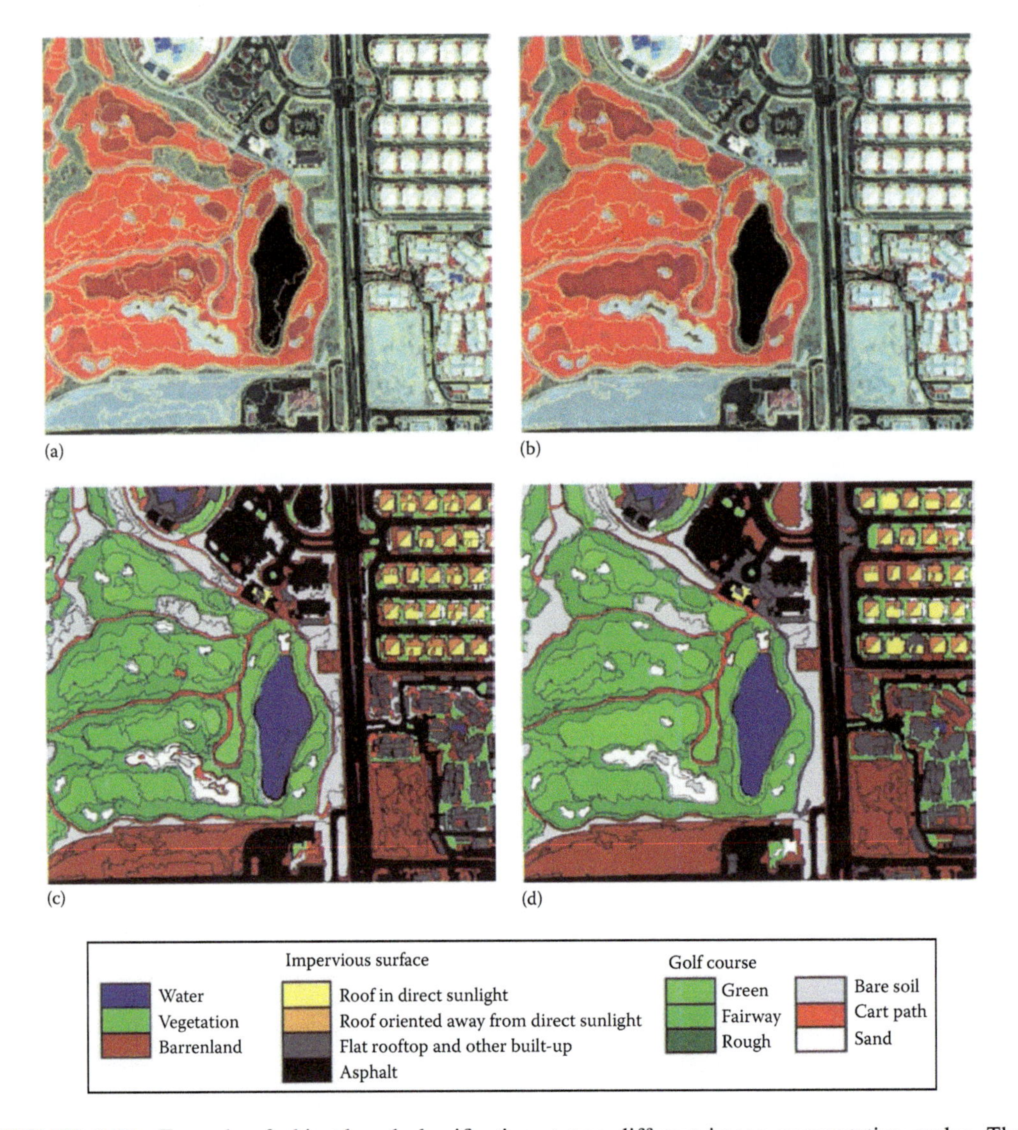

FIGURE 1.15 Example of object-based classification at two different image segmentation scales. The PAN-sharpened QuickBird high spatial resolution (61 × 61 cm) multispectral imagery of Las Vegas, NV was acquired on May 18, 2003. (Source: Jensen et al., 2009.)

developed as technological capabilities have advanced (e.g., Artificial Neural Network classification, Object-oriented classification, etc.) Supervised classification categorizes every image pixel into one of several predefined land-type classes (Jensen et al., 2009). The process requires several steps and includes selecting the land-type categories, pre-processing, defining training data, automated pixel assignment, and accuracy assessment (see Sabins, 1987; Jensen, 2005; Jensen et al., 2009.) Various supervised classification algorithms exist including the maximum-likelihood classifier (Burger and Burge, 2022; Lillesand et al., 2015; Campbell and Wynne, 2011; Strahler, 1980; Foody et al., 1992), nearest-neighbor classification (James, 1985; Hardin, 1994), decision tree classifiers (DeFries and Chan, 2000; Russell and Norvig, 2003; Jensen, 2005), object-oriented classification (Haralick and Shapiro, 1985; Yan et al., 2006) (Figure 1.15), artificial neural networks (ANN) (Elachi and van Zyl, 2021; Chuvieco, 2020; Sabins and Ellis, 2020; Lillesand et al., 2015; Campbell and Wynne, 2011; Gopal and Woodcock, 1996; Hardin, 2000; Jensen et al., 2000).

Unsupervised classification methods generally partition the spectral data of an image into feature space with a minimal input from the analyst. Operating under various constraints specified by the user (e.g., number of clusters, spectral and spatial search radius, bands used, iterations defined, etc.) an unsupervised classification algorithm will search for natural groupings (or clusters) and produce a map of the number of pre-defined clusters. These clusters can subsequently be assigned to previously defined information classes (e.g., land-cover categories) through an iterative process (Figure 1.16). Because there is a high likelihood that there will be clusters of "mixed" categories the user may have to employ other methods to parse such clusters and minimize the undefined areas in an image. For example, "mixed" clusters from the original image may be masked out of the image prior to running an iteration of the unsupervised algorithm using a different combination of inputs (e.g., the number of bands included). The user may also apply supervised classification algorithms to glean information from the "mixed" clusters. Jensen (2005) describes the two common clustering methods including the chain method and the Iterative Self-Organizing Data Analysis (ISODATA). The building of clusters by either of these algorithms does not require any a priori knowledge. However, once the analyst works interactively with the clusters to assign them into the various land-cover categories, considerable familiarity with the study area is needed.

1.9 FUTURE TRENDS

Sensor systems and image processing software/hardware will continue to evolve and improve. Their utility has already been proven by the vast array of applications that image data and its subsequent information extraction have been used in. With the problems of the 21st century being focused on issues such climate change, atmospheric conditions, environmental degradation, natural hazards, population growth, urbanization, resources scarcity and many others, it is inevitable that remote sensing data will aid toward a better understanding of these problems. Future sensors will continue the progression toward more comprehensive and accurate Earth science measurements (Hartley, 2003). Increased spatial, spectral, radiometric and temporal resolutions will provide scientists new levels of detail and new algorithms will be developed to extract the relevant information with improved accuracies (Burger and Burge, 2022; Richards, 2022; Elachi and van Zyl, 2021; Amani et al., 2020; Chuvieco, 2020; Sabins and Ellis, 2020; Liang and Wang, 2019; Nixon and Aguado, 2019; Shao et al., 2019; Sun et al., 2019; Li et al., 2018; Tan and Jiang, 2018; Zhong et al., 2018; Dong and Chen, 2017; Girard, 2017; Shih, 2017; Woodhouse, 2017; Lillesand et al., 2015; Campbell and Wynne, 2011). Furthermore, technological revolutions in the minaturization of electronics, stabilization of optical systems, efficient power sources, minimizing size and weight of senor systems, etc., will lead to changes in the design of the sensors, uninterrupted data collection, very high data quality, and orbital stability of systems.

The fusion of close range and in situ remote sensor data with satellite/airborne imagery, and other geospatially derived information have already transformed analytical capabilities by facilitating multi-scale studies of phenomena. With massive biogeophysical data volumes being generated,

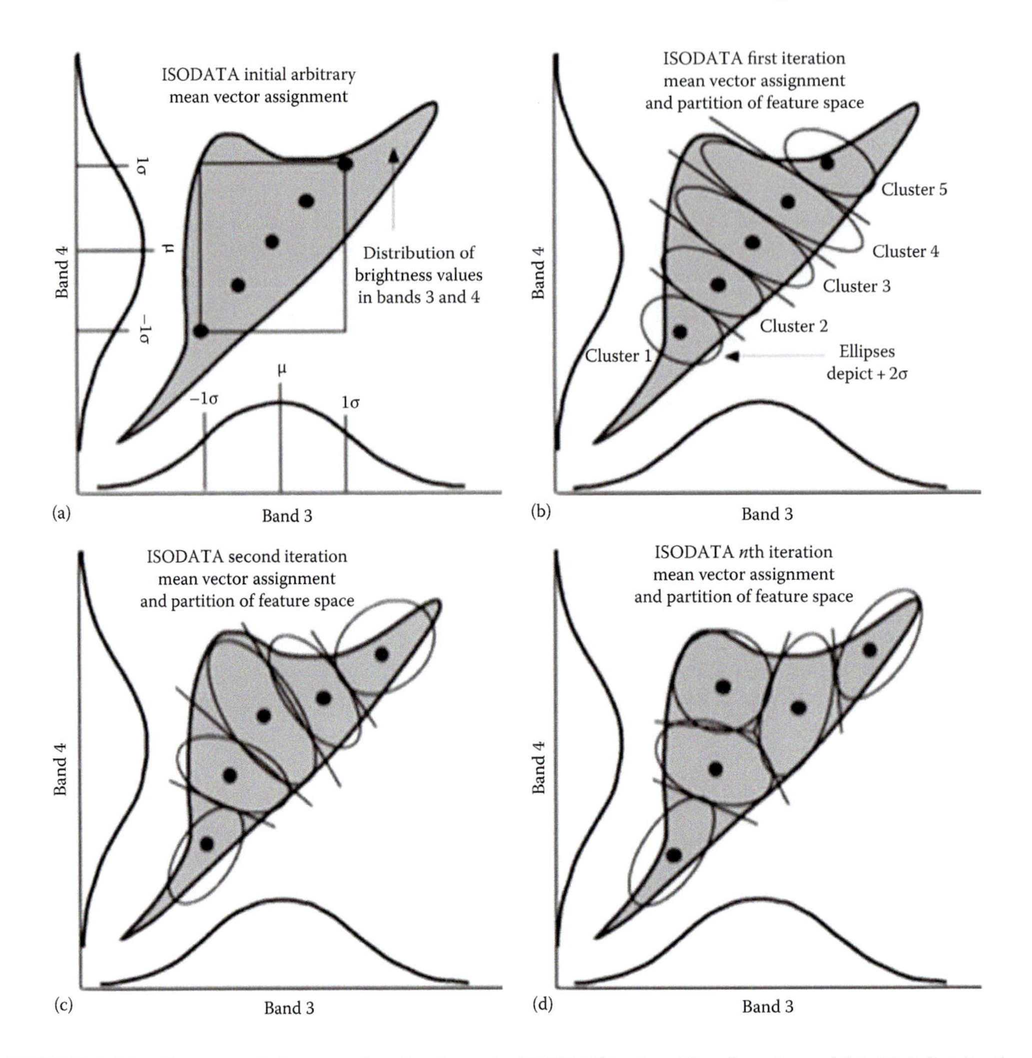

FIGURE 1.16 Conceptual diagram showing how the ISODATA algorithm functions: (a) initial distribution of five hypothetical mean vectors using +1σ in both bands as the beginning and ending points; (b) in the first iteration, each candidate pixel is compared to each cluster mean and assigned to the cluster whose mean is closest in Euclidean distance; (c) during the second iteration, a new mean is calculated for each cluster based on the actual spectral locations of the pixels assigned to each cluster, instead of the initial arbitrary calculation. This involves analysis of several parameters to merge and/or split clusters. After the mean vectors are selected, every pixel in the scene is assigned to one of the new clusters; (d) this split/merge–assign process continues until there is little change in class assignment between iterations (the T threshold is reached) or the maximum number of iterations is reached (M). (Source: Jensen, 2005.)

and the computational challenges for their analyses and visualization, image processing is a major player in the big data arena where new technologies are being sought to process large quantities of data within tolerable timeframes. In addition, there is an urgent need for realistic and computationally simple surface radiation models for inversion of land surface variables from satellite data (Jensen, 2005). Liang (2003) and Liang (2007) have introduced several physically based inversion algorithms and for estimating biogeophysical variables (e.g., LAI, fractional vegetation coverage, broadband albedo, etc.) and further research continues for the derivation of a suite of algorithms that will enable a simple but accurate determination of land surface variables.

REFERENCES

Amani, A., et al. 2020. Google earth engine cloud computing platform for remote sensing big data applications: A comprehensive review. *IEEE J Sel Top Appl Earth Obs Remote Sens*, 13:5326–5350. http://doi.org/10.1109/JSTARS.2020.3021052.

Aronoff, S. 2005. *Remote sensing for GIS managers*. ESRI Press.

Birth, G. S., and G. R. McVey. 1968. Measuring color of growing turf with a reflectance spectrophotometer. *Agro J*, 60:640–649.

Burger, W., and M. J. Burge. 2022. *Digital image processing: An algorithmic introduction* (3rd ed.). Springer Nature Switzerland, p. 653. ISBN 978-3-031-05743-4, ISBN 978-031-05744-1 (eBook). https://doi.org/10.1007/978-3-031-05744-1

Campbell, J. B., and R. H. Wynne. 2011. *Introduction to remote sensing* (5th ed.). The Gilford Press, p. 683. ISBN 978-1-60918-176-5 (hardcover).

Chavez, P. C., Guptill, S. C., and J. A. Bowell. 1984. Image processing techniques for Thematic Mapper Data. *Proc Am Soc of Photogramm*, 2:728–743.

Chen, L., and L. Lee. 1992. Progressive generation of control frameworks for image registration. *Photogramm Eng Rem Sens*, 58(9):1321–1328.

Chuvieco, E. 2020. *Fundamentals of remote sensing: An environmental approach* (3rd ed.), p. 432. eBook ISBM: 9780429506482. https://doi.org/10.1201/9780429506482.

Civco, D. L. 1989. Topographic normalization of Landsat Thematic Mapper digital imagery. *Photogramm Eng Rem Sen*, 55(9):1303–1309.

DeFries, R. S., and J. C. Chan. 2000. Multiple criteria for evaluating machine learning algorithms for land cover classification. *Rem Sens Environ*, 74:503–515.

Dong, P., and Q. Chen. 2017. *LiDAR remote sensing and applications*. CRC Press, Taylor and Frnacis, Inc., p. 199. ISBN: 978-1-4822-4301-7 (Hard cover).

Du, Y., Tiellet, P. M., and J. Cihlar. 2002. Radiometric normalization of multitemporal high-resolution satellite images with quality control for land cover change detection. *Rem Sens Environ*, 82:123.

Eckhardt, D. W., Verdin, J. P., and G. R. Lyford. 1990. Automated update of an irrigated lands GIS using SPOT HRV imagery. *Photogram Eng Rem Sens*, 56(11):1515–1522.

Elachi, C., and J. J. van Zyl. 2021. *Introduction to the physics and techniques of remote sensing*. The John Wiley and Sons, p. 542. ISBN 978-111-952-3086.

Exelis. 2003. *Atmospheric correction module: QUAC and FLAASH user's guide*. https://www.exelisvis.com/portals/0/pdfs/envi/Flaash_Module.pdf (accessed June 8, 2014).

Foody, G. M., Campbell, N. A., Trodd, N. M., and T. F. Wood. 1992. Derivation and applications of probabilistic measures of class membership form the maximum-likelihood classification. *Photogramm Eng Rem Sens*, 58(9):1335–1341.

Fureder, R. 2010. Topographic correction of satellite images for improved LULC classification in alpine areas. *10th International Symposium on High Mountain Remote Sensing Cartography*, Grazer Schriften der Geographie und Raumforschung, Academia.edu. pp. 187–194.

Girard, C. 2017. *Processing of remote sensing data*. Routledge. eBook ISBN: 9780203741917. https://doi.org/10.1201/9780203741917

Gitelson, A. A. 2004. Wide Dynamic Range Vegetation Index for remote quantification of biophysical characteristics of vegetation. *J Plant Phys*, 131:165–173.

Gitelson, A. A., Gritz, Y., and Merzlyak, M. N. 2003. Relationships between leaf chlorophyll content and spectral reflectance and algorithms for non-destructive chlorophyll assessment in higher plant leaves. *J Plant Physiol*, 160(3):271–282. ISSN 0176-1617. https://doi.org/10.1078/0176-1617-00887

Gitelson, A. A., Keydan, G. P., and M. N. Merzlyak. 2006. Three-band model for noninvasive estimation of chlorophyll, carotenoids, and anthocyanin contents in higher plant leaves. *Geophy Res Let*, 33:L11402.

Gonzalez, L. 2014. *Principal component analysis*. http://wiki.landscapetoolbox.org/doku.php/remote_sensing_methods:principal_components_analysis (accessed June 23, 2014).

Gopal, S. and C. E. Woodcock. 1996. Remote sensing of forest change using artificial neural networks. *IEEE Trans Geosci Remote Sens*, 34:398–404.

Gross, M. and V. Klemas. 1988. Remote sensing of biomass of salt marsh vegetation in France. *Int J Rem Sens*, 9(3):397–408.

Hall, F. G., Strebel, D. E., Nickeson, J. E., and S. J. Goetz. 1991. Radiometric rectification: Toward a common radiometric response among multidate, multisenor images. *Rem Sens Env*, 35:11–27.

Haralick, R. M., and L. G. Shapiro. 1985. Image segmentation techniques. *Comp Vis Grap Image Proc*, 29:100–132.

Hardin, P. J. 1994. Parametric and nearest-neighbor methods for hybrid classification. *Photogramm Eng Rem Sens*, 60:1439–1448.

Hardin, P. J. 2000. Neural networks versus nonparametric neighbor-based classifiers for semisupervised classification of Landsat Thematic Mapper imagery. *Opt Eng*, 39:1898–1908.

Hardisky, A., Gross, M., and V. Klemas. 1986. Remote sensing of coastal wetlands. *Biosci*, 37(7):453–460.

Hartley, J. 2003. Earth remote sensing technologies in the twenty-first century. *Proc 2003 IGAARS Symposium,* Volume 1 (IEEE Cat. No.03CH37477), Toulouse, France, 2003, pp. 627–629. https://doi.org/10.1109/IGARSS.2003.1293863

Huete, A., Didan, K., Miura, T., Rodriquez, P., Gao, X., and L. Ferreira. 2002. Overview of the radiometric and biophysical performance of the MODIS vegetation indices. *Rem Sens Env*, 83:195–213.

ImSpec, LLC. 2002. *ACORN 4.0 user's guide*. http://www.aigllc.com/pdf/acorn4_ume.pdf (accessed June 8, 2014).

James, M. 1985. *Classification algorithms*. Wiley-Interscience.

Jensen, J. R. 2005. *Introductory digital image processing: A remote sensing perspective*. Prentice Hall.

Jensen, J. R., Fang, Q., and J. Minhe. 2000. Predictive modeling of coniferous forest age using statistical and neural network approaches applied to remote sensor data. *Int J Rem Sens*, 20:2805–2822.

Jensen, J. R., Im, J., Hardin, P., and R. R. Jensen. 2009. Image classification. In *The SAGE handbook of remote sensing*. Eds. Warner, T. A., Nellis, M. D., Foody, G. M. Sage, 504 pp. ISBN: 978-1412936163 (hardback), pp. 269–281.

Jensen, J. R., Rutchey, K., Koch, M., and S. Narumalani. 1995. Inland wetland change detection in the Everglades Water Conservation Area 2A using a time series of normalized remotely sensed data. *Photogramm Eng Rem Sens*, 61(2):199–209.

Jiang, Z., Juete, A., Didan, R., and T. Miura. 2008. Development of a two-band Enhanced Vegetation Index without a blue band. *Rem Sens Env*, 112(10):3833–3845.

Kawata, Y., Ueno, S., and T. Kusaka. 1988. Radiometric correction for atmospheric and topographic effects on Landsat MSS images. *Int J Rem Sen*, 9(4):729–748.

Li, Y., Zhang, H., Xue, X., Jiang, Y., and S. Qiang. 2018. Deep learning for remote sensing image classification: A survey. *WIREs Data Min Knowl Discov*. 8(6):e1264. https://doi.org/10.1002/widm.1264

Liang, S. 2003. *Quantitative remote sensing of land surfaces*. Wiley.

Liang, S. 2007. Recent developments in estimating land surface biogeophysical variables from optical remote sensing. *Prog Phy Geog*, 31(5):501–516.

Liang, S., and J. Wang. 2019. *Advanced remote sensing: Terrestrial information extraction and applications*. Academic Press, p. 962, imprint of Elsevier. ISBN: 978-0-12-815826-5.

Lillesand, T., Kiefer, R. W., and J. Chipman. 2015. *Remote sensing and image interpretation* (7th ed.). Wiley, p. 531. ISBN 978-1-118-34328-9.

Lu, D., and Q. Weng. 2007. A survey of image classification methods and techniques for improving classification performance. *Int J of Rem Sens*, 28(5):823–870.

Mishra, D., Cho, J., Ghosh, S., Fox, A., Downs, C., Merani, P., Kirui, P., Jackson, N., and S. Mishra. 2012. Post-spill state of the marsh: Impact of the Gulf of Mexico oil spill on the health and productivity of Louisiana salt marshes. *Rem Sens Env*, 118:176–185.

Nixon, M., and A. Aguado. 2019. *Feature extraction and image processing for computer vision* (4th ed.). Academic Press, p. 626. An imprint of Elsevier. ISBN: 978-0-12-814976-8.

Richards, J. A. 2022. *Remote sensing digital image analysis* (6th ed.). Springer, p. 567. Hardcover ISBN: 978-3-030-82326-9. eBook ISBN: 978-3-030-82327-6. https://doi.org/10.1007/978-3-030-82327-6

Rouse, J., Haas, R., Schell, J., and D. Deering. 1974. Monitoring vegetation systems in the Great Plains with ERTS. *3rd Earth Resources Technology Satellite-1 Symposium*, Volume 1. Technical Presentations, NASA SP351, NASA, Washington, DC.

Running, S. W., Justice, C. O., Solomonson, V., Hall, D., Barker, J., Kaufmann, Y. J., Strahler, A. H., Huete, A. R., Muller, J. P., Vanderbilt, V., Wan, Z. M., Teillet, P., and D. Carneggie. 1994. Terrestrial remote sensing science and algorithms planned for EOS/MODIS. *Intl J Rem Sens*, 15(17):3587–3620.

Russell, S. J., and P. Norwig. 2003. *Artificial intelligence: A modern approach*. Prentice-Hall, 1080p.

Sabins, F. F., Jr. 1987. *Remote sensing: Principles and interpretation*. W. H. Freeman.

Sabins, F. F., Jr., and J. M. Ellis. 2020. *Remote sensing: Principles, interpretations, and applications* (4th ed.). Waveland Press Inc., p. 523. ISBN: 978-1-4786-3710-3.

Schowengerdt, R. A. 1997. *Remote sensing: Models and methods for image processing*. Academic Press.

Shao, Z., Pan, Y., Diao, C., and J. Cai. 2019. Cloud detection in remote sensing images based on multiscale features-convolutional neural network, *IEEE Trans Geosci Remote Sens*, 57(6):4062–4076, June. http://doi.org/10.1109/TGRS.2018.2889677.

Sheffield, C. 1985. Selecting band combinations from multispectral data. *Photogramm Eng Rem Sens*, 51(6):681–687.

Shih, F. Y. 2017. *Image processing and mathematical morphology: Fundamentals and applications*. CRC Press, p. 439. ISBN: 978-131-521-8557. https://doi.org/10.1201/9781420089448

Spanglet, H., Ustin, S., and E. Rejmankova. 1998. Spectral reflectance characteristics of California subalpine marsh plant communities. *Wetlands*, 18(3):307–319.

Strahler, A. 1980. The use of prior probabilities in maximum likelihood classification of remotely sensed data. *Rem Sens Env*, 10(2):135–183.

Sun, J., et al. 2019. An efficient and scalable framework for processing remotely sensed big data in cloud computing environments. *IEEE Trans Geosci Remote Sens*, 57(7):4294–4308, July. https://doi.org/10.1109/TGRS.2018.2890513.

Tan, L., and J. Jiang. 2018. *Digital signal processing: Fundamentals and applications*. Academic Press, p. 903, an imprint of Elsevier. ISBN: 978-0-12-815071-9.

Tanre, D., Deroo, C., Duhaut, P., Herman, M., Morcrette, J., Perbos, J., and P. Y. Deschamps. 1986. *Second simulation of the satellite signal in the solar spectrum (6S) user's guide*. U.S.T. de Lille, 59655 Villeneuve d'Ascq. Lab d'Optique Atmospherique.

USGS. 2007. *The power of spatial filters*. http://astrogeology.usgs.gov (accessed June 20, 2014).

van der Meer, F., van der Werff, H. M. A., and S. M. de Jong. 2009. Pre-processing of optical imagery. In *The SAGE handbook of remote sensing*, Eds. Warner, T. A., Nellis, M. D., Foody, G. M. Sage, 504 pp. ISBN: 978-1412936163 (hardback), pp. 229–243.

Viña, A., Gitelson, A., Nguy-Robertson, A., and Y. Peng. 2011. Comparison of different vegetation indices for the remote estimation of green leaf area index of crops. *Rem Sens Env*, 115:3468–3478.

Woodhouse, I. H. 2017. *Introduction to microwave remote sensing*, p. 400. https://doi.org/10.1201/9781315272573.

Yan, G., Mas, J.-F., Maathuis, B. H. P., Xiangmin, Z., and P. M. Van Dijk. 2006. Comparison of pixel-based and object-oriented image classification approaches—a case study in a coal fire area, Wuda, Inner Mongolia, China. *Int J Rem Sens*, 27(18):4039–4055.

Zhang, H., Hu, H., Yao, X., Zheng, K., and Y. Gan. 2009. Estimation of above-ground biomass using HJ-1 hyperspectral images in Hangzhou Bay, China. *Int Conf Info Eng Comp Sci*, 2009.

Zhang, M., Ustin, S., Rejmankova, E., and E. Sanderson. 1997. Monitoring Pacific coast salt marshes using remote sensing. *Eco App*, 7(3):1039–1053.

Zhong, Y., Ma, A., Soon Ong, Y., Zhu, Z., and L. Zhang. 2018. Computational intelligence in optical remote sensing image processing. *Appl Soft Comput*, 64(2018):75–93, ISSN 1568-4946, https://doi.org/10.1016/j.asoc.2017.11.045. https://www.sciencedirect.com/science/article/pii/S1568494617307081

2 Image Classification Methods in Land Cover and Land Use

Mutlu Ozdogan

LIST OF ACRONYMS

ANN	Artificial neural networks
ART	Adaptive resonance theory
AVHRR	Advanced very high resolution radiometer
C-SVC	C-support vector classification
CT	Classification trees
FAO	Food and Agricultural Organization
FROM-GLC	Finer resolution observation and monitoring global land cover
GIS	Geographic information system
KMC	K-means clustering
LACIE	Large area crop inventory experiment
LULC	Land use/land cover
MD	Minimum distance
MLC	Maximum likelihood
MLP	Multilayer perceptron
NDVI	Normalized Difference Vegetation Index
NIR	Near infrared
NLCD	National land cover dataset
OBIA	Object-based image analyses
SVM	Support vector machines
SVR	Support vector regression
TOA	Top of atmosphere
USGS	U.S. Geological Survey
WELD	Web-enabled landsat data

2.1 INTRODUCTION

Today, information on land cover and land use is almost exclusively derived from remotely sensed observations at various spatial and temporal scales. The advantages of these observations include, but not limited to, synoptic view, availability of spectral bands that help distinguish land surface properties, archived temporal record, and digital nature. The principal form of deriving land cover information from remotely sensed images is classification. In the context of remote sensing, classification refers to the process of translating observations into land cover categories with clearly defined biogeophysical function. For example, a typical land cover map may contain categories like forest, water, agriculture, and so on. These maps are then used in a growing number of environmental applications, from resource management to global change studies. To this end, the purpose of this chapter is to review existing and emerging image classification methods applied to remote sensing (Ebenezer and Manohar, 2024; Papoutsis et al., 2023; Nasiri et al., 2022; Prasad et al., 2022; Phan et al., 2020; Carranza-García et al., 2019; Helber et al., 2019; Jozdani et al., 2019; Zhang et al., 2019;

DOI: 10.1201/9781003541158-3

Costa et al., 2018; Goldblatt et al., 2018; Zhang et al., 2018; Ma et al., 2017; Scott et al., 2017; Talukdar et al., 2020).

The chapter has the following outline. First definitions for land cover and land use as used throughout the chapter are given. Then, advantages and disadvantages of remote sensing with respect to land cover mapping are discussed. In the main body of the document, existing as well as emerging methods for land cover mapping are described. The main conclusions and the future of land cover mapping are provided in the last section.

2.1.1 Definitions

Land cover is defined as describing the physical status of the Earth's land surface. Land use, on the other hand, describes the use of that land cover for a particular purpose. For example, in the case of a forested area, the "forest" land-cover label would be used to describe the fact that the land is occupied by a group of trees that constitute a forest stand without any reference to the use of that forest while the land-use label could include its usage status, for example, industrial forest for wood production. In this context, image classification methods applied to remotely sensed data allow classification of Earth's surface into *only* land-cover categories, all of which then can be interpreted into land-use classes. For the rest of this chapter, these definitions will be used and primarily, the land-cover classification with remotely sensed data will be emphasized.

2.1.2 Advantages and Limitations of Remote Sensing for Mapmaking

In the last four decades, both academic and application oriented studies have clearly established both the advantages and limitations of remote sensing for deriving land cover information (Table 2.1) (Ebenezer and Manohar, 2024; Papoutsis et al., 2023; Nasiri et al., 2022; Prasad et al., 2022; Phan et al., 2020; Carranza-García et al., 2019; Helber et al., 2019; Jozdani et al., 2019; Zhang et al., 2019; Costa et al., 2018; Goldblatt et al., 2018; Zhang et al., 2018; Ma et al., 2017; Scott et al., 2017; Talukdar et al., 2020). In terms of advantages, remote sensing offers a relatively cheap and rapid method of acquiring information over a large geographical area. For example, a single Landsat scene covers over 30,000 square kilometers of land surface, producing data at several spectral bands at fairly high spatial resolutions. When subjected to classification, these data could reveal land cover information at the scale of a standard topo sheet (~1:25,000 scale). Even with the cost of ground verification, this is very economical. Second, remote sensing allows unobtrusive means to acquire

TABLE 2.1
Advantages and Limitations of Remote Sensing for Land Cover Studies

Advantages	Disadvantages
Low cost and rapid	Overemphasis on what it can
Up-to-date	Inflated accuracies
Large area coverage	Not a direct sample need to be calibrated
Synoptic view	Large measurement uncertainty
Regular revisit period	Unintended measurements (e.g., clouds)
Digital nature of data for integration in GIS	Lack of observations in certain regions
Data acquisition over inaccessible areas	Spatial/temporal resolution mismatch
Unobtrusive data collection	Difficult to interpret in some cases
Spectral information in the non-visible	Need specific training to process/interpret
Derive a map as the final result	

observations, elucidating the Earth surface discretely by means of electromagnetic radiation. Third, remotely sensed observations offer a synoptic view, which can be described as observations that give a broad view of the Earth's surface at a particular time. As a result, regional phenomena, which are invisible from the ground are clearly visible such as geological structures or forests instead of the trees. Fourth, remote sensing provides one of few means to obtain data from inaccessible areas. Fifth, presence of archival observations allow rapid assessment of land cover change, or cheap ways to update/construct base maps without the need for detailed land surveys. Finally, the (mostly) digital nature of data make it is to use computers for fast processing and combining results with other datasets within a Geographic Information System (GIS).

While there are clear advantages to using remote sensing for extracting land cover information, remote sensing also has important limitations that need to be specified. Perhaps the biggest limitation of remote sensing is that it is oversold (Jensen, 2006). More specifically, remote sensing is often (incorrectly) thought of having the ability to observe/identify/locate objects with excessive or unwarranted enthusiasm. Second, by definition, remotely sensed observations are not direct, and provide only a manifestation of the object of interest, made possible by electromagnetic radiation. This requires that the observations be translated into information of interest and calibrated against reality. However, this calibration is never exact, a classification error of 10% is considered excellent. Third, different land surfaces may yield very similar observations, making them difficult to separate. On the other hand, a single land cover category may be manifested multiple different ways in observations, leading to incorrect classifications. Fourth, depending on the observing platform, observations may contain noise associated with atmosphere and sensor characteristics. In certain cases, these unwanted information must be accounted for. Finally, the spatial resolution of satellite imagery may be too coarse for detailed mapping and for distinguishing small contrasting areas. Rule of thumb: a land use must occupy at least 16 pixels (picture elements, cells) to be reliably identified by automatic methods. However, new satellites are being proposed with 1m resolution, these will have high data volume but will be suitable for land cover mapping at a detailed scale.

Despite these disadvantages remote sensing continues to be the primary data source for deriving land cover and land use information worldwide, from local to global studies. The land cover and land use maps generated from remotely sensed observations also span a wide varieties of themes from crop type identification to forest composition mapping to wetland detection, just to name a few. As our population grows and more pressure is placed on the natural resources, remote sensing will continue to play a significant role to monitor these resources. Both the existing tools and those to be developed in the near future will allow us to achieve this goal more accurately and efficiently (Prasad et al., 2022; Phan et al., 2020; Carranza-García et al., 2019; Costa et al., 2018; Ma et al., 2017; Talukdar et al., 2020).

2.2 IMAGE CLASSIFICATION IN THE CONTEXT OF LAND-USE/LAND-COVER MAPPING

2.2.1 Historical Perspective

Many environmental and natural resource management questions require accurate and timely information on land cover and land use. To this end, many states and agencies today systematically collect, update, and disseminate this most fundamental form of land information. However, it is accurate to suggest that remote sensing has had a much longer history of data collection than we have been making maps using these data. While history of remote sensing is reserved for elsewhere (e.g., Chapter 1, Vol. I), we provide a short history of land cover mapping using remote sensing as it sets the stage for the discussion that follows.

In many ways, the scientific establishment of modern-era land cover and land use mapping with remote sensing can be traced back to early 1970s. In anticipation of civilian space-based remote sensing observations, a number of federal agencies in the United States formed an interagency

TABLE 2.2
Historical Developments in Land Cover/Land Use Mapping from Remote Sensing

1920–1930	Development and early applications of aerial photography
1935–1945	WWII—Extensive applications of non-visible portions of the EMR
1960–1970	First use of the term "remote sensing"
1970–1980	Rapid advances in image processing
1971	Interagency Steering Committee on Land Use information and Classification in US
1974	The Large Area Crop Inventory Experiment (LACIE)
1975	Development of the neural network backpropagation algorithm by Paul Werbos
1976	Publication of James R. Anderson's landmark land use classification system
1970–1980	Launch of early Landsat satellites
1983	Development of the Global vegetation and land-use database by Elaine Matthews
1984	Launch of Landsat 4
1986	Launch of SPOT
1992	Development of the first National Land Cover Dataset (NLCD)
1993	Development of the original SVM algorithm by Vladmir N. Vapnik
1997	Release of the Global Land Cover Characteristics database
1999	Launch of Landsat 7 and MODIS
2000	Development of the FAO Land Cover Classification System (LCCS)
2000	Deep Learning in the neural networks community gets traction
2013	Launch of Landsat 8
2013	Release of the Finer Resolution Observation and Monitoring Global Land Cover (FROM-GLC) database by Chinese researchers

Steering Committee on Land Use information and Classification early in 1971. The formation and subsequent work of this committee resulted in first standards for mapping land cover (Anderson et al., 1976). The land use classification system to be used with remote sensor data, initially proposed by James R. Anderson and published in final form in 1976, was designed to place major reliance on remote sensing and used a system of hierarchically defined categories (Anderson et al., 1976). These developments along with the launch of Landsat I, II, and III are considered to have built the foundation of modern-day land cover mapping with remote sensing.

Several developments occurred from the 1970s to present (Table 2.2). On the technical side, the U.S. Geological Survey (USGS) released its entire satellite data archive at no cost, making it possible to study state's land use/land cover (LULC) on a repeated and low-cost basis. At the same time, new computer-based machine learning algorithms have found their way into regional remote sensing investigations that help provide improved accuracies for map products. On the application side, the environment and the natural resources of our state are under increased pressure from population growth, pollution, and bioenergy prospects.

2.2.2 Methods

Numerous classification algorithms have been developed since the early 1970s when first dedicated land observations became available. These algorithms range from visual interpretation of printed or digital color images to advanced machine learning algorithms that emulate human learning behavior (Ebenezer and Manohar, 2024; Papoutsis et al., 2023; Nasiri et al., 2022; Prasad et al., 2022; Phan et al., 2020; Carranza-García et al., 2019; Helber et al., 2019; Jozdani et al., 2019; Zhang et al., 2019; Costa et al., 2018; Goldblatt et al., 2018; Zhang et al., 2018; Ma et al., 2017; Scott et al., 2017; Talukdar et al., 2020). In this section, we divide the individual methods into two main subcategories:

those that rely on manual interpretation and those rely on automation. Note that the traditional division of classification methods into supervised and unsupervised logic is not used here as that topic is extensively treated in many textbooks and tutorials. The decision to use manual vs. automated categories stems the very purpose of this chapter: to provide a in-depth review of image classification methods. With this in mind, the supervised vs. the unsupervised logic is considered as part of the described classification methods.

2.2.2.1 Approaches That Rely on Manual Interpretation

Overwhelming majority of land cover classification performed on satellite data today are based on some form of automation. However, manual interpretation of satellite data still plays an important role in image categorization, especially in large state and international organizations such as the United Nations Food and Agricultural Organization (FAO). The principle advantage of manual (or visual) interpretation of satellite data for the purpose of land-cover mapping is the accuracy. Humans are extremely good at pattern recognition and categorization, even in very complex landscapes with multiple, perhaps overlapping categories. However, the visual interpretation of satellite images can be a complex process, going beyond what is contained in an image. In fact, successful image interpretation is only possible through the iterative process of recognition and interpretation of objects (Figure 2.1), both of which rely heavily on the knowledge and the expertise of the analyst in charge of the analysis (Albertz, 2007).

In a remotely sensed image, the objects are defined in terms of the way they reflect or emit radiation that give rise to their recorded colors and shapes. Thus, visual elements of tone, shape, size, pattern, texture, and association are the available tools that aid recognition and interpretation of the objects in an image.

Perhaps what separates today's manual interpretation of satellite data for mapping land cover is the medium in which the interpretation is made. Early work with satellite imagery relied on visual interpretation of large hard-copy satellite image prints to identify and map land cover elements. Today's visual image interpretation occurs all in digital format, using the idea of digitizing boundaries of land cover elements directly on a computer using within a GIS system. Regardless of the analysis environment, several known factors significantly influence the outcomes. First, the red and the Near Infrared (NIR) portions of the electromagnetic spectrum provide tremendous information for recognizing land cover elements. Second, the use of multiple images of the same area increases classification accuracy because the phenology of the land cover is captured. Third, studies involving visual interpretation of satellite data benefit from radiometric enhancements and manipulations in the form of spectral indices. For example, compared to automated techniques of image classification, studies involving visual interpretation of satellite data report superior performance, particularly for area estimation problems.

On the down side, manual interpretation of satellite data is expensive, involving thousands of analyst hours. This is one reason, why large organizations such as various state mapping agencies choose manual interpretation as a method of national land cover inventories in many developing

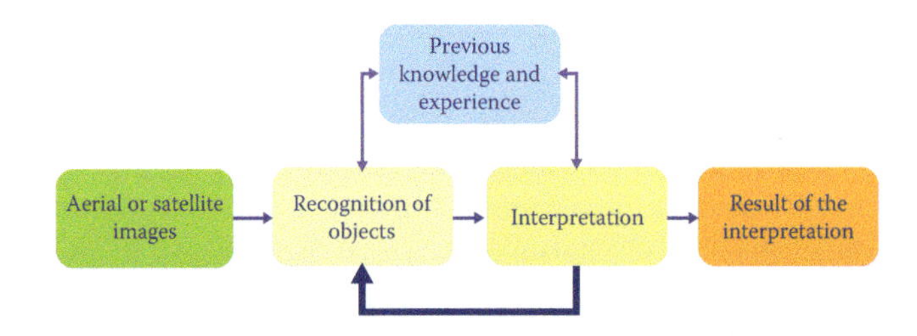

FIGURE 2.1 Schematic Presentation of the Interpretation Process. (Modified from Albertz (2007).)

countries (e.g., Travaglia et al., 2001). Moreover, land cover maps derived from visual interpretation of satellite data may not contain finest levels of categorical detail due to inherent limitations of analyst knowledge and experience. For example, in an agricultural landscape, identification of irrigated vs. non-irrigated fields may be possible by recognizing shapes associated with center pivot irrigation, identification of crop types grown on that field is less straightforward. On the other hand, manual interpretation of image data appears to be more cost effective if using only sampling units and enumeration as the point of area estimates. However, these applications do not provide wall-to-wall maps that are often required for environmental analyses.

2.2.2.2 Approaches That Rely on Automated Classification

Image classification for the purpose of extracting land cover information can also be achieved in an automated way with the help of computers (Prasad et al., 2022; Jozdani et al., 2019; Zhang et al., 2019; Zhang et al., 2018; Ma et al., 2017; Scott et al., 2017). While there are many interpretations of the word "automatic" in the context of classification, what is being referred to here is the idea that only a sample of the landscape need to be known and not the entire population. In other words, unlike manual interpretation in which the knowledge of the whole landscape is needed, automated classification only requires information on a subsample of the landscape. In this context, the idea of automated classification applies to both unsupervised and supervised classification logic. In the unsupervised approach, the relationship between the observed pattern (the feature space) and a predetermined number of statistical clusters (the information classes) is established with the help of an algorithm. It is often applied in situations where prior knowledge of the ground cover is not readily available. Ultimately, the statistical clusters are assigned to land cover labels and it is here where idea of automation comes into play. That is the analyst only uses a sample of known associations between the information clusters and actual land cover labels.

In the case of supervised classification, on the other hand, the user collects a set of learning samples to train a classifier to identify the class label of every pixel in the image. Once again this is what is being referred to automation: only a fraction of the population is needed to estimate the characteristics of every individual in the population: the pixels that constitute the landscape. Another way of looking at this is that there is a large return on investment (training data or known samples): knowledge on less than 5% of the landscape is used to extract information on the entire landscape in question. Note that the accuracy of unsupervised approaches is generally lower than that of supervised methods, especially in complex landscapes where spectral/temporal manifestation of different land cover categories may be similar.

The last two decades has seen a growing number of automated and advanced image classification methods, both supervised and unsupervised, applied to land cover mapping across various spatial and temporal scales (Table 2.3). These methods often belong to the non-parametric statistics domain, defined as a set of statistical tools that are not based on parameterized families of probability distributions, and hence make no assumptions about the probability distributions of the variables being evaluated. The principle advantage of the newer methods is their ability to deal with between-class variability that has limited more traditional pattern recognition methods. To this end, the automated approaches are further categorized into those that rely on parametric and non-parametric statistics.

2.2.2.2.1 Parametric Tools

The image classification tools that are parametric in nature assume that the observation matrix come from a known probability distribution and make inferences about the parameters of the distribution (Ebenezer and Manohar, 2024; Papoutsis et al., 2023; Prasad et al., 2022; Helber et al., 2019; Goldblatt et al., 2018; Scott et al., 2017; Talukdar et al., 2020). Many of the traditional classification algorithms, including the Maximum Likelihood (MLC), Minimum Distance (MD), and to some degree, K-means clustering (KMC) procedure fall into this category of statistical classifiers. Some

TABLE 2.3
Comparison of Automated Image Classification Algorithms Described Here

	Advantages	Limitations	Typical Accuracies	References
Parametric tools	• Long history • Widespread use • Simple formulation • Fast computation • Small set of parameters	• Normality assumption • Lack of ancillary input use • Limited class separation	• 85–95% for small class (< 5) problems • 75–85% for large class problems	Akaike (1973) Loveland et al. (1991) Jia and Richards (1994) Hall and Knapp (2000) Vogelmann et al. (2001) Mondal et al. (2012) Sun et al. (2013)
Non-parametric tools	• Non-distributional requirement • Improved accuracies • Probability estimates • Ability to ignore unimportant features • Ability to estimate feature importance • Many-to-one feature	• Lack of standard software • Can be a black box • Can be computationally expensive • Some optimized for two class problems • Large parameter space • Specialized knowledge may be needed	• 85–98% for small class (< 5) problems • 75–90% for large class (> 5) problems	Benediktsson et al. (1990) Yoshida and Omatu (1994) Gopal and Woodcock (1996) Friedl and Brodley (1997) Carpenter et al. (1997) Pal and Mather (2003) Rodriguez-Galiano et al. (2012)
Object-based tools	• Minimum mapping unit • Natural class boundaries • Improved accuracy	• Lack of standard software • Large parameter space • Computationally expensive	• 85–95% for small class (< 5) problems • 75–90% for large class (> 5) problems	Woodcock and Harward (1992) Blaschke and Hay (2001) Hay et al. (2001) Benz et al. (2004)

of the advantages of parametric algorithms include their simplicity and assumed higher statistical power. From the practical perspective, all off-the-shelf image processing software include these algorithms, thus making them readily available and easy to apply under a wide range of conditions. They also work well, in some cases even better than the new generation of classification techniques as long as the assumptions about the probability distributions of data are valid. However, in many cases the input satellite data do not conform to these assumptions, rendering parametric classifiers less robust in their predictions. An additional limitation occurs when non-traditional inputs, such as topography or other ancillary information are part of the feature space as these variables never conform to known probability distributions.

In the MLC procedure the distribution of each class's learning sample (the training data) is assumed to come from a normal distribution. Using the mean vector and the covariance matrix to characterize a class, the probability that a given pixel belongs to a specific class is calculated (Richards, 2012). Each pixel is then assigned to the class that has the highest probability (hence the maximum likelihood). A prototype land-cover database for the conterminous United States was created by Loveland et al. (1991) by first stratifying vegetated and barren land, then using an unsupervised classification of multitemporal "greenness" data derived from Advanced Very High Resolution Radiometer (AVHRR) imager, and post-classification stratification of classes into homogeneous land-cover regions using ancillary data. Jia and Richards (1994) introduced a simplified MLC technique for handling remotely sensed image data that reduces the processing time and copes with the training of geographically small classes. Hall and Knapp (2000) demonstrated the use of the MLC procedure applied to Landsat data for characterizing the successional and disturbance dynamics of the boreal forest for use in carbon modeling in Canada. Vogelmannn et al. (2001) demonstrated the use of statistical tools to map land cover of the continental U.S. with Landsat data. These earlier studies showed that with smaller number of categories, achieving accuracies better than 80% was possible with the use of parametric classifiers. However, as the number of categories and their complexity increased, these methods did not produce image classification results with better than 75%, particularly for the more refined categories. This notion still holds true today.

In most applications of the MLC, each class is given the same equal likelihood of belonging. However, MLC algorithm allows incorporation of prior probabilities to increase/decrease the likelihood of occurrence for certain classes. For example, Strahler (1980) in an early seminal paper shows how probabilities of occurrence of classes based on separate, independent knowledge such as collateral information datasets (e.g., rock type, soil type, topography) can be used in a Bayesian-type classifier (i.e., MLC) to improve classification accuracies. Strahler (1980) also recognized that using prior probabilities with supervised classification algorithms could also bias the posterior probability of a given class especially if the uncertainty surrounding the information used to prescribe the prior probabilities. Chen et al. (1999) addressed this issue by introducing a subjective confidence parameter (c) that conditions the likelihood estimate for the membership of each pixel in a given class. Ranging from 0.0 to 1.0, the condition parameter controls the influence of the ancillary information on the predictions. Although the choice of a value for c is subjective, McIver and Friedl (2002) show that a simple heuristic based on objective criteria can be used to prescribe a value for c in many situations (Figure 2.2).

The MD procedure is one of the simplest forms of parametric classification algorithms in which the unknown image pixels are categorized by minimizing the distance between the observation matrix of the pixel and the class in multi-feature space (Wacker and Landgrebe, 1972). Variants of the MD procedure use different methods to calculate the distance, which is defined as an index of

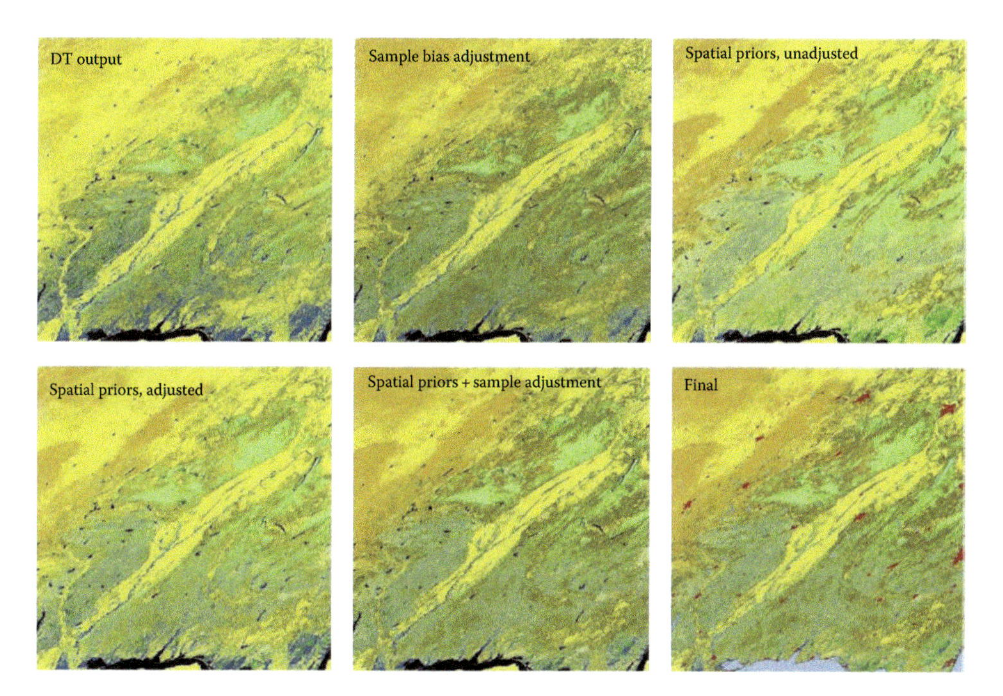

FIGURE 2.2 Classification results for MODIS tile v05h10 (south-central United States) outcomes at each stage of processing. (Adapted from Friedl et al. (2010). Used with permission.)

similarity. For example, the Euclidian distance measure is used in cases where the variances of the classes are very different while the Mahalanobis distance works better in cases where there is correlation between the variables of the feature space (Richards, 2012).

The KMC algorithm, among many, attempts to find a pre-determined number of natural groupings or clusters in the data. It is an iterative approach in which the mean of each information class (the cluster) is initialized in some manner (e.g., randomly) and each pixel in the image matrix is assigned to the cluster that is closest to in the feature space. Then new cluster means are calculated based on the current assignment and an attempt is made to reassign each pixel to its new category based on some similarity index. This procedure is repeated until convergence, which occurs when there is little or no change in assigned cluster means. The KMC method falls into the unsupervised classification domain and is considered to be semi-parametric that lie between fully parametric and fully non-parametric statistical domains (Alpaydin, 2004).

2.2.2.2.2 Non-parametric Tools

The non-parametric methods (also called distribution-free methods) do not require the variables in the image matrix to belong to any particular distribution (Ebenezer and Manohar, 2024; Papoutsis et al., 2023; Prasad et al., 2022; Carranza-García et al., 2019; Zhang et al., 2019; Zhang et al., 2018; Ma et al., 2017). While there is cornucopia of non-parametric classification techniques, the remote sensing community has adapted the artificial neural networks (ANN), classification trees (CT), and Support Vector Machines (SVM) both from the algorithmic and practical (i.e., availability as an image processing software) perspectives (Ebenezer and Manohar, 2024; Papoutsis et al., 2023; Nasiri et al., 2022; Prasad et al., 2022; Phan et al., 2020; Carranza-García et al., 2019; Helber et al., 2019; Jozdani et al., 2019; Zhang et al., 2019; Costa et al., 2018; Goldblatt et al., 2018; Zhang et al., 2018; Ma et al., 2017; Scott et al., 2017; Talukdar et al., 2020; Friedl and Brodley, 1997; Pal and Mather, 2003). These particular methods, sometimes referred to as machine learning classifiers, have been used effectively in a variety of land-cover mapping studies (e.g., DeFries and Chan, 2000; Friedl et al., 2010). With some exceptions, these classifiers have proven superior to parametric classifiers (e.g., maximum likelihood) with accuracy improves on the order of 10–20% (Rogan et al., 2002). Their success can be attributed to (1) ability to incorporate class-relevant categorical and continuous observations into the features space; (2) not being constrained by parametric statistical assumptions so multimodal, noisy, or missing observations are effectively handled; (3) ability to tease apart complex feature spaces; and (4) ability to perform many-to-one classifications where multiple manifestations of the same category are present in the observation matrix.

The ANN technique in remote sensing was introduced over two decades ago by Benediktsson et al. (1990). Since then, a large number of studies have demonstrated their effectiveness in remote sensing image classification. Their potential discriminating power has attracted a great deal of research effort so many types of neural networks have been developed (Lippman, 1987). Among the more popular implementations of ANN are the backpropagation training algorithm and Multilayer Perceptron (MLP) approach (e.g., Carranza-García et al., 2019; Costa et al., 2018; Goldblatt et al., 2018; Talukdar et al., 2020; Atkinson and Tatnall, 1997) and the Adaptive Resonance Theory (ART) approach (Carpenter et al., 1991). MLP networks are architectures in which each node receives inputs from previous layers and information flows in one direction to the output layer (Pratola et al., 2011). The number of nodes in the intermediate layer(s) defines both the complexity and the power of a neural network model to describe underlying relationships and structures inherent in a training dataset (Kavzoglu, 2009). Although ANN classification has been shown to greatly improve accuracy over traditional parametric methods with reduced training sets, the process to implement classification is not straightforward and can be time consuming (Pal and Mather, 2003).

The ART framework describes a number of neural network models, which use supervised and unsupervised learning methods for pattern recognition and prediction. The premise of the ART model is that object identification and recognition generally occur as a result of the interaction of "top-down" observer expectations with "bottom-up" sensory information, in this case information content of satellite and ancillary data. Assuming that the difference between sensation and expectation stays below a set threshold (via the "vigilance parameter"), a pixel will be considered a member of the expected class, thus offering a solution to the problem of plasticity, that is, the problem of acquiring new knowledge without disrupting existing one (Carpenter and Grossberg, 2003). Carpenter et al. (1997) applied the Fuzzy ARTMAP, a version of the ART framework to Landsat data and terrain features for vegetation classification in a challenging environment and reported a fast, reliable, and scaleable algorithm overcoming many limitations of back propagation ANNS, K nearest neighbor algorithms, and MLC. They also report an additional benefit of the ARTMAP method in which a voting strategy improves prediction and assigns confidence estimates by training the system several times on different orderings of an input set.

A CT classifier takes a different approach to land cover classification. It breaks an often very complex classification problem into multiple stages of simpler decision-making processes (Breiman et al., 1984). Depending on the number of variables used at each stage, there are univariate and multivariate decision trees (Friedl and Brodley, 1997). Univariate decision trees have been used to develop land cover classifications at a global scale (DeFries et al., 1998; Hansen et al., 2000). Though multivariate decision trees are often more compact and can be more accurate than univariate decision trees, they involve more complex algorithms and, as a result, are affected by a suite of algorithm-related factors (Friedl and Brodley, 1997). An additional benefit of the CT algorithms is their ease of use and computational efficiency (Pal and Mather, 2003). Several studies have found CTs to be an acceptable classification method (Ebenezer and Manohar, 2024; Papoutsis et al., 2023; Nasiri et al., 2022; Prasad et al., 2022; Phan et al., 2020; Carranza-García et al., 2019; Helber et al., 2019; Jozdani et al., 2019; Zhang et al., 2019; Costa et al., 2018; Goldblatt et al., 2018; Zhang et al., 2018; Ma et al., 2017; Scott et al., 2017; Talukdar et al., 2020) and have shown improvements in accuracy over traditional parametric classifiers (Friedl and Brodley, 1997; Pal and Mather, 2003).

SVMs are a supervised non-parametric statistical learning technique that is increasingly being used by the remote sensing community (Huang et al., 2002; Mantero et al., 2005; Mountrakis et al., 2011). At the heart of an SVM training algorithm lies the concept of a linear *hyperplane*—an optimal boundary found through an iterative learning procedure that separates the training set into a discrete predefined number of classes while minimizing misclassifications errors (Vapnik, 1979; Zhu and Blumberg, 2002). Several approaches have been developed to improve SVM predictive accuracies using multispectral remote sensing data. These include the soft margin approach (Cortes and Vapnik, 1995) and kernel-based learning (Scholkopf and Smola, 2001) that lead to SVM optimization, although the kernel functions often result in more expensive parameterization (Kavzoglu and Colkesen, 2009).

Prior research has identified at least three benefits of SVMs that make them particularly suitable for remote sensing applications. First, regardless of the size of the learning sample, not all the available examples are used in the specification of the hyperplane. This allows SVMs to successfully handle small training datasets because only a subset of points—the support vectors—that lie on the margin are used to define the hyperplane (Mantero et al., 2005). Second, unlike many statistical classifiers, SVMs do not make prior assumptions on the probability distribution of the data, which leads to reduction in classification errors when input data do not conform to a required distribution (e.g., Gaussian). Third, SVM-based classification algorithms have been shown to produce generalizable models from a set of input training data, eliminating the notion of overfitting (Montgomery and Peck, 1992).

One of the more popular implementations of SVMs is software called the LIBSVM implementation that provides linear, polynomial (cubic) and radial-basis kernels (Chang and Lin, 2011). This implementation includes C-support vector classification (C-SVC), ν-support vector classification (ν-SVC), distribution estimation (one-class SVM), ε-support vector regression (ε-SVR), and ν-support vector regression (ν-SVR) formulations. All SVM formulations supported in LIBSVM are quadratic minimization problems. Using the radial-basis kernel classification option, the LIBSVM required only two parameters to be defined: the kernel parameter γ and the cost parameter C (Chang and Lin, 2011). Both of these parameters are data dependent and are identified separately for each footprint/date-pair combination using the grid search option over log-transformed hyper-parameters as suggested by (Hsu et al., 2001).

Note that SVMs have been shown to perform well given a certain level of noise (i.e., mislabeled training data) but they are not completely impervious to outliers (Vapnik, 1995). While a number of methods have been developed to mitigate the effects of outliers on SVMs (Lin and Wang, 2002; Suykens et al., 2002; Tsujinishi and Abe, 2003) they show only incremental improvements over standard SVM methods.

While there is evidence in the literature to show that non-parametric methods perform better than traditional classifiers for mapping land cover, very little attempt has gone into comparing the performance of several non-parametric classifiers. Shao and Lunetta (2012) compared the performance of SVMs, ANNs, and CT for land-cover characterization using MODIS time-series data and investigated the effects of training sample size, sample variability, and landscape homogeneity (purity). Their results indicate a strong relationship between training sample size and classification accuracy but above a set threshold, increasing the number of training data does not leading equal increase in performance. They also show that SVMs had superior generalization capability, particularly with respect to small training sample sizes (Figure 2.3). There was also less variability of SVM performance when classification trials were repeated using different training sets.

A growing number of studies continue to show that the non-parametric image classification tools come with undisputed advantages, not least of the improved classification performance. While the more sophisticated non-parametric algorithms still remain in the research domain and require specialized software environment, the rate at which the remote sensing community take up these methods is high and point to the possibility that they may become the *de facto* choice for image classification.

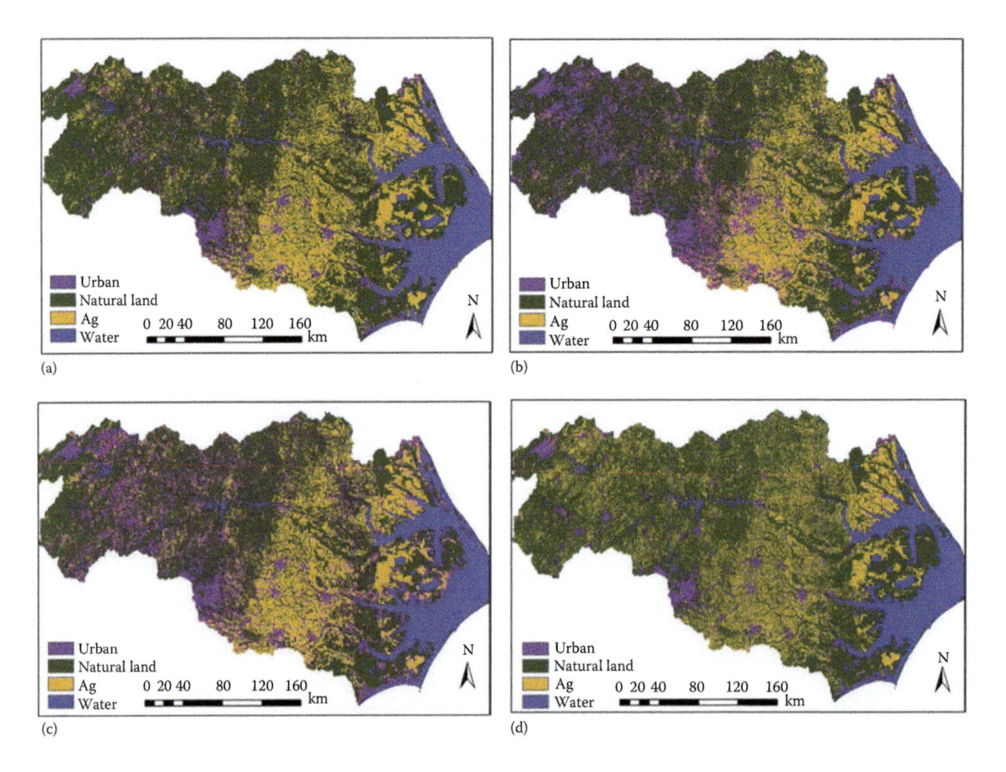

FIGURE 2.3 Comparison of classification results for SVM (a), ANN (b), and CT (c) algorithms. The NLCD 2001 (d) is also included as reference. (Adapted from Shao and Lunetta (2012). Used with permission.)

2.2.2.3 Pixel versus Object-Based Classifications

Individual pixels form the smallest unit of analysis when classifying remotely sensed images for the purpose of land cover mapping. However, it is well known that as arbitrary objects, pixels do not necessarily represent the landscape that is being characterized. Contextual classification methods overcome this issue by incorporating the spatial information into the classification process. In the absence of contextualization, pixel-based classification results tend to contain significant noise as no neighborhood information is being considered (Fisher, 1997). One reason for this is that information content of individual pixels is heavily influenced by the radiance contributed from surrounding pixels, or the MTF effect (Townshend et al., 2000). Moreover, both the landscape structure and the resolving power of the sensor contribute to our ability to identify and map objects of interest in a map. To overcome the errors that result from traditional per-pixel classifiers that ignore contextual properties from surrounding pixels, the remote sensing community has developed a number of image analysis algorithms that go "beyond pixels" and take also into account spatial information (Blaschke, 2010). These new methods include object-based image analyses (OBIA) that have generally yielded higher thematic accuracies than the traditional per-pixel methods (Ebenezer and Manohar, 2024; Papoutsis et al., 2023; Nasiri et al., 2022; Phan et al., 2020; Carranza-García et al., 2019; Costa et al., 2018; Talukdar et al., 2020; Woodcock and Harward, 1992; Blaschke et al., 2006; Platt and Rapoza, 2008; Lizarazo and Elsner, 2011).

In this chapter, we divide the object based image analysis into three categories, primarily distinguished by the order and the purpose by which the objects are utilized. A common denominator of these methods is image segmentation, purpose of which is to produce a set of non-overlapping segments (objects, polygons). Although not new in industrial and medical image processing, image segmentation has only two decades of history in geospatial applications (Ebenezer and Manohar, 2024; Papoutsis et al., 2023; Nasiri et al., 2022; Prasad et al., 2022; Scott et al., 2017; Talukdar et al., 2020; Ryherd and Woodcock, 1996; Blaschke and Hay, 2001; Hay et al., 2005; Esch et al., 2008; Lizarazo and Elsner, 2009) but has been rapidly increasing in the past ten years.

Image segmentation partitions the image into a set of distinct and uniform (homogenous) regions, in which the criteria of homogeneity is defined by one or more dimensions of the feature space (Blaschke, 2010). It is generally understood that these segments represent meaningful landscape objects (e.g., a forest stand or an agricultural field) but this translation is not always clear at the segmentation stage. What is clear however is that different regions are found at different scales of analysis placing image segmentation in the realm of multi-scale landscape analysis (Hay et al., 2001). The image segments also contain additional spectral (e.g., mean values per band, and also median values, minimum and maximum values, mean ratios, variance, etc.) and spatial information compared to single pixels and it is often argued that it is this spatial information that provide greater advantage to OBIA using segments (Blaschke and Strobl, 2001; Flanders et al., 2003; Hay and Castilla, 2008).

One question surrounding the OBIA is the scale at which the analysis is made. More specifically, at what scale (or spatial resolution) does the image under consideration lends itself to image segmentation for the purpose of object-based analysis? It turns out the answer is not related only to spatial resolution but also the size of the objects under consideration on the landscape. From the theoretical perspective, the Shannon-Nyquist sampling theorem suggests that to be able reconstruct the original image (i.e., to reconstruct the landscape objects in the image), the spatial sampling rate (or the spatial resolution) of the image has to be higher than twice the highest spatial frequency of the original image (i.e., spatial resolution must be finer than at least half the size of the smallest objects on the landscape) (Blaschke, 2010). From the practical perspective however, a minimum of six pixels per objects has been found to be necessary in order to accurately identify and map objects.

Returning to different kinds of OBIA, in the first category, pixel-based image classification is performed independent of the image segmentation process. More specifically, first a traditional pixel-based classification is performed using the best possible tools and inputs with the highest

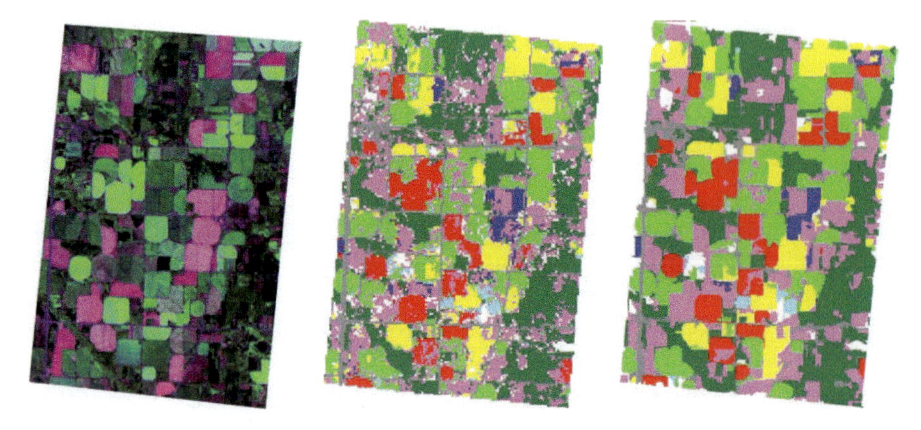

FIGURE 2.4 Object-based analysis example: (a) the original Landsat TM image in true color; (b) the original USDA Cropland Data Layer; (c) object-filled USDA Cropland Data Layer in which the objects (agricultural field boundaries) were used to convert pixel-based classification results to a polygon based map.

possible accuracy. Then image segmentation is performed, either on the original image that went into the classification process or on an image with higher spatial resolution of the same area. The key here is to define the minimum mapping unit, defined as the size of the smallest object to be identified on the landscape, based on the spatial resolution of the image used in classification. Finally, the image segments, either in raster or vector format, are merged with the pixel-based classification using the majority rule, in which the most frequently occurring class label is used to label the entire segment (Figure 2.4). While the majority rule works well, there are other ways to collapse individual pixels into objects (polygons).

2.2.2.4 Emerging Methods

2.2.2.4.1 Cloud and Cloud Shadow Masking

The presence of clouds and cloud shadows affect almost all forms and types of analyses conducted with satellite data and their detection and subsequent removal is often the initial step in most image processing chains (Simpson and Stitt, 1998; Irish, 2000; Arvidson et al., 2001). For example they selectively alter the reflectance and transmission of radiation through the Earth's atmosphere and reduce the accuracy of atmospheric correction of images. Land cover classification involving multiple image composites require clouds and their shadows to be removed from the scene, before the compositing process. In change detection studies, presence of clouds and cloud shadows lead to false detection of land cover change. Calculation of vegetation indices like the Normalized Difference Vegetation Index (NDVI) may be biased in locations where clouds and cloud shadows are present. Archiving processes such as scene selection and scene quality assessment too require cloud information, not only on the total amount but also the location of clouds and cloud shadows in a Landsat scene. In parallel with the growing demand and use of Landsat imagery, there has also been a number of new developments in detecting clouds in Landsat images (e.g., Irish et al., 2006; Oreopoulos et al., 2011; Zhu and Woodcock, 2012; Goodwin et al., 2013).

One of the more robust methods in detecting cloud and cloud shadows was developed by Zhu and Woodcock (2012). This method uses a series of rules based on calculated probabilities of temperature, spectral variability, and brightness, using Top of Atmosphere (TOA) reflectance and Brightness Temperature (BT) as inputs. The clouds and cloud shadows are treated as 3D objects determined via segmentation of the potential cloud layer and an assumption of a constant temperature lapse rate. The solar illumination and sensor view angles are used to predict possible cloud shadow locations and select the one that has the maximum similarity to cloud shape and size.

2.2.2.4.2 Large Area Mapping

Free and open access to most satellite data (e.g., Landsat, Sentinel) are changing the way remotely sensed images are processed. For example, there is increased interest in automatically processing large volumes of imagery covering large areas of the Earth. These methods rely on availability of a large number of images within and across the years and use both the spectral and the temporal information available in satellite data. While the concept of time series observation for large area land cover mapping is not new (Ebenezer and Manohar, 2024; Papoutsis et al., 2023; Nasiri et al., 2022; Prasad et al., 2022; Phan et al., 2020; Carranza-García et al., 2019; Helber et al., 2019; Jozdani et al., 2019; Zhang et al., 2019; Costa et al., 2018; Goldblatt et al., 2018; Zhang et al., 2018; Ma et al., 2017; Scott et al., 2017; Talukdar et al., 2020; Hansen et al., 2000; Loveland et al., 2000; Friedl et al., 2002; Bartholomé and Belward, 2005) the new developments mostly apply to medium resolution (i.e., less than 100-m pixels) images and rely on automated algorithms (Kalensky, 1998; Gong et al., 2013; Yu et al., 2013; Sexton et al., 2013). Also contributing to these developments is the availability of powerful computers and image classification software, mainly based on non-parametric methods that are able to generalize across space and time. For example, development of the Web-Enabled Landsat Data (WELD) (Roy et al., 2010) system has revolutionized the way Landsat data are being processed and temporally aggregated for rapid and efficient large-area applications. However, despite the existence and the availability of maps over large areas, there is still the issue of different land cover definitions as well as the use of different algorithms. In Figure 2.5, we show forest classification results from three large-area classification projects in Kenya. While there is strong correlation between each map product, notable differences remain. For example, the FROM GLC product (Gong et al., 2013) has a large forest area than the other products. Alternatively, in locations where the UMD tree cover map (Sexton et al., 2013) shows small tree cover fractions (i.e., less than 50% threshold), the AfricCover product (Kalensky, 1998) maps them as full forest. Of course this is partly related to class definitions as each of three products use different descriptions of what a forest is. Nevertheless, data in Figure 2.5 simply highlights the difficulty with which large area land cover maps are made using medium- to high-resolution satellite data.

2.2.3 Uncertainty Assessment

Accuracy assessment of classification results developed from satellite data (or other sources) is an important but often neglected step in producing land cover maps. The application of various image classification methods to satellite data produces a map that is hypothesized to represent the land cover in question. Like all hypothesis testing, the next step is to obtain data, preferably

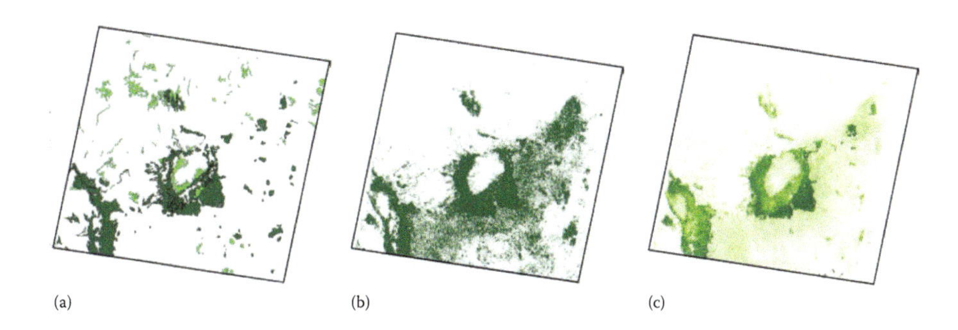

FIGURE 2.5 Comparison of three land cover maps for the forest category in Kenya: (a) the land cover map from the AfriCover project; (b) forest category of a Landsat-based land-cover map as part of the FROM GLC project; (c) tree-cover map derived from multi-temporal Landsat data. The black polygon outlines the boundaries of a single Landsat WRS2 footprint (path:168 row:60).

from an independent sources, and test the validity of the hypothesis in map form. One reason map validation is either neglected or under-estimated stems from the difficulty of which to accuracy assessment task. In general, validation is a complex task as maps made from satellite data cover large areas relative to the spatial sampling unit (i.e., pixel or a polygon). Moreover, collecting the reference data to be used in validation is expensive and labor intensive. For this reason, the sampling design to acquire accurate, representative, statistically sound, and independent reference data must be carefully crafted and evaluated against available resources allocated to the validation procedure. Therefore, the key to an effective accuracy assessment is a sound sampling plan that balances accuracy and pragmatism while including both the common and the rare categories (Zhu et al., 2000).

It is generally accepted that accuracy assessment using an independent, stratified random sample produces the best evaluation results with respect to statistical rigor and representatives (Ebenezer and Manohar, 2024; Papoutsis et al., 2023; Helber et al., 2019; Zhang et al., 2018; Scott et al., 2017; Talukdar et al., 2020; Congalton, 1991; Stehman and Czaplewski, 1998). In a stratified random design, the map categories often constitute the strata so that each class could be evaluated using an independent random sample. The allocation strategy could be proportionate in which known class areas determine the number of samples per class, or disproportionate in which a set number of samples are assigned in category. One disadvantage of the proportionate allocation is that really small classes receive a small, and in some cases unrealistic, number of samples. While the question of sample size, both per class and total, has received due attention in the literature (e.g., see Stehman, 2012 for a comprehensive review), the number of samples to be used in the validation effort ultimately depends on the resources available. Note that with the development of free, high resolution, and accurately located image data, such as those available from Google Earth, some of the cost associated with field work has been shifted to the office in the form of manual interpretation. However, caution must be exercised when using these sources as the independent sample because complex land cover categories do not always lend themselves to visual analysis. Moreover, these online sources often represent a single time period corresponding to current or recent past, rendering them less useful for evaluating maps of past time periods. Finally, maps derived from object-based analysis are evaluated at the object level. That is, accuracy assessment is performed using the objects (e.g., polygons) in the independent sample in which the class label of the entire polygon (not the label of an individual pixel inside the polygon) is used for evaluation. In evaluating pixel-based maps, it is recommended that an area corresponding to at least one pixel shift in all directions be included in the analysis to address the geo-location-based errors inherent in all satellite data. For example, a map made from Landsat data with 30-meter pixels would require a square-like polygon, at least 90 m on a side, centered on the pixel to be evaluated. Information inside this polygon is then used to label the samples in the independent dataset.

Once the independent sample is acquired, development of the confusion matrix (or error matrix) is the standard practice. The confusion matrix is a specific table that allows assessment of both class-specific and overall accuracies rapidly. The table is also used to quantify the omission and commission errors that allow the specific performance of the classifier in use. In general, when reporting map accuracy, it is recommended that the confusion matrix along with all the other accuracy measures be reported. This allows the map user to draw his/her own conclusions when using the land-cover map.

2.3 THE FUTURE

It is the author's opinion that the future of land cover mapping from satellite data is bright. Contributing factors include: (1) free and unlimited access to satellite observations; (2) development of cheaper and better sensors; (3) availability of dense temporal observations, even at high spatial resolutions; (4) advances in computer hardware and software rooted in artificial intelligence community; (5) ease of access using web-enabled services; and (6) availability of ever-growing number of ancillary datasets on environmental and social variables that in turn help determine presence or absence of a specific land cover. As our planet is continually pushed to meeting the demands of a growing and affluent population, identifying and mapping of its resources will be even more paramount. It is hoped that the new developments in image processing and satellite data availability will allow us to make more accurate and refined land cover maps to face this challenge as tremendous recent advances have shown through big-data analytics, machine learning, cloud computing, and artificial intelligence (Ebenezer and Manohar, 2024; Papoutsis et al., 2023; Nasiri et al., 2022; Prasad et al., 2022; Phan et al., 2020; Carranza-García et al., 2019; Helber et al., 2019; Jozdani et al., 2019; Zhang et al., 2019; Costa et al., 2018; Goldblatt et al., 2018; Zhang et al., 2018; Ma et al., 2017; Scott et al., 2017; Talukdar et al., 2020).

REFERENCES

Akaike, H. 1973. Information theory and an extension of the maximum likelihood principle. *Proceedings of the Second International Symposium on Information Theory*, Aakademiai Kidao Budapest, New York: Springer-Verlag.

Albertz, D. 2007. *Einführung in die Fernerkundung*. Darmstadt: Grundlagen der Interpretation von Luft- & Satellitenbildern, 254 pp.

Alpaydin, E. 2004. *Introduction to Machine Learning*. Cambridge, MA: MIT Press.

Anderson, J. R., Hardy, E. E., Roach, J. T., & Witmer, R. E. 1976. A land use and land cover classification system for use with remote sensor data. In *Professional Paper 964. A Revision of the Land Use Classification System in Circular,* Vol. 671, p. 28. https://doi.org/10.3133/pp964

Arvidson, T., Gasch, J., & Goward, S. N. 2001. Landsat-7's long-term acquisition plan—an innovative approach to building a global imagery archive. *Remote Sensing of Environment*, 78(1–2), 13–26.

Atkinson, P. M., & Tatnall, A. R. 1997. Introduction: Neural networks in remote sensing. *International Journal of Remote Sensing*, 18, 699–709.

Bartholomé, E., & Belward, A. S. 2005. GLC2000: A new approach to global land cover mapping from Earth observation data. *International Journal of Remote Sensing*, 26(9), 1959–1977.

Benediktsson, J., Swain, P. H., & Ersoy, O. K. 1990. Neural network approaches versus statistical methods in classification of multisource remote sensing data. *IEEE Transactions on Geoscience and Remote Sensing*, 28(4), 540–552.

Benz, U. C., Hofmann, P., Willhauck, G., Lingenfelder, I., & Heynen, M. 2004. Multi-resolution, object-oriented fuzzy analysis of remote sensing data for GIS-ready information. *ISPRS Journal of Photogrammetry and Remote Sensing*, 58, 239–258.

Blaschke, T. 2010. Object based image analysis for remote sensing. *ISPRS Journal of Photogrammetry and Remote Sensing*, 65, 2–16.

Blaschke, T., Burnett, C., & Pekkarinen, A. 2006. Image segmentation methods for object-based analysis and classification. In S. de Jong & F. van der Meer (eds), *Remote Sensing Image Analysis: Including the Spatial Domain*. New York: Springer-Verlag, pp. 211–236.

Blaschke, T., & Hay, G. J. 2001. Object-oriented image analysis and scale-space: Theory and methods for modeling and evaluating multi-scale landscape structure. *International Archives of Photogrammetry and Remote Sensing*, 34(Part 4/W5), 22–29.

Blaschke, T., & Strobl, J. 2001. What's wrong with pixels? Some recent developments interfacing remote sensing and GIS. *GIS—Zeitschrift für Geoinformationssysteme*, 14(6), 12–17.

Breiman, L., Friedman, J. H., Olshen, R. A., & Stone, C. J. 1984. *Classification and Regression Trees*. Belmont, CA: Wadsworth International Group.

Carpenter, G. A., Gjaja, M. N., Gopal, S., & Woodcock, C. E. 1997. ART neural networks for remote sensing: Vegetation classification from Landsat TM and terrain data. *IEEE Transactions on Geoscience and Remote Sensing*, 35(2), 308–325.

Carpenter, G. A., & Grossberg, S. 2003. Adaptive resonance theory. In M. A. Arbib (ed), *The Handbook of Brain Theory and Neural Networks*, 2nd edition. Cambridge, MA: MIT Press, pp. 87–90.

Carpenter, G. A., Grossberg, S., & Reynolds, J. H. 1991. ARTMAP: Supervised real-time learning and classification of nonstationary data by a self-organizing neural network. *Neural Networks*, 4, 565–588.

Carranza-García, M., García-Gutiérrez, J., & Riquelme, J. C. 2019. A framework for evaluating land use and land cover classification using convolutional neural networks. *Remote Sensing*, 11(3), 274. https://doi.org/10.3390/rs11030274

Chang, C.-C., & Lin, J.-C. 2011. LIBSVM: A library for support vector machines. *ACM Transactions on Intelligent Systems and Technology*, 2, 1–39.

Chen, M.-H., Ibrahim, J. G., & Yianoutsos, C. 1999. Prior elicitation, variable selection, and Bayesian computation for logistics regression models. *Journal of the Royal Statistical Society B*, 61, 223–242.

Congalton, R. G. 1991. A review of assessing the accuracy of classification of remote sensed data. *Remote Sensing of Environment*, 37, 35–46.

Cortes, C., & Vapnik, V. 1995. Support-vector networks. *Machine Learning*, 20, 273–297.

Costa, H., Foody, G. M., & Boyd, D. S. 2018. Supervised methods of image segmentation accuracy assessment in land cover mapping. *Remote Sensing of Environment*, 205(2018), 338–351. ISSN 0034-4257. https://doi.org/10.1016/j.rse.2017.11.024. https://www.sciencedirect.com/science/article/pii/S0034425717305734

DeFries, R. S., & Chan, J. C. 2000. Multiple criteria for evaluating machine learning algorithms for land cover classification from satellite data. *Remote Sensing of Environment*, 74, 503–515.

DeFries, R. S., Hansen, M., Townshend, J. R. G., & Sohlberg, R. 1998. Global land cover classifications at 8 km spatial resolution: The use of training data derived from Landsat imagery in decision tree classifiers. *International Journal of Remote Sensing*, 19(16), 3141–3168.

Ebenezer, P. A., & Manohar, S. 2024. Land use/land cover change classification and prediction using deep learning approaches. *SIViP*, 18, 223–232. https://doi.org/10.1007/s11760-023-02701-0

Esch, T., Thiel, M., Bock, M., Roth, A., & Dech, S. 2008. Improvement of image segmentation accuracy based on multiscale optimization procedure. *IEEE Geosciences and Remote Sensing Letters*, 5(3), 463–467.

Fisher, P. 1997. The pixel: A snare and a delusion. *International Journal of Remote Sensing*, 18, 679–685.

Flanders, D., Hall-Beyer, M., & Pereverzoff, J. 2003. Preliminary evaluation of eCognition object-based software for cut block delineation and feature extraction. *Canadian Journal of Remote Sensing*, 29(4), 441–452.

Friedl, M. A., & Brodley, C. E. 1997. Decision tree classification of land cover from remotely sensed data. *Remote Sensing of Environment*, 61(3), 399–409.

Friedl, M. A., McIver, D. K., Hodges, J. C. F., Zhang, X. Y., Muchoney, D., Strahler, A. H., Woodcock, C. E., Gopal, S., Schneider, A., Cooper, A., Baccini, A., Gao, F., & Schaaf, C. 2002. Global land cover mapping from MODIS: Algorithms and early results. *Remote Sensing of Environment*, 83, 287–302.

Friedl, M. A., Sulla-Menashe, D., Tan, B., Schneider, A., Ramankutty, N., Sibley, A., & Huang, X. 2010. MODIS collection 5 global land cover: Algorithm refinements and characterization of new datasets. *Remote Sensing of Environment*, 114, 168–182.

Goldblatt, R., Stuhlmacher, M. F., Tellman, B., Clinton, N., Hanson, G., Georgescu, M., Wang, C., Serrano-Candela, F., Khandelwal, A. K., Cheng, W., & Balling, R. C. 2018. Using Landsat and nighttime lights for supervised pixel-based image classification of urban land cover. *Remote Sensing of Environment*, 205(2018), 253–275. ISSN 0034-4257. https://doi.org/10.1016/j.rse.2017.11.026. https://www.sciencedirect.com/science/article/pii/S0034425717305758

Gong, P., Wang, J., Yu, L., Zhao, Y. C., Zhao, Y. Y., Liang, L., Niu, Z. G., Huang, X. M., Fu, H. H., Liu, S., Li, C. C., Li, X. Y., Fu, W., Liu, C. X., Xu, Y., Wang, X. Y., Cheng, Q., Hu, L. Y., Yao, W. B., Zhang, H., Zhu, P., Zhao, Z. Y., Zhang, H. Y., Zheng, Y. M., Ji, L. Y., Zhang, Y. W., Chen, H., Yan, A., Guo, J. H., Yu, L., Wang, L., Liu, X. J., Shi, T. T., Zhu, M. H., Chen, Y. L., Yang, G. W., Tang, P., Xu, B., Ciri, C., Clinton, N., Zhu, Z. L., Chen, J., & Chen, J. 2013. Finer resolution observation and monitoring of global land cover: First mapping results with Landsat TM and ETM+ data. *International Journal of Remote Sensing*, 34(7), 2607–2654.

Goodwin, N. R., Collett, L. J., Denham, R. J., Flood, N., & Tindall, D. 2013. Cloud and cloud shadow screening across Queensland, Australia: An automated method for Landsat TM/ETM+ time series. *Remote Sensing of Environment*, 134, 50–65.

Gopal, S., & Woodcock, C. E. 1996. Remote sensing of forest change using artificial neural networks. *IEEE Transactions on Geoscience and Remote Sensing*, 34(2), 398–404.

Hall, F. G., & Knapp, D. 2000. BOREAS TE-18 Landsat TM maximum likelihood classification image of the SSA. *Technical Report Series on the Boreal Ecosystem-Atmosphere Study (BOREAS)*, Volume 175, Greenbelt, MD: NASA.

Hansen, M., DeFries, R., Townshend, J. R. G., & Sohlberg, R. 2000. Global land cover classification at 1km resolution using a decision tree classifier. *International Journal of Remote Sensing*, 21, 1331–1365.

Hay, G. J., & Castilla, G. 2008. Geographic object-based image analysis (GEOBIA): A new name for a new discipline. In T. Blaschke, S. Lang, & G. Hay (eds), *Object Based Image Analysis*. Heidelberg, Berlin, & New York: Springer, pp. 93–112.

Hay, G. J., Castilla, G., Wulder, M. A., & Ruiz, J. R. 2005. An automated object-based approach for the multiscale image segmentation of forest scenes. *International Journal of Applied Earth Observation and Geoinformation*, 7(4), 339–359.

Hay, G. J., Marceau, D. J., Dube, P., & Bouchard, A. 2001. A multiscale framework for landscape analysis: Object-specific analysis and upscaling. *Landscape Ecology*, 16(6), 471–490.

Helber, P., Bischke, B., Dengel, A., & Borth, D. 2019. EuroSAT: A novel dataset and deep learning benchmark for land use and land cover classification. *IEEE Journal of Selected Topics in Applied Earth Observations and Remote Sensing*, 12(7), 2217–2226, July. http://doi.org/10.1109/JSTARS.2019.2918242.

Hsu, C.-W., Chang, C.-C., & Lin C.-J. 2001. *A Practical Guide to Support Vector Classification*. Available at http://www.csie.ntu.edu.tw/~cjlin/

Huang, C., Davis, L. S., & Townshend, J. R. G. 2002. An assessment of support vector machines for land cover classification. *International Journal of Remote Sensing*, 23, 725–749.

Irish, R. 2000. Landsat-7 automatic cloud cover assessment algorithms for multispectral, hyperspectral, and ultraspectral imagery. *The International Society for Optical Engineering*, 4049, 348–355.

Irish, R., Barker, J. L., Goward, S. N., & Arvidson, T. 2006. Characterization of the Landsat-7 ETM+ automated cloud-cover assessment (ACCA) algorithm. *Photogrammetric Engineering and Remote Sensing*, 72(10), 1179–1188.

Jensen, J. R. 2006. *Remote Sensing of the Environment: An Earth Resource Perspective*, 2nd edition. Upper Saddle River, NJ: Prentice Hall, 608 pp.

Jia, X., & Richards, J. A. 1994. Efficient maximum likelihood classification for imaging spectrometer data sets. *IEEE Transactions on Geoscience and Remote Sensing*, 32(2), 274–281.

Jozdani, S. E., Johnson, B. A., & Chen, D. 2019. Comparing deep neural networks, ensemble classifiers, and support vector machine algorithms for object-based urban land use/land cover classification. *Remote Sensing*, 11(14), 1713. https://doi.org/10.3390/rs11141713

Kalensky, Z. D. 1998. AFRICOVER land cover database and map of Africa. *Canadian Journal of Remote Sensing*, 24(3), 292–297.

Kavzoglu, T. 2009. Increasing the accuracy of neural network classification using refined training data. *Environmental Modelling and Software*, 24(7), 850–858.

Kavzoglu, T., & Colkesen, I. 2009. A kernel functions analysis for support vector machines for land cover classification. *International Journal of Applied Earth Observation and Geoinformation*, 11, 352–359.

Lin, C.-F., & Wang, S. D. 2002 Fuzzy support vector machines. *IEEE Transactions on Neural Networks*, 13, 464–471.

Lizarazo, I., & Elsner, P. 2009. Fuzzy segmentation for object-based image classification. *International Journal of Remote Sensing*, 30(6), 1643–1649.

Lizarazo, I., & Elsner, P. 2011. Segmentation of remotely sensed imagery: Moving from sharp objects to fuzzy regions. In Pei-Gee Ho (ed), *Image Segmentation*. InTech. ISBN: 978-953-307-228-9. Available at http://www.intechopen.com/books/image-segmentation/segmentation-of-remotely-sensed-imagery-moving-from-sharp-objects-to-fuzzy-regions

Loveland, T. R., Merchant, J. W., Ohlen, D. O., & Brown, J. F. 1991. Development of a land cover characteristics data base for the conterminous U.S. *Conservation and Survey Division*, 319. Available at https://digitalcommons.unl.edu/conservationsurvey/319

Loveland, T. R., Reed, B. C., Brown, J. F., Ohlen, D. O., Zhu, Z., Yang, L., & Merchant, J. W. 2000. Development of a global land cover characteristics database and IGBP DISCover from 1 km AVHRR data. *International Journal of Remote Sensing*, 21(6–7), 1303–1330.

Ma, L., Li, M., Ma, X., Cheng, L., Du, P., & Liu, Y. 2017. A review of supervised object-based land-cover image classification. *ISPRS Journal of Photogrammetry and Remote Sensing*, 130(2017), 277–293. ISSN 0924-2716. https://doi.org/10.1016/j.isprsjprs.2017.06.001. https://www.sciencedirect.com/science/article/pii/S092427161630661X

Mantero, P., Moser, G., & Serpico, S. B. 2005. Partially supervised classification of remote sensing images through SVM-based probability density estimation. *IEEE Transactions on Geoscience and Remote Sensing*, 43, 559–570.

McIver, D. K., & Friedl, M. A. 2002. Using prior probabilities in decision-tree classification of remotely sensed data. *Remote Sensing of Environment*, 81(2–3), 253–261.

Mondal, A., Kundu, S., Chandniha, S. K., Shukla, R., & Mishr, P. K. 2012. Comparison of support vector machine and maximum likelihood classification technique using satellite imagery. *International Journal of Remote Sensing and GIS*, 1(2), 116–123.

Montgomery, D. C., & Peck, E. A. 1992. *Introduction to Linear Regression Analysis*, 2nd edition. New York: Wiley.

Mountrakis, G., Im, J., & Ogole, C. 2011. Support vector machines in remote sensing: A review. *ISPRS Journal of Photogrammetry and Remote Sensing*, 66, 247–259.

Nasiri, V., Deljouei, A., Moradi, F., Sadeghi, S. M. M., & Borz, S. A. 2022. Land use and land cover mapping using sentinel-2, Landsat-8 satellite images, and google earth engine: A comparison of two composition methods. *Remote Sensing*, 14(9), 1977. https://doi.org/10.3390/rs14091977

Oreopoulos, L., Wilson, M., & Várnai, T. 2011. Implementation on Landsat data of a simple cloud mask algorithm developed for MODIS land bands. *IEEE Transactions on Geoscience and Remote Sensing*, 8(4), 597–601.

Pal, M., & Mather, P. M. 2003. An assessment of the effectiveness of decision tree methods for land cover classification. *Remote Sensing of Environment*, 86, 554–556.

Papoutsis, I., Bountos, N. I., Zavras, A., Michail, D., & Tryfonopoulos, C. 2023. Benchmarking and scaling of deep learning models for land cover image classification. *ISPRS Journal of Photogrammetry and Remote Sensing*, 195(2023), 250–268. ISSN 0924-2716. https://doi.org/10.1016/j.isprsjprs.2022.11.012. https://www.sciencedirect.com/science/article/pii/S0924271622003057

Phan, T. N., Kuch, V., & Lehnert, L. W. 2020. Land cover classification using Google earth engine and random forest classifier—the role of image composition. *Remote Sensing*, 12(15), 2411. https://doi.org/10.3390/rs12152411

Platt, R. V., & Rapoza, L. 2008. An evaluation of an object-oriented paradigm for land use/land cover classification. *The Professional Geographer*, 60(1), 87–100.

Prasad, P., Loveson, V. J., Chandra, P., & Kotha, M. 2022. Evaluation and comparison of the earth observing sensors in land cover/land use studies using machine learning algorithms. *Ecological Informatics*, 68(2022), 101522. ISSN 1574-9541. https://doi.org/10.1016/j.ecoinf.2021.101522. https://www.sciencedirect.com/science/article/pii/S1574954121003137

Pratola, C., Del Frate, F., Schiavon, G., Solimini, D., & Licciardi, G. 2011. Characterizing land cover from X-band COSMO-SkyMed images by neural networks. *Urban Remote Sensing*, 2011, 49–52.

Richards, J. A. 2012. *Remote Sensing Digital Image Analysis: An Introduction*, 5th edition. Heidelberg, Germany: Springer Science & Business Media, 513 pp.

Rogan, J., Franklin, J., & Roberts, D. A. 2002. A comparison of methods for monitoring multitemporal vegetation change using Thematic Mapper imagery. *Remote Sensing of Environment*, 80, 143–156.

Rodriguez-Galiano, V., Ghimire, B., Rogan, J., Chica-Olmo, M., & Rigol-Sanchez, J. 2012. An assessment of the effectiveness of a random forest classifier for land-cover classification. *ISPRS Journal of Photogrammetry and Remote Sensing*, 67, 93–104.

Roy, D. P., Ju, J., Kline, K., Scaramuzza, P. L., Kovalskyy, V., Hansen, M. C., Loveland, T. R., Vermote, E. F., & Zhang, C. 2010. Web-enabled Landsat data (WELD): Landsat ETM+ composited mosaics of the conterminous United States. *Remote Sensing of Environment*, 114, 35–49.

Ryherd, S., & Woodcock, C. E. 1996. Combining spectral and texture data in the segmentation of remotely sensed images. *Photogrammetric Engineering and Remote Sensing*, 62(2), 181–194.

Scholkopf, B., & Smola, A. J. 2001. *Learning with Kernels*. Cambridge, MA: The MIT Press.

Scott, G. J., England, M. R., Starms, W. A., Marcum, R. A., & Davis, C. H. 2017. Training deep convolutional neural networks for land–cover classification of high-resolution imagery. *IEEE Geoscience and Remote Sensing Letters*, 14(4), 549–553, April. http://doi.org/10.1109/LGRS.2017.2657778

Sexton, J. O., Song, X.-P., Feng, M., Noojipady, P., Anand, A., Huang, C., Kim, D.-H., Collins, K. M., Channan, S., DiMiceli, C., & Townshend, J. R. G. 2013. Global, 30-m resolution continuous fields of tree cover: Landsat-based rescaling of MODIS Vegetation Continuous Fields with lidar-based estimates of error. *International Journal of Digital Earth*, 130321031236007. http://doi.org/10.1080/17538947.2013.786146

Shao, Y., & Lunetta, R. S. 2012. Comparison of support vector machine, neural network, and CART algorithms for the land-cover classification using limited training data points, *ISPRS Journal of Photogrammetry and Remote Sensing*, 70, 78–87. ISSN 0924-2716. https://doi.org/10.1016/j.isprsjprs.2012.04.001

Simpson, J. J., & Stitt, J. R. 1998. A procedure for the detection and removal of cloud shadow from AVHRR data over land. *Geoscience and Remote Sensing*, 36(3), 880–890.

Stehman, S. V. 2012. Impact of sample size allocation when using stratified random sampling to estimate accuracy and area of land-cover change. *Remote Sensing Letters*, 3(2), 111–120.

Stehman, S. V., & Czaplewski, R. L. 1998. Design and analysis of thematic map accuracy assessment: Fundamental principles. *Remote Sensing of Environment*, 64, 331–344.

Strahler, H. 1980. The use of prior probabilities in maximum likelihood classification of remotely sensed data. *Remote Sensing of Environment*, 10, 135–163.

Sun, J., Yang, J., Zhang, C., Yun, W., & Qu, J. 2013. Automatic remotely sensed image classification in a grid environment based on the maximum likelihood method. *Mathematical and Computer Modeling*, 58(3–4), 573–581.

Suykens, J. A. K., Brabanter, J. D., Lukas, L., & Vandewalle, J. 2002. Weighted least squares support vector machines: Robustness and sparse approximation. *Neurocomputing*, 48, 85–105.

Talukdar, S., Singha, P., Mahato, S., Shahfahad, S. P., Liou, Y.-A., & Rahman, A. 2020. Land-use land-cover classification by machine learning classifiers for satellite observations—a review. *Remote Sensing*, 12(7), 1135. https://doi.org/10.3390/rs12071135

Townshend, J. R. G., Huang, C., Kalluri, S. N., DeFries, R. S., Liang, S., & Yang, K. 2000. Beware of per-pixel characterization of land cover. *International Journal of Remote Sensing*, 21(4), 839–843.

Travaglia, C., Milenova, L., Nedkov, R., Vassilev, V., Milenov, P., Radkov, R., & Pironkova, Z. 2001. Preparation of land cover database of Bulgaria through remote sensing and GIS. *Environment and Natural Resources Working Paper No. 6*. FAO, Rome, 57pp.

Tsujinishi, D., & Abe, S. 2003. Fuzzy least squares support vector machines for multiclass problems. *Neural Networks*, 16, 785–792.

Vapnik, V. 1979. *Estimation of Dependences Based on Empirical Data*. Nauka Moscow, pp. 5165–5184, 27 (in Russian) (English translation: Springer Verlag, New York, 1982).

Vapnik, V. 1995. *The Nature of Statistical Learning Theory*, 2nd edition. Heidelberg, Germany: Springer.

Vogelmann, J. E., Howard, S. M., Yang, L., Larson, C. R., Wylie, B. K., & Van Driel, J. N. 2001. Completion of the 1990's national land cover data set for the conterminous United States. *Photogrammetric Engineering and Remote Sensing*, 67, 650–662.

Wacker, A. G., & Landgrebe, D. A. 1972. Minimum distance classification in remote sensing. *LARS Technical Reports*. Paper 25. http://docs.lib.purdue.edu/larstech/25

Woodcock, C. E., & Harward, V. J. 1992. Nested-hierarchical scene models and image segmentation. *International Journal of Remote Sensing*, 13, 3167–3187.

Yoshida, T., & Omatu, S. 1994. Neural network approach to land cover mapping. *IEEE Transactions on Geoscience and Remote Sensing*, 32(5), 1103–1109.

Yu, L., Wang, J., & Gong, P. 2013. Improving 30 meter global land cover map FROM-GLC with time series MODIS and auxiliary datasets: A segmentation based approach. *International Journal of Remote Sensing*, 34(16), 5851–5867.

Zhang, C., Sargent, I., Pan, X., Li, H., Gardiner, A., Hare, J., & Atkinson, P. M. 2019. Joint deep learning for land cover and land use classification. *Remote Sensing of Environment*, 221(2019), 173–187. ISSN 0034-4257. https://doi.org/10.1016/j.rse.2018.11.014. https://www.sciencedirect.com/science/article/pii/S0034425718305236

Zhang, P., Ke, Y., Zhang, Z., Wang, M., Li, P., & Zhang, S. 2018. Urban land use and land cover classification using novel deep learning models based on high spatial resolution satellite imagery. *Sensors*, 18(11), 3717. https://doi.org/10.3390/s18113717

Zhu, G., & Blumberg, D. G. 2002. Classification using ASTER data and SVM algorithms; The case study of Beer Sheva, Israel. *Remote Sensing of Environment*, 80, 233–240.

Zhu, Z., & Woodcock, C. E. 2012. Object-based cloud and cloud shadow detection in Landsat imagery. *Remote Sensing of Environment*, 118, 83–94.

Zhu, Z., Yang, L., Stehman, S. V., & Czaplewski, R. L. 2000. Accuracy assessment for the U.S. geological survey regional land-cover mapping program: New York and New Jersey Region. *Photogrammetric Engineering and Remote Sensing*, 66(12), 1425–1435.

3 Urban Image Classification

Per-Pixel Classifiers, Sub-Pixel Analysis, Object-Based Image Analysis, Geospatial Methods, and Machine Learning Approach

Soe W. Myint, Victor Mesev, Dale Quattrochi, and Elizabeth A. Wentz

LIST OF ACRONYMS

CNN	Convolutional neural networks
GEOBIA	Geospatial object based image analysis
GLCM	Gray level co-occurrence matrix
ICAMS	Image characterization and modeling system
K-NN	K-nearest neighbors
MESMA	Multiple endmember spectral mixture analysis
OBIA	Object based image analysis
RF	Random forest
SVM	Support vector machines

3.1 INTRODUCTION

Remote sensing methods used to generate base maps to analyze the urban environment rely predominantly on digital sensor data from space-borne platforms. This is due in part from new sources of high spatial resolution data covering the globe, a variety of multispectral and multitemporal sources, sophisticated statistical and geospatial methods, and compatibility with GIS data sources and methods. The goal of this chapter is to review the four groups of classification methods for digital sensor data from space-borne platforms; per-pixel, sub-pixel, object-based (spatial-based), and geospatial methods. Per-pixel methods are widely used methods that classify pixels into distinct categories based solely on the spectral and ancillary information within that pixel. They are used for simple calculations of environmental indices (e.g., NDVI) to sophisticated expert systems to assign urban land covers (Stefanov et al., 2001). Researchers recognize however, that even with the smallest pixel size the spectral information within a pixel is really a combination of multiple urban surfaces. Sub-pixel classification methods therefore aim to statistically quantify the mixture of surfaces to improve overall classification accuracy (Myint, 2006a). While within pixel variations exist, there is also significant evidence that groups of nearby pixels have similar spectral information and therefore belong to the same classification category. Object-oriented methods have emerged that group pixels prior to classification based on spectral similarity and spatial proximity. Classification accuracy using object-based methods show significant success and promise for numerous urban applications (Myint et al., 2011). Like the object-oriented methods that recognize the importance of spatial proximity, geospatial

DOI: 10.1201/9781003541158-4

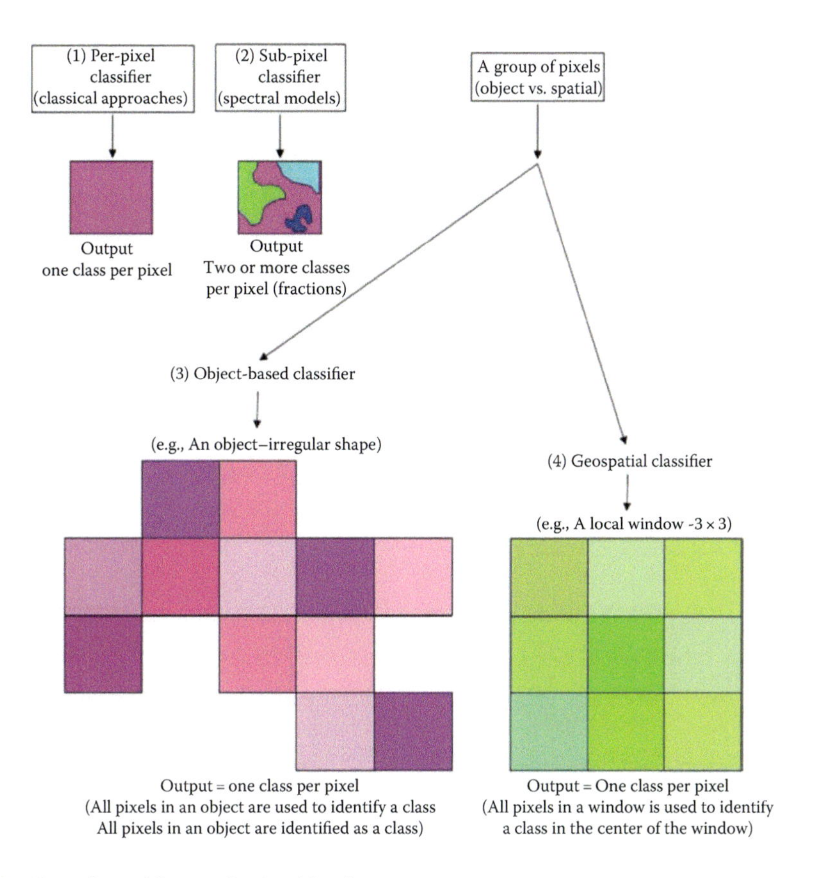

FIGURE 3.1 Overview of four main classification groups.

methods for urban mapping also utilize neighboring pixels in the classification process. The primary difference though is that geostatistical methods (e.g., spatial autocorrelation methods) are utilized during both the pre- and post-classification steps (Myint and Mesev, 2012).

Within this chapter, each of the four approaches is described in terms of scale and accuracy classifying urban land use and urban land cover; and for its range of urban applications. We demonstrate the overview of four main classification groups in Figure 3.1 while Table 3.1 details the approaches with respect to classification requirements and procedures (e.g., reflectance conversion, steps before training sample selection, training samples, spatial approaches commonly used, classifiers, primary inputs for classification, output structures, number of output layers, and accuracy assessment). The chapter concludes with a brief summary of the methods reviewed and the challenges that remain in developing new classification methods for improving the efficiency and accuracy of mapping urban areas.

3.2 REMOTE SENSING METHODS FOR URBAN CLASSIFICATION AND INTERPRETATION

Urban areas are comprised of a heterogeneous patchwork of land covers and land uses that are juxtaposed so that classification of specific classes using remote sensing data can be problematic. Derivation of classification methods for urban landscape features has evolved in tandem with increasing spatial, spectral, and temporal resolutions of remote sensing instruments (e.g., from 90 m Landsat Multispectral Scanner-MSS to 30 m to the Landsat Enhanced Thematic Mapper Plus [ETM+] and Operational Land Imager [OLI] data and progressing to sub-meter spatial resolution products available from commercial systems such as .34 m Geoeye) to achieve more robust digital

TABLE 3.1
Classification Procedures and Characteristics of the Four Main Classification Groups

	Per-pixel	Sub-pixel	Object-based	Geospatial
Reflectance Conversion	Not required	Necessary	Not required	Not required
Additional Step Before Training Sample Selection	No	No	Segment image into objects	No
Training Samples	Irregular polygons that cover multiple pixels representing selected land cover classes	Spectra of selected endmembers	Segmented objects that cover multiple pixels representing selected land cover classes	Square windows that cover multiple pixels representing selected land cover classes
Commonly Used Spatial Approaches	GLCM	No	GLCM	Fractal, Geary's C Moran's I, Getis index Fourier transforms, Lacunarity index, Wavelet transforms
Commonly Used Classifiers	Maximum Likelihood, Mahalanobis Distance, Minimum Distance, Regression Tree, Neural Network Baysian	Linear Spectral Mixture, Multiple Regression, Regression Tree. Neural Network, Baysian MESMA	Nearest Neighbor Decision Rule	Mahalanobis Distance, Minimum Distance
Primary Input for Classification	Regular training samples are used to identify the selected land categories	Endmember spectra are used to quantify fractions of the selected land cover classes	Selected objects that represent the selected land cover classes are used to identify these classes	Pixel values in each of the selected windows are used to identify each land category
No. of Output Layer	One Layer	Multiple Layers	One Layer	One Layer
Output Structure	One class per pixel	One fraction per pixel per class	One class per pixel	One class per pixel
Accuracy Assessment Method	Randomly selected pixels for error matrix	Correlation between predicted and reference fractions	Randomly selected pixels for error matrix (or) object-based accuracy assessment	Randomly selected pixels for error matrix

Note: GLCM = Gray Level (or) Spatial Co-occurrence Matrix; MESMA = Multiple endmember spectral mixture analysis.

classification schemes. This evolution of classification techniques, however, does not imply that one method is better than another. As with the type of satellite remote sensing data that are employed for analyses, the application of a specific algorithm for classification of urban land cover and land use is dependent upon what the user's objectives are, and what level of detail, frequency, and sensors are required for the anticipated or resulting output products. Table 3.2 shows urban remote sensing applications with regards to spatial, temporal, and sensor resolutions.

3.3 PER-PIXEL METHODS

Scale is indelible when conducting per pixel classifications. The spatial resolution of the sensor dictates the classification type, range, and accuracy of urban land use and urban land cover. That is

TABLE 3.2
Urban Remote Sensing Classifications with Regards to Spatial, Temporal, and Sensor Resolutions

	Urban Features	Urban Process	Spatial Resolution	Temporal Resolution	Sensor Resolution
Micro scale: Individual measurements	Building unit (roofs: flat, pitch) (material: tile, natural/metal, synthetic)	Type and architecture Density	1–5 m	1–5 years	Pan–Vis–NIR
	Vegetation unit (tree, shrub)	Type and health Nature	0.25–5 m	1–5 years	Pan–NIR
	Transport unit (width: road lanes, sidewalk) (material: asphalt, concrete, composite)	Infrastructure Mobility and access	0.25–30 m	1–5 years	Pan–Vis–NIR
Macro scale: Aggregation of imperviousness, greenness, soil and water	Residential neighborhood	Suburbanization Gentrification, poverty, crime, racial segregation, etc.	1–5 km	1–10 years	VIS–NIR– TIR
	Industrial/commercial zone	Land use zoning Storm water flow Heat island effect	1–5 km	1–10 years	VIS–NIR– TIR
	Non built urban	Environmental concerns Beautification Public space	1–5 km	1–10 years	VIS–NIR–TIR
	Urban area	Centrality and sprawl Flow and congestion Sustainability	5–100 km	1–10 years	VIS–NIR– TIR– MIR– Radar

because individual urban features are rarely the same size as pixels, nor are they conveniently rectangular in shape. Add temporal scale representing rapid urban activity and per pixel classifications become even more removed from reality. Refining the spatial resolution and reducing the area of the pixel does not necessarily lead to improvements in classification accuracy, and may even introduce additional spectral noise, especially when pixels are smaller than urban features. In all, the ideal situation that each pixel can be identified to represent conclusively one and only one land cover type has now long been abandoned. So, too, the perfect relationship between the pixel and the field-of-view, which assumes reflectance is recorded entirely and uniformly from within the spatial limits of individual pixels (Figure 3.1).

Regardless, the appeal of per-pixel or hard classifications remains; predominantly because they produce crisp and convenient thematic coverages that can be easily integrated with raster-based GIS models (Table 3.1). Composite models and methodologies containing information from remotely sensed sources are critical for revising databases and for producing comprehensive query-based urban applications. To preserve this relationship with GIS, the quality of per-pixel classifications must be monitored not only using conventional determination of accuracy based on comparisons with more reliable reference data, but also in relation to levels of suitability or "scale of appropriateness." Both were evident in the USGS hierarchical scheme (Anderson et al., 1976) using the much-cited 85% as a general guideline for the accuracy of urban features, and which subsequently established a benchmark for researchers to attain and supersede using a variety of statistical and stochastic per-pixel techniques. Some of these focused exclusively on maximizing computational class separability, using the traditional maximum likelihood algorithm (Strahler, 1980) and the more recent support vector machines (Yang, 2011), while others developed methodologies that imported extraneous information when aggregating spectrally similar pixels (Mesev, 1998), by incorporating contextual relationships (Stuckens et al., 2000), or by measuring pixel inter-connectivity (Barr and Barnsley, 1997). In both, classification accuracy typically improves only marginally, simply because there is an inherent numerical limitation to the extent individual pixel values can comprehensively represent the multitude of true urban features within the rigid confines of their regular-sized pixel limits (Fisher, 1997).

However, within these numerical limits per-pixel classification accuracy can be consistently high if the appropriate spatial resolution (i.e., pixel size) is identified with respect to the suitable level of urban detail (Table 3.2). Such ideas of scale appropriateness can be traced back to Welch (1982), and have since been widely accepted as an important part of the class training process. But the decision is far from trivial, and must also consider the appropriate scale of analysis (Mesev, 2012). Consider a continuous scale that can be conceptualized by levels of measurement from remote sensor data; ranging from the representation of atomistic urban features (building, tree, sidewalk, etc.) at the micro scale, to the representation of aggregate urban features (residential neighborhoods, industrial zones, or even complete urban areas) at the macro scale. Micro urban remote sensing by per-pixel classification remains highly tenuous (even using meter and sub meter resolutions from the latest sensors) and any reliable interpretation is extracted directly from the spatial orientation of pixels—in a similar vein to conventional interpretation of aerial photography, but with lower clarity and with limited stereoscopic capabilities. However, the spectral heterogeneity problem is less restrictive at the macro scale of analysis where classified pixels, instead of measuring individual urban objects, can be aggregated to represent a generalized view of urban areas, including total imperviousness, approximate lateral growth, and overall greenness. It is at this scale of analysis that many types of urban processes, such as sprawl, congestion, poverty, land use zoning, storm water flow, and heat islands, can be studied simultaneously across an entire urban area as part of the search for theories of livability and sustainability. In sum, per-pixel classifications produce simple and convenient thematic maps of urban land use and land cover that can be incorporated into GIS models. The spatial resolution of the remote sensor, however, limits their accuracy away from mapping individual urban features with any level of pragmatic precision and towards more traditional macro scales of generalized land cover combinations reminiscent of the timeless V-I-S model (Ridd, 1995).

3.4 SUB-PIXEL METHODS

If locational and thematic accuracy of urban representation from remote sensing is paramount, per-pixel classifications can be modified statistically to measure spectral mixtures representing multiple land cover classes within individual pixels. These are termed sub-pixel algorithms or soft classifications because pixels are no longer constrained to representing single classes, but instead represent various proportions of land cover classes which are conceptually more akin to the spatial and compositional heterogeneity of urban configurations (Ji and Jensen, 1999; Small, 2004). The debate on which approach, per-pixel or sub-pixel, can again be tied to the scale of urban analysis. For example, the measurement of impervious surfaces is particularly amenable to sub-pixel classification because pixels can represent a continuum of imperviousness, from total coverage (downtown areas and industrial estates) to scant dispersion intermingled with bio-physical land covers (city parks). Extensive research has been devoted to more precise quantification of impervious surfaces, and other urban land covers at sub-pixel level, such as linear mixture models (Wu and Murray, 2003; Rashed et al., 2003), background removal spectral mixture analysis (Ji and Jensen, 1999; Myint, 2006a), Bayesian probabilities (Foody et al., 1992; Mesev, 2001; Eastman and Laney, 2002; Hung and Ridd, 2002), artificial neural network (Foody and Aurora, 1996; Zhang and Foody, 2001), normalized spectral mixture analysis (Wu and Yuan, 2007; Yuan and Bauer, 2007), fuzzy c-means methods (Fisher and Pathirana, 1990; Foody, 2000), multivariate statistical analysis (Bauer et al., 2004; Yang and Liu, 2005; Bauer et al., 2007), and regression trees (Yang et al., 2003a, 2003b; Homer et al., 2007).

Among these, linear spectral mixture analysis, regression analysis, and regression trees have had a wider appeal because they are theoretically and computationally simpler, as well as more prevalent in many commercial software packages. However, the success of measuring urban land cover types using linear techniques is dependent on identifying spectrally pure endmembers, preferably using reference samples collected in the field (Adams et al., 1995; Roberts et al., 1998, 2012). Although Weng and Hu (2008) derived moderate accuracy levels from employing linear spectral mixture analysis using ASTER and Landsat ETM+ sensor imagery, they discovered that artificial neural networks were also capable of performing nonlinear mixing of land cover types at the subpixel level (Borel and Gerstl, 1994; Ray and Murray, 1996). Another limitation with linear spectral mixture classifiers is that they do not permit the number of endmembers to be greater than the number of spectral bands (Myint, 2006a). In response, a multiple endmember spectral mixture analysis (MESMA) has been developed to identify many more endmember types to represent the heterogeneous mixture of urban land cover types (Rashed et al., 2003; Powell et al., 2007; Myint and Okin, 2009). Diagrams demonstrating linear spectral mixture analysis and multiple endmember spectral mixture analysis are provided in Figures 3.2 and 3.3, respectively.

Two challenges dominate the research efforts to improve subpixel analysis methods for urban settings. The first challenge is pixel size. Identifying endmembers for all classes in images with large to medium pixels in urban areas is difficult given the heterogeneous nature of urban areas. In small spatial distances (e.g., <30 m), surfaces rapidly change from impervious, to grass, to building. The smaller pixel size (e.g., 1 m or sub-meter), however, is not always the optimal solution. While pixels may not reflect a mixture of the desired endmembers (e.g., a combination of asphalt and grass), reflectance from unwanted features begin to appear that need to be filtered (e.g., oil surfaces and automobiles in asphalt; chimneys and air conditions on rooftops). The second limitation is that it is almost impossible to identify all possible endmembers in a study area and classification accuracy can be degraded by the potential presence of unknown classes or unidentified classes (e.g., the preceding asphalt and rooftop examples). This is because the classifier is based on the assumption that the sum of the fractional proportions of all possible endmembers in a pixel is equal to one. Although this type of modeling is conceptually more representative of urban land cover, from a practical standpoint it nonetheless perpetuates the mixed pixel problem and presents thematic and semantic limitations to urban land classification schemes. In other words, output from sub-pixel analysis produces fractional classes that are more difficult to integrate with GIS data and may even limit their portability for comparisons across space and through time.

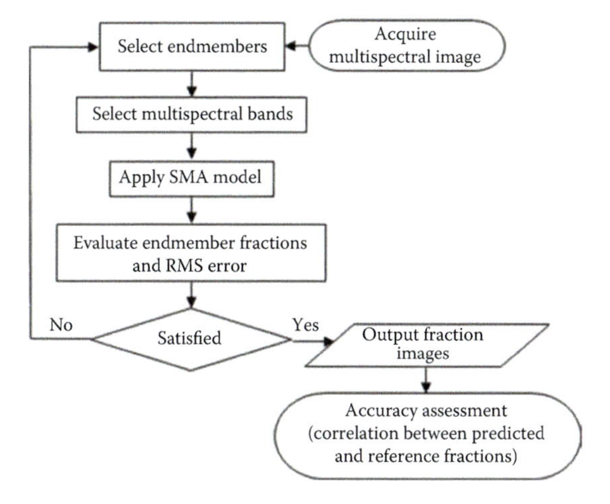

FIGURE 3.2 Spectral mixture analysis.

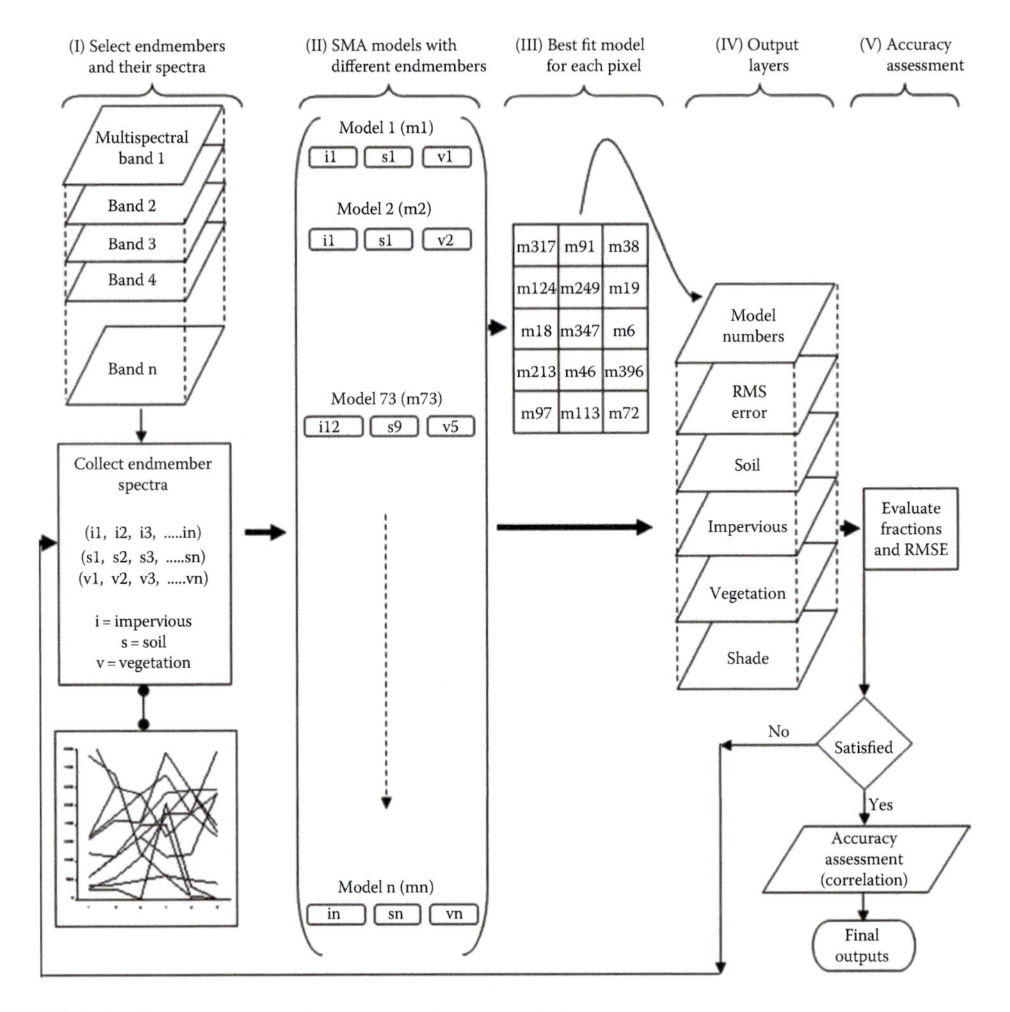

FIGURE 3.3 Multiple endmember spectral mixture analysis.

3.5 OBJECT-BASED METHODS

With the representational limitations of purely spectrally based per-pixel and sub-pixel classifications it was only a matter of time before the shift to the spatial domain gained momentum. Even from a purely intuitive standpoint finer resolution (i.e., smaller pixels or large cartographic-scale) imagery exhibit higher levels of detailed features that mimic the heterogeneous nature of urban areas. This greater level of spatial detail invariably also leads to many more uncertain spectral classes—known as noise—which can be true but potentially unwanted urban features such as chimneys or manhole covers. Assuming spectral noise is reduced, images with spatial resolutions ranging from about 0.25 to 5 m have the potential to help identify urban structures necessary to perform many urban applications, including estimation of population based on the number of dwellings of different housing types, residential water use, predicting energy consumption, urban heat island, outdoor water use, solar energy use, and storm water pollution modeling (Jensen and Cowen, 1999).

Conceptually, spatial or object-based approaches are most applicable to high spatial resolution remote sensing data, where objects of interest are larger than the ground resolution element, or pixel. Urban objects may be vegetated features of urban landscapes (e.g., trees, shrubs, golf course) or anthropogenic features (e.g., buildings, pools, sidewalks, roads, canals). With regards to mapping categorical data or identifying land use land cover classes, remotely sensed image analysis started to shift from pixel-based (per-pixel) to object based image analysis (OBIA) or geospatial object based image analysis (GEOBIA) around the year 2000 (Blaschke, 2010). The object-centered classification prototype starts with the generation of segmented objects at multiple scales (Desclee et al., 2006; Navulur, 2007; Im et al., 2008; Myint et al., 2008). To demonstrate, Walker and Briggs (2007) employed an object-oriented classification procedure to effectively delineate woody vegetation in an arid urban ecosystem using high spatial resolution true-color aerial photography (without the near infrared band) and achieved an overall accuracy of 81%. Hermosilla et al. (2012) developed two object-based approaches for automatic building detection and localization using high spatial resolution imagery and LiDAR data. Stow et al. (2007) further developed object-based classification by taking advantage of the spatial frequency characteristics of multispectral data, and then measuring the proportions of vegetation, imperviousness, and soil sub-objects to identify residential land use in Accra, Ghana (they documented an overall accuracy of 75%). In another study by Zhou et el. (2008), post-classification change detection based on the object-based analysis of multitemporal high spatial resolution produced even higher accuracies of 92% and 94%; while Myint and Stow (2011) demonstrated the effectiveness of object-based strategies based on decision rules (i.e., membership functions) and nearest neighbor classifiers on high spatial resolution Quickbird multispectral satellite data over the city of Phoenix. These are further supported by Myint et al. (2011) who directly compared the accuracy from object-based classifications (90%) with more traditional spectral-based classifications (68%). The land-cover classes that the authors identified for this particular study include buildings, other impervious surfaces (e.g., roads and parking lots), unmanaged soil, trees/shrubs, grass, swimming pools, and lakes/ponds. The study selected 500 samples points that led to approximately 70 points per class (seven total classes) using a stratified random sampling approach for the accuracy assessment of two different subsets of QuickBird over Phoenix. To be consistent and for precise comparison purposes, they applied the same sample points generated for the output generated by the object-based classifier as the output produced by the traditional classification technique (i.e., maximum likelihood).

In general, spectrally similar signatures such as dark/gray soil, dark/gray rooftops, dark/gray roads, swimming pools/blue color rooftops, and red soil/red rooftop remain problematic even with object-based approaches. Furthermore, the most commonly used object-oriented software (Definiens or eCognition) is required to perform a tremendous number of segmentations of objects from all spectral bands using various scale parameters. There is no universally accepted method to determine an optimal level of scale (e.g., object size) to segment objects, and a single scale may not be suitable for all classes. The most feasible approach may be to select the bands for membership functions at the scale that identifies the class with variable options and analyze them heuristically on the display screen. Given that the nearest neighbor classifier and decision rule available in the object-based approach are non-parametric approaches, they are independent of the assumption that data values need to be normally distributed. This is advantageous, because most data are not normally distributed in many real world situations. Another advantage of the object-based approach is that it allows additional selection or modification of new objects (training samples) at iterative stages, until the satisfactory result is obtained. However, the object-based approach has a significant problem when dealing with a remotely sensed data over a fairly large area since computer memory needs to be used extensively to segment tremendous numbers of objects using multispectral bands. This is true even for fine spatial resolution data with fewer bands (e.g., QuickBird) over a small study area when requiring smaller scale parameters (smaller objects). Figure 3.4 shows segmented images at scale level 25, 50, and 100 using a subset of a QuickBird image over Phoenix. Figure 3.5 demonstrates how hierarchical image segmentation delineates image objects at various scales.

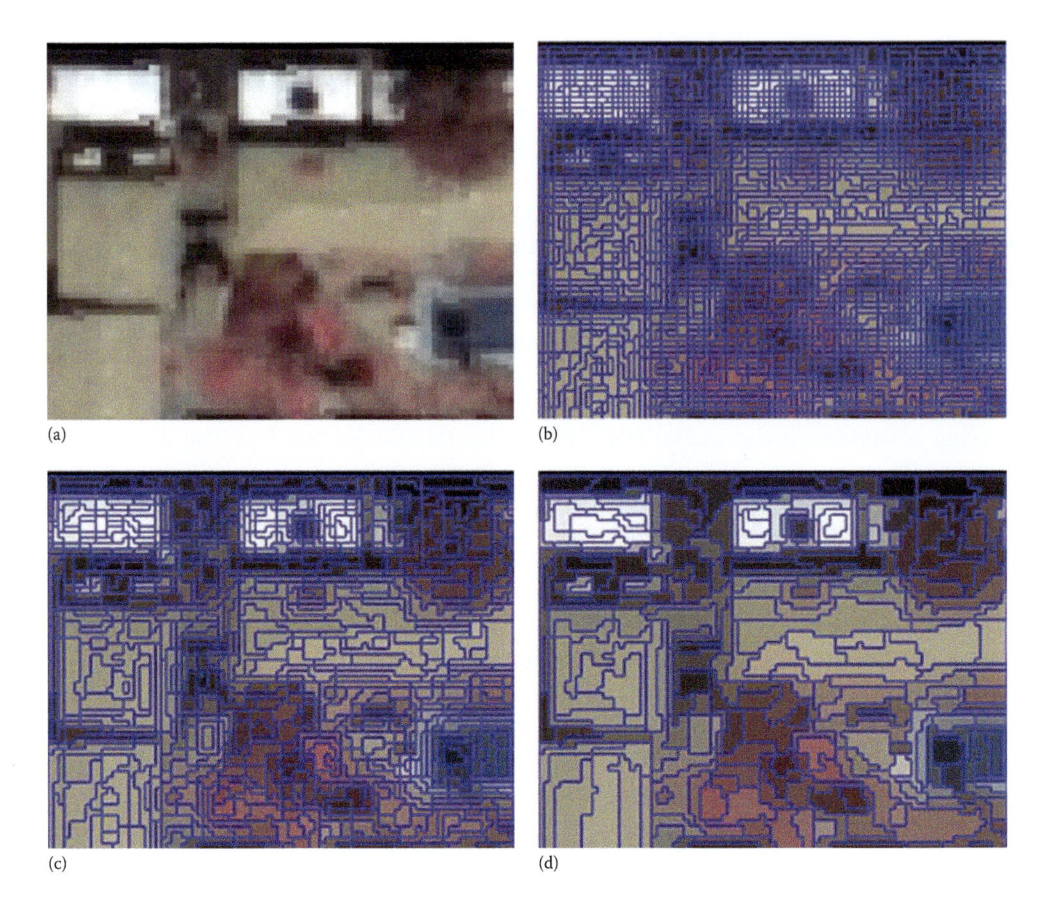

FIGURE 3.4 A subset image and segmented images at different scales: (a) original subset, (b) level 1 (scale parameter 25), (c) level 2 (scale parameters 50), (d) level 3 (scale parameter 100).

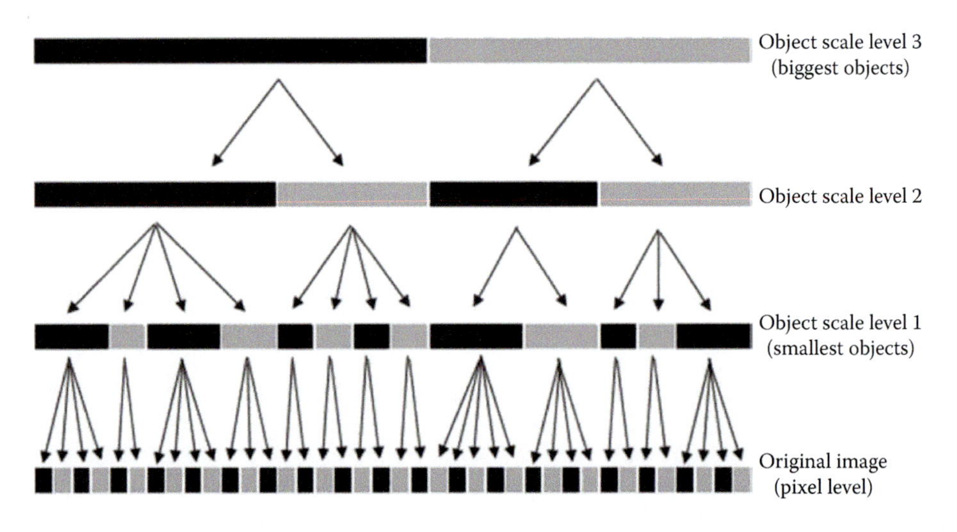

FIGURE 3.5 Image objects at each image scale level. Level 3 = 100, level 2 = 50, level 1 = 25.

3.6 GEOSPATIAL METHODS

Texture plays an important role in the human visual system for pattern recognition and interpretation. For image interpretation, pattern is defined as the overall spatial form of related features, where the repetition of certain forms is a characteristic pattern found in many cultural objects and some natural features. Local variability in remotely sensed data, which is part of texture or pattern analysis, can be characterized by computing the statistics of a group of pixels, for example, standard deviation, coefficient of variance or autocovariance, or by the analysis of fractal similarities or autocorrelation of spatial relationships. There have been some attempts to improve the spectral analysis of remotely sensed data by using texture transforms in which some measure of variability in digital numbers is estimated within local windows; for example the contrast between neighboring pixels (Edwards et al., 1988), standard deviation (Arai, 1993), or local variance (Woodcock and Harward, 1992). One commonly used statistical procedure for interpreting texture uses an image spatial co-occurrence matrix, which is also known as a gray level co-occurrence matrix (GLCM) (Franklin et al., 2000). There are a number of texture measures, which could be applied to spatial co-occurrence matrices for texture analysis (Peddle and Franklin, 1991). For instance, Herold et al. (2003) proposed a method based on using landscape metrics to classify IKONOS sensor images, which in turn is compared to a GLCM. Liu et al. (2006) further contrasted spatial metrics, GLCM, and semi-variograms in terms of urban land use classification.

Lam et al. (1998) demonstrated how fractal dimensions yield quantitative insight into the spatial complexity and information contained in remotely sensed data. Quattrochi et al. (1997) went further and created a software package known as the Image Characterization and Modeling System (ICAMS) to explore how the fractal dimension is related to surface texture. Fractal dimensions were also analyzed by Emerson et al. (1999) who used the isarithm method and Moran's I and Geary's C spatial autocorrelation measures to observe the differing spatial structure of the smooth and rough surfaces in remotely sensed images. In terms of other geospatial techniques, De Jong and Burrough (1995) and Woodcock et al. (1988) implemented variograms to measurements derived from remotely sensed to quantitatively describe urban spatial patterns. Myint and Lam (2005a, 2005b) and Myint et al. (2006) developed a number of lacunarity approaches to characterize urban spatial features with completely different texture appearances that may share the same fractal dimension values. Both studies report that lacunarity can be considered more effective in comparison to fractal approaches for urban mapping.

The geospatial methods described so far may not provide satisfactory accuracies when they are applied to the classification of urban features from fine spatial resolution remotely sensed images. That is mainly because most of them focus primarily on coupling features and objects at a single scale and cannot determine the effective representative value of particular texture features according to their directionality, spatial arrangements, variations, edges, contrasts, and the repetitive nature of object and features. There have been a number of reports in spatial frequency analysis of mathematical transforms, which provide solutions using multi-resolution analysis. Recent developments in spatial/frequency transforms such as the Fourier transform, Wigner distribution, discrete cosine transform, and wavelet transform have all provided sound multi-resolution analytical tools (Bovik et al., 1990; Zhu and Yang, 1998).

Of all transformation approaches, wavelets play the most critical part in texture analysis. Wavelets are part of spatial and frequency based classification approaches, and a local window plays an important role in measuring and characterizing spatial arrangements of objects and features. Homogeneity, size of regions, characteristic scale, directionality, and spatial periodicity are important issues that should be considered to identify local windows when performing wavelet analysis (Myint, 2010). From a computational perspective, the ideal window size is the smallest size that also produces the highest accuracy (Hodgson, 1998). The accuracy should increase with a larger local window size since it contains more information than a smaller window size and therefore provides more complete coverage of spatial variation, directionality and spatial periodicity of a particular texture. However, minimization of local window size is also important in spatial-based urban image classification techniques since a larger window size tends to cover more urban land cover features and consequently creates mixed boundary pixels or mixed land cover problems. However, some spatial and frequency approaches such as wavelet dyadic decomposition approaches require large window sizes to capture spatial information at multiple scales (Myint, 2006b). The potential solution to this problem would be to employ a multi-scale overcomplete wavelet analysis using an infinite scale decomposition procedure. This is because a large spatial coverage or a large local window is not needed to describe a spatial pattern. Furthermore, this approach can measure different directional information of anisotropic features at unlimited scales, and it is designed to normalize and select effective features to identify urban classes. Myint and Mesev (2012) employed a wavelet-based classification method to identify urban land use and land cover classes using different decision rule sets and spatial measures and demonstrated the effectiveness of wavelets. However, the current wavelet-based classification system with the dyadic wavelet approach is limited by the fact that higher-level sub-images are just a quarter of the preceding image. In general, smaller window size is generally thought to yield higher accuracy in geospatial-based image classification because if the window is too large, much spatial information from two or more land cover classes could create a mixed boundary problem. Further research is required to consider an overcomplete wavelet approach that can generate spatial arrangements of objects and features at any scale level for urban mapping. Such an approach could potentially be applicable to any land use/land cover system at any resolution or scale because it can effectively use any window size. Figure 3.6 shows how wavelet approaches work in comparison to other geospatial approaches in urban mapping.

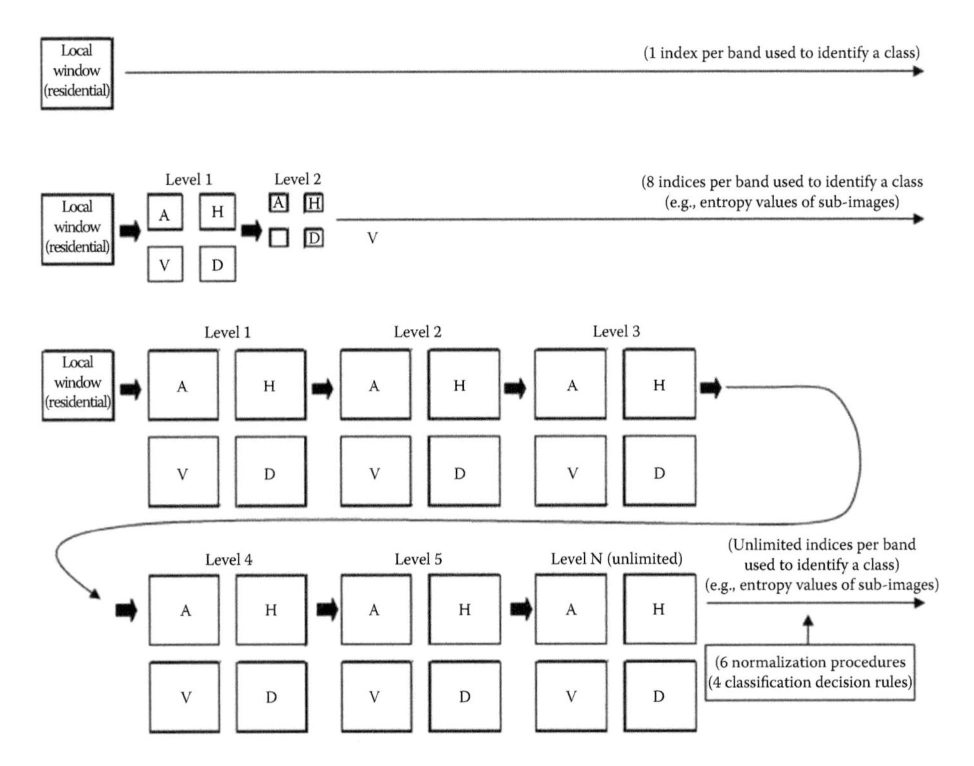

FIGURE 3.6 An example of feature vectors or indices (32 × 32 window or a subset) used to identify an urban class using other geospatial approaches, the dyadic wavelet approach, and the overcomplete wavelet approach. *Note*: Sub-images at level two in the dyadic approach reach the suggested minimum dimension (8 × 8 pixels) since any sub-images smaller than eight pixels may not contain any useful spatial information. A sub-image at a higher level is exactly the same as its original size at the preceding level in the overcomplete approach. It should also be noted that the level of scale with the overcomplete approach is unlimited. A = approximation texture; H = horizontal texture; V = vertical texture; D = diagonal texture.

3.7 MACHINE LEARNING METHODS

Machine learning has been demonstrated as a powerful and versatile technique for the automatic and semi-automatic classification of remote-sensing images (Maxwell et al., 2018). It can leverage the analysis capabilities of various images, including multispectral and hyperspectral images, Radar, and LiDAR data. Combining machine learning and remote-sensing images has been applied to land use and land cover mapping, ecological-environmental monitoring, agriculture, resource planning, disaster response, urban planning, etc. (Lary et al., 2016). It can recognize patterns and features of various images by accurately identifying different land cover classes, such as forests, water bodies, agricultural fields, and urban areas (Myint et al., 2011).

Combining machine learning and remote-sensing images typically involves several key steps, including data collection, preprocessing, feature extraction, labeling, model training, and validation. Machine learning algorithms such as Support Vector Machines (SVM), Random Forest, Convolutional Neural Networks (CNN), and others are commonly employed to perform the classification (Lary et al., 2016). The results are often presented in the form of thematic maps. Here's a detailed explanation of how machine learning applies to remote sensing classification.

Suitable machine learning algorithms for classification should be selected based on the characteristics of data, the specific classification task, and the computational resources available. The algorithm's strengths and weaknesses should also be considered. Common and popular choices include:

- Support Vector Machines (SVM)
- Random Forest (RF)
- Convolutional Neural Networks (CNN), one of the deep learning networks
- K-Nearest Neighbors (K-NN)
- Ensemble learning by cooperating with multiple machine learning models

Once the suitable algorithms are selected, we can perform modeling training by inputting features of remote-sensing images as independent variables and training samples as dependent variables. The model's performance can be assessed using validation samples by metrics, including overall accuracy, precision, recall, F1 score, and confusion matrix.

When machine learning models are built, it is necessary to fine-tune and optimize their key hyperparameters to improve and enhance the robustness of classification tasks. Hyperparameter optimization techniques, such as grid search, random search, or genetic algorithm, can help find the best combination of hyperparameters to maximize model performance. This is particularly important in deep learning, where there may be many hyperparameters to be tuned (Zhang et al., 2016). Optimizing machine learning models in remote sensing is an ongoing process, often requiring a combination of techniques. Additionally, it is crucial to consider the specific challenges posed by the characteristics of the data, such as spatial and spectral resolution, data volume, and data quality. A deep understanding of both the machine learning techniques and the remote sensing domain is essential to achieve accurate and meaningful results.

Once the machine learning models developed by the preceding steps are validated, they can be deployed to perform classification on new or unclassified data, such as satellite images, aerial photographs, radar, or LiDAR point clouds. After classification, the final output of combining machine learning and remote sensing is often classified maps or images that provide valuable information about land cover, objects, or specific features. In general, we also perform post-processing for classified maps to refine the classification results and improve their quality. The common post-processing methods include (1) noise removal by filtering out small isolation regions or outliers, (2) spatial smoothing to create more homogeneous class regions, and (3) edge refinement to enhance the sharpness and accuracy of class boundaries in the classification maps. The process outlined earlier is valuable for various industries and research fields. In summary, machine learning in remote sensing applications can process vast amounts of data efficiently and accurately, thereby being an essential tool for researchers, government agencies, and industries seeking to monitor and understand changes in the Earth's surface.

3.8 CONCLUDING REMARKS

Interpreting urban land cover from data captured by remote sensors remains a conceptual and technical challenge. Accuracy levels are typically lower than the interpretation of more naturally-occurring surfaces. However, huge strides have been made with the formulation of statistical models that help disentangle the spectral and spatial complexity of urban land covers. Whereas per-pixel classification have stood the test of time (primarily for pragmatic reasons, especially when integrated with GIS-handled datasets), developments in sub-pixel, object-based, geospatial, and machine learning techniques have begun, at last, to reproduce the geographical configuration and compositional texture of urban structures. These developments are further tempered by conceptual developments that now consider the "appropriateness" of scale (understanding the level of urban structural measurements) and the "appropriateness" of time (understanding the lag between urban process and urban structure). Both are critical for measuring the rate of urban change; not simply the amount of lateral growth, but also the juxtaposition of land use within existing urban limits. Further research will only improve our use of remote sensor data for measuring urban patterns and in turn will complement our understanding of key urban processes.

REFERENCES

Adams, J. B., Sabol, D. E., Kapos, V., Almeida-Filho, R., Roberts, D. A., Smith, M. O., & Gillespie, A. R. 1995. Classification of multiple images based on fractions of endmembers: Application to landcover change in the Brazilian Amazon. *Remote Sensing of Environment*, 52, 137–154.

Anderson, J. R., Hardy, E. E., Roach, J. T., & Witmer, R. E. 1976. A land use and land cover classification system for use with remote sensor data. *U.S. Geological Survey Professional Paper*, 964. http://landcover.usgs.gov/pdf/anderson.pdf.

Arai, K. 1993. A classification method with a spatial-spectral variability. *International Journal of Remote Sensing*, 14, 699–709.

Barr, S., & Barnsley, M. A. 1997. A region-based, graph-theoretic data model for the inference of second-order thematic information from remotely-sensed images. *International Journal of Geographical Information Science*, 11, 555–576.

Bauer, M. E., Heinert, N. J., Doyle, J. K., & Yuan, F. 2004. Impervious surface mapping and change monitoring using satellite remote sensing. *Proceedings of the ASPRS 2004 Annual Conference*, 24–28 May, Denver, Colorado.

Bauer, M. E., Loeffelholz, B. C., & Wilson, B. 2007. Estimating and mapping impervious surface area by regression analysis of Landsat imagery. In Q. Wang (Ed.), *Remote Sensing of Impervious Surfaces*, pp. 3–20. Boca Raton, FL: CRC Press.

Blaschke, T. 2010. Object-based image analysis for remote sensing. *ISPRS International Journal of Photogrammetry and Remote Sensing*, 65, 2–16.

Borel, C. C., & Gerstl, S. A. W. 1994. Nonlinear spectral mixing models for vegetative and soil surfaces. *Remote Sensing of Environment*, 47, 403–416.

Bovik, A. C., Clark, M., & Geisler, W. S. 1990. Multichannel t exture analysis using localized spatial filters. *IEEE Transactions on Pattern Analysis and Machine Intelligence*, 12, 55–73.

De Jong, S. M., & Burrough, P. A. 1995. A fractal approach to the classification of Mediterranean vegetation types in remotely sensed images. *Photogrammetric Engineering and Remote Sensing*, 61, 1041–1053.

Desclée, B., Bogaert, P., & Defourny, P. 2006. Forest change detection by statistical object-based method. *Remote Sensing of Environment*, 102, 1–11.

Eastman, J. R., & Laney, R. M. 2002. Bayesian soft classification for sub-pixel analysis: A critical evaluation. *Photogrammetric Engineering and Remote Sensing*, 6811, 1149–1154.

Edwards, G., Landry, R., & Thompson, K. P. B. 1988. Texture analysis of forest regeneration sites in high-resolution SAR imagery. *Proceedings of the International Geosciences and Remote Sensing Symposium (IGARSS 88)*, ESA SP-284, pp. 1355–1360. European Space Agency, Paris.

Emerson, C. W., Lam, N. S. N., & Quattrochi, D. A. 1999. Multi-scale fractal analysis of image texture and pattern. *Photogrammetric Engineering and Remote Sensing*, 65, 51–61.

Fisher, P. F. 1997. The pixel: A snare and a delusion. *International Journal of Remote Sensing*, 18, 679–685.

Fisher, P. F., & Pathirana, S. 1990. The evaluation of fuzzy membership of land cover classes in the suburban zone. *Remote Sensing of Environment*, 34, 121–132.

Foody, G. M. 2000. Estimation of sub-pixel land cover composition in the presence of untrained classes. *Computers and Geosciences*, 26, 469–478.

Foody, G. M., & Aurora, M. K. 1996. Incorporating mixed pixels in the training, allocation and testing of supervised classification. *Pattern Recognition Letters*, 17, 1389–1398.

Foody, G. M., Campbell, N. A., Trodd, N. M., & Wood, T. F. 1992. Derivation and applications of probabilistic measures of class membership from the maximum-likelihood classification. *Photogrammetric Engineering and Remote Sensing*, 58, 1335–1341.

Franklin, S. E., Hall, R. J., Moskal, L. M., Maudie, A. J., & Lavigne, M. B. 2000. Incorporating texture into classification of forest species composition from airborne multispectral images. *International Journal of Remote Sensing*, 21, 61–79.

Hermosilla, T., Ruiz, L. A., Recio, J. A., & Cambra-López, M. 2012. Assessing contextual descriptive features for plot-based classification of urban areas. *Landscape and Urban Planning*, 106, 124–137.

Herold, M., Liu, X., & Clarke, K. C. 2003. Spatial metrics and image texture for mapping urban land use. *Photogrammetric Engineering and Remote Sensing*, 69, 991–1001.

Hodgson, M. E. 1998. What size window for image classification? A cognitive perspective. *Photogrammetric Engineering and Remote Sensing*, 64, 797–807.

Homer, C., Dewitz, J., Fry, J., Coan, M., Hossain, N., Larson, C., Herold, N., McKerrow, A., VanDriel, J. N., & Wickham, J. 2007. Completion of the 2001 national land cover database for the conterminous United States. *Photogrammetric Engineering & Remote Sensing*, 73, 337–341.

Hung, M., & Ridd, M. K. 2002. A subpixel classifier for urban land-cover mapping based on a maximum-likelihood approach and expert system rules. *Photogrammetric Engineering and Remote Sensing*, 68, 1173–1180.

Im, J., Jensen, J. R., & Hodgson, M. E. 2008. Object-based land cover classification using high posting density lidar data. *GIScience and Remote Sensing*, 45, 209–228.

Jensen, J. R., & Cowen, D. C. 1999. Remote sensing of urban/suburban infrastructure and socio-economic attributes. *Photogrammetric Engineering and Remote Sensing*, 65, 611–622.

Ji, M., & Jensen, J. R. 1999. Effectiveness of subpixel analysis in detecting and quantifying urban impervious from Landsat Thematic Mapper Imagery. *Geocarto International*, 14, 33–41.

Lam, N. S. N., Quattrochi, D., Qui, H., & Zhao, W. 1998. Environmental assessment and monitoring with image characterization and modeling system using multiscale remote sensing data. *Applied Geographic Studies*, 2, 77–93.

Lary, D. J., Alavi, A. H., Gandomi, A. H., & Walker, A. L. 2016. Machine learning in geosciences and remote sensing. *Geoscience Frontiers*, 7, 3–10.

Liu, X., Clarke, K. C., & Herold, M. 2006. Population density and image texture: A comparison study. *Photogrammetric Engineering and Remote Sensing*, 72, 187–196.

Maxwell, A. E., Warner, T. A., & Fang, F. 2018. Implementation of machine-learning classification in remote sensing: An applied review. *International Journal of Remote Sensing*, 39, 2784–2817.

Mesev, V. 1998. The use of census data in urban image classification. *Photogrammetric Engineering and Remote Sensing*, 64, 431–438.

Mesev, V. 2001. Modified maximum likelihood classifications of urban land use: Spatial segmentation of prior probabilities. *Geocarto International*, 16, 41–48.

Mesev, V. 2012. Multiscale and multitemporal urban remote sensing. *ISPRS International Archives of the Photogrammetry, Remote Sensing & Spatial Information Sciences*, XXXIX-B2, 17–21.

Myint, S. W. 2006a. Urban vegetation mapping using sub-pixel analysis and expert system rules: A critical approach. *International Journal of Remote Sensing*, 27, 2645–2665.

Myint, S. W. 2006b. A new framework for effective urban land use land cover classification: A wavelet approach. *GIScience and Remote Sensing*, 43, 155–178.

Myint, S. W. 2010. Multi-resolution decomposition in relation to characteristic scales and local window sizes using an operational wavelet algorithm. *International Journal of Remote Sensing*, 31, 2551–2572.

Myint, S. W., Giri, C. P., Wang, L., Zhu, Z., & Gillette, S. 2008. Identifying mangrove species and their surrounding land use and land cover classes using an object oriented approach with a lacunarity spatial measure. *GIScience and Remote Sensing*, 45, 188–208.

Myint, S. W., Gober, P., Brazel, A., Grossman-Clarke, S., & Weng, Q. 2011. Per-pixel versus object-based classification of urban land cover extraction using high spatial resolution imagery. *Remote Sensing of Environment*, 115, 1145–1161.
Myint, S. W., & Lam, N. S. N. 2005a. A study of lacunarity-based texture analysis approaches to improve urban image classification. *Computers, Environment, and Urban Systems*, 29, 501–523.
Myint, S. W., & Lam, N. S. N. 2005b. Examining lacunarity approaches in comparison with fractal and spatial autocorrelation techniques for urban mapping. *Photogrammetric Engineering and Remote Sensing*, 71, 927–937.
Myint, S. W., & Mesev, V. 2012. A comparative analysis of spatial indices and wavelet-based classification. *Remote Sensing Letters*, 3, 141–150.
Myint, S. W., Mesev, V., & Lam, N. S. N. 2006. Texture analysis and classification through a modified lacunarity analysis based on differential box counting method. *Geographical Analysis*, 38, 371–390.
Myint, S. W., & Okin, G. S. 2009. Modelling land-cover types using multiple endmember spectral mixture analysis in a desert city. *International Journal of Remote Sensing*, 30, 2237–2257.
Myint, S. W., & Stow, D. 2011. An object-oriented pattern recognition approach for urban classification. In X. Yang (Ed.), *Urban Remote Sensing, Monitoring, Synthesis and Modeling in the Urban Environment*, pp. 129–140. Chichester: John Wiley & Sons, Ltd. http://doi.org/10.1002/9780470979563
Navulur, K. 2007. *Multispectral Image Analysis Using the Object-Oriented Paradigm*. Boca Raton, FL: CRC Press, Taylor and Frances Group.
Peddle, D. R., & Franklin, S. E. 1991. Image texture processing and data integration for surface pattern discrimination. *Photogrammetric Engineering and Remote Sensing*, 57, 413–420.
Powell, R. L., Roberts, D. A., Dennison, P. E., & Hess, L. L. 2007. Sub-pixel mapping of urban land cover using multiple endmember spectral mixture analysis: Manaus, Brazil. *Remote Sensing of Environment*, 106, 253–267.
Quattrochi, D. A., Lam, N. S. N., Qiu, H., & Zhao, W. 1997. Image characterization and modeling system (ICAMS): A geographic information system for the characterization and modeling of multiscale remote sensing data. In D. A. Quattrochi & M. F. Goodchild (Eds.), *Scale in Remote Sensing and GIS*, pp. 295–308. Boca Raton, FL: CRC Press.
Rashed, T., Weeks, J. R., Roberts, D., Rogan, J., & Powell, R. 2003. Measuring the physical composition of urban morphology using multiple endmember spectral mixture models. *Photogrammetric Engineering and Remote Sensing*, 69, 1011–1020.
Ray, T. W., & Murray, B. C. 1996. Nonlinear spectral mixing in desert vegetation. *Remote Sensing of Environment*, 55, 59–64.
Ridd, M. K. 1995. Exploring a V-I-S vegetation-impervious surface-soil model for urban ecosystems analysis through remote sensing: Comparative anatomy for cities. *International Journal of Remote Sensing*, 16, 2165–2186.
Roberts, D. A., Gardner, M., Church, R., Ustin, S., Scheer, G., & Green, R. O. 1998. Mapping chaparral in the Santa Monica Mountains using multiple endmember spectral mixture models. *Remote Sensing of Environment*, 65, 267–279.
Roberts, D. A., Quattrochi, D. A., Hulley, G. C., Hook, S. J., & Green, R. O. 2012. Synergies between VSWIR and TIR data for the urban environment: An evaluation of the potential for the hyperspectral Infrared Imager (HyspIRI) decadal survey mission. *Remote Sensing of Environment*, 117, 83–101.
Small, C. 2004. The landsat ETM+ spectral mixing space. *Remote Sensing of Environment*, 93, 1–17.
Stefanov, W. L., Ramsey, M. S., & Christensen, P. R. 2001. Monitoring urban land cover change: An expert system approach to land cover classification of semiarid to arid urban centers. *Remote Sensing of Environment*, 77(2), 173–185.
Stow, D., Lopez, A., Lippitt, C., Hinton, S., & Weeks, J. 2007. Object-based classification of residential land use within Accra, Ghana based on QuickBird satellite data. *International Journal of Remote Sensing*, 28, 5167–5173.
Strahler, A. H. 1980. The use of prior probabilities in maximum likelihood classification of remotely sensed data. *Remote Sensing of Environment*, 10, 135–163.
Stuckens, J., Coppin, P. R., & Bauer, M. 2000. Integrating contextual information with per-pixel classification for improved land cover classification. *Remote Sensing of Environment*, 71, 282–296.
Walker, J. S., & Briggs, J. M. 2007. An object-oriented approach to urban forest mapping with high-resolution, true-color aerial photography. *Photogrammetric Engineering & Remote Sensing*, 73, 577–583.

Welch, R. A. 1982. Spatial resolution requirements for urban studies. *International Journal of Remote Sensing*, 3, 139–146.

Weng, Q., & Hu, X. 2008. Medium spatial resolution satellite imagery for estimating and mapping urban impervious surfaces using LSMA and ANN. *Transactions on Geoscience and Remote Sensing*, 46, 2387–2406.

Woodcock, C. E., & Harward, V. J. 1992. Nested-hierarchical scene models and image segmentation. *International Journal of Remote Sensing*, 13, 3167–3187.

Woodcock, C. E, Strahler, A. H., & Jupp, D. L. B. 1988. The use of variograms in remote sensing: I. Scene models and simulated images. *Remote Sensing of Environment*, 25, 323–348.

Wu, C., & Murray, A. 2003. Estimating impervious surface distribution by spectral mixture analysis. *Remote Sensing of Environment*, 84, 493–505.

Wu, C., & Yuan, F. 2007. Seasonal sensitivity analysis of impervious surface estimation with satellite imagery. *Photogrammetric Engineering & Remote Sensing*, 73, 1393–1401.

Yang, L., Huang, C., Homer, C. G., Wylie, B. K., & Coan, M. J. 2003a. An approach for mapping large-area impervious surfaces: Synergistic use of Landsat-7 ETM+ and high spatial resolution imagery. *Canadian Journal of Remote Sensing*, 29, 230–240.

Yang, L., Xian, G., Klaver, J. M., & Deal, B. 2003b. Urban land-cover change detection through sub-pixel imperviousness mapping using remotely sensed data. *Photogrammetric Engineering & Remote Sensing*, 69, 1003–1010.

Yang, X. 2011. Parameterizing support vector machines for land cover classification. *Photogrammetric Engineering & Remote Sensing*, 77, 27–37.

Yang, X., & Liu, Z. 2005. Use of satellite-derived landscape imperviousness index to characterize urban spatial growth. *Computers, Environment and Urban Systems*, 29, 524–540.

Yuan, F., & Bauer, M. E. 2007. Comparison of impervious surface area and normalized difference vegetation index as indicators of surface urban heat island effects in Landsat imagery. *Remote Sensing of Environment*, 106, 375–386.

Zhang, J., & Foody, G. M. 2001. Fully-fuzzy supervised classification of sub-urban land cover from remotely sensed imagery: Statistical and neural network approaches. *Photogrammetric Engineering and Remote Sensing*, 22, 615–628.

Zhang, L., Zhang, L., & Du, B. 2016. Deep learning for remote sensing data: A technical tutorial on the state of the art. *IEEE Geoscience and Remote Sensing Magazine*, 4, 22–40.

Zhou, W. Q., Troy, A., & Grove, M. 2008. Object-based land cover classification and change analysis in the Baltimore metropolitan area using multitemporal high resolution remote sensing data. *Sensors*, 8, 1613–1636.

Zhu, C., & Yang, X. 1998. Study of remote sensing image texture analysis and classification using wavelet. *International Journal of Remote Sensing*, 13, 3167–3187.

4 Hyperspectral Image Processing

Methods and Approaches

Jun Li and Antonio Plaza

LIST OF ACRONYMS

LDA	Linear discriminant analysis
QDA	Quadratic discriminant analysis
LogDA	Logarithmic discriminant analysis
SVMs	Support vector machines
TSVMs	Transductive SVMs
MPs	Morphological profiles
PCA	Principal component analysis
EMPs	Extended morphological profiles
MRFs	Markov random fields
RHSEG	Recursive hierarchical segmentation
MLR	Multinomial logistic regression
MLRsubMRF	Subspace-based multinomial logistic regression followed by MRFs
ROSIS	Reflective optics spectrographic imaging system

4.1 INTRODUCTION

Hyperspectral imaging is concerned with the measurement, analysis, and interpretation of spectra acquired from a given scene (or specific object) at a short, medium, or long distance, typically, by an airborne or satellite sensor [1]. The special characteristics of hyperspectral datasets pose different processing problems, which must be necessarily tackled under specific mathematical formalisms [2, 51–71], such as classification and segmentation [3] or spectral mixture analysis [4]. Several machine learning and image processing techniques have been applied to extract relevant information from hyperspectral data during the last decade [5, 6]. Taxonomies of hyperspectral image processing algorithms have been presented in the literature [3, 7, 8, 51–55]. It should be noted, however, that most recently developed hyperspectral image processing techniques focus on analyzing the spectral and the spatial information contained in the hyperspectral data in simultaneous fashion [9]. In other words, the importance of analyzing spatial and spectral information simultaneously has been identified as a desired goal by many scientists devoted to hyperspectral image analysis. This type of processing has been approached in the past from various points of view. For instance, several possibilities are discussed in [10, 58–62] for the refinement of results obtained by spectral-based techniques through a second step based on spatial context. Such contextual classification [11] accounts for the tendency of certain ground cover classes to occur more frequently in some contexts than in others. In certain applications, the integration of high spatial and spectral information is mandatory to achieve sufficiently accurate mapping and/or detection results. For instance, urban area mapping requires sufficient spatial resolution to distinguish small spectral classes, such as trees in a park, or cars on a street [12].

DOI: 10.1201/9781003541158-5

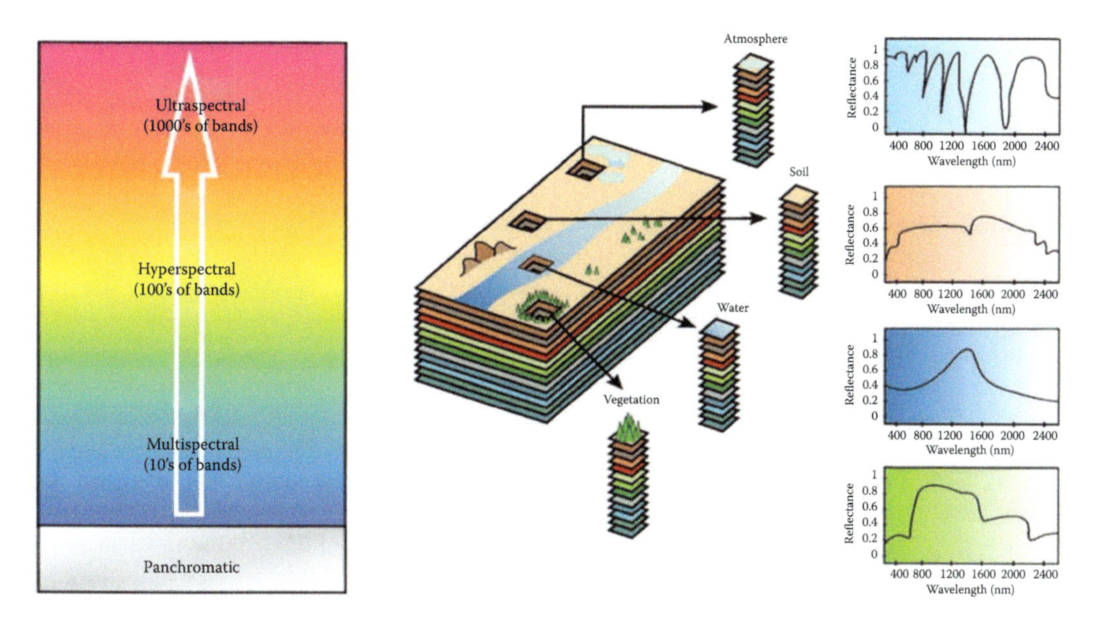

FIGURE 4.1 The challenges of increased dimensionality in remote sensing data interpretation.

However, there are several important challenges when performing hyperspectral image classification. In particular, supervised classification faces challenges related with the unbalance between high dimensionality and the limited number of training samples, or the presence of mixed pixels in the data (which may compromise classification results for coarse spatial resolutions). Specifically, due to the small number of training samples and the high dimensionality of the hyperspectral data, reliable estimation of statistical class parameters is a very challenging goal [13]. As a result, with a limited training set, classification accuracy tends to decrease as the number of features increases. This is known as the Hughes effect [14]. Another relevant challenge is the need to integrate the spatial and spectral information to take advantage of the complementarities that both sources of information can provide. These challenges are quite important for future developments and solutions to some of them have been proposed. Specifically, supervised [15] and semi-supervised [16–18, 51–71] techniques for hyperspectral image classification, strategies for integrating the spatial and the spectral information [19–22], or subspace classifiers [23] that can better exploit the intrinsic nature of hyperspectral data have been quite popular in the recent literature.

Our main goal in this chapter is to provide a seminal view on recent advances in techniques for hyperspectral image analysis that can successfully deal with the dimensionality problem and with the limited availability of training samples *a priori*, while taking into account both the spectral and spatial properties of the data. The remainder of the chapter is structured as follows. Section 4.2 discusses available techniques for hyperspectral image classification, including both supervised and semi-supervised approaches, techniques for integrating spatial and spectral information and subspace-based approaches. Section 4.3 provides an experimental comparison of the techniques discussed in Section 4.2, using a hyperspectral dataset collected by the reflected optics spectrographic imaging system (ROSIS) over the University of Pavia, Italy, which is used here as a common benchmark to outline the properties of the different processing techniques discussed in the chapter. Finally, Section 4.4 concludes the paper with some remarks and hints at the most pressing ongoing research directions in hyperspectral image classification.

4.2 CLASSIFICATION APPROACHES

In this section, we outline some of the main techniques and challenges in hyperspectral image classification. Hyperspectral image classification has been a very active area of research in recent years [3]. Given a set of observations (i.e., pixel vectors in a hyperspectral image), the goal of classification is to assign a unique label to each pixel vector so that it is well-defined by a given class. In Figure 4.2, we provide an overview of a popular strategy to conduct hyperspectral image classification, which is based on the availability of labeled samples. After an optional dimensionality reduction step, a supervised classifier is trained using a set of labeled samples (which are often randomly selected from a larger pool of samples) and then tested with a disjoint set of labeled samples in order to evaluate the classification accuracy of the classifier.

Supervised classification has been widely used in hyperspectral data interpretation [2], but it faces challenges related with the high dimensionality of the data and the limited availability of training samples, which may not be easy to collect in pure form. However, mixed training samples can also offer relevant information about the participating classes [10]. In order to address these issues, subspace-based approaches [23, 24] and semi-supervised learning techniques [25] have been developed. In subspace approaches, the goal is to reduce the dimensionality of the input space in order to better exploit the (limited) training samples available. In semi-supervised learning, the idea is to exploit the information conveyed by additional (unlabeled) samples, which can complement the available labeled samples with a certain degree of confidence. In all cases, there is a clear need to integrate the spatial and spectral information to take advantage of the complementarities that both sources of information can provide [9]. An overview of these different aspects, which are crucial to hyperspectral image classification, is provided in the following subsections.

4.2.1 Supervised Classification

Several techniques have been used to perform supervised classification of hyperspectral data [51–71]. For instance, in discriminant classifiers several types of discriminant functions can be applied: nearest neighbor, decision trees, linear functions or nonlinear functions (see Figure 4.3). In linear discriminant analysis (LDA) [26], a linear function is used in order to maximize the discriminatory power and separate the available classes effectively. However, such a linear function may not be the

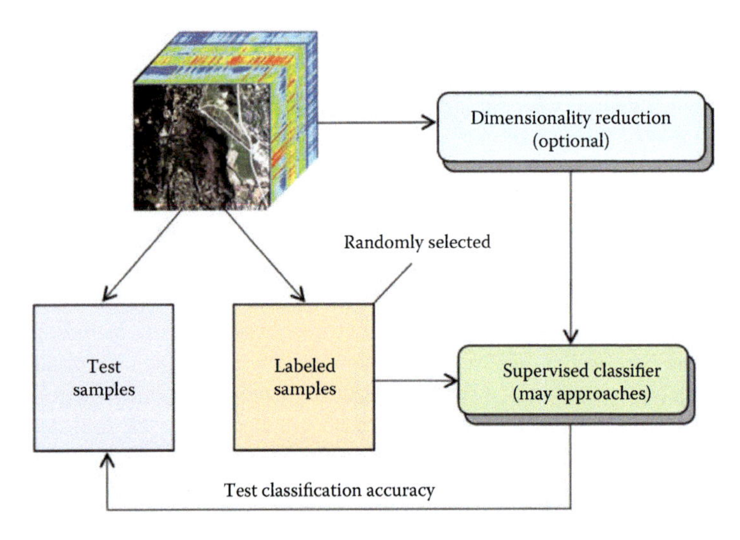

FIGURE 4.2 Standard approach for supervised hyperspectral image classification.

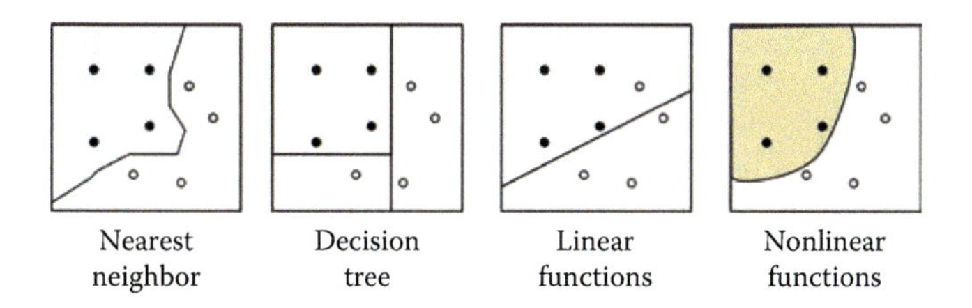

FIGURE 4.3 Typical discriminant functions used in supervised classifiers.

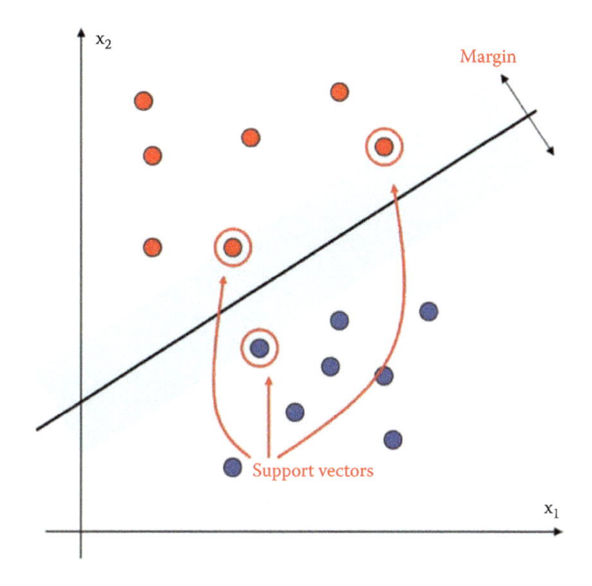

FIGURE 4.4 Soft margin classification with slack variables.

best choice and nonlinear strategies such as quadratic discriminant analysis (QDA) or logarithmic discriminant analysis (LogDA) have also been used. The main problem of these classic supervised classifiers, however, is their sensitivity to the Hughes effect.

In this context, kernel methods such as the support vector machine (SVM) have been widely used in order to deal effectively with the Hughes phenomenon [27, 28]. The SVM was first investigated as a binary classifier [29]. Given a training set mapped into an Hilbert space by some mapping, the SVM separates the data by an optimal hyperplane that maximizes the margin (see Figure 4.4). However, the most widely used approach in hyperspectral classification is to combine soft margin classification with a kernel trick that allows separation of the classes in a higher dimensional space by means of a nonlinear transformation (see Figure 4.5). In other words, the SVM used with a kernel function is a nonlinear classifier, where the nonlinear ability is included in the kernel and different kernels lead to different types of SVMs. The extension of SVM to the multi-class cases is usually done by combining several binary classifiers.

4.2.2 Spectral–Spatial Classification

Several efforts have been performed in the literature in order to integrate spatial-contextual information in spectral-based classifiers for hyperspectral data [3, 9, 51–71]. It is now commonly accepted

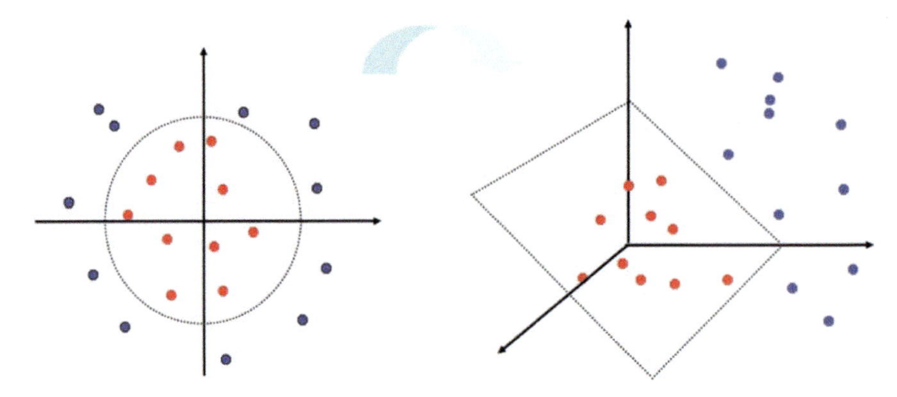

FIGURE 4.5 The kernel trick allows separation of the classes in a higher dimensional space by means of a linear or nonlinear transformation.

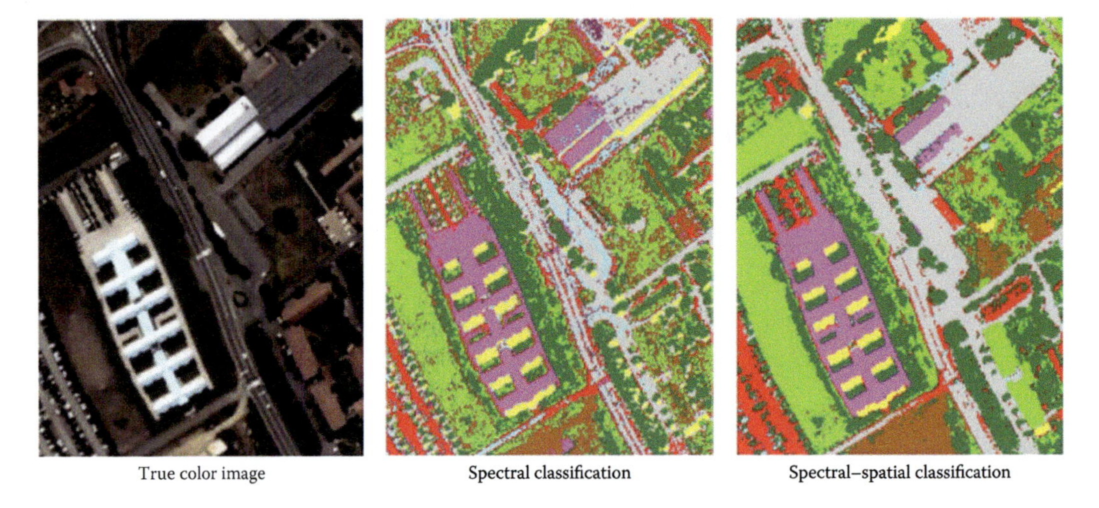

FIGURE 4.6 The importance of using spatial and spectral information in classification.

that using the spatial and the spectral information simultaneously provides significant advantages in terms of improving the performance of classification techniques. An illustration of the importance of integrating spatial and spectral information is given in Figure 4.6. As shown in this figure, spectral-spatial classification (obtained using morphological transformations) provides a better interpretation of classes such as urban features, with a better delineation and characterization of complex urban structures.

Some of the approaches that integrate spatial and spectral information include spatial information prior to the classification, during the feature extraction stage. Mathematical morphology [30] has been particularly successful for this purpose. Morphology is a widely used approach for modeling the spatial characteristics of the objects in remotely sensed images. Advanced morphological techniques such as morphological profiles (MPs) [31] have been successfully used for feature extraction prior to classification of hyperspectral data by extracting the first few principal components of the data using principal component analysis (PCA) [13], and then building so-called extended morphological profiles (EMPs) on the first few components to extract relevant features for classification [32].

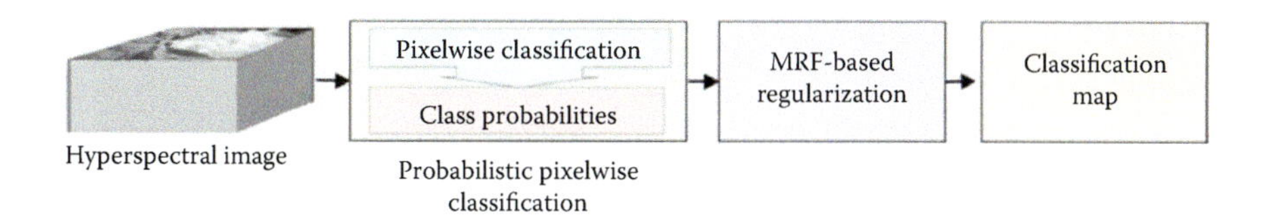

FIGURE 4.7 Standard processing framework using pixelwise probabilistic classification followed by MRF-based spatial postprocessing.

Another strategy in the literature has been to exploit simultaneously the spatial and the spectral information. For instance, in order to incorporate the spatial context into kernel-based classifiers, a pixel entity can be redefined simultaneously both in the spectral domain (using its spectral content) and also in the spatial domain, by applying some feature extraction to its surrounding area which yields spatial (contextual) features, for example, the mean or standard deviation per spectral band. These separated entities lead to two different kernel matrices, which can be easily computed. At this point, one can sum spectral and textural dedicated kernel matrices and introduce the cross-information between textural and spectral features in the formulation. This simple methodology yields a full family of new kernel methods for hyperspectral data classification, defined in [33] and implemented using the SVM classifier, thus providing a composite kernel-based SVM.

Another approach to jointly exploit spatial and spectral information is to use Markov random fields (MRFs) for the characterization of spatial information. MRFs exploit the continuity, in probability sense, of neighboring labels [19, 34]. In this regard, several techniques have exploited an MRF-based regularization procedure which encourages neighboring pixels to have the same label when performing probabilistic classification of hyperspectral datasets. An example of this type of processing is given in Figure 4.7, in which a pixelwise probabilistic classification is followed by an MRF-based spatial postprocessing that refines the initial probabilistic classification output.

Several other approaches include spatial information as a post-processing, that is, after a spectral-based classification has been conducted. One of the first classifiers with spatial post-processing developed in the hyperspectral imaging literature was the well-known ECHO (extraction and classification of homogeneous objects) [10]. Another one is the strategy adopted in [35], which combines the output of a pixel-wise SVM classifier with the morphological watershed transformation [30] in order to provide a more spatially homogeneous classification. A similar strategy is adopted in [36], in which the output of the SVM classifier is combined with the segmentation result provided by the unsupervised recursive hierarchical segmentation (RHSEG)[1] algorithm.

4.2.3 Subspace-Based Approaches

Subspace projection methods [23] have been shown to be a powerful class of statistical pattern classification algorithms. These methods can handle the high dimensionality of hyperspectral data by bringing it to the right subspace without losing the original information that allows for the separation of classes. In this context, subspace projection methods can provide competitive advantages by separating classes which are very similar in spectral sense, thus addressing the limitations in the classification process due to the presence of highly mixed pixels. The idea of applying subspace projection methods to improve classification relies on the basic assumption that the samples within each class can approximately lie in a lower dimensional subspace. Thus, each class may be represented by a subspace spanned by a set of basis vectors, while the classification criterion for a new input sample would be the distance from the class subspace [24].

Recently, several subspace projection methods have been specifically designed for improving hyperspectral data characterization, with successful results. For instance, the subspace-based

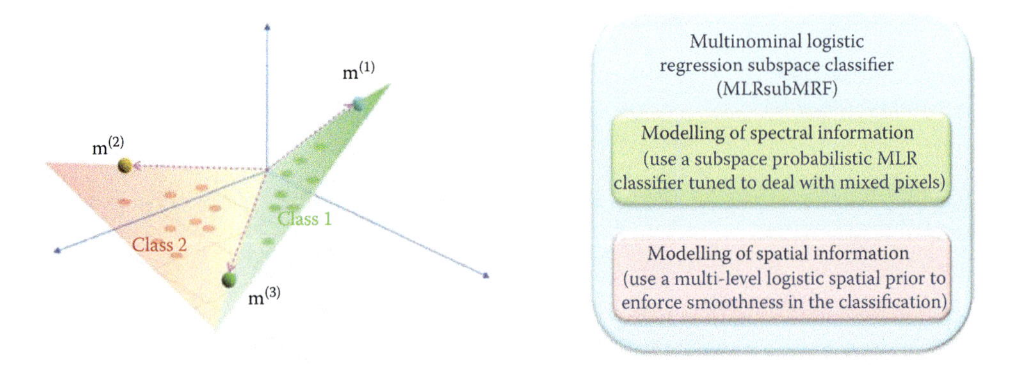

FIGURE 4.8 Multinomial logistic regression subspace classifier.

multinomial logistic regression followed by Markov random fields (MLRsubMRF) method in [23] first performs a learning step in which the posterior probability distributions are modeled by a multinomial logistic regression (MLR) [37] combined with a subspace projection method. Then, the method infers an image of class labels from a posterior distribution built on the learned subspace classifier and on a multilevel logistic spatial prior on the image of labels. This prior is an MRF that exploits the continuity, in probability sense, of neighboring labels. The basic assumption is that, in a hyperspectral image, it is very likely that two neighboring pixels will have the class same label. The main contribution of the MLRsubMRF method is therefore the integration of a subspace projection method with the MLR, which is further combined with spatial–contextual information in order to provide a good characterization of the content of hyperspectral imagery in both the spectral and the spatial domain. As will be shown by our experiments, the accuracies achieved by this approach are competitive with those provided by many other state-of-the-art supervised classifiers for hyperspectral analysis.

4.2.4 Semi-Supervised Classification

A relevant challenge for supervised classification techniques is the limited availability of labeled training samples, since their collection generally involves expensive ground campaigns [38]. While the collection of labeled samples is generally difficult, expensive and time-consuming, unlabeled samples can be generated in a much easier way. This observation has fostered the idea of adopting semi-supervised learning techniques in hyperspectral image classification. The main assumption of such techniques is that new (unlabeled) training samples can be obtained from a (limited) set of available labeled samples without significant effort/cost. This can be simply done by selecting new samples from the spatial neighborhood of available labeled samples, under the principle that it is likely that the new unlabeled samples will have similar class labels as the already available ones.

In contrast to supervised classification, semi-supervised algorithms generally assume that a limited number of labeled samples are available *a priori*, and then enlarge the training set using unlabeled samples, thus allowing these approaches to address ill-posed problems. However, in order for this strategy to work, several requirements need to be met. First and foremost, the new (unlabeled) samples should be obtained without significant cost/effort. Second, the number of unlabeled samples required in order for the semi-supervised classifier to perform properly should not be too high in order to avoid increasing computational complexity in the classification stage. In other words, as the number of unlabeled samples increases, it may be unbearable for the classifier to properly exploit all the available training samples due to computational issues. Further, if the unlabeled samples are not properly selected, these may confuse the classifier, thus introducing significant divergence or even reducing the classification accuracy obtained with the initial set of labeled samples. In order to address these issues, it is very important that the most highly informative unlabeled samples are identified in computationally efficient fashion, so that significant improvements in classification performance can be observed without the need to use a very high number of unlabeled samples.

The area of semi-supervised learning for remote sensing data analysis has experienced a significant evolution in recent years. For instance, looking at machine learning-based approaches, in [39] transductive SVMs (TSVMs) are used to gradually search a reliable separating hyperplane (in the kernel space) with a transductive process that incorporates both labeled and unlabeled samples in the training phase. In [40], a semi-supervised method is presented that exploits the wealth of unlabeled samples in the image, and naturally gives relative importance to the labeled ones through a graph-based methodology. In [41], kernels combining spectral-spatial information are constructed by applying spatial smoothing over the original hyperspectral data and then using composite kernels in graph-based classifiers. In [18], a semi-supervised SVM is presented that exploits the wealth of unlabeled samples for regularizing the training kernel representation locally by means of cluster kernels. In [42], a new semi-supervised approach is presented that exploits unlabeled training samples (selected by means of an active selection strategy based on the entropy of the samples). Here, unlabeled samples are used to improve the estimation of the class distributions, and the obtained classification is refined by using a spatial multi-level logistic prior. In [16], a novel context-sensitive semi-supervised SVM is presented that exploits the contextual information of the pixels belonging to the neighborhood system of each training sample in the learning phase to improve the robustness to possible mislabeled training patterns.

In [43], two semi-supervised one-class (SVM-based) approaches are presented in which the information provided by unlabeled samples present in the scene is used to improve classification accuracy and alleviate the problem of free-parameter selection. The first approach models data marginal distribution with the graph Laplacian built with both labeled and unlabeled samples. The second approach is a modification of the SVM cost function that penalizes more the errors made when classifying samples of the target class. In [44] a new method to combine labeled and unlabeled pixels to increase classification reliability and accuracy, thus addressing the sample selection bias problem, is presented and discussed. In [45], an SVM is trained with the linear combination of two kernels: a base kernel working only with labeled examples is deformed by a likelihood kernel encoding similarities between labeled and unlabeled examples, and then applied in the context of urban hyperspectral image classification. In [46], similar concepts to those addressed before are adopted using a neural network as the baseline classifier. In [17], a semi-automatic procedure to generate land cover maps from remote sensing images using active queries is presented and discussed.

Last but not least, we emphasize that the techniques summarized in this section only represent a small sample (and somehow subjective selection) of the vast collection of approaches presented in recent years for hyperspectral image classification. For a more exhaustive summary of available techniques and future challenges in this area, we point interested readers to [47].

4.3 EXPERIMENTAL COMPARISON

In this section, we illustrate the performance of the techniques described in the previous section by processing a widely used hyperspectral dataset collected by the ROSIS optical sensor over the urban area of the University of Pavia, Italy. The flight was operated by the Deutschen Zentrum for Luftund Raumfahrt (DLR, the German Aerospace Agency) in the framework of the HySens project, managed and sponsored by the European Union. The image size in pixels is 610 × 340, with very high spatial resolution of 1.3 m per pixel. The number of data channels in the acquired image is 103 (with spectral range from 0.43 to 0.86 micrometers. Figure 4.9a shows a false color composite of the image, while Figure 4.9b shows nine reference classes of interest, which comprise urban features, as well as soil and vegetation features. Finally, Figure 4.9c shows a fixed training set available for the scene, which comprises 3,921 training samples (42,776 samples are available for testing). This scene has been widely used in the hyperspectral imaging community to evaluate the performance of processing algorithms [6]. It represents a case study that integrates a challenging urban

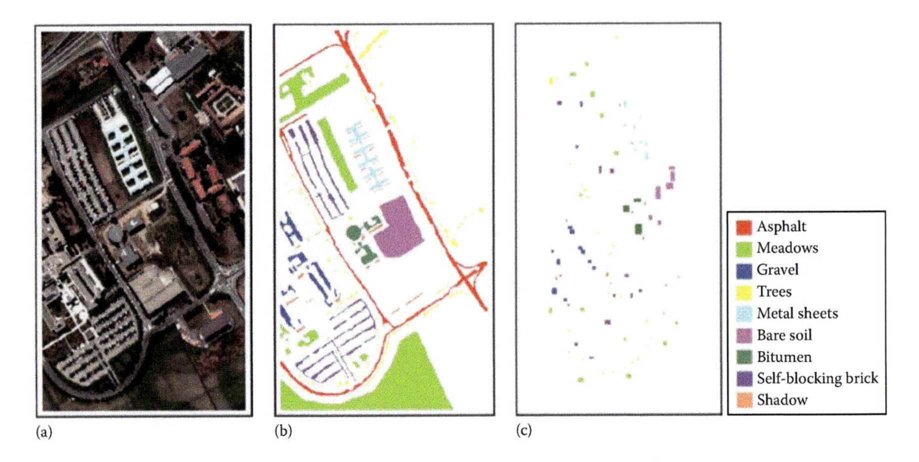

FIGURE 4.9 The ROSIS Pavia University scene used in our experiments. The scene was collected by the ROSIS instrument in the framework of the HySens campaign. It comprises 103 spectral bands between 0.4 and 0.9 microns and was collected over an urban area at the University of Pavia, Italy.

TABLE 4.1
Classification Results Obtained for the ROSIS Pavia University Scene by the SVM Classifier as Compared with Several Discriminant Classifiers

Class	Training	Testing	LDA	QDA	LogDA	SVM
Asphalt (6631)	548	6083	69.45	67.75	70.89	83.71
Meadows (18649)	532	18117	81.92	75.73	76.72	70.25
Gravel (2099)	265	1834	39.11	59.79	55.31	70.32
Trees (3064)	231	2833	95.07	96.64	96.38	97.81
Metal Sheets (1345)	375	970	99.41	99.93	100	99.41
Bare Soil (5029)	540	4489	46.59	73.49	75.06	92.25
Bitumen (1330)	392	938	63.31	93.53	83.98	81.58
Self-Blocking Bricks (3682)	524	3158	88.29	89.52	87.91	92.59
Shadow (947)	514	433	99.79	99.26	99.79	96.62
Overall accuracy	—	—	77.95	77.95	78.41	80.99

Note: LDA stands for linear discriminant analysis. QDA stands for quadratic discriminant analysis. LogDA stands for logarithmic discriminant analysis. SVM stands for support vector machine. The numbers in the parentheses are the total number of available samples.

classification problem, with a dataset comprising high spatial and spectral resolution, and a highly reliable ground-truth, with a well-established training set. All these factors have made the scene a standard and an excellent testbed for evaluation of hyperspectral image classification algorithms, particularly those integrating the spatial and the spectral information.

Table 4.1 illustrates the classification results obtained by different supervised classifiers for the ROSIS University of Pavia scene in Figure 4.9a, using the same training data in Figure 4.9c to train the classifiers and a mutually exclusive set of labeled samples in Figure 4.9b to test the classifiers. As shown by Table 4.1, the SVM classifier obtained comparatively superior performance in terms of the overall classification accuracy (OA) when compared with discriminant classifiers such as LDA, QDA, or LogDA.

In a second experiment, we compared the standard SVM classifier with the composite kernel strategy defined in [33] which combines spatial and spectral information at the kernel level. After carefully evaluating all possible types of composite kernels, the summation kernel provided the best performance in our experiments as reported on Table 4.2. This table suggests the importance of using spatial and spectral information in the analysis of hyperspectral data.

Although the integration of spatial and spectral information carried out by the composite kernel in Table 4.2 is performed at the classification stage, the spatial information can also be included prior to classification. For illustrative purposes, Table 4.3 compares the classification results obtained by

TABLE 4.2

Classification Results Obtained for the ROSIS Pavia University Scene by the SVM Classifier as Compared with a Composite SVM Obtained Using the Summation Kernel in [33]

Class	Training	Testing	SVM	Composite SVM
Asphalt (6631)	548	6083	83.71	79.85
Meadows (18649)	532	18117	70.25	84.76
Gravel (2099)	265	1834	70.32	81.87
Trees (3064)	231	2833	97.81	96.36
Metal Sheets (1345)	375	970	99.41	99.37
Bare Soil (5029)	540	4489	92.25	93.55
Bitumen (1330)	392	938	81.58	90.21
Self-Blocking Bricks (3682)	524	3158	92.59	92.81
Shadow (947)	514	433	96.62	95.35
Overall accuracy	—	—	80.99	87.18

Note: The numbers in the parentheses are the total number of available samples.

TABLE 4.3

Classification Results Obtained for the ROSIS Pavia University Scene by the SVM Classifier as Compared with Those Obtained Using a Combination of the Morphological EMP for Feature Extraction Followed by SVM for Classification (EMP/SVM)

Class	Training	Testing	SVM	EMP/SVM
Asphalt (6631)	548	6083	83.71	95.36
Meadows (18649)	532	18117	70.25	80.33
Gravel (2099)	265	1834	70.32	87.61
Trees (3064)	231	2833	97.81	98.37
Metal Sheets (1345)	375	970	99.41	99.48
Bare Soil (5029)	540	4489	92.25	63.72
Bitumen (1330)	392	938	81.58	98.87
Self-Blocking Bricks (3682)	524	3158	92.59	95.41
Shadow (947)	514	433	96.62	97.68
Overall accuracy	—	—	80.99	85.22

Note: The numbers in the parentheses are the total number of available samples.

the SVM applied on the original hyperspectral image to those obtained using a combination of the morphological EMP for feature extraction followed by SVM for classification (EMP/SVM).

As shown by Table 4.3, the EMP/SVM provides good classification results for the ROSIS University of Pavia scene, which represent a good improvement over the results obtained using the original hyperspectral image as input to the classifier. These results confirm the importance of using spatial and spectral information for classification purposes, as it was already found in the experimental results reported on Table 4.2.

In order to illustrate other approaches that use spatial information as postprocessing, Tables 4.4 and 4.5 respectively compare the classification results obtained by the traditional SVM with those found using the strategy adopted in [35], which combines the output of a pixel-wise SVM classifier with the morphological watershed, and in [36], in which the output of the SVM classifier is combined with the segmentation result provided by the RHSEG segmentation algorithm. As shown by Tables 4.4 and 4.5, these strategies lead to improved classification with regards to the traditional SVM. In fact, Tables 4.2–4.5 illustrate different aspects concerning the integration of spatial and spectral information. The results reported in Table 4.2 are obtained by integrating spatial and spectral information at the classification stage. On the other hand, the results reported in Table 4.3 are obtained by using spatial information at a preprocessing step prior to classification. Finally, the results reported in Tables 4.4 and 4.5 correspond to cases in which spatial information is included at a postprocessing step after conducting spectral-based classification. As a result, the comparison reported in this section illustrates different scenarios in which spatial and spectral information are used in complementary fashion but following different strategies, that is, spatial information is included at different stages of the classification process (preprocessing, postprocessing or kernel level).

After evaluating the importance of including spatial and spectral information, we now discuss the possibility to perform a better modeling of the hyperspectral data by working on a subspace. This is due to the fact that the dimensionality of the hyperspectral data is very high, and often the data lives in a subspace. Hence, if the proper subspace is identified prior to classification, adequate results can be obtained. In order to illustrate this concept, Table 4.6 shows the results obtained after comparing the SVM classifier with a subspace-based classifier such as the MLRsub [23], followed by an MRF-based spatial regularizer.

TABLE 4.4
Classification Results Obtained for the ROSIS Pavia University Scene by the SVM Classifier as Compared with Those Obtained Using the Strategy Adopted in [35], Which Combines the Output of a Pixel-Wise SVM Classifier with the Morphological Watershed

Class	Training	Testing	SVM	SVM+Watershed
Asphalt (6631)	548	6083	83.71	94.28
Meadows (18649)	532	18117	70.25	76.41
Gravel (2099)	265	1834	70.32	69.89
Trees (3064)	231	2833	97.81	98.30
Metal Sheets (1345)	375	970	99.41	99.78
Bare Soil (5029)	540	4489	92.25	97.51
Bitumen (1330)	392	938	81.58	97.14
Self-Blocking Bricks (3682)	524	3158	92.59	98.29
Shadow (947)	514	433	96.62	97.57
Overall accuracy	—	—	80.99	86.64

Note: The numbers in the parentheses are the total number of available samples.

TABLE 4.5

Classification Results Obtained for the ROSIS Pavia University Scene by the SVM Classifier as Compared with Those Obtained Using the Strategy Adopted in [36], in Which the Output of the SVM Classifier Is Combined with the Segmentation Result Provided by the Recursive Hierarchical Segmentation (RHSEG) Algorithm Developed by James C. Tilton at NASA's Goddard Space Flight Center

Class	Training	Testing	SVM	SVM+RHSEG
Asphalt (6631)	548	6083	83.71	94.77
Meadows (18649)	532	18117	70.25	89.32
Gravel (2099)	265	1834	70.32	96.14
Trees (3064)	231	2833	97.81	98.08
Metal Sheets (1345)	375	970	99.41	99.82
Bare Soil (5029)	540	4489	92.25	99.76
Bitumen (1330)	392	938	81.58	100
Self-Blocking Bricks (3682)	524	3158	92.59	99.29
Shadow (947)	514	433	96.62	96.48
Overall accuracy	—	—	80.99	93.85

Note: The numbers in the parentheses are the total number of available samples.

TABLE 4.6

Classification Results Obtained for the ROSIS Pavia University Scene by the SVM Classifier as Compared with a Subspace-Based Classifier Followed by Spatial Post-Processing (MLRsubMLL)

Class	Training	Testing	SVM	MLRsubMRF
Asphalt (6631)	548	6083	83.71	93.83
Meadows (18649)	532	18117	70.25	94.80
Gravel (2099)	265	1834	70.32	71.13
Trees (3064)	231	2833	97.81	92.17
Metal Sheets (1345)	375	970	99.41	100
Bare Soil (5029)	540	4489	92.25	98.43
Bitumen (1330)	392	938	81.58	99.32
Self-Blocking Bricks (3682)	524	3158	92.59	95.19
Shadow (947)	514	433	96.62	96.20
Overall accuracy	—	—	80.99	94.10

Note: The numbers in the parentheses are the total number of available samples.

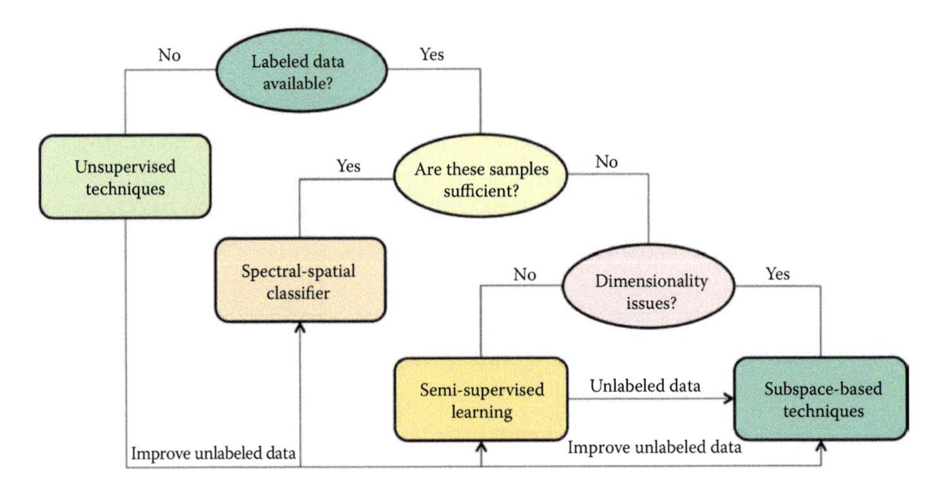

FIGURE 4.10 Summary of contributions in hyperspectral image classification discussed in this chapter.

The idea of applying subspace projection methods relies on the basic assumption that the samples within each class can approximately lie in a lower dimensional subspace. In the experiments reported in Table 4.6 it can be also be seen that spatial information (included as an MRF-based postprocessing) can be greatly beneficial in order to improve classification performance.

Finally, it is worth noting that the results discussed earlier are all based on supervised classifiers that assume the sufficient availability of labeled training samples. In case that no sufficient labeled samples are available, semi-supervised learning techniques can be used to generate additional unlabeled samples (from the initial set of labeled samples) that can complement the available labeled samples. The unlabeled samples can also be used to enhance subspace-based classifiers in case dimensionality issues are found to be relevant in the considered case study. If sufficient labeled samples are available, then the use of a spectral-spatial classifier is generally recommended as spatial information can provide a very important complement to the spectral information. Finally, in case that labeled samples are not available at all, unsupervised techniques need to be used for classification purposes. For instance, a relevant unsupervised method successfully applied to hyperspectral image data is Tilton's RHSEG algorithm,[2] The different analysis scenarios for classification discussed in this chapter are summarized in Figure 4.10.

4.4 CONCLUSIONS AND FUTURE DIRECTIONS

In this chapter, we have provided an overview of recent advances in techniques and methods for hyperspectral image classification [51–71]. The array of techniques particularly illustrated in this chapter comprises supervised and semi-supervised approaches, techniques able to exploit both spatial and spectral information, and techniques able to take advantage of a proper subspace representation of the hyperspectral data before conducting the classification in spatial–spectral terms. These approaches represent a subjective selection of the wide range of techniques currently adopted for hyperspectral image classification [3], and which include other techniques and strategies that have not been covered in detail in this chapter for space considerations. Of particular importance is the recent role of sparse classification methods [2, 20], which are gaining significant popularity and will likely play an important role in this research area in upcoming years.

Despite the wide arrange of techniques and strategies for hyperspectral data interpretation currently available, some unresolved issues still remain. For instance, the geometry of hyperspectral data is quite complex and dominated by nonlinear structures. This issue has undoubtedly an impact in the outcome of the classification techniques discussed in this section. In order to mitigate this, manifold learning has been proposed [48]. An important property of manifold learning is that it can model and characterize the complex nonlinear structure of the data prior to classification [49]. Another remaining issue is the very high computational complexity of some of the techniques available for classification of hyperspectral data [50]. In other words, there is a clear need to develop efficient classification techniques that can deal with the very large dimensionality and complexity of hyperspectral data.

4.5 ACKNOWLEDGMENTS

The authors would like to thank Prof. Paolo Gamba, the University of Pavia and the HySenS project for providing the ROSIS data used in the experiments.

NOTES

1 http://opensource.gsfc.nasa.gov/projects/HSEG/
2 http://opensource.gsfc.nasa.gov/projects/HSEG/

REFERENCES

[1] A. F. H. Goetz, G. Vane, J. E. Solomon and B. N. Rock, "Imaging spectrometry for Earth remote sensing," *Science*, vol. 228, no. 4704, pp. 1147–1153, 1985.
[2] J. M. Bioucas-Dias, A. Plaza, G. Camps-Valls, P. Scheunders, N. Nasrabadi and J. Chanussot, "Hyperspectral remote sensing data analysis and future challenges," *IEEE Geoscience and Remote Sensing Magazine*, vol. 1, no. 2, pp. 6–36, 2013.
[3] M. Fauvel, Y. Tarabalka, J. A. Benediktsson, J. Chanussot and J. C. Tilton, "Advances in spectral-spatial classification of hyperspectral images," *Proceedings of the IEEE*, vol. 101, no. 3, pp. 652–675, 2013.
[4] J. M. Bioucas-Dias, A. Plaza, N. Dobigeon, M. Parente, Q. Du, P. Gader and J. Chanussot, "Hyperspectral unmixing overview: Geometrical, statistical and sparse regression-based approaches," *IEEE Journal of Selected Topics in Applied Earth Observations and Remote Sensing*, vol. 5, no. 2, pp. 354–379, 2012.
[5] J. Chanussot, M. M. Crawford and B.-C. Kuo, "Foreword to the special issue on hyperspectral image and signal processing," *IEEE Transactions on Geoscience and Remote Sensing*, vol. 48, no. 11, pp. 3871–3876, 2010.
[6] A. Plaza, J. M. Bioucas-Dias, A. Simic and W. J. Blackwell, "Foreword to the special issue on hyperspectral image and signal processing," *IEEE Journal of Selected Topics in Applied Earth Observations and Remote Sensing*, vol. 5, no. 2, pp. 347–353, 2012.
[7] N. Keshava and J. Mustard, "Spectral unmixing," *IEEE Signal Processing Magazine*, vol. 19, no. 1, pp. 44–57, 2002.
[8] W.-K. Ma, J. M. Bioucas-Dias, T.-H. Chan, N. Gillis, P. Gader, A. Plaza, A. Ambikapathi and C.-Y. Chi, "A signal processing perspective on hyperspectral unmixing: Insights from remote sensing," *IEEE Signal Processing Magazine*, vol. 31, no. 1, pp. 67–81, 2014.
[9] A. Plaza, J. A. Benediktsson, J. Boardman, J. Brazile, L. Bruzzone, G. Camps-Valls, J. Chanussot, M. Fauvel, P. Gamba, J. A. Gualtieri, M. Marconcini, J. C. Tilton and G. Trianni, "Recent advances in techniques for hyperspectral image processing," *Remote Sensing of Environment*, vol. 113, supplement 1, pp. 110–122, 2009.[10] D. A. Landgrebe, *Signal Theory Methods in Multispectral Remote Sensing*. Hoboken, NJ: Wiley, 2003.
[11] L. O. Jimenez, J. L. Rivera-Medina, E. Rodriguez-Diaz, E. Arzuaga-Cruz and M. Ramirez-Velez, "Integration of spatial and spectral information by means of unsupervised extraction and classification for homogenous objects applied to multispectral and hyperspectral data," *IEEE Transactions on Geoscience and Remote Sensing*, vol. 43, no. 1, pp. 844–851, 2005.
[12] P. Gamba, F. Dell'Acqua, A. Ferrari, J. A. Palmason, J. A. Benediktsson and J. Arnasson, "Exploiting spectral and spatial information in hyperspectral urban data with high resolution," *IEEE Geoscience and Remote Sensing Letters*, vol. 1, no. 3, pp. 322–326, 2004.[13] J. A. Richards and X. Jia, *Remote Sensing Digital Image Analysis: An Introduction*. New York, Berlin and Heidelberg: Springer-Verlag, 2006.

[14] G. F. Hughes, "On the mean accuracy of statistical pattern recognizers," *IEEE Transactions on Information Theory*, vol. 14, no. 1, pp. 55–63, 1968.[15] G. Camps-Valls and L. Bruzzone, *Kernel Methods for Remote Sensing Data Analysis*. Hoboken, NJ: Wiley, 2009.
[16] L. Bruzzone and C. Persello, "A novel context-sensitive semisupervised SVM classifier robust to mislabeled training samples," *IEEE Transactions on Geoscience and Remote Sensing*, vol. 47, no. 7, pp. 2142–2154, 2009.
[17] J. Muñoz-Marí, D. Tuia and G. Camps-Valls, "Semisupervised classification of remote sensing images with active queries," *IEEE Transactions on Geoscience and Remote Sensing*, vol. 50, no. 10, pp. 3751–3763, 2012.
[18] D. Tuia and G. Camps-Valls, "Semisupervised remote sensing image classification with cluster kernels," *IEEE Geoscience and Remote Sensing Letters*, vol. 6, no. 2, pp. 224–228, 2009.
[19] Y. Tarabalka, M. Fauvel, J. Chanussot and J. Benediktsson, "SVM- and MRF-based method for accurate classification of hyperspectral images," *IEEE Geoscience and Remote Sensing Letters*, vol. 7, no. 4, pp. 736–740, 2010.
[20] B. Song, J. Li, M. Dalla Mura, P. Li, A. Plaza, J. M. Bioucas-Dias, J. A. Benediktsson and J. Chanussot, "Remotely sensed image classification using sparse representations of morphological attribute profiles," *IEEE Transactions on Geoscience and Remote Sensing*, vol. 52, no. 8, pp. 5122–5136, 2014.
[21] J. Li, P. R. Marpu, A. Plaza, J. M. Bioucas-Dias and J. A. Benediktsson, "Generalized composite kernel framework for hyperspectral image classification," *IEEE Transactions on Geoscience and Remote Sensing*, vol. 51, no. 9, pp. 4816–4829, 2013.
[22] J. Li, J. M. Bioucas-Dias and A. Plaza, "Spectral-spatial classification of hyperspectral data using loopy belief propagation and active learning," *IEEE Transactions on Geoscience and Remote Sensing*, vol. 51, no. 2, pp. 844–856, 2013.
[23] J. Li, J. Bioucas-Dias and A. Plaza, "Spectral-spatial hyperspectral image segmentation using subspace multinomial logistic regression and Markov random fields," *IEEE Transactions on Geoscience and Remote Sensing*, vol. 50, no. 3, pp. 809–823, 2012.
[24] J. Bioucas-Dias and J. Nascimento, "Hyperspectral subspace identification," *IEEE Transactions on Geoscience and Remote Sensing*, vol. 46, no. 8, pp. 2435–2445, 2008.
[25] I. Dopido, J. Li, P. R. Marpu, A. Plaza, J. M. Bioucas-Dias and J. A. Benediktsson, "Semi-supervised self-learning for hyperspectral image classification," *IEEE Transactions on Geoscience and Remote Sensing*, vol. 51, no. 7, pp. 4032–4044, 2013.
[26] T. V. Bandos, L. Bruzzone and G. Camps-Valls, "Classification of hyperspectral images with regularized linear discriminant analysis," *IEEE Transactions on Geoscience and Remote Sensing*, vol. 47, no. 3, pp. 862–873, 2009.
[27] F. Melgani and L. Bruzzone, "Classification of hyperspectral remote sensing images with support vector machines," *IEEE Transactions on Geoscience and Remote Sensing*, vol. 42, no. 8, pp. 1778–1790, 2004.
[28] G. Camps-Valls and L. Bruzzone, "Kernel-based methods for hyperspectral image classification," *IEEE Transactions on Geoscience and Remote Sensing*, vol. 43, pp. 1351–1362, 2005.[29] B. Schölkopf and A. Smola, *Learning with Kernels: Support Vector Machines, Regularization, Optimization and Beyond.* Cambridge, MA: MIT Press, 2002.[30] P. Soille, *Morphological Image Analysis, Principles and Applications*, 2nd ed. Berlin: Springer-Verlag, 2003.
[31] M. Pesaresi and J. Benediktsson, "A new approach for the morphological segmentation of high-resolution satellite imagery," *IEEE Transactions on Geoscience and Remote Sensing*, vol. 39, no. 2, pp. 309–320, 2001.
[32] J. Benediktsson, J. Palmason and J. Sveinsson, "Classification of hyperspectral data from urban areas based on extended morphological profiles," *IEEE Transactions on Geoscience and Remote Sensing*, vol. 43, no. 3, pp. 480–491, 2005.
[33] G. Camps-Valls, L. Goméz-Chova, J. Muñoz-Marí, J. Vila-Francés and J. Calpe-Maravilla, "Composite kernels for hyperspectral image classification," *IEEE Geoscience and Remote Sensing Letters*, vol. 3, pp. 93–97, 2006.
[34] M. Khodadadzadeh, J. Li, A. Plaza, H. Ghassemian, J. Bioucas-Dias and X. Li, "Spectral–spatial classification of hyperspectral data using local and global probabilities for mixed pixel characterization," *IEEE Transactions on Geoscience and Remote Sensing*, vol. 52, no. 10, pp. 6298–6314, 2014.
[35] Y. Tarabalka, J. Chanussot and J. Benediktsson, "Segmentation and classification of hyperspectral images using watershed transformation," *Pattern Recognition*, vol. 43, pp. 2367–2379, 2010.

[36] Y. Tarabalka, J. A. Benediktsson, J. Chanussot and J. C. Tilton, "Multiple spectral-spatial classification approach for hyperspectral data," *IEEE Transactions on Geoscience and Remote Sensing*, vol. 48, no. 11, pp. 4122–4132, 2011.

[37] D. Böhning, "Multinomial logistic regression algorithm," *Annals of the Institute of Statistics and Mathematics*, vol. 44, pp. 197–200, 1992.

[38] F. Bovolo, L. Bruzzone and L. Carlin, "A novel technique for subpixel image classification based on support vector machine," *IEEE Transactions on Image Processing*, vol. 19, pp. 2983–2999, 2010.

[39] L. Bruzzone, M. Chi and M. Marconcini, "A novel transductive SVM for the semisupervised classification of remote sensing images," *IEEE Transactions on Geoscience and Remote Sensing*, vol. 11, pp. 3363–3373, 2006.

[40] G. Camps-Valls, T. Bandos and D. Zhou, "Semi-supervised graph-based hyperspectral image classification," *IEEE Transactions on Geoscience and Remote Sensing*, vol. 45, pp. 3044–3054, 2007.

[41] S. Velasco-Forero and V. Manian, "Improving hyperspectral image classification using spatial preprocessing," *IEEE Geoscience and Remote Sensing Letters*, vol. 6, pp. 297–301, 2009.

[42] J. Li, J. Bioucas-Dias and A. Plaza, "Semi-supervised hyperspectral image segmentation using multinomial logistic regression with active learning," *IEEE Transactions on Geoscience and Remote Sensing*, vol. 48, no. 11, pp. 4085–4098, 2010.

[43] J. Muñoz Marí, F. Bovolo, L. Gómez-Chova, L. Bruzzone and G. Camp-Valls, "Semisupervised one-class support vector machines for classification of remote sensing data," *IEEE Transactions on Geoscience and Remote Sensing*, vol. 48, no. 8, pp. 3188–3197, 2010.

[44] L. Gómez-Chova, G. Camps-Valls, L. Bruzzone and J. Calpe-Maravilla, "Mean MAP kernel methods for semisupervised cloud classification," *IEEE Transactions on Geoscience and Remote Sensing*, vol. 48, no. 1, pp. 207–220, 2010.

[45] D. Tuia and G. Camps-Valls, "Urban image classification with semisupervised multiscale cluster kernels," *IEEE Journal of Selected Topics in Applied Earth Observations and Remote Sensing*, vol. 4, no. 1, pp. 65–74, 2011.

[46] F. Ratle, G. Camps-Valls and J. Weston, "Semisupervised neural networks for efficient hyperspectral image classification," *IEEE Transactions on Geoscience and Remote Sensing*, vol. 48, no. 5, pp. 2271–2282, 2010.[47] G. Camps-Valls, D. Tuia, L. Gómez-Chova, S. Jiménez and J. Malo, *Remote Sensing Image Processing*. San Rafael, CA: Morgan and Claypool, 2011.

[48] L. Ma, M. Crawford and J. Tian, "Local manifold learning based k-nearest-neighbor for hyperspectral image classification," *IEEE Transactions on Geoscience and Remote Sensing*, vol. 48, no. 11, pp. 4099–4109, 2010.

[49] W. Kim and M. Crawford, "Adaptive classification for hyperspectral image data using manifold regularization kernel machines," *IEEE Transactions on Geoscience Remote Sensing*, vol. 48, no. 11, pp. 4110–4121, 2010.

[50] A. Plaza, J. Plaza, A. Paz and S. Sanchez, "Parallel hyperspectral image and signal processing," *IEEE Signal Processing Magazine*, vol. 28, no. 3, pp. 119–126, 2011.

[51] N. Audebert, B. Le Saux and S. Lefevre, "Deep learning for classification of hyperspectral data: A comparative review," *IEEE Geoscience and Remote Sensing Magazine*, vol. 7, no. 2, pp. 159–173, 2019. http://doi.org/10.1109/MGRS.2019.2912563.

[52] J. Bai, W. Shi, Z. Xiao, T. A. A. Ali, F. Ye and L. Jiao, "Achieving better category separability for hyperspectral image classification: A spatial–spectral approach," *IEEE Transactions on Neural Networks and Learning Systems*, 2023. http://doi.org/10.1109/TNNLS.2023.3235711.

[53] X. Cao, J. Yao, Z. Xu and D. Meng, "Hyperspectral image classification with convolutional neural network and active learning," *IEEE Transactions on Geoscience and Remote Sensing*, vol. 58, no. 7, pp. 4604–4616, 2020. http://doi.org/10.1109/TGRS.2020.2964627.

[54] P. Ghamisi, et al., "New frontiers in spectral-spatial hyperspectral image classification: The latest advances based on mathematical morphology, markov random fields, segmentation, sparse representation, and deep learning," *IEEE Geoscience and Remote Sensing Magazine*, vol. 6, no. 3, pp. 10–43, 2018. http://doi.org/10.1109/MGRS.2018.2854840.

[55] P. Ghamisi, J. Plaza, Y. Chen, J. Li and A. J. Plaza, "Advanced spectral classifiers for hyperspectral images: A review," *IEEE Geoscience and Remote Sensing Magazine*, vol. 5, no. 1, pp. 8–32, 2017. http://doi.org/10.1109/MGRS.2016.2616418.

[56] Reaya Grewal, Singara Singh Kasana and Geeta Kasana, "Machine learning and deep learning techniques for spectral spatial classification of hyperspectral images: A comprehensive survey," *Electronics*, vol. 12, no. 3, p. 488, 2023. https://doi.org/10.3390/electronics12030488

[57] Y. Gu, J. Chanussot, X. Jia and J. A. Benediktsson, "Multiple kernel learning for hyperspectral image classification: A review," *IEEE Transactions on Geoscience and Remote Sensing*, vol. 55, no. 11, pp. 6547–6565, 2017. http://doi.org/10.1109/TGRS.2017.2729882.

[58] N. He, et al., "Feature extraction with multiscale covariance maps for hyperspectral image classification," *IEEE Transactions on Geoscience and Remote Sensing*, vol. 57, no. 2, pp. 755–769, 2019. http://doi.org/10.1109/TGRS.2018.2860464.

[59] M. Imani and H. Ghassemian, "An overview on spectral and spatial information fusion for hyperspectral image classification: Current trends and challenges," *Information Fusion*, vol. 59, no. 2020, pp. 59–83, 2020. ISSN 1566-2535. https://doi.org/10.1016/j.inffus.2020.01.007. (https://www.sciencedirect.com/science/article/pii/S1566253519307857)

[60] T. Ishida, J. Kurihara, F. A. Viray, S. B. Namuco, E. C. Paringit, G. J. Perez, Y. Takahashi and J. J. Marciano, "A novel approach for vegetation classification using UAV-based hyperspectral imaging," *Computers and Electronics in Agriculture*, vol. 144, no. 2018, pp. 80–85, 2018. ISSN 0168-1699. https://doi.org/10.1016/j.compag.2017.11.027. https://www.sciencedirect.com/science/article/pii/S0168169917310499

[61] A. Khan, A. D. Vibhute, S. Mali and C. H. Patil, "A systematic review on hyperspectral imaging technology with a machine and deep learning methodology for agricultural applications," *Ecological Informatics*, vol. 69, no. 2022, p. 101678, 2022. ISSN 1574-9541. https://doi.org/10.1016/j.ecoinf.2022.101678. https://www.sciencedirect.com/science/article/pii/S1574954122001285

[62] S. Liu, D. Marinelli, L. Bruzzone and F. Bovolo, "A review of change detection in multitemporal hyperspectral images: Current techniques, applications, and challenges," *IEEE Geoscience and Remote Sensing Magazine*, vol. 7, no. 2, pp. 140–158, 2019. http://doi.org/10.1109/MGRS.2019.2898520.

[63] M. J. Khan, H. S. Khan, A. Yousaf, K. Khurshid and A. Abbas, "Modern trends in hyperspectral image analysis: A review," *IEEE Access*, vol. 6, pp. 14118–14129, 2018. http://doi.org/10.1109/ACCESS.2018.2812999.

[64] M. E. Paoletti, et al., "Capsule networks for hyperspectral image classification," *IEEE Transactions on Geoscience and Remote Sensing*, vol. 57, no. 4, pp. 2145–2160, 2019. http://doi.org/10.1109/TGRS.2018.2871782.

[65] M. E. Paoletti, J. M. Haut, J. Plaza and A. Plaza, "A new deep convolutional neural network for fast hyperspectral image classification," *ISPRS Journal of Photogrammetry and Remote Sensing*, vol. 145, Part A, pp. 120–147, 2018. ISSN 0924-2716. https://doi.org/10.1016/j.isprsjprs.2017.11.021. https://www.sciencedirect.com/science/article/pii/S0924271617303660

[66] B. Rasti, et al., "Feature extraction for hyperspectral imagery: The evolution from shallow to deep: Overview and toolbox," *IEEE Geoscience and Remote Sensing Magazine*, vol. 8, no. 4, pp. 60–88, 2020. http://doi.org/10.1109/MGRS.2020.2979764.

[67] A. Sellami and S. Tabbone, "Deep neural networks-based relevant latent representation learning for hyperspectral image classification," *Pattern Recognition*, vol. 121, p. 108224, 2022. ISSN 0031-3203. https://doi.org/10.1016/j.patcog.2021.108224. https://www.sciencedirect.com/science/article/pii/S0031320321004052

[68] L. Shuai, Z. Li, Z. Chen, D. Luo and J. Mu, "A research review on deep learning combined with hyperspectral Imaging in multiscale agricultural sensing," *Computers and Electronics in Agriculture*, vol. 217, p. 108577, 2024. ISSN 0168-1699. https://doi.org/10.1016/j.compag.2023.108577. https://www.sciencedirect.com/science/article/pii/S0168169923009651

[69] A. Signoroni, M. Savardi, A. Baronio and S. Benini, "Deep learning meets hyperspectral image analysis: A multidisciplinary review," *Journal of Imaging*, vol. 5, no. 5, p. 52, 2019. https://doi.org/10.3390/jimaging5050052

[70] F. Ullah, I. Ullah, R. U. Khan, S. Khan, K. Khan and G. Pau, "Conventional to deep ensemble methods for hyperspectral image classification: A comprehensive survey," *IEEE Journal of Selected Topics in Applied Earth Observations and Remote Sensing*, vol. 17, pp. 3878–3916, 2024. http://doi.org/10.1109/JSTARS.2024.3353551.

[71] G. Vivone, "Multispectral and hyperspectral image fusion in remote sensing: A survey," *Information Fusion*, vol. 89, pp. 405–417, 2023. ISSN 1566-2535. https://doi.org/10.1016/j.inffus.2022.08.032. https://www.sciencedirect.com/science/article/pii/S1566253522001312

5 Collaborative Representation for Hyperspectral Image Classification and Detection

Qian Du, Chiranjibi Shah, Hongjun Su, and Wei Li

LIST OF ACRONYMS

AA	Average accuracy
CDSRC	Class-dependent SRC
CR	Collaborative representation
CRC	Collaborative Representation-Based Classifier
CRD	Collaborative representation-based detector
CRT	Collaborative Representation with Tikhonov
EMAP	Extended multi-attribute profile
HSI	Hyperspectral image
JCRC	Joint within-class CRC
MRF	Markov random field
OA	Overall accuracy
PCA	Principal component analysis
ProCRC	Probabilistic CRC
ProCRT	Probabilistic CRT
ROC	Receiver operating characteristic
ROSIS	Reflective operating system imaging spectrometer
SaCRC	Spatial-aware CRC
SaCRT	Spatial-aware CRT
SRC	Sparse representation-based classification

5.1 INTRODUCTION

A hyperspectral imaging sensor collects hundreds of narrow spectral bands for an imaged scene. Due to high spectral resolution, a hyperspectral image (HSI) taken from an airborne or spaceborne sensor can be used to measure, analyze, and interpret spectra for a specific object or target [1–2]. Commercial hyperspectral sensors have many applications, such as food safety, precision farming, geological exploration, environmental monitoring, etc. [3–4].

Many pattern classification schemes have been proposed for accurate classification of HSI. Representation-based classification and detection are of great interest due to the paradigm of compressed sensing. For instance, sparse representation-based classification (SRC) [5–8] has been investigated in many applications, such as face recognition. It has also been applied to remote sensing images, such as optical hyperspectral images [9–12]. In SRC, a testing pixel is sparsely represented by the labeled data via an l_0 or l_1 -norm regularization, and its class label is determined to be that of the class whose labeled samples provides the smallest approximation error.

DOI: 10.1201/9781003541158-6

However, it has been argued that it is the "collaborative" nature of the approximation instead of "competitive" nature imposed by sparseness constraint that actually improves classification accuracy [13–15]. Collaborative representation (CR) means all the atoms "collaborate" on the representation of a single pixel, and each atom has an equal chance to participate in the representation. It is solved as an l_2 -norm regularized least squares problem [16]. The CRC provides a closed-form solution and is more efficient in comparison to the SRC [17].

In addition, the idea of collaborative representation has been employed for hyperspectral anomaly detection [18], where an anomaly (often corresponding to a man-made target) is a pixel whose spectral signature is different from its surroundings, and it can be effectively detected by collaborative representation of background pixels. It has been demonstrated that the collaborative representation-based detector (CRD) can offer better detection performance than the classical ones such as the RX algorithm [19].

This chapter reviews various forms of CR in hyperspectral image classification and anomaly detection that can provide better performance in comparison to its original versions.

5.2 COLLABORATIVE REPRESENTATION-BASED CLASSIFICATION

5.2.1 Collaborative Representation-Based Classifier (CRC)

Let an HSI dataset with n labeled samples be denoted as $\mathbf{X} = \{\mathbf{x}_i\}_{i=1}^{n} \in \Re^d$, where d is the number of spectral bands. Let M be the total number of classes. The weight vector $\mathbf{a}$ in CRC for a testing pixel $\mathbf{y}$ can be estimated based on the following objective function [17]:

$$\arg\min \lambda \arg\min_{\mathbf{a}} \|\mathbf{y} - \mathbf{Xa}\|_2^2 + \lambda \|\mathbf{a}\|_2^2 \tag{5.1}$$

where λ is the regularization term. Equation 5.1 has a closed-form solution as:

$$\mathbf{a} = \left(\mathbf{X}^T\mathbf{X} + \lambda\mathbf{I}\right)^{-1} \mathbf{X}^T\mathbf{y} \tag{5.2}$$

Then, the class label of $\mathbf{y}$ can be determined by minimizing class-specific representation residuals as:

$$r_m = \|\mathbf{y} - \mathbf{X}_m\mathbf{a}_m\|_2^2 \tag{5.3}$$

where $\mathbf{r}m$ is the residual for the m-th class, $\mathbf{X}m$ includes the labeled samples belonging to the m-th class, and $\mathbf{a}m$ is the coefficients in $\mathbf{a}$ corresponding to $\mathbf{X}m$. Then rm is used to predict the class label of $\mathbf{y}$ as:

$$class(\mathbf{y}) = \arg\min_{m} \left(r_m(\mathbf{y})\right) \tag{5.4}$$

5.2.2 Collaborative Representation with Tikhonov (CRT) Regularization

Some samples in $\mathbf{X}$ may be different from the testing pixel $\mathbf{y}$. In this case, allowing all samples to equally contribute in representation is problematic. Distance weighted Tikhonov regularization is implemented in CRT such that the objective function is formulated as:

$$\arg\min_{\mathbf{a}} \|\mathbf{y} - \mathbf{Xa}\|_2^2 + \lambda \|\Gamma\mathbf{a}\|_2^2 \tag{5.5}$$

where $\mathbf{\Gamma}$ is the Tikhonov regularization term that defines the distance between training samples and the testing sample, and it can be represented in a diagonal matrix as [20]:

$$\Gamma = \begin{bmatrix} \|\mathbf{y}-\mathbf{x}_1\|_2 & & 0 \\ & \ddots & \\ 0 & & \|\mathbf{y}-\mathbf{x}_n\|_2 \end{bmatrix} \tag{5.6}$$

For a sample dissimilar to $\mathbf{y}$, its Euclidean distance with $\mathbf{y}$ is large, and the overall penalty in Equation 5.5 becomes large, preventing a large coefficient for representation. The closed-form solution to Equation 5.5 is:

$$\mathbf{a} = \left(\mathbf{X}^T\mathbf{X} + \lambda\mathbf{\Gamma}^2\right)^{-1}\mathbf{X}^T\mathbf{y} \tag{5.7}$$

Similarly, Equations 5.3 and 5.4 can be used to predict the class label of $\mathbf{y}$.

5.2.3 Probabilistic CRC (ProCRC) and Probabilistic CRT (ProCRT)

The ProCRC is proposed with the following objective function [3, 21]:

$$\arg\min_{\mathbf{a}} \|\mathbf{y}-\mathbf{Xa}\|_2^2 + \lambda\|\mathbf{a}\|_2^2 + \gamma\sum_{m=1}^{M}\|\mathbf{Xa}-\mathbf{X}_m\mathbf{a}_m\|_2^2 \tag{5.8}$$

where the third term is added to minimize the inconsistency between global representation and class-specific representations, and γ is a controlling parameter. Then, $\mathbf{a}$ can be determined by finding $\mathbf{Xa}$ that is closest to $\mathbf{y}$ as:

$$\mathbf{a}\left(\mathbf{X}^T\mathbf{X} + \frac{\gamma}{M}\sum_{m=1}^{M}\bar{\mathbf{X}}'^T_m\bar{\mathbf{X}}'_m + \lambda\mathbf{I}\right)^{-1}\mathbf{X}^T\mathbf{y} \tag{5.9}$$

where $\bar{\mathbf{X}}'_m = \mathbf{X} - \mathbf{X}'_m$ and $\mathbf{X}'_m = [0, \ldots, \mathbf{X}_m, \ldots, 0]$ **have the same size of X**. Then $\mathbf{a}$ is used to predict class label of $\mathbf{y}$ as:

$$class(\mathbf{y}) = \arg\min_{m}\|\mathbf{Xa}-\mathbf{X}_m\mathbf{a}_m\|_2^2 \tag{5.10}$$

The Tikhonov regularization can be imposed in the objective function of ProCRC as follows:

$$\arg\min_{\mathbf{a}} \|\mathbf{y}-\mathbf{Xa}\|_2^2 + \lambda\|\mathbf{\Gamma a}\|_2^2 + \gamma\sum_{m=1}^{M}\|\mathbf{Xa}-\mathbf{X}_m\mathbf{a}_m\|_2^2 \tag{5.11}$$

resulting in ProCRT. The coefficient $\mathbf{a}$ in Equation 5.11 can be found as follows:

$$\mathbf{a} = \left(\mathbf{X}^T\mathbf{X} + \frac{\gamma}{M}\sum_{m=1}^{M}\bar{\mathbf{X}}'^T_M\bar{\mathbf{X}}'_m + \lambda\mathbf{\Gamma}^2\right)^{-1}\mathbf{X}^T\mathbf{y} \tag{5.12}$$

Then the class label of $\mathbf{y}$ can be obtained using Equation 5.10.

5.2.4 Spatial-Aware CRC (SaCRC) and Spatial-Aware CRT (SaCRT)

The following objective function is formed in spatial-aware collaborative representation (SaCRC) that consists of spatial regularization term [22]:

$$\underset{\mathbf{a}}{\arg\min}\|\mathbf{y}-\mathbf{Xa}\|_2^2+\lambda\|\mathbf{a}\|_2^2+\beta\|diag(\mathbf{s})\|_2^2 \tag{5.13}$$

where $\mathbf{s} = (s_1, s_2 \ldots, s_n)^T$and

$$s_i = [dist((a_i,b_i),(a_y,b_y))]^p \tag{5.14}$$

and β is an additional regularization parameter. Here, dist(.) denotes the Euclidean distance, *si* is the distance between the *i*-th training sample **x***i* and **y** at locations (a_i,b_i) and (a_y,b_y), respectively, and p is a smoothing parameter for spatial coherence. Adding the spatial distance constraint means training samples located in a local spatial neighbor is given a high opportunity to contribute in the representation, because local pixels are often similar. The solution of coefficient **a** can be estimated as:

$$\mathbf{a}=\left(\mathbf{X}^T\mathbf{X}+\lambda\mathbf{I}+\beta diag(\mathbf{s})\right)^{-1}\mathbf{X}^T\mathbf{y} \tag{5.15}$$

Similarly, the Tikhonov regularized spatial aware collaborative representation (SaCRT) can be estimated as:

$$\underset{\mathbf{a}}{\arg\min}\|\mathbf{y}-\mathbf{Xa}\|_2^2+\lambda\|\boldsymbol{\Gamma}\mathbf{a}\|_2^2+\beta\|diag(\mathbf{s})\|_2^2 \tag{5.16}$$

and its solution is:

$$\mathbf{a}=\left(\mathbf{X}^T\mathbf{X}+\lambda\boldsymbol{\Gamma}^2+\beta diag(\mathbf{s})\right)^{-1}\mathbf{X}^T\mathbf{y} \tag{5.17}$$

5.2.5 SaProCRC and SaProCRT

Spatial feature induced regularization term can be used in the objective function of SaProCRC as [23]:

$$\underset{\mathbf{a}}{\arg\min}\|\mathbf{y}-\mathbf{Xa}\|_2^2+\lambda\|\mathbf{a}\|_2^2+\gamma\sum_{m=1}^{M}\|\mathbf{Xa}-\mathbf{X}_m\mathbf{a}_m\|_2^2+\beta\|duag(\mathbf{s})\|_2^2 \tag{5.18}$$

The solution of **a** is given by

$$\mathbf{a}\left(\mathbf{X}^T\mathbf{X}+\frac{\gamma}{M}\sum_{m=1}^{M}\bar{\mathbf{X}}'^T_m\bar{\mathbf{X}}'_m+\lambda\mathbf{I}+\beta diag(\mathbf{s})\right)^{-1}\mathbf{X}^T\mathbf{y} \tag{5.19}$$

The objective function of SaProCRT is

$$\underset{\mathbf{a}}{\arg\min}\|\mathbf{y}-\mathbf{Xa}\|_2^2+\lambda\|\boldsymbol{\Gamma}\mathbf{a}\|_2^2+\gamma\sum_{m=1}^{M}\|\mathbf{Xa}-\mathbf{X}_m\mathbf{a}_m\|_2^2+\beta\|diag(\mathbf{s})\|_2^2 \tag{5.20}$$

and the solution is

$$\mathbf{a}\left(\mathbf{X}^T\mathbf{X}+\frac{\gamma}{M}\sum_{m=1}^{M}\bar{\mathbf{X}}'^T_M\bar{\mathbf{X}}'_m+\lambda\mathbf{\Gamma}^2+\beta diag(\mathbf{s})\right)^{-1}\mathbf{X}^T\mathbf{y} \tag{5.21}$$

5.2.6 Representation-Based Classification with Dictionary Partition

The original CRC and CRT use all the labeled samples belonging to different classes in **X** for representation. An alternative way is to estimate testing samples using within-class samples to generate class-specific representations and residuals. In [24–25], class-dependent SRC (CDSRC) and nearest subspace classifier (NSC) were presented. The CDSRC improves the traditional SRC by exploiting both correlation and inter-sample distance relationship between testing and training samples. The NSC was designed to employ an l_2-norm penalty to measure the similarity between the testing pixel and within-class labeled samples, which is denoted as within-class CRC. In the NRS [20], the objective function is

$$\arg\min_{\mathbf{a}_m}\left\|\mathbf{y}-\mathbf{X}_m\mathbf{a}_m\right\|_2^2+\lambda\left\|\mathbf{\Gamma}_{m,\mathbf{y}}\mathbf{a}_m\right\|_2^2 \tag{5.22}$$

where $\mathbf{X}m$ includes the samples in the m-th class, $\mathbf{\Gamma}_{m,\mathbf{y}}$ is a diagonal matrix specific to the samples in $\mathbf{X}m$ for the testing sample **y**, and the weight vector $\mathbf{a}m$ corresponding to $\mathbf{X}m$ can be calculated as

$$\mathbf{a}_m=\left(\mathbf{X}_M^T\mathbf{X}_m+\lambda\mathbf{\Gamma}_{m,\mathbf{y}}^2\right)^{-1}\mathbf{X}_m^T\mathbf{y} \tag{5.23}$$

The class label of the testing sample **y** is then determined according to the class which minimizes the representation residual. According to previous study, the class-specific CRC and CRT outperform their global counterparts [24].

5.2.7 Joint Representation-Based Classification

Incorporating contextual information is another important strategy to improve the traditional representation-based classification, with the assumption that neighboring pixels tend to belong to the same class. In [26], a joint within-class CRC (JCRC) was presented, where neighboring pixels near the testing pixel are simultaneously represented via a joint collaborative model with class-specific labeled samples. The JCRC [26] is a spatially-joint extension of the previously-introduced NRS where both the testing and training data are spatially averaged with their nearest neighbors before representation. Take the testing pixel **y** centered at a 3 × 3 window as an example, the averaged signature is obtained using its 8 neighbors and itself, that is, $\tilde{\mathbf{y}}=\frac{1}{9}\sum_{i=1}^{9}\mathbf{y}_i$. Similarly, each atom in the m-th sub-dictionary $\mathbf{X}m$ is replaced with its spatial average, resulting in $\tilde{\mathbf{X}}m$, which is used for the representation of $\tilde{\mathbf{y}}$.

The JCRC considers all surrounding pixels equally important, which may be problematic, especially in heterogeneous regions where the central pixel and neighboring pixels may not belong to the same class. Under this case, only the neighboring pixels that are similar to the central pixel should be taken into consideration or given higher weights when producing a smoothed version of $\tilde{\mathbf{y}}$. The Gaussian kernel function was employed to generate the weights or importance of a neighboring pixel in the proposed weighted JCRC [27]. It turns out that the weighted version outperforms the original JCRC.

5.2.8 Kernel CRC and CRT

In a kernel method, data are projected into a high-dimensional kernel-induced feature space by an implicit nonlinear mapping function, that is, $\mathbf{y} \rightarrow \Phi(\mathbf{y}) \in \Re^{D\times 1}$ ($D>>d$ is the dimension of kernel feature space). Usually, the explicit mapping $\Phi(\cdot)$ is unknown. According to the kernel trick, the inner product of two vectors, $\mathbf{y}$ and $\mathbf{y}'$, in the feature space is equal to the output of a kernel function, that is, $k(\mathbf{y}, \mathbf{y}') = \Phi(\mathbf{y})^T \Phi(\mathbf{y}')$ [28]. Thus, the explicit mapping $\Phi(\cdot)$ can be avoided. If a kernel function $k(\cdot,\cdot)$ with appropriate parameters is chosen, the testing pixel can be represented more accurately than in the original space. It was demonstrated that the aforementioned kernelized representation-based algorithms can provide higher classification accuracy than the original counterparts, and have been applied to various applications (e.g., face recognition [29–30]). In [31], kernel CRC (KCRC) and kernel collaborative representation with Tikhonov regularization (KCRT) were proposed. In [32], kernel nonlocal joint CRC based on column-generation was introduced.

In KCRC [31], the weight vector $\mathbf{a}^{(K)}$ is estimated as

$$\arg\min_{\mathbf{a}^{(K)}} \left\|\Phi(\mathbf{y}) - \mathbf{\Phi a}^{(K)}\right\|_2^2 + \lambda \left\|\mathbf{a}^{(K)}\right\|_2^2 \tag{5.24}$$

where $\mathbf{\Phi} = \Phi(\mathbf{X}) \in \Re^{D\times n}$. The closed-form solution is

$$\mathbf{a}^{(K)} = (\mathbf{K} + \lambda\mathbf{I})^{-1} k(\cdot, \mathbf{y}) \tag{5.25}$$

where $\mathbf{K} = \mathbf{\Phi}^T \mathbf{\Phi} \in \Re^{n\times n}$ is the Gram matrix with the (i,j)-th element being $\mathbf{K}_{i,j} = k(\mathbf{x}_i, \mathbf{x}_j)$, and $k(\cdot, \mathbf{y}) = [k(\mathbf{x}_1, \mathbf{y}), k(\mathbf{x}_2, \mathbf{y}), \ldots, k(\mathbf{x}_n, \mathbf{y})]^T \in \Re^{n\times 1}$. In KCRT, the weight vector $\mathbf{a}^{(K)}$ is estimated by minimizing

$$\arg\min_{\mathbf{a}^{K}} \left\|\Phi(\mathbf{y}) - \mathbf{\Phi a}^{(K)}\right\|_2^2 + \lambda \left\|\mathbf{\Gamma}_{\Phi(\mathbf{y})} \mathbf{a}^{(K)}\right\|_2^2 \tag{5.26}$$

and the Tikhonov matrix $\mathbf{\Gamma}_{\Phi(\mathbf{y})}$ in the kernel space becomes

$$\mathbf{\Gamma}_{\Phi(\mathbf{y})} = \begin{bmatrix} \left\|\Phi(\mathbf{y}) - \Phi(\mathbf{x}_1)\right\|_2 & & 0 \\ & \ddots & \\ 0 & & \left\|\Phi(\mathbf{y}) - \Phi(\mathbf{x}_n)\right\|_2 \end{bmatrix} \tag{5.27}$$

where $\|\Phi(y) - \Phi(x_i)\|_2$, $i = [k(y,y) + k(x_i, x_i) - 2k(y,x_i)]^{1/2}$, $i = 1, 2, \ldots, n$. The solution of $\mathbf{a}^{(K)}$ is $\mathbf{a}^{(K)} = (\mathbf{K} + \lambda\mathbf{\Gamma}^2_{\Phi(y)})^{-1} k(\cdot,y)$.

5.2.9 Representation in Feature Spaces

Spatial-spectral classifiers, considering spatial features and spectral signatures simultaneously, have been recently developed (e.g., [33]). Representation based classifiers can be applied in a feature space. In [34–35], extended multi-attribute profile (EMAP) was employed to exploit spatial features, followed by the traditional SRC. In [36–37], a Gabor filter was used to capture spatial structures of objects, such as orientation information. The other strategy is to incorporate spatial-context information at postprocessing level. In [38–39], Markov random field (MRF) was combined with SRC or NRS to produce the final output. The benefit of MRF is that it can work with different probability distributions and extract spatial correlation among pixels. In [34, 36, 37],

EMAP or Gabor features were extracted in a preprocessing step. EMAP features are generated by the morphological attribute profiles (APs) obtained using different types of attributes (e.g., thinning, thickening, etc.). Note that a Gabor filter [40] is a sinusoidal function modulated by a Gaussian envelope. In [41], 3-D Gabor feature-based collaborative representation was proposed. It stated that spatial features extracted by 3-D Gabor transformation could potentially increase the discrimination power of material features.

Before the operators, for example, EMAP and Gabor, are employed, a dimensionality reduction technique is usually applied to select a representative subset. It is to reduce redundance of spectral bands and avoid tremendous extracted spatial features. For example, a band selection technique, such as the one using linear prediction error [42], has been proposed for unsupervised band selection based on band similarity assessment. Compared to the widely used principal component analysis (PCA), band selection techniques can keep the clear physical meaning of the low-dimensional bands.

5.3 COLLABORATIVE REPRESENTATION-BASED ANOMALY DETECTION

Anomaly detection has attracted extensive attention, as anomalies are often related to man-made targets of interest whose spectral signatures are unknown but different from the surroundings. The Reed–Xiaoli (RX) algorithm proposed by Reed and Xiaoli Yu is one of the most widely used anomaly detection algorithms [19], and it is a constant false alarm detection algorithm with maximum likelihood detection. There are two versions of the RX: global and local. It is based on a hypothesis that the background pixels around an anomalous pixels obey the multivariate Gaussian model, and the parameters of the model are obtained by estimating the mean and covariance of the background. Therefore, accurate global/local background statistics is crucial. The RX detector for the testing pixel $\mathbf{y}$ has the form of

$$d_{RX}(\mathbf{y}) = (\mathbf{y} - \mathbf{m})^T \mathbf{\Sigma}^{-1} (\mathbf{y} - \mathbf{m}) \tag{5.28}$$

where $\mathbf{m}$ and $\mathbf{\Sigma}$ are the mean vector and covariance matrix of background.

Li and Du proposed a hyperspectral anomaly detection algorithm based on collaborative representation (CRD) [18], where a central pixel in a dual-window is collaboratively represented by pixels in the frame of dual-window. Anomalies have much lower occurrence other objects or background pixels. If the central pixel is an anomaly, then the representation error by local background pixels is large. The CRD can be expressed as

$$d_{CRD}(\mathbf{y}) = \left\| \mathbf{y} - \mathbf{X}_b \mathbf{a}_b \right\|_2^2 \tag{5.29}$$

where $\mathbf{X}b$ include the background pixels and $\mathbf{a}b$ is the coefficient vector estimated using Equations 5.7 and 5.6.

Obviously, the CRD can be extended to the kernel version (KCRD). According to Li and Du [18], CRD algorithm is simpler and more efficient than other detectors. However, the performance of CRD is not optimal. If the testing pixel is an abnormal pixel and several pixels from surrounding pixels are similarly anomalous, the judge error may occur. Vafadar and Ghassemian [43] proposed CR with outlier removal anomaly detector (CRBORAD) method to improve the accuracy. This method removes outlier by disregarding the small probability pixels in the Gaussian distribution. However, the actual situation does not necessarily conform to the statistical law, the improvement of detection accuracy may be limited. Recently, the morphology based collaborative representation detector (MCRD) was proposed in [44] to utilize the spatial information for anomaly detection, but it does not consider the pollution of abnormal pixels.

Principal component analysis (PCA) is often used to reduce dimension to extract useful information. When PCA is applied in the spatial domain, it can maintain the major spatial information while removing minor variations including anomalous pixels. Ref. [45] proposed an algorithm that incorporates CRD with PCA for removing outliers (PCAroCRD), thereby eliminating the influence of outliers in collaborative representation. Two versions of the PCAroCRD are proposed: global and local. For the global PCAroCRD detector, the PCA of the whole image is conducted, and then the extracted pixel information is used to represent each testing pixel. For the local PCAroCRD detector, a sliding dual window is adopted, and then PCA is used to extract the major background pixel information as samples to represent the testing pixel. If the testing pixel can be well represented, it is a normal pixel; otherwise it is claimed to be an abnormal pixel.

There exist other related works in the literature. For instance, Ref. [46] introduced a combined sparse and collaborative representation-based (CSCR) algorithm for target detection. The basic idea in the CSCR is that a testing sample is sparsely represented by target atoms because it can include only one target; meanwhile, it is collaboratively represented by background atoms because multiple background atoms may appear in a single pixel area. So for each testing pixel, sparse representation of known target signatures was estimated via an l_1-regularization, while collaborative representation of background atoms was obtained using an l_2-regularization. The final target detection was achieved by evaluating the difference of these two representation residuals.

5.4 EXPERIMENTS

Experimental results are presented later in the chapter to illustrate the performance of some CRC and CRD variants.

5.4.1 Classification

The dataset used in classification experiment is about University of Pavia dataset collected by the reflective operating system imaging spectrometer (ROSIS) sensor (Figure 5.1). It consists spatial size of 610 × 340 with 115 spectral bands in wavelengths from 0.43 to 0.86 μm. In the experiment, 103 channels were used after noisy bands being removed. It includes nine different classes and 43,316 labeled samples. One hundred samples of each class are taken as training samples and the remaining is taken as testing samples. To reduce ambiguity, each case is repeated ten times, and average performance is presented. Performance of CRC, CRT, ProCRC, ProCRT, SaCRC, SaCRT, SaProCRC, and SaProCRT on different parameters is investigated. Leave-one-out cross validation is performed to estimate optimal parameters, as listed in Table 5.1.

Table 5.2 shows Overall accuracy (OA), Average accuracy (AA), and Kappa coefficient (Kappa). It shows that CRT is better than the original CRC, ProCRT is better than the original ProCRC in terms of classification accuracies, and SaProCRC can further enhance the performance of ProCRC. In addition, SaProCRT provides better performance than ProCRT, and it can be observed that SaProCRC and SaProCRT provides competitive results as that of SaCRC and SaCRT. Table 5.2 also lists the execution time of different algorithms. It can be observed that ProCRT is slower than ProCRC. Moreover, SaProCRC and SaProCRT are slightly slower than ProCRT. It is because ProCRT consists of Tikhonov regularization, and spatial term regularization is incorporated in SaProCRC in addition to Tikhonov regularization in SaProCRT. SaProCRT is slightly slower than SaProCRC and SaCRT is slightly more computationally complex than SaCRC.

5.4.2 Anomaly Detection

The dataset for anomaly detection was acquired by the HyMap airborne hyperspectral imaging sensor covering one area of Cooke City, MT, USA on July 4, 2006, with the spatial size of 200 × 800 and 126 spectral bands spanning the wavelength interval of 0.4–2.5 μm. The spatial resolution is

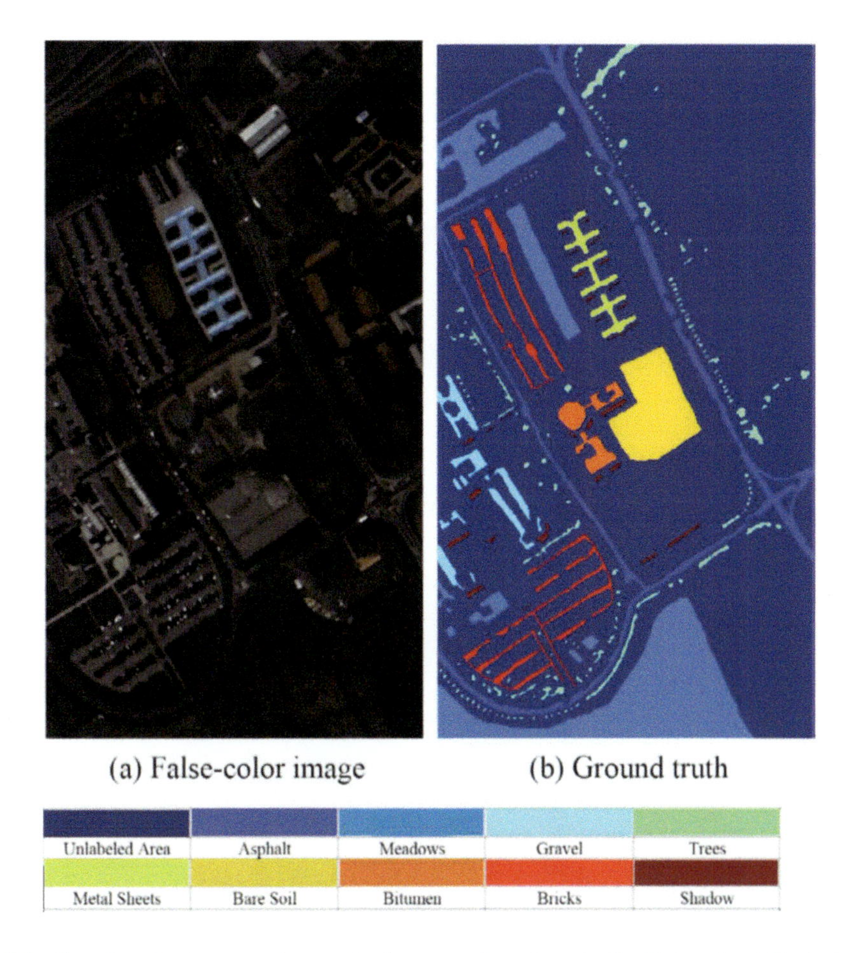

FIGURE 5.1 Pseudo-color image and ground reference information for the University of Pavia dataset.

TABLE 5.1
Parameters for Algorithms in University of Pavia Dataset

	λ	β	Γ	p
CRC	1e-4	–	–	–
CRT	1e-4	–	–	–
ProCRC	1e-0	–	1e-1	–
ProCRT	1e-0	–	1e-1	–
SaCRC	1e-2	1e4	-	6
SaCRT	1e-2	1e6	-	8
SaProCRC	1e-0	1e-0	1e-1	8
SaProCRT	1e6	1e-1	1e-0	6

approximately 3 m. This image has seven types of targets that are four fabric panel targets and three vehicle targets. In the experiment, a subimage of 68 × 238 was cropped, including all the targets as presented in Figure 5.2.

The performance of detection can be evaluated with receiver operating characteristic (ROC) curves and the area under the ROC curve (AUC). The larger the area under the ROC curve, the

TABLE 5.2
Classification Performance on University of Pavia Dataset

	CRC	CRT	ProCRC	ProCRT	SaCRC	SaCRT	SaProCRC	SaProCRT
1	77.92	96.86	89.95	97.63	99.21	99.48	89.13	99.45
2	87.71	94.69	91.05	93.17	99.36	99.78	99.48	99.30
3	36.31	54.43	73.55	65.15	94.34	88.99	98.18	91.82
4	64.77	80.85	21.90	34.68	89.39	88.57	64.21	91.58
5	99.40	99.76	95.60	97.94	99.12	99.52	93.15	99.20
6	47.32	73.23	92.95	97.81	97.06	98.46	100	97.75
7	32.24	44.83	98.39	85.41	76.75	81.59	99.27	91.92
8	42.67	74.54	72.30	87.66	66.98	82.54	84.04	95.61
9	100	100	99.26	99.54	100	99.76	76.30	90.77
OA	**66.07**	**83.37**	**66.46**	**80.20**	**93.11**	**95.74**	**93.00**	**97.44**
AA	**65.37**	**79.91**	**81.67**	**84.34**	**91.36**	**93.19**	**89.31**	**95.27**
Kappa	**0.58**	**0.77**	**0.59**	**0.74**	**0.90**	**0.94**	**0.90**	**0.96**
Time(s)	**1013**	**1029**	**6099**	**6418**	**637**	**641**	**6397**	**6502**

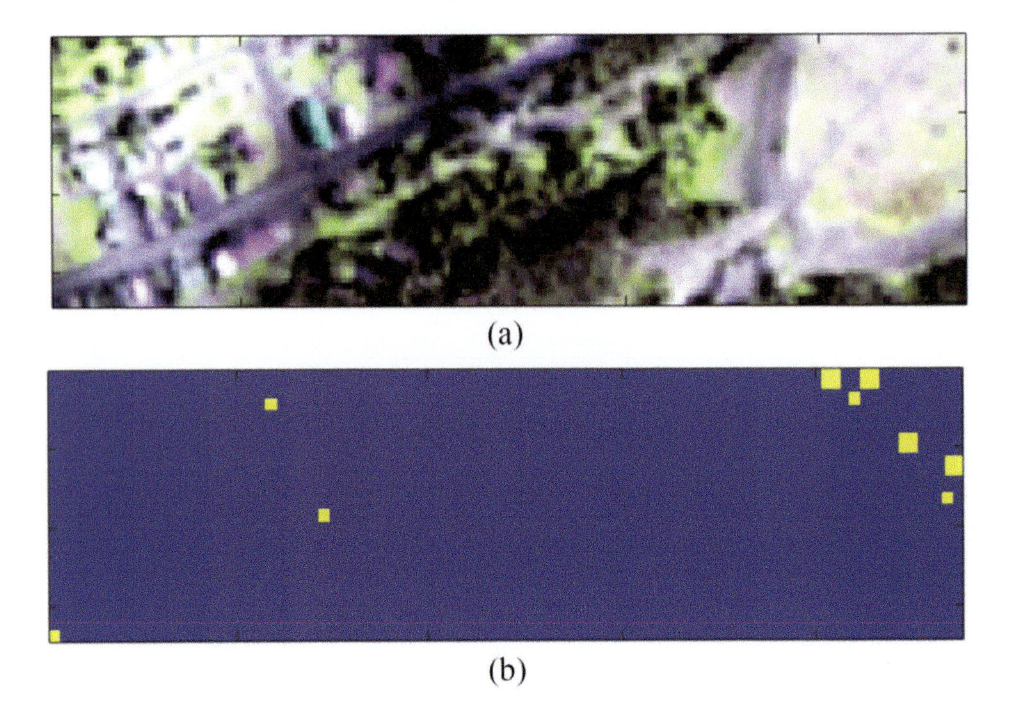

FIGURE 5.2 (a) Pseudocolor image of the HyMap scene. (b) Ground-truth map of anomalous pixels.

better the algorithm. AUC (%) is adopted to measure the detection accuracy. In order to compare the maximum detection performance of algorithms, the optimized parameters such as λ and window size ($w$out, $w$in) are adopted. They were obtained when other parameters remain unchanged and the AUC is the maximum. Table 5.3 lists the parameters of seven algorithms and their performance. Obviously, CRD offers better performance than the RX algorithms, and local versions outperform

TABLE 5.3
Anomaly Detection Performance for the HyMap Dataset with Optimal Parameter Settings

Detectors	Window Size	Parameter λ	AUC (%)
Global RX	-		64.58
Local RX	(11,3)		67.81
CRD	(15,11)	1e–7	75.56
KCRD	(13,11)	1e–5	74.09
CRBORAD	(15,13)	1e–8	76.30
MCRD	(13,11)	1e–6	78.07
Global PCAroCRD	–	1e–2	79.23
Local PCAroCRD	(11,9)	1e1	90.57

global versions. CRBORAD can further improve the performance of CRD by eliminating anomalous pixels in the background. LocalPCAroCRD outperforms by using major background pixels for representation.

5.5 CONCLUSION

During the past few years, a variety of CR-based classification and detection algorithms have been proposed for hyperspectral image analysis. They become popular due to simple implementation, efficient calculation, and excellent performance. In particular, the CRC does not have the traditional training-testing pattern, while the concept of CRD well fits the idea of anomaly detection without the prior knowledge of background and target.

In this chapter, we have reviewed the CRC and the variants to improving the performance including: dictionary partition, weighted regularization, joint representation, representation using spatial features, and kernel extensions. The common objective is to increase representation accuracy. Particularly, joint representation-based classification and spatial-spectral representation-based classification tackle the same problem (i.e., how to utilize the spatial contextual information) but from different perspectives. The former attempts to incorporate information on neighboring locations by reorganizing the structure of representation (e.g., JCRC), whereas the latter mainly employs a spatial feature extraction algorithm (i.e., Gabor, AP, etc.) as preprocessing before implementing representation-based classifier. This chapter also reviews the CRD related detectors. Improvement can be resulted in by removing the background contamination through techniques such as PCAroCRD.

Certainly, the best performance of representation-based algorithms depends on the choice of parameters to some extent. As a global regularization parameter, the adjustment of λ affects the balance between the residual and weight vector norm, and eventually classification and detection accuracy. Current works employ cross validation for parameter tuning, which can offer satisfactory performance. Approaches with less human intervention could be investigated. Overall, CRC and CRD offer additional options for hyperspectral image analysis and can be good candidates for ensemble learning based techniques.

REFERENCES

[1] Plaza, A., Benediktsson, J. A., Boardman, J. W., Brazile, J., Bruzzone, L., Camps-Valls, G., Chanussot, J., Fauvel, M., Gamba, P., et al., "Recent advances in techniques for hyperspectral image processing," *Remote Sens. Environ.* **113**(2009).

[2] Fauvel, M., Tarabalka, Y., Benediktsson, J. A., Chanussot, J., Tilton, J. C., "Advances in spectral-spatial classification of hyperspectral images," *Proc. IEEE* **101**(3) 652–675 (2013).

[3] Xu, Y., Du, Q., Li, W., Younan, N., "Gabor-filtering-based probabilistic collaborative representation for hyperspectral image classification," *Proc. IEEE Int. Geosci. Remote Sens. Symp.*, IGARSS 2018, Valencia, Spain (2018).

[4] Sun, W., Zhang, L., Zhang, L., Lai, Y. M., "A dissimilarity-weighted sparse self-representation method for band selection in hyperspectral imagery classification," *IEEE J. Select. Top. Appl. Earth Observ. Remote Sens.* **9**(9) 4374–4388 (2016).

[5] Wright, J., Yang, A. Y., Ganesh, A., Sastry, S. S., Ma, Y., "Robust face recognition via sparse representation," *IEEE Trans. Pattern Anal. Mach. Intell.* **31**(2) 210–227 (2009).

[6] Sami ul Haq, Q., Tao, L., Sun, F., Yang, S., "A fast and robust sparse approach for hyperspectral data classification using a few labeled samples," *IEEE Trans. Geosci. Remote Sens.* **50**(6) 2287–2302 (2012).

[7] Shrivastava, A., Patel, V. M., Chellappa, R., "Multiple kernel learning for sparse representation-based classification," *IEEE Trans. Image Process.* **23**(7) 3013–3024 (2014).

[8] Wang, J., Lu, C., Wang, M., Li, P., Yan, S., Hu, X., "Robust face recognition via adaptive sparse representation," *IEEE Trans. Cybern.* **44**(12) 2368–2378 (2014).

[9] Castrodad, A., Xing, Z., Greer, J. B., Bosch, E., Carin, L., Sapiro, G., "Learning discriminative sparse representations for modeling, source separation, and mapping of hyperspectral imagery," *IEEE Trans. Geosci. Remote Sens.* **49**(11) 4263–4281 (2011).

[10] Wu, Z., Wang, Q., Plaza, A., Li, J., Wei, Z., "Parallel implementation of sparse representation classifiers for hyperspectral imagery on GPUs," *IEEE J. Select. Top. Appl. Earth Observ. Remote Sens.* **8**(6) 2912–2925 (2015).

[11] Wu, Z., Wang, Q., Plaza, A., Li, J., Sun, L., Wei, Z., "Parallel spatial-spectral hyperspectral image classification with sparse representation and Markov random fields on GPUs," *IEEE J. Select. Top. Appl. Earth Observ. Remote Sens.* **8**(6) 2926–2938 (2015).

[12] Yuan, H., Tang, Y. Y., "Sparse representation based on set-to-set distance for hyperspectral image classification," *IEEE J. Select. Top. Appl. Earth Observ. Remote Sens.* **8**(6) 2464–2472 (2015).

[13] Zhang, L., Yang, M., Feng, X., "Sparse representation or collaborative representation: Which helps face recognition?" *Proc. Int. Conf. Comput. Vis.*, pp. 471–478, ICCV 2011, Barcelona, Spain (2011).

[14] Yang, J., Zhang, L., Xu, Y., Yang, J. "Beyond sparsity: The role of $L1$-optimizer in pattern classification," *Pattern Recogn.* **45**(3) 1104–1118 (2012).

[15] Waqas, J., Yi, Z., Zhang, L., "Collaborative neighbor representation based classification using $l2$ -minimization approach," *Pattern Recogn. Lett.* **34** 201–208 (2013).

[16] Chen, J., Jiao, L., "Hyperspectral imagery classification using local collaborative representation," *Int. J. Remote Sens.* **36**(3) 734–748 (2015).

[17] Li, W., Du, Q., "A survey on representation-based classification and detection in hyperspectral remote sensing imagery," *Pattern Recogn. Lett.* **83**, 115–123 (2016).

[18] Li, W., Du, Q., "Collaborative representation for hyperspectral anomaly detection," *IEEE Trans. Geosci. Remote Sens.* **53**(3) 1463–1474 (2015).

[19] Reed, I. S., Yu, X., "Adaptive multiple-band CFAR detection of an optical pattern with unknown spectral distribution," *IEEE Trans. Acoust. Speech Signal Process.* **38**(10) 1760– 1770 (1990).

[20] Li, W., Tramel, E. W., Prasad, S., Fowler, J. E., "Nearest regularized subspace for hyperspectral classification," *IEEE Trans. Geosci. Remote Sens.* **52**(1) 477–489 (2014).

[21] Cai, S., Zhang, L., Zuo, W., Feng, X., "A probabilistic collaborative representation based approach for pattern classification," *Proc. IEEE Conf. Comput. Vis. Pattern Recognition (CVPR)*, IEEE CVPR 2016, Las Vegas, USA (2016).

[22] Jiang, J., Chen, C., Yu, Y., Jiang, X., Ma, J., "Spatial-aware collaborative representation for hyperspectral remote sensing image classification," *IEEE Geosci. Remote Sens. Lett.* **14**(3) 404–408 (2017).

[23] Shah, C., Du, Q., "Spatial-aware probabilistic collaborative representation for hyperspectral image classification," *Proc. SPIE* **4256** (2020).

[24] Cui, M., Prasad, S., "Class-dependent sparse representation classifier for robust hyperspectral image classification," *IEEE Trans. Geosci. Remote Sens.* **53**(5) 2683–2695 (2015).

[25] Zhang, L., Zhou, W., Liu, B., "Nonlinear nearest subspace classifier," *Proc. 18th Int. Conf. Neural Inf. Process.*, pp. 638–645, ICONIP 2011, Shanghai, China (2011).

[26] Li, W., Du, Q., "Joint within-class collaborative representation for hyperspectral image classification," *IEEE J. Select. Top. Appl. Earth Observ. Remote Sens.* **7**(6) 2200–2208 (2014).

[27] Xiong, M., Ran, Q., Li, W., Zou, J., Du, Q., "Hyperspectral image classification using weighted joint collaborative representation," *IEEE Geosci. Remote Sens. Lett.* **12**(1) 48–52 (2015).

[28] Fauvel, M., Chanussot, J., Benediktsson, J. A., "A spatial-spectral kernel-based approach for the classification of remote-sensing images," *Pattern Recogn.* **45**(1) 381–392 (2012).
[29] Wang, B., Li, W., Poh, N., Liao, Q., "Kernel collaborative representation-based classifier for face recognition," *Proc. Int. Conf. Acoust. Speech Signal Process.*, pp. 2877–2881, ICASSP 2013, Vancouver, Canada (2013).
[30] Wang, D., Lu, H., Yang, M. H., "Kernel collaborative face recognition," *Pattern Recogn.* **48**(10) 3025–3037 (2015).
[31] Li, J., Zhang, H., Zhang, L., "Column-generation kernel nonlocal joint collaborative representation for hyperspectral image classification," *ISPRS J. Photogram. Remote Sens.* **94** 25–36 (2014).
[32] Li, W., Du, Q., "Kernel collaborative representation with Tikhonov regularization for hyperspectral image classification," *IEEE Geosci. Remote Sens. Lett.* **12**(1) 1–5 (2015).
[33] Li, W., Chen, C., Su, H., Du, Q., "Local binary patterns and extreme learning machine for hyperspectral imagery classification," *IEEE Trans. Geosci. Remote Sens.* **53**(7) 3681– 3693 (2015).
[34] Song, B., Li, J., Mura, M. D., Li, P., Plaza, A., Benediktsson, J. A., Chanussot, J., "Remotely sensed image classification using sparse representations of morphological attribute profiles," *IEEE Trans. Geosci. Remote Sens.* **52**(8) 5122–5136 (2014).
[35] Yang, J., Qian, J., "Joint collaborative representation with shape adaptive region and locally adaptive dictionary for hyperspectral image classification," *IEEE Geosci. Remote Sens. Lett.* **17**(4) 671–675 (2020).
[36] Li, W., Du, Q., "Gabor-filtering based nearest regularized subspace for hyperspectral image classification," *IEEE J. Select. Top. Appl. Earth Observ. Remote Sens.* **7**(4) 1012–1022 (2014).
[37] Yang, M., Zhang, L., Shiu, S., Zhang, D., "Gabor feature based robust representation and classification for face recognition with Gabor occlusion dictionary," *Pattern Recogn.* **46**(7) 1868–1878 (2013).
[38] Xu, L., Li, J., "Bayesian classification of hyperspectral imagery based on probabilistic sparse representation and Markov random field," *IEEE Geosci. Remote Sens. Lett.* **11**(4) 823–827 (2014).
[39] Xiong, M., Zhang, F., Ran, Q., Hu, W., Li, W., "Representation-based classifications with Markov random field model for hyperspectral urban data," *J. Appl. Remote Sens.* **8**, 085097 (2014).
[40] Clausi, D. A., Jernigan, M. E., "Designing Gabor filters for optimal texture separability," *Pattern Recogn.* **33**(11) 1835–1849 (2000).
[41] Jia, S., Shen, L., Li, Q., "Gabor feature-based collaborative representation for hyperspectral imagery classification," *IEEE Trans. Geosci. Remote Sens.* **53**(2) 1118–1129 (2015).
[42] Du, Q., Yang, H., "Similarity-based unsupervised band selection for hyperspectral image analysis," *IEEE Geosci. Remote Sens. Lett.* **5**(4) 564–568 (2008).
[43] Vafadar, M., Ghassemian, H., "Hyperspectral anomaly detection using outlier removal from collaboration representation," *Proc. 3rd Int. Conf. Pattern Recognit. Image Anal.*, pp. 13–19, IPRIA 2017, Shahrekord, Iran (2017).
[44] Imani, M. "Anomaly detection using morphology-based collaborative representation in hyperspectral imagery," *Eur. J. Remote Sens.* **51**(1) 457–471 (2018).
[45] Su, H., Wu, Z., Du, Q., Du, P. "Hyperspectral anomaly detection using collaborative representation with outlier removal," *IEEE J. Select. Top. Appl. Earth Observ. Remote Sens.* **11**(12) 5029–5038 (2018).
[46] Li, W., Du, Q., Zhang, B. "Combined sparse and collaborative representation for hyperspectral target detection," *Pattern Recogn.* **48**, 3904–3916 (2015).

6 Thermal Infrared Remote Sensing

Principles and Theoretical Background

Claudia Kuenzer, Philipp Reiners, Jianzhong Zhang, and Stefan Dech

LIST OF ACRONYMS

ASTER	Advanced spaceborne thermal emission and reflection radiometer
AVHRR	Advanced very high resolution radiometer
HyTES	Hyperspectral thermal emission spectrometer
IPCC	Intergovernmental panel on climate change
MODIS	Moderate resolution imaging spectrometers
NOAA-AVHRR	National oceanic and atmospheric administration—Advanced very high resolution radiometer
UAV	Unmanned aerial vehicle
UHI	Urban heat island

6.1 INTRODUCTION

It is temperature, which defines the habitat boundaries of plants, animals, and humans. Temperature—among other variables—defines global, regional, and local climate and weather, the rate of sea level rise, the magnitude of forest fire risk, the reproduction conditions for bacteria and viruses, and it even defines our cultural preferences. Housing style, preferred means of transport, customs of eating and free-time behavior focusing on the inside or the outside, as well as the attractiveness of tourist destinations are all defined by temperature. Temperature strongly impacts our well- or mal-being, and extreme cold spills or heat waves have proven to negatively impact local economies and even lead to fatalities.

Although the number of studies using remote sensing in the thermal infrared wavelength increased rapidly in the last ten years (Reiners, Sobrino et al., 2023), the domain is still an often neglected sub-discipline within the field of remote sensing. Reasons for this range from a limited availability of space borne sensors acquiring data in the thermal infrared, TIR, domain to the—generally—lower spatial resolution of this data, when compared with other—for example, optical—data, to last but not least different data analyses approaches, which are mandatory when dealing with TIR data. Thus, it is the goal of this chapter to provide a comprehensive overview of the principles of remote sensing in the thermal infrared. We address the theoretic background of the TIR domain, including important physical laws, address parameters such as kinetic and radiance temperature, emissivity and thermal inertia, discuss the pre-processing, the analyses, as well as the validation of thermal data, and present a range of illustrative application examples.

 DOI: 10.1201/9781003541158-7

Much of this chapter has been reformulated based on Kuenzer and Dech (2013).

In general, electromagnetic radiation is emitted by all objects that have a temperature above 0 Kelvin (equals –273°C). Depending on the temperature of the object it has its peak of electromagnetic emittance in a certain wavelength domain. The peak of emission of our planet, which has an average temperature of about 300 K, is located in the TIR domain at a wavelength of about 9.7 μm (Sabins, 1996; Tipler, 2000). Objects on Earth absorb the Sun's incoming radiation and emit a corresponding amount at longer wavelengths in the TIR domain.

Several sensors, such as Landsat-9 TIRS-2, contain bands (detectors), which are responsive in the thermal domain and record TIR radiation emitted by objects (see Figure 6.1). The thermal imagery then represents the kinetic temperature of objects at a certain spatial resolution. We can differentiate between different categories of temperature. Land Surface Temperature, LST, describes the temperature at the land surface. It represents the temperature of object's surfaces—be it meadows, house roofs, forest canopies, or inland water bodies. Typical LST based thermal analyses either focus on spatial LST patterns or spatio-temporal analysis of LST time series at local, continental or global scale (Reiners, Sobrino et al., 2023). Spatial anomalies are used for Urban Heat Island (UHI) studies, assessments of forest fires, grassland fires, gas flares, underground coal fires, or geothermal phenomena, the observation of nuclear accident sites (e.g., Chernobyl, Fukushima), or the monitoring of industry. LST time series analysis can help to analyze vegetation health, land

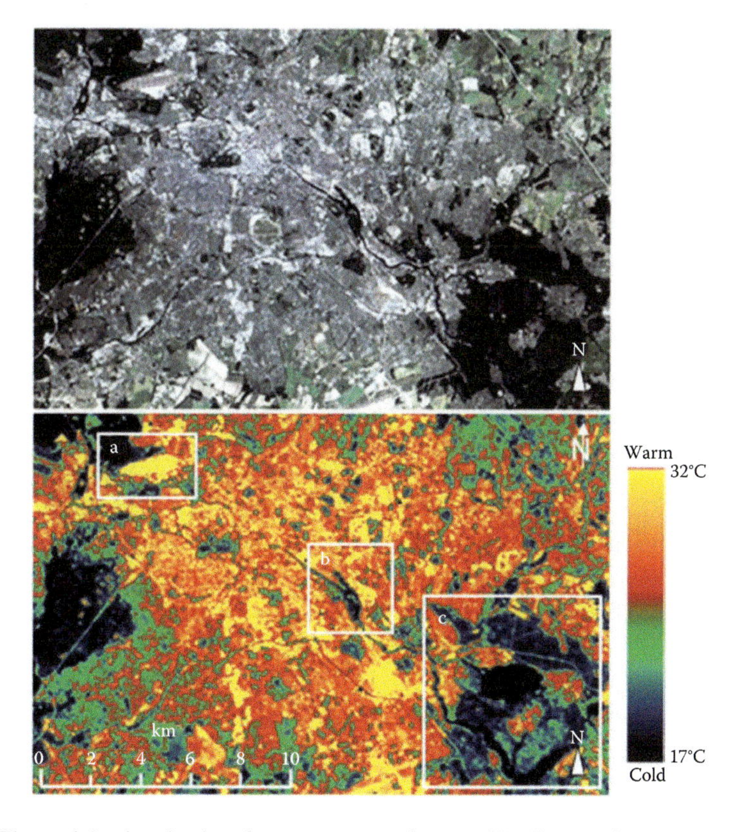

FIGURE 6.1 Thermal daytime land surface temperature image of Berlin, acquired on June 24, 2022 by the Landsat-9 TIRS-2 sensor. The subset areas depict the Berlin airport "Tegel" (a), the river "Spree" (b), as well as a large lake named "Müggelsee" and it surrounding forest (c). Water surfaces and densely vegetated areas make up the cooler places within this daytime thermal image, while areas with a high percentage of artificial surfaces appear warmest.

Source: Coordinates: UL: 52.64°N, 13.06°E, LR: 52.38°N, 13.74°E.

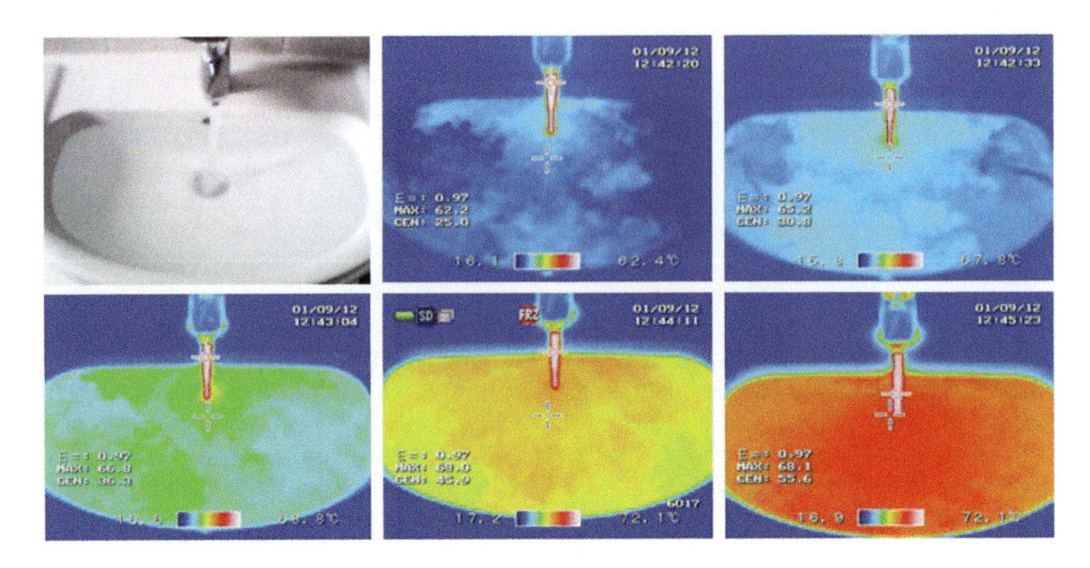

FIGURE 6.2 Temperature dynamics: A sink is filled with cold water, and successively hot water is added. The temperature of the cold water is 22°C, while the hot water is around 70°C. Thermal camera images were taken over the course of three minutes. As water in Munich, Germany, is extremely hard, at over >2.5 mmol/L $CaCO_3$, emissivity is set to 0.97. (Source: Photographs: C. Kuenzer, 2012, illustration: Kuenzer and Dech, 2013, modified.)

cover transformations, soil moisture dynamics and drought events. Contrary to LST, Sea Surface Temperature—SST—is the temperature of the upper water layer of an ocean or large inland sea (Dech et al., 1998). The SST datasets are usually exploited for an improved understanding of global circulations, for analyses in the context of algae blooms, or for example thermal water pollution (see Figure 6.2 for a handheld camera example). Satellite-derived LST and SST furthermore play a crucial role for monitoring the current state of the global climate and quantify global warming, as it can be seen in the recent reports of the intergovernmental panel on climate change (IPCC) (Rama et al., 2022).

However, thermal sensor data is also employed outside of the geosciences. In medical imaging, inflamed areas can be detected due to their higher temperature compared with their surroundings. In industry, machine performance is monitored with thermal cameras. Architects use thermal imagery to detect energy leaks in buildings, and also police and the military use thermal cameras for object detection. Several examples for the use of thermal camera imagery are illustrated by Figure 6.3.

A large variety of TIR sensors data is available in data archives, as well as for current tasking (Kuenzer et al., 2013a). Figure 6.4 presents all platforms and instruments that allow for data acquisition in the TIR domain. We can see that especially the fleet of AVHRR (Advanced Very High Resolution Radiometer), as well as Landsat sensors, enable a long-term monitoring of our planet in the TIR domain (Frey et al., 2012). Most TIR data is available from American sensors, and good access is especially granted to data of AVHRR, MODIS (Moderate Resolution Imaging Spectrometers), Landsat, and ASTER (Advanced Spaceborne Thermal Emission and Reflection Radiometer). However, since the last years, also European missions offer operational LST and SST services, as for example the Sentinel satellites or the geostationary Meteosat satellites with their sensor SEVIRI. Furthermore, numerous Chinese sensors record data in the TIR domain—however, here data access is not as easy. A detailed overview of TIR related instruments and their preferred application domain can be found in Kuenzer et al. (2013a). Abbreviations can be found in the abbreviations section of this book volume.

FIGURE 6.3 Thermal camera images. **Upper left (a):** day-time image of a high-rise residential area in Beijing, China, acquired during a cold wave in the winter of 2012. While the outside air and background indicates temperatures of well below 0°C (up to –20°C) it can be seen that heat from apartments inside penetrates through the windows, so that temperatures at the outside of the building reach about –3°C. **Upper right (b):** daytime image of a skyscraper façade acquired in the same residential area: the surface temperature of the façade is well below 0°C (down to –11°C), however, from two windows heat penetrates to the outside. This heat originates from air condition systems people install privately above their windows (inside), which are used as cooling devices in summer and heating devices in winter. Temperatures of up to 9.8°C occur outside. **Lower left (c):** picture of an opened freezer. Temperatures in the freezer go down to –20°C. The lower part of the freezer is colder than the upper part (cold air sinks down). Outside of the freezer temperatures of up to 21.5°C are reached. **Lower right (d):** tree lined pathway in the village of Gilching, near Munich. In the evening around sunset cemented surfaces, as well as vegetation appear warmest. (Source: photographs: C. Kuenzer, 2012.)

6.2 PRINCIPLES, THEORETICAL BACKGROUND, IMPORTANT LAWS

6.2.1 The Thermal Infrared Domain

Different authors define the thermal infrared domain differently, so an overall valid strict, physical definition does not exist. Sabins (1996) defines the TIR range from about 3 μm to 13 μm—a wavelength range, in which two important atmospheric windows are located (see Figure 6.5). In the 8–14 μm window only ozone absorption occurs, which is omitted by most sensors. In the 3–5 μm range reflected sunlight can still slightly contaminate the emitted thermal signal, so the data has to be interpreted with care. However, according to Lillesand and Kiefer (1994) and Löffler (1994) the TIR

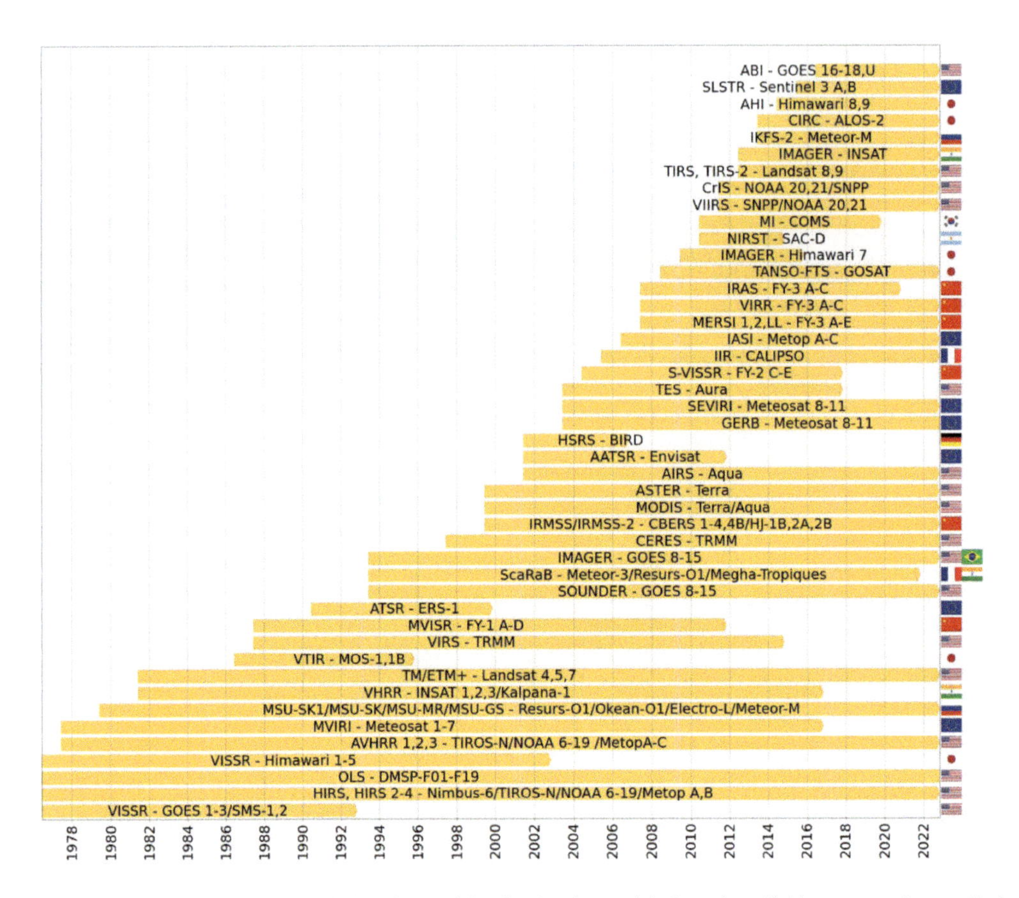

FIGURE 6.4 Satellite sensors with bands sensitive in the thermal infrared until (documentation until the end of 2022). The chart depicts the sensor's start and ending dates, name of the platform and the instrument, as well as the country of origin.

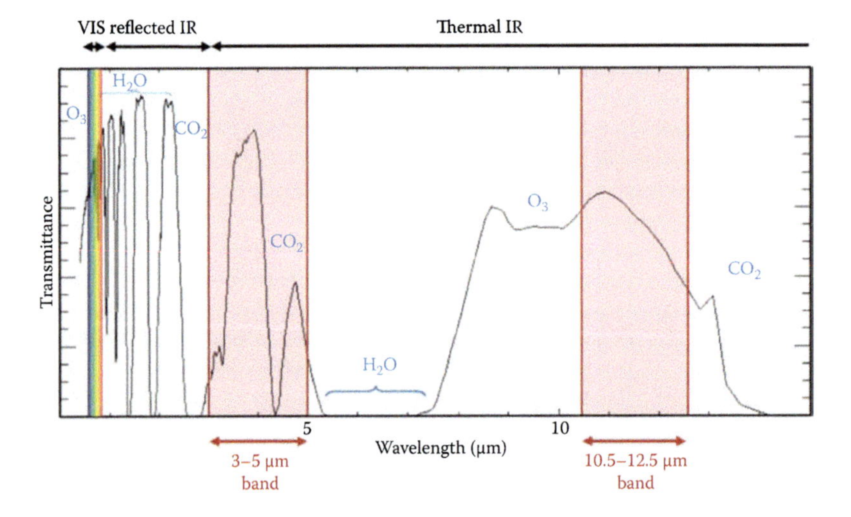

FIGURE 6.5 The diagram depicts the thermal infrared wavelength domain, typical absorption bands induced by gasses and water, as well as atmospheric transmittance (atmospheric windows). (Source: Kuenzer and Dech, 2013, modified.)

domain ranges from 3 µm to 1,000 µm. Common to all authors is that they define thermal infrared remote sensing as the field that deals with emitted radiation, whereas multispectral remote sensing in the visible (VIS) and near infrared (NIR) domain records reflected radiation.

6.2.2 Important Laws: Planck

Planck's law, or Planck's blackbody radiation law as it is officially termed, describes the electromagnetic radiation emitted by a blackbody at a given wavelength, M, as a function of the blackbody's temperature (Planck, 1900). A blackbody is a theoretical concept and does not exist in reality. It is an ideal radiator, which re-emits all energy it absorbs. However, there are surfaces on the Earth that show "near blackbody like" behavior for certain wavelengths. The Planck formula allows to calculate the emitted radiation, M, by inserting a certain wavelength as well as the body's temperature. It also allows deriving a blackbody's temperature, if M and the wavelength are known (Equation 6.1).

$$M_\lambda = \frac{2\pi hc^2}{\lambda^5\left(e^{hc/1kT} - 1\right)} \tag{6.1}$$

where

M_λ = spectral radiant exitance [W m^{-2} µm^{-1}]
h = Planck's constant [6.626×10^{-34} J s]
c = speed of light [2.9979246×10^{8} m s^{-1}]
k = Boltzmann constant [1.3806×10^{-23} J K^{-1}]
T = absolute temperature [K]
λ = wavelength [µm]

The Stefan–Boltzmann law (following) as well as Wien's law (following as well) allow calculation of the total energy a theoretical blackbody radiates, as well as its wavelength of maximum emittance (Tipler, 2000; Walker, 2008).

6.2.3 Important Laws: Stefan–Boltzmann

The Stefan–Boltzmann enables to calculate the total energy a theoretical blackbody radiates, as a function of its temperature. As depicted in Figure 6.6 the emitted radiation is described by the area under the radiation curve. The overall energy an object radiates is the larger, the higher its temperature. As Equation 6.2 elucidates this relationship between temperature and energy is not linear, but is described by a 4th power relationship (see Equation 6.2).

$$T_{RadBB} = \sigma T_{kin}^{\ 4} \tag{6.2}$$

where

T_{RadBB} = radiant flux of a blackbody [W/m²],
T = absolute kinetic temperature [K]
σ = Stefan-Boltzmann constant [5.6697×10^{-8} W m^{-2} K^{-4}]

6.2.4 Important Laws: Wien

The law of Wien, also sometimes called Wien's displacement law, describes the wavelength of maximum spectral exitance (in µm) as a function of an object's temperature (see Equation 6.3). The

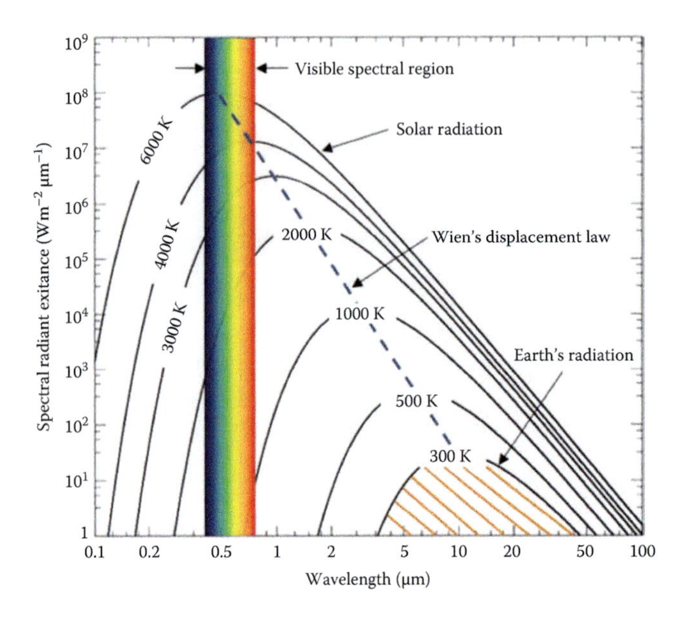

FIGURE 6.6 Blackbody radiation curves for blackbodies with different temperatures, as derived from Planck's equation. Stefan–Boltzmann's equation describes the area under the curves. The rainbow colored bar marks the VIS region. (Source: Kuenzer and Dech, 2013, modified.)

hotter an object, the further shifts its maximum exitance towards shorter wavelengths (Heald, 2003; Walker, 2008; Tipler, 2000). This has also already been demonstrated by Figure 6.6. The Sun has an average temperature of 5505°C (or 5778 K) and it peak of emission is located in the visible (VIS) domain of the electromagnetic spectrum at 0.55 μm. Colder objects, such as the earth, have their peak of emission in the TIR.

$$\lambda_{max} = \frac{A}{T} \tag{6.3}$$

where

λ_{max} = wavelength of maximum spectral radiant exitance [μm]
A = Wien's constant [2897.8 μm K];
T = absolute kinetic temperature [K]

Wien's displacement law can be well demonstrated on multispectral remote sensing imagery depicting areas of different temperature, including extreme hot spots. Figure 6.7 presents several bands of a Landsat-5 TM image of the Etna volcano in Italy, acquired on July 29, 2001. In the upper-left true color composite we can see the volcano and its main chimney on the top. Clouds of smoke are emanating from the volcano. We can see areas of bare soil on the left flank of the peak, as well as some vegetated patches. Lava flows cannot be observed. In the temperature image displayed on the lower right (band 6 of Landsat 5 TM, representing the TIR domain from 10.4 to 12.5 μm) we can see that the smoke clouds are warm, whereas the flanks of the volcano are much colder. Vegetated areas appear coldest. At the same time, we can observe some straight structures in the upper center of this thermal image, which do not seem to belong to the smoke itself, but cannot further differentiate this. Here now band 7—the short wave infrared, SWIR, band 7 (2.09–2.35 μm) comes into play. It is depicted at the lower left of Figure 6.7.

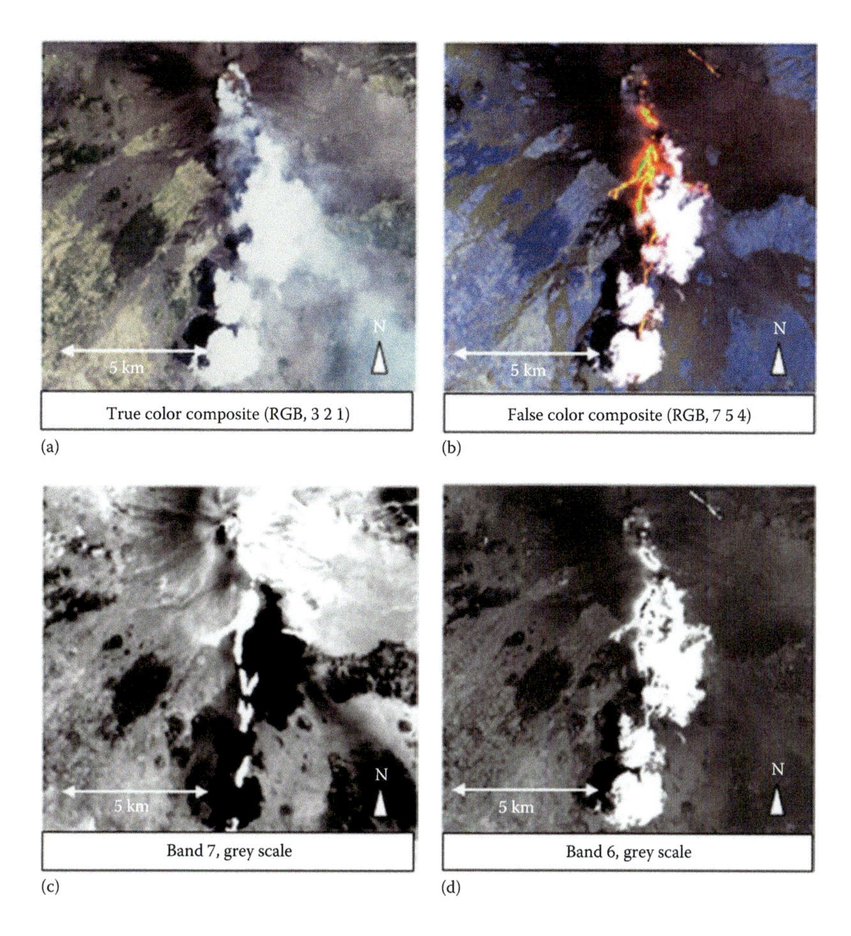

FIGURE 6.7 Landsat-5 TM data of Etna volcano, Italy, acquired on July 29, 2001. Upper left: true color composite with the red, green and blue bands displayed in RGB; upper right: false: color composite with the two SWIR bands 7 and 5 and NIR band 4 displayed in RGB; lower left: grayscale image of the shortwave infrared band 7 (2.09–2.35 μm); lower right: grayscale image of the thermal infrared, TIR, band 6 (10.40–12.50 μm). (Source: Coordinates: UL: 37°50'32N, 14°52'11E; LR: 37°39'30N, 15°05'07E.)

In this image we can see that lava flows of extreme temperature lead to an elevated signal in this band 7 (Wien's law). The material is so hot that its peak of emission does not occur in the TIR domain from 10.4 to 12.5 μm, but rather in the SWIR. At the same time, the clearly warm clouds (lower right) appear dark in this lower left image. Reason is the high water vapor content of the clouds. Furthermore, they are not hot enough to elevate the signal in the SWIR: they do have the peak of emittance in the TIR. Creating a color composite of bands 7, 5, and 4 now allows to beautifully depict hottest lava (orange), warm smoke (white), former cooled lava flows (black), and sparse and dense vegetation (blue).

6.2.5 Important Laws: Kirchhoff and the Role of Emissivity

Planck's law describes blackbody radiation, but it has already been stated that blackbodies are fictive objects, as most objects on our planet emit less energy than would be predicted based on their kinetic temperature. The so-called emissivity coefficient, ($\varepsilon_{(\lambda)}$), is taking this fact into account and is defined as the radiant flux of an object at a given temperature over the radiant flux of a blackbody at the same temperature. A perfect blackbody would emit all radiation it absorbed and

Kirchhoff's law states that—for a blackbody—emittance and absorbance at a given wavelength are equal (Kirchhoff, 1860).

$$\varepsilon_{(\lambda)} = \alpha_{(\lambda)} \tag{6.4}$$

As—according to energy conservation—the sum of absorption (α), reflection (ρ), and transmission (τ) equals 1, and considering Kirchoff's law, we can postulate that:

$$\varepsilon_{(\lambda)} + \rho_{(\lambda)} + \tau_{(\lambda)} = 1 \tag{6.5}$$

However, most solid objects (except for, e.g., water, leafs, etc.) do not transmit radiation. They are opaque, so Equation 6.5 can be reformulated to:

$$\varepsilon_{(\lambda)} + \rho_{(\lambda)} = 1 \tag{6.6}$$

This means that an object's (blackbody's) reflectance allows to calculate its emittance and vice versa. Materials with low ε absorb and radiate lower amounts of energy, whereas materials with a high ε absorb large amounts of incident energy and radiate large quantities of energy, (Kirchhoff, 1860; Sabins, 1996). Emissivity varies depending on surface type and wavelength, but is not temperature dependent (Becker, 1987). Table 6.1 illustrates the varying emissivities of common surfaces for the wavelength region 8–14 µm (averaged).

As can be seen, for all real materials emissivity is below 1. This means that the radiance temperature $T_{(rad)}$, measured by a thermal sensor (e.g., handheld thermal camera, airborne thermal scanner, thermal detector on-board a satellite) is always lower than the real kinetic (surface) temperature, $T_{(kin)}$, (which one would measure with a contact thermometer) of the object. This also means that objects with exactly the same kinetic temperature on the ground can exhibit a very different radiant temperature. If we consider a typical land surface with different geologic surfaces, different types of vegetation, differing moisture conditions and maybe even containing a large variety of construction materials (e.g., urban areas) it is obvious that a remotely acquired thermal image does not represent

TABLE 6.1
Emissivity of Different Surfaces in the 8–14 µm Wavelength Range as Compiled from Different Sources (Own Measurements, and Lillesand et al., 2008; Sabins, 1996; Hulley et al., 2009; Source: Kuenzer and Dech, 2013.)

Surface	Emissivity at 8–14 µm
Carbon powder	0.98–0.99
Water	0.98
Ice	0.97–0.98
Plant leaves, healthy	0.96–0.99
Plant leaves, dry	0.88–0.94
Asphalt	0.96
Sand	0.93
Basalt	0.92
Wood	0.87
Granite	0.83–0.87
Polished metals, averaged	0.02–0.21
Aluminium foil	0.036

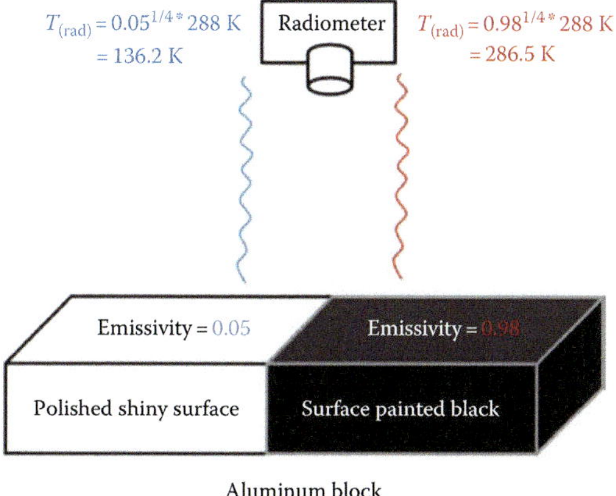

FIGURE 6.8 Emissivity strongly impacts the temperature recorded at a (remote, non-contact) sensor. In this figure modified after Sabins (1996) a block of aluminum with a homogeneous kinetic temperature and a very low emissivity is covered with carbon-rich dark paint (on its right side). This paint has a very different emissivity than the polished aluminum block. Although the object is 15°C warm, it appears as −136.9°C on the uncovered side, and as 13.3°C on the painted side (calculation based on Equations 6.2 and 6.7).

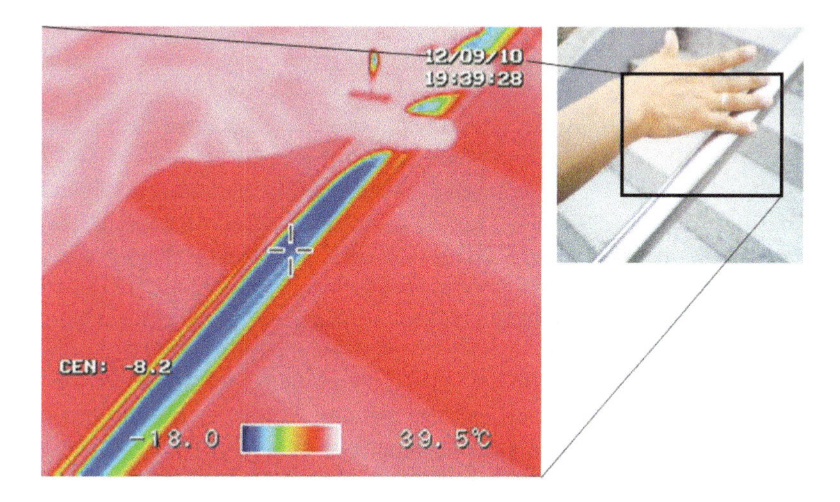

FIGURE 6.9 Impact of emissivity differences on radiance temperature recorded at the sensor of a handheld thermal camera. Picture taken by C. Kuenzer in September, 2012. (Source: Kuenzer and Dech, 2013, modified.)

the true kinetic temperatures of these objects unless the data is corrected for emissivity effects. This is especially crucial for all thermal infrared applications in urban areas as the emissivities of different construction materials vary considerably. Emissivities of metals for example (aluminium, tin, copper, in some areas used as roof materials) are extremely low. This will lead to the fact that temperatures appear much lower than the sensed temperatures of the surrounding objects of similar kinetic temperature (see also Figures 6.8 and 6.9). Contrary, objects with a very high emissivity, such as water and vegetated surfaces allow for a pretty exact assessment of their kinetic temperature. The conversion of radiance temperature to kinetic temperature can be undertaken as follows:

$$T_{(rad)} = \varepsilon^{(1/4)} * T_{(kin)} \tag{6.7}$$

where

$T_{(rad)}$: Radiance Temperature;
$T_{(kin)}$: Kinetic Temperature
ε: Emissivity

Also Figure 6.9 visualized this phenomenon very well, based on an imminent example. A handheld thermal camera photograph is taken. The photograph shows a concrete stairs with a polished metal railing and a male hand with a wedding ring on one finger. The picture was acquired on a warm day in late summer, with an ambient outside air temperature of 22°C. The human hand depicts temperature values of 37°C (as would be expected). Note that even the veins within the hand can be seen (white, hot lines). Also the stairs in the background appears warm. However, the handrail appears at a radiance temperature of −8°C (cross at image center) although it is definitely not that cold. However, as this picture was acquired with a standard pre-set emissivity value of 1 (so no emissivity correction) the handrail, as well as the gold ring on the person's hand, appear much colder than they actually are.

Now imagine a remotely sensed thermal image of an area with many metal roofs. The temperatures in the image will all appear much cooler than reality, if not corrected for this emissivity effect. Therefore, thermal imagery in artificial environments, as well as in areas with a large variety of exposed rocks and minerals has to be handled with care.

6.3 POTENTIAL OF DIURNAL AND TIME SERIES OF THERMAL INFRARED REMOTE SENSING DATA

Thermal infrared data has an enormous advantage to other multispectral data: it can be acquired independent of the Sun as an illumination source. This means that TIR data can be acquired during the daytime, as well as during the nighttime. Many multispectral sensors either automatically acquire data during the night, or can be tasked to acquire nighttime data. MODIS (Moderate Resolution Imaging Spectroradiometer), NOAA-AVHRR (National Oceanic and Atmospheric Administration—Advanced Very High Resolution Radiometer), or MSG SEVIRI (Meteosat Second Generation—Spinning Enhanced Visible and Infrared Imager) for example all acquire TIR data in an automated mode all day long. MSG Seviri delivers geostationary data for every 15 to 30 minutes at 3 km spatial resolution in the TIR. MODIS (as on-board the platforms TERRA and AQUA) for each spot on Earth delivers up to four acquisitions per day at 1 km spatial resolution, which usually cover the morning, the afternoon, an early nighttime images, as well as a pre-dawn image. Therefore, this sensor holds a large potential for diurnal thermal mapping (Kuenzer, Hecker et al., 2008). AVHRR also delivers a 1km nighttime LST image for each spot on the earth. Nighttime LST from Landsat and ASTER can be downloaded from the USGS Earth Explorer website. Several other of the thermal infrared sensors depicted in Figure 6.4 can acquire nighttime data upon request with data providers.

Nighttime data has the advantage that influences of solar uneven heating due to topography solar are minimized. This data is therefore especially suitable to detect thermal anomalies, such as hot spots induced by forest fires, coal fires (Kuenzer et al., 2007; Zhang et al., 2007; Kuenzer et al., 2008; Kuenzer et al., 2013), peat fires, industry related hot spots or geothermal phenomena. Daytime data usually reflects uneven solar heating due to varying Sun-sensor object geometry, topography, thermal inertia) and therefore often complicates the extraction of anomalies, or the analyses of time series of thermal data. Furthermore, nighttime LST trends are more consistent with air temperature trends than daytime LST trends (Pepin et al., 2016; Zhou and Wang, 2016; Abera et al., 2020; NourEldeen et al., 2020) and are therefore a more reliable source to study global warming. Especially in pre-dawn thermal nighttime data the solar component is least accentuated.

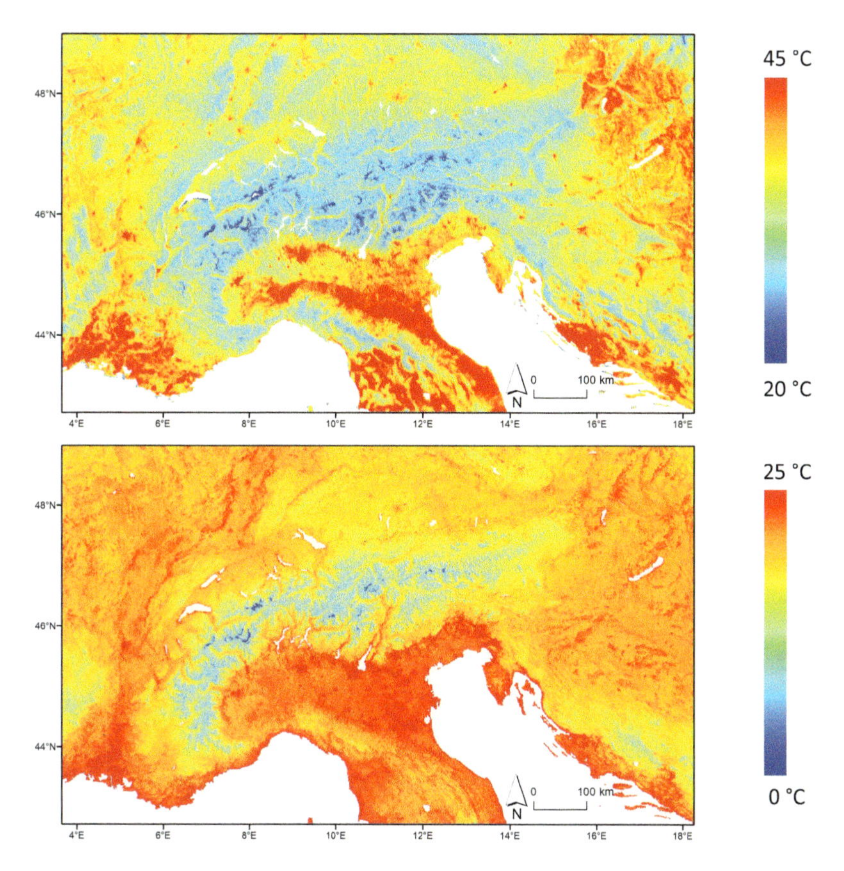

FIGURE 6.10 Long term maximum LST in July for daytime (up, around 1 p.m.) and nighttime (bottom, around 4 a.m.) derived from the TIMELINE AVHRR Level 3 products for the years 2006–2017. (Source: Coordinates: UL: 48.68°N, 3.92°E; LR: 42.81°N, 18.24°E.)

Figure 6.10 illustrates the differences in thermal daytime and nighttime data based on two composites of AVHRR LST over the Alps and its surrounding regions. In the upper daytime image relief as well as land cover show high effects on LST, where settled and low elevated areas heat up during the day, while vegetated and higher elevated areas stay relatively cool. In the bottom nighttime image these effects are weaker, while coastal areas appear relatively warm due to the heat storage of the Mediterranean Sea. Permanent features like the Alpine glaciers in dark blue appear on both images.

All objects on the Earth's surface have their own characteristic diurnal temperature curve (Göttsche and Olesen, 2001). This curve represents the temperature behavior of the object within a 24-hour cycle and describes how much and how fast an object heats up and cools down. The diurnal curve of an object depends and material of the object in the first place—defined by its so called thermal inertia. But also season (Sun-object geometry characterizing strength of illumination), atmospheric disturbances, and—complicating the matter for land surfaces—its exposure (aspect, slope) influence those diurrnal curves. Figure 6.11 depicts diurnal temperature curves of dry soil/rock and water. Differences in this example are purely based on material related properties, described by the thermal inertia. This parameter (measured in $J/m^2/K/sec^{0.5}$) is defined as the resistance of a material to heating and is the product of three factors: the product of three factors: the energy needed to raise the temperature of a material by 1°C (heat capacity c) per mass unit of the substance, the density of a material, p, and the thermal conductivity, k, of the object:

$$I = \sqrt{c \times p \times k} \tag{6.8}$$

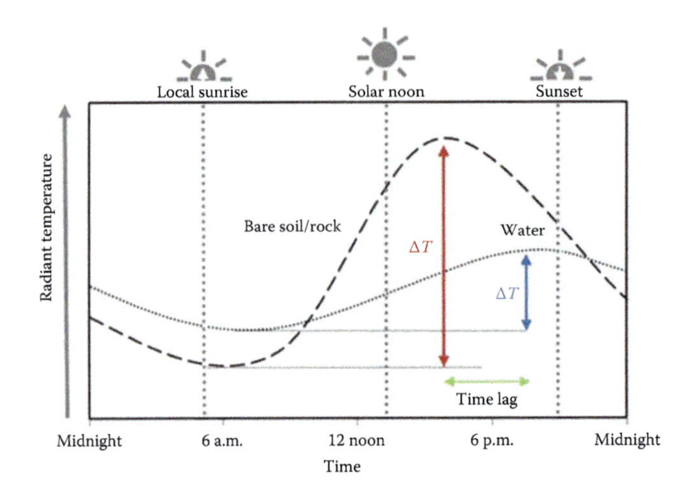

FIGURE 6.11 Diurnal temperature variation of water and dry soil/rock. Each object shows a distinct diurnal temperature cycle determined by the thermal inertia of the object and the history of the incoming solar radiation. (Modified from Lillesand et al., 2008.)

The difference between the highest and lowest temperature in a typical diurnal temperature curve is called ΔT (Lillesand et al., 2008). Variations in the thermal inertia (e.g., due to moisture for example) lead to changes in ΔT. Small ΔTs indicate a high thermal inertia: so a high resistance of the object against temperature changes. Water is a good representative for such a surface. High ΔTs indicate a low thermal inertia—so little resistance of the material against temperature change (Cracknell and Xue, 1996).

Thermal inertia cannot be derived from remote sensing data directly, but the idea behind it still allows to be employed with remote sensing data, as minimum and maximum temperature of a surface within the diurnal cycle can be derived from daytime and nighttime thermal images, such as the ones presented in Figure 6.10. ΔT can then be calculated subtracting the nighttime from the daytime temperature. In soils ΔT for example changes under different moisture conditions. Dry soils have a larger ΔT than wet soils. Thus, already in the 1970s did Idso et al., (1975) and Schmugge et al. (1991) employ synthetic ΔT images for surface soil moisture retrieval. The concept of ΔT over time has been extended to the concept of the so-called Apparent Thermal Inertia, ATI. The ATI is defined as:

$$\mathrm{ATI} = (1 - \mathrm{A})/\Delta \mathrm{T} \quad (6.9)$$

Here A represents the albedo of the pixel in the visible band. Albedo is included as dark materials absorb more sunlight than light materials—so by including it here the impact of this effect is compensated for. As one example, (Yuan et al., 2020) established an empirical relationship between ATI and the volumetric soil moisture content for the Chinese Loess Plateau. However, working with ΔT and ATI is a difficult task as they can vary depending on an objects bidirectional reflectance distribution function, BRDF, relief induced variations, shadowing, and impacts of wind. To compensate for relief induced variations in ΔT topographic data and information on solar elevation and azimuth need to be integrated into approximated corrections.

Diurnal effects of solar uneven heating were observed in situ on the ground by Zhang and Kuenzer (2007) as depicted in Figure 6.12. Temperatures were measured with a handheld radiometer at a very dense temporal interval of 10 minutes on a sand dune with slopes to the North, East, South, and West.

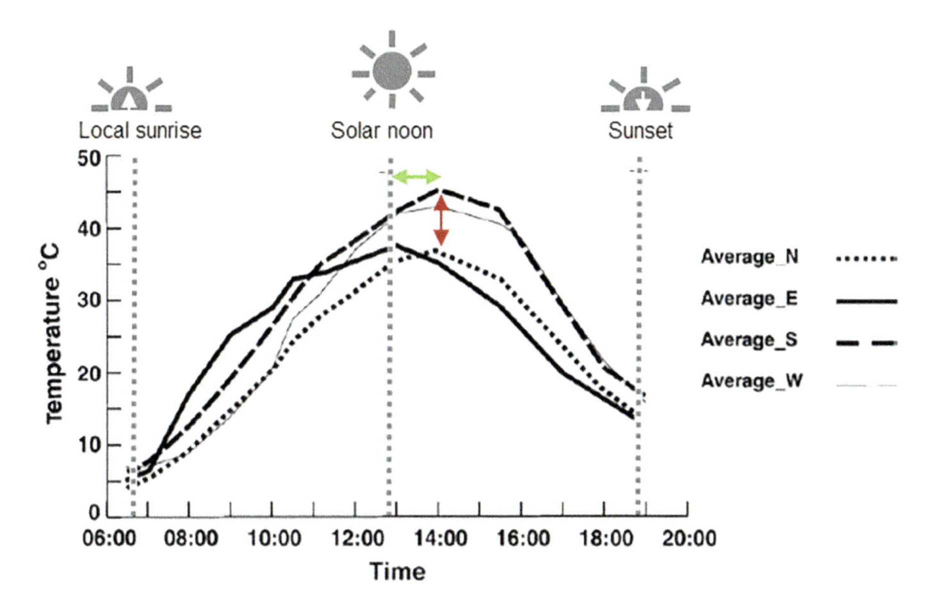

FIGURE 6.12 Diurnal temperature variation as measured on a sand dune, where one and the same material occurs at different aspects (thermal anisotropy). (Source: Zhang and Kuenzer, 2007.)

As expected, we can see that the East exposed slope heats up earliest and fastest in the morning (as the Sun rises in the East), and also the peak temperature here is reached earliest. Highest temperatures are reached on the southwards facing slope, and peak temperature here occurs more than one hour later than on the eastern slope. While peak temperature on the eastern and northern slope reach around 35°C, on the southern slope 45°C is reached. This demonstrated that one and the same surface material can—at one time step—exhibit completely different temperature depending on aspect and slope. At local Landsat overpass time of 10:30 for example, temperature of the same object can differ up to 10°C. The sand dune exhibits temperature ranging from 5°C to 45°C over the course of the day.

Solar uneven heating effects, as well as effects of differing Sun-sensor-object geometries in LST imagery thus have to be corrected for—especially when a study focuses on thermal change detection or even time series analyses (Warner and Chen, 2001).

Prior to these more complex correction TIR data of course has to be pre-processed like other multispectral data also. Just like optical imagery, data has to be geo-corrected, sensor calibration needs to be undertaken with constantly updated calibration coefficients (establishing a constant relationship between the radiation received at the sensor and the DN), DNs than have to be transferred in object radiance, considering atmospheric effects as well, which is usually undertaken in atmospheric correction software tools (Vidal, 1991).

But not only diurnal variation is interesting to exploit in multiple TIR acquisitions, but also annual temperature curves of objects can be analyzed (see Figure 6.13). Annual ΔT can be calculated from the temperature difference of an object between the coldest (winter) and the warmest (summer) season.

To define, at which point in time an object is the warmest and the coldest, usually gap-free or gap-filtered annual time series of daily available data are needed. In higher resolution data (e.g., Landsat data, which is only acquired every 16 days in the best case—meaning cloud free conditions) the search for the highest and lowest annual temperature would be a large challenge. However, even here can images acquired within only a few weeks difference illustrate the strong variability of temperature—with season, but also with season related land use changes. In Figure 6.13 we can clearly see how agricultural areas turn cooler with the expansion of vegetation covering the underlying soil (Kant et al., 2009). Urban areas usually appear warmest.

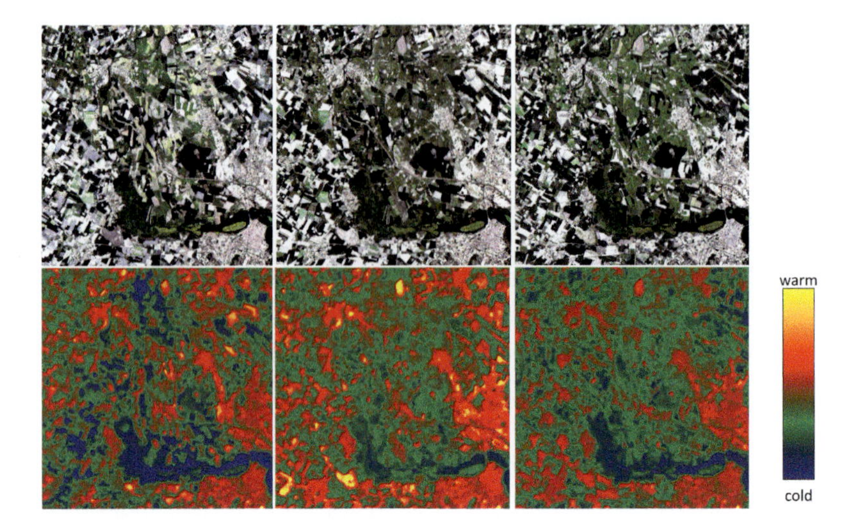

FIGURE 6.13 Landsat 8 TIRS daytime image subsets depicting an agricultural area in northern Italy, acquired on 16.06.2022 (left), 02.07.2022 (middle), 06.08.2022 (right). The upper row presents true color composites in RGB, whereas the lower row presents the color coded TIR band 6. UL: 45.28°N, 10.62°E, LR: 45.14°N, 10.81°E.

The strong variability of thermal data with acquisition data and time is often not considered—not even in studies, where it would be most crucial. Numerous authors study so called "urban heat islands" (Schwarz et al., 2011, Tiangco et al., 2008, Streutker, 2003), to assess, if a city—due to increased surface sealing—is getting hotter over time due, but do not really consider that actually a large amount of data would be needed, to eliminate the effects of different acquisition dates and times. A comprehensive time series of scenes acquired during the same date and time need to be analyzed to derive a real trend (e.g., possible with MODIS or AVHRR data).

6.4 APPLICATION EXAMPLES OF THERMAL INFRARED DATA ANALYSES

In the field of TIR analyses, SST analyses is the furthest advanced (see Figure 6.14) (Iwasaki et al., 2008), and one of the few domains in TIR remote sensing, where operational services are already offered—for example, by the European Space Agency, ESA. Figure 6.14 depicts an SST product derived from NOAA-19 and Metop-A data for the 18th of July 2016. Poles are not covered due to the near polar orbit of the satellites. Time series of these products are freely available form ESA and can be analyzed with respect to SST averages, minima, mean, variability, anomalies, and the representation of occurrences of, for example, El Niño in TIR derived SST data. TIR data acquired over the ocean has the big advantage that scientists do not have to deal with topographic effects of solar uneven heating. Furthermore, the water surface is more or less homogeneous—at least when compared with the patchy mosaics of the land surface. Furthermore, BRDF effects are minimized, and the thermal inertia of water is high and therefore temperature changes take place relatively slowly.

Figure 6.15 presents a typical application example of TIR data analyses on land. Here, hot spots have been derived over the area of the Niger Delta, Nigeria, Africa, during three time steps, and covering the time span from 1986–1987 until 2013. Oil industry in the Niger Delta flares enormous amounts of natural gas. This practice brings with it the harmful release of climate relevant and toxic gasses and substances, contributing to the severe environmental degradation in the area. Different approaches for thermal anomaly extraction exist, and here an automated, moving window based approach (Kuenzer et al., 2007) was applied to extract local hot spots of different temperature—an

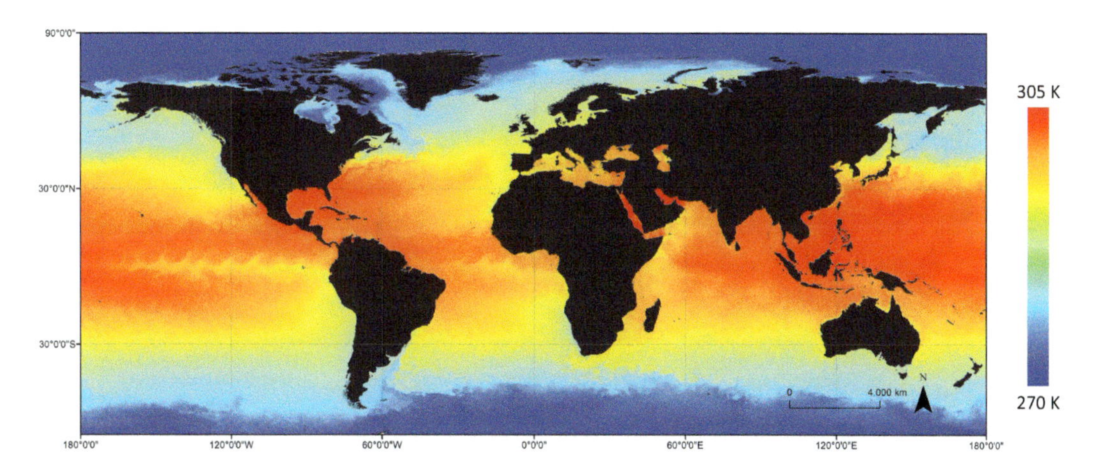

FIGURE 6.14 Sea Surface Temperature, SST, on 18th of July 2016 from the ESA CCI Level 4 SST product derived from NOAA-19 and Metop-A data at 0.05° resolution.

approach clearly superior to simple empirical thresholding. The illustration of hot spot occurrences over time (in this case gas flares) can depict the dynamics of oil exploiting industry development in the region, and can support the designation of especially threatened natural resource or communities.

The range of TIR based application studies is very broad, and published studies are numerous. A comprehensive overview of thermal infrared remote sensing is provided in the book "Thermal Infrared Remote Sensing: Sensors, Method, Applications" (Kuenzer and Dech, 2013), to which numerous authors contributed, and which contains three parts. Sensor related chapters focus, amongst others, on the geometric calibration of thermographic cameras, thermal infrared spectroscopy, challenges and opportunities for UAV (Unmanned Aerial Vehicle)-borne thermal imaging, or planned new thermal missions, such as NASA's Hyperspectral Thermal Emission Spectrometer (HyTES), NASA's Hyperspectral Infrared Imager (HyspIRI), or thermal remote sensing with small satellites such as BIRD, TET and the next generation BIROS. Method oriented chapters present cross-comparisons of daily LST products from NOAA-AVHRR and MODIS, compare the advantages and shortcomings of the thermal sensors of SEVIRI and MODIS for LST mapping, discuss methods for improving atmospheric correction of TIR data, or for time series corrections and analyses in TIR data, or novel concepts for the derivation of SST products. Application chapters address approaches to derive urban structure types, address TIR based mineral mapping, soil moisture derivation, the assessment of vegetation fires, analyses of lava flows, thermal analyses of volcanoes, investigations in underground coal fire regions, as well as the analyses of geothermal systems.

6.5 GROUND DATA AND VALIDATING THERMAL INFRARED DATA

To validate temperatures derived from thermal infrared data, or to confirm phenomena or patterns observed in this data validation with ground data is a common procedure undertaken (Coll et al., 2005). Validation on past data is of course not possible, as LST varies from hour to hour and from day to day. Therefore, ground data collection activities should be performed during the satellite's overpass. As temperatures change within minutes there is little time to measure the on ground surface temperatures of several objects within the scenes footprint. Usually, several people with inter-calibrated radiometers, contact thermometers, as well as thermal cameras are on the ground to measure the kinetic and radiance temperatures of hottest and coldest objects (see Figure 6.16). Water surfaces are a good target to establish a relationship between ground-measured temperatures and satellite imagery derived temperatures, as they have a high thermal inertia and do exhibit fast or accentuated temperature changes over time.

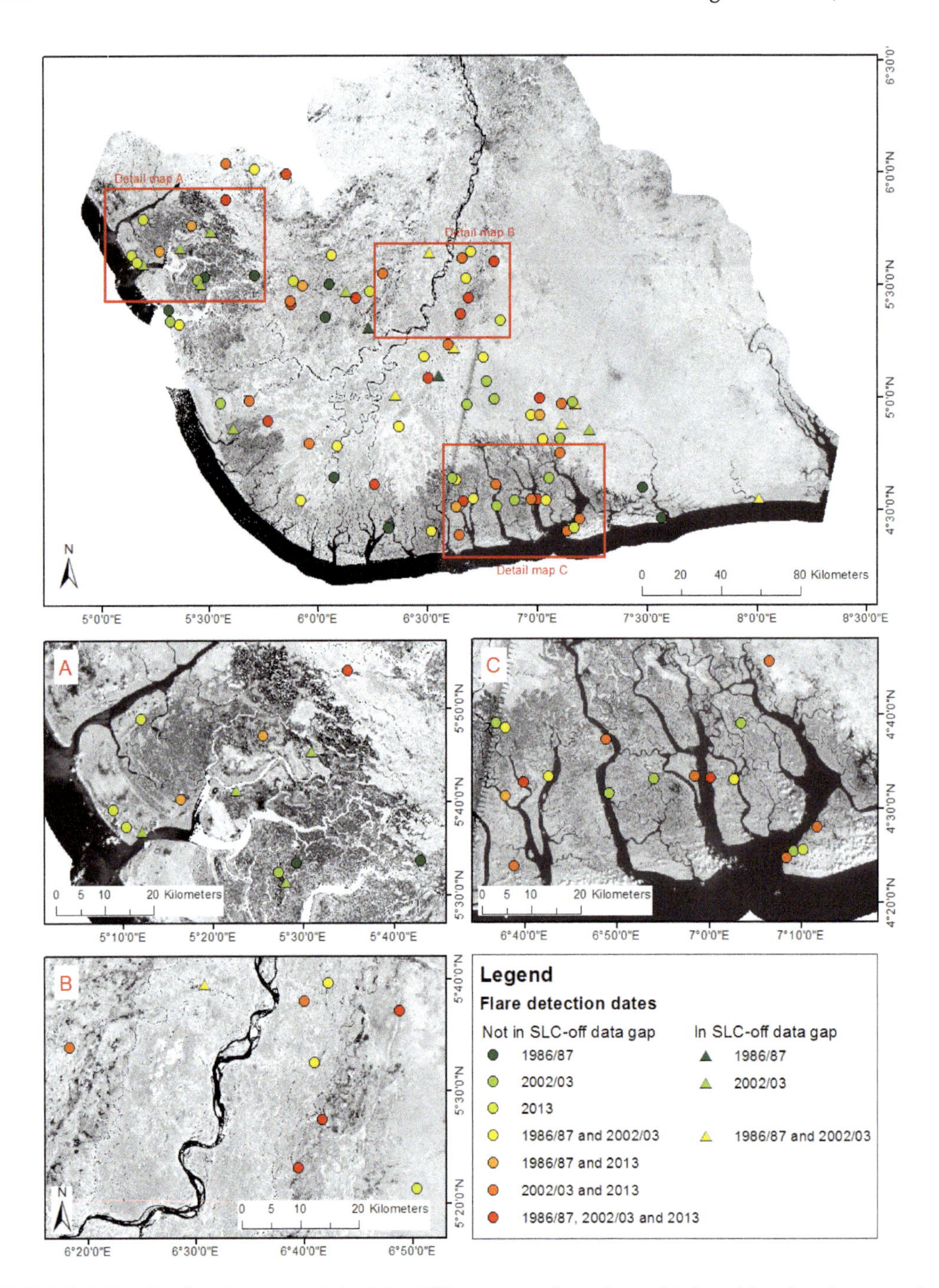

FIGURE 6.15 Gas flare hot spots derived for different years from thermal infrared Landsat data over the course of 27 years (1986–1987 until 2013). Oil industry in the Niger Delta flares enormous amounts of natural gas. This practice brings with it the harmful release of climate relevant and toxic gasses and substances, contributing to the severe environmental degradation in the area. (Source: Kuenzer et al., 2014.)

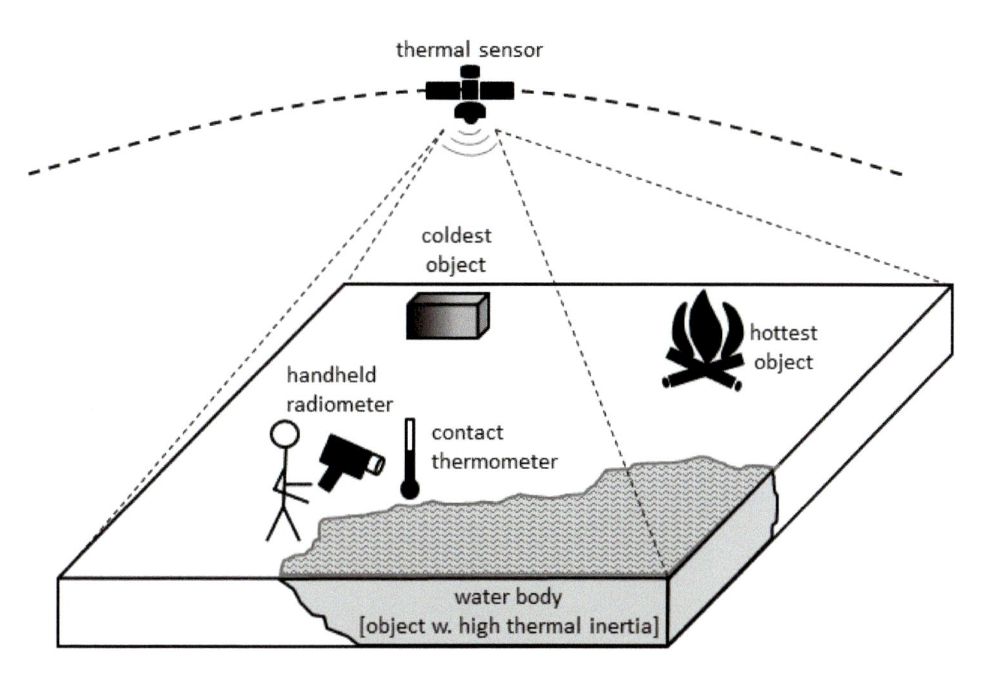

FIGURE 6.16 Ground data collection/validation of thermal infrared data and observed phenomena undertaken during satellite overpass. (Source: Kuenzer and Dech, 2013, modified.)

Indirect validation of via Google Earth can be undertaken for selected applications. For example, the Niger Delta gas flare locations, which were presented in Figure 6.14 were partially validated via checks in high resolution data Google Earth. The high resolution data available for most parts of the planet allows to put thermal anomalies detected into the right context. Gas flares—for example—are clearly visible also in high-resolution optical data. If a flame is obscured by smoke other indirect indicators are burnt vegetation, soot around the flare site, and hot spots can sometimes also be validated by the numerous photographs available in Google Earth. The same applies for forest fire occurrences or grassland fires.

6.6 DISCUSSION AND CONCLUSION

Thermal remote sensing data is used for a variety of applications such as the assessment of Land Surface Temperature, Sea Surface Temperature, inland water body temperature, the analyses of land cover related thermal patterns, such as urban heat islands, the detection of hot spots and thermal anomies resulting from forest fires, coal fires, and industry related activities, as well as the derivation of moisture conditions via the investigation of diurnal temperature dynamics. While the fleet of satellites with a thermal sensor on board is larger than often assumed, there are only few sensors available, which offer free and easy access to thermal data. However, thermal infrared data offers several advantages. Although thermal infrared data—just like multispectral data—is affected by cloud cover an active illumination source is not needed. Therefore, the data can be acquired during the daytime as well as the nighttime. Thermal data depicts a direct physical quantity (in K or °C) and is thus easy to interpret without too much bias. Furthermore, long term data archives of common medium resolution sensors with daily acquisition coverage allow for the analyses as decadal, annual and monthly means, deviations, variability, and trends. Disadvantages of thermal infrared remote sensing are the relatively low spatial resolution of the data. Currently, the highest resolution space born data has a 90 m pixel size. Thermal sensors—be it airborne or spaceborne—are more costly than, for example, visible sensors. Thus, the TIR bandwidth is usually the domain which is usually discarded if budget cuts have to be made. At the same time the thermal infrared community is relatively small, and lobby voices for thermal sensors are thus not so powerful.

This chapter presented an overview of the principles and theoretical background of remote sensing in the thermal infrared domain. We addressed data characteristics and important laws of physics, presented common past and up to date thermal infrared instruments, discussed approaches to analyze thermal data, and presented selected application examples. Remote Sensing of the thermal infrared is an often neglected discipline of remote sensing. Thermal sensors are very expensive, and due to the longer wavelength compare to optical and near infrared data, spatial resolution is always inferior to shorter wavelength multispectral data. Astonishingly, while the ETM+ sensor on Landsat 7 offered 60m resolution, its successors TIRS and TIRS-2 on Landsat 8 and 9 only provide TIR data in 100m resolution. Today MODIS is the most commonly used sensor for LST time series studies with high temporal resolution, while Landsat's TIRS and TIRS-2 sensors are popular for studies, where high spatial resolution is required (Reiners et al., 2023). Climate-change related studies, which require a time series of 30 years or longer have to rely on AVHRR or Landsat data. However, with the upcoming decommission of MODIS announced by NASA, the TIR LST community is looking for alternatives, namely VIIRS and SLSTR.

REFERENCES

Abera, T. A., J. Heiskanen, E. E. Maeda and P. K. E. Pellikka (2020). "Land Surface Temperature Trend and Its Drivers in East Africa." *Journal of Geophysical Research: Atmospheres* 125.

Becker, F. (1987). "The Impact of Emissivity on the Measurement of Land Surface Temperature from a Satellite." *International Journal of Remote Sensing* 8(10):1509–1522.

Coll, C, V. Caselles, J. M. Galve, E. Valor, R. Niclòs, J. M. Sánchez and R. Rivas (2005). "Ground Measurements for the Validation of Land Surface Temperatures Derived from AATSR and MODIS Data." *Remote Sensing of Environment* 97:288–300.

Cracknell, A. P. and Y. Xue (1996). "Thermal Inertia Determination from Space— A Tutorial Review." *International Journal of Remote Sensing* 17(3):431–461.

Dech, S. W., P. Tungalagsaikhan, C. Preusser and R. E. Meisner (1998). "Operational Value-Adding to AVHRR Data Over Europe: Methods, Results, and Prospects." *Aerospace Science and Technology* 2:335–346.

Frey, C., C. Kuenzer and S. Dech (2012). "Quantitative Comparison of the Operational NOAA AVHRR LST Product of DLR and the MODIS LST Product V005." *International Journal of Remote Sensing* 33(22):7165–7183.

Göttsche, F.-M. and F. S. Olesen (2001). "Modelling of Diurnal Cycles of Brightness Temperature Extracted from METEOSAT Data." *Remote Sensing of Environment* 76(3):337–348.

Heald, M. A. (2003). "Where Is the 'Wien Peak?'" *American Journal of Physics* 71(12):1322–1323.

Hulley, G. C., S. J. Hook and A. M. Baldridge (2009). "Validation of the North American ASTER Land Surface Emissivity Database (NAALSED) Version 2.0 Using Pseudo-Invariant Sand Dune Sites." *Remote Sensing of Environment* 113(10):2224–2233.

Idso, S. B., R. D. Jackson and R. J. Reginato (1975). "Detection of Soil Moisture by Remote Surveillance." *American Scientist* 63:549–557.

Iwasaki, S., M. Kubota and H. Tomita (2008). "Inter-Comparison and Evaluation of Global Sea Surface Temperature Products." *International Journal of Remote Sensing* 29(21):6263–6280.

Kant, Y., B. D. Bharath, J. Mallick, C. Atzberger and N. Kerle (2009). "Satellite-based Analysis of the Role of Land Use/Land Cover and Vegetation Density on Surface Temperature Regime of Delhi, India." *Journal of the Indian Society of Remote Sensing* 37(2):201–214.

Kirchhoff, G. (1860). "Ueber das Verhältniss zwischen dem Emissionsvermögen und dem Absorptionsvermögen der Körper für Wärme und Licht." *Annalen der Physik und Chemie (Leipzig)* 109:275–301.

Kuenzer, C. and S. Dech (2013). "Theoretical Background of Thermal Infrared Remote Sensing." In C. Kuenzer and S. Dech (eds.) *Thermal Infrared Remote Sensing—Sensors, Methods, Applications*. Remote Sensing and Digital Image Processing Series, Volume 17. Springer, Dordrecht, The Netherlands, 572 pp., pp. 1–26, ISBN 978-94-007-6638-9.

Kuenzer, C., U. Gessner and W. Wagner (2013b). "Deriving Soil Moisture from Thermal Infrared Satellite Data—Synergies with Microwave Data." In C. Kuenzer and S. Dech (eds.) *Thermal Infrared Remote Sensing—Sensors, Methods, Applications*. Remote Sensing and Digital Image Processing Series, Volume 17. Springer, Dordrecht, The Netherlands, 572 pp., pp. 315–330, ISBN 978-94-007-6638-9.

Kuenzer, C., H. Guo, M. Ottinger and S. Dech (2013a). "Spaceborne Thermal Infrared Observation – An Overview of Most Frequently Used Sensors for Applied Research." In C. Kuenzer and S. Dech (eds.) *Thermal Infrared Remote Sensing—Sensors, Methods, Applications*. Remote Sensing and Digital Image Processing Series, Volume 17. Springer, Dordrecht, The Netherlands, 572 pp., pp. 131–148, ISBN 978-94-007-6638-9.

Kuenzer, C., C. Hecker, J. Zhang, S. Wessling and W. Wagner (2008). "The Potential of Multi-Diurnal MODIS Thermal Bands Data for Coal Fire Detection." *International Journal of Remote Sensing* 29:923–944.

Kuenzer, C., S. van Beijma, U. Gessner and S. Dech (2014). *Land Surface Dynamics and Environmental Challenges of the Niger Delta, Africa: Remote Sensing based Analyses Spanning Three Decades (1986–2013)*. Accepted at Applied Geography.

Kuenzer, C., J. Zhang, L. Jing, G. Huadong and S. Dech (2013). "Thermal Infrared Remote Sensing of Surface and Underground Coal Fires." In C. Kuenzer and S. Dech (eds.) *Thermal Infrared Remote Sensing—Sensors, Methods, Applications*. Remote Sensing and Digital Image Processing Series, Volume 17. Springer, Dordrecht, The Netherlands, 572 pp., pp. 429–451, ISBN 978-94-007-6638-9.

Kuenzer, C., J. Zhang, J. Li, S. Voigt, H. Mehl and W. Wagner (2007). "Detection of Unknown Coal Fires: Synergy of Coal Fire Risk Area Delineation and Improved Thermal Anomaly Extraction." *International Journal of Remote Sensing* 28:4561–4585.

Lillesand, T. M. and R. W. Kiefer (1994). *Remote Sensing and Image Interpretation*. Third Edition. John Wiley & Sons, New York, 748pp.

Lillesand, T. M., R. W. Kiefer and J. W. Chipman (2008). *Remote Sensing and Image Interpretation*. Sixth Edition. John Wiley & Sons, New York, 768pp, ISBN-10: 0470052457.

Löffler, E. (1994). *Geographie und Fernerkundung*, Third Edition, Teubner, Stuttgart, 251p., ISBN 3-519-13423-3.

NourEldeen, N., K. Mao, Z. Yuan, X. Shen, T. Xu and Z. Qin (2020). "Analysis of the Spatiotemporal Change in Land Surface Temperature for a Long-Term Sequence in Africa (2003–2017)." *Remote Sensing* 12(3).

Pepin, N. C., E. E. Maeda and R. Williams (2016). "Use of Remotely Sensed Land Surface Temperature as a Proxy for Air Temperatures at High Elevations: Findings from a 5000 m Elevational Transect Across Kilimanjaro." *Journal of Geophysical Research: Atmospheres* 121(17).

Planck, M. (1900). "Entropie und Temperatur strahlender Wärme." *Annalen der Physik* 306(4):719–737.

Pörtner, H.-O., D. Roberts, M. Tignor, E. Poloczanska, K. Mintenbeck, A. Alegría, M. Craig, S. Langsdorf, S. Löschke, V. Möller, A. Okem, B. Rama, D. Belling, W. Dieck, S. Götze, T. Kersher, P. Mangele, B. Maus, A. Mühle and N. Weyer. (2022). *Climate Change 2022: Impacts, Adaptation and Vulnerability Working Group II Contribution to the Sixth Assessment Report of the Intergovernmental Panel on Climate Change*. http://doi.org/10.1017/9781009325844.

Reiners, P., J. Sobrino and C. Kuenzer (2023). "Satellite-Derived Land Surface Temperature Dynamics in the Context of Global Change—A Review." *Remote Sensing* 15. http://doi.org/10.3390/rs15071857

Sabins, F. F. (1996). *Remote Sensing: Principles and Interpretation*. Third Edition, W.H. Freeman and Company, New York, 450pp.

Schmugge, T. J., F. Becker and Z. L. Li (1991). "Spectral Emissivity Variations Observed in Airborne Surface Temperature Measurements." *Remote Sensing of Environment* 35(2):95–104.

Schwarz, N., S. Lautenbach and R. Seppelt (2011). "Exploring Indicators for Quantifying Surface Urban Heat Islands of European Cities with MODIS Land Surface Temperatures." *Remote Sensing of Environment* 115(12):3175–3186.

Streutker, D. (2003). "Satellite-Measured Growth of the Urban Heat Island of Houston, Texas." *Remote Sensing of Environment* 85:282–289.

Tiangco, M., A. M. F. Lagmay and J. Argete (2008). "ASTER-Based Study of the Night-Time Urban Heat Island Effect in Metro Manila." *International Journal of Remote Sensing* 29(10):2799–2818.

Tipler, P. A. (2000). *Physik*. Third Edition. Spektrum Akademischer Verlag, Heidelberg, Germany, 1520pp.

Vidal, A. (1991). "Atmospheric and Emissivity Correction of Land Surface Temperature Measured from Satellite Using Ground Measurements or Satellite Data." *International Journal of Remote Sensing* 12(12):2449–2460.

Walker, J. (2008). *Fundamentals of Physics*. Eighth Edition. John Wiley and Sons, New York, 891pp. ISBN 9780471758013.

Warner, T. A. and X. Chen (2001). "Normalization of Landsat Thermal Imagery for the Effect of Solar Heating and Topography." *International Journal of Remote Sensing* 22(5):773–788.

Yuan, L., L. Li, T. Zhang, L. Chen, J. Zhao, S. Hu, L. Cheng and W. Liu (2020). "Soil Moisture Estimation for the Chinese Loess Plateau Using MODIS-derived ATI and TVDI." *Remote Sensing* 12. http://doi.org/10.3390/rs12183040

Zhang, J. and C. Kuenzer (2007). "Thermal Surface Characteristics of Coal Fires 1: Results of In-situ Measurements." *Journal of Applied Geophysics* 63:117–134.

Zhang, J., C. Kuenzer, A. Tetzlaff, D. Oertel, B. Zhukov and W. Wagner (2007). "Thermal Characteristics of Coal Fires 2: Results of Measurements on Simulated Coal Fires." *Journal of Applied Geophysics* 63:135–147.

Zhou, C. and K. Wang (2016). "Land Surface Temperature Over Global Deserts: Means, Variability, and Trends." *Journal of Geophysical Research: Atmospheres* 121(24):14, 344–314, 357.

7 Remote Sensing Image Segmentation

Methods, Approaches, and Advances

Mohammad D. Hossain and Dongmei Chen

LIST OF ACRONYMS

AI Artificial intelligence
ASPP Atrous spatial pyramid pooling
CNN Convolutional neural network
CRF Conditional random field
DEM Digital elevation models
DFN Deep fusion networks
DL Deep learning
DSM Digital surface models
DTM Digital terrain models
ERS Entropy rate superpixel
FCN Fully convolution network
FNEA Fractal net evolution approach
GAN Generative adversarial network
HED Holistically-nested edge detection
HRM Hybrid region merging
HSI Hyperspectral images
HSWO Hierarchical stepwise optimization
LiDAR Light detection and ranging
LRW Lazy random walk
MCL Markov clustering
MCNN Multi-scale convolutional neural network
MRS Multi-resolution segmentation
PSO Particle swarm optimization
ReLU Rectified linear unit
RHSeg Recursive hierarchical segmentation
RISA Region-based image segmentation algorithm
SLIC Simple linear iterative clustering
UAVs Unmanned aerial vehicles

7.1 INTRODUCTION

Following the introduction of the IKONOS (IK) satellite in 1999, there was a revolutionary enhancement in spatial resolution, with subsequent even more significant improvements in satellites such

DOI: 10.1201/9781003541158-8

as QuickBird (QB), WorldView1 (WV-1), WorldView2 (WV-2), WorldView3 (WV-3), WorldView4 (WV-4), GeoEye-1, Pleiades, OrbView-3, SPOT-5, SPOT-6, SPOT-7, Ziyuan-3 (ZY-3), KOMPSAT, Gaofen-1 (GF-1), Gaofen-2 (GF-2), SuperView (Super), Planet Labs, and Unmanned Aerial Vehicles (UAVs). High-resolution images, in contrast to moderate- or low-resolution ones, exhibit a profound distinction: clusters of pixels form discernible land covers representing real-world features, denoted as objects in (GEographic) Object-Based Image Analysis (GEOBIA or OBIA) (Blaschke, 2010). GEOBIA tackles intricate class definitions dependent on complex spatial and hierarchical relationships throughout the classification process (Lang, 2008). Segmentation is often considered the linchpin of GEOBIA and can significantly influence the accuracy of subsequent processes in object-based feature extraction and classification (Su & Zhang, 2018).

Image segmentation, as defined by Pal and Pal (1993), is the process of partitioning an image into coherent and homogeneous regions. These regions correspond to distinct land cover elements such as buildings, trees, water bodies, and grasslands, collectively called "image objects" within the GEOBIA framework (Costa et al., 2018). Image segmentation aims to simplify an image's representation into meaningful objects that are easier to analyze. Image segmentation groups neighboring pixels into objects with boundaries based on certain characteristics, such as color, intensity, geometry, texture, etc. Besides its application in remote sensing, image segmentation has many other applications across various domains, such as computer vision, medical imaging, autonomous driving, object detection, face recognition, etc.

The landscape of segmentation algorithms is vast, with numerous approaches conceived and applied in different contexts. Broadly speaking, there are two classes of segmentation techniques in remote sensing image analysis: traditional segmentation approaches and semantic segmentation. As portrayed in Figure 7.1, traditional segmentation methods group pixels in an image into individual objects and assign generic labels (e.g., region A, region B) to each object. However, they do not assign the class to the objects. After segmentation, an additional classification procedure will be needed to label objects into corresponding classes (e.g., building, tree, road). Semantic segmentation takes traditional segmentation further by incorporating object classification. They treat multiple objects of the same class as a single entity and output both the boundary and class of objects. Modern semantic segmentation is based on deep learning (DL) methods that assign specific

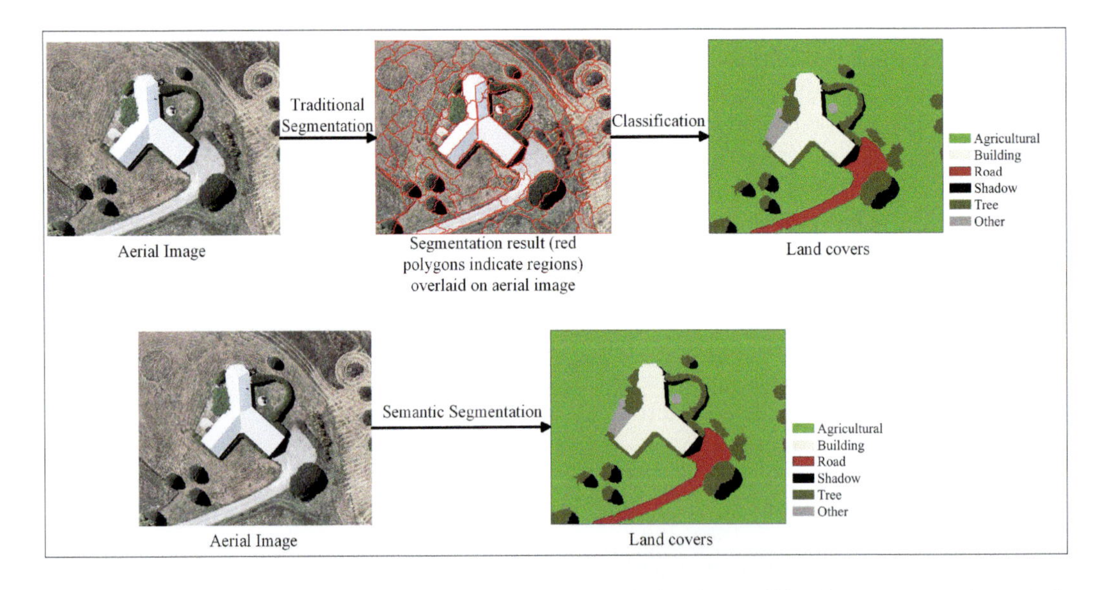

FIGURE 7.1 A typical workflow of remote sensing image analysis using traditional versus semantic method.

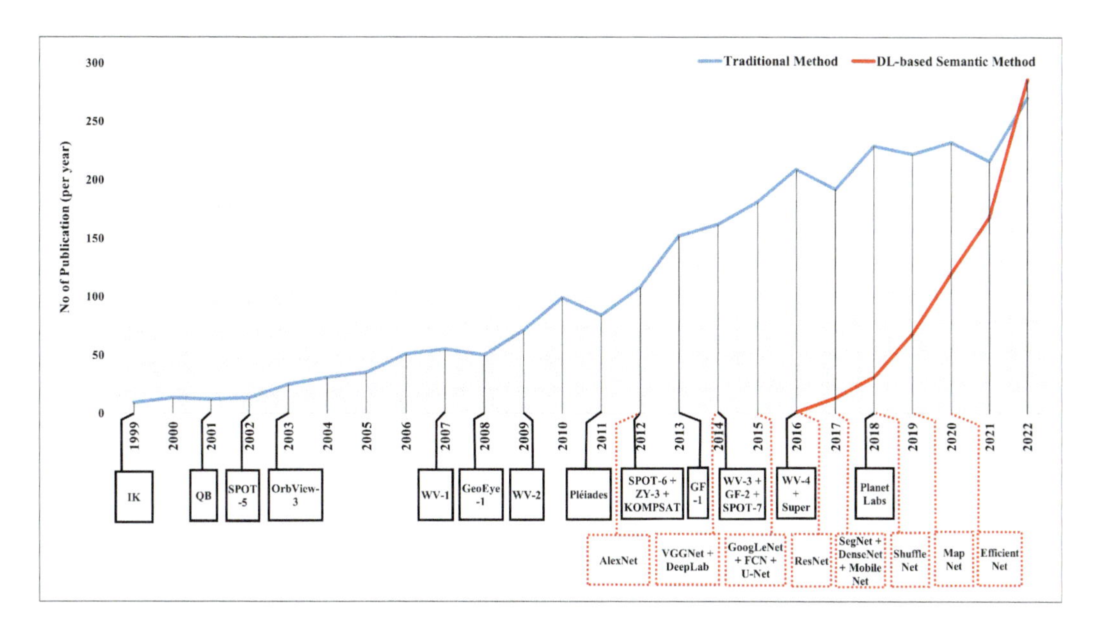

FIGURE 7.2 Evolution of publications over time: a comparative analysis of traditional and DL-based semantic segmentation methods, highlighting influential trends and triggers.

labels (e.g., building, road, tree, waterbody) to individual pixels by leveraging spatial and spectral information.

This chapter conducts a concise literature survey using the Web of Science, employing keywords like OBIA, GEOBIA, segmentation, remote sensing, semantic segmentation, deep learning, and variations. Figure 7.2 illustrates a notable surge in articles utilizing traditional segmentation methods, particularly with the advent of high-resolution images. Intriguingly, following the introduction of DL-based semantic segmentation methods, their application witnessed an exponential rise, eventually surpassing traditional methods. It is noteworthy that image segmentation is not confined to multispectral images but extends to hyperspectral and Light Detection And Ranging (LiDAR) images. This chapter is dedicated to unveiling recent advancements in remote sensing image segmentation, focusing on DL-based semantic methods. It is an exhaustive exploration of cutting-edge image segmentation algorithms not covered in the preceding chapters. The aim is to delve into their concepts and critically analyze their merits, limitations, and application domains. Additionally, this chapter presents recent and captivating applications of methods already discussed in previous chapters.

7.2 TRADITIONAL SEGMENTATION METHODS

7.2.1 Spectrally Based Methods

The primary focus of spectrally based, or pixel-based, image segmentation techniques is examining individual pixels (Kotaridis & Lazaridou, 2021). These techniques can be further separated into two categories: clustering-based and thresholding-based techniques. Among the earliest image segmentation methods is thresholding. Typically, clustering techniques entail partitioning the spectral feature space, assigning each pixel to a cluster, and combining adjacent pixels that have the same cluster value to form segments. Spectrally based methods often encounter salt-and-pepper effects due to atmospheric disturbances and noise (Subudhi et al., 2021). Superpixel segmentation provides a solution to this issue and has garnered a lot of attention in the field of computer vision (M. Wang et al., 2017).

For the purpose of creating superpixels, Achanta et al. (2012) presented Simple Linear Iterative Clustering (SLIC), a modification of the k-means clustering algorithm. The input multispectral image is represented as $MSI \equiv \{s_1^b, s_2^b, s_3^b, ..s_n^b\}$ where n denotes the number of pixels and $\{s_i^b\}$ signifies the value of ith pixel for the bth band with i ranging from 1 to n and b from 1 to B. Here B symbolizes the set of spectral bands. In the SLIC method, distance is computed within a $2S\,X\,2S$ window around the cluster center where $S = \sqrt{\frac{n}{K}}$, where K represents the desired number of superpixels. The cluster centers are initially placed randomly on a regular grid space and are iteratively relocated to the position of the lowest gradient within a 3 × 3 kernel. Each pixel is linked to the nearest cluster center during the assignment phase using the distance measure D, as depicted in Equation 7.1. This departs from the traditional k-means clustering approach, where each pixel is compared with all cluster centers. The equation for D is as follows:

$$D = \sqrt{\left(d_{spectral}\right)^2 + \left(\frac{d_{spatial}}{S}\right)^2 m^2} \tag{7.1}$$

In the equation, m serves as the weighting factor between spatial and spectral features. The spectral distance between pixel i and pixel j are calculated as follows:

$$d_{spectral} = \sqrt{\sum_{i=1}^{B}\left(s_i^b - s_j^b\right)^2} \tag{7.2}$$

Here, $d_{spectral}$ represents the homogeneity within superpixels. The spatial distance between pixels within superpixels is calculated as follows:

$$d_{spatial} = \sqrt{\left(x_j - x_i\right)^2 + \left(y_j - y_i\right)^2} \tag{7.3}$$

The cluster centers are then updated to reflect the mean vector of each pixel in the cluster through an update process. These assignment and update stages are repeated until the error converges or hits the input iteration threshold. In order to guarantee connectivity, a post-processing phase reassigns disjunct pixels to adjacent superpixels. The same authors also suggested SLICO, a parameter-free version of SLIC that creates uniformly shaped superpixels throughout the scene regardless of whether the image's regions are textured or not. On the other hand, SLIC is sensitive to texture; in non-textured areas, it produces superpixels that are smooth and regular in size, whereas in textured areas, it produces very irregular superpixels. SLIC does not capture global image attributes, despite its efficiency and good boundary adherence (M. Wang et al., 2017). However, while creating superpixels with more regular shapes, the Linear Spectral (Z. Li & Chen, 2015) method maintains the global features of images.

In an intriguing development, Y. J. Liu et al. (2016) expanded the capabilities of SLIC to compute content-sensitive superpixels. This was achieved by mapping the image onto a 2-dimensional manifold. This innovative extension facilitates the creation of smaller superpixels in regions dense with content and larger superpixels in sparse regions. Boundaries are also followed by Levinshtein et al. (2009)'s TurboPixel. This technique divides an image into compact sections that form a lattice-like structure. Dilating seeds enables them to adjust to the local image structure. Using a level-set-based geometric flow that depends on the local image gradient to produce regular superpixels, TurboPixel determines a set of seeds. In Vedaldi and Soatto's (2008) Quick Shift, it is not possible to control the number and size of superpixels, but it demonstrates strong

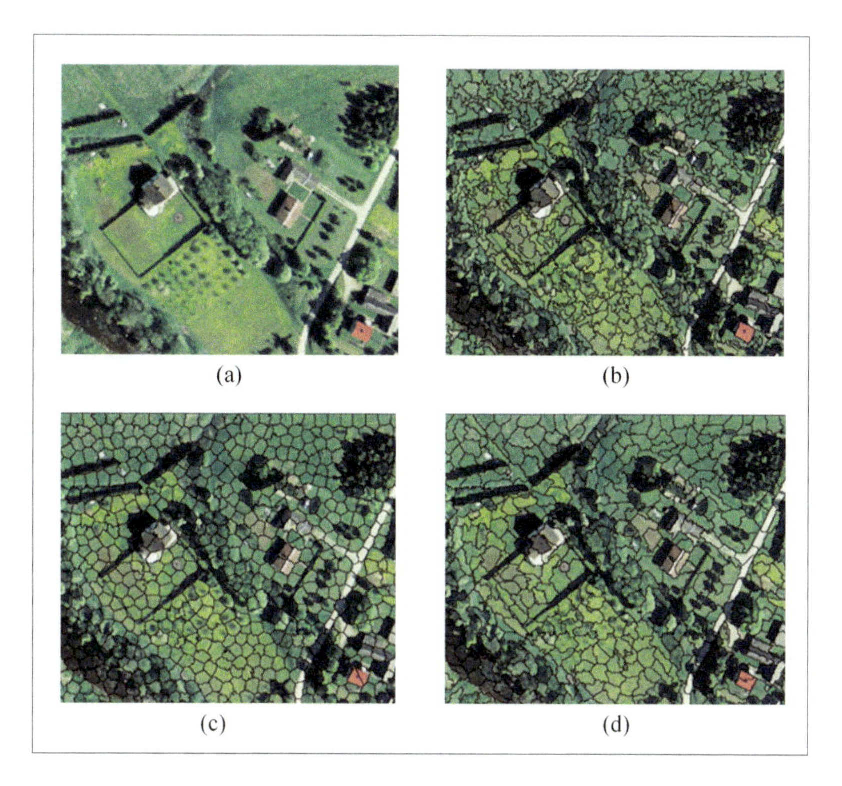

FIGURE 7.3 Example of segments generated by superpixel algorithms. (a) An example image, (b) segments generated by SLIC and overlaid on the image, (c) segments generated by SLICO and overlaid on the image, and (d) segments generated by SEEDS and overlaid on the image.

boundary adherence. On the other hand, Moore et al. (2008)'s Superpixel Lattices do not adhere well to image borders. Van den Bergh et al. (2015) proposed SEEDS, which has substantial form irregularity and difficulty controlling the number of superpixels. Superpixels have been criticized for generating over-segmentation (Csillik, 2017) even though they can produce a coherent grouping of pixels, as seen in Figure 7.3.

7.2.2 Spatially Based Methods

7.2.2.1 Edge-Based Method

Edge-based techniques have long been a cornerstone in image segmentation, offering valuable insights into the boundaries and contours of objects within an image. These techniques fundamentally rely on identifying edges, which are then meticulously closed using contouring algorithms. The fundamental assumption driving edge-based techniques is that pixel properties undergo abrupt changes between edges. In this context, edges are conceptualized as the precise boundaries where these changes occur, effectively demarcating the transition between distinct objects within the image. Diverse algorithms have been developed to identify object edges, each capturing unique geometrical and physical characteristics of image objects.

Jain et al. (1995) divided edge detection into three steps: filtering, enhancement, and detection. Filtering methods are used to remove noise available in the images. The noise content of a color image usually has the same characteristics in each channel. Those noises contribute to the identification of false object edges. Typical noise removal algorithms such as Gaussian and median filters modify even those pixels that are not impacted by noise. To address this issue, Z. Chen et al.

(2006) proposed a peer-group filtering algorithm to remove impulse noise. Besides, Nikolaou and Papamarkos (2009) proposed the edge-preserving smoothing filter to remove noise and preserve object boundaries. This method applied a 3 × 3 convolution mask with a Manhattan color distance between the central pixel and the eight neighboring pixels. In the kernel, they used 0 for the central pixel to remove impulse noise. Kerem and Ulusoy (2013) pointed out that high-spatial-resolution images are mostly free of impulse noise. Thus, they chose 1 for the central pixel in the edge-preserving smoothing filter.

The enhancement makes pixels more noticeable where the brightness level varies significantly in the nearby area. Enhanced data are used for detecting actual edges. Edge-detection techniques were divided into two categories by Jing et al. (2022): hand-crafted and machine-learning-based. The collection of manual edge detection techniques uses one or more scales to extract first- or second-order derivatives of an image's light intensity in order to identify edge outlines. These edge detection techniques include, for example, Sobel (Sobel, 1970), Laplace-of-Gaussian (Marr & Hildreth, 1980), Canny (Canny, 1987), Prewitt (Prewitt, 1970), Roberts (Roberts, 1963), and others. Since edges in remote sensing images typically span multiple scales, multi-scale edge detection techniques are widely used. These techniques combine several edge strength maps that were acquired at various scales. For example, Shui and Zhang (2012) fused edge strength by multiplying the gradients at two separate scales. Nonetheless, since edges are not well defined, fuzzy theory has been employed by academics to identify edge structures. For instance, Melin et al. (2014) used the morphological gradient method and type-2 fuzzy logic to identify edges. Furthermore, hand-crafted based approaches include spatial-frequency-based methods (Yi et al., 2009), sub-pixel localization methods (Seo, 2018), anisotropic diffusion-based methods (Tsiotsios & Petrou, 2013), active contour methods (C. Liu et al., 2017).

DL techniques have also been widely used for edge detection in recent years. The edge detection problem was defined by Dollár and Zitnick (2015) as a random decision forest-based prediction of image local segmentation masks. The nearest neighbor technique matched the extracted features from image patches to detect edge contours from images, and Ganin and Lempitsky (2014) created neural network nearest neighbor fields, the first deep convolutional neural network (CNN) based edge detection method. A CNN was extensively utilized in the holistically-nested edge detection (HED) approach presented by Xie and Tu (2015). In order to recognize object edges, Yang, Price, et al. (2016) created a completely convolutional encoder-decoder network based on a fully convolution network (FCN). Even though the aforementioned edge-detecting algorithms provided promising results, most of them were tested only on natural images. Researchers are still looking for an edge detection algorithm that can achieve high precision and efficiency (Jing et al., 2022) when dealing with remote sensing images.

Converting edges into closed borders comes next once edges have been identified. This stage typically entails removing noise-generated edges, filling in gaps where edges aren't found, and choosing which edge segments to link together to form a single object. Various edge-linking techniques have been proposed to make up for incompletely linked edges (Dronova et al., 2012; Lu & Chen, 2008). The Hough transform (Ballard, 1981; Kiryati & Eldar, 1991) is another tool researchers use to find the optimal edges that best fit the partial edges. But for basic parametric shapes, this approach performs admirably (Maintz, 2005). Neighborhood search was another tool used by researchers to identify a potential connection point for the edge pixels (Ghita & Whelan, 2002). Graph theory was used by Flores et al. (2013) and Sappa and Vintimilla (2008) to construct closed contour. Ji et al. (2013) linked the edges using a heuristic A* search, gradient, and directions. Guan et al. (2015), on the other hand, suggested a partial differential equation-based technique to connect the edges. Furthermore, a predictive edge-linking method was suggested by Akinlar and Chome (2016), which traverses the edge map by generating predictions from its historical movements. Finding the correct edges to construct image objects is difficult, even though numerous algorithms have been developed to locate and connect actual edges to generate objects (Hossain & Chen, 2019).

7.2.2.2 Region-Based Method

Edge-based methodologies commence by delineating the boundaries of an object, and subsequently, the interior is filled to define the object. In contrast, region-based techniques take a divergent route, initiating from the object's interior and expanding outward until the boundaries are encountered. While edge-based and region-based approaches theoretically describe different facets of the same object, outcomes from region-based methods can markedly differ from those derived through edge-based methodologies (Kavzoglu & Tonbul, 2017). Region-based methods operate on the premise that adjacent pixels within the same object manifest similar values. These methods predominantly involve two operations: merging and splitting. Following the foundational approach to region-based image segmentation entails the following steps: (1) Attain an initial segmentation of the image, which may be either over or under-segmented. (2) Merge or split adjacent segments based on their similarity or dissimilarity. (3) Repeat this process until no segments remain that necessitate merging or splitting. This iterative procedure enhances the precision of the segmentation.

Region growing, a widely recognized method for region-based segmentation, encounters two principal challenges: the selection of the seed region and defining the similarity criteria. An innovative variation, termed seeded region growing, was proposed by Deng and Manjunath (2001). This variant introduces two internal pixel order dependencies, resulting in distinct segments. The first-order dependency occurs when multiple pixels exhibit the same difference measure to their neighboring regions. The second-order dependency arises when a single pixel displays the same variation measure in multiple regions. Despite its simplicity, the seed selection process amplifies computational costs and execution time. Verma et al. (2011) introduced a single-seeded region-growing technique to mitigate this. In response to localization issues, Wang et al. (2010) creatively used the K-means clustering algorithm for seed generation in the Region-based Image Segmentation Algorithm (RISA), while Mirghasemi et al. (2013) used Particle Swarm Optimization (PSO). In order to segment high-resolution remote sensing images, X. Zhang et al. (2014) presented the Hybrid Region Merging (HRM) technique, which combines both local- and global-oriented region merging algorithms into a single, coherent framework. Byun et al. (2011), in contrast, presented a technique that combined a block-based seed selection method with a modified form of seeded region expanding and region merging. Despite these advancements, there is still a quest for improved techniques that function without seeds (L. Wu et al., 2015) or that, even when seeded, are impartial toward neighbors.

Following seed selection, the region undergoes expansion by incorporating adjacent pixels exhibiting similarity based on a designated homogeneity criterion. This gradual process enlarges the region, with the homogeneity criterion being a critical factor in determining pixel inclusion. The decision to merge is influenced solely by the contrast between the current pixel and the region. The segmentation process encounters limitations in identifying objects smaller than the image's spatial resolution, while larger objects face pixelation. In contrast to region-growing techniques, region-merging methods commence from an initial region, fostering the development of Multi-Resolution Segmentation (MRS) (Hay et al., 2003). A noteworthy contribution to this field is the Fractal Net Evolution Approach (FNEA), a multi-resolution method introduced by Baatz and Schäpe (2000), widely employed in various studies (Hossain & Chen, 2019; J. Liu et al., 2021; Ninsawat & Hossain, 2016; Som-ard et al., 2018). FNEA, functioning as a region-merging hierarchical segmentation, originates from a single pixel (Blaschke et al., 2004). Each coarser level integrates information from the finer level, ensuring that identified objects at finer levels persist in representations at coarser levels. Beyond MRS, diverse region-merging approaches, including Mean-Shift (MS) (Comaniciu & Meer, 2002), Hierarchical Stepwise Optimization (HSWO) (Beaulieu & Goldberg, 1989), and Recursive Hierarchical Segmentation (RHSeg) (Tilton et al., 2012), have garnered attention from researchers, broadening the spectrum of segmentation methodologies.

To address the performance imbalances stemming from the use of global measures in region merging, researchers have introduced split and merge techniques as local measures, aiming to enhance segmentation outcomes. The split process initiates with the entire image and, guided by criteria like grey values, texture, internal edges, or other indicators of inhomogeneity, partitions the image into segments. Region merging and splitting are often coupled to consolidate similar regions and create maximally homogeneous entities. The tendency of the resultant image to reflect the data structure, giving it a square aspect, is a disadvantage of area splitting. Numerous improvements and modifications to these methods have been suggested. Split-and-merge segmentation using an improved quadtree approach was first presented by Kelkar and Gupta (2008). Manousakas et al. (1998) improved on conventional split and merge algorithms by combining the concepts of boundary removal and simulated annealing. In order to extract roads from images of metropolitan areas, Alshehhi and Marpu (2017) used hierarchical merging and splitting image segmentation, utilizing color and shape data.

7.2.3 Graph-Based Method

Image segmentation is formulated as a graph partitioning problem within the context of a graph-based methodology, where nodes are often individual pixels or regions connected by edges that reflect spatial adjacency. These edges have weights that represent the dissimilarity of pixels, with the goal of locating subgraphs that match imaging clusters (Alshehhi & Marpu, 2017). The surge in interest surrounding graph-based algorithms for image segmentation underscores their potency. Superpixels are also produced as part of this procedure by minimizing a cost function that is established throughout the graph. To compute superpixels, for example, the Homogeneous Superpixels (HS) (Perbet & Maki, 2011) method uses a graph-based algorithm called Markov clustering (MCL), which uses stochastic flow circulation. However, superpixel segmentation is framed as a multi-label assignment issue in the Superpixels via Pseudo-Boolean (PB) (Y. Zhang et al., 2011) technique, which produces superpixels with consistent size and shape. A derivation of the RW technique called Lazy Random Walk (LRW) (Shen et al., 2014) converts the input image into a graph and performs exceptionally well at respecting object boundaries. Superpixel segmentation is approached as a graph maximizing problem in Entropy Rate Superpixel (ERS) (M. Y. Liu et al., 2011), which introduces a novel objective function based on graph topology. By cherry-picking a subset of edges from the resulting graph, superpixel segmentation seeks to create precisely K-connected sub-graphs, where K is the required number of superpixels. Graph-based superpixels produce over-segmented regions even though they outperform cluster-based superpixels (M. Wang et al., 2017).

7.2.4 Hybrid Methods (HMs)

To address the limitations inherent in both edge- and region-based methods, researchers have undertaken an integrative approach that combines the strengths of both paradigms, offering the potential for enhanced segmentation outcomes. While edge-based methods excel in edge detection precision, they often struggle to form closed segments. In contrast, region-based methods are adept at creating closed regions but may sacrifice precision in defining segment boundaries (M. Wang & Li, 2014). In HMs (Gaetano et al., 2015; X. Li et al., 2014), initial segments are delineated using edge-based approaches and then consolidated using region-based methods. This integration leverages boundary pixels for initial segment outlines and interior pixels for segment merging (X. Zhang et al., 2014). Mueller et al. (2004) successfully employed a hybrid of edge and region-based techniques for extracting large objects, such as agricultural fields, demonstrating the potential of combining shape information and edge maps to guide region growing algorithms. Gambotto (1993) suggested incorporating edge information to halt the growing process. N. Li et al. (2010) proposed texture clustering as a constraint in Hybrid Spatial Watershed Optimization (HSWO), incorporating region adjacency and neighbor graph-based merging criteria.

While many region-merging methods rely on a single global parameter to govern the iterative merging process, offering user control over under- and over-segmentation, issues arise as the same threshold is applied uniformly across all segments. To overcome this, Johnson and Xie (2011) and

Chen et al. (2014) introduced local measures to identify segments under- and over-segmented at an optimal scale parameter, refining them through appropriate splitting and merging. This local refinement strategy effectively improves segmentation quality by addressing under- and over-segmentation problems (Yang et al., 2017). However, executing further splitting and merging steps presents challenges in an operational context (Yang, He, et al., 2016). Furthermore, considering both homogeneities within segments and heterogeneities between segments is crucial. Y. Wang et al. (2018) proposed a HM that integrates the objectives of heterogeneity and relative homogeneity during the merging process. Hossain and Chen (2022) introduced a donut-filling technique to alleviate over-segmentation caused by roof elements and used illumination differences to restrict merging with shadows. Commonly, studies initiate segmentation from the edge-based method, resulting in an over-segmented image. Subsequently, the region-based method is employed to merge similar segments based on either homogeneity or heterogeneity.

7.3 DL-BASED SEMANTIC SEGMENTATION METHODS

Machine learning algorithms have played a pivotal role in remote sensing applications, particularly in the classification of individual pixels within an image. This involves training models to assign specific labels to pixels based on their distinctive features. However, the evolution of semantic segmentation transcends mere pixel classification and delves into the intricate spatial connections between pixels. Here, the core aim is to attain a profound understanding of the detailed spatial arrangement of diverse objects present in the image. Unlike traditional pixel-wise classification, semantic segmentation provides a holistic perspective, capturing the nuanced interplay and layout of objects within the image canvas. Semantic segmentation of remote sensing images aims at precision in obtaining pixel-level spatial accuracy, especially along the boundaries of unique objects (Marmanis et al., 2018). Unlike traditional segmentation methods, semantic segmentation assigns a label to each pixel. Key architectural contributors to the evolution of deep neural networks include AlexNet (Krizhevsky et al., 2012), VGGNet (Simonyan & Zisserman, 2014), and GoogLeNet (Szegedy et al., 2015). Their significance lies in enhancing the capabilities of DL models for intricate tasks like semantic segmentation in remote sensing (Yuan et al., 2021).

7.3.1 ALGORITHMS' BASIC

AlexNet boasts five convolutional layers, denoted as feature extraction layers, and three fully connected layers, as depicted in Figure 7.4. Interposed between consecutive convolutional layers are pooling layers designed to reduce dimensionality, lowering computational complexity. AlexNet employs max pooling, wherein the largest value within the filter-covered image region is retained,

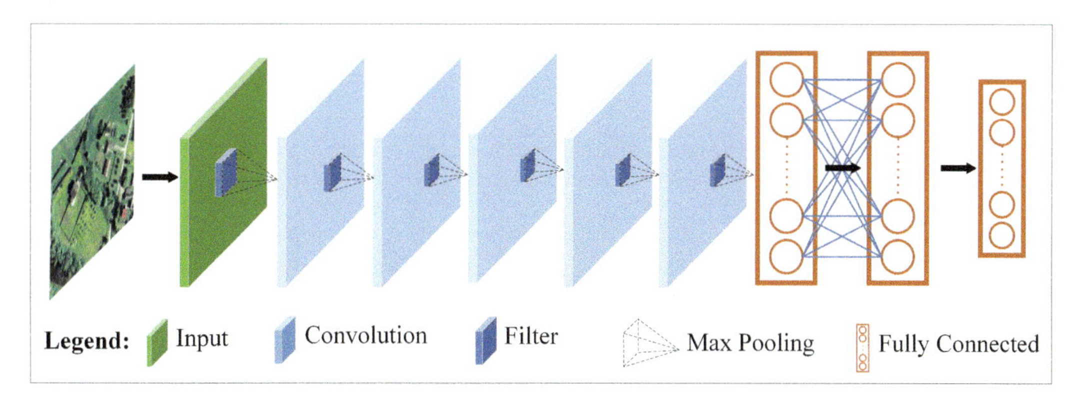

FIGURE 7.4 The schematic network architecture of AlexNet.

effectively eliminating noise. For feature extraction, AlexNet's first and second convolutional layers use filters with sizes of 11 × 11 and 5 × 5, respectively, while the next three convolutional layers use smaller filters, 3 × 3. Objects of different sizes can be accommodated by this various filter method. To train a classification function, the fully connected layers process feature vectors that have been flattened. In three important aspects, AlexNet has been a trailblazer in the development of CNNs: (1) It uses the Rectified Linear Unit (ReLU), which is a non-saturating activation function and is defined as f(x) = max(x, 0). All that is needed for this computationally efficient function is a comparison operation. (2) Overlapping max pooling is used, in which each filtering operation's stride is smaller than the filter's size. (3) In order to prevent overfitting, the dropout approach is introduced in the fully connected layers. With a probability of 0.5, this method randomly assigns zero to the output.

VGGNet presents a compelling architecture featuring multiple convolutional layers and three fully connected layers. The schematic structure of the VGG-16 network is illustrated in Figure 7.5, offering a flexible design that allows for the creation of various VGGNets, including VGG-11, VGG-16, and VGG19, by adjusting the number of convolutional layers. A notable departure from AlexNet is VGGNet's utilization of 3 × 3 filters in its convolutional layers, and the convolution stride is set at one pixel. The training procedure is made simpler by using smaller filters, which effectively lower the number of network weights. To preserve spatial resolution, one-pixel padding is used for the three-by-three convolutional layers. Max-pooling adds even more distinctive features to the network. It is implemented over a 2 × 2 window with a stride of two pixels. Similar to AlexNet, VGGNet makes use of each hidden layer neuron's ReLU activation function. The features that were collected from the convolutional layers create a hierarchy of scales that show how well the network performs a variety of tasks, including target detection and semantic segmentation.

GoogLeNet is different from other CNN versions since it uses a single fully connected layer, integrates auxiliary classifiers during training, and has an inception module. GoogLeNet's inception module combines the output with max-pooling results by applying filters with different sizes (1 × 1, 3 × 3, and 5 × 5) to the input. Between inception modules, max pooling is used, and after the last inception module, a dropout-enhanced average pooling technique is employed. The architecture of GoogLeNet consists of three convolutional layers and nine inception modules. As illustrated in Figure 7.6, it adds auxiliary classifiers in intermediate layers that use the output of the inception modules. These classifiers' losses are applied to the overall network loss during training. This design enhances the network's training stability and gradient flow through the deep layers.

The FCN (Long et al., 2015) was developed especially for image semantic segmentation, building upon the foundation of those three networks. Multi-layer convolution, deconvolution, and fusion are the three main phases of FCN. Figure 7.7 illustrates how convolutional layers are used in FCN in place of traditional fully connected layers. Interestingly, a 1 × 1 convolution—also known as pixel-wise convolution—calculates scores for every class in an image. The resulting image is smaller

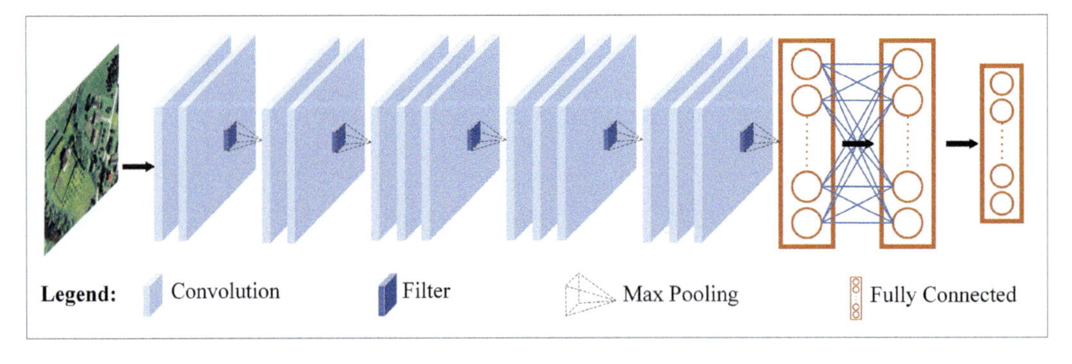

FIGURE 7.5 The schematic network architecture of VGG-16.

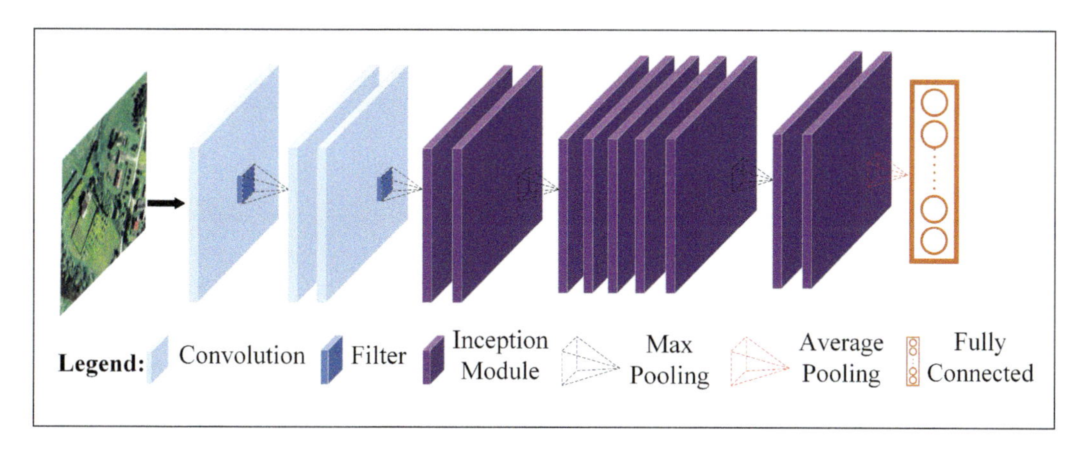

FIGURE 7.6 The schematic network architecture of GoogLeNet.

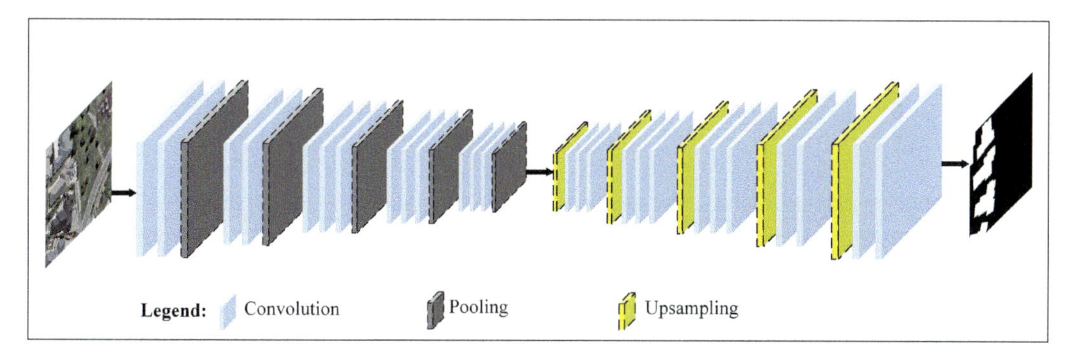

FIGURE 7.7 The schematic network architecture of FCN.

than the original image because of the pooling processes that followed. For bilinear upsampling of these coarse outputs, deconvolution is used to overcome this and satisfy the segmentation requirements. While deconvolution is similar to convolution, it serves to "expand" the input by combining parts within a deconvolution filter and padding the matrix. A skip architecture combines location information from previous levels with semantic information from deep layers to produce the final segmentation.

One of the challenges faced by semantic segmentation algorithms is their requirement for large volumes of training data. U-Net (Ronneberger et al., 2015) was introduced to perform image segmentation using a smaller training dataset to tackle this issue. The U-Net architecture comprises convolution and deconvolution layers. The graphic network architecture of U-Net is portrayed in Figure 7.8. Two 3 × 3 filters are applied by the convolution layers, and the outputs are then processed using ReLU. Next, downsampled outputs are produced by max pooling with a stride of two. It is noteworthy that in the convolutional layers, the quantity of feature channels doubles with each step. A 2 × 2 convolution is used to minimize the amount of features after the feature map has been upsampled in the deconvolution layers. Cropped to fit the input size are the feature maps produced by the convolutional layers. In order to make sure that the dimensions of the convolution and deconvolution results match, this cropping step compensates for the loss of boundary pixels during the convolution period. Through shortcut connections, the deconvolution results are stacked with the cropped feature maps. After labeling pixels in the feature map using a 1 × 1 convolution, the network produces the segmentation result.

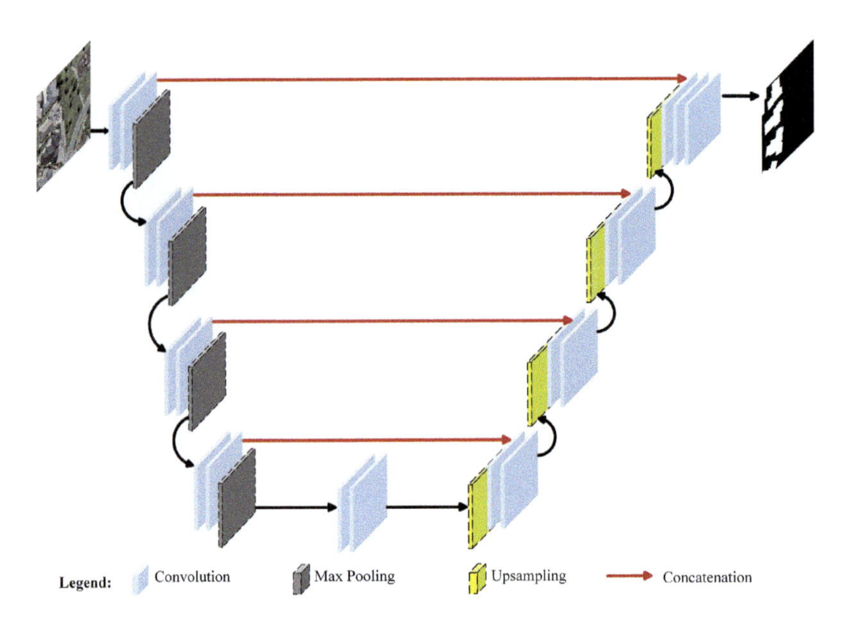

FIGURE 7.8 The schematic network architecture of U-Net.

In addressing a persistent challenge encountered by many CNNs—the extensive tuning of thousands of parameters—Howard et al. (2017) introduced MobileNet. This innovative architecture offers an efficient solution, providing a pair of hyperparameters that facilitate the creation of compact, low-latency models. This adaptability is particularly advantageous for meeting the specific demands of mobile and embedded vision applications. MobileNet differs from traditional CNNs in that it does not rely on a single 3 × 3 convolution layer, instead using batch normalization and ReLU. Rather than using a conventional method, MobileNet divides the convolution into two distinct types: a 1 × 1 pointwise convolution and a 3 × 3 depthwise convolution. This small design change greatly improves the network's efficiency. MobileNet's two global hyperparameters, which provide a dynamic trade-off between accuracy and latency, are what make it so brilliant. While the resolution multiplier controls the network's input image size, the width multiplier allows for control over the number of channels. This flexibility empowers users to tailor the network precisely to their requirements, making MobileNet a groundbreaking solution in mobile and embedded vision applications. EfficientNet (Tan & Le, 2019) also has fewer parameters yet provides better accuracy when compared to previous models in terms of accuracy and the number of parameters (Alhichri et al., 2021).

SegNet (Badrinarayanan et al., 2017) comprises two integral components: an encoder leveraging a pre-trained VGG16 model and a decoder network. The encoder network adopts a structure aligned with FCN, incorporating multiple convolution and max-pooling operations for effective feature extraction. While the deeper layers of this network excel at extracting features with enhanced semantic meanings, they tend to introduce ambiguity in spatial information within the output. SegNet offers a novel solution to this problem by keeping the element index, which indicates where an element is located in the filter window. The decoder network's upsampling procedure then heavily relies on this index. The decoder network mimics the architecture of the encoder network, directing convolutions and guided upsampling to convert low-quality features to higher resolution. In this process, the encoder network's pooling index comes in rather handy. A 2 × 2 low-resolution feature map, for example, is scaled to a 4 × 4 matrix that is entirely zeroed out. The 2 × 2 map's material is arranged in a deliberate manner at the spots where it was pooled in the matching encoder layer. This strategic reuse of the pooling index significantly contributes to retrieving spatial information and augments boundary accuracy. SegNet exhibits a more memory-efficient approach than U-Net, transferring only the pooling indices from the encoder to the decoder instead of the entire feature maps.

DeepLab (L.-C. Chen et al., 2014) is an FCN extension with a unique focus on atrous convolution. This innovation significantly broadens the filter field, encompassing a larger image context and offering specific control over the resolution of feature responses. Atrous convolution, represented as $y(a)=\sum_{p=1}^{K} w(p)x(a+rp)$, integrates the element index (a), filter window offset (p), weight to the element p (w(p)), and sampling rate to the input x (r). When r = 1, atrous convolution aligns with conventional convolution. With different sampling rates, several atrous convolutions with the same kernel are utilized in Atrous Spatial Pyramid Pooling (ASPP). Addition is used to combine the results of these convolutions. DeepLab incorporates Conditional Random Field (CRF) to improve spatial localization, especially in boundary details. After bi-linear interpolation, a fully connected CRF is applied to the network output.

The method is improved by using successive atrous convolutions with different sampling rates in DeepLab V3 (L.-C. Chen et al., 2017). For the final feature map in the model, ASPP applies global average pooling. Bilinear upsampling is used to recover the desired spatial dimensions after the features undergo a 1 × 1 convolution with 256 filters. By adding a decoder module to enhance boundary details in the DeepLabV3 network, DeepLab V3+ (L.-C. Chen et al., 2018) goes one step further. To do this, the ASPP pooling and the decoder modules must perform separable, depth-wise convolution. Depth-wise spatial convolutions are performed for every input channel, and a 1 × 1 convolution is used to harmonize the outputs.

ResNet, a network structure similar to VGGNet but different in layer composition, was presented by He et al. (2016). ResNet integrates residual modules with two 3 × 3 filters and a shortcut link, in contrast to VGGNet, which primarily uses 3 × 3 filters. This novel method makes residual mapping with shortcut connections easier to understand and makes it possible to build incredibly deep networks. DenseNet (Huang et al., 2017) builds on ResNet by introducing connections from one layer to every layer below, encouraging greater information flow and feature reuse. DenseNet's central component, the dense block, is made up of layers stacked with two filters: a 1 × 1 filter in front of a 3 × 3 filter. Every layer that comes before it, even the input layer, contributes to each layer. By encouraging feature reuse and mitigating the vanishing gradient issue, this approach reduces the number of features per layer and, as a result, the number of learning parameters.

ShuffleNet (X. Zhang et al., 2018) uses group convolutions to lower the complexity of 1 × 1 convolutions, improving computing efficiency. It also makes use of channel shuffling to improve information flow between feature channels. In order to maximize GPU processing, the group convolution splits computations into distinct portions that are done in parallel. In a ShuffleNet Unit, each channel's 3 × 3 convolution is swapped out for a depth-wise convolution, and the channel dimension is restored for concatenation with the residual by the second group convolution. This novel method greatly increases the computational efficiency of DL networks.

7.3.2 Application Examples

Combining convolution layers with max-pooling in semantic segmentation algorithms often leads to losing spatial details, making pixel-level accuracy a crucial requirement. Dilated convolutions, also known as atrous convolutions, are frequently utilized to address this challenge. This technique integrates elements at noncontiguous positions in a kernel to enhance spatial context. Zhao and Du (2016) presented a multi-scale convolutional neural network (MCNN) that is intended to understand intricate details of spatial interactions. MCNN extracts spatial characteristics at multiple scales by building a pyramid structure from the image. After that, spectral features and high-level spatial features are combined to provide a dataset that may be used to train a logistic regression model. The ultimate outcomes are determined by means of a majority vote process.

Through the use of three parallel convolution layers with different kernel sizes (3 × 3, 5 × 5, and 7 × 7), Audebert et al. (2016) accomplished a revolutionary extension of SegNet. Through the clever averaging of the responsive cells over several scales, the predictions from these layers were combined. Four CNNs, each concentrating on a different contextual scale, were used in a simultaneous fashion by Längkvist et al. (2016). Fine-tuning of fully linked layers was accomplished via backpropagation, and mid-level features having semantic implications were derived from low-level features. A further degree of sophistication was introduced by CNN's use of pre-trained filters. Conversely, G. Wu et al. (2018) introduced a brand-new method for building segmentation using a FCN with many constraints. These constraints were generated based on predictions at various decoder layers and their accompanying ground truth, integrating the U-Net architecture with scale restrictions for intermediate layers. In parallel, Diakogiannis et al. (2020) presented ResUNet-a, an advanced technique based on a U-Net encoder/decoder backbone that combines pyramid scene parsing pooling, residual connections, atrous convolutions, and multi-tasking inference to successfully infer object boundaries, distance transform, segmentation mask, and input reconstruction.

By extending the U-Net architecture, R. Li et al. (2018) included UpBlocks, which are two convolutional layers followed by an upsampling layer, in the expansion path and DownBlocks, which are two convolutional layers concatenated through a ReLU layer in the contracting path. Two convolution layers with a set number of channels make up these units, which are similar to the residual units in the deep Residual U-Net (Z. Zhang et al., 2018). A convolutional encoder neural network with two layers was also developed by C. Zhang et al. (2018). Farmland and woodland characteristics are extracted by the first layer using two sets of convolutional kernels. In the second layer, the learned features are encoded by two encoders using nonlinear functions, and the output is mapped to the appropriate category. The network can now identify features and classify a wider range of landscapes because of this clever architecture.

A further method for enhancing precision is using edge maps in the segmentation procedure. By building an edge detection and segmentation network, D. Cheng et al. (2017) expanded SegNet and suggested an edge-aware convolutional network. The edge detection network is trained using the segmentation network, which extracts semantic information at various sizes. The network is adjusted using the edge map of the edge detection network. The edge detection network's edge map is used to fine-tune the network. To address the segmentation of manmade structures and intricate objects concurrently, Y. Liu et al. (2018) introduced a self-cascaded network. This network incorporates dilated convolutions in the final layer of the encoder to achieve a multi-scale representation. Beyond encompassing a broader contextual scope, the multi-scale representation incorporates hierarchical dependencies within the context. Different dilation rates are employed to capture features at various scales, and the outcomes are consolidated in a coarse-to-fine fashion through skip connections from the encoder, resulting in an enhanced delineation of the target object. CNNs were shuffled further by G. Chen et al. (2018) in order to segment aerial photos. Low-resolution feature maps are downsampled and then converted to a higher-resolution version using a shuffling operator. Pixels in an overlapping region are predicted using the closest patches, and the final segmentation is achieved by averaging the score maps.

Due to the growing availability of heterogeneous, geo-registered data, integrating geometry and spectral information to improve segmentation accuracy in remote sensing is becoming more and more popular. Marmanis et al. (2016) presented feature-level fusion using an ensemble of FCNs based on the VGG-16 architecture. Feature-level fusion is usually carried out when input data share comparable patterns. These FCNs are blended by averaging predictions after being trained with various initializations from the ImageNet, Pascal VOC, and Places datasets. In order to preserve the fine boundary details from the early layers, a skip connection is an essential component. At the same time, intermediate features were utilized at different resolutions by Maggiori et al. (2017a). When incoming data has multiple structures, a different network typically handles each data type, and fusion happens during the classification phase. The method was expanded by Audebert et al. (2018) by using Digital Surface Models (DSM) produced from LiDAR point clouds and using FCNs to extract semantic feature maps at various resolutions. In order to handle topological inconsistencies, fusion techniques included late fusion employing residual correction and early fusion utilizing FuseNet at the encoder stages. For point classification, Yousefhussien et al. (2018) proposed a modified FCN handling normalized point clouds and related spectral data. W. Sun and Wang (2018) introduce the idea of maximum fusion in a novel way. This technique uses color images to create an initial segmentation and then uses a DSM backend to correct segmentation faults that an FCN produces. The elevation difference between each pixel and the surrounding ground is computed by the DSM backend, which effectively removes fake top hats and incorrect ground pixels. This all-inclusive strategy demonstrates the adaptability of combining different data sources and utilizing cutting-edge processes for precise segmentation in intricate settings.

In semantic segmentation methods, hybrid approaches are pivotal in refining segmentation outcomes (Kemker, Salvaggio, et al., 2018). The core concept behind this lies in recognizing the strong correlation among labels of adjacent pixels due to the spatial continuity of objects. This inherent correlation implies that neighboring segments are likely to be merged into a unified object when they share the same classification. Using pre-generated superpixels from SLIC, Längkvist et al. (2016) expanded on this idea and introduced a refining technique. In order to achieve this, the segmentation map has to be smoothed by adjusting the class of each pixel by taking into account the RGB channels and the average classifications of pixels inside each superpixel. Similar to this, Alshehhi et al. (2017) deftly employed spatial proximity and color similarity, integrating pixel coordinates into an initial segmentation of an image produced by SLIC. Their method combined segments of the same object that were not coherently segmented by using shape parameters such as density of neighboring superpixels, elongation, asymmetry, and compactness. Convolutional Neural Networks (CNN) and MRS were combined by Y. Sun et al. (2018) in an innovative approach. Superior semantic tagging was achieved by means of a rigorous data fusion procedure that combined point clouds with high-resolution images through a hybrid approach. The tactical incorporation greatly reduced the salt-and-pepper artifacts, improving object boundary accuracy.

7.3.3 Publicly Available Data for Semantic Segmentation of Remote Sensing Images

Semantic segmentation methods of high-resolution imagery often involve deep neural networks with millions of parameters, demanding a substantial volume of labeled examples for practical training. In computer vision, expansive datasets, such as ImageNet (Krizhevsky et al., 2012) with 1.28 million training images, have been instrumental. The conventional strategy involves constructing a deep network using a comprehensive dataset and fine-tuning it with a smaller dataset specific to the target problem (Pan et al., 2018). However, the challenge arises in remote sensing imagery, where the availability of large labeled datasets is limited. Unlike scenic images, the intricacies of remote sensing imagery necessitate specialized expertise for accurate delineation, making the crowdsourcing strategy less applicable. Popular public datasets utilized in existing studies, as presented in Table 7.1, highlight properties like the number of channels, classes, and spatial resolution. Remote sensing datasets manifest as singular images or tiles with reported sizes.

A practical solution to the lack of training examples is to create artificial images to supplement the training collection. For example, Kemker, Luu, et al. (2018) generated artificial multispectral images and labels using tools for digital imaging and remote sensing image generation. For training, each annotated image was divided into patches measuring 160 by 160 pixels. Although there are benefits to producing synthetic images, matching feature-space distributions between synthetic and actual images is hampered by the synthetic gap. In order to address this, Feng et al. (2019) carefully considered spectral similarity when choosing unlabeled data, adding high-similarity samples to the training set. The Generative Adversarial Network (GAN) model was utilized by Gao et al. (2019) in order to bridge the synthetic gap. Both labeled examples and unlabeled samples could be classified by the network thanks to the discriminator of the GAN, which produced label categories. In order to

TABLE 7.1
Available Public Datasets for Semantic Segmentation of Remote Sensing Images

Dataset	Source and Reference	Image Size (Channel)	# of Classes	Spatial Resolution	# of Tiles
Building and road	https://www.cs.toronto.edu/~vmnih/data/ (Mnih, 2013)	1500 x 1500 (3)	2	1 m	1352
Botswana	http://aviris.jpl.nasa.gov/data/free_data.html (Houshyaripour & Momeni, 2020)	1476 × 256 (145)	14	30 m	
ISPRS 2D Semantic Labeling—Vaihingen	https://www.isprs.org/education/benchmarks/UrbanSemLab/2d-sem-label-vaihingen.aspx (Zhu et al., 2021)	2000 × 2500 (3)	6	9 cm	33
ISPRS 2D Semantic Labeling—Potsdam	https://www.isprs.org/education/benchmarks/UrbanSemLab/2d-sem-label-potsdam.aspx (G. Yang et al., 2020)	6000 × 6000 (4)	6	5 cm	38
ISPRS 3D Semantic Labeling	https://www.isprs.org/education/benchmarks/UrbanSemLab/3d-semantic-labeling.aspx (Niemeyer et al., 2014)	753,859 pts	9	4 pst/m^2	–
Inria Aerial Image	https://project.inria.fr/aerialimagelabeling/ (Maggiori et al., 2017b)	1500 × 1500 (3)	2	30 cm	360
Indian Pines	https://purr.purdue.edu/publications/1947/1 (Baumgardner et al., 2015)	145 × 145 (200)	16	20 m	–
Salinas Scene	https://www.ehu.eus/ccwintco/index.php/Hyperspectral_Remote_Sensing_Scenes (Kemker & Kanan, 2017)	512 × 217 (204)	16	3.7 m	–
Pavia Center	https://www.ehu.eus/ccwintco/index.php/Hyperspectral_Remote_Sensing_Scenes (Barbato et al., 2022)	1096 × 715 (102)	9	1.3 m	–
Pavia University	https://www.ehu.eus/ccwintco/index.php/Hyperspectral_Remote_Sensing_Scenes (S. Liu et al., 2019)	1096 × 715 (102)	9	1.3 m	–
WHU Dataset	http://gpcv.whu.edu.cn/data/building_dataset.html (Hossain & Chen, 2022)	512 × 512 (3)	2	0.075 m	8,189

identify bands from hyperspectral images, Ghamisi et al. (2016) presented a self-improving convolutional neural network that uses particle swarm optimization. This optimization technique, based on concepts from fractional-order Darwinian theory, identified bands that are essential for network training. These novel approaches lessen the difficulty of hardly available labeled datasets when used to remote sensing photography. ResDilation and ResNet were used by MapNet (Gorbatsevich et al., 2020) to restore 3D models using GAN. Besides, ResiDualGAN (Y. Zhao et al., 2023) leveraged an in-network resizer module to tackle scale discrepancies within remote sensing image datasets. Furthermore, a strategic integration of a residual connection was employed to enhance the stability of real-to-real image translation.

7.3.4 Pros and Cons of Semantic Methods

In contrast to conventional methods, DL-based semantic approaches exhibit promising efficacy, especially as the resolution of remote sensing images continues to improve. This progress enables semantic methods to capture better the characteristics of high spatial resolution remote sensing images, enhancing feature recognition accuracy and reducing processing time. Standard semantic segmentation methods for remote sensing images are often optimized based on CNNs such as VGG, GoogLeNet, and ResNet. As satellites capture images with ever-increasing resolutions, the semantic segmentation of high-resolution remote sensing images becomes a focal point, albeit with substantial challenges. Maintaining consistency between segmentation results and semantic segmentation targets poses a challenge, as portrayed in Figure 7.9, urging researchers to enhance accuracy. Strategies include integrating deconvolution fusion structures, constructing deep fusion networks (DFN), and fusing shallow detailed information with deep semantic information. Incorporating spatial context information, as demonstrated in FCN with fully conditional random fields (CRF), aids boundary localization and improves the accuracy of the semantic method.

Balancing segmentation's effectiveness with time complexity remains another challenge, especially when leveraging rich feature and context information. Network modifications, like those proposed by L.-C. Chen et al. (2018), reduce complexity without compromising feature extraction capability, thus improving training speed and accuracy. Addressing the need for labeled training

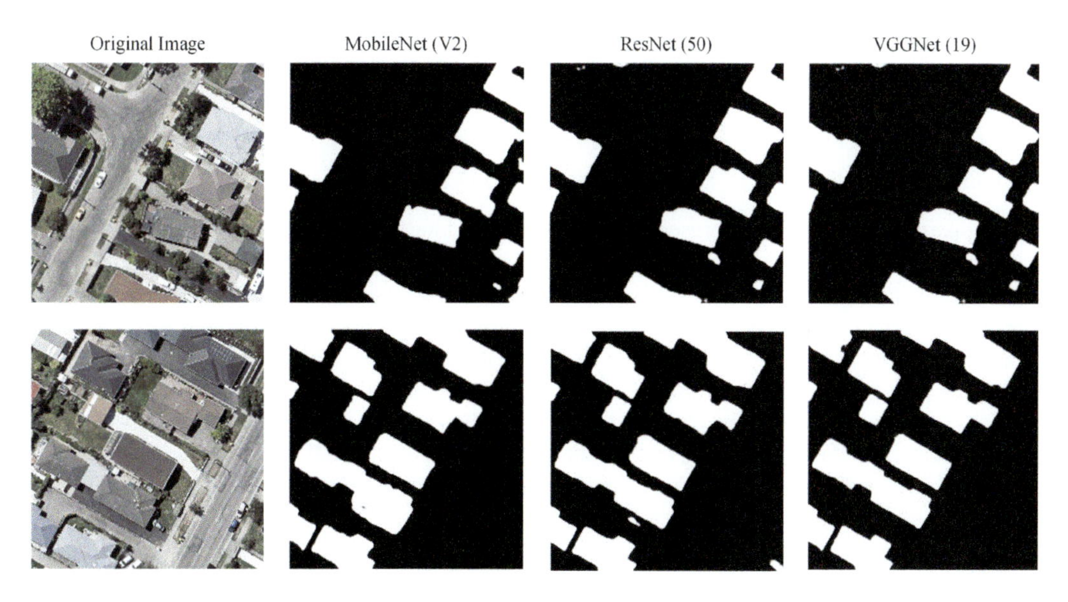

FIGURE 7.9 Semantic segmentation results with black representing the background and white representing the buildings.

samples, researchers explore weak supervised, semi-supervised, and unsupervised conditions. This shift reduces the laborious process of manual labeling while maintaining semantic segmentation accuracy comparable to fully supervised methods. Due to limited datasets for semantic segmentation of remote sensing images, studies often utilize existing datasets and fine-tune parameters during training for improved efficiency and model effectiveness. In the context of complex images, extracting features at different scales becomes crucial for DL-based semantic segmentation. Innovative encoder designs, such as those proposed by J. Zhang et al. (2019), leverage multi-scale features and convolution kernels of varying sizes, resulting in models that adeptly capture semantic information and exhibit high precision in segmentation and building extraction.

7.4 SEGMENTATION OF LiDAR DATA

LiDAR remote sensing provides a quick and affordable way to record and visualize the three-dimensional (3D) structure of surface objects using x, y, and z coordinates. Different 3D segmentation algorithms have been developed to identify features of the terrain from complex 3D point clouds. LiDAR remote sensing is becoming more popular due to the growing need for a 3D perspective of urban landscapes and the limitations of conventional photogrammetric methods, which need stereo images and labor-intensive processes for 3D surface synthesis. A work by Ramiya et al. (2017) on 3D segmentation-based building extraction from LiDAR point clouds serves as an example of image segmentation on LiDAR data. Outside of buildings, LiDAR can create 3D point clouds that accurately depict wooded landscapes. In order to classify trees as coniferous or deciduous based on segmentation on aerial LiDAR data, Hamraz et al. (2019) explore the use of deep learning. For image segmentation research, the raw LiDAR data can be transformed into DSM, Digital Elevation Models (DEM), and Digital Terrain Models (DTM). These include using LiDAR data to identify individual trees in intricate landscapes and mapping geomorphological features (Anders et al., 2011; Jakubowski et al., 2013).

LiDAR point clouds offer an extra layer of information and are widely used in semantic segmentation applications (W. Sun & Wang, 2018). In order to use CNNs to process LiDAR point clouds, points are usually gridding into raster images (Zhou & Gong, 2018). In particular, Arief et al. (2018) extracted the normalized elevation from the LiDAR point cloud, generating a two-dimensional elevation matrix, and suggested a network design that combined the stochastic depth approach with an atrous network to enforce regularization. Point clouds were converted into grayscale images by Zhou and Gong (2018), with each pixel denoting the quantized elevation of related points that were relevant to the ground.

7.5 SEGMENTATION OF HYPERSPECTRAL DATA

Hyperspectral Images (HSI) represent a distinctive data modality, boasting hundreds, if not thousands, of channels capturing intricate spectral information. According to Ball et al. (2017), there are difficulties using traditional DL frameworks for semantic segmentation of HSI because of this complexity. In order to address this, Geng et al. (2015) created a deep convolutional autoencoder using creative layer integration that included scale transformation, convolutional, sparse autoencoder-based, and post-processing layers. Reducing spectral bands and filling the gap with tri-color images is an obvious way to modify deep networks for HSI. Zhao and Du (2016) used two CNNs, one for spectral characteristics (a 1D CNN) and another for spatial features (a 2D CNN), to address spectral and spatial features independently. Yu et al. (2017) used higher dropout rates and average pooling to extract features across bands using 1 × 1 convolutional kernels. Y. Li et al. (2017) processed several bands using 3D convolutions by utilizing 3D CNNs for hyperspectral cubes. A 3D CNN network that manages spectral and spatial features simultaneously was developed by Paoletti et al. (2018). Fang et al. (2019) added 3D dilated convolutions to DenseNet, extending it with a spectrum attention method for HSI image categorization. A stacked convolutional auto-encoder model for HSI classification was presented by Kemker and Kanan (2017). This model uses unsupervised learning to extract spatial-spectral features. With no need for training, Xu et al. (2018) suggested a Random Patches Network that uses image patches as convolution kernels. A divide-and-conquer strategy was used by Feng et al. (2019) to develop a dual-architecture CNN that processes homogeneous and heterogeneous regions independently. In the former case, a fine-grained CNN learns hierarchical spectral characteristics, whereas a multi-scale CNN learns joint spatial-spectral features.

7.6 SEGMENTATION EVALUATION

Based on the assessment unit, evaluation techniques can be roughly divided into two categories: per-pixel and per-polygon methods (Stehman & Wickham, 2011). The goal of per-pixel techniques is to describe accuracy in terms of areas that are detected properly and inaccurately. Although map users are interested in area-based accuracy, the per-pixel technique has been criticized because of its vulnerability to positional inaccuracies and lack of a meaningful association with Earth's characteristics (Whiteside et al., 2014). Notwithstanding these concerns, many deep learning semantic segmentation methods (Audebert et al., 2016; Boonpook et al., 2021; G. Chen et al., 2018; Cheng et al., 2017; Diakogiannis et al., 2020; Papadomanolaki et al., 2019; Pedrayes et al., 2021; W. Sun & Wang, 2018; Y. Sun et al., 2018) rely on per-pixel accuracy assessments.

Per-polygon-based approaches further branch into supervised and unsupervised evaluations. Supervised methods, termed empirical discrepancy methods, measure the variance between computer-generated segments and reference polygons/objects of interest to determine dissimilarity (Corcoran et al., 2010). This entails collecting a set of ground-truth polygons, either through manual digitization or field measurements. Subsequently, multiple segmentations with varying parameter combinations are applied to the image. Similarity metrics, based on color, location, and geometric differences, quantify the discrepancy between each ground-truth polygon and its corresponding image segment(s) (Clinton et al., 2010; Ghorbanzadeh et al., 2020; Neubert et al., 2008; Tian & Chen, 2007). Determining the optimal segmentation involves selecting the one most similar to the reference digitization. While different similarity measures may yield varied "best" segmentations, considering results from multiple measures is often necessary (Jozdani & Chen, 2020). Notably, most studies employing supervised methods evaluate single-scale segmentations, with some exceptions exploring hierarchical, multi-scale segmentations (Johnson & Xie, 2011). However, the supervised approach has a significant drawback: creating a reference digitization can be challenging, subjective, and time-consuming (H. Zhang et al., 2008). It becomes impractical for large images or numerous datasets.

On the other hand, unsupervised methods, or empirical goodness methods, offer a more efficient and objective alternative for assessing segmentation quality. These methods permit quantitative evaluation without requiring reference digitization or intricate visual comparisons. Unsupervised evaluation involves scoring and ranking multiple image segmentations based on predetermined quality criteria, often aligned with human perception of a good segmentation (Chabrier et al., 2006). The definition of good segmentation includes criteria such as uniformity, distinctiveness, simplicity, and spatial accuracy of regions. While these criteria apply well to uniform and homogeneous regions, for textured or natural images, only the first two criteria are realistically applicable (H. Zhang et al., 2008). In remote sensing, an ideal segmentation should maximize intra-segment homogeneity and inter-segment heterogeneity. Most unsupervised methods involve calculating these measures for each segment and then aggregating them into a global value, determining an overall "goodness" score for the segmentation (Belgiu & Drăguţ, 2014). Yet, defining criteria for quantifying intra-segment homogeneity and inter-segment heterogeneity poses a challenge. Thus, supervised methods are widely used as a per-polygon approach.

Consideration of the application's purpose is vital. For applications focused on specific classes, such as building and tree extraction, geometric characteristics like position and shape are crucial (Belgiu & Drăguţ, 2014). In contrast, maximizing the correctly represented area in the final map becomes paramount for wall-to-wall land cover classification and mapping. Geometric and area-based methods can be employed, aiming to optimize the segmentation output that represents the largest amount of area for the corresponding polygon, enhancing overall map accuracy (Costa et al., 2018).

7.7 CONCLUDING REMARKS

Image segmentation is critical in GEOBIA and is a pivotal procedure in various remote sensing applications. The emergence of "big earth data" presents a challenge due to the sheer size of earth observation datasets, requiring integration with GEOBIA for effective analysis. When dealing with big earth data, it is evident that these concerns will remain at the forefront of the remote sensing community for years to come. Integrating semantic segmentation methodologies, mainly through Artificial Intelligence (AI) and DL, is a current focal point in the remote sensing community. The promising nature of AI, especially DL, in handling vast earth observation datasets and identifying patterns holds significant potential. However, challenges persist, such as the black-box nature of AI, reliance on labeled datasets, and the need for human expertise to ensure meaningful outcomes. Deep CNNs are potent tools for semantic segmentation, but reducing confusion among predicted classes without sacrificing spatial information remains a critical hurdle. Considering their more complex implementation compared to traditional approaches, a thorough understanding of image segmentation fundamentals is essential for successful application.

REFERENCES

Achanta, R., Shaji, A., Smith, K., Lucchi, A., Fua, P., & Süsstrunk, S. (2012). SLIC superpixels compared to state-of-the-art superpixel methods. *IEEE Transactions on Pattern Analysis and Machine Intelligence*, *34*(11), 2274–2281. https://doi.org/10.1109/TPAMI.2012.120

Akinlar, C., & Chome, E. (2016). PEL: A predictive edge linking algorithm. *Journal of Visual Communication and Image Representation*, *36*, 159–171. https://doi.org/10.1016/j.jvcir.2016.01.017

Alhichri, H., Alswayed, A. S., Bazi, Y., Ammour, N., & Alajlan, N. A. (2021). Classification of remote sensing images using efficientnet-B3 CNN model with attention. *IEEE Access*, *9*, 14078–14094. https://doi.org/10.1109/ACCESS.2021.3051085

Alshehhi, R., & Marpu, R. P. (2017). Hierarchical graph-based segmentation for extracting road networks from high-resolution satellite images. *ISPRS Journal of Photogrammetry and Remote Sensing*, *126*, 245–260. https://doi.org/10.1016/j.isprsjprs.2017.02.008

Alshehhi, R., Marpu, P. R., Woon, W. L., & Mura, M. D. (2017). Simultaneous extraction of roads and buildings in remote sensing imagery with convolutional neural networks. *ISPRS Journal of Photogrammetry and Remote Sensing*, *130*, 139–149. https://doi.org/10.1016/j.isprsjprs.2017.05.002

Anders, N. S., Seijmonsbergen, A. C., & Bouten, W. (2011). Segmentation optimization and stratified object-based analysis for semi-automated geomorphological mapping. *Remote Sensing of Environment*, *115*(12), 2976–2985. https://doi.org/10.1016/J.RSE.2011.05.007

Arief, H. A., Strand, G. H., Tveite, H., & Indahl, U. G. (2018). Land cover segmentation of airborne LiDAR data using stochastic atrous network. *Remote Sensing*, *10*, 973. https://doi.org/10.3390/RS10060973

Audebert, N., Le Saux, B., & Lefèvre, S. (2016). Semantic segmentation of earth observation data using multimodal and multi-scale deep networks. In S. Lai, V. Lepetit, K. Nishino & Y. Sato (Eds.), *Asian conference on computer vision* (pp. 180–196). Springer. https://doi.org/10.1007/978-3-319-54181-5_12

Audebert, N., Le Saux, B., & Lefèvre, S. (2018). Beyond RGB: Very high resolution urban remote sensing with multimodal deep networks. *ISPRS Journal of Photogrammetry and Remote Sensing*, *140*, 20–32. https://doi.org/10.1016/J.ISPRSJPRS.2017.11.011

Baatz, M., & Schäpe, A. (2000). Multi-resolution segmentation: An optimization approach for high quality multi-scale image segmentation. *Angewandte Geographische Informationsverarbeitung XII*, *58*, 12–23.

Badrinarayanan, V., Kendall, A., & Cipolla, R. (2017). SegNet: A deep convolutional encoder-decoder architecture for image segmentation. *IEEE Transactions on Pattern Analysis and Machine Intelligence*, *39*(12), 2481–2495. https://doi.org/10.1109/TPAMI.2016.2644615

Ball, J. E., Anderson, D. T., & Chan, C. S. (2017). Comprehensive survey of deep learning in remote sensing: Theories, tools, and challenges for the community. *Journal of Applied Remote Sensing*, *11*(4), 1. https://doi.org/10.1117/1.JRS.11.042609

Ballard, D. H. (1981). Generalizing the Hough transform to detect arbitrary shapes. *Pattern Recognition*, *13*(2), 111–122. https://doi.org/10.1016/0031-3203(81)90009-1

Barbato, M. P., Napoletano, P., Piccoli, F., & Schettini, R. (2022). Unsupervised segmentation of hyperspectral remote sensing images with superpixels. *Remote Sensing Applications: Society and Environment*, *28*, 100823. https://doi.org/10.1016/J.RSASE.2022.100823

Baumgardner, M. F., Biehl, L. L., & Landgrebe, D. A. (2015). 220 Band AVIRIS hyperspectral image data set: June 12, 1992 Indian Pine Test Site 3. https://doi.org/doi:/10.4231/R7RX991C

Beaulieu, J. M., & Goldberg, M. (1989). Hierarchy in picture segmentation: A stepwise optimization approach. *IEEE Transactions on Pattern Analysis and Machine Intelligence*, *11*(2), 150–163. https://doi.org/10.1109/34.16711

Belgiu, M., & Drăguţ, L. (2014). Comparing supervised and unsupervised multi-resolution segmentation approaches for extracting buildings from very high resolution imagery. *ISPRS Journal of Photogrammetry & Remote Sensing*, *96*, 67–75. https://doi.org/10.1016/j.isprsjprs.2014.07.002

Blaschke, T. (2010). Object based image analysis for remote sensing. *ISPRS Journal of Photogrammetry and Remote Sensing*, *65*(1), 2–16. https://doi.org/10.1016/j.isprsjprs.2009.06.004

Blaschke, T., Charles, B., & Pekkarinen, A. (2004). Remote sensing image analysis: Including the spatial domain. In F. D. de Jong & S. M. van der Meer (Eds.), *Remote sensing image analysis: Including the spatial domain* (pp. 211–236). Springer. https://doi.org/10.1017/S0032247400010123

Boonpook, W., Tan, Y., & Xu, B. (2021). Deep learning-based multi-feature semantic segmentation in building extraction from images of UAV photogrammetry. *International Journal of Remote Sensing*, *42*(1), 1–19. https://doi.org/10.1080/01431161.2020.1788742/FORMAT/EPUB

Byun, Y., Kim, D., Lee, J., & Kim, Y. (2011). A framework for the segmentation of high-resolution satellite imagery using modified seeded-region growing and region merging. *International Journal of Remote Sensing*, *32*(16), 4589–4609. https://doi.org/10.1080/01431161.2010.489066

Canny, J. (1987). A computational approach to edge detection. In *Readings in computer vision* (pp. 184–203). Elsevier. https://doi.org/10.1016/B978-0-08-051581-6.50024-6

Chabrier, S., Emile, B., Rosenberger, C., & Laurent, H. (2006). Unsupervised performance evaluation of image segmentation. *EURASIP Journal on Applied Signal Processing*, *2006*, 1–12. https://doi.org/10.1155/ASP/2006/96306

Chen, G., Zhang, X., Wang, Q., Dai, F., Gong, Y., & Zhu, K. (2018). Symmetrical dense-shortcut deep fully convolutional networks for semantic segmentation of very-high-resolution remote sensing images. *IEEE Journal of Selected Topics in Applied Earth Observations and Remote Sensing*, *11*(5), 1633–1644. https://doi.org/10.1109/JSTARS.2018.2810320

Chen, J., Deng, M., Mei, X., Chen, T., Shao, Q., & Hong, L. (2014). Optimal segmentation of a high-resolution remote-sensing image guided by area and boundary. *International Journal of Remote Sensing*, *35*(19), 6914–6939. https://doi.org/10.1080/01431161.2014.960617

Chen, L.-C., Papandreou, G., Kokkinos, I., Murphy, K., & Yuille, A. L. (2014). Semantic image segmentation with deep convolutional nets and fully connected CRFs. *ArXiv*, 1–12. https://arxiv.org/abs/1412.7062v4

Chen, L.-C., Papandreou, G., Schroff, F., & Adam, H. (2017). Rethinking atrous convolution for semantic image segmentation. *ArXiv*. http://arxiv.org/abs/1706.05587

Chen, L.-C., Zhu, Y., Papandreou, G., Schroff, F., & Adam, H. (2018). Encoder-decoder with atrous separable convolution for semantic image segmentation. *European Conference on Computer Vision (ECCV)*, *34*, 801–818.

Chen, Z., Zhao, Z., Gong, P., & Zeng, B. (2006). A new process for the segmentation of high resolution remote sensing imagery. *International Journal of Remote Sensing*, *27*(22), 4991–5001. https://doi.org/10.1080/01431160600658131

Cheng, D., Meng, G., Xiang, S., & Pan, C. (2017). FusionNet: Edge aware deep convolutional networks for semantic segmentation of remote sensing harbor images. *IEEE Journal of Selected Topics in Applied Earth Observations and Remote Sensing*, *10*(12), 5769–5783. https://doi.org/10.1109/JSTARS.2017.2747599

Clinton, N., Holt, A., Scarborough, J., Yan, L., & Gong, P. (2010). Accuracy assessment measures for object-based image segmentation goodness. *Photogrammetric Engineering & Remote Sensing*, *76*(3), 289–299. https://doi.org/10.14358/PERS.76.3.289

Comaniciu, D., & Meer, P. (2002). Mean shift: A robust approach toward feature space analysis. *IEEE Transactions on Pattern Analysis and Machine Intelligence*, *24*(5), 603–619. https://doi.org/10.1109/34.1000236

Corcoran, P., Winstanley, A., & Mooney, P. (2010). Segmentation performance evaluation for object-based remotely sensed image analysis. *International Journal of Remote Sensing*, *31*(3), 617–645. https://doi.org/10.1080/01431160902894475

Costa, H., Foody, G. M., & Boyd, D. S. (2018). Supervised methods of image segmentation accuracy assessment in land cover mapping. *Remote Sensing of Environment*, *205*(December 2016), 338–351. https://doi.org/10.1016/j.rse.2017.11.024

Csillik, O. (2017). Fast segmentation and classification of very high resolution remote sensing data using SLIC superpixels. *Remote Sensing*, *9*(243), 19. https://doi.org/10.3390/rs9030243

Deng, Y., & Manjunath, B. S. (2001). Unsupervised segmentation of color-texture regions in images and video. *IEEE Transactions on Pattern Analysis and Machine Intelligence*, *23*(8), 800–810. https://doi.org/10.1109/34.946985

Diakogiannis, F. I., Waldner, F., Caccetta, P., & Wu, C. (2020). ResUNet-a: A deep learning framework for semantic segmentation of remotely sensed data. *ISPRS Journal of Photogrammetry and Remote Sensing*, *162*, 94–114. https://doi.org/10.1016/J.ISPRSJPRS.2020.01.013

Dollár, P., & Zitnick, C. L. (2015). Fast edge detection using structured forests. *IEEE Transactions on Pattern Analysis and Machine Intelligence*, *37*(8), 1558–1570. https://doi.org/10.1109/TPAMI.2014.2377715

Dronova, I., Gong, P., Clinton, N. E., Wang, L., Fu, W., Qi, S., & Liu, Y. (2012). Landscape analysis of wetland plant functional types: The effects of image segmentation scale, vegetation classes and classification methods. *Remote Sensing of Environment*, *127*, 357–369. https://doi.org/10.1016/j.rse.2012.09.018

Fang, B., Li, Y., Zhang, H., & Chan, J. C. W. (2019). Hyperspectral images classification based on dense convolutional networks with spectral-wise attention mechanism. *Remote Sensing*, *11*, 159. https://doi.org/10.3390/RS11020159

Feng, J., Wang, L., Yu, H., Jiao, L., & Zhang, X. (2019). Divide-and-conquer dual-architecture convolutional neural network for classification of hyperspectral images. *Remote Sensing*, *11*, 484. https://doi.org/10.3390/RS11050484

Flores, J. L., Ayubi, G. A., Alonso, J. R., Fernández, A., Di Martino, J. M., & Ferrari, J. A. (2013). Edge linking and image segmentation by combining optical and digital methods. *Optik – International Journal for Light and Electron Optics*, *124*(18), 3260–3264. https://doi.org/10.1016/J.IJLEO.2012.10.036

Gaetano, R., Masi, G., Poggi, G., Verdoliva, L., & Scarpa, G. (2015). Marker-controlled watershed-based segmentation of multi-resolution remote sensing images. *IEEE Transactions on Geoscience and Remote Sensing*, *53*(6), 2987–3004. https://doi.org/10.1109/TGRS.2014.2367129

Gambotto, J.-P. (1993). A new approach to combining region growing and edge detection. *Pattern Recognition Letters*, *14*(11), 869–875. http://www.sciencedirect.com/science/article/B6V15-48M2KTS-4/2/80f077ecef7c33e7770e356be7e6ef54

Ganin, Y., & Lempitsky, V. (2014). N4-fields: Neural network nearest neighbor fields for image transforms. *Asian Conference on Computer Vision*, 536–551. https://doi.org/10.1007/978-3-319-16808-1_36

Gao, H., Yao, D., Wang, M., Li, C., Liu, H., Hua, Z., & Wang, J. (2019). A hyperspectral image classification method based on multi-discriminator generative adversarial networks. *Sensors*, *19*, 3269. https://doi.org/10.3390/S19153269

Geng, J., Fan, J., Wang, H., Ma, X., Li, B., & Chen, F. (2015). High-resolution SAR image classification via deep convolutional autoencoders. *IEEE Geoscience and Remote Sensing Letters*, *12*(11), 2351–2355. https://doi.org/10.1109/LGRS.2015.2478256

Ghamisi, P., Chen, Y., & Zhu, X. X. (2016). A self-improving convolution neural network for the classification of hyperspectral data. *IEEE Geoscience and Remote Sensing Letters*, *13*(10), 1537–1541. https://doi.org/10.1109/LGRS.2016.2595108

Ghita, O., & Whelan, P. F. (2002). Computational approach for edge linking. *Journal of Electronic Imaging*, *11*(4), 479. https://doi.org/10.1117/1.1501574

Ghorbanzadeh, O., Tiede, D., Wendt, L., Sudmanns, M., & Lang, S. (2020). Transferable instance segmentation of dwellings in a refugee camp – integrating CNN and OBIA. *European Journal of Remote Sensing*, *54*(sup1), 127–140. https://doi.org/10.1080/22797254.2020.1759456

Gorbatsevich, V., Kulgildin, B., Melnichenko, M., Vygolov, O., & Vizilter, Y. (2020). Semi-automatic cityscape 3D model restoration using generative adversarial network. *International Archives of the Photogrammetry, Remote Sensing and Spatial Information Sciences – ISPRS Archives*, *43*(B2), 415–420. https://doi.org/10.5194/isprs-archives-XLIII-B2-2020-415-2020

Guan, T., Zhou, D., Peng, K., & Liu, Y. (2015). A novel contour closure method using ending point restrained gradient vector flow field. *Journal of Information Science and Engineering*, *31*(1), 43–58.

Hamraz, H., Jacobs, N. B., Contreras, M. A., & Clark, C. H. (2019). Deep learning for conifer/deciduous classification of airborne LiDAR 3D point clouds representing individual trees. *ISPRS Journal of Photogrammetry and Remote Sensing*, *158*, 219–230. https://doi.org/10.1016/J.ISPRSJPRS.2019.10.011

Hay, G. J., Blaschke, T., Marceau, D. J., & Bouchard, A. (2003). A comparison of three image-object methods for the multi-scale analysis of landscape structure. *ISPRS Journal of Photogrammetry and Remote Sensing*, *57*(5–6), 327–345. https://doi.org/10.1016/S0924-2716(02)00162-4

He, K., Zhang, X., Ren, S., & Sun, J. (2016). Identity mappings in deep residual networks. In B. Leibe, J. Matas, N. Sebe & M. Welling (Eds.), *European conference on computer vision-ECCV 2016* (pp. 630–645). Springer International Publishing.

Hossain, M. D., & Chen, D. (2019). Segmentation for object-based image analysis (OBIA): A review of algorithms and challenges from remote sensing perspective. *ISPRS Journal of Photogrammetry and Remote Sensing*, *150*, 115–134. https://doi.org/10.1016/j.isprsjprs.2019.02.009

Hossain, M. D., & Chen, D. (2022). A hybrid image segmentation method for building extraction from high-resolution RGB images. *ISPRS Journal of Photogrammetry and Remote Sensing*, *192*, 299–314. https://doi.org/10.1016/j.isprsjprs.2022.08.024

Houshyaripour, A. H., & Momeni, M. (2020). Target spectrum based feature selection (TSFS): A new method based on chain coding for target detection problems. *Infrared Physics & Technology*, *109*, 103429. https://doi.org/10.1016/J.INFRARED.2020.103429

Howard, A. G., Zhu, M., Chen, B., Kalenichenko, D., Wang, W., Weyand, T., Andreetto, M., & Adam, H. (2017). MobileNets: Efficient convolutional neural networks for mobile vision applications. *ArXiv*, 1704.04861. https://doi.org/10.48550/arXiv.1704.04861

Huang, G., Liu, Z., Van Der Maaten, L., & Weinberger, K. Q. (2017). Densely connected convolutional networks. *2017 IEEE conference on computer vision and pattern recognition (CVPR)*, 2261–2269. https://doi.org/10.1109/CVPR.2017.243

Jain, R., Kasturi, R., & Schunck, B. G. (1995). *Machine vision*. McGraw-Hill. https://pdfs.semanticscholar.org/b471/764b7bb35abbcacb3e9f585d2031f4fddff9.pdf

Jakubowski, M. K., Li, W., Guo, Q., & Kelly, M. (2013). Delineating individual trees from lidar data: A comparison of vector- and raster-based segmentation approaches. *Remote Sensing*, *5*, 4163–4186. https://doi.org/10.3390/RS5094163

Ji, X., Zhang, X., & Zhang, L. (2013). Sequential edge linking method for segmentation of remotely sensed imagery based on heuristic search. *International Conference on Geoinformatics*, 1–5. https://doi.org/10.1109/Geoinformatics.2013.6626164

Jing, J., Liu, S., Wang, G., Zhang, W., & Sun, C. (2022). Recent advances on image edge detection: A comprehensive review. *Neurocomputing*, *503*, 259–271. https://doi.org/10.1016/j.neucom.2022.06.083

Johnson, B., & Xie, Z. (2011). Unsupervised image segmentation evaluation and refinement using a multi-scale approach. *ISPRS Journal of Photogrammetry and Remote Sensing*, *66*(4), 473–483. https://doi.org/10.1016/j.isprsjprs.2011.02.006

Jozdani, S., & Chen, D. (2020). On the versatility of popular and recently proposed supervised evaluation metrics for segmentation quality of remotely sensed images: An experimental case study of building extraction. *ISPRS Journal of Photogrammetry and Remote Sensing*, *160*, 275–290. https://doi.org/10.1016/j.isprsjprs.2020.01.002

Kavzoglu, T., & Tonbul, H. (2017). A comparative study of segmentation quality for multi-resolution segmentation and watershed transform. *Proceedings of 8th international conference on recent advances in space technologies, RAST 2017*, 113–117. https://doi.org/10.1109/RAST.2017.8002984

Kelkar, D., & Gupta, S. (2008). Improved quadtree method for split merge image segmentation. *2008 first international conference on emerging trends in engineering and technology*, 44–47. https://doi.org/10.1109/ICETET.2008.145

Kemker, R., & Kanan, C. (2017). Self-taught feature learning for hyperspectral image classification. *IEEE Transactions on Geoscience and Remote Sensing*, *55*(5), 2693–2705. https://doi.org/10.1109/TGRS.2017.2651639

Kemker, R., Luu, R., & Kanan, C. (2018). Low-shot learning for the semantic segmentation of remote sensing imagery. *IEEE Transactions on Geoscience and Remote Sensing*, *56*(10), 6214–6223. https://doi.org/10.1109/TGRS.2018.2833808

Kemker, R., Salvaggio, C., & Kanan, C. (2018). Algorithms for semantic segmentation of multi-spectral remote sensing imagery using deep learning. *ISPRS Journal of Photogrammetry and Remote Sensing*, *145*(A), 60–77. https://doi.org/10.1016/j.isprsjprs.2018.04.014

Kerem, S., & Ulusoy, I. (2013). Automatic multi-scale segmentation of high spatial resolution satellite images using watersheds. *Geoscience and Remote Sensing Symposium (IGARSS)*, 2505–2508.

Kiryati, N., & Eldar, Y. (1991). A probabilistic hough transform. *Pattern Recognition*, *24*(4), 303–316.

Kotaridis, I., & Lazaridou, M. (2021). Remote sensing image segmentation advances: A meta-analysis. *ISPRS Journal of Photogrammetry and Remote Sensing*, *173*, 309–322. https://doi.org/10.1016/j.isprsjprs.2021.01.020

Krizhevsky, A., Sutskever, I., & Hinton, G. E. (2012). ImageNet classification with deep convolutional neural networks. In F. Pereira, C. J. Burges, L. Bottou & K. Q. Weinberger (Eds.), *Advances in neural information processing systems* (pp. 1097–1105). Curran Associates. https://doi.org/10.1145/3065386

Lang, S. (2008). Object-based image analysis for remote sensing applications: Modeling reality—dealing with complexity. In *Object-based image analysis: Spatial concepts for knowledge-driven remote sensing applications* (pp. 3–27). Springer.

Längkvist, M., Kiselev, A., Alirezaie, M., Loutfi, A., Li, X., Saleh, R. A., & Thenkabail, P. S. (2016). Classification and segmentation of satellite orthoimagery using convolutional neural networks. *Remote Sensing*, *8*, 329. https://doi.org/10.3390/RS8040329

Levinshtein, A., Stere, A., Kutulakos, K. N., Fleet, D. J., Dickinson, S. J., & Siddiqi, K. (2009). TurboPixels: Fast superpixels using geometric flows. *IEEE Transactions on Pattern Analysis and Machine Intelligence*, *31*(12), 2290–2297. https://doi.org/10.1109/TPAMI.2009.96

Li, N., Huo, H., & Fang, T. (2010). A novel texture-preceded segmentation algorithm for high-resolution imagery. *IEEE Transactions on Geoscience and Remote Sensing*, *48*(7), 2818–2828.

Li, R., Liu, W., Yang, L., Sun, S., Hu, W., Zhang, F., & Li, W. (2018). DeepUNet: A deep fully convolutional network for pixel-level sea-land segmentation. *IEEE Journal of Selected Topics in Applied Earth Observations and Remote Sensing*, *11*(11), 3954–3962. https://doi.org/10.1109/JSTARS.2018.2833382

Li, X., Myint, S. W., Zhang, Y., Galletti, C., Zhang, X., & Turner, B. L. (2014). Object-based land-cover classification for metropolitan Phoenix, Arizona, using aerial photography. *International Journal of Applied Earth Observation and Geoinformation*, *33*, 321–330. https://doi.org/10.1016/J.JAG.2014.04.018

Li, Y., Zhang, H., & Shen, Q. (2017). Spectral–spatial classification of hyperspectral imagery with 3D convolutional neural network. *Remote Sensing*, *9*, 67. https://doi.org/10.3390/RS9010067

Li, Z., & Chen, J. (2015). Superpixel segmentation using Linear Spectral Clustering. *2015 IEEE conference on computer vision and pattern recognition (CVPR)*, 1356–1363. https://doi.org/10.1109/CVPR.2015.7298741

Liu, C., Xiao, Y., & Yang, J. (2017). A coastline detection method in polarimetric SAR images mixing the region-based and edge-based active contour models. *IEEE Transactions on Geoscience and Remote Sensing*, *55*(7), 3735–3747. https://doi.org/10.1109/TGRS.2017.2679112

Liu, J., Hossain, M. D., & Chen, D. (2021). A procedure for identifying invasive wild parsnip plants based on visible bands from UAV images. *The International Archives of the Photogrammetry, Remote Sensing and Spatial Information Sciences*, *XLIII*, 173–181. https://doi.org/10.5194/isprs-archives-XLIII-B1-2021-173-2021

Liu, M. Y., Tuzel, O., Ramalingam, S., & Chellappa, R. (2011). Entropy rate superpixel segmentation. *Proceedings of the IEEE computer society conference on computer vision and pattern recognition*, 2097–2104. https://doi.org/10.1109/CVPR.2011.5995323

Liu, S., Qi, Z., Li, X., & Yeh, A. G.-O. (2019). Integration of convolutional neural networks and object-based post-classification refinement for land use and land cover mapping with optical and SAR data. *Remote Sensing*, *11*(6), 690. https://doi.org/10.3390/RS11060690

Liu, Y., Ren, Q., Geng, J., Ding, M., & Li, J. (2018). Efficient patch-wise semantic segmentation for large-scale remote sensing images. *Sensors*, *18*, 3232. https://doi.org/10.3390/S18103232

Liu, Y. J., Yu, C. C., Yu, M. J., & He, Y. (2016). Manifold SLIC: A fast method to compute content-sensitive superpixels. *IEEE computer society conference on computer vision and pattern recognition*, 651–659. https://doi.org/10.1109/CVPR.2016.77

Long, J., Shelhamer, E., & Darrell, T. (2015). Fully convolutional networks for semantic segmentation. *IEEE conference on computer vision and pattern recognition (CVPR)*, 3431–3440. https://doi.org/10.1109/ICCVW.2019.00113

Lu, D.-S., & Chen, C.-C. (2008). Edge detection improvement by ant colony optimization. *Pattern Recognition Letters*, *29*(4), 416–425. https://doi.org/10.1016/J.PATREC.2007.10.021

Maggiori, E., Tarabalka, Y., Charpiat, G., & Alliez, P. (2017a). High-resolution aerial image labeling with convolutional neural networks. *IEEE Transactions on Geoscience and Remote Sensing*, *55*(12), 7092–7103. https://doi.org/10.1109/TGRS.2017.2740362

Maggiori, E., Tarabalka, Y., Charpiat, G., & Alliez, P. (2017b). Can semantic labeling methods generalize to any city? The inria aerial image labeling benchmark. *International geoscience and remote sensing symposium (IGARSS)*, 2017-July, 3226–3229. https://doi.org/10.1109/IGARSS.2017.8127684

Maintz, T. (2005). Segmentation. In *Digital and medical image processing*. Universiteit Utrecht. http://www.cs.uu.nl/docs/vakken/ibv/reader/chapter10.pdf

Manousakas, I. N., Undrill, P. E., Cameron, G. G., & Redpath, T. W. (1998). Split-and-merge segmentation of magnetic resonance medical images: Performance evaluation and extension to three dimensions. *Computers and Biomedical Research*, *31*(6), 393–412. https://doi.org/10.1006/CBMR.1998.1489

Marmanis, D., Schindler, K., Wegner, J. D., Galliani, S., Datcu, M., & Stilla, U. (2018). Classification with an edge: Improving semantic image segmentation with boundary detection. *ISPRS Journal of Photogrammetry and Remote Sensing*, *135*, 158–172. https://doi.org/10.1016/J.ISPRSJPRS.2017.11.009

Marmanis, D., Wegner, J. D., Galliani, S., Schindler, K., Datcu, M., & Stilla, U. (2016). Semantic segmentation of aerial images with an ensemble of CNNs. *ISPRS Annals of Photogrammetry, Remote Sensing and Spatial Information Sciences*, *III– 3*, 473–480. https://doi.org/10.5194/isprsannals-iii-3-473-2016

Marr, D., & Hildreth, E. (1980). Theory of edge detection. In *Proceedings of the royal society of London – biological sciences* (Vol. 207, Issue 1167, pp. 187–217). The Royal Society. https://doi.org/10.1098/rspb.1980.0020

Melin, P., Gonzalez, C. I., Castro, J. R., Mendoza, O., & Castillo, O. (2014). Edge-detection method for image processing based on generalized type-2 fuzzy logic. *IEEE Transactions on Fuzzy Systems*, *22*(6), 1515–1525. https://doi.org/10.1109/TFUZZ.2013.2297159

Mirghasemi, S., Rayudu, R., & Zhang, M. (2013). A new image segmentation algorithm based on modified seeded region growing and particle swarm optimization. *Image and Vision Computing*, *28*, 382–387.

Mnih, V. (2013). *Machine learning for aerial image labeling*. PhD Thesis. University of Toronto.

Moore, A. P., Prince, S. J. D., Warrell, J., Mohammed, U., & Jones, G. (2008). Superpixel lattices. *IEEE conference on computer vision and pattern recognition, CVPR*, 1–8. https://doi.org/10.1109/CVPR.2008.4587471

Mueller, M., Segl, K., & Kaufmann, H. (2004). Edge-and region-based segmentation technique for the extraction of large, manmade objects in high-resolution satellite imagery. *Pattern Recognition*, *37*, 1619–1628. https://doi.org/10.1016/j.patcog.2004.03.001

Neubert, M., Herold, H., & Meinel, G. (2008). Assessing image segmentation quality—concepts, methods and application. In G. Hay, S. Lang & G. H. T. Blaschke (Eds.), *Object-based image analysis: Spatial concepts for knowledge-driven remote sensing applications* (pp. 769–784). Springer.

Niemeyer, J., Rottensteiner, F., & Soergel, U. (2014). Contextual classification of lidar data and building object detection in urban areas. *ISPRS Journal of Photogrammetry and Remote Sensing*, *87*, 152–165. https://doi.org/10.1016/J.ISPRSJPRS.2013.11.001

Nikolaou, N., & Papamarkos, N. (2009). Color reduction for complex document images. *International Journal of Imaging Systems and Technology*, *19*, 14–26. https://doi.org/10.1002/ima.20174

Ninsawat, S., & Hossain, M. D. (2016). Identifying potential area and financial prospects of rooftop solar photovoltaics (PV). *Sustainability*, *8*(10), 1068. https://doi.org/10.3390/su8101068

Pal, N. R., & Pal, S. K. (1993). A review on image segmentation techniques. *Pattern Recognition*, *26*(9), 1277–1294. https://doi.org/10.1016/0031-3203(93)90135-J

Pan, B., Shi, Z., & Xu, X. (2018). MugNet: Deep learning for hyperspectral image classification using limited samples. *ISPRS Journal of Photogrammetry and Remote Sensing*, *145*, 108–119. https://doi.org/10.1016/J.ISPRSJPRS.2017.11.003

Paoletti, M. E., Haut, J. M., Plaza, J., & Plaza, A. (2018). A new deep convolutional neural network for fast hyperspectral image classification. *ISPRS Journal of Photogrammetry and Remote Sensing*, *145*, 120–147. https://doi.org/10.1016/J.ISPRSJPRS.2017.11.021

Papadomanolaki, M., Vakalopoulou, M., & Karantzalos, K. (2019). A novel object-based deep learning framework for semantic segmentation of very high-resolution remote sensing data: Comparison with convolutional and fully convolutional networks. *Remote Sensing*, *11*(6), 684. https://doi.org/10.3390/RS11060684

Pedrayes, O. D., Lema, D. G., García, D. F., Usamentiaga, R., & Alonso, Á. (2021). Evaluation of semantic segmentation methods for land use with spectral imaging using sentinel-2 and PNOA imagery. *Remote Sensing*, *13*, 2292. https://doi.org/10.3390/RS13122292

Perbet, F., & Maki, A. (2011). Homogeneous superpixels from random walks. *12th IAPR conference on machine vision applications, MVA 2011*, 26–30.

Prewitt, J. M. S. (1970). *Picture processing and psychopictorics* (B. S. Lipkin & A. Rosenfeld, Eds.). Academic Press Inc., Elsevier Science. https://books.google.ca/books?hl=en&lr=&id=vp-w_pC9JBAC&oi=fnd&pg=PA75&dq=Object+enhancement+and+extraction&ots=szCd-poxJb&sig=Dpy4Lxoen-6t_uI9BQSTWQZQT4M#v=onepage&q=Object enhancement and extraction&f=false

Ramiya, A. M., Nidamanuri, R. R., & Krishnan, R. (2017). Segmentation based building detection approach from LiDAR point cloud. *The Egyptian Journal of Remote Sensing and Space Science*, *20*(1), 71–77. https://doi.org/10.1016/J.EJRS.2016.04.001

Roberts, L. G. (1963). *Machine perception of three-dimensional solids*. Massachusetts Institute of Technology. http://dspace.mit.edu/handle/1721.1/11589

Ronneberger, O., Fischer, P., & Brox, T. (2015). U-net: Convolutional networks for biomedical image segmentation. In N. Navab, J. Hornegger, W. M. Wells & A. F. Frangi (Eds.), *Lecture notes in computer science* (pp. 234–241). Springer. https://doi.org/10.1007/978-3-319-24574-4_28

Sappa, A. D., & Vintimilla, B. X. (2008). Edge point linking by means of global and local schemes. In *Signal processing for image enhancement and multimedia processing* (pp. 115–125). Springer US. https://doi.org/10.1007/978-0-387-72500-0_11

Seo, S. (2018). Subpixel edge localization based on adaptive weighting of gradients. *IEEE Transactions on Image Processing*, *27*(11), 5501–5513. https://doi.org/10.1109/TIP.2018.2860241

Shen, J., Du, Y., Wang, W., & Li, X. (2014). Lazy random walks for superpixel segmentation. *IEEE Transactions on Image Processing*, *23*(4), 1451–1462. https://doi.org/10.1109/TIP.2014.2302892

Shui, P. L., & Zhang, W. C. (2012). Noise-robust edge detector combining isotropic and anisotropic Gaussian kernels. *Pattern Recognition*, *45*(2), 806–820. https://doi.org/10.1016/J.PATCOG.2011.07.020

Simonyan, K., & Zisserman, A. (2014). *Very deep convolutional networks for large-scale image recognition*. http://arxiv.org/abs/1409.1556

Sobel, I. E. (1970). *Camera models and machine perception*. Stanford University.

Som-ard, J., Hossain, M. D., Ninsawat, S., & Veerachitt, V. (2018). Pre-harvest sugarcane yield estimation using UAV-based RGB images and ground observation. *Sugar Tech*. https://doi.org/10.1007/s12355-018-0601-7

Stehman, S. V., & Wickham, J. D. (2011). Pixels, blocks of pixels, and polygons: Choosing a spatial unit for thematic accuracy assessment. *Remote Sensing of Environment*, *115*(12), 3044–3055. https://doi.org/10.1016/J.RSE.2011.06.007

Su, T., & Zhang, S. (2018). Multi-scale segmentation method based on binary merge tree and class label information. *IEEE Access*, *6*, 17801–17816. https://doi.org/10.1109/ACCESS.2018.2819988

Subudhi, S., Patro, R. N., Biswal, P. K., & Dell'acqua, F. (2021). A survey on superpixel segmentation as a preprocessing step in hyperspectral image analysis. *IEEE Journal of Selected Topics in Applied Earth Observations and Remote Sensing*, *14*, 5015–5035. https://doi.org/10.1109/JSTARS.2021.3076005

Sun, W., & Wang, R. (2018). Fully convolutional networks for semantic segmentation of very high resolution remotely sensed images combined with DSM. *IEEE Geoscience and Remote Sensing Letters*, *15*(3), 474–478. https://doi.org/10.1109/LGRS.2018.2795531

Sun, Y., Zhang, X., Xin, Q., & Huang, J. (2018). Developing a multi-filter convolutional neural network for semantic segmentation using high-resolution aerial imagery and LiDAR data. *ISPRS Journal of Photogrammetry and Remote Sensing*, *143*, 3–14. https://doi.org/10.1016/J.ISPRSJPRS.2018.06.005

Szegedy, C., Liu, W., Jia, Y., Sermanet, P., Reed, S., Anguelov, D., Erhan, D., Vanhoucke, V., & Rabinovich, A. (2015). Going deeper with convolutions. *2015 IEEE conference on computer vision and pattern recognition (CVPR)*, 1–9. https://doi.org/10.1109/CVPR.2015.7298594

Tan, M., & Le, Q. (2019). EfficientNet: Rethinking model scaling for convolutional neural networks. In K. Chaudhuri & R. Salakhutdinov (Eds.), *Proceedings of the 36th international conference on machine learning* (Vol. 97, pp. 6105–6114). PMLR. https://proceedings.mlr.press/v97/tan19a.html

Tian, J., & Chen, D.-M. (2007). Optimization in multi-scale segmentation of high-resolution satellite images for artificial feature recognition. *International Journal of Remote Sensing*, *28*(20), 4625–4644. https://doi.org/10.1080/01431160701241746

Tilton, J. C., Tarabalka, Y., Montesano, P. M., & Gofman, E. (2012). Best merge region-growing segmentation with integrated nonadjacent region object aggregation. *IEEE Transactions on Geoscience and Remote Sensing*, *50*(11 PART1), 4454–4467. https://doi.org/10.1109/TGRS.2012.2190079

Tsiotsios, C., & Petrou, M. (2013). On the choice of the parameters for anisotropic diffusion in image processing. *Pattern Recognition*, *46*(5), 1369–1381. https://doi.org/10.1016/J.PATCOG.2012.11.012

Van den Bergh, M., Boix, X., Roig, G., & Van Gool, L. (2015). SEEDS: Superpixels extracted via energy-driven sampling. *International Journal of Computer Vision*, *111*(3), 298–314. https://doi.org/10.1007/S11263-014-0744-2/FIGURES/15

Vedaldi, A., & Soatto, S. (2008). Quick shift and kernel methods for mode seeking. In D. Forsyth, P. Torr & A. Zisserman (Eds.), *European conference on computer vision—ECCV 2008* (pp. 705–718). Springer. https://doi.org/10.1007/978-3-540-88693-8_52

Verma, O. P., Hanmandlu, M., Susan, S., Kulkarni, M., & Jain, P. K. (2011). A simple single seeded region growing algorithm for color image segmentation using adaptive thresholding. *2011 international conference on communication systems and network technologies*, 500–503. https://doi.org/10.1109/CSNT.2011.107

Wang, M., & Li, R. (2014). Segmentation of high spatial resolution remote sensing imagery based on hard-boundary constraint and two-stage merging. *IEEE Transactions on Geoscience and Remote Sensing*, *52*(9), 5712–5725. https://doi.org/10.1109/TGRS.2013.2292053

Wang, M., Liu, X., Gao, Y., Ma, X., & Soomro, N. Q. (2017). Superpixel segmentation: A benchmark. *Signal Processing: Image Communication*, *56*, 28–39. https://doi.org/10.1016/J.IMAGE.2017.04.007

Wang, Y., Meng, Q., Qi, Q., Yang, J., & Liu, Y. (2018). Region merging considering within- and between-segment heterogeneity: An improved hybrid remote-sensing image segmentation method. *Remote Sensing*, *10*(5), 781. https://doi.org/10.3390/rs10050781

Wang, Z., Jensen, J. R., & Im, J. (2010). An automatic region-based image segmentation algorithm for remote sensing applications. *Environmental Modelling and Software*, *25*(10), 1149–1165. https://doi.org/10.1016/j.envsoft.2010.03.019

Whiteside, T. G., Maier, S. W., & Boggs, G. S. (2014). Area-based and location-based validation of classified image objects. *International Journal of Applied Earth Observation and Geoinformation*, *28*, 117–130. https://doi.org/10.1016/J.JAG.2013.11.009

Wu, G., Shao, X., Guo, Z., Chen, Q., Yuan, W., Shi, X., Xu, Y., & Shibasaki, R. (2018). Automatic building segmentation of aerial imagery using multi-constraint fully convolutional networks. *Remote Sensing*, *10*, 407. https://doi.org/10.3390/RS10030407

Wu, L., Wang, Y., Long, J., & Liu, Z. (2015). A non-seed-based region growing algorithm for high resolution remote sensing image segmentation. In Y. J. Zhang (Ed.), *Image and graphics. ICIG 2015*. Lecture Notes in Computer Science (pp. 263–277). Springer.

Xie, S., & Tu, Z. (2015). Holistically-nested edge detection. *International Journal of Computer Vision*, 1395–1403. https://doi.org/10.1007/s11263-017-1004-z

Xu, Y., Du, B., Zhang, F., & Zhang, L. (2018). Hyperspectral image classification via a random patches network. *ISPRS Journal of Photogrammetry and Remote Sensing*, *142*, 344–357. https://doi.org/10.1016/J.ISPRSJPRS.2018.05.014

Yang, G., Zhang, Q., & Zhang, G. (2020). EANet: Edge-aware network for the extraction of buildings from aerial images. *Remote Sensing*, *12*(13), 2161. https://doi.org/10.3390/rs12132161

Yang, J., He, Y., & Caspersen, J. (2016). A self-adapted threshold-based region merging method for remote sensing image segmentation. *International Geoscience and Remote Sensing Symposium (IGARSS)*, 6320–6323.

Yang, J., He, Y., & Caspersen, J. (2017). Region merging using local spectral angle thresholds: A more accurate method for hybrid segmentation of remote sensing images. *Remote Sensing of Environment*, *190*, 137–148. https://doi.org/10.1016/j.rse.2016.12.011

Yang, J., Price, B., Cohen, S., Lee, H., & Yang, M. H. (2016). Object contour detection with a fully convolutional encoder-decoder network. *IEEE computer society conference on computer vision and pattern recognition*, 193–202. https://doi.org/10.1109/CVPR.2016.28

Yi, S., Labate, D., Easley, G. R., & Krim, H. (2009). A shearlet approach to edge analysis and detection. *IEEE Transactions on Image Processing*, *18*(5), 929–941. https://doi.org/10.1109/TIP.2009.2013082

Yousefhussien, M., Kelbe, D. J., Ientilucci, E. J., & Salvaggio, C. (2018). A multi-scale fully convolutional network for semantic labeling of 3D point clouds. *ISPRS Journal of Photogrammetry and Remote Sensing*, *143*, 191–204. https://doi.org/10.1016/J.ISPRSJPRS.2018.03.018

Yu, S., Jia, S., & Xu, C. (2017). Convolutional neural networks for hyperspectral image classification. *Neurocomputing*, *219*, 88–98. https://doi.org/10.1016/J.NEUCOM.2016.09.010

Yuan, X., Shi, J., & Gu, L. (2021). A review of deep learning methods for semantic segmentation of remote sensing imagery. *Expert Systems with Applications*, *169*(June 2020), 114417. https://doi.org/10.1016/j.eswa.2020.114417

Zhang, C., Liu, J., & Yu, F. (2018). Segmentation model based on convolutional neural networks for extracting vegetation from Gaofen-2 images. *Journal of Applied Remote Sensing*, *12*(4), 1. https://doi.org/10.1117/1.jrs.12.042804

Zhang, H., Fritts, J. E., & Goldman, S. A. (2008). Image segmentation evaluation: A survey of unsupervised methods. *Computer Vision and Image Understanding*, *110*(2), 260–280. https://doi.org/10.1016/j.cviu.2007.08.003

Zhang, J., Jin, Q., Wang, H., Da, C., Xiang, S., & Pan, C. (2019). Semantic segmentation on remote sensing images with multi-scale feature fusion. *Journal of Computer-Aided Design and Computer Graphics*, *31*(9), 1509–1517. https://doi.org/10.3724/SP.J.1089.2019.17645

Zhang, X., Xiao, P., Feng, X., Wang, J., & Wang, Z. (2014). Hybrid region merging method for segmentation of high-resolution remote sensing images. *ISPRS Journal of Photogrammetry & Remote Sensing*, *98*, 19–28. https://doi.org/10.1016/j.isprsjprs.2014.09.011

Zhang, X., Zhou, X., Lin, M., & Sun, J. (2018). ShuffleNet: An extremely efficient convolutional neural network for mobile devices. *2018 IEEE/CVF conference on computer vision and pattern recognition*, 6848–6856. https://doi.org/10.1109/CVPR.2018.00716

Zhang, Y., Hartley, R., Mashford, J., & Burn, S. (2011). Superpixels via pseudo-Boolean optimization. *Proceedings of the IEEE international conference on computer vision*, 1387–1394. https://doi.org/10.1109/ICCV.2011.6126393

Zhang, Z., Liu, Q., & Wang, Y. (2018). Road extraction by deep residual U-net. *IEEE Geoscience and Remote Sensing Letters*, *15*(5), 749–753. https://doi.org/10.1109/LGRS.2018.2802944

Zhao, W., & Du, S. (2016). Spectral-spatial feature extraction for hyperspectral image classification: A dimension reduction and deep learning approach. *IEEE Transactions on Geoscience and Remote Sensing*, *54*(8), 4544–4554. https://doi.org/10.1109/TGRS.2016.2543748

Zhao, Y., Guo, P., Sun, Z., Chen, X., & Gao, H. (2023). ResiDualGAN: Resize-residual DualGAN for cross-domain remote sensing images semantic segmentation. *Remote Sensing 2023*, *15*(5), 1428. https://doi.org/10.3390/RS15051428

Zhou, Z., & Gong, J. (2018). Automated residential building detection from airborne LiDAR data with deep neural networks. *Advanced Engineering Informatics*, *36*, 229–241. https://doi.org/10.1016/J.AEI.2018.04.002

Zhu, Y., Liang, Z., Yan, J., Chen, G., & Wang, X. (2021). E-D-Net: Automatic building extraction from high-resolution aerial images with boundary information. *IEEE Journal of Selected Topics in Applied Earth Observations and Remote Sensing*, *14*, 4595–4606. https://doi.org/10.1109/JSTARS.2021.3073994

8 Object-Based Image Analysis
Evolution, History, State of the Art, and Future Vision

Thomas Blaschke, Maggi Kelly, and Helena Merschdorf

LIST OF ACRONYMS

CART	Classification and regression trees
DAIS	Digital airborne imaging system
GEOBIA	Geographic object based image analysis
GEOSS	Global Earth observing system of systems
GIS	Geographic information systems
GMES	Global monitoring of environment and security
IT	Information technology
LiDAR	Light detection and ranging
OBIA	Object based image analysis
UAS	Unmanned aerial systems
UAV	Unmanned aerial vehicle
WoS	Web of science

8.1 INTRODUCTION

Remote Sensing, what it is and what it be can used for is laid out in various chapters of this comprehensive book. We may only state here that remote sensing has a short history—when compared to traditional disciplines such as mathematics or physics. Contrarily, we may state that it has a long history when we compare it to recent internet-based technology like social media or, closer to our field, the tracking of people and moving objects by means of cell phone signals. Remote sensing has been a domain for specialists for many years and to some degree it still is. Similarly, Geographic Information Systems (GIS) has for years been a field where professionals worked on designated workstations while not being fully integrated in standard corporate Information Technology (IT) infrastructures. The latter changed more than a decade ago while for remote sensing only recently, and one may still witness remnants of historical developments of RS specific hardware and software. The dominant concept in remote sensing has been the pixel, while GIS functionality has always been somehow splintered into the raster and vector domains. Blaschke and Strobl (2001) provocatively raised the question "What's wrong with pixels?" having identified an increasing dissatisfaction with pixel-by-pixel image analysis. Although this critique was not new (Gui et al., 2024; Jiang et al., 2023; Putra and Wijayanto, 2023; Ye et al., 2023; Li et al., 2022; Liu et al., 2018; Hossain and Chen, 2019; Belgiu and Csillik, 2018; Chen et al., 2018; Deng et al., 2018; Ye et al., 2018; Zhang et al., 2018; Leichtle et al., 2017; Long et al., 2017; Zhao et al., 2017; Cracknell, 1998; see also Blaschke and Strobl, 2001; Burnett and Blaschke, 2003; Blaschke, 2010; Blaschke et al., 2014 for a more thorough discussion) these authors described a need for applications "beyond pixels" and for specific methods and methodologies which support this.

DOI: 10.1201/9781003541158-9

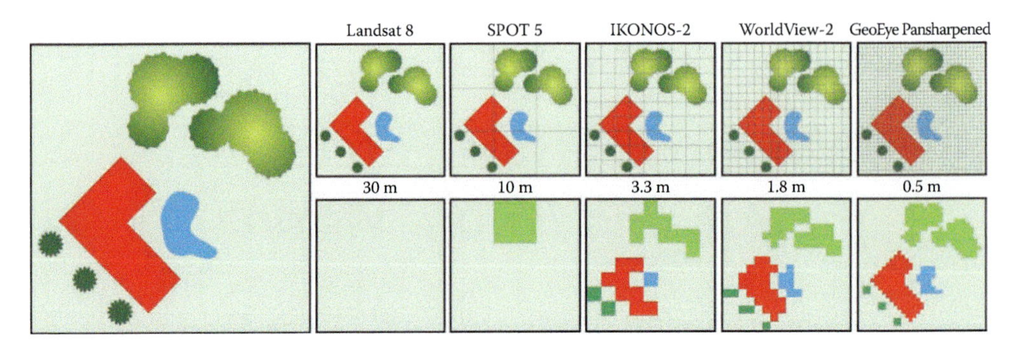

FIGURE 8.1 Objects and resolutions: OBIA methods are associated with the notion of high resolution—whereby "high" has always to be seen in context.

Over the last years the number of applications which conceptually aim for objects—still built on the information of the underlying pixels—rose quickly. Blaschke et al. (2014) identified a high number of relevant publications which use—with some degree of fuzziness in their terminology—the concept of object based image analysis (OBIA). They even claim that this concept and its instantiation to a particular order of scale—the "Geographic" as opposed to applications in medical imaging or cell biology—is a new paradigm in remote sensing. For this level of scale and the "geo-domain" this paradigm is then referred to by some scholars as Geographic Object Based Image Analysis (GEOBIA), while the generic principles—the multi-scale segmentation and object handling may generically be called OBIA. Other sources use the more generic term of OBIA when referring to the geospatial domain also, and Blaschke et al. refer to Kuhn (1962) stating that an inconsistent use of terminology can be expected for a new paradigm. Nevertheless, it is high time to consolidate this terminology and to support a coherent usage of terms and naming conventions—after having agreed upon the concepts and the conception of the overall approach.

This chapter therefore briefly explains OBIA methods as used in the geospatial domain and elsewhere. We will start from the quest to partitioning geospatial data into meaningful image-objects, and the needs and possibilities to assessing their characteristics through spatial, spectral and temporal scale. At its most fundamental level, OBIA requires image segmentation, attribution, classification and the ability to query and link individual objects (a.k.a. segments) in space and time. We will elucidate the evolution of this approach, its relatively short history and its older origins. Instead of a comprehensive state-of-the-art analysis we refer to the key literature and try to summarize the core concepts for the reader in an understandable way, with a particular emphasis on a common nomenclature, definitions, and reporting procedures. Ultimately, we will ask where this development will lead to in terms of applications, research questions and needs in education, training and professional workforce development, and we conclude with the main advances and recommendations for future work.

8.2 HISTORY OF OBIA

8.2.1 Intellectual Roots

8.2.1.1 Conceptual Foundations

The conceptual foundations of OBIA are rooted in the 1960s with pre-digital aerial photography. The spatial information found in digital imagery that is harnessed in the object-based approach, for example: image texture, contextual information, pixel proximity, and geometric attributes of features were discussed in the 1960s as possible components to yet possible automation of photo interpretation. In his seminal work on aerial photography and early remote sensing applications, Colwell (1965) describes the photo interpretation process as the act of examining photographic images for the purpose of identifying objects and judging their significance. He said that photo interpretation involves the observation of "size, shape, shadow, tone, texture, pattern and location" of features, as well as the significance of the features, based largely on their "interrelationships or association" (Colwell, 1965). His assessment of the potential for automation of an object recognition process depended on the capacities of a digital scanner and the ability of an algorithm to assess the differences, in photographic tone, between a "blob" and its surroundings (Colwell, 1964, 1965). Colwell was an important advisor on the Landsat 1 mission, and his ideas on extraction of meaningful features transferred to his ambitions for the satellite missions (Colwell, 1973).

8.4.1.2 Image Segmentation

Image segmentation is the division of an image into different regions, each having certain properties, and it provides the building blocks of object based image analysis (Blaschke, 2010). The desire expressed by Colwell and others in the 1960s to more automatically delineate meaningful features, objects, or "blobs" in his early terminology launched numerous approaches to image segmentation that rapidly advanced in the 1980s. It is widely agreed that the segmentation algorithms implemented in the OBIA software of today owe a debt to theoretical and applied work in the 1970s and 1980s that developed and refined numerous methods for image segmentation (Blaschke et al., 2004; Blaschke, 2010). Early key papers for the remote sensing field include Kettig and Landgrebe (1976) who presented experimental results in segmentation of Landsat-1 (ERTS-1) imagery, and McKeown et al. (1989) who developed a knowledge-based system with image segmentation and classification tools designed for semi-automated photo interpretation of aerial photographs. Key reviews are provided in numerous papers (Gui et al., 2024; Jiang et al., 2023; Putra and Wijayanto, 2023; Ye et al., 2023; Li et al., 2022; Liu et al., 2018; Hossain and Chen, 2019; Belgiu and Csillik, 2018; Chen et al., 2018;

TABLE 8.1
Overview of Major Groups of Image Segmentation Techniques

	Main Issues	Strengths	Weaknesses
Thresholding and Clustering	Threshold values are applied globally (to the whole image), or locally (applied to sub-regions)	Most thresholding algorithms are computationally simple. Clustering an image or a raster may be intuitive for a given number of clusters.	The results depend on the initial set of clusters and user values, or thresholds, respectively.
Edge-detection	Boundaries of object or regions under consideration and edges are assumed to be closely related, since there are often sharp differences in intensity at the region boundaries.	Discontinuities are identified across the array of values studied. Particularly suited for internally relatively homogeneous objects such as buildings, roads or water bodies.	Edges identified by edge detection are often disconnected. To segment an object from an image however, one needs closed region boundaries. Typically problematic in objects with high internal heterogeneity such as forests.
Region growing	Starting from the assumption that the neighboring pixels within one region have similar values a similarity criterion is defined and applied in to neighboring pixels		Selection of the similarity criterion significantly influences the results.

Deng et al., 2018; Ye et al., 2018; Zhang et al., 2018; Leichtle et al., 2017; Long et al., 2017; Zhao et al., 2017; Fu and Mui, 1981; Haralick and Shapiro, 1985; Pal and Pal, 1993). Building on that work, image segmentation techniques implemented today include those focused on thresholding or clustering, edge detection, region extraction and growing, and some combination of these have been explored since the 1970s (Fu and Mui, 1981; Blaschke, 2010).

8.2.2 Needs and Driving Forces

With a focus on geospatial data, OBIA has particular needs that were not anticipated by its antecedents. The OBIA methods were driven first by the need to more accurately map multi-scaled Earth features with high spatial-resolution imagery such as the tree, the building, and the field. Following that, the spatial dimension of objects (distances, pattern, neighborhoods, topologies) was mined for classification accuracy (e.g., Guo et al., 2007). Most recently, the OBIA field has been characterized by discussions of object semantics within fixed or emergent ontologies (Ye et al., 2023; Li et al., 2022; Belgiu and Csillik, 2018; Leichtle et al., 2017; Long et al., 2017; Zhao et al., 2017; Yue et al., 2013; Arvor et al., 2013), and by the need for interoperability between OBIA and GIS and spatial modeling frameworks (Harvey and Ruskin, 2011; Yue et al., 2013). The OBIA approach has evolved from a method of convenience to what has been called a new paradigm in remote sensing and spatial analysis (Blaschke et al., 2014).

8.2.3 GEOBIA Developments

8.2.3.1 Emergence (1999–2003/04)

The emergence of OBIA has been written about extensively elsewhere (e.g., Blaschke, 2010; Blaschke et al., 2014), and had it largest boost from the availability of satellite imagery of increasing spatial resolution such as IKONOS (1–4 m), QuickBird (resolution), and OrbView (resolution) sensors (launched in 1999, 2001, and 2003 respectively) (Blaschke, 2010). This ready availability of high-resolution multi-band imagery coincided with increasing awareness in the remote sensing literature that novel methods to extract meaningful and more accurate results were critically needed. The "business-as-usual" pixel-based algorithms were not reliable with imagery exhibiting high local variability and obvious spatial context (Cracknell, 1998; Townshend et al., 2000; Blaschke and Strobl, 2001).

Importantly, the software package called "eCognition" from the company "Definiens" (subsequently called "Definiens Earth Science") became commercially available in 2000. This event is marked as a milestone in the emergence of a body of work on OBIA, as it was the first commercially available, object based, image analysis software (Flanders et al., 2003; Benz et al., 2004; Blaschke, 2010), and many peer-reviewed papers during this phase relied on the software. The eCognition software is built on to the approach originally known as Fractal Net Evolution (Blaschke, 2010) that is not easily, nor often described in detail in early papers that relied on the software.

Representative papers demonstrating the utility of the newly released software from this timeframe include the following. Flanders et al. (2003) evaluated the object-based approach from eCognition software and classified forest clearings and forest structure elements in British Columbia, Canada, using a Landsat enhanced thematic mapper plus (ETM+) image. They found that forest clearings as well as forest growth stage, water, and urban features were classified with significantly higher accuracy than using a traditional pixel-based method. With slightly different results, Dorren et al. (2003) also compared pixel- and object-based classification of forest stands using Landsat imagery in Austria. They used eCogntion for the object-based approach and found that while the pixel-based method provided slightly better accuracies, the object-based approach was more realistic, and better served the needs of local foresters. Benz et al. (2004) used eCognition to update urban maps (buildings, roofs, etc.) from high-resolution (0.5 m) RGB aerial orthoimages in Austria. Theirs was an early and comprehensive examination of the use of the software, and they discussed

TABLE 8.2
Overview of Early Application Fields

Application	Images	Comparison to Pixel-Based/Findings	Software
Forest clearings and forest structure (Flanders et al., 2003)	Landsat enhanced thematic mapper plus (ETM+) image	Significantly higher accuracy of OBIA compared to pixel-based	eCognition
Forests Dorren at al. (2003)	Landsat	Pixel-based method provided slightly better accuracies, but the object-based approach was more realistic, and better served the needs of local foresters	eCognition
Update urban maps Benz et al. (2004)	high-resolution (0.5 m) RGB aerial ortho-images	Comprehensive examination of the use of the software. Discussed—for example the importance of semantic features, and uncertainties in representation	eCognition
Shrub cover and rangeland characteristics Laliberte et al. (2004)	Historic (1937–1996) scanned aerial photos and a contemporary QuickBird satellite image		eCognition
Correlate field-derived forest inventory parameters and image objects Chubey et al. (2006)	IKONOS-2 imagery	The strongest relationships were found for discrete land-cover types, species composition, and crown closure	eCognition
Vegetation inventory Yu et al. (2006)	Digital Airborne Imaging System (DAIS) imagery	The object-based approach outperformed the pixel-based approach	
Map surface coal fires Yan et al. (2006)	ASTER image (15m resolution)	The OBIA approach yielded classifications of marked improvement over the pixel-based approach	

numerous aspects of the OBIA approach that are still actively discussed today—for example the importance of semantic features, and uncertainties in representation. Laliberte et al. (2004) used a combination of historic (1937–1996) scanned aerial photos and a contemporary QuickBird satellite image to map shrub cover and rangeland characteristics over time. Ecognition was critical in their workflow. Chubey et al. (2006) used eCognition to segment IKONOS-2 imagery, and decision tree analysis to correlate field-derived forest inventory parameters and image objects for forests in Alberta, Canada. They found that the strongest relationships were found for discrete land-cover types, species composition, and crown closure. While much work focused on the use of eCognition for high resolution imagery, not all work in this phase did. Many papers explored the method using Landsat imagery (e.g., Putra and Wijayanto, 2023; Hossain and Chen, 2019; Belgiu and Csillik, 2018; Long et al., 2017; Zhao et al., 2017; Dorren et al., 2003).

8.2.3.2 Establishment (2005–2010)

Accuracy. Many papers during this timeframe focused on proving the utility of the new approach, and provided comparisons between OBIA and pixel-based classifiers (Li et al., 2022; Liu et al., 2018; Chen et al., 2018; Deng et al., 2018; Zhao et al., 2017; Yan et al., 2006; Cleve et al., 2008; Maxwell, 2010). For example, Yu et al. (2006) used high spatial resolution airborne Digital Airborne Imaging System (DAIS) imagery and associated topographic data the Point Reyes National Seashore in California USA for a comprehensive and detailed vegetation inventory at the alliance level. The object-based approach outperformed the pixel-based approach. Yan et al. (2006) compared

TABLE 8.3
Development of OBIA Application Fields

Application Area	Examples
Forests	Feng et al. (2023)
Individual trees	Guo et al. (2007) Yang et al. (2022)
Forest stands	Filippi et al. (2022)
Parklands	Yu et al. (2006)
Rangelands	Laliberte et al. (2007)
Wetlands and other critical habitat	
Urban areas	Cleve et al. (2008)
Land use and land cover	Maxwell (2010)
Public health	Mudau and Mhangara (2023)
Disease vector habitats	Troyo et al. (2009)
Public health infrastructure (e.g., refugee camps)	Gao et al. (2022)
Hazard vulnerability and disaster aftermath	Belcore et al. (2022)

pixel- and object-based classification of an ASTER image (15 m resolution) to map surface coal fires and coal piles. The OBIA approach yielded classifications of marked improvement over the pixel-based approach. Similar results were shown using high resolution aerial imagery for urban features (Cleve et al., 2008), Landsat imagery and land cover (Maxwell, 2010).

Applications. From 2005—2010 there was a wide net cast around OBIA application areas. The following table provides an overview of the various application areas, which emerged over these years.

Capturing, attributing, and understanding changing landscapes continues to be a primary research area in remote sensing, and the use of OBIA methods for studying and understanding change were increasingly popular during this period. In a comprehensive review article, Chen et al. (2012) presented a timely overview of the main issues in remote sensing change detection, and suggested reasons for favoring object-based change detection over pixel-based approaches. They suggested that an object-based approach to change detection allows for multiscale analysis to optimize the delineation of individual landscape features, it reduces spurious changes due to high spectral variability in high-spatial-resolution imagery, and the approach also allows for more meaningful ways to evaluate change.

Data Fusion. Data fusion became increasingly common during this phase. The utility of light detection and ranging (LiDAR) data for capturing height and that could be used in both segmentation and classification was recognized soon after LiDAR became somewhat operational. Pascual et al. (2008) incorporated LiDAR data to help characterize forest stands and structure using OBIA in a complex *Pinus sylvestris* dominated forest in central Spain. Zhou and Troy (2008) used LiDAR with high-resolution digital aerial imagery to analyze and characterize the urban landscape structure of Baltimore at the parcel level. Ebert et al. (2009) used optical, LiDAR and digital elevation models to estimate social vulnerability indicators through the use of physical characteristics and hazard potential. Tullis et al. (2010) found that certain land covers (e.g., forest and herbaceous cover, rather than impervious surface) benefited more from a synergy between LIDAR and optical imagery. Image fusion has also involved multiple spatial and spectral resolutions. For example, Aneece et al. (2022) used both Quickbird and Hyperion hyper spectral imagery to map an invasive plant species in the Galapagos Islands. The fusion of multi- and hyper-spectral imagery was beneficial.

Software. During this timeframe, papers evolved from naive and sometimes simplistic use of complicated software (e.g., “we used eCognition to segment and classify our imagery”) to more

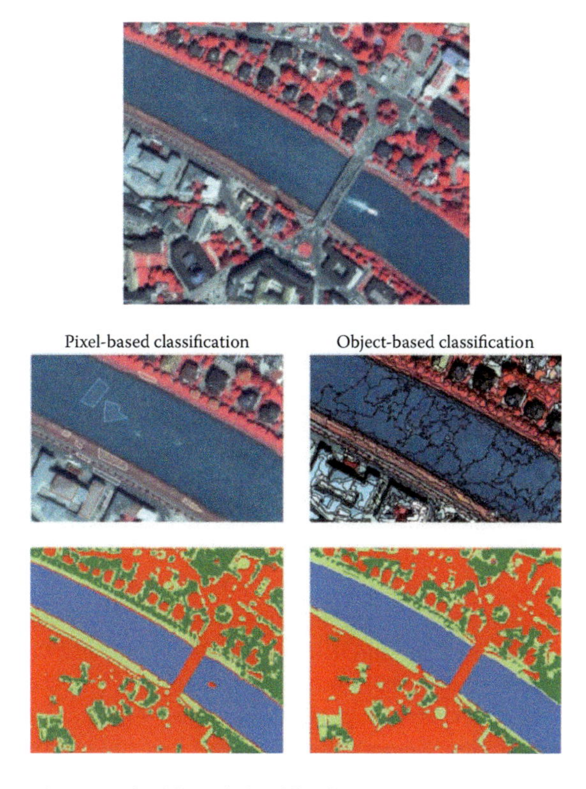

FIGURE 8.2 Object-based versus pixel-based classification.

nuanced descriptions of methodology. Editorial boards of journals with higher impact factors (e.g., Remote Sensing of Environment) began demanding more explanation in a Methods section than the use of a software package. The success of the suite of eCognition/Definiens software packages likely prompted rapid development of alternative software for the OBIA workflow. Berkeley ImageSeg (Clinton et al., 2010), Visual Learning Systems' Feature Analyst extension for ArcGIS (Metaferia et al., 2023), SAGA (Pawar et al., 2024), ENVI Feature Extraction (Hölbling and Neubert, 2008), and ERDAS IMAGINE's Objective module (Jawak et al., 2022) appeared between 2006 and 2024. Use of additional, external software, particularly for the classification step of the OBIA workflow became increasingly common. For example, many papers discuss the use of decision trees such as classification and regression trees (CART), usually run externally to a software package such as eCognition, in R (http://www.r-project.org/) or See5 (http://www.rulequest.com/see5-info.html), to classify objects. Yu et al. (2006) used this approach to map vegetation alliances in a California reserve, Laliberte et al. (2007) did so with high resolution data over rangelands, as did Chubey et al. (2006) for forest inventory mapping. Green and Lopez (2007) used CART to label polygons created in eCognition for benthic habitat in Texas, and Stow et al. (2007) used a similar combined approach to map urban areas in Accra, Ghana. Since then, eCognition has implemented a decision tree algorithm for classification.

Next to the commercial software mentioned, several open source software products have been developed. While earlier attempts may be considered to be more of an academic, not very user-friendly and not well documented prototypical software such as GeoAida (Bückner et al., 2001), recent open source developments aim to compete with commercial software such as eCognition, Erdas or ENVI in respect to a modern user-friendly GUI and software documentation. InterIMAGE is an open source and free access framework for knowledge-based image classification. It is based on algorithms from GeoAida and provides a capacity for customization and extension tools. Costa et al. (2010), describe the InterIMAGE system as a multi-platform framework, implemented for LINUX and Windows operational systems (http://www.lvc.ele.puc-rio.br/projects/interimage/).

A more recent development is the Geographic Data Mining Analyst, GeoDMA. It bridges GIS and image processing functionality and includes algorithms for segmentation, feature extraction, feature selection, classification, landscape metrics and multitemporal methods for change detection and analysis (Körting et al., 2013).

Bunting et al. (2014) developed the open source platform RSGISLib for data processing techniques. Users interact with the software through an XML script, where XML tags and attributes are used to parameterize 300 available commands. The developers claim that command options are easily recognizable to the user because of their logical and descriptive names. Through the XML interface, processing chains and batch processing are supported. More recently a Python binding has been added to RSGISLib allowing individual XML commands to be called as Python functions. The software has been released under a GPL3 license and makes use of a number of other open source software libraries (e.g., GDAL/OGR), a user guide and the source code are available at http://www.rsgislib.org.

8.2.3.3 Consolidation (Since around 2010)

Since around 2010 the field has emerged from its earlier stages, and is displaying more maturity. Blaschke et al. (2014) raise the discussion that in some ways this maturity suggests a label of new "paradigm." From a workshop on OBIA convened at the 2012 GIScience Conference in Columbus, OH to discuss key theoretical and applied aspects of the approach emerged several important topics for the next decade: integration with GIS, semantics, accuracy, change, standards, and learning from the past. These themes are born out in the literature. There have also been some important developments on the software front. For example, in 2010 Trimble (a company expert in field and mobile technology, and one of the leading manufacturers of research and survey grade GPS systems) purchased Definiens Earth Science ("Trimble Acquires Definiens' Earth Sciences Business to Expand its GeoSpatial Portfolio": https://www.trimble.com/news/release.aspx?id=061110a), with expectations that the OBIA workflow would be of particular use to mobile mapping, survey and urban environment reconnaissance. Additionally, there has been increasing use in the remote sensing world of unmanned aerial systems (UAS) or drones, which provide small footprint, very high resolution imagery (cm to meter pixel size). Once geometric and radiometric corrections and mosaicking have been applied, these images are routinely being approached with the OBIA workflow. UAS provide the ability for repeated deployment for acquisition of multispectral imagery at high temporal resolution data at very high spatial resolution. For example, Laliberte et al. (2011) acquired multispectral imagery using UAS and obtained orthorectified, radiometrically calibrated image mosaics for the purpose of rangeland vegetation classification. They relied heavily on an OBIA approach for classification of rangeland classes and achieved relatively high accuracies. Castro et al. (2013) were able to generate weed maps early in the growing season for maize fields by using an unmanned aerial vehicle (UAV) and object-based image analysis.

The current global explosion of imagery resources at high temporal and spatial resolution is actively changing all aspects of the geospatial enterprise. The ways in which we acquire, store, serve, and generate information from an increasing supply of imagery across domains necessitates the continued development of streamlined OBIA workflows that render imagery useful through geospatial semantics and shared knowledge (Harvey and Raskin, 2011; Blaschke et al., 2014). The time-sensitive decision support tasks found in disaster response, for example, which typically make use of rapidly acquired imagery to find targets are often facilitated currently by human volunteers or "distributed thinking" (Zook et al., 2010). These tasks in the future might be supported by OBIA workflows. And the accelerated pace of geospatial work that accompanies disaster response is increasingly characteristic of science in general than it has ever been in the past. Decisions that routinely waited for annual, seasonal or monthly data (e.g., forest loss, peak greenness, soil water

TABLE 8.4
Summary of Historic Effects and OBIA Developments

External Effects/Triggers	OBIA Developments
1972: Landsat 1 and its multispectral sensor set the standard for civilian remote sensing applications for the next decades	
Late 1970s: image segmentation techniques are developed and are subsequently being used in image processing but not much in geospatial applications	Kettig and Landgrebe (1976) developed the first hybrid classification approach which included neighborhood aspects.
Late 1999 and 2000: advent of the first two civilian 1m resolution satellites mark a new area of high resolution spaceborn imaging	1999/2000: commercialization of Definiens company and eCognition software
1998/99: commercial LiDAR systems available June 2003: Orbview-3 High resolution digital airborne cameras such as the Ultracam (Leberl and Gruber, 2003)	July 2001 first scientific workshop on OBIA methods: FE/GIS'2001: Remote Sensing: New sensors—innovative methods, Salzburg, Austria (German language) 2002: first book on OBIA in German language based on the 2001 workshop (Blaschke, 2002) 2001–2003: first dozen papers in peer-reviewed journals
2004 onward: more high resolution satellites, decreasing prices of data, higher accessibility 2005: Google Earth raised public awareness about remote sensing imagery and subsequently increased demand for information products	2005: First OBIA-related book for the fast developing Brazilian market (Blaschke and Kux, 2005). OBIA workshop at the XII Brazilian Remote Sensing Symposium, June 2005, Goinania, Brazil 2006: First OBIA conference in Salzburg, Austria. 2007: OBIA workshop at UC Berkeley 2008: GEOBIA int. conference in Calgary, Canada 2009: Object based landscape analysis workshop at the University of Nottingham, UK 2010: GEOBIA int. conference in Ghent, Belgium. 2012: GEOBIA int. conference in Rio de Janeiro, Brazil 2014: GEOBIA int. conference in Thessaloniki, Greece

deficits, etc.) can now be made based on data at finer spatial and temporal resolutions (e.g., Ye et al., 2023; Li et al., 2022; Liu et al., 2018; Belgiu and Csillik, 2018; Zhang et al., 2018; Leichtle et al., 2017; Long et al., 2017). Doubtlessly, future research within OBIA will focus on transferring imagery quickly into comprehensive and web-enabled geographic knowledge-bases to be used for decision making.

8.3 OBIA—A SHORT SUMMARY OF THE STATE OF THE ART

This section is kept very short and aims to succinctly summarize the main findings from other state of the art reviews, particularly Blaschke (2010) and Blaschke et al. (2014).

8.3.1 Segmentation Is Part of OBIA—But Not Married to It

A common denominator of OBIA applications was, and still is, that they are built on image segmentation (see also Gui et al., 2024; Jiang et al., 2023; Putra and Wijayanto, 2023; Ye et al., 2023; Li et al., 2022; Liu et al., 2018; Hossain and Chen, 2019; Belgiu and Csillik, 2018; Chen et al., 2018; Deng et al., 2018; Ye et al., 2018; Zhang et al., 2018; Leichtle et al., 2017; Long et al., 2017; Zhao et al., 2017; Burnett and Blaschke, 2003; Lang, 2008; Hay and Castilla, 2008). Image segmentation is not at all new (Haralick and Shapiro, 1985; Pal and Pal, 1993), but has its roots in industrial image processing and was not used extensively in geospatial applications throughout the 1980s and 1990s (Blaschke et al., 2004).

Interestingly, independent from most of the OBIA-related developments described in Blaschke (2010) but also triggered by the advent of high resolution satellite imagery, Aplin et al. (1999) and Aplin and Atkinson (2001) developed an approach to segment image pixels using vector field boundaries and to assign sub-pixel land cover labels to the pixel segments. Subsequently, hard per-field classification, the assignment of land cover classes to fields (land cover parcels) rather than pixels (Aplin et al., 1999), was achieved by grouping and analyzing all land cover labels for all pixels and pixel segments within each individual field. Their approach was somewhat different in a sense that they aimed to classify pre-defined objects, namely fields. These developments coincided later with the "OBIA-community" when Paul Aplin and Geoff Smith organized a symposium on "Object based Landscape Analysis" in 2009 in Nottingham, UK, and edited a special issue in International Journal of Geographical Information Science (Aplin and Smith, 2011).

Although most scientists would associate OBIA with segmentation, recent work has shown that some segmentations steps typically involved in OBIA research don't necessarily play a major role, as sometimes postulated in the earlier development of OBIA. See particularly the discussion of Tiede (2014) who in essence decouples OBIA from image processing, Lang et al. (2010, 2014) and their work on concept-related fiat objects, geons, and on "latent phenomena."

8.3.2 Classification

Blaschke and Strobl (2001) have posed the question "What's wrong with pixels?" and elucidated some short-comings of a pure per-pixel approach. This was certainly not the first time to highlight the limitations of treating pixels individually based on multivariate statistics. In fact, Kettig and Landgrebe (1976) developed the first algorithm called ECHO which at least partially utilizes contextual information. Based on the short history of OBIA in the section before, we may argue that around the turn of the Millennium the quest for objects reached a new dimension. Particularly for high resolution image it seems to make much sense to classify segments—rather than pixels. The segments may or may not correspond exactly to the objects of desire. Burnett and Blaschke (2003) called such segments from initial delimitation steps objects candidates. They already offer parameters such as size, shape, relative/absolute location, boundary conditions and topological relationships which can be used within the classification process in addition to their associated spectral information.

There is increasing awareness that object-based methods make better use of—often neglected—spatial information implicit within REMOTE SENSING images. Such approaches allow for a tightly coupled or even full integration with both vector and raster based GIS. In fact, when studying the early OBIA literature for the geospatial domain, it may be concluded that many applications were driven by the demand for classifications which incorporate structural and functional aspects.

There are many examples of comprehensive reviews and studies (Gui et al., 2024; Jiang et al., 2023; Putra and Wijayanto, 2023; Ye et al., 2023; Li et al., 2022; Liu et al., 2018; Hossain and Chen, 2019; Belgiu and Csillik, 2018; Chen et al., 2018; Deng et al., 2018; Ye et al., 2018; Zhang et al., 2018; Leichtle et al., 2017; Long et al., 2017; Zhao et al., 2017). One good example of a comprehensive review is the paper by Salehi et al. (2012). They conducted of recent literature and evaluated performances in urban land cover classifications using high resolution imagery. They analyzed the classification results for both pixel-based and object-based classifications. In general, object-based classification outperformed pixel-based approaches. These authors reason that the cause for the superiority was the use of spatial measures and that utilizing spatial measures significantly improved the classification performance particularly for impervious land cover types.

8.3.3 Complex "Geo-Intelligence" Tasks

Increasingly, OBIA is used beyond simple image analysis tasks such as image classification and feature extraction from one image or a series of images from the same sensor.

Today, terabytes of data are acquired from space- and air-borne platforms, resulting in massive archives with incredible information potential. As Hay and Blaschke (2010) argue, it is only recently that we have begun to mine the spatial wealth of these archives. These authors claim that, in essence, we are data rich, but geospatial information poor. In most cases, data/image access is constrained by technological, national, and security barriers, and tools for analyzing, visualizing, comparing, and sharing these data and their extracted information are still in their infancy. In the few years since this publication "big data" has fully arrived in many sciences and this debate seems not to be OBIA specific from today's point of view.

Furthermore, policy, legal, and remuneration issues related to who owns (and are responsible for) value-added products resulting from the original data sources, or from products that represent the culmination of many different users input (i.e., citizen sensors) are not well understood and still developing. Thus, myriad opportunities exist for improved geospatial information generation and exploitation.

OBIA has been claimed to be a sub-discipline of GIScience devoted to developing automated methods to partition remote sensing imagery into meaningful image-objects, and assessing their characteristics through scale (Hay and Castilla, 2008). Its primary objective is the generation of geographic information (in GIS-ready format) from which new geo-intelligence can be obtained. Based on this argument, Hay and Blaschke (2010) have defined geo-intelligence as geospatial content in context.

The final theme is intelligence—referring to geo-intelligence—which denotes the "right (geographically referenced) information" (i.e., the content) in the "right situation" so as to satisfy a specific query or queries within user specified constraints (i.e., the context).

Moreno et al. (2010) describe a geographic object-based vector approach for cellular automata modeling to simulate land-use change that incorporates the concept of a dynamic neighborhood. This represents a very different approach for partitioning a scene, compared to the commonly used OBIA segmentation techniques, while producing a form of temporal geospatial information with a unique heritage and attributes.

Lang (2008) provided a more holistic perspective on an image analysis and the extraction of geospatial information or what he called at this time an upcoming paradigm. He started from a review of requirements from international initiatives like GMES (Global Monitoring of Environment and Security, now: "Copernicus") and he discussed in details the concept of "class modeling." Also, such methods may need further advancement of the required adaptation of standard methods of accuracy assessment and change detection. He introduced the term "conditioned information." With this term he addresses processes which entail the creation of new geographies as a flexible, yet statistically robust and (user-) validated unitization of space.

Lang et al. (2014) developed the concept of geons as a strategy to represent and analyze latent spatial phenomena across different geographical scales (local, national, regional) incorporating domain-specific expert knowledge. The authors exemplified how geons are generated and explored. So-called composite geons represent functional land-use classes, required for regional planning purposes. They are created via class modeling to translate interpretation schemes from mapping keys. Integrated geons, on the other hand, address abstract, yet policy-relevant phenomena such as societal vulnerability to hazards. They are delineated by regionalizing continuous geospatial datasets representing relevant indicators in a multidimensional variable space. In fact, the geon approach creates spatially exhaustive sets of units, scalable to the level of policy intervention, homogenous in their domain-specific response, and independent from any predefined boundaries. Despite its validity for decision making and its transferability across scales and application fields, the delineation of geons requires further methodological research to assess their statistical and conceptual robustness.

8.4 ONGOING DEVELOPMENTS: INFLUENCES OF OBIA TO OTHER FIELDS AND VICE VERSA

8.4.1 GIScience and Remote Sensing

OBIA arguably has its roots firmly in the field of remote sensing. Developments in remote sensing through the decades of the 2000–2010s—including most importantly the widespread availability of high resolution imagery globally, but also from LiDAR, and novel methods of data fusion—have continued this alliance. However, this early grounding of OBIA in theoretical and practical aspects of remote sensing is recently being enhanced through multiple novel interactions with aspects of the GIScience field, and OBIA is poised to develop further from new trends in GIScience.

Since Goodchild (1992) first coined the term GIScience, suggesting it as a manner of dealing with the issues raised by GIS technology by focusing on the unaddressed theoretical shortcomings of conventional GIS, the contents and borders have constantly shifted, especially in light of recent advances in geospatial technologies, including remote sensing (Blaschke and Merschdorf, 2014). In order to deal with the special properties of spatial information in an era of Web 2.0 technologies, the field of GIScience has embraced not only classic geographical knowledge and concepts, but also increasingly incorporated approaches from other disciplines such as computer science and cognitive sciences (Jiang et al., 2023; Belgiu and Csillik, 2018; Ye et al., 2018; Long et al., 2017; Zhao et al., 2017; Blaschke and Merschdorf, 2014). In turn, other disciplines have recently discovered the potential of GIScience, utilizing its tools and methodologies to serve their own needs, and to drastically advance the knowledge base in their own respective fields. Such is not least the case for remote sensing, which has experienced a drastic shift from purely pixel based methods of image interpretation, to the identification of "objects" in remotely sensed imagery by means of OBIA. Hay and Castilla (2006) propose that OBIA is a sub-discipline of GIScience, combining a "unique focus on remote sensing and GI" (Hay and Castilla, 2006:1). In this sense, OBIA may be seen as the first in a string of developments leading to the consolidation of GIS and remote sensing, facilitated through the common denominator of GIScience. This implies that current and ongoing developments in the discipline of GIScience may bare a significant impact on the field of remote sensing. Such developments include, but are not limited to volunteered geographic information, ubiquitous sensing, indoor sensing and the integration of in situ measurements with classic remote sensing datasets.

Web 2.0 technologies have had a significant impact on GIScience, as they have enabled the bi-directional and participatory use of the internet (Blaschke and Merschdorf, 2014). These technologies go beyond "GIS-centered" assemblages of hardware, software, and functionalities. Wiki-like collective mapping environments, geovisualization APIs, and geo-tagging may either be based on GIS or they have common denominators in the digital storage, retrieval, and visualization of information based upon its geographic content (Sheppard, 2006).

These developments have led to an influx of spatial content, contributed by individual users or groups of users, which nowadays composes a valuable data source in GIS. Such content has been termed as "volunteered geographic information" (VGI), by Goodchild (2007), and Atzmanstorfer and Blaschke (2013) claim that its full realm of possibilities, in terms of citizens partaking in planning initiatives, yet remains unknown. VGI is not limited to online applications such as the provision of geotagged photographs on the photo management service Flickr, or geo-located messages on the online messaging portal Twitter, but also includes the information collected by wireless sensors on common mobile devices. Due to the proliferation of wireless sensors in all sorts of mobile devices, sensory data collection is no longer constrained to few experts equipped with expensive sensors, but rather has shifted more into the lay domain. In GIScience, this notion is referred to as ubiquitous sensing, and can be used for monitoring activities and locations of users, or groups of users, in near real time. The near real time capabilities of ubiquitous sensing can assist decision makers in a variety of applications, such as emergency response, public safety, traffic management, environmental monitoring or public health (Resch, 2013). For example, Sagl et al. (2012) utilize the movements of cell-phones between pairs of radio cells—termed as handovers—in order to analyze spatio-temporal urban mobility patterns, and demonstrate how mobile phone data can be utilized to analyze patterns of real-world events using the example of a soccer match, while Zook et al. (2010) present how a mash-up of various data sources, including both government data and VGI significantly contributed to disaster relief in Haiti, following the earthquake in 2010.

While VGI is oftentimes a passive by-product, resulting from the use of Web 2.0 technologies and mobile computing devices, millions of internet users can nowadays choose to actively utilize GIS methodologies and applications by means of "public participation geographic information system" (PPGIS). Manifestations of such participation can, for instance, be found in the widespread community of users contributing to virtual globes and maps by superimposing new layers, such as street networks or landmarks, or even in disaster relief efforts such as the recent search for the debris of the missing Malaysian Airline flight MH370, which was assisted by tens of thousands of internet users, who helped in sifting through the vast magnitude of satellite data recorded during the time frame in question.

The contribution of the general public, be it actively by uploading data to virtual globes or maps, or passively by utilizing social media platforms such as Twitter or Flickr, has also fuelled the collection of in situ data, such as photos taken at a certain location, values measured there, etc. Such data is particularly valuable in the era of very high resolution satellite imagery, as well as the subsequent surge of urban remote sensing applications, such as the mapping of mega-cities, the monitoring of fast expanding settlements in developing countries, or the routine monitoring of informal settlements, conducted either by public administration or by commercial companies, as outlined by Blaschke et al. (2011). Based on an extensive literature review, Blaschke et al. (2011) conclude that the increased availability of high resolution satellite imagery has resulted in a greater demand for timely urban mapping and monitoring. However, remotely sensed imagery, which provides the basis for urban mapping applications, can only provide the bird's-eye view of a given location, neglecting ground information such as the building facades or interiors. With the advent of widely applied OGC standards, in situ measurement data recorded at ground locations can be integrated with the remote sensing imagery, providing a more holistic approach to urban mapping applications (Blaschke et al., 2011). Blaschke et al. (2011) note that although remote sensing and in situ measurements are currently two separate technologies, the strengths of both can be combined by means of sensor webs and OGC standards, potentially producing new and meaningful information (Blaschke et al., 2011). They conclude that "while available information will always be incomplete, decision makers can be better informed through such technology integration, even if loosely coupled" (Blaschke et al., 2011:1768).

Another trend enabled by the recent advances in mobile technology is the concept of indoor sensing, sometimes referred to as indoor geography (Blaschke and Merschdorf, 2014). Naturally, remote sensing imagery can only provide a planar view of the Earth's surface, including natural features, as well as human infrastructure. While LiDAR technology complements the classic two-dimensional imagery with the added dimension of depth, it still doesn't provide any insight as to the contents of buildings. In this sense, indoor sensing may be a future trend in indoor positioning and mapping, whereby sensor fusion will evolve to support indoor locations, paving the way for geo-enabled manufacturing (Blaschke and Merschdorf, 2014).

8.4.2 The Changing Workplace

In the past, Remote Sensing and GIS were distinctly separated disciplines, whereby remotely sensed imagery was primarily considered as a data source for GIS (Jensen, 1996). However, in light of more recent technical and theoretical advancements, these disciplines have begun to consolidate, not least attributed to the quest for tangible objects. The emergence of OBIA as a sub-discipline of GIScience (Gui et al., 2024; Jiang et al., 2023; Putra and Wijayanto, 2023; Ye et al., 2023; Li et al., 2022; Liu et al., 2018; Hossain and Chen, 2019; Belgiu and Csillik, 2018; Chen et al., 2018; Deng et al., 2018;

Ye et al., 2018; Zhang et al., 2018; Leichtle et al., 2017; Long et al., 2017; Zhao et al., 2017) laid a foundation for the use of shared methodologies, and remote sensing was recognized as "one element of an integrated GIS environment, rather than simply an important data source" (Malczewski, 1999:20). The bi-directional nature of the relationship between remote sensing and GIS implies that not only advances in remote sensing technology influence the GIS environment, but also vice versa. In this sense we can witness the impact of recent trends in GIScience, described in section 4.2, on the remote sensing discipline. Especially the technological advances brought about by the Web 2.0, such as VGI, ubiquitous sensing, or PPGIS, call for new approaches of data integration, with the primary aim of developing more comprehensive and accurate datasets. Such integration can complement the birds-eye view perspective offered by remotely sensed imagery, with in situ information, which in turn can more efficiently represent dynamic urban environments (Blaschke et al., 2011). To this end, OGC standards can provide the necessary interface for data integration, as is the case for the Global Earth Observing System of Systems (GEOSS), which seamlessly integrates remotely sensed imagery with in situ measurements.

One particular example of an OBIA application as a substitute for GIS-overlay is provided by Tiede (2014). GIS-overlay routines usually build on relatively simple data models. Topology is—if at all—calculated on the fly for very specific tasks only. If, for example, a change comparison is conducted between two or more polygon layers, the result leads mostly to a complete and also very complex from–to class intersection. Additional processing steps need to be performed to arrive at aggregated and meaningful results. To overcome this problem Tiede (2014) presented an automated geospatial overlay method in a topologically enabled (multi-scale) framework. The implementation works with polygon and raster layers and uses a multi-scale vector/raster data model developed in the object-based image analysis software eCognition. Advantages are the use of the software inherent topological relationships in an object-by-object comparison, addressing some of the basic concepts of object-oriented data modeling such as classification, generalization, and aggregation. Results can easily be aggregated to a change-detection layer; change dependencies and the definition of different change classes are interactively possible through the use of a class hierarchy and its inheritance (parent–child class relationships. The author demonstrates the flexibility and transferability of change comparison for CORINE Land Cover datasets. This is only one example where OBIA and GIS are fully integrated and, although this case maybe being an exception so far, one field may jeopardize the other field if the fields are seen isolated.

8.4.3 Who Uses OBIA?

In a recent publications (Gui et al., 2024; Jiang et al., 2023; Putra and Wijayanto, 2023; Ye et al., 2023; Li et al., 2022; Liu et al., 2018; Hossain and Chen, 2019; Belgiu and Csillik, 2018; Chen et al., 2018; Deng et al., 2018; Ye et al., 2018; Zhang et al., 2018; Leichtle et al., 2017; Long et al., 2017; Zhao et al., 2017 these have been widely discussed and implemented. For example, Blaschke et al., 2014) found an increasing number of publications concerned with OBIA in peer reviewed journals, special issues, books, and book chapters, and concluded that OBIA is a new evolving paradigm in remote sensing and to some degree in GIScience also. However, they also noted that the exact terminology used within these publications is distinctly ambiguous, as is characteristic for an emerging multidisciplinary field (Blaschke et al., 2014). Therefore, we herein aim to review the literature databases of the ISI's Web of Science (WoS), as well as Scopus, in an attempt to quantify who uses OBIA, both in terms of countries of origin, as well as contributing field, and to track its presence in literature over the past years.

A search in the Web of Science database for the phrases "Object based image analysis", or "Object oriented image analysis," or "OBIA," or "Geographic object based image analysis," or "GEOBIA," contained in the title, abstract or keywords, returns a total of 451 articles (17.04.2014). When analyzing which countries the publications primarily come from, we determined that the highest number of publications is contributed by the USA, accounting for 24% of all publications, followed by the People's Republic of China with a 14% contribution, Germany contributing 12%, Austria 8%, Canada 7%, Australia, Brazil and the Netherlands 6%, respectively, and Italy and Spain with 4% each, just to name the top ten contributing countries. This shows that while the USA is the main contributor, accounting for nearly a quarter of all publications returned in the search, many other smaller countries also make a noteworthy contribution. In particularly remarkable is the 8% contribution made by Austria, which has only a fraction of the population (approx. 8.5 million) compared to most other countries represented within the top 10. Compared to the leading country—the United States—Austria has merely 2.7% of the population, however, has 33% as many publications. Such a comparison becomes even more extreme when made with China, the second largest contributor, whereby Austria has only 0.6% as many inhabitants, but accounts for 57% as many publications. This shows that there may be certain research clusters in certain countries, which largely contribute to OBIA/GEOBIA research, rather than all countries contributing relatively to their population.

A further analysis consisting of the research areas contributing to OBIA/GEOBIA reveals that the largest contribution is made by Remote Sensing, accounting for 61% of all publications. The second largest contributor, namely Imaging Science, accounts for only 31%, followed by Geology with a share of 27%. A full chart of the top 10 contributing fields is depicted in Figure 8.4.

When assessing the publication years, it is notable that the number of publications on the topic of OBIA/GEOBIA has drastically increased over the last five years, whereby 22% were written in 2013 alone, as compared to >16% prior to 2008. The first OBIA publication indexed in ISI's Web of Science database dates back to 1985, preceding the second OBIA publication by 10 years, and at least 20 years prior to a steady incline in the number of publications.

When the same search is conducted in the Scopus database (same phrases searched for in Title, Abstract and Keywords), a total of 586 publications are returned (July 17, 2014). The discrepancy in terms of numbers of publications as compared to the Web of Science database can be attributed to the fact that Scopus contains a broader range of document types, such as notes, short surveys, in press articles, etc., while the WoS database only contains peer reviewed journal articles, conference proceedings, reviews and editorials, all of which are additionally included in Scopus.

Although including a slightly greater number in overall publications, the trends revealed in the Scopus data are largely in line with those depicted in the WoS data. Some discrepancies were

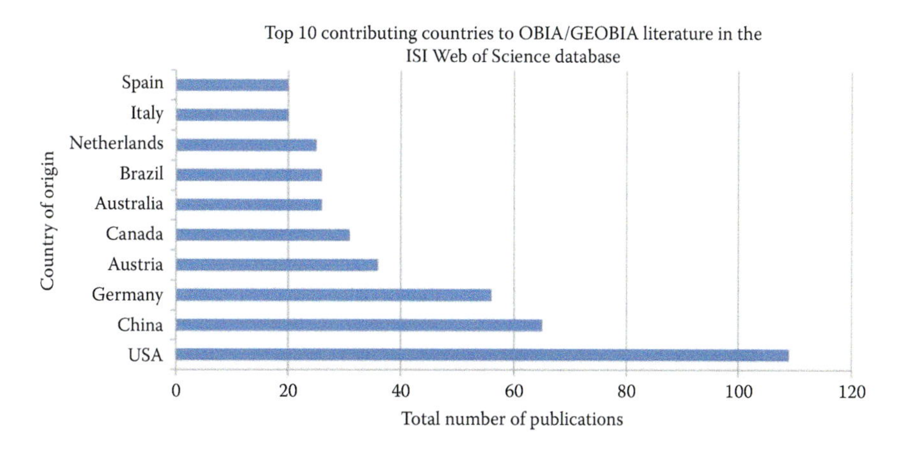

FIGURE 8.3 Top 10 contributing countries to the OBIA/GEOBIA Literature in the Web of Science Database and their Respective Contributions.

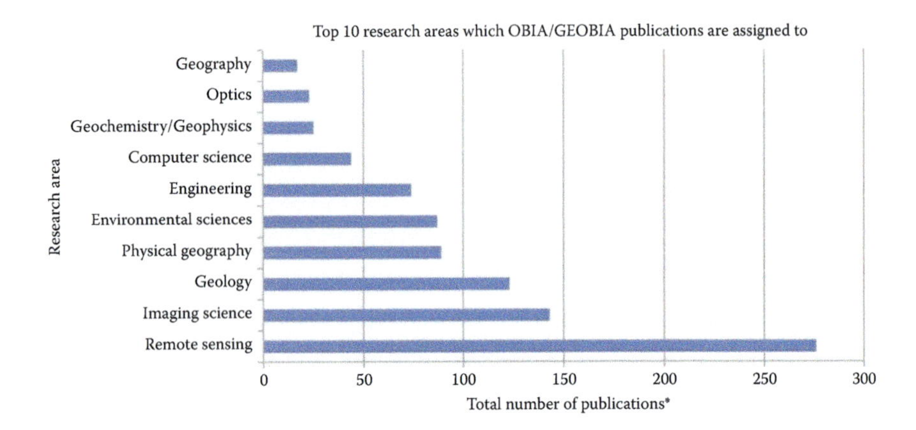

FIGURE 8.4 Main research areas for OBIA/GEOBIA publications in ISI's Web of Science database*.

*The total numbers add up to more than the total of 451 publications due to the fact that some multidisciplinary publications may have been assigned to more than one research area.

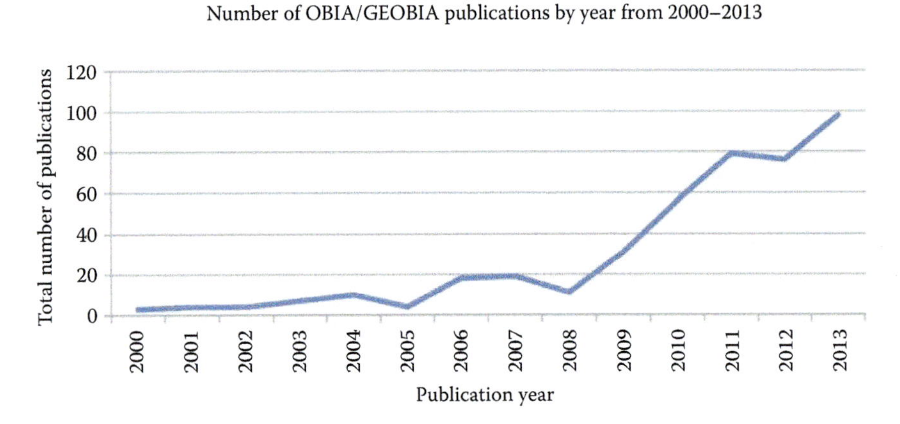

FIGURE 8.5 Number of OBIA/GEOBIA Publications by Publication Year from 2000–2014, as indexed in ISI's Web of Knowledge Database.

found in terms of research areas, which, however, may largely be down to the different naming conventions utilized by both databases (e.g., the top contributing discipline to OBIA/GEOBIA in the Scopus database is "Earth and Planetary Sciences," with a total of 315 publications or 54%, respectively, which corresponds to the largest WoS contributor of "Remote Sensing"). Furthermore, both the publication year timeline and the contributing countries roughly correspond to the results obtained from the analyses of the WoS data.

In conclusion, when analyzing the literature, and some key milestone events and publications, the rise of OBIA/GEOBIA can be clearly traced through the course of the last decade and a half. This is depicted in the timeline shown in Figure 8.6, which exemplifies how both technological, as well as methodological advances gave birth to object oriented approaches and, according to Blaschke et al. (2014) to a new paradigm in Remote Sensing although it must be clearly stated that in absolute terms "classic" per-pixel methods are represented way more in publications at the moment.

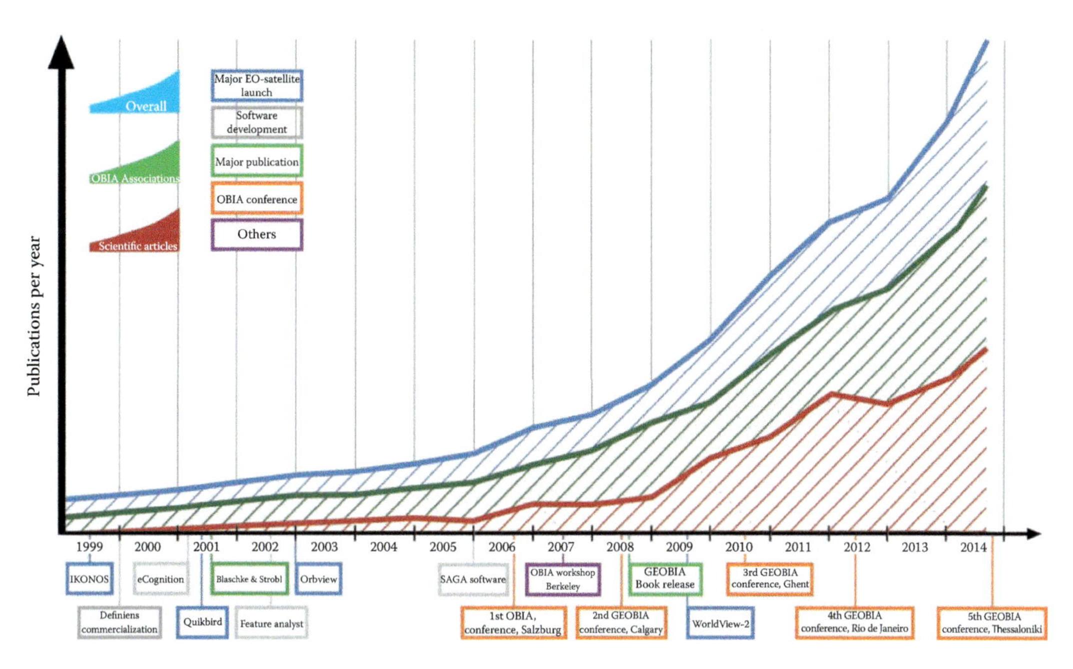

FIGURE 8.6 Milestone timeline of OBIA/GEOBIA development from the late 1990s until today.

8.5 CONCLUDING PERSPECTIVES

8.5.1 New Paradigm: The Need for a Common Nomenclature and Sound Methodologies

OBIA has certainly arrived at the sciences (Gui et al., 2024; Jiang et al., 2023; Putra and Wijayanto, 2023; Ye et al., 2023; Li et al., 2022; Liu et al., 2018; Hossain and Chen, 2019; Belgiu and Csillik, 2018; Chen et al., 2018; Deng et al., 2018; Ye et al., 2018; Zhang et al., 2018; Leichtle et al., 2017; Long et al., 2017; Zhao et al., 2017). While the first years of the development were characterized by a lack of high quality peer-reviewed scientific publications, the last few years witnessed a sharp increase in such articles. Some of them are remarkably highly cited such as the review paper by Blaschke (2010). Much of the excitement about this new methodology or paradigm has to do with the increasing availability of high resolution datasets, which can now be used to produce information and, in particular, information on demand or "conditioned information." Some predict that researchers, policymakers, citizen groups, and private institutions might use information contributed by ordinary people for any number of purposes, including emergency response, mobilizing activist efforts, monitoring environmental change, filling gaps in existing spatial databases, or identifying and addressing needs and problems in urban neighborhoods

OBIA has developed a rich array of approaches for grappling with the challenges associated with high resolution data. One remaining task is to standardize terms across for methods and methodologies being used. While Blaschke et al. (2014) argue that this is very common for a new paradigm it is nevertheless troublesome. OBIA needs to urgently harmonize and streamline the terms being used. Otherwise, a widespread recognition from other fields may be hindered.

8.5.2 Toward a Civilian Geo-Intelligence

We don't exactly know how the future will look like. One possible development can best be illustrated by the power and the innovative potential of the "object-by-object change comparison framework"

(Tiede, 2014) described before. This framework yields flexible, transferable, and highly complex change comparisons that can be visualized or calculated and aggregated to higher level composite objects. Here, the Geon-concept of Lang et al. (2014) comes into play: as briefly described it also allows for the creation of more conceptual objects which may represent latent phenomena which are not directly mappable.

Hay and Blaschke (2010) suggested the term (civilian) geo-intelligence. Since then, the number of technical developments and the number of documented applications which may support the hypothesis of locational intelligence has clearly grown. As discussed earlier, Lang (2008) laid some theoretical foundations for the concept of "conditioned information" and Lang et al. (2014) developed the concept of Geons which may also serve as units to characterize and delimitate latent phenomena.

An area for future research emerges from a wider set of organizational changes within the software industry such as the "software as a service" paradigm. This is a significant development in the organization and deployment of remote sensing image analysis for the professional and advanced users. It may also open opportunities for non-expert users in remote sensing in general and for OBIA in particular. Tiede et al. (2012) presented an OBIA geoprocessing service which integrates OBIA methods into a geoprocessing service. This development was—to our best knowledge—the first integration of an eCognition based OBIA application into an interactive WebGIS geoprocessing environment.

Interestingly, the emergence of OBIA has not been generating a substantial quantity of critical reflection about the technology as such nor about the wider scientific and technological implications of this paradigm for various user groups, both geographically as well as seen along an educational ladder (students—graduates—professionals in private industry and academia).

Another future research area concerns the remote sensing and GIScience practices of non-professional actors, not as outsourced operatives for research institutions, but as private actors or NGOs following their own agendas. In remote sensing in general—much more than in GIScience—the vast majority of existing literature investigates widely agreed scientific or commercially interesting problems, and reflects both the focus of an Anglophone research community looking primarily back in time and a focus primarily on activities in the global North by state actors. Although we did not carry out a severe literature study we may speculate that OBIA researchers may be a little bit less Anglophone dominated than the general remote sensing community.

8.5.3 Epistemological and Ontological Challenges

We may claim here for the field of remote sensing that some long-known principles about technological determinism (McLuhan, 1964, who basically claimed that humans shape their tools and they in turn shape humans) may become more obvious today because its practical and theoretical implications are now much faster discovered. Nevertheless, the process of the social shaping of technology can be a long-term, interactive, and sometimes conflict-ridden (Rohracher, 2003).

Like GIS which has for some years been decried as "ontologically shallow" and insufficient to the task of comprehending the many epistemological points of difference among users (Schuurman, 1999) remote sensing literature offers very little in regard to its ontological and epistemological foundation. Without doubt, remote sensing principles have solid foundations in Physics. Only through the amalgamation with GIS-principles with OBIA the need for a theoretical, that is: epistemological and ontological establishment increases. As long as the pixel is more or less the only subject of studies and, more importantly, as long as objects of interest are smaller than or similar in size compared to the pixels, such questions may not be urgent. With the advent of high resolution imagery the question "what's wrong with pixels?" (Blaschke and Strobl, 2001) is valid to be asked. In fact, concerns about the appropriate use of technology in the application of remote sensing data suggest that non-expert users involved in interpretation tasks may gain a relatively sophisticated understanding not just of what the technology can do, but of the processes involved in visualizing and disseminating findings via interactive representations and Web-GIS.

We refer to Pickles (2004) who contends that the contingent nature of technical outcomes from GIS use is often overlooked, and the exploitation of some groups, particularly those with less access to technology, becomes a real possibility. He also emphasizes how important it is "to study maps in human terms, to unmask their hidden agendas, to describe and account for their social embeddedness and the way they function as microphysics" (ibid. p. 181).

Lastly, we may call for a relaxation of a potential friction between OBIA and per-pixel approaches (Gui et al., 2024; Jiang et al., 2023; Putra and Wijayanto, 2023; Ye et al., 2023; Li et al., 2022; Liu et al., 2018; Hossain and Chen, 2019; Belgiu and Csillik, 2018; Chen et al., 2018; Deng et al., 2018; Ye et al., 2018; Zhang et al., 2018; Leichtle et al., 2017; Long et al., 2017; Zhao et al., 2017). There are dozens, most likely more than a hundred of scientific papers which compare both methods. Nevertheless, the future may not be dominated by a "either—or" question. Rather, we should be cautious about abandoning too hastily the concepts and terminologies of the "old" paradigm with reference to the its dazzling object of cognition in this debate—the pixel. The pixel is a technical construct which may be useful in many cases from a technical, that is: data acquisition point of view but sometimes also as a cognitional prerogative. In this sense, the aforementioned question "what's wrong with pixels" (Blaschke and Strobl, 2001) may appear in a less unfavorable light—for the latter, the pixels.

REFERENCES

Aneece, D. F., Thenkabail, P., Oliphant, A., & Teluguntla, P. (2022). New generation hyperspectral data from DESIS compared to high spatial resolution PlanetScope data for crop type classification. *IEEE Journal of Selected Topics in Applied Earth Observations and Remote Sensing*, 15, 7846–7858. https://doi.org/10.1109/JSTARS.2022.3204223

Aplin, P., & Atkinson, P. M. (2001). Sub-pixel land cover mapping for per-field classification. *International Journal of Remote Sensing*, 22(14), 2853–2858.

Aplin, P., Atkinson, P. M., & Curran, P. J. (1999). Fine spatial resolution simulated satellite sensor imagery for land cover mapping in the UK. *Remote Sensing of Environment*, 68, 206–216.

Aplin, P., & Smith, G. M. (2011). Introduction to object-based landscape analysis. *International Journal of Geographical Information Science*, 25(6), 869–875.

Arvor, D., Durieux, L., Andrés, S., & Laporte, M. A. (2013). Advances in geographic object-based image analysis with ontologies: A review of main contributions and limitations from a remote sensing perspective. *ISPRS Journal of Photogrammetry and Remote Sensing*, 82, 125–137.

Atzmanstorfer, K., & Blaschke, T. (2013). Geospatial web: A tool to support the empowerment of citizens through e-participation? In C. Nunes Silva (Ed.), *Handbook of Research on E-Planning: ICTs for Urban Development and Monitoring* (pp. 144–171). Hershey, PA: IGI-Global.

Belcore, E., Piras, M., & Pezzoli, A. (2022). Land cover classification from very high-resolution UAS data for flood risk mapping. *Sensors*, 22, 5622. https://doi.org/10.3390/s22155622

Belgiu, M., & Csillik, O. (2018). Sentinel-2 cropland mapping using pixel-based and object-based time-weighted dynamic time warping analysis. *Remote Sensing of Environment*, 204(2018), 509–523. ISSN 0034-4257. https://doi.org/10.1016/j.rse.2017.10.005. https://www.sciencedirect.com/science/article/pii/S0034425717304686

Benz, U. C., Hofmann, P., Willhauck, G., Lingenfelder, I., & Heynen, M. (2004). Multi-resolution, object-oriented fuzzy analysis of remote sensing data for GIS-ready information. *ISPRS Journal of Photogrammetry and Remote Sensing*, 58(3–4), 239–258. ISSN 0924-2716. https://doi.org/10.1016/j.isprsjprs.2003.10.002.

Blaschke, T. (Ed.). (2002). *Fernerkundung und GIS: Neue Sensoren—Innovative Methoden*. Karlsruhe: Wichmann Verlag.

Blaschke, T. (2010). Object based image analysis for remote sensing. *ISPRS International Journal of Photogrammetry and Remote Sensing*, 65(1), 2–16.

Blaschke, T., Burnett, C., & Pekkarinen, A. (2004). New contextual approaches using image segmentation for object-based classification. In F. De Meer & S. de Jong (Eds.), *Remote Sensing Image Analysis: Including the Spatial Domain* (pp. 211–236). Dordrecht: Kluwer Academic Publishers.

Blaschke, T., Hay, G. J., Kelly, M., Lang, S., Hofmann, P., Addink, E., Feitosa, R., van der Meer, F., van der Werff, H., Van Coillie, F., & Tiede, D. (2014). Geographic object-based image analysis: A new paradigm in remote sensing and geographic information science. *ISPRS International Journal of Photogrammetry and Remote Sensing*, 87(1), 180–191.

Blaschke, T., Hay, G. J., Weng, Q., & Resch, B. (2011). Collective sensing: Integrating geospatial technologies to understand urban systems – an overview. *Remote Sensing*, 3(8), 1743–1776.

Blaschke, T., Lang, S., & Hay, G. J. (Eds.). (2008). *Object-based image analysis, spatial concepts for knowledge-drives remote sensing applications*, Lecture Notes in Geoinformation and Cartography. Springer-Verlag.

Blaschke, T., & Merschdorf, H. (2014). Geographic information science as a multidisciplinary and multi-paradigmatic field. *Cartography and Geographic Information Science*, 41(3), 196–213.

Blaschke, T., & Strobl, J. (2001). What's wrong with pixels? Some recent developments interfacing remote sensing and GIS. *GIS—Zeitschrift für Geoinformationssysteme*, 14(6), 12–17.

Bückner, J., Pahl, M., Stahlhut, O., & Liedtke, C.-E. (2001). GEOAIDA a knowledge based automatic image data analyser for remote sensing data. *Paper Presented at the ICSC Congress on Computational Intelligence Methods and Applications—CIMA*, Bangor, UK.

Bunting, P., Clewley, D., Lucas, R. M., & Gillingham, S. (2014). The remote sensing and GIS software library (RSGISLib). *Computers & Geosciences*, 62, 216–226.

Burnett, C., & Blaschke, T. (2003). A multi-scale segmentation/object relationship modelling methodology for landscape analysis. *Ecological Modelling*, 168(3), 233–249.

Chen, G., Hay, G. J., Carvalho, L. M., & Wulder, M. A. (2012). Object-based change detection. *International Journal of Remote Sensing*, 33(14), 4434–4457.

Chen, G., Weng, Q., Hay, G. J., & He, Y. (2018). Geographic object-based image analysis (GEOBIA): Emerging trends and future opportunities. *GIScience & Remote Sensing*, 55(2), 159–182. http://doi.org/10.1080/15481603.2018.1426092

Chubey, M. S., Franklin, S. E., & Wulder, M. A. (2006). Object-based analysis of Ikonos-2 imagery for extraction of forest inventory parameters. *Photogrammetric Engineering and Remote Sensing*, 72(4), 383–394.

Cleve, C., Kelly, M., Kearns, F. R., & Moritz, M. (2008). Classification of the wildland–urban interface: A comparison of pixel-and object-based classifications using high-resolution aerial photography. *Computers, Environment and Urban Systems*, 32(4), 317–326.

Clinton, N., Holt, A., Scarborough, J., Yan, L., & Gong, P. (2010). Accuracy assessment measures for object-based image segmentation goodness. *Photogrammetric Engineering and Remote Sensing*, 76, 289–299.

Colwell, R. N. (1965). The extraction of data from aerial photographs by human and mechanical means. *Photogrammetria*, 20(6), 211–228.

Colwell, R. N. (1973). Remote sensing as an aid to the management of earth resources. *American Scientist*, 61(2), 175–183.

Costa, G. A. O. P., Feitosa, R. Q., Fonseca, L. M. G., Oliveira, D. A. B., Ferreira, R. S., & Castejon, E. F. (2010). Knowledge-based interpretation of remote sensing data with the InterIMAGE system: Major characteristics and recent developments. *Paper presented at the 3rd Int. Conference on Geographic Object-Based Image Analysis (GEOBIA 2010)*, Ghent, Belgium.

Cracknell, A. P. (1998). Synergy in remote sensing – what's in a pixel? *International Journal of Remote Sensing*, 19(11), 2025–2047.

Deng, Z., Sun, H., Zhou, H., Zhao, J., Lei, L., & Zou, H. (2018). Multi-scale object detection in remote sensing imagery with convolutional neural networks. *ISPRS Journal of Photogrammetry and Remote Sensing*, 145(Part A), 3–22. ISSN 0924-2716. https://doi.org/10.1016/j.isprsjprs.2018.04.003. https://www.sciencedirect.com/science/article/pii/S0924271618301096

Dorren, L. K., Maier, B., & Seijmonsbergen, A. C. (2003). Improved Landsat-based forest mapping in steep mountainous terrain using object-based classification. *Forest Ecology and Management*, 183(1–3), 31–46.

Ebert, A., Kerle, N., & Stein, A. (2009). Urban social vulnerability assessment with physical proxies and spatial metrics derived from air-and spaceborne imagery and GIS data. *Natural Hazards*, 48(2), 275–294.

Feng, C., Zhang, W., Deng, H., Dong, L., Zhang, H., Tang, L., Zheng, Y., & Zhao, Z. (2023). A combination of OBIA and random forest based on visible UAV remote sensing for accurately extracted information about weeds in areas with different weed densities in farmland. *Remote Sensing*, 15, 4696. https://doi.org/10.3390/rs15194696

Filippi, A. M., Güneralp, İ., Castillo, C. R., Ma, A., Paulus, G., & Anders, K.-H. (2022). Comparison of image endmember- and object-based classification of very-high-spatial-resolution unmanned aircraft system (UAS) narrow-band images for mapping riparian forests and other land covers. *Land*, 11, 246. https://doi.org/10.3390/land11020246

Flanders, D., Hall-Beyer, M., & Pereverzoff, J. (2003). Preliminary evaluation of eCognition object-based software for cut block delineation and feature extraction. *Canadian Journal of Remote Sensing*, 29(4), 441–452.

Fu, K.-S., & Mui, J. (1981). A survey on image segmentation. *Pattern Recognition*, 13(1), 3–16.

Gao, Y., Lang, S., Tiede, D., Gella, G. W., & Wendt, L. (2022). Comparing OBIA-generated labels and manually annotated labels for semantic segmentation in extracting refugee-dwelling footprints. *Applied Science*, 12, 11226. https://doi.org/10.3390/app122111226

Goodchild, M. F. (1992). Geographical information science. *International Journal of Geographic Information Systems*, 6, 31–45.

Goodchild, M. F. (2007). Citizens as sensors: The world of volunteered geography. *GeoJournal*, 69(4), 211–221.

Green, K., & Lopez, C. (2007). Using object-oriented classification of ADS40 data to map the benthic habitats of the state of Texas. *Photogrammetric Engineering and Remote Sensing*, 73(8), 861.

Gui, S., Song, S., Qin, R., & Tang, Y. (2024). Remote sensing object detection in the deep learning era—a review. *Remote Sensing*, 16(2), 327. https://doi.org/10.3390/rs16020327

Guo, Q., Kelly, M., Gong, P., & Liu, D. (2007). An object-based classification approach in mapping tree mortality using high spatial resolution imagery. *GIScience & Remote Sensing*, 44(1), 24–47.

Haralick, R. M., & Shapiro, L. (1985). Survey: Image segmentation techniques. *Computer Vision, Graphics, and Image Processing*, 29, 100–132.

Harvey, F., & Raskin, R. G. (2011). Spatial cyberinfrastructure: Building new pathways for geospatial semantics on existing infrastructures. *Geospatial Semantics and the Semantic Web* (pp. 87–96). Springer US.

Hay, G. J., & Blaschke, T. (2010). Special issue: Geographic object-based image analysis (GEOBIA). *Photogrammetric Engineering and Remote Sensing*, 76(2), 121–122.

Hay, G. J., & Castilla, G. (2006). Object-based image analysis: Strengths, weaknesses, opportunities and threats (SWOT). *International Archives of Photogrammetry, Remote Sensing and Spatial Information Sciences*, 36(4).

Hay, G. J., & Castilla, G. (2008). Geographic object-based image analysis (GEOBIA): A new name for a new discipline. In T. Blaschke, S. Lang & G. J. Hay (Eds.), *Object Based Image Analysis* (pp. 93–112). Heidelberg, Berlin and New York: Springer.

Hölbling, D., & Neubert, M. (2008). ENVI feature extraction 4.5. Snapshot. *GIS Business*, 48–51.

Hossain, M. D., & Chen, D. (2019). Segmentation for object-based image analysis (OBIA): A review of algorithms and challenges from remote sensing perspective. *ISPRS Journal of Photogrammetry and Remote Sensing*, 150(2019), 115–134. ISSN 0924-2716. https://doi.org/10.1016/j.isprsjprs.2019.02.009. https://www.sciencedirect.com/science/article/pii/S0924271619300425

Jawak, S. D., Wankhede, S. F., Luis, A. J., & Balakrishna, K. (2022). Effect of image-processing routines on geographic object-based image analysis for mapping glacier surface facies from Svalbard and the Himalayas. *Remote Sensing*, 14, 4403. https://doi.org/10.3390/rs14174403

Jensen, J. R. (1996). *Introductory Digital Image Processing: A Remote Sensing Perspective*. Prentice-Hall Inc.

Jiang, B., An, X., Xu, S. et al. (2023). Intelligent image semantic segmentation: A review through deep learning techniques for remote sensing image analysis. *Journal of the Indian Society of Remote Sensing*, 51, 1865–1878. https://doi.org/10.1007/s12524-022-01496-w

Kettig, R. L., & Landgrebe, D. A. (1976). Classification of multispectral image data by extraction and classification of homogeneous objects. *IEEE Transactions on Geoscience and Remote Sensing*, 14(1), 19–26.

Körting, T. S., Garcia Fonseca, L. M., & Câmara, G. (2013). GeoDMA—geographic data mining analyst. *Computers & Geosciences*, 57, 133–145.

Kuhn, T. S. (1962). *The Structure of Scientific Revolutions*. Chicago: The Chicago University Press.

Laliberte, A. S., Fredrickson, E. L., & Rango, A. (2007). Combining decision trees with hierarchical object-oriented analysis for mapping arid rangelands. *Photogrametric Engineering & Remote Sensing*, 73(2), 197–207.

Laliberte, A. S., Goforth, M. A., Steele, C. M., & Rango, A. (2011). Multispectral remote sensing from unmanned aircraft: Image processing workflows and applications for rangeland environments. *Remote Sensing*, 3(11), 2529–2551.

Laliberte, A. S., Rango, A., Havstad, K. M., Paris, J. F., Beck, R. F., McNeely, R., & Gonzalez, A. L. (2004). Object-oriented image analysis for mapping shrub encroachment from 1937 to 2003 in southern New Mexico. *Remote Sensing of Environment*, 93(1–2), 198–210.

Lang, S. (2008). Object-based image analysis for remote sensing applications: Modeling reality—dealing with complexity. In T. Blaschke, S. Lang & G. J. Hay (Eds.), *Object-based Image Analysis* (pp. 1–25). Heidelberg, Berlin & New York: Springer.

Lang, S., & Blaschke, T. (2006). Bridging remote sensing and GIS—what are the main supporting pillars? *International Archives of Photogrammetry, Remote Sensing and Spatial Information Sciences*, XXXVI-4/C42.

Lang, S., Kienberger, S., Tiede, D., Hagenlocher, M., & Pernkopf, L. (2014). Geons–domain-specific regionalization of space. *Cartography and Geographic Information Science*, 41(3), 214–226.

Leberl, F., & Gruber, M. (2003). Flying the new large format digital aerial camera Ultracam. *Photogrammetric Week*, 3, 67–76.

Leichtle, T., Geiß, C., Wurm, M., Lakes, T., & Taubenböck, H. (2017). Unsupervised change detection in VHR remote sensing imagery—an object-based clustering approach in a dynamic urban environment. *International Journal of Applied Earth Observation and Geoinformation*, 54(2017), 15–27. ISSN 1569-8432. https://doi.org/10.1016/j.jag.2016.08.010. https://www.sciencedirect.com/science/article/pii/S0303243416301490

Li, Z., Wang, Y., Zhang, N., Zhang, Y., Zhao, Z., Xu, D., Ben, G., & Gao, Y. (2022). Deep learning-based object detection techniques for remote sensing images: A survey. *Remote Sensing*, 14(10), 2385. https://doi.org/10.3390/rs14102385

Liu, T., Abd-Elrahman, A., Morton, J., & Wilhelm, V. L. (2018). Comparing fully convolutional networks, random forest, support vector machine, and patch-based deep convolutional neural networks for object-based wetland mapping using images from small unmanned aircraft system. *GIScience & Remote Sensing*, 55(2), 243–264. http://doi.org/10.1080/15481603.2018.1426091

Long, Y., Gong, Y., Xiao, Z., & Liu, Q. (2017). Accurate object localization in remote sensing images based on convolutional neural networks. *IEEE Transactions on Geoscience and Remote Sensing*, 55(5), 2486–2498, May. http://doi.org/10.1109/TGRS.2016.2645610

Maxwell, S. K. (2010). Generating land cover boundaries from remotely sensed data using object-based image analysis: Overview and epidemiological application. *Spatial and Spatio-Temporal Epidemiology*, 1(4), 231–237.

McKeown Jr, D. M., Harvey, W. A., & Wixson, L. E. (1989). Automating knowledge acquisition for aerial image interpretation. *Computer Vision, Graphics, and Image Processing*, 46(1), 37–81.

McLuhan, M. (1964). *Understanding Media*. New York: McGraw-Hill.

Metaferia, M. T., Bennett, R. M., Alemie, B. K., & Koeva, M. (2023). Furthering automatic feature extraction for fit-for-purpose cadastral updating: Cases from peri-urban Addis Ababa, Ethiopia. *Remote Sensing*, 15, 4155. https://doi.org/10.3390/rs15174155

Moreno, N., Wang, F., & Marceau, D. J. (2010). A geographic object-based approach in cellular automata modeling. *Photgrammetric Engineering and Remote Sensing*, 76(2), 183–191.

Mudau, N., & Mhangara, P. (2023). Mapping and assessment of housing informality using object-based image analysis: A review. *Urban Science*, 7, 98. https://doi.org/10.3390/urbansci7030098

Pal, R., & Pal, K. (1993). A review on image segmentation techniques. *Pattern Recognition*, 26(9), 1277–1294.

Pawar, B., Prakash, V., Garg, L., Galdies, C., Buttigieg, S., & Calleja, N. (2024). A review of satellite image analysis tools. In V. K. Gunjan & J. M. Zurada (Eds.), *Proceedings of 4th International Conference on Recent Trends in Machine Learning, IoT, Smart Cities and Applications*. ICMISC 2023. Lecture Notes in Networks and Systems, vol 873. Singapore: Springer. https://doi.org/10.1007/978-981-99-9442-7_65

Pascual, C., García-Abril, A., García-Montero, L. G., Martín-Fernández, S., & Cohen, W. (2008). Object-based semi-automatic approach for forest structure characterization using LiDAR data in heterogeneous *Pinus sylvestris* stands. *Forest Ecology and Management*, 255(11), 3677–3685.

Putra, O. C., & Wijayanto, A. W. (2023). Automatic detection and counting of oil palm trees using remote sensing and object-based deep learning. *Remote Sensing Applications: Society and Environment*, 29(2023), 100914. ISSN 2352-9385. https://doi.org/10.1016/j.rsase.2022.100914. https://www.sciencedirect.com/science/article/pii/S2352938522002221

Resch, B. (2013). People as sensors and collective sensing-contextual observations complementing geo-sensor network measurements. In *Progress in Location-Based Services* (pp. 391–406). Berlin & Heidelberg: Springer.

Rohracher, H. (2003). The role of users in the social shaping of environmental technologies. *Innovation*, 16(2), 177–192.

Sagl, G., Loidl, M., & Beinat, E. (2012). A visual analytics approach for extracting spatio-temporal urban mobility information from mobile network traffic. *ISPRS International Journal of Geo-Information*, 1(3), 256–271.

Salehi, B., Ming Zhong, Y., & Dey, V. (2012). A review of the effectiveness of spatial information used in urban land cover classification of VHR imagery. *International Journal of Geoinformatics*, 8(2), 35–51.

Schuurman, N. (1999). Critical GIS: Theorizing an emerging science. *Cartographica*, 36(4), 1–101.

Sheppard, E. (2006). Knowledge production through critical GIS: Genealogy and prospects. *Cartographica*, 40, 5–21.

Stow, D., Lopez, A., Lippitt, C., Hinton, S., & Weeks, J. (2007). Object-based classification of residential land use within Accra, Ghana based on QuickBird satellite data. *International Journal of Remote Sensing*, 28(22), 5167.

Tiede, D. (2014). A new geospatial overlay method for the analysis and visualization of spatial change patterns using object-oriented data modeling concepts. *Cartography and Geographic Information Science*, 41(3), 227–234.

Tiede, D., Huber, J., & Kienberger, S. (2012). Implementation of an interactive WebGIS-based OBIA geoprocessing service. In *International Conference on Geographic Object-Based Image Analysis, 4 (GEOBIA), May 7–9, 2012* (pp. 402–406).

Townshend, J. R. G., Huang, C., Kalluri, S. N. V., Defries, R. S., Liang, S., & Yang, K. (2000). Beware of per-pixel characterization of land cover. *International Journal of Remote Sensing*, 21(4), 839–843.

Troyo, A., Fuller, D. O., Calderon Arguedas, O., Solano, M. E., & Beier, J. C. (2009). Urban structure and dengue incidence in Puntarenas, Costa Rica. *Singapore Journal of Tropical Geography*, 30(2), 265–282.

Tullis, J. A., Jensen, J. R., Raber, G. T., & Filippi, A. M. (2010). Spatial scale management experiments using optical aerial imagery and LiDAR data synergy. *GIScience & Remote Sensing*, 47, 338–359.

Yan, G., Mas, J.-F., Maathuis, B. H. P., Xiangmin, Z., & Van Dijk, P. M. (2006). Comparison of pixel-based and object-oriented image classification approaches—a case study in a coal fire area, Wuda, Inner Mongolia, China. *International Journal of Remote Sensing*, 27(18), 4039–4055.

Yang, K., Zhang, H., Wang, F., & Lai, R. (2022). Extraction of broad-leaved tree crown based on UAV visible images and OBIA-RF model: A case study for Chinese olive trees. *Remote Sensing*, 14, 2469. https://doi.org/10.3390/rs14102469

Ye, S., Pontius, R. G., & Rakshit, R. (2018). A review of accuracy assessment for object-based image analysis: From per-pixel to per-polygon approaches. *ISPRS Journal of Photogrammetry and Remote Sensing*, 141(2018), 137–147. ISSN 0924-2716. https://doi.org/10.1016/j.isprsjprs.2018.04.002. https://www.sciencedirect.com/science/article/pii/S0924271618300947

Ye, S., Zhu, Z., & Cao, G. (2023). Object-based continuous monitoring of land disturbances from dense Landsat time series. *Remote Sensing of Environment*, 287(2023), 113462. ISSN 0034-4257. https://doi.org/10.1016/j.rse.2023.113462. https://www.sciencedirect.com/science/article/pii/S0034425723000135

Yu, Q., Gong, P., Clinton, N., Kelly, M., & Schirokauer, D. (2006). Object-based detailed vegetation classification with airborne high spatial resolution remote sensing imagery. *Photogrammetric Engineering and Remote Sensing*, 72(7), 799–811.

Yue, P., Di, L., Wei, Y., & Han, W. (2013). Intelligent services for discovery of complex geospatial features from remote sensing imagery. *ISPRS Journal of Photogrammetry and Remote Sensing*, 83, 151–164.

Zhang, C., Sargent, I., Pan, X., Li, H., Gardiner, A., Hare, J., & Atkinson, P. M. (2018). An object-based convolutional neural network (OCNN) for urban land use classification. *Remote Sensing of Environment*, 216(2018), 57–70. ISSN 0034-4257. https://doi.org/10.1016/j.rse.2018.06.034

Zhao, W., Du, S., & Emery, W. J. (2017). Object-based convolutional neural network for high-resolution imagery classification. *IEEE Journal of Selected Topics in Applied Earth Observations and Remote Sensing*, 10(7), 3386–3396, July. http://doi.org/10.1109/JSTARS.2017.2680324.

Zhou, W., & Troy, A. (2008). An object-oriented approach for analysing and characterizing urban landscape at the parcel level. *International Journal of Remote Sensing*, 29(11), 3119–3135.

Zook, M., Graham, M., Shelton, T., & Gorman, S. (2010). Volunteered geographic information and crowdsourcing disaster relief: a case study of the Haitian earthquake. *World Medical & Health Policy*, 2(2), 7–33.

9 Geospatial Data Integration in OBIA—Implications of Accuracy and Validity

Stefan Lang and Dirk Tiede

LIST OF ACRONYMS

ARD	Analysis ready data
EO	Earth observation
EU	European Union
FCA	Formal concept analysis
IUS	Image understanding system
KOS	Knowledge organizing system
LCCS	Land cover classification system
NIR	Near infrared
OBAA	Object-based accuracy assessment
OBCD	Object-based change detection
OBIA	Object-based image analyses
OFA	Object fate analysis
OWL	Ontology Web Language
SDGs	Sustainable development goals
SDI	Spatial data infrastructures
SWEET	Semantic web for Earth and environmental terminology
VHR	Very high resolution
WFD	Water framework directive

9.1 CONDITIONED INFORMATION

The advancement of feature recognition, scene reconstruction and advanced image analysis techniques facilitates the extraction of thematic information, for policy making support and informed decisions (Mollick et al. 2023; Nasiri et al. 2023; Kotaridis and Lazaridou 2022; Vizzari 2022; Zaabar et al. 2022; Lang et al. 2019; Luciano et al. 2019; Yurtseven et al. 2019; Chen et al. 2018; Labib and Harris 2018; Ye et al. 2018; Shahi et al. 2017). As a strong driver, the ubiquitous availability of image data meets an ever-increasing need for updated geo-information. In response, we strive for methods that exploit image information comprehensively and likewise more effectively. Object-based image analysis (OBIA), with its explicit focus on objects and their relationships, enables to address and model composite and spatially defined classes, enriching information extracted from image data (Mollick et al. 2023; Vizzari 2022; Luciano et al. 2019; Labib and Harris 2018; Ye et al. 2018). The OBIA approach in this context is inherently linked to an expert-based strategy that focuses on the handling, (hierarchical) modeling, and validation of image objects irrespective of the initial object generation process.

 DOI: 10.1201/9781003541158-10

9.1.1 OBIA in Support to Geospatial Information Needs

Object-based image analysis, as the name indicates, operates on objects that represent "real-world" entities as the constituents of our geographical reality. Spatial concepts (basically size, shape, and distance) are intrinsic parameters to be considered in the information extraction process, making spatial characteristics an additional feature domain in classification. Using regionalization techniques for image segmentation, there is a complementarity (say, a trade-off) between spectral and spatial similarity, that is, closeness in color and neighborhood. In other words, the spatial constraint balances the spectral behavior, which leads to (scalable) generalization and a reduction of the so-called salt-and-pepper effect (Bischof et al. 1992; Blaschke and Strobl 2001). In OBIA, the class that a group of picture elements is assigned to, depends on both spatial and spectral characteristics. This may sound like a restriction narrowing down classification power, but it actually *extends* the set of target classes that can be addressed. More precisely, it opens another dimension of potential target classes (Lang 2008), on two additional semantic levels next to spectral classes (see Figure 9.1). This entails, at first, all subcategories or instances of a spectral class that are defined by shape features on individual object level: a class <water> may be further specified as <lake | river> depending on its length/width ratio. A class <built-up> may split into several subcategories of a village typology based on footprint physiognomy. Second, and even more crucial, we find spatial properties in terms of relations among objects. We will come back to this in Section 9.2, when discussing the term "class modeling." In a nutshell, the integration of object relationships (including relative coverage, distance, and in particular topological features), allows for addressing complex, composite target classes on high semantic level.

While "object-based" suggests objects to be the fundamental elements to base the analysis on, OBIA goes beyond objects as isolated items distributed over space (Nasiri et al. 2023; Kotaridis and Lazaridou 2022). In capturing and utilizing the manifold relationships among these objects, we can use additional features in classification that exceed classical quantitative measures. Topological and hierarchical relationships are more qualitative features that complement geometrical standard features such as object size, or spectral or spatial distance. The trend set by OBIA—some may call it a paradigm shift,

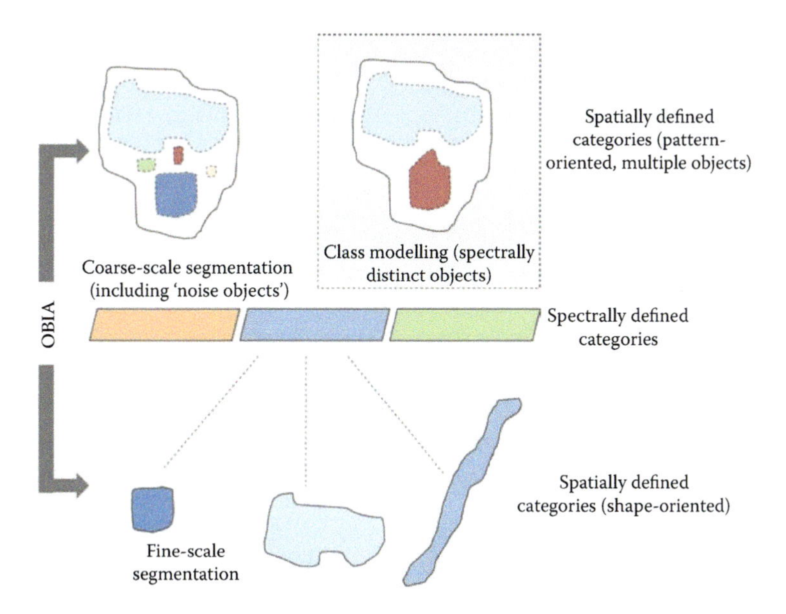

FIGURE 9.1 Extended set of categories (target classes) using features in addition to spectral features (see text for further explanations).

cf. Blaschke et al. (2014)—from pixels (as technically defined units), via objects (as spatially manifested and conceptually ascertained entities) towards relational patterns (as multi-scalar qualities of objects) opens a plethora of possible new ways of flexible spatial representations.

The explicit focus on spatial properties and relationships of OBIA[1] allows us to address higher-level semantic classes, including spatial composite objects that "emerge" (in terms of relevance) on a certain phenomenon scale. Here we clearly meet the ever increasing demand for geospatial information from the perspective of various space-related policies (Lang et al. 2019; Lang et al. 2009b). The term "space-related" is used in a broad sense to comprise all policies with a spatial component that are—directly or indirectly—relying on critical geospatial information updates. Such policies entail societal and environmental domains across geographical scale domains (i.e., from local to continental), for example a local policy for climate change adaption, a regional policy of sustainable urban development, the EU (European Union) policy on integrated water management (Water Framework Directive, WFD), or the globally adopted sustainable development goals (SDGs). The primary data source is increasingly imagery with respective information extracted for a specific purpose. Earth observation (EO) is a term that entails societal relevance (Lang et al. 2019; Lang 2008) and thus highlights the capacity of satellite (and other) sensor technology for the purpose of updated information provision in civil application domains. EO comprising space and ground infrastructure for satellite imagery, sensor networks, and the analysis capabilities supports societal benefits and the political ambition of such policies. Recent examples of international endeavors is the European initiative Copernicus as a large-scale contribution to the intergovernmental Group on Earth Observation. In order to reach a consistent approach to monitoring manifold environmental and societal conditions, considerable efforts are made to investigate opportunities EO may provide to facilitate the reporting requirements and the evaluation of intervention options (Lang et al. 2019; Lang et al. 2009b). EO technology is increasingly the tool of choice for achieving the objectives of multilateral agreements by (1) providing contextual spatial knowledge about the underlying processes; (2) supporting the efficient management of monitoring tasks; and (3) contributing overall to more effectiveness of conventions or treaties. Review articles and white papers document the information needs in the various societal domains, as well as the phenomenon scale and respective policy scale, and link these to existing Earth observation capacity (for an overview see, Zeil et al. (2008)). The scoping on the demand side is done regularly when planning new satellite missions and the suite of onboard instruments. The Copernicus Sentinel fleet, with its various sensor types and instruments, is designed to be as flexible and generic in application as possible.

A widespread demand requires, next to the adequate observation systems, highly capable analysis tools and methods to transform the enormous amount of raw data into—first—analysis ready data (ARD) as a midstream step (Baraldi et al., 2022) and then meaningful and ready-to-use information. OBIA may prove this capability in the following (non-exclusive) areas:

- OBIA, through its multi-scale option (Mollick et al. 2023; Chen et al. 2018; Labib and Harris 2018; Ye et al. 2018; Shahi et al. 2017; Marceau 1999; Hay et al. 2001), enables to match the scale inherent in the source data to the scale of observation or intervention. With an appropriate resolution level of the imagery, we can work flexibly towards a commensurate analysis scale.
- The object-oriented data model inherent to the OBIA approach supports complex and hierarchical class models, transferable and adjustable between application domains and geographical areas (Tiede 2014). This is not to say that OBIA entails fully automated procedures, but the adjustment of once established class model architectures is a minor effort as compared to manual grouping and reclassifications (Zaabar et al. 2022; Lang et al. 2019; Labib and Harris 2018; Ye et al. 2018; Shahi et al. 2017; Tiede et al. 2010).
- OBIA not only produces objects but also integrates existing geographical features in the analysis process. In many domains, the update of (existing) geospatial information with existing geometries is more critical than the provision of new, unprecedented units (Lang

et al. 2019; Ye et al. 2018; Shahi et al. 2017; Tiede et al. 2007). Geodata from existing spatial data infrastructures may be used as a reference dataset, or potential boundaries to report results on. Remote sensing products used to be generated outside authoritative workflows and a match with existing geodata repositories remains a challenging task (see Section 9.3). A missing agreement may occur in terms of nomenclature (thematic congruence) and cartographical generalization (geometrical match). Interestingly, by addressing the spatial component in the analysis process we can improve both, as spatial detail goes hand in hand with thematic depth.

Spatial data policies like the European INSPIRE directive (2007/2/EC) seeks to provide a common baseline for interoperable data usage, to overcome technological constraints via standardization and opening the way to a content-driven debate. Information provision, whether spatial or nonspatial requires compatibility to existing data pools. Spatial data infrastructures (SDI) and large-scale data repositories or geospatial data ecosystems store data in a given level-of-detail explicitly specified by cartographic scale, which needs remote sensing products to be aligned accordingly. This is not only limited to spatial or thematic (dis-)agreement but also relates to the fitness of information products derived from different sources. Decision makers should be able to interpret the information unambiguously, but this key requirement of usability and fitness-for-purpose need to be specifically addressed by remote sensing products so that decision makers and stakeholders clearly see a benefit in enriching their daily routines and established workflows (Lang et al. 2009b).

Let's consider an example (see Figure 9.2) that will be discussed in greater detail in Section 9.3: the representation of different agricultural parcels derived by image segmentation from SPOT-5 data might not suffice for users because the analysis scale does not match the scale of intervention. The regional development plan in the Stuttgart Metropolitan area in Southwest Germany requires a baseline geometry representing so-called biotope complexes (Schumacher et al. 2007). These are functional, composed spatial units whose outlines should resemble, where applicable, the existing digital cadaster boundaries. The automated segmentation shown in Figure 9.2a provides neither of these. By means of a strategy called "spatially constraint class modeling" (Tiede et al. 2010), the authors were able to deliver the appropriate geospatial information (Figure 9.2b).

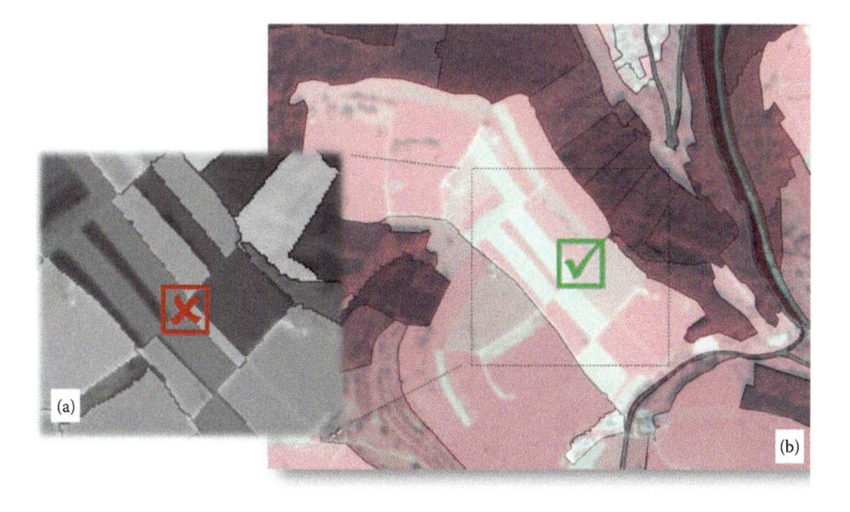

FIGURE 9.2 Object understanding as a key to an adequate representation: (a) unconditioned information, simple delineation of land cover objects; (b) conditioned information, aggregated biotope complex.

9.1.2 Enriched Information through OBIA

As pointed out by several authors (Kotaridis and Lazaridou 2022; Vizzari 2022; Blaschke et al. 2007) at the launching phase of the Copernicus initiative some 20 years ago, EO-based monitoring systems were expected to once be capable of transforming complex scene content into ready-to-use information, in a reliable, transferable, effective, and—desirably—cost-efficient manner. Global commitments, directives, and policies with their demand for timely, accurate, and conditioned geospatial information (Lang et al. 2019; Lang et al. 2008), valuate and valorize the load of raw data collected from various monitoring systems. With a technically and spatially more literate user community asking for advanced geospatial products, and expressing their needs accordingly, the overall consciousness level pertains to prevailing societal challenges (Lang et al. 2009a).

The advancement in feature recognition and image analysis techniques facilitates the extraction of rich thematic information, irrespective of a particular application domain. The term "rich" relates to semantic content and requirements from particular application fields or policy contexts, beyond generic land cover classification schemes (Augustin et al. 2019; Lang et al. 2009a). With increasing capacities both on the imaging and the processing side of observation systems, high expectations (Lang 2008) were raised from all sides, including academia, service providers, and users at the dawning of very high resolution (VHR) data (Aschbacher 2002), the such as WorldView, Pléiades, etc. Despite some shortcomings may exist in data provision and handling, data acquired through EO missions are abundant, so that users as well as service providers increasingly challenged by integrating and valorizing, rather than lacking, these data. Lang et al. (2009b) use the term "information conditioning" for the act of turning unstructured data into ready-to-use information, thus turning big Earth data into true value (Sudmanns et al., 2019). OBIA, also called GEOBIA (Hay and Castilla 2008) to highlight geographic application context, provides the toolbox, capable of treating users' needs by combining intelligent tools and algorithms for automatically delineating, extracting, and categorizing geographical objects from remotely sensed images (Lang et al. 2019; Shahi et al. 2017; Lang et al. 2010a). Methodologically, OBIA seeks to accommodate various geospatial concepts, remote sensing principles (spectral behavior) and the wealth of experience from visual interpretation and manual delineation (Hay and Castilla 2008).

Decision makers and other policy implementation bodies pursue their missions; in other words: informed decisions are mostly embedded in any kind of operational workflow, with each decision underlying a spatial constraint, often linked to political boundaries. This kind of dual constraint (both space- and mandate/task-related) is called the *policy scope* (Lang et al. 2008). "Conditioned information is the result of a process to fulfill the user demand in technological and conceptual sense and has undergone an evaluation for its operational use ('user validation')" (Lang et al. 2010a: 3). Conditioned information thus means a full match between geospatial information provided and the accommodation of this information in established workflows.

Another key aspect is the issue of scale. According to hierarchy theorists (Allen and Starr 1982), scale is defined as the "period of time or space over which signals are integrated to give message." Converting imagery to information ideally follows the principles of communication models. The role of image interpretation or classification is a kind of translation channeling the oft-overwhelming (unstructured) information residing in images. Scale is an intrinsic aspect in communication, when signals need to be filtered and integrated to give message. Here we need to match the phenomenon scale in general and the policy scale in particular. The flexibility of current digital representations turns a fixed cartographic scale on maps to a visualization scale confining dynamically the level of detail of displayed information (Montello 2001). Whether in traditional paper map production or dynamic on-screen visualizations, we rely on the cartographic principle of generalization. This refers to the reduction of redundant detail with coarser scale, a filtering effect that makes representations commensurate to the scale of interest (simplification, enhancement, selection). In digital representations, generalization can be achieved by (1) algorithmic smoothing of lines or polygon outlines by dropping a selected number of vertices; (2) selective display of raster cells or pixels with decreasing zoom level, (3) manual, context-specific omissions or exaggerations to suppress or emphasize certain cartographic aspects.

OBIA supports the idea of multi-scale representation (Mollick et al. 2023; Zaabar et al. 2022; Yurtseven et al. 2019; Chen et al. 2018; Hay et al. 2001) by enabling nested hierarchies of image objects in several scales. In this process, generalization is a particular challenge, balancing between spatial embeddedness and appropriate scale. Either the object hierarchy is strictly fitting, that is, outlines of objects are not generalized but fully nested (scale-adaptive strategy); or boundaries are generalized according to the scale level and the object hierarchy is not spatial explicit any longer discussed (scale-specific strategy).

The challenges extend to concepts and methods of assessing and validating the results. This includes specific object-based approaches to accuracy assessment and change detection (Lang 2008). "Object-based" implies that accuracy or changes are not only assessed thematically but also spatially and on object level: this refers to assessing specificities in object delineation, that is, their spatial representation (position and precision of boundary), and to what degree these representations change over time.

9.2 OBJECT VALIDITY

The transformation and reconstruction of an image scene into ready-to-use geospatial information implies that image objects may be correctly classified but not "valid." This may sound contradictory, but the power to delineate spatially flexible units challenges the classical concept of binary accuracy assessment (being *correctly* classified or not). In addition to the assigned label or category, we need to evaluate the way units are delineated. This implies issues of scale, generalization, as well as representation (including visualization). We subsume the entirety of aspects considered in this process under the term "object validity."

9.2.1 Color and Form: Elements of Image Understanding

The term "image understanding" (Pinz 1994; Zlatof et al. 2004) refers to the reconstruction of the imaged scene as a complete process of transforming reflectance values into symbolic representations meaning ultimately, to valid information. Ideally, the target symbology is a fixed set of meaningful patterns, which are detectable in an image. This applies, for example, to optical character recognition where a scanned document is analyzed for the occurrence of a given set of characters (letters, numbers, and special characters). Their appearance may vary within certain ranges, but in each language or script there is a fixed and limited set of characters to target on. We can objectively assess the success rate of this image reconstruction by counting the correctly identified characters. Once characters are rightly extracted, a word processor is able to interpret their positions relative to each other and eventually identify words, which can be further "understood" by a spellchecker or even be read out by the machine. A visually impaired person, not able to read the content of a letter, may consider such image-understanding capabilities as a "full match" between the information provided and his or her need (while any interpretation error can be clearly spotted out as well). Characters to represent letters, words, etc., are standardized representations (just as notes are for musical events) as a means to encode and decode information in a (thereby regulated) communication process. In other words, the process of information exchange uses unequivocal, even standardized, symbology ("fonts") and is thereby well-posed and automatable.

Returning to remotely sensed images, there is no such fixed target symbology through which we could communicate content and exchange information in a standardized way. As close as possible we get to this is applying discrete color levels, which are intersubjectively comprehensible and used consensually (Baraldi, 2011). Still, even if color categories exist, color remains a continuous phenomenon whose discretization is per intercultural convention and not standardized. A pixel (picture element, i.e., the smallest, technically defined unit) appears in a certain color, which results from an integrated signal of the spectral reflectance of a deliberately small fraction of the Earth's surface. The spectral resolution (number of bands, *n*) determines the *n*-dimensionality of the feature space, which controls—along with the radiometric resolution (number of quantization levels, a.k.a. "bit-depth")—the number of discernible colors.

Yet, what do the colors mean? How do we reach from a sub-symbolic level (color impression) to a symbolic one (the actual class)? According to the radiometric principles of remote sensing, the integrated signal stored in a pixel represents physical conditions of the ground ("land cover"), so we can link color with corresponding physical models using spectral signatures. Globally applicable classification schemes such as the LCCS (Land Cover Classification System) (Di Gregorio and Jansen 2005) use spectral signatures to convert reflectance values (via colors) into nominal classes with a semantic meaning, that is, (land cover) categories. With higher spectral sensitivity, that is, increasing number of bands, we can incorporate more and better physical models in a classification scheme, when analyzing, for example, vegetation health versus stress in the near infrared (NIR) spectrum or the presence of snow against clouds in the short-wave IR spectrum. An error matrix is used to check the respective assignment of a pixel to a nominal class in a procedure called site-specific (meaning pixel-by-pixel) accuracy assessment (Congalton and Green 1998). The semantic ambiguity in the classification process (e.g., a mixed forest pixel can have a similar color as a pure deciduous forest under shady conditions) can be tackled by applying fuzzy logic (Zadeh 1965), which uses class membership probabilities rather than crisp class assignments. Since a pixel is a technically defined unit, which has no immediate correspondence with real-world features, the assessment focuses on the color-based class assignment of this particular pixel. For the actual assessment, independent reference data need to be gathered from the ground or from any other data source that qualifies as a source of "truth." This implies a certain level of detail and timeliness, which often challenges an ideal assessment due to limitations imposed by inaccessibility, security, and time/costs. Here we encounter a certain paradoxon: the main reason for *remote* sensing is inaccessibility of an area and, consequently, a lack of actual ground information; while, for a proper assessment, we would assume such perfect information is available as a reference. To cope with that, visual inspection of the classification is an alternative (Mollick et al. 2023; Zaabar et al. 2022; Lang et al. 2019; Campbell 2002), which assumes there is an agreement between the color impression on screen and the evaluator's experience of physical models and the spectral behavior of various land cover types in color space. Usually, due to our everyday experience, a visual inspector considers spatial context as additional cues in taking the decision (see Figure 9.3).

Ideally, physical conditions on the ground are the only influence of a pixel's integrated signal, and different signals would only indicate a change in these conditions. In reality, however, various disturbances occur such as atmospheric influences and topographic effects (Richter and Schläpfer 2002), viewing angle of a rotatable VHR sensor, etc., that make the same land cover captured recursively appear differently. While color is the most direct and unambiguous property in image classification at a per-pixel basis, other features of geographic objects, like size, form, orientation, etc., are likewise important. Such geospatial features cannot be established from single pixels alone (Blaschke and Strobl 2001), since they depend on spatial context and a grouping of neighboring pixels into image regions. Geometric measures such as area, perimeter, shape, texture, etc., of such pixel aggregates (segments) characterize these features, while a standardized scheme is hard to establish, due to the variability of geographic features. Imagine a river that has a distinct, elongated shape, yet there is no absolute spatial feature to describe its appearance. Still, geospatial features allow for additional differentiation among target classes on a (higher) symbolic level: a river is an elongated water body, a lake a more compact water body. Such features can be obtained only once neighboring pixels are organized in segments or image objects (Benz et al. 2004). When co-assessing color and form simultaneously, subtle variances in color might be sacrificed for the overall uniform spatial appearance of the aggregate. In other words, a lake is recognized by a human interpreter due to its rather compact form, even if there are some subtle variances in color tones due to varying water depths or the presence of underwater vegetation, etc. Image objects, in

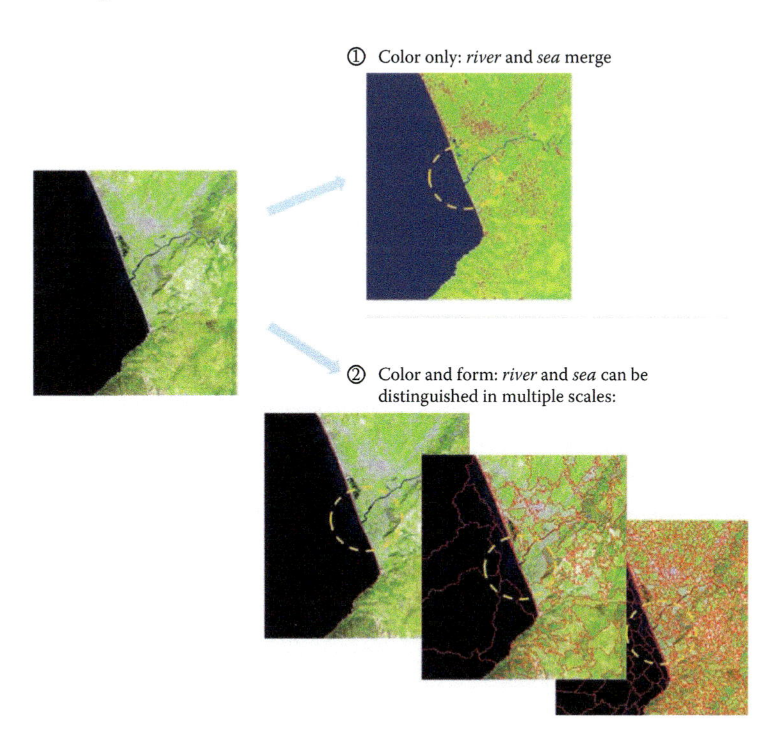

FIGURE 9.3 Shape is a highly distinguishable feature of land cover classes similar in color (here large river and sea).

contrast to pixels, have irregular shapes, which allows for the delineation of meaningful boundaries. A human interpreter applies generalization principles and adapts the delineation of objects according to the mapping scale and the level of detail foreseen by the classification scheme, while higher-level semantics usually imply a coarser scale (Belgiu et al. 2014). In OBIA, often multi-resolution segmentation is a strategy to perform image segmentation at several scale levels to represent the respective multi-level content of a scene scale-adaptively (i.e., without generalization) or scale-specifically (i.e., with generalization) (Luciano et al. 2019; Shahi et al. 2017). Pixel-based assessment methods fail to assess the spatial dimension of object delineations. Lucieer and Stein (2002) distinguish in this context between existential and extensional uncertainty of delineated objects.

To treat color and form separately suggests independence of both aspects. Yet an image understanding system (IUS) in general needs to integrate both aspects, according to models of human vision. An IUS that follows the multi-stage paradigm of human vision by Marr (1982), starts with pre-attentive vision as a first stage, and then works along these building blocks to compose more complex target classes. There are several strategies to accomplish this first stage; two of which are shortly discuss here. The first starts with a symbolic (pre-)classification reflecting pre-attentive vision (Baraldi and Broschetti 2012) according to color similarity based on physical models of spectral behavior as implemented in the SIAM™ (Satellite Image Automatic Mapper) expert system (Baraldi 2011), where pixels are pre-classified by rule-based clustering in color space. Based on that we can form segments by grouping neighboring pixels exhibiting the same symbolic color level. The initial pre-classification generates a semantic level of (pre-)classes, while leaving the spectral distance as high as possible: depending on the aggregation level, an 8-bit 6-band Landsat-7 imagery with 256^6 quantization levels can be meaningfully transformed in a categorization layer ranging from 18 up to 96 spectral categories. While a decreased number of quantization levels is desirable from the perspective of deductive reasoning, challenges arise from grouping similar pixels into a region. Here the second strategy comes into play: multi-scale segmentation provides image objects by grouping neighboring pixels based on the geographic key principles of spatial autocorrelation, a.k.a. regionalization (Wise et al. 2001). By classifying these sub-symbolic segments according to their spectral and spatial behavior, we arrive at a level of (higher) semantic classes. The building blocks used in the first method are pre-classified pixel aggregates—no matter what spatial properties; while in the second we use image regions—fairly homogenous and spatially contiguous—as an input to the IUS, which enables class modeling (Lang 2005). The first strategy rests on heavy use of preexisting knowledge on physical principles and spectral signatures, while the second requires empirically determined parameterization and heuristics about class compositions.

9.2.2 Human versus Machine Vision

Great advances in computer vision over the last decades still do not achieve the power of human vision. While most of the physiological principles like retinal structure, and functioning of singular processes such as the cerebral reaction, are analytically known, we still lack the bigger "picture" on human perception as a whole (Blaschke et al. 2014). In contrast to recent deep learning approaches, where the understanding is outsourced to convolutional and other neural networks architectures relying on huge amount of training samples to learn spatial patterns and relations in varying local neighborhoods (Yuan et al. 2021), OBIA tries to make explicit basic principles of human perception (Lang et al. 2019; Ye et al. 2018; Shahi et al. 2017; Lang 2008), so it is important to understand how we deal with imaged information in various scales, how we relate recognized objects in patterns, and how we understand complex scene contents. "The ultimate benchmark of OBIA is human perception" (Lang 2008), but in fact it remains hard to describe what exactly happens when we look at an image and suddenly "see" something (Blaschke et al. 2014).

Pattern recognition—not necessarily the interpretation of it—works without major effort (Eysenck and Keane 1995; Tarr and Cheng 2003). Human perception is a complex matter of filtering relevant signals from noise (Lang et al. 2019; Lang 2008), a selective process of information processing, and, finally, experience. With respect to visual information processing, Marr (1982) provides a conceptual framework of a three-leveled structure. Hence, we distinguish the following levels: (1) the computational level, related to the purpose, logic, and strategy of perception; (2) the algorithmic level dealing with the transfer functions between input and output; and (3) the hardware level that ensures a physical implementation. The computational level is characterized according to (Marr 1982) by several stages of the visual representation of an image's content. These stages (or "sketches") progress in a subsequent manner revealing gradually more detailed information on the scene under concern. The primal sketch involves a 2D description of the radiometric properties of pixels and pixel aggregates and their geometrical characteristics. The pure spectral differentiation of gray shades and color tones makes up the basic level (raw primal sketch), whereas the grouping of spectrally like pixels into geometrical units represents the full primal sketch. The latter is built upon quasi-homogenous regions (blobs) or bounding zones of high contrast (contours or edges). The raw primal sketch marks the stage where image primitives such as blobs, edge segments, and their low-level descriptions are produced. They are then ordered and organized into higher-level place tokens (Marr 1982) or perceptual chunks (Bruce and Green 1990). This is aligned with perceptual organization that logically groups a pattern and transfers it to meaningful symbolic representation, while an arrangement of small place tokens may cause a certain effect of granularity (Julesz 1975).

When moving from image perception to image interpretation, experience gains more attention. Findings in neuro-psychology (Spitzer 2000) suggest that signal processing by our senses uses vector coding in a high-dimensional feature space. "Experience" relates to a certain cluster or nexus in the feature space. When signals are received, they are compared to this experience nexus, to reach from (1) raw data through (2) patterns or aggregates of color and form structured in various levels to (3) relationships between object-concepts in an image. Still, human perception is likely far from mechanistic or linear sequential (Gorte 1998), and often more than one model is used to construct meaning from an image (Vizzari 2022; Labib and Harris 2018; Lang et al. 2004). Image interpretation, in particular when dealing with unfamiliar scenes, requires "multi-object recognition" in

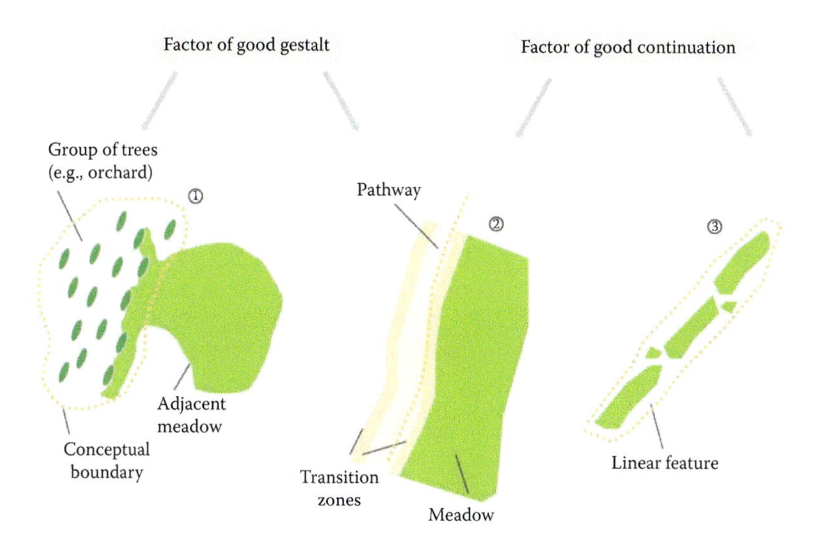

FIGURE 9.4 Gestalt theorems applicable to image analysis problems.

a rather abstracted mode. Values and meanings of objects are attributed via (object) "affordance" (Gibson 1979). The skilled visual interpreter may match features of the visual impression against experience or examples listed in an interpretation key (Blaschke et al. 2014).

The *gestalt* approach (Wertheimer 1925) has established general principles (factors, "laws") that apply when certain patterns or figures are examined. Following the Ehrenfels criterion, a *gestalt* shows emergent properties resulting in a strong predictive (although not an explanatory) power of how we perceive structures and patterns. Based on the factor of *proximity* single elements within a scene are grouped together once they are close enough to another, depending on the scale of observation. The factor of *good gestalt* assumes that simple, well-shaped geometric figures are more likely to be perceived than more complex ones. Both factors suggest that interpreters tend to group similar elements into larger aggregates and to continue drawing lines across gaps, something that is also reflected by the factor of *good continuation* (Lang 2005).

9.2.3 Class Modeling

Multi-scale segmentation, for example implemented by the "multi-resolution segmentation" algorithm (Baatz and Schäpe 2000), generates homogenous image objects in a nested hierarchy. Spectral information is aggregated while loss of detail is minimized (Drăgut et al. 2014), leading to more functional homogeneity in the sense of Spence (1961). Certain composite classes (e.g., forest stands composed of tree species with similar spectral behavior) can be directly delineated by multi-resolution segmentation (Strasser and Lang 2015). More often, however, composite classes such as orchard fields, mixed arable land, riparian forests, residential areas, informal settlements, and the like, show internal heterogeneity that exceeds the capability of state-of-the-art segmentation algorithms (Mollick et al. 2023; Lang et al. 2019; Luciano et al. 2019; Lang et al. 2010a). The internal building blocks may consist of different land cover, being spectrally diverse (e.g., grassland and crop patches) and there is no *a priori* "true" delineation. Lang et al. (2014) use the term "composite geon" for such target classes, whose composition is scale-depended. Body plans represent the composite objects inter-subjectively perceived by experts as functional *bona fide* units (Smith 1994). Class modeling (Zaabar et al. 2022; Chen et al. 2018; Shahi et al. 2017; Tiede et al. 2010) can be utilized to topologically describe spatial constellations of a set of subunits in a way that image information is structured into hierarchical divisions based on ontology-like rule sets that employ relational features.

Tiede et al. (2010) used a class modeling approach based on spectrally homogeneous landscape units applying a supervised regionalization technique with iterative segmentation and classification steps to model composite classes. The initial segmentation and preliminary classification of the basic units is the first step in the modeling process. Additional expert knowledge with respect to the overall mapping key is formalized in rule sets, including the use of auxiliary datasets, such as specific biotope mapping. The result is a cyclic process for a generation of composite objects (functionally homogenous, structurally heterogeneous; see Figure 9.5a) using region-based

(a)

(b)

FIGURE 9.5 (a) SPOT derived landuse units to fully validated composite geons; (b) initial forest patch/generalized forest patch.

segmentation algorithms. In the case of well distinguishable (sub-)units, class modeling aims at assembling expert-known larger units, according to the aforementioned principle of *good gestalt*. Again, this is a matter of scale. When looking at patches of (natural) forest, quasi-homogenous forest stands may appear according to their specific composition of prevailing and accompanying tree species. This latter example is an intermediary stage between a pure texture-based impression (where individual elements are not clearly distinguishable) and a building-block composition (where elements are few in number and inter-subjectively perceivable). This is what *gestalt* theorists would call the factor of *granularity*. When fine granularity enables the segmentation algorithm to find higher hierarchical boundaries (Strasser and Lang 2015), composite classes may be delineated directly (Figure 9.5b).

Experts design interpretation keys for mapping and survey based on experience and convention (Lang et al. 2004), hence it is hard to automate and force into rule-based systems the intrinsic and intuitive knowledge. The criterion of being "relatively homogeneous" on the level of composite classes remains a relative measure and a matter of disposition depending on the target scale dimension.

9.2.4 Validity of Object Delineation and Classification

Object identification implies—next to the cognitive skills of interpreting color and form and linking them to biophysical properties—a grasp of the relevance of the extracted information for a certain purpose. "Object validity" (Mollick et al. 2023; Lang et al. 2019; Luciano et al. 2019; Yurtseven et al. 2019; Lang et al. 2010a) has been defined as the degree of fitness of object delineations for operational tasks in a policy context. Comprising object representations in both the *bona fide* as well as the *fiat* domain, OBIA poses new challenges for evaluating object validity in an operational context ("geon" concept). According to Smith (1994), objects can be differentiated in (1) concrete, tangible objects with a visible physical boundary (*bona fide*); (2) objects that lack a physical border, thus not directly visible in the landscape (*fiat*). Examples are administrative boundaries or, as discussed in Section 9.2.3, expert-driven composite classes. In remote sensing, depending on scale and resolution, we treat a spectrally homogeneous image segment as a *bona fide* object. Examples are a distinct deciduous forest patch in an agricultural landscape, a lake or pond, or a sealed surface such as a large parking lot. As we leave such "crisp" *bona fide* objects behind, the binary decision between "correct" and "false" labeling gets blurred. Fuzzy rule sets (Zadeh 1965), as mentioned taking into consideration a probability measure for the class membership, only partly solve this. For more concept-related, composite classes, a binary assessment will fail. Thus, Lang et al. (2010a, 2019) and Shahi et al. (2017) propose to use the term "validity" to underline the challenge of binary decisions in object assessments.

Object recognition has a long tradition in computer vision, including theoretical concepts of template matching theories, feature analysis, and structural descriptions (Bruce and Green 1990) to describe appearing patterns more explicitly and tackle the components of a configuration to understand the aggregates. The term "perceptual classification" (Eysenck and Keane 1995) denotes the matching of extracted visual information with our stored structural description. Semantic classification and labeling (or naming) are subsequent stages and involve retrieval of functions and object associates (Lang et al. 2009a). One of the striking capabilities of human perception is to tell signal from noise, in other words to distinguish information against a "simplified" environment (Bruce and Green 1990). By experience and training, we continuously feed our implicit knowledge with explicit knowledge derived from formal learning situations (e.g., spectral behavior of stressed vegetation). We may thereby differentiate between procedural knowledge and structural knowledge. Procedural knowledge is concerned with specific computational functions and can be translated into a set of rules. Structural knowledge implies the way concepts of a domain are ontologically interrelated: in our case that means, in how far links between image objects and "real-world" geographical features (Hay and Castilla 2008) are established. It is characterized by high semantic contents and can be organized in knowledge organizing systems (KOSs) such as semantic networks (Liedtke et al. 1997), graph representations, or formal concept analysis (FCA) (Ganter and Wille 1996). Semantic nets and frames (Pinz 1994) used to act as a formal framework for semantic knowledge representation in image analysis using an inheritance concept (*is part of, is more specific than, is instance of*). Ontologies are a general and domain-agnostic approach to formalize knowledge, for example, for including geospatial knowledge representation (De Martino and Albertoni 2011), geographic information retrieval (Lutz and Klien 2006), image processing and analysis (Tönjes et al. 1999), semantic sensor networks (Kuhn 2009), and geodata interoperability (Reitsma et al. 2009). The Semantic Web for Earth and Environmental Terminology (SWEET) ontologies provided an upper-level ontology for Earth system science comprising thousands of terms from the Earth system science domain using OWL (Ontology Web Language) (Raskin and Pan 2003).

While in visual interpretation cognitive processes run more or less intuitively, within OBIA, we try to organize the components along a procedural processing line, which can be set up, controlled, and maintained individually (Vizzari 2022; Zaabar et al. 2022; Lang et al. 2019; Yurtseven et al. 2019; Shahi et al. 2017; Lang et al. 2010) (see Figure 9.6).

Machine-based knowledge representation and hierarchy theory (i.e., the principle understanding of decomposability of complex scene content) complement each other by enriching automated image analysis with geospatial scale concepts. The rule-based intelligence of a production system can be enriched by learning algorithms, empowering the classification system to improve itself (with increasing number of classification tasks), but to some degree limiting the transparency (and potentially the transferability) of the rule-base as such. The spatial relationships among (elementary) objects are useful in modeling higher-level (i.e., composed) object classes. These have to match with the conceptual reality that underlies the classification problem. Simple quality criteria (like any

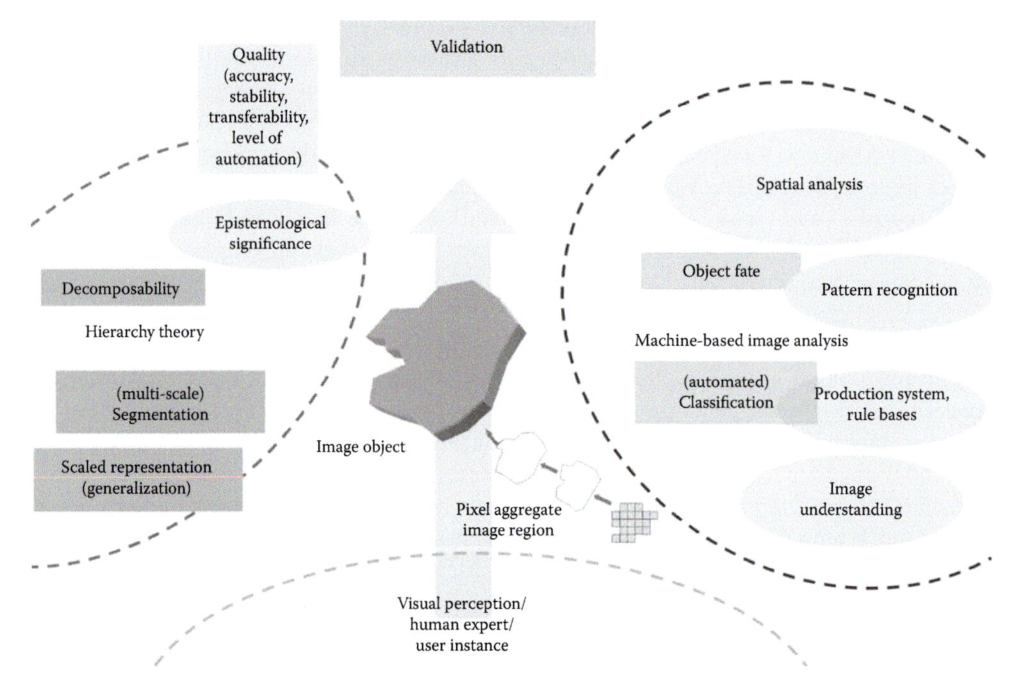

FIGURE 9.6 Object validity in object-based image analysis (Lang et al. 2010). Starting from pixel aggregates (image regions) as initial products of image portioning, a set of routines may turn suitable segments into image objects which (1) are considered appropriate for the respective context of use and (2) are adhering to general principles of image segmentation such as scale domain, spatial coherence, non-overlap, and the like (Blaschke et al. 2014).

TABLE 9.1
Aspects of Object Validity

	Bona Fide Objects	Modeled Composite Objects
Example(s)	Land cover classes, for example, meadow, field, forest	Mixed arable land and grassland area, residential area, wetlands
Main methodological challenge	Scale-specific delineation, boundary generalization	Meet requirements in terms of appropriateness and mapping key
Real-world reference	Existing and recordable in field	Driven by expert vision and understanding, agreement with mapping key and existing boundaries
Label verification	Binary or fuzzified	In stages, validation a concept to be operationalized
Boundary validation	Scale-dependent, low degree of freedom	Expert-based, by convention with a high degree of freedom
Policy-driven	On an elementary object level: yes	Yes, EO-based techniques foreseen in policy documents but not detailed
Delineation	Automated	Partly automated, class modeling, expert regrouping

Source: From Lang et al. (2010a), modified.

accuracy measure) should consider the process lineage of the object provision and a general understanding of the epistemological significance of the target objects (Lang et al. 2010b).

Table 9.1 summarizes particularities of validation tasks within OBIA for *bona fide* objects and composite. Specific details are discussed by Lang et al. (2019, 2010a). The main methodological challenges are the ones attached to optimizing the process of OBIA-specific automation and machine-assistance within each category. The issue of reference data varies between the different types of object classes. Label verification refers to the external process of confirming the assigned category. The validation of boundary delineation as such is scale-dependent, as well as policy-related. It includes the issue of degree of freedom in delineation when it comes to grouping of elementary units. Lastly, the automation of the delineation as such is characterized.

9.2.5 Multiple-Stage Validation

Semi-automated classification of complex land use/cover units faces a high number of degrees of freedom (Lang et al. 2019, 2010a). Therefore, the validation of modeled composite classes uses both traditional methods of quantifying accuracy and more qualitative validation measures (matching scale and relationship patterns, etc.). The quality assessment of OBIA class modeling needs to determine which step in the modeling approach to validate. The validation sequence, as proposed by Tiede et al. (2010), includes verification, improvement, quality assurance, reliability, and even usability to ensure the product fully satisfies the user requirements. That results in a four-stage validation procedure (Figure 9.7):

- Statistical point-based accuracy assessment based on stratified random sampling (Congalton and Green 1998) to validate the elementary (*bona fide*) units based on spectral homogeneous units.
- First expert-based validation cycle on both the composite classes (in this case: biotope complexes) and their dependent basic units, as a part of a hybrid modeling approach (on-screen expert assessment).
- Second expert-based validation cycle on the quality of the modeling approach regarding the given mapping key as compared to field mappings by domain experts, but also considering geometric accuracy of the delineated boundaries.

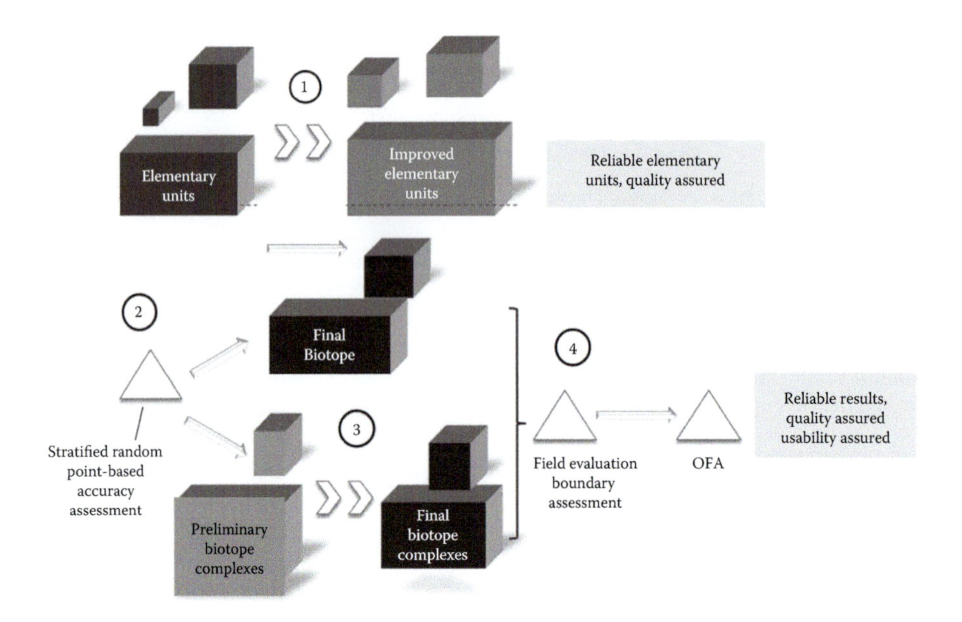

FIGURE 9.7 Multi-stage validation of composite classes (after Tiede et al. (2010), adapted).

- Evaluation of inaccuracies introduced by the policy-demanded matching of delineated biotope complex boundaries to cadastral data (scale gap). Object-based fate analysis (see Section 9.3) has been applied on individual object level.

9.3 OBJECT FATE

Image objects as real-world objects are subject to change. Changes may occur more gradually (varying quality) or more abruptly (altering category or label). In parallel, image objects may change their extent, location, or composition and even topological relationships. Object-based change detection (OBCD) strives to assess such changes in terms of type, intensity, size and shape, as well as horizontal and vertical relations of image objects. Thereby, we face the challenge to tell real changes from generalization effects or any kind of "noise" occurring in the imagery. A similar problem arises in object-based accuracy assessment (OBAA) when evaluating the quality (accuracy, robustness, etc.) of OBIA classifications, when a reference data layer is difficult to compare. The concept of object fate analysis (OFA) provides a strategy for how to deal with such challenges.

9.3.1 From Static to Dynamic Change Concepts

Geographical features, when observed over time, are subject to change—and so are image objects that represent them. It seems a straightforward task to extend established change detection methods to the object-based domain. However, there are some additional challenges and specifics to what we may call object-based change detection (OBCD) (Lang et al. 2019; Lang 2008; Chen et al. 2012). "Change," in fact, is a more challenging issue in OBIA than in pixel-based analysis, due to the reasons already discussed: an image object can change both in label and extent. The fact that a particular geographical feature (a lake, a forest, a town) changes its category within exactly the same boundary is a rare case. There is not the single pixel converted from meadow to non-meadows, but the expanding settlement that eats up grassland patches in the surrounding. Blaschke (2005) has proposed a typology of object geometry changes considering different categories (existence-related, size- and shape-related, and location-related changes). In reality, we are facing a combination of all of these basic categories of geometric changes (Lang et al. 2019, 2009a). In addition, when an object is shrinking or expanding, the surrounding objects are inevitably affected. Raza and Kainz (2001) include subdivision and amalgamation of objects in their list of spatiotemporal characteristics of parcels. Other studies specifically investigate topological relations to mathematically evaluate the identity of two object hypotheses when comparing two different datasets (Straub and Heipke 2004). They discuss key topological relations being reducible to the principal relationships: disjoint, equal, overlap, and containment.

Practically, when changes are detected in multi-temporal imagery, such spatial implications are often neglected so that established methods for spatially explicit object-based change analysis are still rare (Schöpfer et al. 2008). Traditional techniques of map-to-map comparisons refer to changes in pixels but not of objects. Vector overlays with physical intersections of the boundaries produce complex geometry with sliver polygons (Chen et al. 2012). Visual comparisons, on the other hand, are powerful but rather subjective and time-consuming (Schöpfer et al. 2008).

Next to the simultaneity of change in label and extent, there are other reasons why it is hard to find straightforward solutions for object-based change. Image objects may vary due to different representations, while those geographical features they represent remain constant. A forest patch may be represented in different scales and generalization levels (Weinke et al. 2008) and the same applies to other features like settlements, wetlands, and the like. The question arises whether a spatial mismatch of corresponding image objects are indeed caused by data mismatches, different representations, or real change. More specific, there are several reasons for spatially inconsistent features, which are inherent to the representation itself (Lang et al. 2019, 2009a) and not related to the geographical features under concern ("pseudo-changes") (Mollick et al. 2023; Nasiri et al. 2023; Kotaridis and Lazaridou 2022; Vizzari 2022; Zaabar et al. 2022; Lang et al. 2019; Luciano et al. 2019; Yurtseven et al. 2019; Chen et al. 2018; Labib and Harris 2018; Ye et al. 2018; Shahi et al. 2017):

- Image mis-registrations due to poor spatial referencing may cause shifts in corresponding objects and pseudo-boundaries when overlaying those.
- Differing image characteristics of multi-temporal or multi-seasonal datasets in terms of viewing angles, sun illumination, etc.
- Combing image data of different spatial resolutions and/or different segmentation algorithms.
- Applying different scales and outline complexities when comparing segmentation results to visual interpretations and manual delineations.

This has implications on the assessment of OBIA outcomes, for example when comparing a segmented and classified result with a visual interpretation map. Spatial disagreement or "spatial error" (Radoux and Defourny 2008) arises, as object boundaries may mismatch by simply applying a different generalization strategy, or none at all. The quality of the segmentation, "segmentation goodness" (Clinton et al. 2010; Hernando et al. 2012), is a critical factor in this process. Van Coillie et al. (2008) proposed a Purity Index measuring the common area between segmentation objects and reference objects. Vector-based measures (area-based, location-based, and a combination of both) were analyzed by Clinton et al. (2010), calculating the similarities between segments and training objects. Two goodness measures for optimal segmentation were proposed by Johnson and Zhixiao (2011): global intra-segment homogeneity (the variance of each object) and inter-segment heterogeneity (how similar a region is to its neighbors).

In the following section, we demonstrate how to analyze differences in objects by investigating spatial relationships among corresponding objects. In doing so, we abstract from the issue whether this is a product of object transition (change over time) or an outcome of different object representations or delineations. Since spatial relations are various and appear in reality in different combinations, there is a demand for ready-to-use solutions that are able to structure and categorize these.

9.3.2 Application Scenario 1—Object-Based Information Update

The need for updating existing information through remote sensing based technology is undoubted and still increases with the availability of especially high spatial resolution data. In operations workflows the update of information from image data is hampered by the lack of full integration into existing geospatial infrastructures. This issue has been discussed for decades (Ehlers 1990), and the integration of GIS and remote sensing is progressing (Nasiri et al. 2023; Kotaridis and Lazaridou 2022; Labib and Harris 2018; Ye et al. 2018; Shahi et al. 2017). But some problems are still not solved for many cases. For example, the operational workflow for a planning authority to integrate a—even very accurate and based on VHR data—land cover classification in existing vector datasets (e.g., cadastral data) is still limited, if the boundaries of the land cover classes do not match the spatial characteristics of the geo data in use. The geometric matching of the different datasets is an issue of scale/resolution, the related issue of boundary complexity but also a principle problem of conversion between different data models (usually raster to vector) (Merchant and Narumalani 2009). To overcome the latter, expectations have been placed on the ability of segmentation-based approaches to overcome the problem of mismatches that are due to the use of different data models (raster vs. vector) by delivering "GIS-ready information" (Benz et al. 2004) from remote sensing data. However, as Tiede et al. (2010) pointed out, such a full integration of image-derived spatial information is not a trivial task: the problem is residing lesser in the raster-vector conversion itself and more in the integration of information from different (spatial) resolutions. Coarser image resolution than the existing geo data to be updated results in objects usually not matching boundaries compared to the existing ones. Also the contrary, very high spatial resolution imagery as information source to update lower-resolution (i.e., coarser scale, generalized) vector data results in objects often too complex regarding shape and boundaries. Smoothening and generalization routines need to be applied, which can lead—according to their non-deterministic operation—to unsatisfactory matching with existing boundaries. Refer to Walter and Fritsch (1999) and Buthenuth et al. (2007) for a discussion about matching of different (vector) data models and the integration of raster and vector data.

In a study for a regional planning body (*Verband Region Stuttgart*, an association of local authorities in the Stuttgart region, Germany) both problems were addressed (Tiede et al. 2010). The requirements to support regional planning based on EO data encompassed the delineation of biotope complexes as target units. To be fed into the existing data structure of the planning authority ("cadaster-conform"), the generated biotope complexes should preserve boundaries of the digital cadaster map in cases where a change in biotope complex type is also reflected in the cadastral data, but remove boundaries within the same biotope complex. In addition, biotope boundaries not reflected in the cadastral data need to be introduced.

As a trade-off between costs and quality image data of 5 m ground sample distance was used for the analysis, which introduced a certain scale-gap, especially for newly introduced boundaries not yet reflected in the cadastral map. A twofold strategy was used to tackle the problem, by using (1) the digital cadastral data as spatial constraints in the initial object building step (using an adaptive per-parcel approach), and (2) the adjustment and validation of newly introduced rasterized boundaries by comparison with the target vector geometry. In the first step, it was possible to flag either boundaries of the digital cadastral map to be preserved if they were also representing biotope complex boundaries or to flag boundaries to be removed if they are within the same biotope complex. By this, the scale-gap of the image data compared to the existing data is avoided. In the second step—to integrate newly introduced boundaries from the image based biotope complex delineation—it was necessary to perform an adjustment in regard to the same generalization/smoothness level of the existing data (instead of a rasterized boundary). The latter was achieved using a combination of established GI-tools and additionally programmed solutions. Figure 9.8 shows schematically the three occurring problems that were addressed to reach the requirements for an accurate and compatible dataset for administrative purposes: (1) the replacement of biotope complex boundaries with corresponding

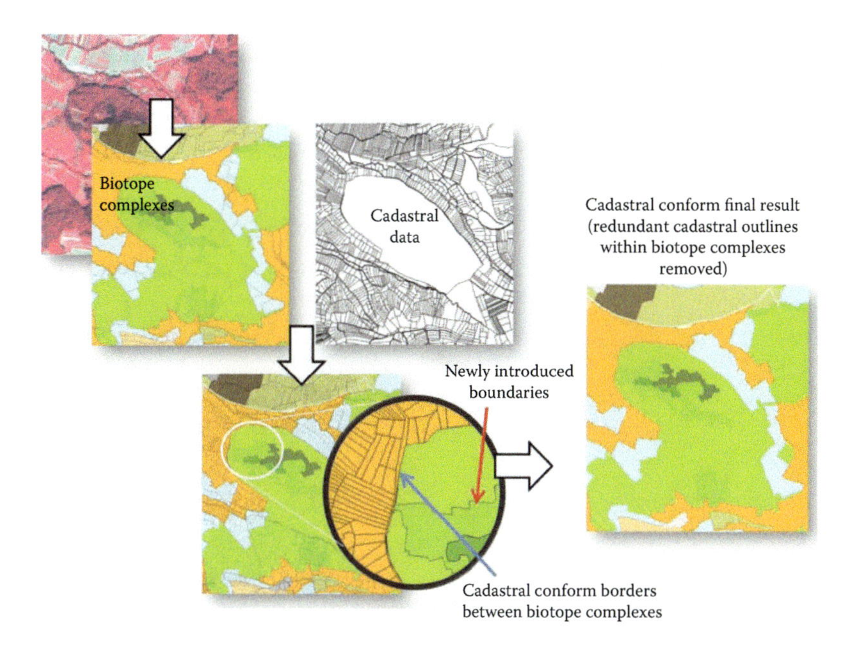

FIGURE 9.8 Cadaster-conform extraction of biotope complex units from EO data. Cadastral data is retained, where biotope complexes are changing, redundant cadastral boundaries within the same biotope complex are eliminated.

cadastral boundaries by considering a spatial displacement tolerance ("scale-gap") according to the pixel size of the image data; (2) merging of cadastral polygons within a biotope complex (removing of cadastral boundaries); and (3) the introduction of new boundaries (based on an OBIA based biotope complex delineation) not reflected in the cadastral dataset but which represent changes of biotope complexes. The newly introduced boundaries were finally smoothed and generalized using standard GI procedures (Lang et al. 2019; Luciano et al. 2019; Yurtseven et al. 2019; Douglas and Peucker 1973; Bodansky et al. 2002).

Specific challenges are faced when existing boundaries (e.g., digital cadastral data) are considered in a process of information update. We shortly present a collection of instances on how to perform OBIA for monitoring purposes, aiming at the provision of up-to-date information under the explicit consideration of existing geospatial data. Both illustrations and real examples from project-related work are shown, and the different cases are grouped into three categories. "Boundaries" refers to existing polygon or line data which are used for adaptive parcel-based segmentation. A specific case of using cadaster information for generating polygon datasets has been dubbed "adaptive per parcel segmentation" (APPS) by Tiede et al. (2007) (see Figure 9.9).

9.3.3 Object Fate Analysis

Object Fate Analysis (OFA) is a method proposed by Lang (Lang et al. 2019; Lang 2005) and developed by Schöpfer et al. (2008) for investigating spatial relationships between corresponding objects in two different representations. Rephrased and put from an individual image object's perspective this would reflect the "fate" of this particular object and its representation (see Figure 9.10). Object fate may be caused by real change captured in data from different points of time. Or otherwise it may root in differences in object generation by using segmentation algorithms or visual analysis, heterogeneous data reference datasets from other sources (Schöpfer et al. 2008), etc. In reality we often face a combination of change and representation-induced divergences.

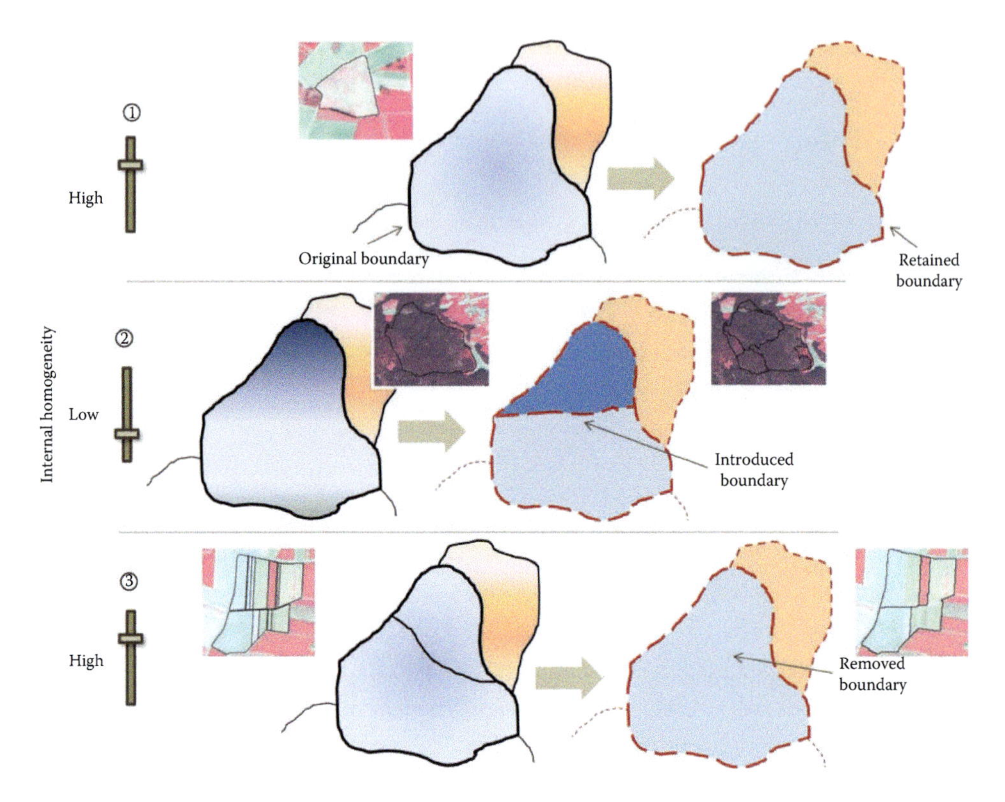

FIGURE 9.9 Illustration of differentiated boundary treatment. Additional cases (not shown): boundaries are generalized, fractalized, shifted, found in the surroundings. Case 1: existing boundaries are retained (limited internal heterogeneity); case 2: new boundaries are introduced—internal variance larger than (given) threshold; case 3: boundaries are removed—internal variance larger than (given) threshold.

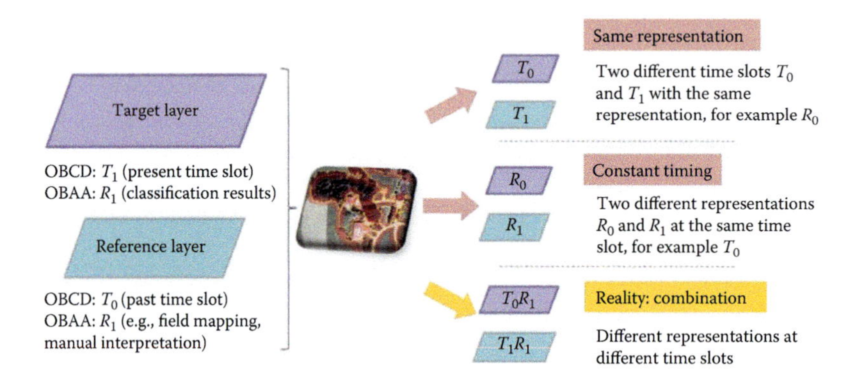

FIGURE 9.10 Object fate considering aspects of different status (change) and representation.

OFA has been implemented for example in a tool called LIST (Landscape Interpretation Support Tool) (Weinke et al. 2007; Lang et al. 2019, 2009a), as an extension for ESRI's ArcGIS software. Following the concept of parent and child relationships, two vector layers are used to represent the specific "fate" of corresponding objects. Thereby, only the geometry is considered, assuming that corresponding objects retain their label (see more on an extension of OFA to the thematic dimension later). To overcome spatial uncertainty in image object, an error band (positive and negative buffer around the object boundary) is used. The size of the buffer is either specified manually or

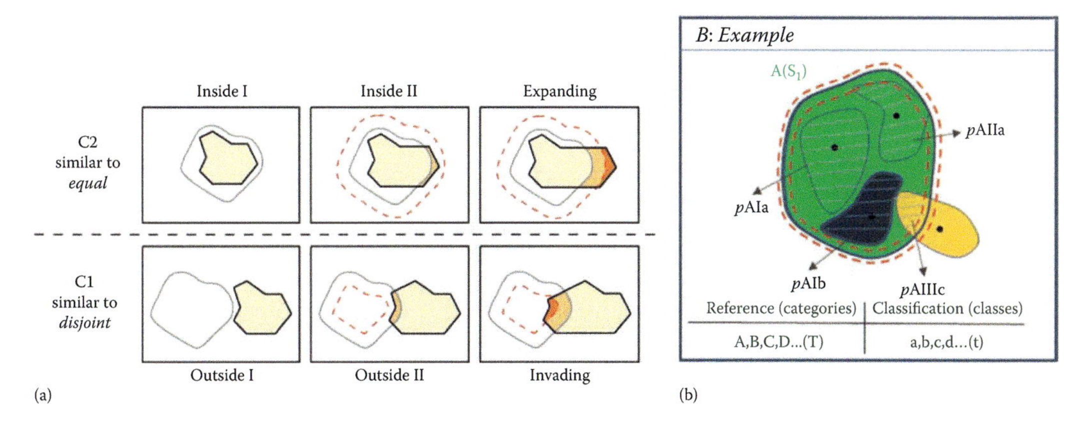

FIGURE 9.11 Spatial and thematic assessment in extended OFA. (From Hernando et al. (2012), modified.)

as function relative to object size. A "virtual" overlay characterizes spatial relationships via the relative positions of centroids, retaining the original object boundaries. This distinguishes between weak overlap (*similar to disjoint*) and strong overlap (*similar to identical*). The resulting categories of object relationships are discussed by Schöpfer et al. (2008) and shown in a simplified version in Figure 9.11. Object relationships are divided into either C1 cluster or C2 cluster by evaluating whether the centroid of the classification object fell inside the reference object (upper row: "*inside*" or "*good I*," "*inside II*," "*expanding*") or not (lower row: "*outside* or *not interfering I*," "*outside II*," "*invading*"). There are six relationships; five of them are relevant for further analysis when two corresponding objects at least partly overlap. The relationship "*not interfering I*" assumes that objects do not "correspond" any longer. Two stability measures, *object (or "offspring") loyalty* and *interference* (Lang et al. 2009a) were derived based on the relationships between objects.

Hernando et al. (2012) extended OFA to the thematic dimension by assessing results from a detailed OBIA classification of forest stands as compared to an existing forest stand mapping derived from visual air-photo interpretation. By combining spatial and thematic aspects (see Figure 9.11, right) an OFA matrix was established. First, each classified object (classes a, b, c, d, . . . t) was compared to the corresponding reference object with S_T the overlapping area for each of the reference categories (A, B, C, D . . . T) and S_{Total} the total reference area for all *w* reference categories (*A, B, C . . . w . . . T*). The matrix is established by listing all relationships between all category/class shares for each of the five relationship type. Each cell of this matrix is filled by a value r_{TCt} as the sum of the relative areas of the *t* classified objects considering an OFA-type.

9.3.4 Object Linking

Traditional automated change detection approaches are limited if, for example, co-registration errors are present or if an object changed over time (changes especially in size/form or it disappeared). Such change detection approaches, looking into pixel value changes rather than into the—also geometrical—change of objects, usually need a very precise co-registration between multi-temporal images (Lu et al. 2004). Following a study about the robustness of different pixel-based change detection algorithms by Sundersan et al. (2007), the registration error between the images should not exceed 0.2 pixels (RMS) for satisfactory performance of the algorithms.

If we compare objects either created from different datasets (different sensors) or different time intervals, we are able to compensate to some degree these high demands for image co-registration (generalization issue in the object generation process, comparing of statistical values per object instead of single pixels, comparing also geometric properties). Nevertheless, a strict vertical object hierarchy, as often used in OBIA frameworks, is hampering sophisticated analysis, because of the

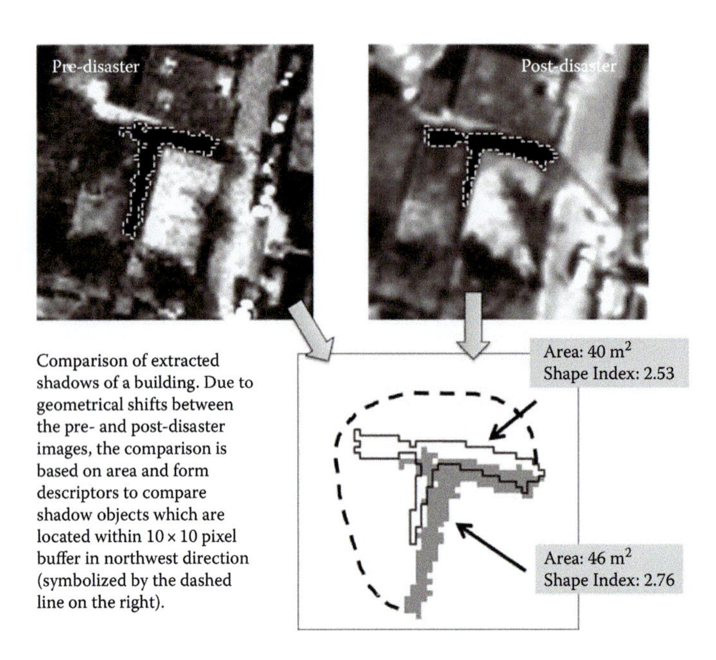

FIGURE 9.12 Conceptual illustration of the object-linking approach. Here: comparing extracted shadow objects from an earthquake pre- and post-disaster images—despite a geometrical shift between the images—to indicate if the shadow is considered to be still in existence (here: no indication of destruction of the building since the area and form descriptors are quite similar), from Tiede et al. (2011).

intersection of boundaries in such hierarchy treating multi-temporal data similar to a multi-scale hierarchy.

If we look at human observers, they usually can compensate positional shifts between datasets much easier for recognizing similar objects. Such a "loose" coupling can be mimicked to some degree using an object-linking approach to track objects over time or to analyze changes per object instead of, for example, intersections of objects (Tiede et al. 2011; Hofmann and Blaschke 2012). Object-linking relies on GI-methods adopted in an topologically enabled OBIA framework and is establishing spatial relationships between objects incorporating shifts introduced through positional errors using (multi-)directional buffering and can be enriched by the comparison of geometric object properties (like form and size).

Figure 9.12 shows the implementation of an object-linking approach for shadows casted by buildings, before and after an earthquake (Tiede et al. 2011). The class-constrained object-by-object comparison is able to compare objects of different classifications including geometrical buffering in x or y directions, that is, a geometrical shift between the dates can be compensated (or to track moving objects). Since such a linking of objects—including the neighborhood and not only the direct spatial overlap—can lead to 1:n or even n:n matches, not only class-constraints (matching of objects of the same classes only) are important but also the comparison of geometrical object properties (e.g., form and size). By this, the degree of change or the degree of "movement" can be estimated. In the following an application scenario is described using an object-linking approach in the case of automated rapid information extraction after a disaster (here: damage indication after an earthquake).

9.3.5 Application Scenario 2—Rapid Information Extraction

Specific automated image analysis techniques based on very high spatial resolution (VHR) images, have reached a status of maturity to be utilized in application domains crucially depending on reliability and timeliness (Lang et al. 2019, 2010a). Several studies (Mollick et al. 2023; Nasiri et al. 2023; Kotaridis and Lazaridou 2022; Vizzari 2022; Zaabar et al. 2022; Lang et al. 2019; Luciano et al. 2019; Yurtseven et al. 2019; Chen et al. 2018; Labib and Harris 2018; Ye et al. 2018; Shahi et al. 2017) have shown that different approaches can lead to successful applications in the field of damage assessment using change detection methodologies, for example comparing of pre- and post-event images. Gusella et al. (2005) demonstrated the use of QuickBird data for quantifying the number of buildings that collapsed after the Bam earthquake in Iran, 2003. Using QuickBird data, Pesaresi et al. (2007) showed that very high accuracies allow for rapid damage assessment of built-up structures in tsunami-affected areas based on a multi-criteria recognition system within a restricted set of general assumptions (e.g., built-up structures cast a shadow and collapsed built-up structures no longer cast a shadow, and they leave debris on the ground). Vu and Ban (2010) developed a context-based automated approach for earthquake damage mapping relying on debris areas identification. In addition the calculation has been speed optimized by parallel processing implementation showing good results for the test area, after the Sichuan earthquake in China, 2008. Ehrlich et al. (2008) and Brunner et al. (2010) demonstrate the usefulness of applications with the additional use of synthetic aperture radar (SAR) data for damaged building assessment also for the Sichuan earthquake area. 3D information has been taken into account to detect damaged built-up structures after earthquakes by Turker and Cetinkaya (2005) with the help of stereo-images, but for rapid damage assessment shortly after an earthquake, the necessity of capturing a pair of fitting stereo-images is usually not feasible. A broad overview of automated techniques for earthquake damage detection is given by Chini (2009) and Rathje and Adams (2008).

However, these approaches have rarely been applied on an operational level in the relief phase shortly after a disaster. One reason might be that the process critically depends on a good quality of the pre-processed data, that is, mainly a high reliability of the image (co-)registration. For most of the automated change detection algorithms, this is a crucial point (Lu et al. 2004; Sundersan et al. 2007). Right after a disaster there is often a lack of high quality pre- and post-disaster VHR imagery (Rathje and Adams 2008). After the Haiti earthquake in January 2010, the availability of VHR satellite data was outstanding compared to previous disasters. Data have even been provided to the public for free (also outside the International Charter Space and Major Disasters) by companies like Google or Digital Globe / Maxar. One of the effects was that a tremendous amount of "crowd-sourced" information was provided in the aftermath of the earthquake. Nevertheless, these datasets are often not well suited for automated analysis methods, due to reasons such as lacking metadata, nondocumented or insufficient pre-processing, and missing spectral NIR bands.

A strategy for automated extraction of damage indication from very high spatial resolution satellite imagery was presented for the Haitian town of Carrefour after the January 2010 earthquake (Tiede et al. 2011). Damaged buildings are identified by changes in their shadows compared in pre- and post-event data. The approach builds on object-based image analysis concepts to extract the relevant information about damage distribution (collapsed buildings). The lacking quality of the pre- and post-disaster imagery usually available directly after an event (in this case the shift between the multi-temporal images was up to 5 m and varied throughout the images) was bypassed using the described object-linking approach between objects extracted from the different images. This approach allows to overcome sub-pixel co-registration of images (especially if the image source and the conducted pre-processing steps are not knows) in rapid information extraction workflows as in the context of disaster events (see Figure 9.13). For the area of Carrefour the new methodology produced positively validated results in acceptable processing time and was, to our best knowledge, the only automated damage assessment method that delivered relevant results to requesting relief organizations in the first days after the Haiti earthquake.

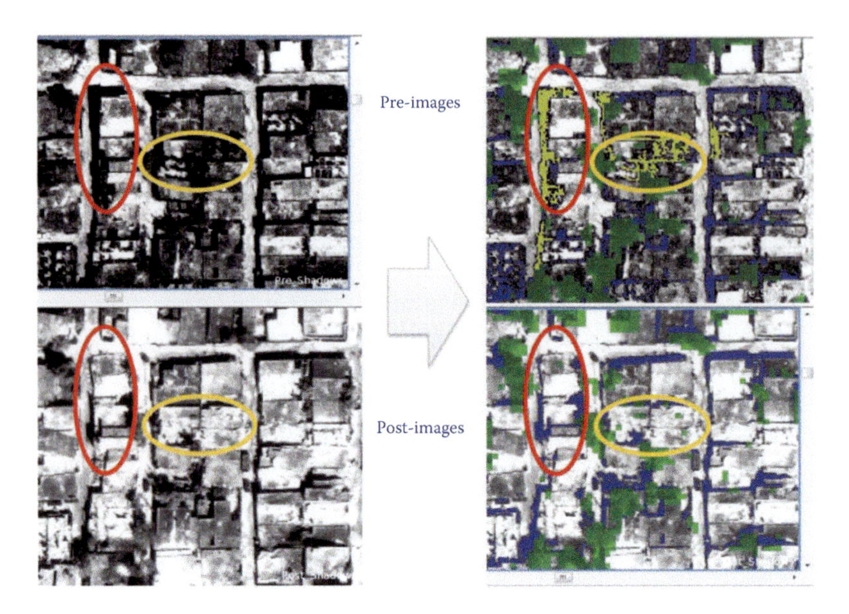

FIGURE 9.13 Object-linking for relating collapsed buildings.

9.4 CONCLUSIONS

In this chapter we have discussed implications of geodata integration in the context of OBIA and geospatial information update from an interdisciplinary, applied perspective (Mollick et al. 2023; Nasiri et al. 2023; Kotaridis and Lazaridou 2022; Vizzari 2022; Zaabar et al. 2022; Lang et al. 2019; Luciano et al. 2019; Yurtseven et al. 2019; Chen et al. 2018; Labib and Harris 2018; Ye et al. 2018; Shahi et al. 2017). We have shown how the advancement of feature recognition and advanced image analysis techniques facilitates the extraction of thematic information for decision support. OBIA techniques were discussed in the light of transforming and reconstructing image scenes into ready-to-use geospatial information matching the conceptual understanding and the practical needs of users. This implies issues of scale, generalization, as well as representation (including visualization), thus we subsumed the entirety of aspects to be considered in this process under the term "object validity." Since image objects mimicking real-world objects are subject to change, OBCD was reviewed in terms of type, intensity, size and shape, as well as horizontal and vertical relations of image objects. Object-based accuracy assessment faces similar challenges when evaluating the quality (accuracy, robustness, etc.) of OBIA classifications. The concept of "object fate analysis" (OFA) provides a strategy for how to deal with such challenges. We have illustrated the chapter with two representative application scenarios to illustrate the discussed conceptual items in a practical context.

NOTE

1 In order to highlight the relationships between objects rather than emphasizing objects as such, we may use a more generic term: *spatial image analysis*. Still, since OBIA is an established term in the community, we use it throughout the chapter.

REFERENCES

Allen, T. F. H. and T. B. Starr. 1982. *Hierarchy*. Chicago, University of Chicago Press.

Aschbacher, J. 2002. Monitoring environmental treaties using earth observation. In *VERTIC Verification Yearbook*: 171–186.
Augustin, H., M. Sudmanns, D. Tiede, S. Lang and A. Baraldi. 2019. Semantic Earth observation data cubes. *Data* 4(3): 102–115.
Baatz, M. and A. Schäpe. 2000. *Multiresolution Segmentation: An Optimization Approach for High Quality Multi-Scale Image Segmentation*. Salzburg, Wichmann Verlag.
Baraldi, A. 2011. Satellite Image Automatic Mapper™ (SIAM™). A turnkey software button for automatic near-real-time multi-sensor multi-resolution spectral rule-based preliminary classification of spaceborne multi-spectral images. *Recent Patents on Space Technology* 2011(1): 81–106.
Baraldi, A. and L. Boschetti. 2012. Operational automatic remote sensing image understanding systems: Beyond geographic object-based and object-oriented image analysis (GEOBIA/GEOOIA). Part 1: Introduction. *Remote Sensing* 4: 2694–2735.
Baraldi, A., L. D. Sapia, D. Tiede, M. Sudmanns, H. Augustin and S. Lang. 2022. Innovative analysis ready data (ARD) product and process requirements, software system design, algorithms and implementation at the midstream as *necessary-but-not-sufficient* precondition of the downstream in a new notion of space economy 4.0—Part 1 and part 2. *Big Earth Data* 7(3): 455–811.
Belgiu, M., B. Hofer and P. Hofmann. 2014. Coupling formalized knowledge bases with object-based image analysis. *Remote Sensing Letters* 5(6): 530–538.
Benz, U., P. Hofmann, G. Willhauck, I. Lingenfelder and M. Heynen. 2004. Multi-resolution, object-oriented fuzzy analysis of remote sensing data for GIS-ready information. *ISPRS Journal of Photogrammetry* 58: 239–258.
Bischof, H., W. Schneider and A. Pinz. 1992. Multispectral classification of Landsat-images using neural networks. *IEEE Transactions on Geoscience and Remote Sensing* 30: 482–490.
Blaschke, T. 2005. Towards a framework for change cetection based on image objects. Göttinger Geographische Abhandlungen. In *Remote Sensing and GIS for Environmental Studies*. Eds. S. Erasmi, B. Cyffka and M. Kappas. Göttingen, 113.
Blaschke, T. and J. Strobl. 2001. What's wrong with pixels? Some recent developments interfacing remote sensing and GIS. *Zeitschrift für Geoinformationssysteme* 14(6): 12–17.
Blaschke, T., et al. 2007. GMES: From research projects to operational environmental monitoring services. *ISPRS Workshop on High-Resolution Earth Imaging for Geospatial Information*. Hannover.
Blaschke, T., et al. 2014. Geographic object-based image analysis: A new paradigm in remote sensing and geographic information science. *International Journal of Photogrammetry and Remote Sensing* 87(1): 180–191.
Bodansky, E., A. Griboy and M. Pilouck. 2002. Smoothing and compression of lines obtained by raster-to-vector; conversion. *Graphics Recognition: Algorithms and Applications* 2390: 256–265.
Bruce, V. and P. R. Green. 1990. *Visual Perception*. East Sussex, Lawrence Erlbaum Associates.
Brunner, D., G. Lemoine and L. Bruzzone. 2010. Earthquake damage assessment of buildings using VHR optical and SAR imagery. *IEEE Transactions on Geoscience and Remote Sensing* 48(5): 2403–2420.
Buthenuth, M., G. Gösseln, M. Tiedge, C. Heipke, U. Lipeck and M. Sester. 2007. Integration of heterogeneous geospatial data in a federated database. *ISPRS Journal of Photogrammetry* 62(5): 328–346.
Campbell, J. B. 2002. *Introduction to Remote Sensing*. New York, The Guilford Press.
Chen, G., G. J. Hay and B. St-Onge. 2012. A GEOBIA framework to estimate forest parameters from lidar transects, Quickbird imagery and machine learning: A case study in Quebec, Canada. *The International Journal of Applied Earth Observation and Geoinformation* 15: 28–37.
Chen, G., Q. Weng, G. J. Hay and Y. He. 2018. Geographic object-based image analysis (GEOBIA): Emerging trends and future opportunities. *GIScience & Remote Sensing*, 55(2): 159–182. http://doi.org/10.1080/15481603.2018.1426092
Chini, M. 2009. Earthquake damage mapping techniques using SAR and optical remote sensing satellite data. In *Advances in Geoscience and Remote Sensing*. Ed. G. Jedlovec. InTech—Open Access Publisher: 269–278.
Clinton, N., A. Holt, J. Scarborough, L. Yan and P. Gong. 2010. Accuracy assessment measures for object-based image segmentation goodness. *Photogrammetric Engineering and Remote Sensing*: 289–299.
Congalton, R. G. and K. Green. 1998. *Assessing the Accuracy of Remotely Sensed Data: Principles and Practices*. Boca Raton, Lewis Publishers.

De Martino, M. and R. Albertoni. 2011. A multilingual/multicultural semantic-based approach to improve data sharing in an SDI for nature conservation. *International Journal of Spatial Data Infrastructures Research* 6: 206–233.
Di Gregorio, A. and L. J. M. Jansen. 2005. *Land Cover Classification System (LCCS): Classification Concepts and User Manual*. Rome, Food and Agriculture Organization of the United Nations.
Douglas, D. H. and T. K. Peucker. 1973. Algorithms for the reduction of the number of points required to represent a digitized line or its caricature. *Cartographica* 10(2): 112–122.
Drăgut, L., O. Csillik, C. Eisank and D. Tiede. 2014. Automated parameterisation for multi-scale image segmentation on multiple layers. *ISPRS Journal of Photogrammetry* 88: 119–127.
Ehlers, M. 1990. Remote sensing and geographic information systems: Towards integrated spatial information processing. *IEEE Transactions on Geoscience and Remote Sensing* 28(4): 763–766.
Ehrlich, D., H. D. Guo, K. Molch, J. W. Ma and M. Pesaresi. 2008. Identifying damage caused by the 2008 Wenchuan earthquake from VHR remote sensing data. *International Journal of Digital Earth* 4: 309–326.
Eysenck, M. W. and M. T. Keane. 1995. *Cognitive Psychology. A Student's Handbook*. East Sussex, Psychology Press.
Ganter, B. and R. Wille. 1996. *Die formale Begriffsanalyse*. Berlin, Springer.
Gibson, J. J. 1979. *The Ecological Approach to Visual Perception*. Boston, Houghton Mifflin.
Gorte, B. 1998. *Probabilistic Segmentation of Remotely Sensed Images*. ITC Publication Series.
Gusella, L., B. J. Adams, G. Bitelli, C. K. Huyck and A. Mognol. 2005. Object-oriented image understanding and post-earthquake damage assessment for the 2003 Bam, Iran, earthquake. *Earthquake Spectra* 21(S1): 225–238.
Hay, G. J. and G. Castilla. 2008. Geographic object-based image analysis (GEOBIA): A new name for a new discipline. In *Object-based Image Analysis: Spatial Concepts for Knowledge-driven Remote Sensing Applications*. Eds. T. Blaschke, S. Lang and G. J. Hay. Berlin, Springer.
Hay, G. J., D. J. Marceau, P. Dubé and A. Buchard. 2001. A multiscale framework for landscape analysis: Object-specific analysis and upscaling. *Landscape Ecology* 16(6): 471–490.
Hernando, A., D. Tiede, F. Albrecht and S. Lang. 2012. Spatial and thematic assessment of object-based forest stand delineation using an OFA-matrix. *The International Journal of Applied Earth Observation and Geoinformation* 19: 214–225.
Hofmann, P. and T. Blaschke. 2012. Object based change detection using temporal linkages. *Proceedings of the 4th GEOBIA Conference 2012*. Rio de Janeiro, INPE.
Johnson, B. and X. Zhixiao. 2011. Unsupervised image segmentation evaluation and refinement using a multi-scale approach. *ISPRS Journal of Photogrammetry* 66: 473–483.
Julesz, B. 1975. Experiments in the visual perception of texture. *Scientific American* 212: 38–48.
Kotaridis, I. and M. Lazaridou. 2022. Integrating object-based image analysis and geographic information systems for waterbodies delineation on synthetic aperture radar data. *Geocarto International* 37(16): 4655–4670. http://doi.org/10.1080/10106049.2021.1892213
Kuhn, W. 2009. A functional ontology of observation and measurement. In *GeoSpatial Semantics*. Eds. K. Janowicz, M. Raubal and S. Levashkin. Berlin, Springer: 26–43.
Labib, S. M. and A. Harris. 2018. The potentials of Sentinel-2 and LandSat-8 data in green infrastructure extraction, using object based image analysis (OBIA) method. *European Journal of Remote Sensing* 51(1): 231–240. http://doi.org/10.1080/22797254.2017.1419441
Lang, S. 2005. *Image Objects and Landscape Objects*. Salzburg, Interpretation, Hierarchical Representation And Significance.
Lang, S. 2008. Object-based image analysis for remote sensing applications: Modeling reality—dealing with complexity. In *Object-Based Image Analysis – Spatial Concepts for Knowledge-Driven Remote Sensing Applications*. Eds. T. Blaschke, S. Lang and G. J. Hay. Berlin, Springer: 3–28.
Lang, S., F. Albrecht, S. Kienberger and D. Tiede. 2010a. Object validity for operational tasks in a policy context. *Journal for Spatial Science* 55(1): 9–22.
Lang, S., C. Burnett and T. Blaschke. 2004. Multi-scale object-based image analysis: A key to the hierarchical organisation of landscapes. *Ekologia* 23(Supplement): 1–9.
Lang, S., G. J. Hay, A. Baraldi, D. Tiede and T. Blaschke. 2019. GEOBIA achievements and spatial opportunities in the era of big earth observation data. *ISPRS International Journal of Geo-Information* 8(11): 474. https://doi.org/10.3390/ijgi8110474
Lang, S., S. Kienberger, D. Tiede, M. Hagenlocher and L. Pernkopf. 2014. Geons—domain-specific regionalization of space. *Cartography and Geographic Information Science* 41(3): 214–226.

Lang, S., E. Schöpfer and T. Langanke. 2009a. Combined object-based classification and manual interpretation – Synergies for a quantitative assessment of parcels and biotopes. *Geocarto International* 24(2): 99–114.

Lang, S., D. Tiede, F. Albrecht and P. Füreder. 2010b. Automated techniques in rapid geospatial reporting – issues of object validity. In *VALgEO*. Eds. C. Corbane, D. Carrion, M. Broglia and M. Pesaresi. Ispra: 65–75.

Lang, S., D. Tiede, D. Hölbling, P. Füreder and P. Zeil. 2009b. Conditioning land-use information across scales and borders. *Geospatial Crossroads @ GI_Forum' 09*. Adrijana Car, Gerald Griesebner, Josef Strobl. Heidelberg, Wichmann: 100–109.

Lang, S., P. Zeil, S. Kienberger and D. Tiede. 2008. *Geons—Policy-Relevant Geo-Objects for Monitoring High-Level Indicators*. GI Forum. Adrijana Car and Josef Strobl. Salzburg, Geospatial Crossroads @ GI_Forum' 08: 180–186.

Liedtke, C. E., J. Bückner, O. Grau, S. Growe and R. Tönjes. 1997. AIDA: A system for the knowledge based interpretation of remote sensing data. *Third International Airborne Remote Sensing Conference*. Copenhagen.

Lu, D., P. Mausel and E. Moran. 2004. Change detection techniques. *International Journal of Remote Sensing* 25: 2365–2401.

Luciano, A. C. D. S., M. C. A. Picoli, J. V. Rocha, D. G. Duft, R. A. Camargo, R. S. C. Lamparelli, M. R. L. V. Leal and G. L. Maire. 2019. A generalized space-time OBIA classification scheme to map sugarcane areas at regional scale, using Landsat images time-series and the random forest algorithm. *International Journal of Applied Earth Observation and Geoinformation* 80(2019): 127–136. ISSN 1569-8432. https://doi.org/10.1016/j.jag.2019.04.013. https://www.sciencedirect.com/science/article/pii/S0303243418311917

Lucieer, A. and A. Stein. 2002. Existential uncertainty of spatial objects segmented from satellite sensor imagery. *IEEE Transactions on Geoscience and Remote Sensing* 40(11): 2518–2521.

Lutz, M. and E. Klien. 2006. Ontology based retrieval of geographic information. *International Journal of Geographical Information Science* 20(3): 2006.

Marceau, D. J. 1999. The scale issue in the social and natural sciences. *Canadian Journal of Remote Sensing* 25: 347–356.

Marr, D. 1982. *Vision*. New York, W.H. Freeman.

Merchant, J. W. and S. Narumalani. 2009. Integrating remote sensing and geographic information system. In *The SAGE handbook of remote sensing*. Eds. T. A. Warner, M. D. Nellis and G. M. Foody. London, SAGE Publications Ltd: 257–268.

Mollick, T., M. G. Azam and S. Karim. 2023. Geospatial-based machine learning techniques for land use and land cover mapping using a high-resolution unmanned aerial vehicle image. *Remote Sensing Applications: Society and Environment* 29(2023), 100859. ISSN 2352-9385. https://doi.org/10.1016/j.rsase.2022.100859. https://www.sciencedirect.com/science/article/pii/S2352938522001677

Montello, D. R. 2001. Scale in Geography. In *International encyclopedia of the social & behavioural sciences*. Oxford, Pergamon Press: 13501–13504.

Nasiri, V., P. Hawryło, P. Janiec and J. Socha. 2023. Comparing object-based and pixel-based machine learning models for tree-cutting detection with planetscope satellite images: Exploring model generalization. *International Journal of Applied Earth Observation and Geoinformation* 125(2023), 103555. ISSN 1569-8432. https://doi.org/10.1016/j.jag.2023.103555. https://www.sciencedirect.com/science/article/pii/S1569843223003795

Pesaresi, M., A. Gerhardinger and F. Haag. 2007. Rapid damage assessment of built-up structures using VHR satellite data in tsunami-affected areas. *International Journal of Remote Sensing* 28(13): 3013–3036.

Pinz, A. 1994. *Bildverstehen*. Vienna, Springer.

Radoux, J. and P. Defourny. 2008. A framework for the quality assessment of object-based classification. In *Object-Based Image Analysis – Spatial Concepts for Knowledge-Driven Remote Sensing Applications*. Eds. T. Blaschke, S. Lang and G. J. Hay. Berlin, Springer: 257–271.

Raskin, R. and M. Pan. 2003. Semantic web for earth and environmental terminology (sweet). *Proc. of the Workshop on Semantic Web Technologies for Searching and Retrieving Scientific Data*.

Rathje, E. M. and B. J. Adams. 2008. The role of remote sensing in earthquake science and engineering: Opportunities and challenges. *Earthquake Spectra* 24(2): 471–492.

Raza, A. and W. Kainz. 2001. An object-oriented approach for modeling urban land-use changes. *Journal of the Urban and Regional Information Association* 14(1): 37–55.

Reitsma, F., J. Laxton, S. Ballard, W. Kuhn and A. Abdelmoty. 2009. Semantics, ontologies and eScience for the geosciences. *Computers and Geosciences* 35: 706–709.

Richter, R. and D. Schläpfer. 2002. Geo-atmospheric processing of airborne imaging spectrometry data, part 2: Atmospheric/topographic correction. *International Journal of Remote Sensing*: 2631–2649.

Schöpfer, E., S. Lang and F. Albrecht. 2008. Object-fate analysis – spatial relationships for the assessment of object transition and correspondence. In *Object-Based Image Analysis – Spatial Concepts for Knowledge-Driven Remote Sensing Applications*. Eds. T. Blaschke, S. Lang and G. J. Hay. Berlin, Springer: 785–801.

Schumacher, J., S. Lang, D. Tiede, D. Hölbling, J. Rietzke and J. Trautner. 2007. Einsatz von GIS und objekt-basierter Analyse von Fernerkundungsdaten in der regionalen Planung Methoden und erste Erfahrungen aus dem Biotopinformations- und Managementsystem (BIMS) Region Stuttgart. In *Angewandte Geoinformatik 2007*. Eds. J. Strobl, T. Blaschke and G. Griesebner. Heidelberg, Wichmann: 703–708.

Shahi, K., H. Zulhaidi, M. Shafri and A. Hamedianfar. 2017. Road condition assessment by OBIA and feature selection techniques using very high-resolution WorldView-2 imagery. *Geocarto International* 32(12): 1389–1406. http://doi.org/10.1080/10106049.2016.1213888

Smith, B. 1994. Fiat objects. In *11th European Conference on Artificial Intelligence*. Eds. N. Guarino, L. Vieu and S. Pribbenow. Amsterdam.

Spence, N. A. 1961. A multifactor uniform regionalization of British Countries on the basis of employment data for 1961. *Regional Studies* II: 87–104.

Spitzer, M. 2000. *Geist im Netz. Modelle für Lernen, Denken und Handeln*. Heidelberg, Spektrum Akademischer Verlag.

Strasser, T. and S. Lang. 2015. Object-based class modelling for multi-scale riparian forest habitat mapping. *The International Journal of Applied Earth Observation and Geoinformation* (pages pending).

Straub, B. M. and C. Heipke. 2004. Concepts for internal and external evaluation of automatically delineated tree tops. *IntArchPhRS* 26(8): 62–65.

Sudmanns, M., D. Tiede, S. Lang, H. Bergstedt, G. Trost, H. Augustin, A. Baraldi and T. Blaschke. 2019. Big Earth data: disruptive changes in Earth observation data management and analysis? *International Journal of Digital Earth* 13(7): 1–19.

Sundersan, A., P. K. Varshney and M. K. Arora. 2007. Robustness of change detection algorithms in the presence of registration errors. *Photogrammetric Engineering & Remote Sensing* 73: 165–174.

Tarr, M. J. and Y. D. Cheng. 2003. Learning to see faces and objects. *Trends in Cognitive Science* 7(1): 23–30.

Tiede, D. 2014. A new geospatial overlay method for the analysis and visualization of spatial change patterns using object-oriented data modeling concept. *Cartography and Geographic Information Science* 41: 227–234.

Tiede, D., S. Lang, F. Albrecht and D. Hölbling. 2010. Object-based class modeling for cadastre constrained delineation of geo-objects. *Photogrammetric Engineering and Remote Sensing*: 193–202.

Tiede, D., S. Lang, P. Füreder, D. Hölbling, C. Hoffmann and P. Zeil. 2011. Automated damage indication for rapid geospatial reporting. An operational object-based approach to damage density mapping following the 2010 Haiti earthquake. *Photogrammetric Engineering and Remote Sensing* 9: 933–942.

Tiede, D., M. Möller, S. Lang and D. Hölbling. 2007. Adapting, splitting and merging cadastral boundaries according to homogenous LULC types derived from Spot 5 dat. *ISPRS Journal of Photogrammetry* 36(3): 99–104.

Tönjes, R., S. Growe, J. Bückner and C. E. Liedtke. 1999. Knowledge-based Interpretation of remote sensing images using semantic nets *Photogramm. Engineering Remote Sensing* 65: 811–821.

Turker, M. and B. Cetinkaya. 2005. Automatic detection of earthquake-damaged buildings using DEMs created from pre- and post-earthquake stereo aerial photographs. *International Journal of Remote Sensing* 26(4): 823–832.

Van Coillie, F. M. B., R. P. C. Verbeke and R. R. De Wulff. 2008. Semi-automated forest stand delineation using wavelet based segmentation of very high resolution optical imagery. In *Object-Based Image Analysis for Remote Sensing Applications: Modeling Reality-Dealing with Complexity*. Eds. T. Blaschke, S. Lang and G. J. Hay. Berlin, Springer: 237–256.

Vizzari, Marco. 2022. PlanetScope, Sentinel-2, and Sentinel-1 data integration for object-based land cover classification in Google earth engine. *Remote Sensing* 14(11): 2628. https://doi.org/10.3390/rs14112628

Vu, T. T. and Y. Ban. 2010. Context-based mapping of damaged buildings from high-resolution optical satellite images. *International Journal of Remote Sensing* 31(13): 3411–3425.

Walter, V. and D. Fritsch. 1999. Matching spatial data sets: A statistical approac. *International Journal of Geographical Information Science* 13(5): 445–473.

Weinke, E., S. Lang and M. Preiner. 2008. Strategies for semi-automated habitat delineation and spatial change assessment in an Alpine environment. In *Object-Based Image Analysis – Spatial Concepts for Knowledge-Driven Remote Sensing Applications*. Eds. T. Blaschke, S. Lang and G. J. Hay. Berlin, Springer: 711–732.

Weinke, E., S. Lang and D. Tiede. 2007. *Landscape Interpretation Support Tool (LIST)*. Shaker Verlag.

Wertheimer, M. 1925. *Drei Abhandlungen zur Gestalttheorie (in German)*. Erlangen, Palm & Enke.

Wise, S., R. Haining and J. Ma. 2001. Providing spatial statistical data analysis functionality for the GIS user: The SAGE project. *International Journal of Geographical Information Science* 3(1): 239–254.

Ye, S., R. G. Pontius and R. Rakshit. 2018. A review of accuracy assessment for object-based image analysis: From per-pixel to per-polygon approaches. *ISPRS Journal of Photogrammetry and Remote Sensing* 141(2018): 137–147. ISSN 0924-2716. https://doi.org/10.1016/j.isprsjprs.2018.04.002. https://www.sciencedirect.com/science/article/pii/S0924271618300947

Yuan, X., J. Shi and L. Gu. 2021. A review of deep learning methods for semantic segmentation of remote sensing imagery. *Expert Systems with Applications* 169: 114417.

Yurtseven, H., M. Akgul, S. Coban and S. Gulci. 2019. Determination and accuracy analysis of individual tree crown parameters using UAV based imagery and OBIA techniques. *Measurement* 145(2019), 651–664. ISSN 0263-2241. https://doi.org/10.1016/j.measurement.2019.05.092. https://www.sciencedirect.com/science/article/pii/S0263224119305421

Zaabar, N., S. Niculescu and M. M. Kamel. 2022. Application of convolutional neural networks with object-based image analysis for land cover and land use mapping in coastal areas: A case study in ain témouchent, Algeria. *IEEE Journal of Selected Topics in Applied Earth Observations and Remote Sensing* 15: 5177–5189. http://doi.org/10.1109/JSTARS.2022.3185185.

Zadeh, L. A. 1965. Fuzzy sets. *Information and Control* 8(3): 338–353.

Zeil, P., H. Klug and I. Niemeyer. 2008. GIS and remote sensing: Monitoring Environmental conventions and agreements. *BICC Brief.* Bonn. 37: 50–56.

Zlatof, N., B. Tellez and A. Baskurt. 2004. Image understanding and scene models: A generic framework integrating domain knowledge and Gestalt theory. *ICIP '04. 2004 International Conference on Image Processing*.

10 Image Segmentation Algorithms for Land Categorization

James C. Tilton, Selim Aksoy, and Yuliya Tarabalka

LIST OF ACRONYMS

GEOBIA	Geographic object-based image analysis
HSWO	Hierarchical step-wise optimization
ICM	Iterated conditional modes
IOER	Institute for Ecological and Regional Development
NDVI	Normalized difference vegetation index
ROSIS	Reflective optics system imaging spectrometer
SAM	Spectral angle mapper
SAR	Synthetic aperture radar
SVM	Support vector machines

10.1 INTRODUCTION

Image segmentation is the partitioning of an image into related meaningful sections or regions. Segmentation is a key first step for a number of image analysis approaches. The nature and quality of the image segmentation result is a critical factor in determining the level of performance of these image analysis approaches. It is expected that an appropriately designed image segmentation approach will provide a better understanding of a landscape, and/or significantly increase the accuracies of a landscape classification. An image can be partitioned in several ways, based on numerous criteria. Whether or not a particular image partitioning is useful depends on the goal of the image analysis application that is fed by the image segmentation result.

The focus of this chapter is on image segmentation algorithms for land categorization (Jia et al., 2024; Martinez-Sanchez et al., 2024; Lilay and Taye, 2023; Liu et al., 2023; Yu et al., 2023; Raei et al., 2022; Sertel et al., 2022; Li et al., 2022; Pare et al., 2020; Phan et al., 2020; Tassi and Vizzari, 2020; Jin et al., 2019; Liu et al., 2019; Zhang et al., 2019; Shan, 2018; Yin et al., 2018; Zhang et al., 2018; Csillik, 2017; Maulik and Chakraborty, 2017). Our image analysis goal will generally be to appropriately partition an image obtained from a remote sensing instrument on-board a high flying aircraft or a satellite circling the earth or other planet. An example of an earth remote sensing application might be to produce a labeled map that divides the image into areas covered by distinct earth surface covers such as water, snow, types of natural vegetation, types of rock formations, types of agricultural crops and types of other man created development. Alternatively, one can segment the land based on climate (e.g., temperature, precipitation) and elevation zones. However, most image segmentation approaches do not directly provide such meaningful labels to image partitions. Instead, most approaches produce image partitions with generic labels such as region 1, region 2, and so on, which need to be converted into meaningful labels by a post-segmentation analysis.

DOI: 10.1201/9781003541158-11

An early survey on image segmentation grouped image segmentation approaches into three categories (Fu and Mui, 1981): (1) characteristic feature thresholding or clustering, (2) boundary detection, and (3) region extraction. Another early survey (Haralick and Shapiro, 1985) divides region extraction into several region growing and region split and merge schemes. Both of these surveys note that there is no general theory of image segmentation, most image segmentation approaches are *ad hoc* in nature, and there is no general algorithm that will work well for all images. This is still the case even today.

We start our image segmentation discussion with spectrally based approaches, corresponding to Fu and Mui's characteristic feature thresholding or clustering category. We include here a description of support vector machines, as a supervised spectral classification approach that has been a popular choice for analyzing multispectral and hyperspectral images (images with several tens or even hundreds of spectral bands). We then go on to describe a number of spatially based image segmentation approaches that could be appropriate for land categorization applications, generally going from the simpler approaches to the more complicated and more recently developed approaches. Here our emphasis is guided by the prevalence of reported use in land categorization studies. We then take a brief look at various approaches to image segmentation quality evaluation, and include a closer look at a particular empirical discrepancy approach with example quality evaluations for a particular remotely sensed hyperspectral dataset and selected image segmentation approaches. We wrap up with some concluding comments and discussion.

10.2 SPECTRALLY BASED SEGMENTATION APPROACHES

The focus of this section is on approaches that are mainly based on analyses of individual pixels. These approaches use an initial labeling of pixels using unsupervised or supervised classification methods, and then try to group neighboring pixels with similar labels using some form of post-processing to produce segmentation results.

10.2.1 Thresholding-Based Algorithms

Thresholding has been one of the oldest and most widely used techniques for image segmentation (Martinez-Sanchez et al., 2024; Liu et al., 2023; Yu et al., 2023; Raei et al., 2022; Phan et al., 2020; Tassi and Vizzari, 2020; Jin et al., 2019; Shan, 2018; Yin et al., 2018; Zhang et al., 2018; Maulik and Chakraborty, 2017). Thresholding algorithms used for segmentation assume that the pixels that belong to the objects of interest have a property whose values are substantially different from those of the background, and aim to find a good set of thresholds that partition the histogram of this property into two or more non-overlapping regions (Sezgin and Sankur, 2004). While the spectral channels can be directly used for thresholding, other derived properties of the pixels are also commonly used in the literature. For example, Akcay and Aksoy (2011) used thresholding of the red band to identify buildings with red roofs, Bruzzone and Prieto (2005) performed change detection by thresholding the difference image, Rosin and Hervas (2005) used thresholding of the difference image for determining landslide activity, Aksoy et al. (2010) applied thresholding to the normalized difference vegetation index (NDVI) for segmenting vegetation areas, and Unsalan and Boyer (2005) combined thresholding of NDVI and a shadow-water index to identify potential building and street pixels in residential regions.

Selection of the threshold values is often done in an ad hoc manner usually when a single property is involved, but optimal values can also be found by employing exhaustive or stochastic search procedures that look for the values that optimize some criteria on the shape or the statistics of the histogram such as minimization of the within-class variance and maximization of the between-class variance (Otsu, 1979). A stochastic search procedure is particularly needed for finding multiple thresholds where an exhaustive search is not computationally feasible due to the combinatorial increase in the number of candidate values. For example, a recent use of multilevel thresholding for the segmentation of Earth observation data is described in (Lilay and Taye, 2023; Ghamisi et al., 2014) where a particle swarm optimization based stochastic search algorithm was used to obtain a multilevel thresholding of each spectral channel independently by maximizing the corresponding between-class variance.

Even when the selected thresholds are obtained by optimizing some well-defined criteria on the distributions of the properties of the pixels, they do not necessarily produce operational image segmentation results because they suffer from the lack of the use of spatial information as the decisions are independently made on individual pixels. Thus, thresholding is usually applied as a pre-processing algorithm, and various post-processing methods such as morphological operations are often applied to the results of pixel-based thresholding algorithms as discussed in the following section.

10.2.2 Clustering-Based Algorithms

The clustering-based approaches to image segmentation aim to make use of the rich literature on data grouping and/or partitioning techniques for pattern recognition (Yu et al., 2023; Phan et al., 2020; Zhang et al., 2019; Duda et al., 2001). It is intuitive to pose the image segmentation problem as the clustering of pixels, and thus, pixel-based image analysis techniques in the remote sensing literature have found natural extensions to image segmentation. In the most widely used methodology, first, the spectral feature space is partitioned and the individual pixels are grouped into clusters without regard to their neighbors, and then, a post-processing step is applied to form regions by merging neighboring pixels having the same cluster label by using a connected components labeling algorithm.

The initial clustering stage commonly employs well-known techniques such as k-means (Aksoy and Akcay, 2005), fuzzy c-means (Shankar, 2007), and their probabilistic extension using the Gaussian mixture model estimated via expectation-maximization (Fauvel et al., 2013). Since no spatial information is used during the clustering procedure, pixels with the same cluster label can either form a single connected spatial region, or can belong to multiple disjoint regions that are assigned different labels by the connected components labeling algorithm. This reduces the significance of the difficulty of the user's *a priori* selection of the number of clusters in many popular clustering algorithms as there is no strict correspondence between the initial number of clusters and the final number of image regions. However, it still has a high potential of producing an oversegmentation consisting of noisy results with isolated pixels having labels different from those of their neighbors due to the lack of the use of spatial data.

Therefore, a following post-processing step aims to produce a smoother and spatially consistent segmentation by converting the pixel-based clustering results into contiguous regions. A popular approach is to use an additional segmentation result (often also an oversegmentation), and to use a majority voting procedure for spatial regularization by assigning each region in the oversegmentation a single label that is determined according to the most frequent cluster label among the pixels in that region (Li et al., 2022; Pare et al., 2020; Zhang et al., 2019; Fauvel et al., 2013). An alternative approach is to use an iterative split-and-merge procedure as follows (Aksoy et al., 2005; Aksoy, 2006):

1. Merge pixels with identical labels to find the initial set of regions and mark these regions as foreground,
2. Mark regions with areas smaller than a threshold as background using connected components analysis,
3. Use region growing to iteratively assign background pixels to the foreground regions by placing a window at each background pixel and assigning it to the class that occurs the most in its neighborhood.

This procedure corresponds to a spatial smoothing of the clustering results. The resulting regions can be further processed using mathematical morphology operators to automatically divide large regions into more compact sub-regions as follows:

1. Find individual regions using connected components analysis for each cluster,
2. For all regions, compute the erosion transform and repeat:
   a. Threshold erosion transform at steps of 3 pixels in every iteration,
   b. Find connected components of the thresholded image,
   c. Select sub-regions that have an area smaller than a threshold,
   d. Dilate these sub-regions to restore the effects of erosion,
   e. Mark these sub-regions in the output image by masking the dilation using the original image, until no more sub-regions are found.
3. Merge the residues of previous iterations to their smallest neighbors.

Even though we focused on producing segmentations using clustering algorithms in this section, similar post-processing techniques for converting the pixel-based decisions into contiguous regions can also be used with the outputs of pixel-based thresholding (Section 10.2.1) and classification (Section 10.2.3) procedures (Aksoy et al., 2005).

It is also possible to pose clustering, and the corresponding segmentation, as a density estimation problem. A commonly used algorithm that combines clustering with density estimation and segmentation is the mean shift algorithm (Comaniciu and Meer, 2002). Mean shift is based on non-parametric density estimation where the local maxima (i.e., modes) of the density can be assumed to correspond to clusters. The algorithm does not require *a priori* knowledge of the number of clusters in the data, and can identify the locations of the local maxima by a set of iterations. These iterations can be interpreted as the shifting of points toward the modes where convergence is achieved when a point reaches a particular mode. The shifting procedure uses a kernel with a scale parameter that determines the amount of local smoothing performed during density estimation. The application of the mean shift procedure to image segmentation uses a separate kernel for the feature (i.e., spectral) domain and another kernel for the spatial (i.e., pixel) domain. The scale parameter for the spectral domain can be estimated by maximizing the average likelihood of held-out data. The scale parameter for the spatial domain can be selected according to the amount of compactness or over-segmentation desired in the image, or can be determined by using geospatial statistics (e.g., by using semivariogram-based estimates) (Dongping et al., 2012). Furthermore, agglomerative clustering of the mode estimates can be used to obtain a multi-scale segmentation (Sertel et al., 2022; Tassi and Vizzari, 2020; Jin et al., 2019; Shan, 2018; Csillik, 2017; Maulik and Chakraborty, 2017).

10.2.3 Support Vector Machines

Output of supervised classification of pixels can also be used as input for segmentation techniques. In recent years, support vector machines (SVM) and the use of kernels to transform data into a new feature space where linear separability can be exploited have been proposed. The SVM method attempts to separate training samples belonging to different classes by tracing maximum margin hyperplanes in the space where the samples are mapped. SVM have shown to be particularly well suited to classify high-dimensional data (e.g., hyperspectral images) when a limited number of training samples is available (Camps-Valls, 2005; Vapnik, 1998). The success of SVM for pixel-based classification has led to its subsequent use as part of image segmentation methods. Thus we discuss the SVM approach in detail later.

SVM are primarily designed to solve binary tasks, where the class labels take only two values: 1 or –1. Let us consider a binary classification problem in a B-dimensional space $\mathbb{R}^B$, with N training samples, $x_i \in \mathbb{R}^B$, and their corresponding class labels $y_i = \pm 1$ available. The SVM technique consists in finding the hyperplane that maximizes the margin, that is, the distance to the closest training

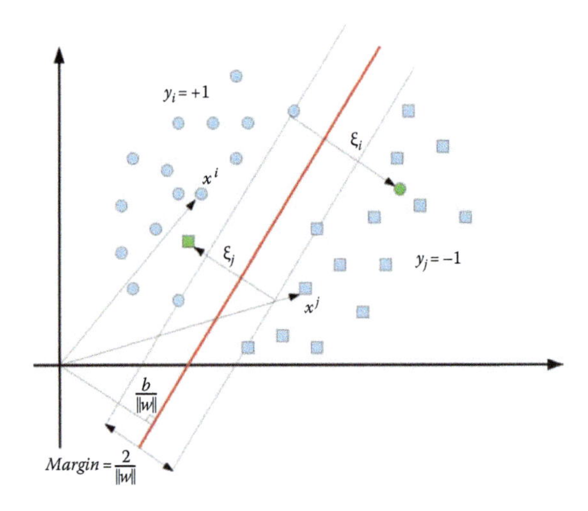

FIGURE 10.1 Schematic illustration of the SVM binary classification method. There is one nonlinearly separable sample in each class.

data points in both classes (see Figure 10.1). Noting $\mathbf{w} \in \mathbb{R}^B$ as the vector normal to the hyperplane and $b \in \mathbb{R}$ as the bias, the hyperplane H is defined as:

$$\mathbf{w} \cdot x + b = 0, \ \forall x \in H$$

If $x \notin H$ then

$$f(x) = \frac{|\mathbf{w} \cdot x + b|}{\| \mathbf{w} \|}$$

defines the distance of the sample x to H. In the linearly separable case, such a hyperplane must satisfy:

$$y_i (\mathbf{w} \cdot x_i + b) > 1, \forall i \in [1, N]. \tag{10.1}$$

The optimal hyperplane is the one that maximizes the margin $2/\| \mathbf{w} \|$. This is equivalent to minimizing $\| \mathbf{w} \| /2$ and leads to the quadratic optimization problem:

$$\min \left[\frac{\| \mathbf{w} \|^2}{2} \right], \text{ subject to (10.1)} \tag{10.2}$$

To take into account nonlinearly separable data, slack variables ξ are introduced to deal with misclassified samples (see Figure 10.1). Equation (10.1) becomes

$$y_i (\mathbf{w} \cdot x_i + b) > 1 - \xi_i, \xi_i \geq 0, \ \forall i \in [1, N]. \tag{10.3}$$

The final optimization problem is formulated as

$$\min \left[\frac{\| \mathbf{w} \|^2}{2} + C \sum_{i=1}^{N} \xi_i \right], \text{ subject to (10.3)} \tag{10.4}$$

where the constant C is a regularization parameter that controls the amount of penalty. This optimization problem is typically solved by quadratic programming (Vapnik, 1998). The classification is further performed by computing $y_u = sign(\mathbf{w} \cdot x_u + b)$, where $(\mathbf{w}, \mathrm{b})$ are the hyperplane parameters found during the training process and x_u is an unseen sample.

One can notice that the pixel vectors in the optimization and decision rule equations always appear in pairs related through a scalar product. These products can be replaced by nonlinear functions of the pairs of vectors, essentially projecting the pixel vectors in a higher dimensional space $\mathbb{H}$ and thus improving linear separability of data:

$$\mathbb{R}^B \rightarrow \mathbb{H},$$

$$x \rightarrow \Phi(x) \tag{10.5}$$

$$x_i \cdot x_j \rightarrow \Phi(x_i) \cdot \Phi(x_j) = K(x_i, x_j).$$

Here, $\Phi(\cdot)$ is a nonlinear function to project feature vectors into a new space, $K(\cdot)$ is a *kernel* function, which allows one to avoid the computation of scalar products in the transformed space $[\Phi(x_i) \cdot \Phi(x_j)]$ and thus reduces the computational complexity of the algorithm. The kernel K must satisfy Mercer's condition (Burges, 1998). The Gaussian Radial Basis Function (RBF) kernel is the most widely used for remote sensing image classification:

$$K_{Gaussian}(x_i, x_j) = \exp\left[-\gamma \| x_i - x_j \|^2\right], \tag{10.6}$$

where γ is the spread of the RBF kernel.

To solve the K-class problem, various approaches have been proposed. Two main approaches combining a set of binary classifiers are defined as (Scholkopf and Smola, 2002):

- One versus all: K binary classifiers are applied on each class against the others. Each pixel x_i is assigned to the class with the maximum output $f(x_i)$.
- One versus one: $K(K-1)/2$ binary classifiers are applied on each pair of classes. Each pixel is assigned to the class winning the maximum number of binary classification procedures.

As a conclusion, SVM directly exploit the geometrical properties of data, without involving a density estimation procedure. This method has proven to be more effective than other nonparametric classifiers (such as neural networks or the k-Nearest Neighbor classifier (Duda et al., 2001)) in terms of classification accuracies, computational complexity and robustness to parameter setting. SVM can efficiently handle high-dimensional data, exhibiting low sensitivity to the Hughes phenomenon (Hughes, 1968). Finally, it exhibits good generalization capability, fully exploiting the discrimination capability of available training samples.

Pixel-based supervised classification results, such as those obtained using an SVM classifier, are often given as input to segmentation procedures that aim to group the pixels to form contiguous regions (Jia et al., 2024; Sertel et al., 2022; Li et al., 2022; Zhang et al., 2019; Shan, 2018; Csillik, 2017; Maulik and Chakraborty, 2017) as discussed in the following sections.

10.3 SPATIALLY BASED SEGMENTATION APPROACHES

We cannot possibly discuss myriad of spatially-based image segmentation approaches that have been proposed and developed over the years. Instead we will focus on approaches that have achieved demonstrated success in remote sensing land categorization applications. A compilation of such approaches can be found in a series of papers published by a research group based at the Leibniz Institute for Ecological and Regional Development (IOER) that present comparative evaluations of image segmentation approaches implemented in various image analysis packages (Meinel and Neubert, 2004; Neubert et al., 2006; Neubert et al., 2008; Marpu et al., 2010). Table 10.1 provides a summary listing of most of the remote sensing oriented image analysis packages whose image segmentation approach was evaluated in these papers, plus image segmentation approaches from three additional notable remote sensing oriented image analysis packages (GRASS GIS, IDRISI and the Orfeo toolbox).

We note from Table 10.1 that region growing is the most frequent image segmentation approach utilized by these remote sensing oriented image analysis software packages. Further, several packages combine region growing with other techniques (RHSeg with spectral clustering, SCRM and GRASS GIS with watershed, SegSAR with edge detection). Watershed segmentation (an approach based on region boundary detection) is the next most popular approach. Simulated annealing is often utilized in analysis packages oriented towards analyzing Synthetic Aperture Radar (SAR) imagery data (Ceasar and InfoPACK).

The next several sections describe various spatially based image segmentation approaches, starting with region growing algorithms, and continuing with texture-based algorithms, morphological

TABLE 10.1
The Algorithmic Basis of Image Segmentation Approaches in Remote Sensing Oriented Image Analysis Packages

Name	Website or Reference	Algorithmic Basis
BerkeleyImgseg	http://www.imageseg.com/	Region-growing, region-merging
Ceasar	(Cook et al., 1996)	Simulated annealing
eCognition Developer[1]	http://www.ecognition.com/products/ecognition-developer and (Baatz and Schape, 2000)	Region growing
ENVI Feature Extraction	http://www.exelisvis.com/docs/using_envi_Home.html and (Robinson et al., 2002)	Edge-based (Full Lambda-Schedule algorithm for region merging)
Extended Watershed EWS	(Li and Xiao, 2007)	Multi-channel watershed transformation
Image WS for Erdas Imagine	(Sramek and Wrbka, 1997)	Hierarchical watershed
InfoPACK	(Cook et al., 1996)	Simulated annealing
PARBAT	http://parbat.lucieer.net/ and (Lucieer, 2004)	Region growing
RHSeg	(Tilton, 1998) and (Tilton et al., 2012)	Region growing and spectral clustering
SCRM	(Castilla et al., 2008)	Watershed and region merging
SegSAR	http://www.segsar.googlepages.com/	Hybrid (edge/region oriented)
SEGEN	(Gofman, 2006)	Region growing
GRASS GIS	http://grasswiki.osgeo.org/wiki/GRASS-Wiki	Region growing and watershed
IDRISI	http://www.clarklabs.org/applications/upload/Segmentation-IDRISI-Focus-Paper.pdf	Watershed
Orfeo Toolbox	http://www.orfeo-toolbox.org/otb/	Region growing, watershed, level sets, mean-shift

Note: (1) Was Definiens Developer—but the remote sensing package is now marketed by Trimble, and the Definiens product is now oriented to biomedical image analysis. (2) Most of these image segmentation approaches were evaluated in a series of papers by the Leibnitz IOER group.

algorithms, graph-based algorithms, and MRF-based algorithms (Jia et al., 2024; Martinez-Sanchez et al., 2024; Lilay and Taye, 2023; Liu et al., 2023; Yu et al., 2023; Raei et al., 2022; Sertel et al., 2022; Li et al., 2022; Pare et al., 2020; Phan et al., 2020; Tassi and Vizzari, 2020; Jin et al., 2019; Liu et al., 2019; Zhang et al., 2019; Shan, 2018; Yin et al., 2018; Zhang et al., 2018; Csillik, 2017; Maulik and Chakraborty, 2017).

10.3.1 Region Growing Algorithms

In the region growing approach to image segmentation an image is initially partitioned into small region objects. These initial small region objects are often single image pixels, but can also be $n \times n$ blocks of pixels or another partitioning of the image into small spatially connected region objects. Then pairs of spatially adjacent region objects are compared and merged together if they are found to be similar enough according to some comparison criterion. The underlying assumption is that region objects of interest are several image pixels in size and relatively homogeneous in value. Most region growing approaches can operate on either grey scale, multispectral or hyperspectral image data, depending on the criterion used to determine the similarity between neighboring region objects (Martinez-Sanchez et al., 2024; Raei et al., 2022; Phan et al., 2020; Tassi and Vizzari, 2020; Shan, 2018; Yin et al., 2018; Zhang et al., 2018; Csillik, 2017; Maulik and Chakraborty, 2017).

A very early example of region growing was described in (Muerle and Allen, 1968). Muerle and Allen experimented with initializing their region growing process with region objects consisting of 2×2 up to 8×8 blocks of pixels. After initialization, they started with the region object at the upper left corner of the image and compared this region object with the neighboring region objects. If a neighboring region object was found to be similar enough, the region objects were merged together. This process was continued until no neighboring region objects could be found that were similar enough to be merged into the region object that was being grown. Then the image was scanned (left-to-right, top-to-bottom) to find an unprocessed region object, that is, a region object that not yet been considered as an initial object for region growing or merged into a neighboring region. If an unprocessed region object was found, they conducted their region growing process from that region object. This continued until no further unprocessed region objects could be found, upon which point the region growing segmentation process was considered completed.

Many early schemes for region growing, such as Muerle and Allen's, can be formulated as logical predicate segmentation, defined in (Zucker, 1976) as:

A segmentation of an image X can be defined as a partition of X into R disjoint subsets $X_1, X_2, \ldots, X_R$, such that the following conditions hold:

1. $\cup_{i=1}^{R} X_i = X$,
2. X_i, $i = 1, 2, \ldots, R$ are connected,
3. $P(X_i) = \text{TRUE}$ for $i = 1, 2, \ldots, R$, and
4. $P(X_i \cup X_j) = \text{FALSE}$ for $i \neq j$, where X_i and X_j are adjacent.

$P(X_i)$ is a logical predicate that assigns the value TRUE or FALSE to X_i, depending on the image data values in X_i.

These conditions are summarized in (Zucker, 1976) as follows: the first condition requires that every picture element (pixel) must be in a region (subset). The second condition requires that each region must be connected, that is, composed of contiguous image pixels. The third condition determines what kind of properties each region must satisfy, that is, what properties the image pixels must satisfy to be considered similar enough to be in the same region. Finally, the fourth condition specifies that any merging of adjacent regions would violate the third condition in the final segmentation result.

Several researchers in this early era of image segmentation research, including Muerle and Allen, noted some problems with logical predicate segmentation. For one, the results were very dependent on the order in which the image data was scanned. Also, the statistics of a region object can change quite dramatically as the region is grown, making it possible that many adjacent cells that were rejected for merging early in the region growing process would have been accepted in later stages based on the changed statistics of the region object. The reverse was also possible, where adjacent cells that were accepted for merging early in the region growing process would have been rejected in later stages.

Subsequently, an alternate approach to region growing was developed that avoids the aforementioned problems, and approach that eventually became to be referred to as best merge region growing. An early version of best merge region growing, hierarchical step-wise optimization (HSWO), is an iterative form of region growing, in which the iterations consist of finding the most optimal or best segmentation with one region less than the current segmentation (Beaulieu and Goldberg, 1989). The HSWO approach can be summarized as follows:

1. Initialize the segmentation by assigning each image pixel a region label. If a pre-segmentation is provided, label each image pixel according to the pre-segmentation. Otherwise, label each image pixel as a separate region.
2. Calculate the dissimilarity criterion value, d, between all pairs of spatially adjacent regions, find the smallest dissimilarity criterion value, T_{merge}, and merge all pairs of regions with $d = T_{merge}$.
3. Stop if no more merges are required. Otherwise, return to step 2.

HSWO naturally produces a segmentation hierarchy consisting of the entire sequence of segmentations from initialization down to the final trivial one region segmentation (if allowed to proceed that far). For practical applications, however, a subset of segmentations needs to be selected out from this exhaustive segmentation hierarchy. At a minimum such a subset can be defined by storing the results only after a preselected number of regions is reached, and then storing selected iterations after that such that no region is involved in more than one merge between stored iteration until a two region segmentation is reached. A portion of such a segmentation hierarchy is illustrated in Figure 10.2. (The selection of a single segmentation from a segmentation hierarchy is discussed in Section 10.3.1.5.)

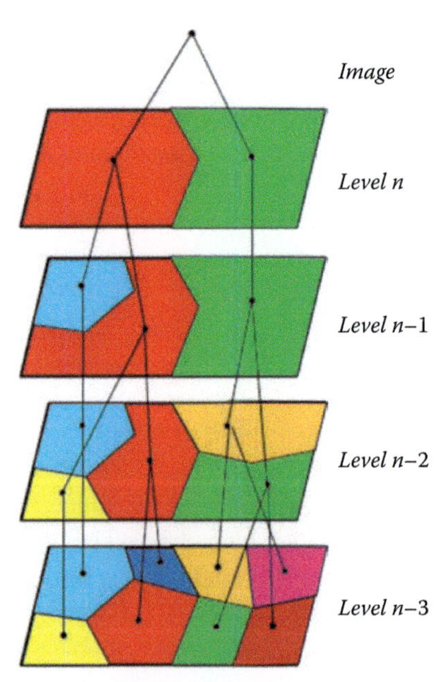

FIGURE 10.2 The last four levels of an n level segmentation hierarchy produced by a region growing segmentation process. Note that when depicted in this manner, the region growing process is a "bottom-up" approach.

A unique feature of the segmentation hierarchy produced by HSWO and related region growing segmentation approaches is that the segment or region boundaries are maintained at the full image spatial resolution for all levels of the segmentation hierarchy. The region boundaries are coarsened in many other multilevel representations.

Many variations on best merge region growing have been described in the literature. As early as 1994, Kurita (1994) described an implementation of the HSWO form of best merge region growing that utilized a heap data structure (Williams, 1964) for efficient determination of best merges and a dissimilarity criterion based on minimizing the mean squared error between the region mean image and original image. Further, several of the image segmentation approaches studied in the previously referenced series of papers published by the Liebniz Institute are based on best merge region growing (see Table 10.1). As we will discuss in more detail later, the main differences between most of these region growing approaches are the dissimilarity criterion employed and, perhaps, some control logic designed to remove small regions or otherwise tailor the segmentation output.

10.3.1.1 Seeded Region Growing

Seeded region growing is a variant on best merge region growing in which regions are grown from pre-selected seed pixels or seed regions. Adams and Bischof (1994) is an early example of this approach. In seeded region growing, the best merges are found by examining the pixels adjacent to each seed pixel of region formed around each seed pixel. As described by Adams and Bischof (1994), the region growing process continues until each image pixel is associated with one of the pre-selected seed pixels or regions.

10.3.1.2 Split and Merge Region Growing

In the split and merge approach, the image is repeatedly subdivided until each resulting region has a sufficiently high homogeneity (Sertel et al., 2022; Li et al., 2022; Zhang et al., 2019; Shan, 2018; Maulik and Chakraborty, 2017). Examples of measures of homogeneity are the mean squared error between the region mean and the image data values, or the region standard deviation. After the region-splitting process converges, the regions are grown using one of the previously described region growing approaches. This approach is more efficient when large homogeneous regions are present. However, some segmentation detail may be lost. See (Horowitz and Pavlidis, 1974; Cross et al., 1988; Strasters and Gerbrands, 1991) for examples of this approach.

10.3.1.3 Hybrid of Spectral Clustering and Region Growing

Tilton (1998) and Tilton et al. (2012) describe a hybridization of HSWO best merge region growing with spectral clustering, called HSeg (for Hierarchical Segmentation). We remind the reader that HSWO is performed by finding a threshold value, T_{merge}, equal to the value of a dissimilarity criterion of the most similar pair of spatially adjacent regions, and then merging all pairs of regions that have dissimilarity equal to T_{merge}. HSeg adds to HSWO a step following each step of adjacent region merges in which all pairs of spatially non-adjacent regions are merged that have dissimilarity $<= S_w T_{merge}$, where $0.0 <= S_w <= 1.0$ is a factor that sets the priority between spatially adjacent and non-adjacent region merges. Note that when $S_w = 0.0$, HSeg reduces to HSWO.

Unfortunately, the inclusion of the step in HSeg of merging spatially non-adjacent regions adds significantly to the computational requirements of this image segmentation approach. This is because comparisons must now be made between all pairs of regions instead of just between pairs of spatially adjacent regions. A recursive divide-and-conquer approximation of HSeg (called RHSeg), with a straightforward parallel implementation, was developed to help overcome this problem (Tilton, 2007). The computational requirements of this approach were further reduced by a refinement in which the non-adjacent region merging was limited to between regions of a minimum size, P_{min} (Tilton et al., 2012). In this refinement the value of P_{min} is dynamically adjusted to keep the number of "large regions" (those of at least P_{min} in size) to a range that substantially reduces the computational requirements without out significantly affecting the image segmentation results.

10.3.1.4 Dissimilarity Criterion and Specialized Control Logic

Muerle and Allen (1968) experimented with various criteria for determining whether or not pairs of region objects were similar enough to be merged together. They concluded that an optimal criterion would be a threshold of a function of the mean and standard deviation of the gray levels of the pixels contained in the compared region objects.

In our image segmentation research, we have implemented and studied several region merging criteria in the form of dissimilarity criteria (Tilton, 2013). Included among these dissimilarity criteria are criteria based on vector norms (the 1-, 2-, and ∞–norms), criteria based on minimizing the increase of mean squared error between the region mean image and the original image data, and a criterion based on the Spectral Angle Mapper (SAM) criterion (Kruse et al., 1993). We briefly describe these dissimilarity criteria here.

The dissimilarity criterion based on the 1-Norm of the difference between the region mean vectors, u_i and u_j, of regions X_i and X_j, each with B spectral bands, is:

$$\mathrm{d}_{\text{1-Norm}}\left(X_i, X_j\right) = \left\|u_i - u_j\right\|_1 = \sum_{b=1}^{B} \left|\mu_{ib} - \mu_{jb}\right|, \tag{10.7}$$

where μ_{ib} and μ_{jb} are the mean values for regions i and j, respectively, in spectral band b, that is, $u_i = (\mu_{i1}, \mu_{i2}, \ldots, \mu_{iB})^T$ and $u_j = (\mu_{j1}, \mu_{j2}, \ldots, \mu_{jB})^T$.

The dissimilarity criterion based on the 2-Norm is:

$$\mathrm{d}_{\text{2-Norm}}\left(X_i, X_j\right) = \left\|u_i - u_j\right\|_2 = \left[\sum_{b=1}^{B} \left(\mu_{ib} - \mu_{jb}\right)^2\right]^{1/2}, \tag{10.8}$$

and the dissimilarity criterion based on the ∞-Norm is:

$$\mathrm{d}_{\infty\text{-Norm}}\left(X_i, X_j\right) = \left\|u_i - u_j\right\|_\infty = \max\left(\left|\mu_{ib} - \mu_{jb}\right|, b = 1, 2, \cdots, B\right). \tag{10.9}$$

As noted earlier, a criterion based on mean squared error minimizes the increase of mean squared error between the region mean image and the original image data as regions are grown. The sample estimate of the mean squared error for the segmentation of band b of the image X into R disjoint subsets $X_1, X_2, \ldots, X_R$ is given by:

$$MSE_b\left(X\right) = \frac{1}{N-1} \sum_{i=1}^{R} MSE_b\left(X_i\right), \tag{10.10a}$$

where N is the total number of pixels in the image data and

$$MSE_b\left(X_i\right) = \sum_{x_p \in X_i} \left(\chi_{pb} - \mu_{ib}\right)^2 \tag{10.10b}$$

is the mean squared error contribution for band b from segment X_i. Here, x_p is a pixel vector (in this case, a pixel vector in data subset X_i), and χ_{pb} is the image data value for the bth spectral band of the pixel vector, x_p. The dissimilarity function based on a measure of the increase in mean squared error due to the merge of regions X_i and X_j is given by:

$$d_{BSMSE}\left(X_i,X_j\right)=\sum_{b=1}^{B}\Delta MSE_b\left(X_i,X_j\right), \tag{10.11a}$$

where

$$\Delta MSE_b(X_i,X_j) = MSE_b(X_i \cup X_j) — MSE_b(X_i) — MSE_b(X_j) \tag{10.11b}$$

BSMSE refers to "band sum *MSE*." Using (10.10b) and exchanging the order of summation, (10.11b) can be manipulated to produce an efficient dissimilarity function based on aggregated region features (for the details, see (Tilton, 2013)):

$$d_{BSMSE}\left(X_i,X_j\right)=\frac{n_i n_j}{\left(n_i+n_j\right)}\sum_{b=1}^{B}\left(\mu_{ib}-\mu_{jb}\right)^2. \tag{10.12}$$

The dimensionality of the d_{BSMSE} dissimilarity criteria is equal to the square of the dimensionality of the image pixel values, while the dimensionality of the vector norm based dissimilarity criteria is equal to the dimensionality of the image pixel values. To keep the dissimilarity criteria dimensionalities consistent, the square root of d_{BSMSE} is often used.

The Spectral Angle Mapper (SAM) criterion is widely used in hyperspectral image analysis (Kruse et al., 1993). This criterion determines the spectral similarity between two spectral vectors by calculating the "angle" between the two spectral vectors. An important property of the SAM criterion is that poorly illuminated and more brightly illuminated pixels of the same color will be mapped to the same spectral angle despite the difference in illumination. The spectral angle θ between the region mean vectors, u_i and u_j, of regions X_i and X_j is given by:

$$\theta\left(u_i,u_j\right)=\arccos\left(\frac{u_i \circ u_j}{\left\|u_i\right\|_2\left\|u_j\right\|_2}\right)=\arccos\left(\frac{\sum_{b=1}^{B}\mu_{ib}\mu_{jb}}{\left(\sum_{b=1}^{B}\mu_{ib}^2\right)^{1/2}\left(\sum_{b=1}^{B}\mu_{jb}^2\right)^{1/2}}\right). \tag{10.13a}$$

where μ_{ib} and μ_{jb} are the mean values for regions i and j, respectively, in spectral band b, that is, $u_i = (\mu_{i1}, \mu_{i2}, \ldots, \mu_{iB})^T$ and $u_j = (\mu_{j1}, \mu_{j2}, \ldots, \mu_{jB})^T$. The dissimilarity function for regions X_i and X_j, based on the SAM distance vector measure, is given by:

$$d_{SAM}\left(X_i,X_j\right)=\theta\left(u_i,u_j\right). \tag{10.13b}$$

Note that the value of d_{SAM} ranges from 0.0 for similar vectors up to $\pi/2$ for the most dissimilar vectors.

A problem that can often occur with basic best merge region growing approaches is that the segmentation results contain many small regions. We have found this to be the case when employing dissimilarity criteria based on vector norms or SAM, but not a problem for dissimilarity criteria based on minimizing the increase of mean squared error. This is because the mean squared error criterion has a factor, $n_i n_j / \left(n_i - n_j\right)$, where n_i and n_j are the number of pixels in the two compared regions, that biases toward merging small regions into larger ones. We have found it useful to add on a similar "small region merge acceleration factor" to the vector norm and SAM based criterion when one of the compared regions is smaller than a certain size. See Tilton et al. (2012) for more details.

Implementations of best merge region growing often add special control logic to reduce the number of small regions or otherwise improve the final classification result. An example of this is SEGEN (Gofman, 2006), which uses the vector 2-norm (otherwise known as Euclidean distance) for the dissimilarity criterion. As noted in Tilton et al. (2012), SEGEN is a relatively pure implementation of best merge region growing, optimized for efficiency in performance, memory utilization, and image segmentation quality. SEGEN adds a number of (optional) procedures to best merge region growing, among them a low-pass filter to be applied on the first stage of the segmentation and outlier dispatching on the last stage. The latter removes outlier pixels and small segments by imbedding them in neighborhood segments with the smallest dissimilarity. SEGEN also provides several parameters to control the segmentation process. A set of "good in average" control values is suggested in (Gofman, 2006).

The best merge region growing segmentation approach employed in eCognition Developer utilizes a dissimilarity function that balances minimizing the increase of heterogeneity, f, in both color and shape (Baatz and Schape, 2000; Benz et al., 2004):

$$f = w_{color} \cdot \Delta h_{color} + w_{shape} \cdot \Delta h_{shape} \tag{10.14}$$

where w_{color} and w_{shape} are weights that range in value from zero to one and must mutually sum up to one.

The color component of heterogeneity, Δh_{color}, is defined as follows:

$$\Delta h_{color} = \sum_{c} w_c \left(n_{merge} \cdot \sigma_{c,merge} - \left(n_1 \cdot \sigma_{c,1} + n_2 \cdot \sigma_{c,2} \right) \right) \tag{10.15}$$

where n_x is the number of pixels, with the subscript x=*merge* referring to the merged object, and subscripts $x = 1$ or 2 referring to the first and second objects considered for merging. σ_c refers to the standard deviation in channel (spectral band) c, with the same additional subscripting denoting the merged or pair of considered objects. w_c is a channel weighting factor.

The shape component of heterogeneity, Δh_{shape}, is defined as follows:

$$\Delta h_{shape} = w_{compt} \cdot \Delta h_{compt} + w_{smooth} \cdot \Delta h_{smooth} \tag{10.16}$$

where

$$\Delta h_{compt} = n_{merge} \cdot \frac{l_{merge}}{\sqrt{n_{merge}}} - \left(n_1 \cdot \frac{l_1}{\sqrt{n_1}} + n_2 \cdot \frac{l_2}{\sqrt{n_2}} \right) \tag{10.17}$$

$$\Delta h_{smooth} = n_{merge} \cdot \frac{l_{merge}}{b_{merge}} - \left(n_1 \cdot \frac{l_1}{b_1} + n_2 \cdot \frac{l_2}{b_2} \right) \tag{10.18}$$

where l is the perimeter of the object, and b is the perimeter of the object's bounding box. The weights w_c, w_{color}, w_{shape}, w_{smooth} and w_{compt} can be selected to best suit a particular application.

The level of detail, or scale, of the segmentation is set by stopping the best merge region growing process when the increase if heterogeneity, f, for the best merge reaches a predefined threshold value. A multiresolution segmentation can be created by performing this process with a set of increasing thresholds.

In eCognition Developer, other segmentation procedures can be combined with the multiresolution approach. For example, spectral difference segmentation merges neighbor objects that fall within a user-defined maximum spectral difference. This procedure can be used to merge spectrally similar objects from the segmentation produced by the multiresolution approach (Yu et al., 2023; Phan et al., 2020; Zhang et al., 2018; Csillik, 2017).

10.3.1.5 Selection of a Single Segmentation from a Segmentation Hierarchy

Some best merge region growing approaches such as HSWO and HSeg do not produce a single segmentation result, but instead produce a segmentation hierarchy. The best merge region growing segmentation approach employed in eCognition Developer can also produce a segmentation hierarchy. A segmentation hierarchy is a set of several image segmentations of the same image at different levels of detail in which the segmentations at coarser levels of detail can be produced from simple merges of regions at finer levels of detail. In such a structure, an object of interest may be represented by multiple segments in finer levels of detail, and may be merged into a surrounding region at coarser levels of detail. A single segmentation level can be selected out of the segmentation hierarchy by analyzing the spatial and spectral characteristics of the individual regions, and by tracking the behavior of these characteristics throughout different levels of detail. A manual approach for doing this using a graphical user interface is described in (Tilton, 2003; Tilton, 2013). A preliminary study on automating this approach is described in (Plaza and Tilton, 2005) where it was proposed to automate this approach using joint spectral/spatial homogeneity scores computed from segmented regions. An alternate approach for making spatially localized selections of segmentation detail based on matching region boundaries with edges produced by an edge detector is described in (Le Moigne and Tilton, 1995).

Tarabalka et al. (2012) proposed a modification of HSeg through which a single segmentation output is automatically selected for output from the usual segmentation hierarchy. The idea is similar to the previously described seeded region growing. The main idea behind the marker-based HSeg algorithm consists in automatically selecting seeds, or *markers*, for image regions, and then performing best merge region growing with an additional condition: two regions with different marker labels cannot be merged together (see Figure 10.3). The authors proposed to choose markers of spatial regions by analyzing results of a probabilistic supervised classification of each pixel and by retaining the most reliably classified pixels as region seeds (Jia et al., 2024; Yu et al., 2023; Tassi and Vizzari, 2020).

An alternative algorithm that produces a final segmentation by automatically selecting subsets of regions appearing in different levels in the segmentation hierarchy is described in Section 10.3.3.2.

10.3.2 Texture-Based Algorithms

Along with spectral information, textural features have also been heavily used for various image analysis tasks, including their use as features in thresholding-based, clustering-based, and region growing-based segmentation in the literature. In this section, we focus on the use of texture for *unsupervised* image segmentation (Jia et al., 2024; Martinez-Sanchez et al., 2024; Lilay and Taye, 2023; Liu et al., 2023; Yu et al., 2023). In the following, first, we discuss some particular examples that involve texture modeling for segmenting natural landscapes and man-made structures using local image properties, and then, present recent work on generalized texture algorithms that aim to model complex image structures in terms of the statistics and spatial arrangements of simpler image primitives.

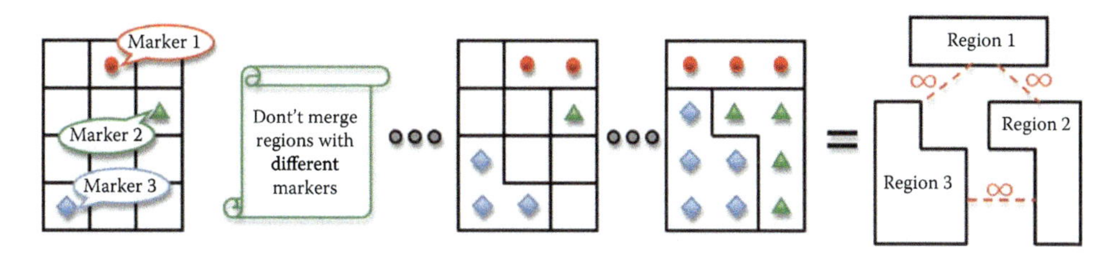

FIGURE 10.3 Scheme illustrating the marker-based HSeg algorithm.

One of the common uses of texture in remote sensing image analysis is the identification of natural landscapes. For example, Epifanio and Soille (2007) used morphological transformations such as white and black top-hat by reconstruction and thresholded image gradients with an unsupervised clustering procedure where the texture prototypes were automatically selected based on the dissimilarities between the feature vectors of neighboring image windows to segment vegetation zones and forest stands, and Wang and Boesch (2007) combined an initial color-based oversegmentation step with a threshold-based region merging procedure that used wavelet feature statistics inside candidate image regions to delineate forest boundaries.

Man-made structures also exhibit particular characteristics that can be modeled using textural features. For example, Pesaresi et al. (2008) proposed a rotation invariant anisotropic texture model that used contrast features computed from gray level co-occurrence matrices, and used this model to produce a built-up presence index. The motivation behind the use of the contrast-based features was to exploit the relationships between the buildings and their shadows. Sirmacek and Unsalan (2010) and Gueguen et al. (2012) used a similar idea and modeled urban areas using spatial voting (smoothing) of local feature points extracted from the local maxima of Gabor filtering results and corner detection results, respectively. All of these results can be converted to a segmentation output by thresholding the corresponding urban area estimates.

An open problem in image segmentation is to identify the boundaries of regions that are not necessarily homogeneous with respect to low-level features such as color and texture that are extracted from individual pixels or from small local pixel neighborhoods (Raei et al., 2022; Sertel et al., 2022; Zhang et al., 2019; Shan, 2018; Yin et al., 2018; Zhang et al., 2018; Csillik, 2017). These recent literature includes several examples that can be considered as generalized texture algorithms that aim to model heterogeneous image content in terms of the spatial arrangements of relatively homogeneous image primitives. Some of this work has considered the segmentation of particular structures that have specific textural properties. For example, Dogrusoz and Aksoy (2007) aimed the segmentation of regular and irregular urban structures by modeling the image content using a graph where building objects and the Voronoi tessellation of their locations formed the vertices and the edges, respectively, and the graph was clustered by thresholding its minimum spanning tree so that organized (formal) and unorganized (informal) settlement patterns were extracted to model urban development. Some agricultural structures such as permanent crops also exhibit specific textural properties that can be useful for segmentation. For example, Aksoy et al. (2012) proposed a texture model that is based on the idea that textures are made up of primitives (trees) appearing in a near-regular repetitive arrangement (planting patterns), and used this model to compute a regularity score for different scales and orientations by using projection profiles of multi-scale isotropic filter responses at multiple orientations. Then, they illustrated the use of this model for segmenting orchards by iteratively merging neighboring pixels that have similar regularity scores at similar scales and orientations.

More generic approaches have also been proposed for segmenting heterogeneous structures (Jia et al., 2024; Martinez-Sanchez et al., 2024; Pare et al., 2020; Shan, 2018). For example, Gaetano et al. (2009) started with an oversegmentation of atomic image regions, and then performed hierarchical texture segmentation by assuming that frequent neighboring regions are strongly related. These relations were represented using Markov chain models computed from quantized region labels, and the image regions that exhibit similar transition probabilities were clustered to construct a hierarchical set of segmentations. Zamalieva et al. (2009) used a similar frequency-based approach by finding the significant relations between neighboring regions as the modes of a probability distribution estimated using the continuous features of region co-occurrences. Then, the resulting modes were used to construct the edges of a graph where a graph mining algorithm was used to find subgraphs that may correspond to atomic texture primitives that form the heterogeneous structures. The final segmentation was obtained by using the histograms of these subgraphs inside sliding windows centered at individual pixels, and by clustering the pixels according to these histograms. As an alternative to graph-based grouping, Akcay et al. (2010) performed Gaussian mixture-based clustering of the region co-occurrence features to identify frequent region pairs that are merged in each iteration of a hierarchical texture segmentation procedure.

In addition to using co-occurrence properties of neighboring regions to exploit statistical information, structural features can also be extracted to represent the spatial layout for texture modeling. For example, Akcay and Aksoy (2011) described a procedure for finding groups of aligned objects by performing a depth-first search on a graph representation of neighboring primitive objects. After the search procedure identified aligned groups of three or more objects that have centroids lying on a straight line with uniform spacing, an agglomerative hierarchical clustering algorithm was used to find larger groups of primitive objects that have similar spatial layouts. The approach was illustrated in the finding of groups of buildings that have different statistical and spatial characteristics that cannot be modeled using traditional segmentation methods. Another approach for modeling urban patterns using hierarchical segmentations extracted from multiple images of the same scene at various resolutions was described by (Kurtz et al., 2012) where binary partition trees were used to model image data and tree cuts were learned from user-defined segmentation examples for interactive partitioning of images into semantic heterogeneous regions (Pare et al., 2020; Phan et al., 2020; Zhang et al., 2018; Csillik, 2017; Maulik and Chakraborty, 2017).

10.3.3 Morphological Algorithms

Mathematical morphology has been successfully used for various tasks such as image filtering for smoothing or enhancement, texture analysis, feature extraction, and detecting objects with certain shapes in the remote sensing literature (Soille and Pesaresi, 2002; Soille, 2003). Morphological algorithms have also been one of the most widely used techniques for segmenting remotely sensed images. These approaches view the two-dimensional image data that consist of the spectral channels or some other property of the pixels as an imaginary topographic relief where higher pixel values map to higher imaginary elevation levels (see Figure 10.4). Consequently, differences in the elevations of the pixels in a spatial neighborhood can be exploited to partition those pixels into different regions. Two morphological approaches for segmentation have found common use in the literature: watershed algorithms and morphological profiles. These approaches are described in the following sections. Other approaches using mathematical morphology for image segmentation, and particularly for producing segmentation hierarchies, can be found in (Soille, 2008; Soille and Najman, 2010; Ouzounis and Soille, 2012; Perret et al., 2012).

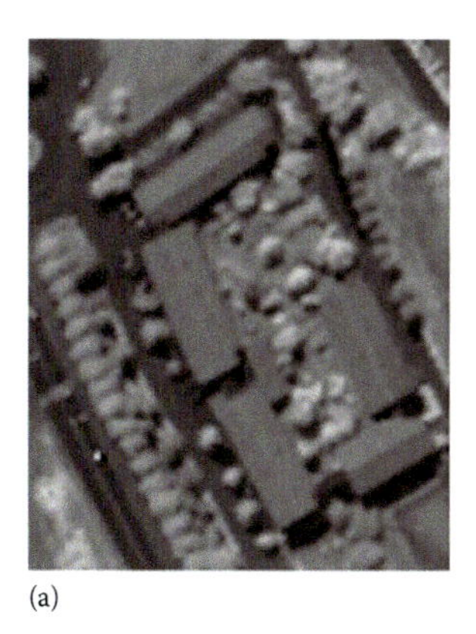
(a)

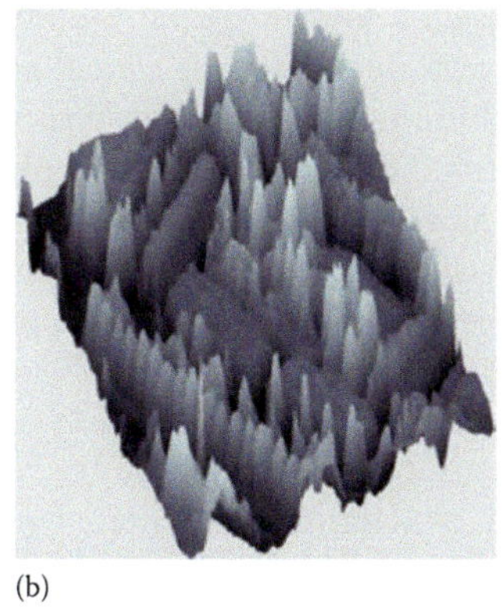
(b)

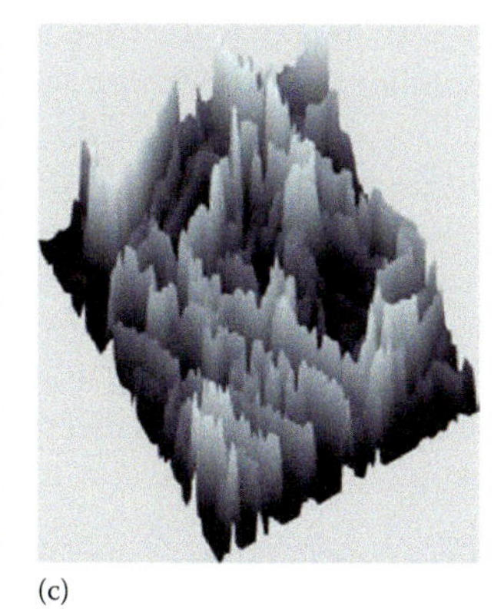
(c)

FIGURE 10.4 Illustration of mapping of the two-dimensional image data that consist of the spectral channels or some other property of the pixels as an imaginary topographic relief so that higher pixel values map to higher imaginary elevation levels. (a) An example spectral band. (b) The spectral values viewed as a three-dimensional topographic relief. (c) Gradient of the spectral data at each pixel viewed as a three-dimensional topographic relief.

10.3.3.1 Watershed Algorithms

The watershed algorithm divides the imaginary topographic relief into catchment basins so that each basin is associated with one local minimum in the image (i.e., individual segments) and the watershed lines correspond to the pixel locations that separate the catchment basins (i.e., segment boundaries). Watershed segmentation can be simulated by an immersion process (Vincent and Soille, 1991). If we immerse the topographic surface in water, the water rises through the holes at the regional minima with a uniform rate. When two volumes of water coming from two different minima are about to merge, a dam is built at each point of contact. Following the immersion process, the union of all those dams constitutes the watershed lines. A graph-theoretical interpretation of the watershed algorithm can be found in (Meyer, 2001). Couprie et al. (2011) describes a common framework that unifies watershed segmentation and some other graph-based segmentation algorithms that are described in Section 10.3.4.

The most commonly used method for constructing the topographic relief from the image data to be segmented is to use the gradient function at each pixel (Sertel et al., 2022; Liu et al., 2019; Zhang et al., 2019; Shan, 2018; Yin et al., 2018). This approach incorporates edge information in the segmentation process, and maps homogeneous image regions with low gradient values into the catchment basins and the pixels in high-contrast neighborhoods with high gradient values into the peaks in the elevation function. The gradient function for single-channel images can easily be computed using derivative filters. Multivariate extensions of the gradient function can be used to apply watershed segmentation to multispectral and hyperspectral images (Aptoula and Lefevre, 2007; Noyel et al., 2007; Li and Xiao, 2007; Fauvel et al., 2013).

A potential problem in the application of watershed segmentation to images with high levels of detail is oversegmentation when the watersheds are computed from raw image gradient where an individual segment is produced for each local minimum of the topographic relief. Pre-processing or post-processing methods can be used to reduce oversegmentation. For example, smoothing filters such as the mean or median filters can be applied to the original image data as a pre-processing step. Alternatively, the oversegmentation produced by the watershed algorithm can be given as input to a region merging procedure for post-processing (Haris et al., 1998).

Another commonly used alternative to reduce the oversegmentation is to use the concept of dynamics that are related to the regional minima of the image gradient. A regional minimum is composed of a group of neighboring pixels with the same value where the pixels on the external boundary of this group have a greater value. When we consider the image gradient as a topographic surface, the dynamic of a regional minimum can be defined as the minimum height that a point in the minimum has to climb to reach a lower regional minimum (Najman and Schmitt, 1996). The h-minima transform can be used to suppress the regional minima with dynamics less than or equal to a particular value h by performing geodesic reconstruction by erosion of the input image f from $f + h$ (Soille, 2003). When it is difficult to select a single h value, it is common to create a multi-scale segmentation by using an increasing sequence of h values. The multi-scale watershed segmentation generates a set of nested partitions where the partition at scale s is obtained as the watershed segmentation of the image gradient whose regional minima with dynamics less than or equal to s are eliminated by using the h-minima transform. First, the initial partition is calculated as the classical watershed corresponding to all local minima. Next, the two catchment basins having a dynamic of 1 are merged with their neighbor catchment basins at scale 1. Then, at each scale s, the minima with dynamics less than or equal to s are filtered whereas the minima with dynamics greater than s remain the same or are extended. This continues until the last scale corresponding to the largest dynamic in the gradient image. Figure 10.5 illustrates the use of the h-minima transform for suppressing regional minima for obtaining a multi-scale watershed segmentation.

Yet another popular approach for computing the watershed segmentation without a significant amount of oversegmentation is to use markers (Meyer and Beucher, 1990). Marker controlled watershed segmentation can be defined as the watershed of an input image transformed to have regional minima only at the marker locations. Possible methods for identifying the markers include manual

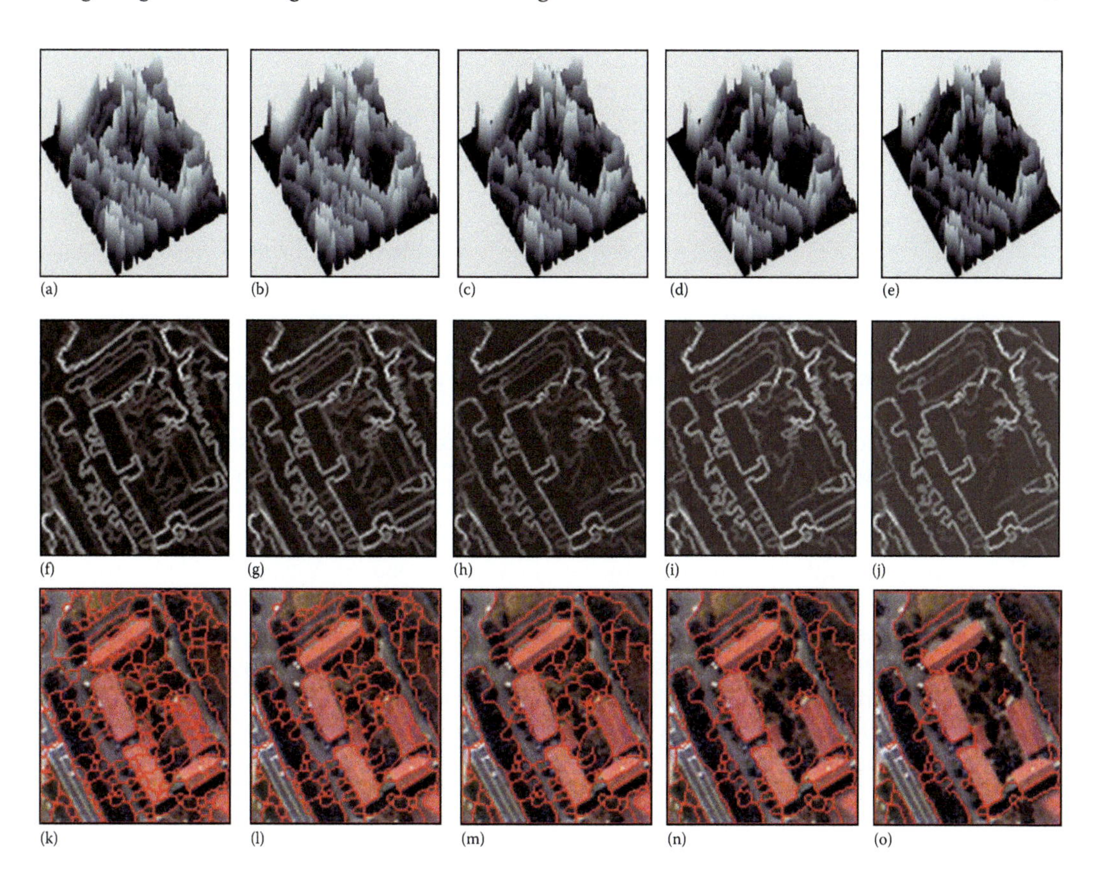

FIGURE 10.5 Illustration of the h-minima transform for suppressing regional minima for obtaining a multi-scale watershed segmentation. The columns represent increasing values of h, corresponding to decreasing amount of detail in the gradient data. The first row shows the gradient information at each pixel as a topographic relief. The second row shows the gradient data as an image. Brighter values represent higher gradient. The third row shows the segmentation boundaries obtained by the watershed algorithm in red.

selection or selection of the pixels with high confidence values at the end of pixel-based supervised classification (Tarabalka et al., 2010a). Given a marker image f_m that consists of pixels whose value is 0 at the marker locations and a very large value in the rest of the image, the minima in the input image f can be rearranged by using minima imposition. First, minima can be created only at the locations of the markers by taking the point-wise minimum between $f+1$ and f_m. Note that the resulting image is lower than or equal to the marker image. The second step of the minima imposition is the morphological reconstruction by erosion of the resulting image from the marker image f_m. Finally, watershed segmentation is applied to the resulting image. It is also possible to produce a multi-scale segmentation by applying marker controlled watershed segmentation to the input image by using a decreasing set of markers. Marker selection is also discussed in Section 10.3.4.

10.3.3.2 Morphological Profiles

The image representation called morphological profiles was popularized in the remote sensing literature by Pesaresi and Benediktsson (2001). The representation uses the morphological residuals between the original image function and the composition of a granulometry constructed at multiple scales. The proposed approach makes use of both classical morphological operators such as opening and closing, and recent theoretical advances such as leveling and morphological spectrum to build the morphological profile (Yu et al., 2023; Tassi and Vizzari, 2020; Shan, 2018).

The fundamental operators in mathematical morphology are erosion and dilation (Soille, 2003). Both of these operators use the definition of a pixel neighborhood with a particular shape called a structuring element (SE) (e.g., a disk of radius of 3 pixels). The erosion operator can be used to identify the image locations where the SE fits the objects in the image, and is defined as the infimum of the values of the image function in the neighborhood defined by SE. The dilation operator can be used to identify the pixels where the SE hits the objects in the image, and is defined as the supremum of the image values in the neighborhood defined by SE. These two operators can be combined to define other operators. For example, the opening operator, which is defined as the result of erosion followed by dilation using the same SE, can be used to cut the peaks of the topographic relief that are smaller than the SE. On the other hand, the closing operator, which is defined as the result of dilation followed by erosion using the same SE, can be used to fill the valleys that are smaller than the SE.

The morphological operations are often used with the non-Euclidean geodesic metric instead of the classical Euclidean metric (Pesaresi and Benediktsson, 2001). The elementary geodesic dilation of f (called the marker) under g (called the mask) based on SE is the infimum of the elementary dilation of f (with SE) and g. Similarly, the elementary geodesic erosion of f under g based on SE is the supremum of the elementary erosion of f (with SE) and g. A geodesic dilation (respectively, erosion) of size k can also be obtained by performing k successive elementary geodesic dilations (respectively, erosions). Next, the reconstruction by dilation (respectively, erosion) of f under g is obtained by the iterative use of an elementary geodesic dilation (respectively, erosion) of f under g until idempotence is achieved. Then, the opening by reconstruction of an image f can be defined as the reconstruction by dilation of the erosion under the original image. Similarly, the closing by reconstruction of the image f can be defined as the dual reconstruction by erosion of the dilation above the original image.

The advantage of the reconstruction filters is that they do not introduce discontinuities, and therefore, preserve the shapes observed in the input images. Hence, the opening and closing by reconstruction operators can be used to identify the sizes and shapes of different objects present in the image such that opening (respectively, closing) by reconstruction preserves the shapes of the structures that are not removed by erosion (respectively, dilation), and the residual between the original image and the result of opening (respectively, closing) by reconstruction, called the top-hat (respectively, inverse top-hat, or bot-hat) transform, can be used to isolate the structures that are brighter (respectively, darker) than their surroundings.

However, to determine the shapes and sizes of all objects present in the image, it is necessary to use a range of different SE sizes. This concept is called granulometry. The morphological profile (MP) of size $(2k+1)$ can be defined as the composition of a granulometry of size k constructed with opening by reconstruction (opening profile), the original image, and an antigranulometry of size k constructed with closing by reconstruction (closing profile) using a sequence of k SEs with increasing sizes. Then, the derivative of the morphological profile (DMP) is defined as a vector where the measure of the slope of the opening-closing profile is stored for every step of an increasing SE series (see Figures 10.6 and 10.7 for the illustration of opening and closing profiles, respectively).

Pesaresi and Benediktsson (2001) used DMP for image segmentation. They defined the size of each pixel as the SE size at which the maximum derivative of the morphological profile is achieved. Then, they defined an image segment as a set of connected pixels showing the greatest value of the DMP for the same SE size. That is, the segment label of each pixel is assigned according to the scale corresponding to the largest derivative of its profile. This scheme works well in images where the structures are mostly flat so that all pixels in a structure have only one derivative maximum. A potential drawback of this scheme is that neighborhood information is not used at the final step of assigning segment labels to pixels. This may result in an over-segmentation consisting of small noisy segments in very high spatial resolution images with non-flat structures where the scale with the largest value of the DMP may not correspond to the true structure.

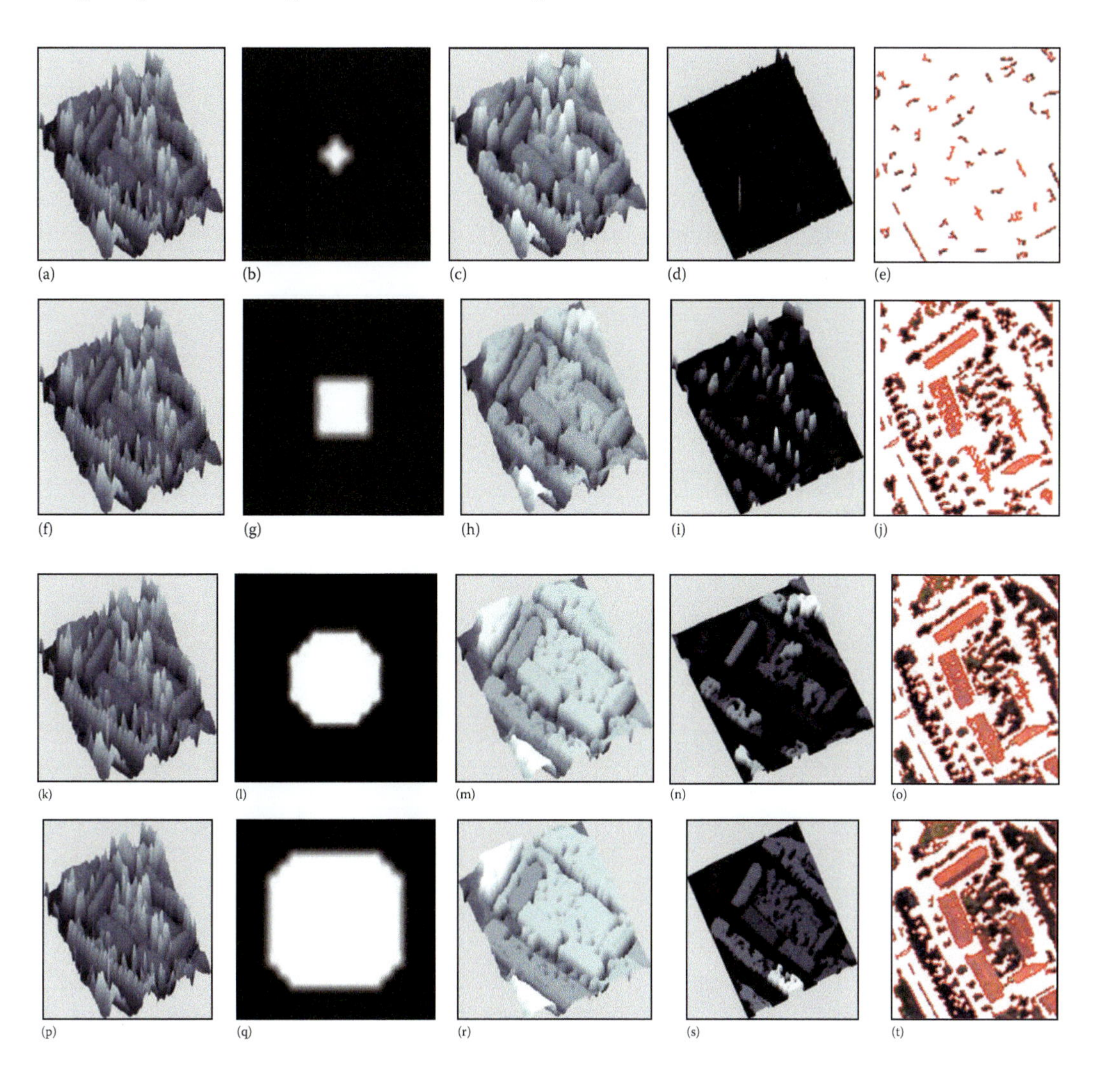

FIGURE 10.6 Illustration of the opening profile obtained using increasing SE sizes. Each row shows the results for an increasing SE series. The first column shows the input spectral data as a topographic relief. The second column shows the SEs used. The third column shows the result of opening by reconstruction of the topographic relief with the corresponding SEs. The fourth column shows the derivative of the opening morphological profile. The fifth column shows the boundaries of the connected components having a non-zero derivative profile for the corresponding SE for a multi-scale segmentation.

Akcay and Aksoy (2008) proposed to consider the behavior of the neighbors of a pixel while assigning the segment label for that pixel. The method assumes that pixels with a positive DMP value at a particular SE size face a change with respect to their neighborhoods at that scale. As opposed to Pesaresi and Benediktsson (2001) where only the scale corresponding to the greatest DMP is used, the main idea is that a neighboring group of pixels that have a similar change for any particular SE size is a candidate segment for the final segmentation. These groups can be found by applying connected components analysis to the DMP at each scale. For each opening and closing profile, through increasing SE sizes from 1 to m, each morphological operation reveals connected components that are contained within each other in a hierarchical manner where a pixel may be assigned to more than one connected component appearing at different SE sizes. Each component is treated as a candidate meaningful segment (see Figures 10.6 and 10.7). Using these segments, a

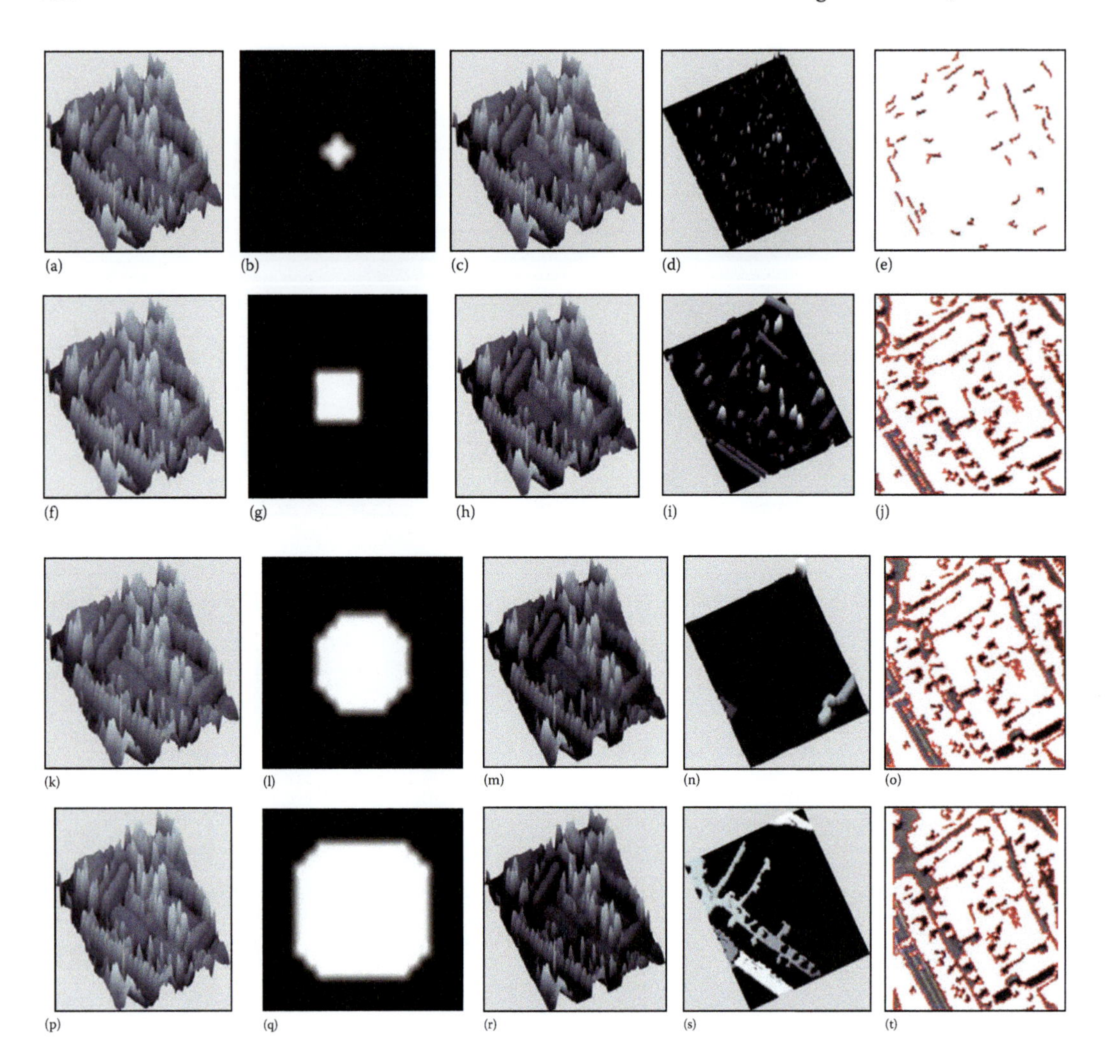

FIGURE 10.7 Illustration of the closing profile obtained using increasing SE sizes. Each row shows the results for an increasing SE series. The first column shows the input spectral data as a topographic relief. The second column shows the SEs used. The third column shows the result of closing by reconstruction of the topographic relief with the corresponding SEs. The fourth column shows the derivative of the closing morphological profile. The fifth column shows the boundaries of the connected components having a non-zero derivative profile for the corresponding SE for a multi-scale segmentation.

tree is constructed where each connected component is a node and there is an edge between two nodes corresponding to two consecutive scales if one node is contained within the other. Leaf nodes represent the components that appear for SE size 1. Root nodes represent the components that exist for SE size m.

After forming a tree for each opening and closing profile, the goal is to search for the most meaningful connected components among those appearing at different scales in the segmentation hierarchy. Ideally, a meaningful segment is expected to be spectrally as homogeneous as possible. However, in the extreme case, a single pixel is the most homogeneous. Hence, a segment is also desired to be as large as possible. In general, a segment stays almost the same (both in spectral homogeneity and size) for some number of SEs, and then faces a large change at a particular scale either because it merges with its surroundings to make a new structure or because it is completely lost. Consequently, the size of interest corresponds to the scale right before this change. In other words, if the nodes on a path in the tree stay homogeneous until some node n, and then the homogeneity is lost in the next level, it can be said that n corresponds to a meaningful segment in the hierarchy. With this motivation, to check the meaningfulness of a node, Akcay and Aksoy (2008) defined a measure consisting of two factors: spectral homogeneity, which is calculated in terms of the difference of the standard deviation of the spectral features of the node and its parent, and neighborhood connectivity, which is calculated using sizes of connected components. Then, starting from the leaf nodes (level 1) up to the root node (level m), this measure is computed at each node, and a node is selected as a meaningful segment if it is highly homogeneous and large enough on its path in the hierarchy (a path corresponds to the set of nodes from a leaf to the root).

After the tree is finalized, each node is regarded as a candidate segment for the final segmentation. Given the goodness measure of each node in the hierarchy, the segments that optimize this measure are selected by using a two-pass algorithm that satisfies the following conditions. Given N as the set of all nodes and P as the set of all paths in the tree, the algorithm selects $N^* \subseteq N$ as the final segmentation such that any node in N^* must have a measure greater than all of its descendants, any two nodes in N^* cannot be on the same path (i.e., the corresponding segments cannot overlap in the hierarchical segmentation), and every path must include a node that is in N^* (i.e., the segmentation must cover the whole image). The first pass finds the nodes having a measure greater than all of their descendants in a bottom-up traversal. The second pass selects the most meaningful nodes having the largest measure on their corresponding paths of the tree in a top-down traversal. The details of the algorithm can be found in (Akcay and Aksoy, 2008). Even though the algorithm was illustrated using a tree constructed from a DMP, it is a generic selection algorithm in the sense that it can be used with other hierarchical image partitions, such as the ones described in Section 10.3.1, and can be applied to specific applications by defining different goodness measures for desired image segments (e.g., see (Genctav et al., 2012) for an application of this selection algorithm to a hierarchical segmentation produced by a multi-scale watershed procedure (Yu et al., 2023; Tassi and Vizzari, 2020).

10.3.4 Graph-Based Algorithms

Graph-based segmentation techniques gained popularity in recent years. In the graph-based framework, the image is modeled by a graph, where nodes typically represent individual pixels or regions, while edges connect spatially adjacent nodes. The weights of the edges reflect the (dis)similarity between the neighboring pixels/regions linked by the edge. The general idea is then to find subgraphs in this graph, which correspond to regions in the image scene. The early graph-theoretic approaches for image segmentation were described in (Zahn, 1971), where a minimum spanning tree was used to produce connected groups of vertices, and (Narendra and Goldberg, 1977), where directed graphs have been employed to define regions in edge-detected images. In this section we will review two algorithms that have been successfully applied for remote sensing applications: optimal spanning forests and normalized cuts.

10.3.4.1 Optimal Spanning Forests

The optimal spanning forest segmentation is based on the minimum spanning tree algorithm introduced by Kruskal (1956) and Prim (1957). It was employed for segmentation of remote sensing images in (Tarabalka et al., 2009; Skurikhin, 2010).

We denote an image undirected graph as $G=(V,E,W)$, where each pixel is considered as a vertex $v \in V$, each edge $e_{i,j} \in E$ connects a couple of vertices i and j corresponding to the neighboring pixels. Furthermore, a weight $w_{i,j}$ is assigned to each edge $e_{i,j}$, which indicates the degree of dissimilarity between two pixels connected by this edge. Different dissimilarity measures can be used to compute weights of edges, such as vector norms, SAM, or spectral information divergence (Tarabalka et al., 2010a).

Given a connected graph $G=(V,E)$, a spanning tree $T=(V,E_T)$ of G is a connected graph without cycles such that $E_T \subset E$. A spanning forest $F=(V,E_F)$ of G is a non-connected graph without cycles such that $E_F \subset E$. Given a graph $G=(V,E,W)$, the *minimum spanning tree* is defined as a spanning tree $T^*=(V,E_{T^*})$ of G such that the sum of the edge weights of T^* is minimal among all the possible spanning trees of G. The *minimum spanning forest* (MSF) rooted on a set of m distinct vertices $\{t_1,...,t_m\}$ consists in finding a spanning forest $F^*=(V,E_{F^*})$ of G, such that each distinct tree of F^* is grown from one root t_i, and the sum of the edge weights of F^* is minimal among all the spanning forests of G rooted on $\{t_1,...,t_m\}$.

The MSF-based segmentation typically consists of two steps:

1. The objective of this step is to select a *marker*, or region seed, for each spatial object in the image. Such region seeds $\{t_1,...,t_m\}$ can be manually selected from image pixels via interactive image analysis software, however automatic marker selection is highly desirable. Markers are often defined by automatically searching flat zones (i.e., connected components of pixels of constant intensity value), zones of homogeneous texture, or image extrema (Soille, 2003). Tarabalka et al. (2010a) proposed to perform a supervised probabilistic classification of each pixel (i.e., compute probabilities for each pixel to belong to each of the land categories of interest), and to choose the most reliably classified pixels as markers of spatial regions.
2. Image pixels are grouped into an MSF, where each tree is rooted on a marker. To compute an MSF, an additional root vertex r is added and is connected by the null-weight edges to the marker vertices t_i. The minimum spanning tree of the constructed graph induces an MSF in G, where each tree is grown on a marker vertex t_i; the MSF is obtained after removing the vertex r. The two most commonly used algorithms for computing a minimum spanning tree are Prim's (1957) and Kruskal's (1956) algorithms.

The watershed transform described in the previous section can be efficiently built by computing an MSF rooted on the image minima (Cousty et al., 2009). For this purpose, an ultrametric flooding distance has to be used to compute weights of edges (Meyer, 2005). This distance is defined as the minimal level of flooding for which two pixels belong to the same lake.

Meyer (2005) showed that an MSF can also be efficiently computed from a minimum spanning tree of image pixels, without introducing an additional root vertex r, as depicted in Figure 10.4. This algorithm is useful if the initial markers can be modified (e.g., suppression and addition of markers during interactive segmentation). Given the minimum spanning tree T of a graph $G=(V,E,W)$ and a set of markers $\{t_1,...,t_m\}$, the edges of T are first sorted in the order of their decreasing weights and are considered one after another. Suppose that e is the edge currently under consideration. The edge e belongs to a sub-tree of T. Suppressing e will cut this tree into two smaller sub-trees; if each of them contains at least one marker, then the suppression of e is validated (this is the case in Figure 10.8c–e, where an edge has been suppressed each time); if at least one of the sub-trees does not contain a marker, then the edge e is reintroduced. The process stops when each of the created sub-trees contains one and only one marker. This algorithm outputs an MSF with one tree rooted in each marker. It was applied in (Bernard et al., 2012) for segmentation of hyperspectral remote sensing images. If the markers of the image regions cannot be reliably found, a similar algorithm can be iteratively applied, by suppressing the edge of the minimum spanning tree with the highest weight at each iteration until convergence. A threshold for the edge weight can control in this case the convergence.

10.3.4.2 Normalized Cuts

The normalized cuts segmentation method introduced by (Shi and Malik, 2000) aims at partitioning the image in the way to minimize the similarity between adjacent regions while maximizing

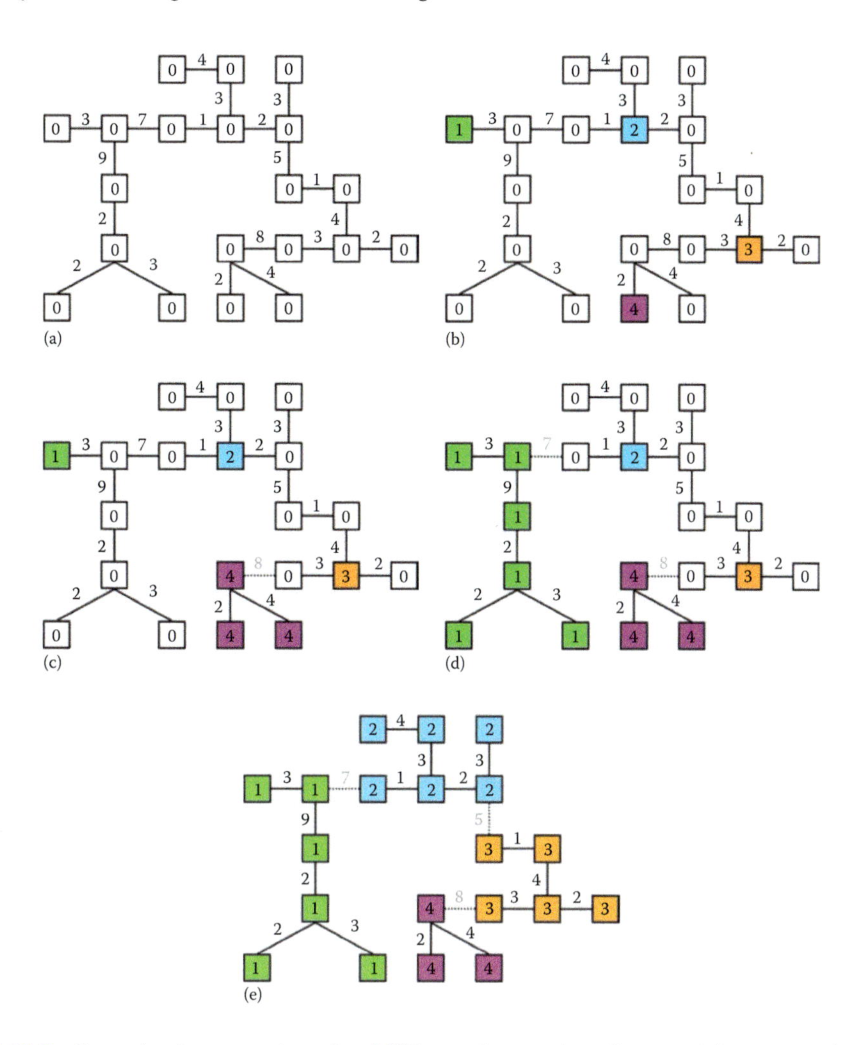

FIGURE 10.8 Example of construction of an MSF rooted on markers from a minimum spanning tree: (a) the initial minimum spanning tree; (b) four markers defined by the colored nodes; (c)–(d) illustration of the construction of the MSF from the four markers by highest weight edge suppression; and (e) final MSF, where each tree has the color of its marker.

the similarity within the regions. Figure 10.9 shows an example graph, where the pixels in group X are strongly connected with high similarities between adjacent pixels, shown as thick red lines, as are the pixels in group Y. The connections between groups X and Y, shown as blue lines, are much weaker. A normalized cut between these two groups separates them into two clusters.

The *cut* between two groups X and Y is computed as the sum of the weights of all edges being cut:

$$cut\left(X,Y\right)=\sum\nolimits_{i\in X,j\in Y} w_{i,j}, \tag{10.19}$$

where the weights of the edges between two vertices i and j measure the similarity between the corresponding pixels (or regions). The optimal bipartitioning of the graph is the one that minimizes this cut value, that is, the finds the *minimum cut* of a graph. However, because the value of the cut computed by Equations 10.16–10.19 increases with the number of edges separating two partitions,

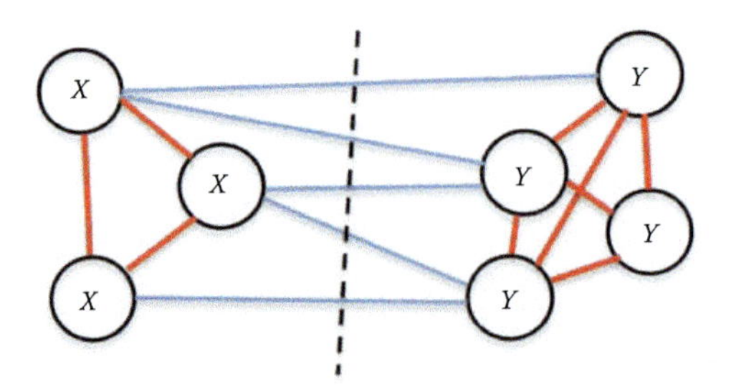

FIGURE 10.9 Example of a weighted graph and its normalized cut (shown as a dashed line).

using a minimum cut as a segmentation criterion favors keeping in one group isolated pixels or small sets of isolated nodes in the graph.

To avoid such partitioning of small sets of nodes into separate groups, (Shi and Malik, 2000) proposed a new measure of disassociation between two groups. The new measure, called a *normalized cut*, computes the cut cost as a fraction of the total edge connections to all the nodes in the graph:

$$Ncut(X,Y) = \frac{cut(X,Y)}{assoc(X,V)} + \frac{cut(X,\ Y)}{assoc(Y,V)} \tag{10.20}$$

where $assoc(X,V) = \sum_{i \in X, v \in V} w_{i,v}$ is the sum of all weights from nodes in X to all nodes in the graph, and $assoc(Y,V)$ is similarly defined. The normalized cut defined by Equations 10.16–10.20 better reflects the fitness of a particular segmentation, because it seeks to cut a collection of edges that are weak relatively to all of the edges both inside and emanating from each of the regions. Shi and Malik (2000) proposed an efficient algorithm based on a generalized eigenvalue problem to optimize the normalized cut criterion.

Numerous works employed a normalized cut algorithm for segmentation of remote sensing images, such as (Grote et al., 2007) and (Sung and He, 2009). Normalized cut were also applied in combination with other techniques. For instance, Jing et al. (2010) first used watershed algorithm to find initial segments, and then the normalized cuts technique grouped these segments into the final regions.

10.3.5 MRF-Based Algorithms

Markov random fields (MRFs) are probabilistic graphical models that conceptually generalize the notion of Markov chain (Wang et al., 2013; Moser et al., 2013). They provide a flexible tool to include spatial context into image-analysis schemes in terms of minimization of suitable energy functions. While earlier algorithms for optimizing MRF energy, such as iterated conditional modes (ICM) and simulated annealing (Solberg et al., 1996; Tarabalka et al., 2010b) were time consuming, more advanced methods, such as graph cuts (Boykov et al., 2001; Li et al., 2012) provided powerful alternatives from both theoretical and computational viewpoints, resulting in a growing use of the MRF-based segmentation techniques.

For land categorization applications, MRFs are usually applied in the framework of image classification, where the output is a landcover map, with every region assigned to one of the thematic classes. The commonly used MRF energy function in this case is computed as a linear combination of two terms:

$$E(L) = E_{data} + E_{smooth} \tag{10.21}$$

The first term $E_{data} = \sum_{i=1}^{n} V_i(L_i)$ is related to pixelwise information and it measures for each pixel the disagreement between a prior probabilistic model and the observed data. Thus, individual potentials $V_i(L_i)$ measure a penalty for a pixel i $(i = 1, 2, ..., n)$ to have a label L_i. This term is often formulated in terms of the probability density function of feature vectors conditioned to the related class label:

$$V_i(L_i) = -ln(p(x_i \mid L_i)) \tag{10.22}$$

This probability density can be estimated based on the available training samples for each landcover class. A discussion of the main approaches for this estimation problem can be found in (Duda et al., 2001).

The second contribution $E_{smooth} = \sum_{i \sim j} W_{i,j}(L_i, L_j)$ expresses interaction between neighboring pixels, thus exploiting image spatial context. $i \sim j$ denotes a pair of spatially adjacent pixels, and $W_{i,j}(L_i, L_j)$ is an interaction term for these pixels. Most works in remote sensing image classification use a Potts model (Tarabalka et al., 2010b; Li et al., 2012) to compute this spatial term, which favors spatially adjacent pixels to belong to the same class (or spatial region):

$$W_{i,j}(L_i, L_j) = \beta\left(1 - \delta(L_i, L_j)\right) \tag{10.23}$$

where $\delta(\cdot)$ is the Kronecker function ($\delta(a,b) = 1$ for $a = b$ and $\delta(a,b) = 0$ otherwise) and β is a positive constant parameter that controls the importance of spatial smoothing. This model tends to deteriorate classification results at the edges between land-cover classes and near small-scale details. In order to preserve the border in the output thematic map, "edge" functions have been proposed and integrated in the spatial energy term, such as (Tarabalka et al., 2010b):

$$W_{i,j}(L_i, L_j) = \beta\left(1 - \delta(L_i, L_j)\right) \frac{t}{t + \left|\rho_{i|j}\right|} \tag{10.24}$$

where $\rho_{i|j}$ is the gradient value of the pixel i in the direction of j, and t is a parameter controlling the fuzzy edge threshold.

To optimize the MRF energy, different methods were proposed and applied for remote sensing applications, as described in the following subsections.

10.3.5.1 Simulated Annealing and Iterated Conditional Modes

The main idea of simulated annealing described in (Kirkpatrick et al., 1983) is to iteratively propose a change from the current configuration of pixel labels (region or class labels) L to the new randomly generated configuration of labels, and to probabilistically decide is this change is accepted or not. This procedure yields configurations of labels with the lower MRF energy.

Stewart et al. (2000) applied MRF-based method with simulated annealing optimization for segmentation and classification of synthetic aperture radar (SAR) data. They used Gamma distribution to model the data energy term and a Potts model to compute the spatial term. Tarabalka et al. (2010b) applied simulated annealing technique for classification of hyperspectral remote sensing images. They performed probabilistic support vector machines (SVM) classification to derive the data term, and employed the spatial energy contribution with the fuzzy edge function described previously. At each iteration, a new class label L_i^{new} was randomly selected for a randomly

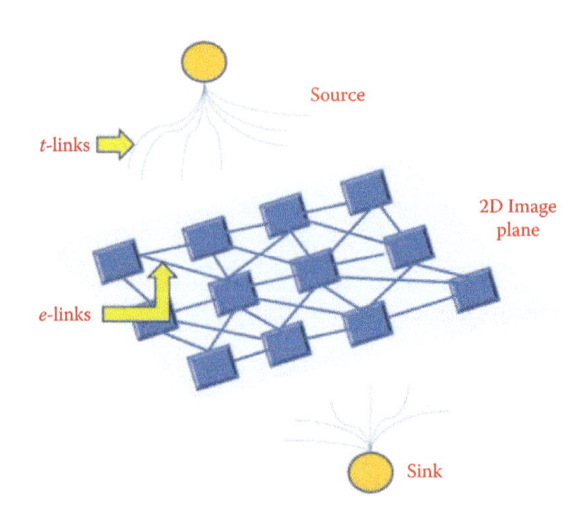

FIGURE 10.10 Mapping of an image to the graph for the graph-cut optimization.

selected pixel x_i, and a new local energy $E^{new}(x_i)$ was computed. If the variation of the energy $\Delta E = E^{new}(x_i) - E(x_i) < 0$, the new class label is accepted. Otherwise, the new class assignment is accepted with the probability $p = exp(-\Delta E / T)$, where T is a global parameter called temperature. The optimization begins at a high temperature, which is gradually lowered as the iteration procedure proceeds. This helps to avoid converging to local minima.

The iterated conditional modes (ICM) optimization algorithm proposed in (Besag, 1986) is conceptually simpler and computationally less expensive. It iteratively considers every pixel x_i, and assigns a (region or class) label to this pixel, which minimizes the local energy centered at x_i. Solberg et al. (1996) applied the ICM optimization for multisource classification of optical, SAR and geographic information systems images. Farag et al. (2005) employed the ICM method for hyperspectral image classification. The density functions in this case were estimated by mean field-based SVM regression. The drawback of the ICM algorithm is that it is suboptimal and converges only to a local minimum of the energy function.

10.3.5.2 Graph Cuts

Boykov and Jolly (2001) were the first to apply the graph-cut optimization technique proposed by (Greig et al., 1989) to binary image segmentation. This algorithm is computationally efficient and it gives a globally-optimal solution to the binary segmentation problem. In the approach of (Boykov and Jolly, 2001), all pixels are connected by t-links to the additional two nodes, called source and sink, respectively, as depicted in Figure 10.10. The weights of these edges are computed as the potentials $V_i(L_i)$ (see Equations 10.16–10.22), so that pixels that are more compatible with the foreground or background region get stronger connections to the respective source or sink. The e-links between adjacent pixels are assigned weights $W_{i,j}(L_i, L_j)$. The resulting MRF energy (see Equations 10.16–10.21) is optimized by solving the minimum-cut/maximum-flow problem, yielding a binary segmentation map.

Boykov et al. (2001) proposed two efficient approximation algorithms based on graph-cuts (α-expansion and α-β-swap) to solve multi-label classification problems. In remote sensing, Denis et al. (2009) applied a graph-cut-based algorithm for regularization of SAR images. Li et al. (2012) and Tarabalka and Rana (2014) used an α-expansion optimization of the MRF energy for segmentation of hyperspectral images.

10.4 IMAGE SEGMENTATION QUALITY EVALUATION

We noted in the introduction to this chapter that there is no general theory of image segmentation, and most image segmentation approaches are *ad hoc* in nature. This is also the case with image segmentation quality evaluation. In practice, image segmentation quality is best evaluated by how well the goals of the analysis utilizing the segmentation are served (Jia et al., 2024; Li et al., 2022; Pare et al., 2020; Phan et al., 2020).

Several general approaches to quantitative segmentation quality evaluation are listed and described in the series of papers published by the research group based at the Leibniz Institute for Ecological and Regional Development (IOER) mentioned in the introduction to this chapter, in particular (Neubert et al., 2006). Among the evaluation approaches discussed and employed are:

1. Fragmentation and Area-Fit-Index (Strasters and Gerbrands, 1991; Lucieer, 2004): These measures address over- and under-segmentation by analyzing the number of segmented and reference regions.
2. Geometric features Circularity and Geometric features Shape Index (Yang et al., 1995; Neubert and Meinel, 2003): These measures address the shape conformity between segmentation and reference regions.
3. Empirical Evaluation Function (Borsotti et al., 1998): This measure addresses how uniform a feature is within segmented regions.
4. Entropy-based evaluation function and a weighted disorder function (Zhang and Gerbrands, 1994): This measure addresses the uniformity within segmented regions using entropy as a criterion of disorder.
5. Fitness function (Everingham et al., 2002): This measure addresses multiple criteria and parameterizations of algorithms by a probabilistic Fitness/Cost Analysis.

Neubert et al. (2006) also note that the vast majority of quantitative image segmentation approaches are empirical discrepancy methods that analyze the number of misclassified pixels in relation to a reference classification. In the remainder of this section we describe and demonstrate such an approach to image segmentation quality evaluation. The first step in this evaluation approach is to perform a pixelwise classification of an image dataset. In our example we create our pixelwise classification using the support vector machine (SVM) classifier. Then a region classification is obtained by assigning each spatially connected region from the segmentation result to the most frequently occurring class within the region. The SVM classifier and this "plurality vote" (PV) region-based classification approach are described in more detail in (Tilton et al., 2012).

Our test dataset is the University of Pavia dataset was recorded by the Reflective Optics System Imaging Spectrometer (ROSIS) over the University of Pavia, Pavia, Italy. The image is 610 × 340 pixels in size, with a spatial resolution of 1.3 m. The ROSIS sensor has 115 spectral channels, with a spectral range of 0.43–0.86 μm. The 12 noisiest channels were removed, and the remaining 103 spectral bands were used in this experiment. See Tarabalka et al. (2010) for more details on this dataset. The reference data contain nine ground cover classes: asphalt, meadows, gravel, trees, metal sheets, bare soil, bitumen, bricks and shadows. A three-band false color image of this dataset and the ground reference data are shown in Figure 10.11.

We note that this dataset is not rigorously geo-referenced. North is roughly towards the bottom of the image and the area covered is 793 m by 442 m.

The classification accuracy results are listed in Table 10.2 for the pixelwise SVM classification and the PV region-based classification approach for several region growing approaches. Figure 10.12 shows the corresponding classification maps. The classification maps for the PV region-based classification approach appear smoother than the pixelwise SVM classification, with the D8 and SEGEN based classification appearing the smoothest. All of the PV region-based classification accuracies as substantially higher that the accuracy of the SVM pixelwise classification, with the D8, SEGEN and HSeg PV region-based classifications notably more accurate than the PV region-based

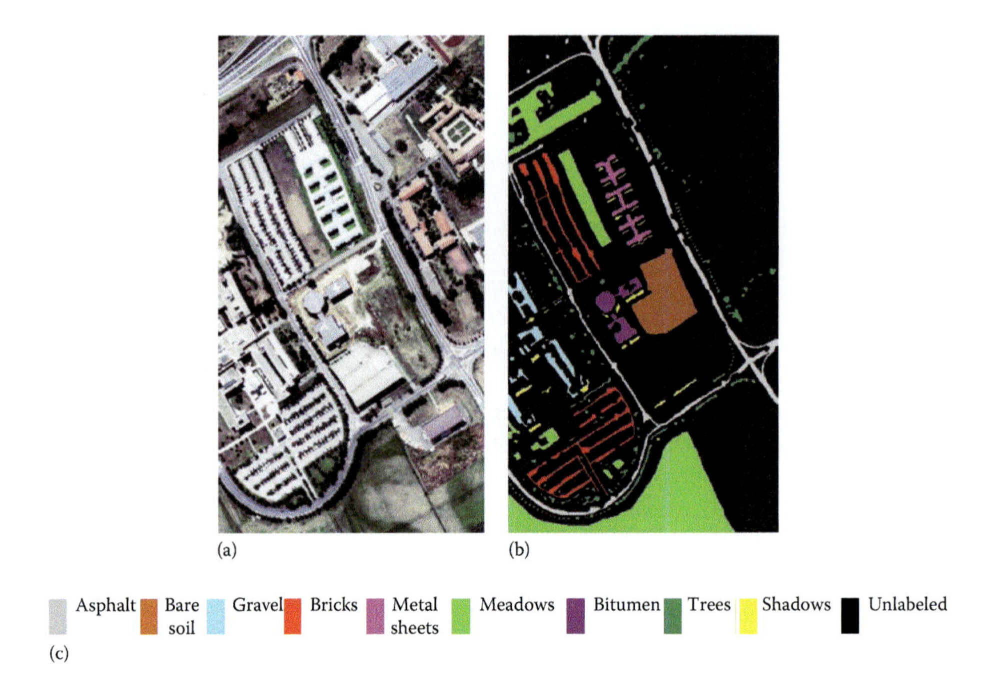

FIGURE 10.11 (a) Three-band false color image of the University of Pavia hyperspectral dataset (RGB = bands 56, 33 and 13). (b) Reference data. (c) Color key. The training data for the SVM classifier and segmentation optimization are two separate randomly selected subsets of the reference data.

TABLE 10.2
Comparison of Classification Accuracies on the University of Pavia Hyperspectral Dataset for Per-Pixel SVM and with the Plurality Vote (PV) Region-Based classification method for the region Muerle Allen, HSWO, Definiens 8.0 (D8), SEGEN, and HSeg.

	SVM	Muerle Allen	HSWO +PV	D8+PV	SEGEN +PV	HSeg+PV $S_w = 0.3$	MSF+PV	Graph-cut +PV
OA	89.03	95.35	95.38	97.54	98.09	98.35	96.99	98.49
AA	89.56	95.26	95.50	97.26	97.95	98.15	97.01	97.73
κ	85.46	93.78	93.83	96.71	97.45	97.79	95.99	98.47

HSeg was performed with small region merge acceleration. Results are also included for two graph-based approaches—the minimum spanning forest (MSF) with plurality vote (PV) method and MRF-based method using the α-expansion graph-cut algorithm. Percentage classification accuracies in terms of OA, AA, and Kappa Coefficient (κ).

classifications based on Muerle Allen or HSWO segmentation. The HSeg based classification has a marginally higher classification than the other region growing approaches.

Table 10.2 and Figure 10.6 also show classification results obtained by using the graph-based segmentation approaches. The MSF segmentation with plurality vote was performed as described in (Tarabalka et al., 2010a), by using the *L1* norm between spectral vectors to compute weights of edges between adjacent pixels. For the graph-cut method, probabilistic SVM classification was applied to derive the data energy term, and a Potts model was used to express the spatial term. We fixed the parameter $\beta=1.5$ as suggested in (Tarabalka et al., 2010b). The MRF-based method using the α-expansion optimization yields the highest overall classification accuracy.

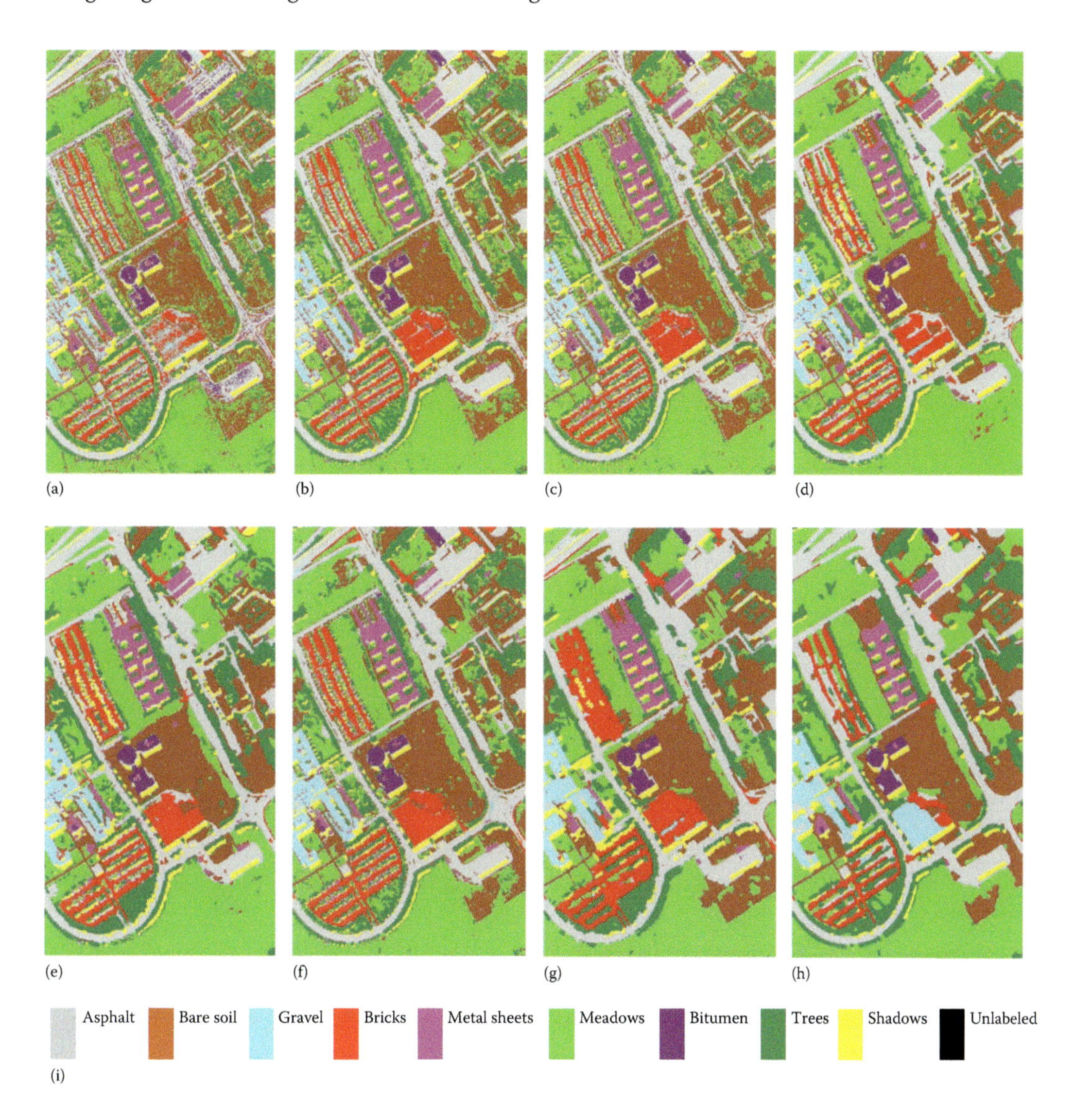

FIGURE 10.12 (a) SVM classification. (b) PV classification with Muerle Allen segmentation. (c) PV classification with HSWO segmentation. (d) PV classification with D8. (e) PV classification with SEGEN segmentation. (f) PV classification with HSeg segmentation ($S_w = 0.3$). (g) Minimum spanning forest (MSF) with plurality vote (PV) classification map. (h) Classification map of the α-expansion graph-cut algorithm. (i) Color key.

The classifications are evaluated in terms of Overall Accuracy (OA), Average Accuracy (AA) and the kappa coefficient (κ). OA is the percentage of correctly classified pixels, AA is the mean of the class-specific accuracies, and κ is the percentage of agreement (correctly classified pixels) corrected by the number of agreements that would be expected purely by chance (e.g., see Richards and Jia, 1999).

10.5 CONCLUDING REMARKS

Land categorization applications are quite varied. They range from land cover monitoring (e.g., croplands, forests, wetlands, and urbanization), snow and ice mapping, geology and mineral exploration, and even wildfire and agricultural burn monitoring. Each land categorization will have its particular image analysis needs that will be best served by a particular class of image segmentation algorithm. Generally, region-growing approaches may be best suited for finding homogenous regions, while if locating region borders are more important, watershed approaches may be more effective. Region-growing and morphological approaches that produce segmentation hierarchies would be most appropriate if analysis of the image data at different scales is desired.

We have demonstrated through a simple plurality vote classification approach that utilizing image segmentation can dramatically improve image classification results both qualitatively and quantitatively. Another advantage of image segmentation is that image segmentation defines spatial objects with particular shapes and sizes that can be utilized to perform a deeper, object-based analysis of the image scene than possible with pixel-based analysis. In the context of land categorization, this form of analysis has led to the establishment of the new sub-discipline of Geographic Object-Based Image Analysis (GEOBIA) and a popular series of biennial conferences starting with GEOBIA 2006 in Salzburg, Austria—and most recently with GEOBIA 2014 in Thessaloniki, Greece (Gitas et al., 2014).

We have described and discussed a wide range of image segmentation approaches developed over the past century, focusing on those approaches most applicable to the analysis of remotely sensed imagery data for land categorization (Jia et al., 2024; Martinez-Sanchez et al., 2024; Lilay and Taye, 2023; Liu et al., 2023; Yu et al., 2023; Raei et al., 2022; Sertel et al., 2022; Li et al., 2022; Pare et al., 2020; Phan et al., 2020; Tassi and Vizzari, 2020; Jin et al., 2019; Liu et al., 2019; Zhang et al., 2019; Shan, 2018; Yin et al., 2018; Zhang et al., 2018; Csillik, 2017; Maulik and Chakraborty, 2017). The wide range of approaches included in our discussion and the ongoing active research are a clear indication that there still is no general theory of segmentation and that most of the successful image segmentation approaches are rather *ad hoc* in nature. But the wide range of approaches provide a rich menu to choose from for image analysis practitioners in tailoring their image analysis approach to their particular application.

REFERENCES

Adams, R., and L. Bischof. 1994. Seeded region growing. *IEEE Trans. Pattern Anal. Mach. Intell.*, 16(6):641–646.

Akcay, H. G., and S. Aksoy. 2008. Automatic detection of geospatial objects using multiple hierarchical segmentations. *IEEE Trans. Geosci. Remote Sens.*, 46(7):2097–2111.

Akcay, H. G., and S. Aksoy. 2011. Detection of compound structures using hierarchical clustering of statistical and structural features. *Proc. of IGARSS*, Vancouver, Canada.

Akcay, H. G., S. Aksoy, and P. Soille. 2010. Hierarchical segmentation of complex structures. *Proc. of ICPR*, Istanbul, Turkey.

Aksoy, S. 2006. Spatial techniques for image classification. In: C. H. Chen, ed. *Signal and Image Processing for Remote Sensing*. Boca Raton, FL: CRC Press, pp. 491–513.

Aksoy, S., and H. G. Akcay. 2005. Multi-resolution segmentation and shape analysis for remote sensing image classification. *Proc. of 2nd Int'l Conf. Recent Advances in Space Technologies*, Istanbul, Turkey.

Aksoy, S., H. G. Akcay, and T. Wassenaar. 2010. Automatic mapping of linear woody vegetation features in agricultural landscapes using very-high resolution imagery. *IEEE Trans. Geosci. Remote Sens.*, 48(1):511–522.

Aksoy, S., K. Koperski, C. Tusk, G. Marchisio, and J. C. Tilton. 2005. Learning Bayesian classifiers for scene classification with a visual grammar. *IEEE Trans. Geosci. Remote Sens.*, 43(3):581–589.

Aksoy, S., I. Z. Yalniz, and K. Tasdemir. 2012. Automatic detection and segmentation of orchards using very high-resolution imagery. *IEEE Trans. Geosci. Remote Sens.*, 50(8):3117–3131.

Aptoula, E., and S. Lefevre. 2007. A comparative study on multivariate mathematical morphology. *Pattern Recognit.*, 40(11):2914–2929.

Baatz, M., and A. Schape. 2000. Multiresolution segmentation: An optimizing approach for high quality multiscale segmentation. In: J. Strobl and T. Blaschke, eds. *Angewandte Geographich Informationsverarbeitung, XII*. Heidelberg, Germany: Wichmann, pp. 12–23.

Beaulieu, J.-M., and M. Goldberg. 1989. Hierarchy in picture segmentation: A stepwise optimal approach. *IEEE Trans. Pattern Anal. Mach. Intell.*, 11:150–163.
Benz, U., P. Hofmann, G. Wilhauck, I. Lingenfelder, and M. Heynen. 2004. Multi-resolution, objected-oriented fuzzy analysis of remote sensing data for GIS-ready information. *ISPRS J. Photogramm. Remote Sens.*, 58(3/4):239–258.
Bernard, K., Y. Tarabalka, J. Angulo, J. Chanussot, and J. A. Benediktsson. 2012. Spectral-spatial classification of hyperspectral data based on a stochastic minimum spanning forest approach. *IEEE Trans. Image Proc.*, 21(4):2008–2021.
Besag, J. 1986. On the statistical analysis of dirty pictures. *J. R. Statist. Soc.*, 48(3):259–302.
Borsotti, M., P. Campadelli, and R. Schettini. 1998. Quantitative evaluation of color image segmentation results. *Pattern Recognit. Lett.*, 19(8):741–747.
Boykov, Y., and M.-P. Jolly. 2001. Interactive graph cuts for optimal boundary and region segmentation of objects in N-D images. *Proc. of ICCV*, Vancouver, Canada, 1:105–112.
Boykov, Y., O. Veksler, and R. Zabih. 2001. Fast approximate energy minimization via graph cuts. *IEEE Trans. Pattern Anal. Mach. Intell.*, 23(11):1222–1239.
Bruzzone, L., and D. F. Prieto. 2005. Automatic analysis of the difference image for unsupervised change detection. *IEEE Trans. Geosci. Remote Sens.*, 38(3):1171–1182.
Burges, C. J. 1998. A Tutorial on Support Vector Machines for Pattern Recognition. *Data Mining and Knowledge Discovery*, 2:121–167.
Camps-Valls, G. and Bruzzone, L. 2005. Kernel-based methods for hyperspectral image classification. *IEEE Trans. Geosci. Remove Sens.,* 43(6):1351-1362.
Castilla, G., G. J. Hay, and J. R. Ruiz 2008. Size-constrained region merging (SCRM): An automated delineation tool for assisted photointerpretation. *PE&RS*, 74(4):409–419.
Comaniciu, D., and P. Meer. 2002. Mean shift: A robust approach toward feature space analysis. *IEEE Trans. Pattern Anal. Mach. Intell.*, 24(5):603–619.
Cook, R., I. McConnell, D. Stewart, and C. J. Oliver. 1996. Segmentation and simulated annealing. In: G. Franceschetti, F. S. Rubertone, C. J. Oliver, and S. Tajbakhsh, eds. *Microwave Sensing and Synthetic Aperture Radar, Proc. SPIE* 2958:30–35.
Couprie, C., L. Grady, L. Najman, and H. Talbot. 2011. Power watershed: A unifying graph-based optimization framework. *IEEE Trans. Pattern Anal. Mach. Intell.*, 33(7):1384–1399.
Cousty, J., G. Bertrand, L. Najman, and M. Couprie. 2009. Watershed cuts: Minimum spanning forests and the drop of water principle. *IEEE Trans. Pattern Anal. Mach. Intell.*, 31(8):1362–1374.
Cross, A. M., D. C. Mason, and S. J. Dury. 1988. Segmentation of remotely sensed image by a split-and-merge process. *Int. J. Remote Sens.*, 9:1329–1345.
Csillik, Ovidiu. 2017. Fast segmentation and classification of very high resolution remote sensing data using SLIC superpixels. *Remote Sens.* 9(3):243. https://doi.org/10.3390/rs9030243
Denis, L., F. Tupin, J. Darbon, and M. Sigelle. 2009. SAR image regularization with fast approximate discrete minimization. *IEEE Trans. Image Proc.*, 18(7):1588–1600.
Dogrusoz, E., and S. Aksoy. 2007. Modeling urban structures using graph-based spatial patterns. *Proc. of IGARSS*, Barcelona, Spain, pp. 4826–4829.
Dongping, M., C. Tianyu, C. Hongyue, L. Longxiang, Q. Cheng, and D. Jinyang. 2012. Semivariogram-based spatial bandwidth selection for remote sensing image segmentation with mean-shift algorithm. *IEEE Geosci. Remote Sens. Lett.*, 9(5):813–817.
Duda, R. O., P. E. Hart, and D. G. Stork. 2001. *Pattern Classification*, 2nd ed. New York: Wiley.
Epifanio, I., and P. Soille. 2007. Morphological texture features for unsupervised and supervised segmentations of natural landscapes. *IEEE Trans. Geosci. Remote Sens.*, 45(4):1074–1083.
Everingham, M., H. Muller, and B. Thomas. 2002. Evaluating image segmentation algorithms using monotonic hulls in fitness/cost space. In: T. Cootes and C. Taylor, eds. *Proc. 12th British Machine Vision Conference*, pp. 363–372.
Farag, A. A., R. M. Mohamed, and A. El-Baz. 2005. A unified framework for MAP estimation in remote sensing image segmentation. *IEEE Trans. Geosci. Remote Sens.*, 43(7):1617–1634.
Fauvel, M., Y. Tarabalka, J. A. Benediktsson, J. Chanussot, and J. C. Tilton. 2013. Advances in spectral-spatial classification of hyperspectral images. *Proceedings of the IEEE*, 101(3):652–675.
Fu, K. S., and J. K. Mui. 1981. A survey on image segmentation. *Pattern Recognit.*, 13:3–16.
Gaetano, R., G. Scarpa, and G. Poggi. 2009. Hierarchical texture-based segmentation of multiresolution remote-sensing images. *IEEE Trans. Geosci. Remote Sens.*, 47(7):2129–2141.

Genctav, A., S. Aksoy, and S. Onder. 2012. Unsupervised segmentation and classification of cervical cell images. *Pattern Recognit.*, 45(12):4151–4168.

Ghamisi, P., M. S. Couceiro, F. M. L. Martins, and J. A. Benediktsson. 2014. Multilevel image segmentation based on fractional-order Darwinian particle swarm optimization. *IEEE Trans. Geosci. Remote Sens.*, 52(5):2382–2394.

Gitas, I., G. Mallinis, P. Patias, D. Stathakis, and G. Zalidis, guest editors. 2014. GEOBIA 2014, advancements, trends and challenges, 5th geographic object-based image analysis conference, Thessaloniki, Greece, May 21–24. Special issue of the *South-Eastern Eur. J. Earth Observ. Geomat.*, 3(2S):1–768.

Gofman, E. 2006. Developing an efficient region growing engine for image segmentation. *Proc. of ICPR*, Hong Kong, China, pp. 2413–1416.

Greig, D., B. Porteous, and A. Seheult. 1989. Exact maximum a posteriori estimation for binary images. *Journal of the Royal Statistical Society, Series B*, 51(2):271–279.

Grote, A., M. Betenuth, M. Gerke, and C. Heipke. 2007. Segmentation based on normalized cuts for the detection of suburban roads in aerial imagery. *Urban Remote Sensing Joint Event*, Paris, France, pp. 1–5.

Gueguen, L., P. Soille, and M. Pesaresi. 2012. A new built-up presence index based on density of corners. *Proc. of IGARSS*, Munich, Germany.

Haralick, R. M., and L. G. Shapiro. 1985. Survey: Image segmentation techniques. *Comput. Vis. Graph. Image Process.*, 29(1):100–132.

Haris, K., S. N. Efstratiadis, N. Maglaveras, and A. Katsaggelos. 1998. Hybrid image segmentation using watersheds and fast region merging. *IEEE Trans. Image Proc.*, 7(12):1684–1699.

Horowitz, S. L., and T. Pavlidis. 1974. Picture segmentation by a directed split-and-merge procedure. *Proc. of the 2nd Int'l Joint Conf. Pattern Recognit.*, pp. 424–433.

Hughes, G. 1968. On the mean accuracy of statistical pattern recognizers. *IEEE Trans. Information Theory*, 14(1):55-63.

Jia, P., C. Chen, D. Zhang, Y. Sang, and L. Zhang. 2024. Semantic segmentation of deep learning remote sensing images based on band combination principle: Application in urban planning and land use. *Comput. Commun.*, 217(2024):97–106. ISSN 0140-3664. https://doi.org/10.1016/j.comcom.2024.01.032. https://www.sciencedirect.com/science/article/pii/S014036642400032X

Jin, B., P. Ye, X. Zhang, et al. 2019. Object-oriented method combined with deep convolutional neural networks for land-use-type classification of remote sensing images. *J. Indian Soc. Remote Sens.* 47:951–965. https://doi.org/10.1007/s12524-019-00945-3

Jing, W., J. Hua, and W. Yubin. 2010. Normalized cut as basic tool for remote sensing image. *Proc. of Int'l Conf. on ICISS*, Gandhinagar, India, pp. 247–249.

Kirkpatrick, S., C. D. Gelatt Jr, and M. P. Vecchi. 1983. Optimization by simulated annealing. *Science*, 220(4598):671–680.

Kruse, F. A., A. B. Lefkoff, J. W. Boardman, K. B. Heidebrecht, A. T. Shapiro, P. J. Barloon, and A. F. H. Goetz. 1993. The spectral image processing system (SIPS) – Interactive visualization and analysis of imaging spectrometer data. *Remote Sens. Environ.*, 44(2/3):145–163.

Kruskal, J. 1956. On the shortest spanning tree of a graph and the traveling salesman problem. *Proc. Am. Math. Soc.*, 7:48–50.

Kurita, T. 1994. An efficient agglomerative clustering algorithm for region growing. *Proc. of MVA, IAPR Workshop on Mach. Vis. Appl.*, Kawasaki, Japan, pp. 210–213.

Kurtz, C., N. Passat, P. Gancarski, and A. Puissant. 2012. Extraction of complex patterns from multiresolution remote sensing images: A hierarchical top-down methodology. *Pattern Recognit.*, 45(2):685–706.

Le Moigne, J., and J. C. Tilton. 1995. Refining image segmentation by integration of edge and region data. *IEEE Trans. Geosci. Remote Sens.*, 33(3):605–615.

Li, J., J. M. Bioucas-Dias, and A. Plaza. 2012. Spectral-spatial hyperspectral image segmentation using subspace multinomial logistic regression and markov random fields. *IEEE Trans. Geosci. Remote Sens.*, 50(3):809–823.

Li, P., and B. Xiao. 2007. Multispectral image segmentation by a multichannel watershed-based approach. *Int'l Jour. Remote Sens.*, 28(19):4429–4452.

Li, X., G. Zhang, H. Cui, S. Hou, S. Wang, X. Li, Y. Chen, Z. Li, and L. Zhang. 2022. MCANet: A joint semantic segmentation framework of optical and SAR images for land use classification. *Int. J. Appl. Earth Obs. Geoinf.*, 106(2022):102638. ISSN 1569-8432. https://doi.org/10.1016/j.jag.2021.102638. https://www.sciencedirect.com/science/article/pii/S0303243421003457

Lilay, M. Y., and G. D. Taye. 2023. Semantic segmentation model for land cover classification from satellite images in Gambella National Park, Ethiopia. *SN Appl. Sci.*, 5:76. https://doi.org/10.1007/s42452-023-05280-4

Liu, Li, Emad Mahrous Awwad, Yasser A. Ali, Muna Al-Razgan, Ali Maarouf, Laith Abualigah, and Azadeh Noori Hoshyar. 2023. Multi-dataset hyper-CNN for hyperspectral image segmentation of remote sensing images. *Processes*, 11(2):435. https://doi.org/10.3390/pr11020435

Liu, X., Z. Deng, and Y. Yang. 2019. Recent progress in semantic image segmentation. *Artif Intell Rev.*, 52:1089–1106. https://doi.org/10.1007/s10462-018-9641-3

Lucieer, A. 2004. *Uncertainties in Segmentation and Their Visualisation*, PhD Thesis, International Institute for Geo-Information Science and Earth Observation (ITC) and University of Utrecht, The Netherlands.

Marpu, P. R., M. Neubert, H. Herold, and I. Niemeyer. 2010. Enhanced evaluation of image segmentation results. *J. Spatial Sci.*, 55(1):55–68.

Martinez-Sanchez, L., L. See, M. Yordanov, A. Verhegghen, N. Elvekjaer, D. Muraro, R. d'Andrimont, and M. van der Velde. 2024. Automatic classification of land cover from LUCAS in-situ landscape photos using semantic segmentation and a Random Forest model. *Environmental Modelling & Software*, 172(2024):105931. ISSN 1364–8152. https://doi.org/10.1016/j.envsoft.2023.105931. https://www.sciencedirect.com/science/article/pii/S1364815223003171

Maulik, U., and D. Chakraborty. 2017. Remote sensing image classification: A survey of support-vector-machine-based advanced techniques. *IEEE Geosci. Remote Sens. Mag.*, 5(1):33–52, March. http://doi.org/10.1109/MGRS.2016.2641240.

Meinel, G., and M. Neubert. 2004. A comparison of segmentation programs for high resolution remote sensing data. *Proc. of Commission IV, XXth ISPRS Congr.*, Istanbul, Turkey, pp. 1097–1102.

Meyer, F. 2001. An overview of morphological segmentation. *Int'l J. Pattern Recognit. Artif. Intell.*, 15(7):1089–1118.

Meyer, F. 2005. Grey-weighted, ultrametric and lexicographic distances. In: C. Ronse, L. Najman and Etienne Decenciere, eds. *Mathematical Morphology: 40 Years On.* Dordrecht, Netherlands: Springer, pp. 289–298.

Meyer, F., and S. Beucher. 1990. Morphological segmentation. *Journal of Visual Communication and Image Representation*, 1(1):21–46.

Moser, G., S. B. Serpico, and J. A. Benediktsson. 2013. Landcover mapping by Markov modeling of spatial-contextual information in very-high-resolution remote sensing images. *Proc. IEEE*, 101(3):631–651.

Muerle, J. L., and D. C. Allen. 1968. Experimental evaluation of techniques for automatic segmentation of objects in a complex scene. In: G. C. Cheng, ed. *Pictorial Pattern Recognition.* Washington, DC: Thompson, pp. 3–13.

Najman, L., and M. Schmitt. 1996. Geodesic saliency of watershed contours and hierarchical segmentation. *IEEE Trans. Pattern Anal. Mach. Intell.*, 18(12):1163–1173.

Narendra, P. M., and M. Goldberg. 1977. A graph-theoretic approach to image segmentation. *Proc. IEEE Comp. Soc. Conf. Pattern Recognit. Image Proc.*, pp. 248–256.

Neubert, M., H. Herold, and G. Meinel. 2006. Evaluation of remote sensing image segmentation quality—Further results and concepts. *Proc. 1st Int. Conf. OBIA*, Salzburg, Austria.

Neubert, M., H. Herold, and G. Meinel. 2008. Assessing image segmentation quality – Concepts, methods and application. In: T. Blaschke, S. Lang and G. J. Hay, eds. *Object-Base Image Analysis: Spatial Concepts for Knowledge-Driven Remote Sensing Applications.* Berlin, Germany: Springer-Verlag.

Neubert, M., and G. Meinel. 2003. Evaluation of segmentation programs for high resolution remote sensing applications. In: *Proc. Joint ISPRS/EARSel. Workshop "High Resolution Mapping from Space 2003."* Hannover, Germany.

Noyel, G., J. Angulo, and D. Jeulin. 2007. Morphological segmentation of hyperspectral images. *Image Anal. Stereol.*, 26(3):101–109.

Otsu, N. 1979. A threshold selection method from gray-level histograms. *IEEE Trans. Syst. Man Cybernet.*, SMC-9(1):62–66.

Ouzounis, G. and Soille, P. 2012. The alpha-tree algorithm. Joint Research Centre Technical Reports, Italy.

Pare, S., A. Kumar, G. K. Singh, et al. 2020. Image segmentation using multilevel thresholding: A research review. *Iran. J. Sci. Technol. Trans. Electr. Eng.* 44:1–29. https://doi.org/10.1007/s40998-019-00251-1

Perret, B., S. Lefevre, C. Collet, and E. Slezak. 2012. Hyperconnections and hierarchical representations for grayscale and multiband image processing. *IEEE Trans. Image Proc.*, 30(7):14–27.

Pesaresi, M., and J. A. Benediktsson. 2001. A new approach for the morphological segmentation of high-resolution satellite imagery. *IEEE Trans. Geosci. Remote Sens.*, 39(2):309–320.

Pesaresi, M., A. Gerhardinger, and F. Kayitakire. 2008. A robust built-up area presence index by anisotropic rotation-invariant textural measure. *IEEE JSTARS*, 1(3):180–192.

Phan, Thanh Noi, Verena Kuch, and Lukas W. Lehnert. 2020. Land cover classification using google earth engine and random forest classifier—the role of image composition. *Remote Sens.* 12(15):2411. https://doi.org/10.3390/rs12152411

Plaza, A. J., and J. C. Tilton. 2005. Automated selection of results in hierarchical segmentations of remotely sensed hyperspectral images. *Proc. of IGARSS*, Seoul, Korea, 7:4946–4949.

Prim, R. 1957. Shortest connection networks and some generalizations. *Bell Syst. Tech. J.*, 36:1389–1401.

Raei, E., A. A. Asanjan, M. Z. Nikoo, M. Sadegh, S. Pourshahabi, and J. F. Adamowski. 2022. A deep learning image segmentation model for agricultural irrigation system classification. Comput. Electron. Agric., 198(2022):106977. ISSN 0168-1699. https://doi.org/10.1016/j.compag.2022.106977. https://www.sciencedirect.com/science/article/pii/S0168169922002940

Richards, J. A., and X. Jia. 1999. *Remote Sensing Digital Image Analysis: An Introduction.* New York: Springer-Verlag.

Robinson, D. J., N. J. Redding, and D. J. Crisp. 2002. Implementation of a fast algorithm for segmenting SAR imagery. In: *Scientific and Technical Report.* Defence Science and Technology Organization, Edinburgh, South Australia, Australia.

Rosin, P. L., and J. Hervas. 2005. Remote sensing image thresholding methods for determining landslide activity. *Int'l J. Remote Sens.*, 26(6):1075–1092.

Scholkopf, B., and Smola, A. J. 2002. Learning with Kernels. MIT Press, Cambridge, MA.

Sertel, Elif, Burak Ekim, Paria Ettehadi Osgouei, and M. Erdem Kabadayi. 2022. Land use and land cover mapping using deep learning based segmentation approaches and VHR worldview-3 images. *Remote Sens.*, 14(18):4558. https://doi.org/10.3390/rs14184558

Sezgin, M., and B. Sankur. 2004. Survey over image thresholding techniques and quantitative performance evaluation. *J. Electron. Imaging*, 13(1):146–168.

Shan, P. 2018. Image segmentation method based on K-mean algorithm. *J Image Video Proc.*, 2018(81). https://doi.org/10.1186/s13640-018-0322-6

Shankar, B. U. (2007). Novel classification and segmentation techniques with application to remotely sensed images. In: V. W. Marek, E. Orlowska, R. Slowinski, and W. Ziarko, eds. *Transactions on Rough Sets VII.* New York: Springer-Verlag, pp. 295–380.

Shi, J., and J. Malik. 2000. Normalized cuts and image segmentation. *IEEE Trans. Pattern Anal. Mach. Intell.*, 22(8):888–905.

Sirmacek, B., and C. Unsalan. 2010. Urban area detection using local feature points and spatial voting. *IEEE Geosci. Remote Sens. Lett.*, 7(1):146–150.

Skurikhin, A. N. 2010. Patch-based image segmentation of satellite imagery using minimum spanning tree construction. *Proc. GEOBIA*, Ghent, Belgium.

Soille, P. 2003. *Morphological Image Analysis.*New York: Springer-Verlag.

Soille, P. 2008. Constrained connectivity for hierarchical image partitioning and simplification. *IEEE Trans. Pattern Anal. Mach. Intell.*, 30(7):1132–1145.

Soille, P., and L. Najman. 2010. On morphological hierarchical representations for image processing and spatial data clustering. *Int'l Workshop on Applications of Discrete Geometry and Mathematical Morphology*, Istanbul, Turkey, pp. 43–67.

Soille, P., and M. Pesaresi. 2002. Advances in mathematical morphology applied to geoscience and remote sensing. *IEEE Trans. Geosci. Remote Sens.*, 40(9):2042–2055.

Solberg, A. H. S., T. Taxt, and A. K. Jain. 1996. A Markov random field model for classification of multisource satellite imagery. *IEEE Trans. Geosci. Remote Sens.*, 34(1):100–113.

Sramek, M., and T. Wrbka. 1997. Watershed based image segmentation – an effective tool for detecting landscape structure. *Digital Image Processing and Computer Graphics* (DIP'97), Proc. SPIE 3346, pp. 227–235.

Stewart, D., B. Blacknell, A. Blake, R. Cook, and C. Oliver. 2000. Optimal approach to SAR segmentation and classification. *IEE Proc. Radar Sonar Navig.*, 147(3):134–142.

Strasters, K., and J. Gerbrands. 1991. Three-dimensional segmentation using a split, merge and group approach. *Pattern Recognit. Lett.*, 12:307–325.

Sung, F., and J. He. 2009. The remote-sensing image segmentation using textons in the Normalized Cuts framework. *Mechatronics and Automation: International Conference on ICMA*, Changchun, China, pp. 1877–1881.

Tarabalka, Y., J. A. Benediktsson, J. Chanussot, and J. C. Tilton. 2010. Multiple spectral-spatial classification approach for hyperspectral data. *IEEE Trans. Geosci. Remote Sens.*, 48(11):4122–4132.

Tarabalka, Y., J. Chanussot, and J. A. Benediktsson. 2009. Classification of hyperspectral images using automatic marker selection and minimum spanning forest. *Proc. of IEEE WHISPERS*, Grenoble, France, pp. 1–4.

Tarabalka, Y., J. Chanussot, and J. A. Benediktsson. 2010a. Segmentation and classification of hyperspectral images using minimum spanning forest grown from automatically selected markers. *IEEE Trans. Syst. Man Cybern. Part B: Cybern.*, 40(5):1267–1279.

Tarabalka, Y., M. Fauvel, J. Chanussot, and J. A. Benediktsson. 2010b. SVM- and MRF-based method for accurate classification of hyperspectral images. *IEEE Geosc. Remote Sens. Lett.*, 7(4):736–740.

Tarabalka, Y., and A. Rana. 2014. Graph-cut-based model for spectral-spatial classification of hyperspectral images. *Proc. of IGARSS*, Quebec City, Quebec, Canada.

Tarabalka, Y., J. C. Tilton, J. A. Benediktsson, and J. Chanussot. 2012. A marker-based approach for the automated selection of a single segmentation from a hierarchical set of image segmentations. *IEEE JSTARS*, 5(1):262–272.

Tassi, Andrea, and Marco Vizzari. 2020. Object-oriented LULC classification in Google Earth engine combining SNIC, GLCM, and machine learning algorithms. *Remote Sens.*, 12(22):3776. https://doi.org/10.3390/rs12223776

Tilton, J. C. 1998. Image segmentation by region growing and spectral clustering with a natural convergence criterion. *Proc. of IGARSS*, Seattle, WA, USA.

Tilton, J. C. 2003. Analysis of hierarchically related image segmentations. *Proc. of IEEE Workshop on Advances in Techniques for Analysis of Remotely Sensed Data*, Greenbelt, MD, pp. 60–69.

Tilton, J. C. 2007. Parallel implementation of the recursive approximation of an unsupervised hierarchical segmentation algorithm. In: A. J. Plaza and C. Chang, eds. *High Performance Computing in Remote Sensing*.New York: Chapman and Hall, pp. 97–107.

Tilton, J. C. 2013. *RHSeg User's Manual: Including HSWO, HSeg, HSegExtract, HSegReader, HSegViewer and HSegLearn*, version 1.59: available via email request to James.C.Tilton@nasa.gov.

Tilton, J. C., Y. Tarabalka, P. Montesano, and E. Gofman. 2012. Best merge region-growing segmentation with integrated nonadjacent region-object aggregation. *IEEE Trans. Geosci. Remote Sens.*, 50(11):4454–4467.

Unsalan, C., and K. L. Boyer. 2005. A system to detect houses and residential street networks in multispectral satellite images. *Comput. Vis. Image Underst.*, 98(3):423–461.

Vapnik, V. 1998. *Statistical Learning Theory*. John Wiley & Sons.

Vincent, L., and P. Soille. 1991. Watersheds in digital spaces: An efficient algorithm based on immersion simulations. *IEEE Trans. Pattern Anal. Mach. Intell.*, 13(6):583–598.

Wang, C., N. Komodakis, and N. Paragios. 2013. Markov Random field modeling, inference and learning in computer vision and image understanding: A survey. *Comput. Vis. Image Underst.*, 117:1610–1627.

Wang, Z., and R. Boesch. 2007. Color- and texture-based image segmentation for improved forest delineation. *IEEE Trans. Geosci. Remote Sens.*, 45(10):3055–3062.

Williams, J. W. 1964. Heapsort. *Commun. ACM*, 7(12):347–348.

Yang, L., F. Albregsten, T. Lonnestad, and P. Grottum. 1995. A supervised approach to the evaluation of image segmentation methods. In: *Proc. CAIP, Lect. Notes Comput. Sc. 970*, Prague, Czech Republic, pp. 759–765.

Yin, H., A. V. Prishchepov, T. Kuemmerle, B. Bleyhl, J. Buchner, and V. C. Radeloff. 2018. Mapping agricultural land abandonment from spatial and temporal segmentation of Landsat time series. *Remote Sens. Environ.*, 210(2018):12–24. ISSN 0034-4257. https://doi.org/10.1016/j.rse.2018.02.050. https://www.sciencedirect.com/science/article/pii/S0034425718300622

Yu, Ying, Chunping Wang, Qiang Fu, Renke Kou, Fuyu Huang, Boxiong Yang, Tingting Yang, and Mingliang Gao. 2023. Techniques and challenges of image segmentation: A review. *Electronics*, 12(5):1199. https://doi.org/10.3390/electronics12051199

Zahn, C. T. 1971. Graph-theoretic methods detecting and describing gestalt clusters. *IEEE Trans. Comput.*, C-20(1):68–86.

Zamalieva, D., S. Aksoy, and J. C. Tilton. 2009. Finding compound structures in images using image segmentation and graph-based knowledge discovery. *Proc. of IGARSS*, Cape Town, South Africa.

Zhang, C., I. Sargent, X. Pan, H. Li, A. Gardiner, J. Hare, and P. M. Atkinson. 2018. An object-based convolutional neural network (OCNN) for urban land use classification. *Remote Sens. Environ.*, 216(2018):57–70. ISSN 0034-4257. https://doi.org/10.1016/j.rse.2018.06.034. https://www.sciencedirect.com/science/article/pii/S0034425718303122

Zhang, C., I. Sargent, X. Pan, H. Li, A. Gardiner, J. Hare, and P. M. Atkinson. 2019. Joint Deep learning for land cover and land use classification. *Remote Sens. Environ.*, 221(2019):173–187. ISSN 0034-4257. https://doi.org/10.1016/j.rse.2018.11.014. https://www.sciencedirect.com/science/article/pii/S0034425718305236

Zhang, Y. J., and J. J. Gerbrands. 1994. Objective and quantitative segmentation evaluation and comparison. *Signal Process.*, 39:3–54.

Zucker, S. W. 1976. Region growing: Childhood and adolescence. *Comput. Graph. Image Process.*, 5:382–399.

11 LiDAR Data Processing and Applications

Shih-Hong Chio, Tzu-Yi Chuang, Pai-Hui Hsu, Jen-Jer Jaw, Shih-Yuan Lin, Yu-Ching Lin, Tee-Ann Teo, Fuan Tsai, Yi-Hsing Tseng, Cheng-Kai Wang, Chi-Kuei Wang, Miao Wang, and Ming-Der Yang

LIST OF ACRONYMS

3D	Three-dimensional
ALS	Airborne laser scanner
ASPRS	American Society of Photogrammetry and Remote Sensing
CBH	Crown base height
CF	Constant fraction
CG	Center of gravity
DBH	Diameter at breast height
DBMS	DataBase management system
DEM	Digital elevation model
DG	Direct geo-referencing
DSM	Digital surface model
DTM	Digital terrain model
FDNs	Fixed distance neighbors
GCPs	Ground control points
GIS	Geographical information system
GLAS	Geoscience laser altimeter system
GNSS	Global navigation satellite system
GPS	Global positioning system
GRCS	Ground reference coordinate system
ICE	SatIce cloud and land elevation satellite
ICP	Iterative closest point
IMU	Inertial measurement unit
INS	Inertial navigation system
ITS	Intelligent transportation system
k-NNs	*k*-nearest neighbors
LiDAR	Light detection and ranging
LS3D	Least squares 3D surface matching
MA	Maximum
OGC	Open geospatial consortium
PCA	Principal component analysis
POS	Positioning and orientation systems
QA/QC	Quality assessment and quality control
RANSAC	RANdom SAmple Consensus
RMSD	Root mean square difference

DOI: 10.1201/9781003541158-12

RMSE	Root mean square error
SDMBS	Spatial DBMS
TH	Threshold
TIN	Triangulated irregular networks
TLS	Terrestrial laser scanner
VLR	Variable length record
WB	Wavelet-based
VMNS	Voxel-based marked neighborhood searching
Voxel	Volume element
ZC	Zero crossing of the first deviation

11.1 INTRODUCTION

LiDAR (Light Detection and Ranging), also called laser scanning, is an effective remote sensing system for acquiring three-dimensional (3D) information about scanned objects, which has been widely applied in a broad range of disciplines since it was first developed less than two decades ago (Dong and Chen, 2017; Baltsavias, 1999a; Krabill et al., 2000). A LiDAR system integrates several accurate optical, electronic, mechanic, timing and geo-referencing units that make it enable to acquire 3D point measurements rapidly. LiDAR systems are mainly classified into two types based on the motions that occur when they collect data. The first category includes static terrestrial (or ground-based) LiDAR systems (usually mounted at a tripod) with a variety of scanning ranges, which are also called 3D laser scanners. The primary component of a terrestrial LiDAR system is a laser ranging unit to obtain the distance to the target. Integrating with one or two rotating mirror or prism which changes the direction of emitting laser pulses, the LiDAR system can scan and measure the distances to the surrounding objects. By calculating the obtained range and the scanning angle, the 3D coordinates of the scanned target are obtained. Because the mirror rotates in rapid speed with small angle variation when the LiDAR operates, a huge number of dense and accurate point measurements, often called point cloud, of the surfaces of scanned objects are obtained (Guo et al., 2023; Dong and Chen, 2017; Petrie and Toth, 2008b).

The second category includes airborne LiDAR systems (also called Airborne Laser Scanning, ALS), mobile terrestrial LiDAR systems or other LiDAR systems boarded on moving platforms (Lohani and Ghosh, 2017; Baltsavias, 1999b; Petrie and Toth, 2008a; Petrie and Toth, 2008b). These systems offer the capability of acquiring 3D information about the scanned objects at the mapping coordinate system by means of direct geo-referencing, which is enabled by the combination

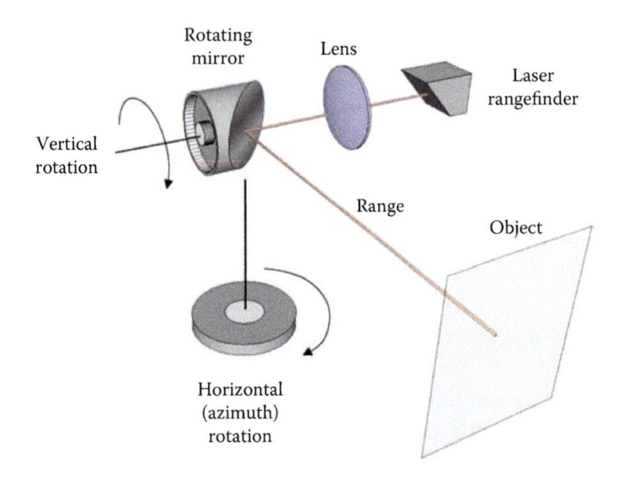

FIGURE 11.1 The measurement mechanism of a terrestrial laser scanner.

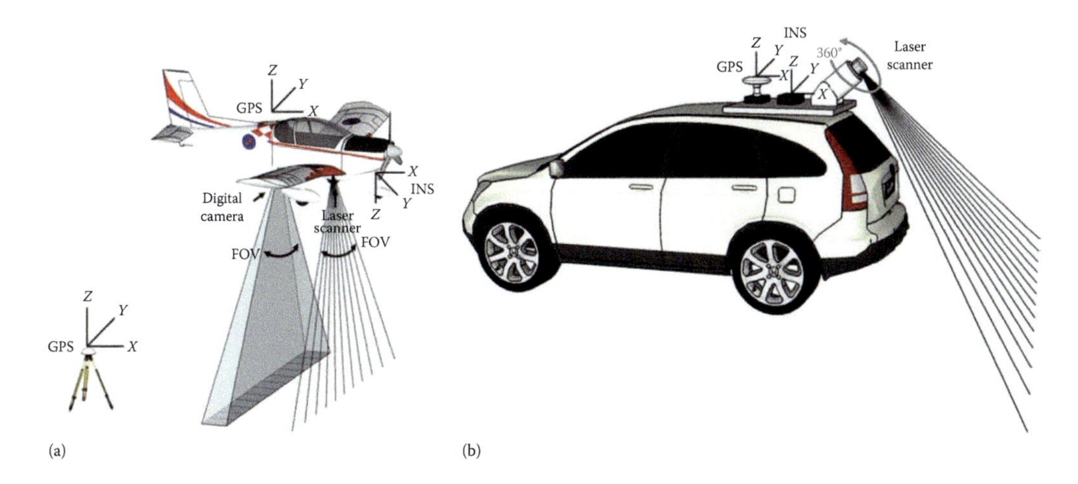

FIGURE 11.2 (a) Airborne and (b) vehicle-borne LiDAR systems.

of a GPS unit and an IMU unit (together these are also known as Positioning and Orientation Systems, or POS) (Baltsavias, 1999a; Krabill et al., 2000), as shown at Figure 11.2. The spaceborne LiDAR systems can be regarded in the second category. However, data processing consideration for such systems is somewhat different. One of the reasons is that the footprint size and the sampling interval of the spaceborne systems are significantly larger, for which they have been called large footprint systems. For example, the Geoscience Laser Altimeter System (GLAS) aboard the Ice Cloud and Land Elevation Satellite (ICESat) has the footprint size of 60 m and sampling interval of 172 m (Zwally et al., 2002). The geometry details obtained from spaceborne LiDAR systems have smoother appearance, which prevents many of the processing logic designed for the small footprint systems to be directly applied. For brevity and conciseness, these large footprint systems are not included in this chapter.

LiDAR measures the surfaces of objects by accurately recording the time of flight of the laser pulse, which is transmitted from the laser ranging system and then returned back as it reflects off the scanned surface, and the scan angle of the laser pulse, which the laser is directed to, with respect to the laser ranging system. The distance between the laser ranging system and the scanned object can be obtained by halving the multiplication of the speed of light and the round trip time of the laser pulse, and when this is combined with the scan angle of the laser pulse the 3D coordinates of the point measurement at a local coordinate system can be obtained. For a LiDAR system installed on a moving platform, the POS provides highly accurate data on the position and attitude of the platform for direct geo-referencing. The resulting 3D information about the scanned objects is then registered in the same mapping coordinate system as the POS. The choice of the mapping coordinate system depends on the scope of utility for the LiDAR data (Guo et al., 2023; Dong and Chen, 2017; Fornaciai et al., 2010). Figure 11.3a and b show the point clouds of a city area and a forest area, and Table 11.1 list the characteristics of ALS, TLS, and mobile TLS. Table 11.2 lists the main applications, strengths and limitations of ALS, TLS and mobile TLS (Debnath et al., 2023; Guo et al., 2023; Kovanič et al., 2023; Rivera et al., 2023; Mirzaei et al., 2022; Roriz et al., 2022; Xu et al., 2021; Li and Ibanez-Guzman, 2020; Muhadi et al., 2020; Dong and Chen, 2017; Gargoum and El-Basyouny, 2017; Lohani and Ghosh, 2017; Zhenlong et al., 2018; Cao et al., 2015).

Airborne LiDAR systems have been widely employed for large area or nationwide applications, in many cases replacing photogrammetric approaches, as the process flow of LiDAR data is relatively straightforward and less time-consuming (Guo et al., 2023; Dong and Chen, 2017; Lohani and Ghosh, 2017). There are two major technological benefits of airborne LiDAR that have contributed to its popularity. The first is that LiDAR can acquire 3D information of homogeneous surfaces,

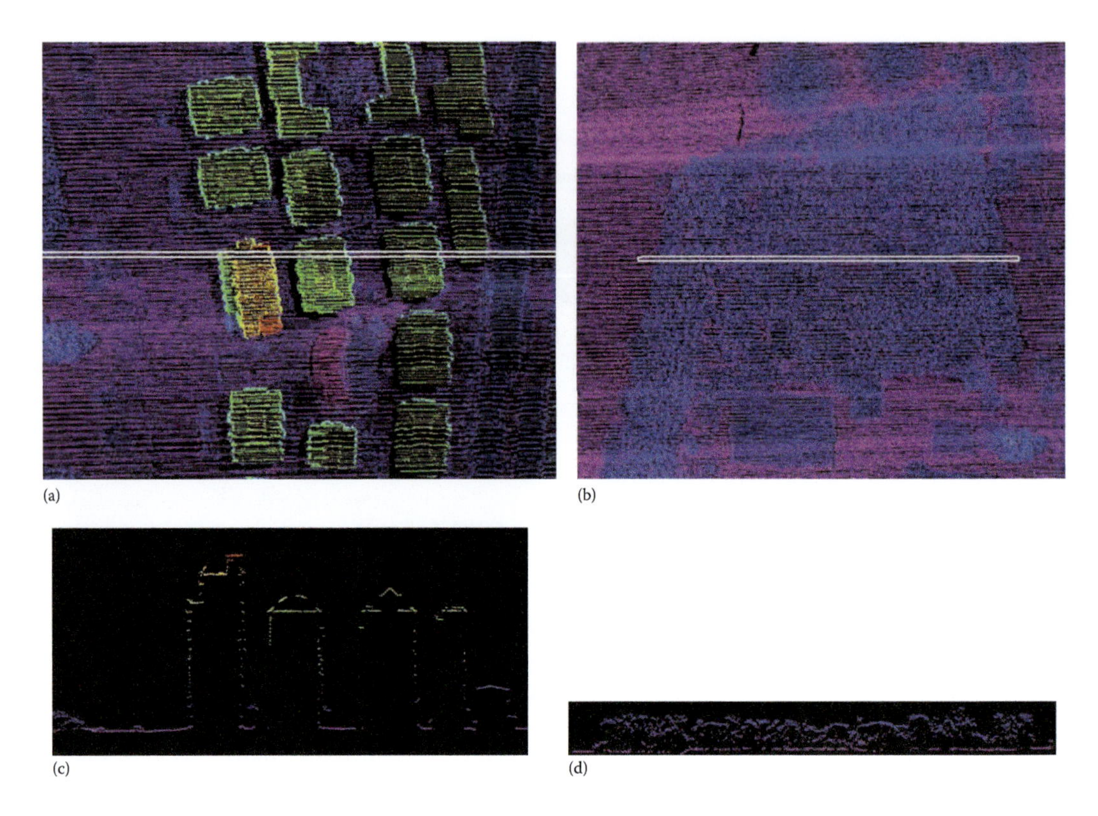

FIGURE 11.3 (Up) Point clouds and (down) profile of (a) a city area and (b) a forest areas. Color represent the height of the points.

TABLE 11.1
Characteristics and Applications of ALS, TLS and Mobile TLS

System	ALS	TLS	Mobile TLS
Footprint size	Small (~20–40 cm in diameter)	Extreme small (~2 cm in diameter)	Extreme small (~2 cm in diameter)
Waveform data	Available from more recent systems	Available from limited systems	Available from limited systems
Geo-referencing	Direct	In-direct	Direct
Measuring distance	200–3000 m	< 1 km	< 1 km
Data characteristic	The sunlit surfaces are scanned, where most of the point measurements are distributed on the op surfaces of scanned objects.	Point measurements are distributed on the side of the scanned objects. Usually extremely high detail of the surface can be obtained within the scan range of the static station	Point measurements are distributed on the side of the scanned objects, which are close to the path of the moving platform
Applications	DEM, DSM, DBM, CHM, Land-cover classification, City modeling, Road network, Biomass	DBH, building modeling	DBH, city modeling, road network
Survey Area	Nation-wide	Single or multiple objects with small vicinity	City-wide

which is a difficult task for photogrammetry due to the lack of textures for successful stereo matching. The other is that it is capable of acquiring multiple measurements at different distances with only one pulse. For forested areas it is very useful to be able to acquire 3D information about not only the top of the canopy, but also the ground surfaces beneath it and the structures within it. As a comparison, photogrammetry tends to be used only to measure the top canopy, because the ground surfaces are only partially visible or completely invisible from aerial photographs.

Careful data processing must be carried out in order to fully exploit the point measurements obtained from the scanned objects. This chapter covers the issues of data quality assessment and control (QA/QC), data management, point cloud feature extraction, and full-waveform data processing of the LiDAR data. In addition, applications for digital elevation model (DEM) and digital surface model (DSM) generation and 3D city modeling are also described.

In Section 11.2 the concept of error budget is introduced and implemented as the main tool for the QA/QC of the LiDAR point measurements. Internal and externals controls are then suggested to ensure that the data derived from the LiDAR system meets the requirement of a specific application.

The number of point measurements resulting from a normal LiDAR scan project, either airborne or terrestrial, can easily exceed several tens of millions. Due to the nature of these 3D point measurements, which lack an efficient spatial index to easily identify neighboring points and quickly access any points of interest, the LiDAR data are difficult to visualize, edit, and process. Section 11.3 thus presents some concepts that can be used to establish a spatial index for point clouds.

Assigning meaningful attribute to the point clouds can significantly increase the usage of the LiDAR data (Debnath et al., 2023; Guo et al., 2023),. In an urban setting, extracting spatial features from the point clouds, such as points, lines, and surfaces, can be quite useful as the results can be readily adopted by 3D city modeling. Section 11.4 gives detailed information regarding this kind of feature extraction from LiDAR data. Furthermore, several examples of using LiDAR point clouds for 3D city modeling are presented in Section 11.5.

The complete temporal history of the laser return signal is called the full waveform. In practice, the waveform is processed on-the-fly and only a few meaningful signals (i.e., returns) are extracted while the waveform data are then discarded. Some of the early LiDAR systems were only able to extract two returns for each waveform. More r systems normally provide four returns, that is, the first, second, third, and last, for each laser pulse. An unlimited number of returns for each laser pulse is possible, provided the full waveform data and a decent post-processing algorithm are available. However, handling such a massive amount of data is rather difficult, and thus this approach has not been widely embraced by the community (e.g., manufacturers, service providers and end users). Nevertheless, processing the full waveform data is suggested for areas where the penetration of LiDAR is unsatisfactory, and Section 11.6, explains the benefits provided by the full waveform data.

Generating DEM using LiDAR point clouds is the main purpose of obtaining airborne LiDAR data (Lohani and Ghosh, 2017). To ensure a good quality DEM, the ground points must be identified from the whole LiDAR point clouds, and several algorithms that have been designed to achieve this. The methodology used can be as simple as treating all of the last returns as ground points. While this simple solution may be sufficient for a bare surface or in a city setting, it is likely to fail in forested areas, since many of the last returns may not represent the ground surface (being merely close to it). To obtain good quality DEM in an efficient manner, the LiDAR points are filtered by a morphology-based algorithm, which considers the connectivity of the nearby points, and this is then followed by manual inspection/editing to ensure all ground points are reasonably identified. The production of DEM is finally realized by interpolating the identified ground points. Section 11.7 provides a detailed explanation of DEM generation from LiDAR data.

The characteristics of the point clouds obtained from the ground-based and mobile LiDAR systems, for example, the pattern of point distribution, the variation of point density within the LiDAR data, and lack of echo information, are different from those from airborne LiDAR systems (Guo et al., 2023; Dong and Chen, 2017; Lohani and Ghosh, 2017). The processing strategy used for such data should take into account these differences and make any necessary adjustments. The processing of the ground-based and mobile LiDAR point clouds is thus described in Section 11.8.

11.2 LiDAR DATA QUALITY ASSESSMENT AND CONTROL

Due to the nature of LiDAR systems, every single point that is generated usually contains no redundant measurements, and there is no associated measure that can be used to evaluate the quality of point clouds, except for the nominal precision that manufacturers claim. Therefore, a well-defined set of quality assessment and quality control procedures is needed before embarking on any aspect of LiDAR data processing. This chapter thus introduces basic components of LiDAR systems and relevant error budgets, and this is followed by a discussion of common approaches to the quality assessment and control of LiDAR data.

11.2.1 System Components

The components of LiDAR systems differ with regard to different LiDAR platforms, which are classified into two categories, namely mobile platforms, for example, airborne or vehicle-based LiDAR systems, and static platforms, for example, terrestrial LiDAR mounted on a tripod. The system structure of static LiDAR platforms, which integrate state-of-the-art laser ranging and scanning for the rapid, highly dense, and precise acquisition of 3D point clouds (Petrie and Toth, 2008a), is relatively simple as compared to mobile ones. Mobile LiDAR systems are based on the combination of a laser scanning system and a direct geo-referencing (DG) system. The laser scanning system can be subdivided into three key units: the opto-mechanical scanner, the ranging unit, and the control processing unit; The direct geo-referencing system composed of a Global Positioning System (GPS)/Global Navigation Satellite System (GNSS) and an Inertial Navigation System (INS) measures the sensor's position and orientation directly with respect to a referenced coordinate system. The combination of GPS/INS technologies may also be referred to as a GPS-Aided INS or a Position and Orientation System (POS). Each of these technologies alone has limitations, but the integration of GPS and INS is a powerful solution for direct geo-referencing. The DG system and scanning sensor must be in a rigidly fixed position with respect to each other and calibrated within the reference frame of the platform for meaningful results. During a scanning task, the DG system records position and orientation data, and also records a corresponding time tag for each laser scan. The DG post-processing software interpolates the position and orientation of the laser reference point at each time tag. The 3D ground coordinates of every laser return can then be computed using this data and the range measured by the laser.

11.2.2 LiDAR Error Budget

The LiDAR systems introduced earlier face problems with regard to the random and systematic errors that may occur in each component, as well as due their integration (Filin, 2001). Random errors include those related to position and orientation measurements from the direct geo-referencing system, mirror angles, and ranges, and these are based on the precision of the instrumental measurements within a mobile system. On the other hand, systematic errors are mainly caused by biases in the mounting parameters related to the system components, as well as those in the system measurements resulting from biases related to the range and mirror angle in a laser scanner system. Moreover, systematic errors are also caused by differences in the troposphere and ionosphere, as well as multi-path, INS initialization and misalignment errors, and gyro drifts in a DG system. Lichti and Licht (2006) presented a systematic error model which consisted of 19 coefficients grouped into two categories: physical and empirical. The physical group comprises known error sources such as rangefinder offset, cyclic errors, collimation axis error, trunnion axis error, vertical circle index error and others. The empirical terms lack ready physical explanation but nonetheless model significant errors. Typically, errors of a static LiDAR platform come principally from the laser scanning system, while those of a mobile LiDAR platform are the combined effects of the laser scanning system and DG system. A detailed description of random and systematic errors of LiDAR systems is given in several publications such as Guo et al. (2023), Xu et al. (2021), Gargoum and El-Basyouny (2017), Lohani and Ghosh (2017), Zhenlong et al. (2018), Cao et al. (2015), Huising and Pereira (1998), Baltsavias (1999b), Schenk (2001), Latypov and Zosse (2002), Habib et al. (2008), and Habib et al. (2009). A summary of laser scanning and direct geo-referencing errors is given in the following paragraphs.

Laser scanner errors: There are a number of factors affecting the accuracy of a laser scanner. Reshetyuk (2006) classified scanning errors into three major categories, which highlight the influence of the scanned object on the accuracy of the related point clouds. The first category is instrumental system errors, based on the range and angular measurements, and these vary with different scanning devices. The second category is object related errors, and these are related to the reflectance properties of the object's surface, due to several factors such as material properties, laser wavelength, polarization, surface color, moisture, roughness, and temperature. The last category is environmental errors, which affect laser beam propagation in the atmosphere, causing both distortion and attenuation of the returned signal. The degree of attenuation depends on the wavelength, temperature, pressure, microscopic particles in the air and weather conditions. Other factors influencing laser beam propagation are reflection and atmospheric turbulence, caused by the beam wandering from its initial direction and Gaussian wave-front distortion, called beam intensity fluctuation. Based on these characteristics, it can be understood that the longer the range, the greater the expected errors.

Direct geo-referencing errors: With regard to positioning performance, most errors in a GPS/GNSS system are dependent on the operating conditions and set-up (Morin, 2002). These errors, such as atmospheric errors, multipath effects, poor satellite geometry, baseline length, and loss of lock, have a direct impact on the resulting positioning accuracy of a GPS/GNSS system, and most of the related factors are difficult to predict and describe via a mathematic model. Positioning noise leads to similar amounts of noise in the derived point cloud. As for orientation errors, the overall accuracy of navigation attitude depends on the quality of IMU. Errors in the LiDAR return position due to attitude errors are directly proportional to the range from the scanner to target. As a result, an IMU with higher accuracy is normally required for fixed wing operations compared to helicopter or ground-based data collection, due to the increased target range (Glennie, 2007). When taking the integration of the various sensor components into consideration, the bore-sighting angles and the physical mounting angles between an IMU and a laser scanning system, are usually the main sources of systematic errors, and thus careful calibration of these is required. Moreover, due to the fact that the center of observations from the laser scanner and the origin of the navigation system cannot be co-located, the precise offset or lever-arm between the two centers must be known in order to accurately determine the laser scanner measurements. Since the physical measurement origin of the navigation system or laser scanner assembly cannot be directly observed, the lever-arm offset must be obtained indirectly by a calibration procedure. Biases in the lever-arm offsets can then lead to constant shifts in the derived point cloud.

11.2.3 Quality Assessment

Quality assessment encompasses management activities that are carried out prior to data collection to ensure that the derived data is of the quality demanded by the user (Debnath et al., 2023; Guo et al., 2023; Kovanič et al., 2023; Rivera et al., 2023; Cao et al., 2015). These management controls cover the calibration, planning, implementation, and review of data collection activities. In consideration of the potential errors mentioned earlier, calibration of LiDAR systems is prerequisite for reducing the effects of these on the accuracy of acquired point clouds. Ground-based LiDAR systems are calibrated by scanning a field of distributed control targets, while airborne LiDAR systems are calibrated by flying over a set of ground calibration targets which have been precisely surveyed and well-mapped. However, here is currently no standard procedure for a calibration task, and thus each LiDAR manufacturer may have its own calibration scheme for it own products, based on specific parameters. Further discussions on LiDAR system calibration can be found in Schenk (2001), Burman (2002), Filin (2003), Kager (2004), and Skaloud and Lichti (2006). One problems is that careful calibrations require some raw measurements, such as navigation data, mirror angles, and ranges, and in general LiDAR systems do not supply such raw measurements, and thus quality control procedures are needed to assess system performance after the data has been collected.

11.2.4 Quality Control

Quality control procedures can be realized by internal and external checks. The internal measures are used to check the relative consistency of the LiDAR data, while the external measures verify the absolute quality of the LiDAR data by checking its compatibility with an independently collected and more accurate dataset.

Internal quality control: LiDAR data is usually acquired from different scanning viewpoints or different strips, so that the relative consistency of point clouds within the overlapping areas can be assessed through the correspondence of conjugate features (Xu et al., 2021; Li and Ibanez-Guzman, 2020; Zhenlong et al., 2018; Cao et al., 2015). This check is commonly conducted by comparing interpolated range or intensity images derived from the overlapping areas, or by comparing the conjugate features extracted from corresponding datasets. The degree of coincidence of the extracted features can be used as a measure of the data quality and to detect the presence of systematic biases. A well-presented demonstration in internal quality control can be found in Habib et al. (2010).

External quality control: A common approach to external checks involves check-point analysis using specially designed LiDAR targets. The targets are then extracted from the range and intensity LiDAR imagery using a segmentation procedure. The coordinates of the extracted targets are then compared with the surveyed coordinates using an RMSE (Root Mean Square Error) or RMSD (Root Mean Square Difference) indicator. The former usually refers to comparing the estimated results with error-free data, while the latter compares the estimated results with erroneous reference data, and both share the same expression (Mirzaei et al., 2022; Roriz et al., 2022; Gargoum and El-Basyouny, 2017; Cao et al., 2015). For more details regarding this approach, one can refer to Csanyi and Toth (2007). On the other hand, in addition to employing special control targets, features within point clouds can also be used as a measure of external quality control. In such case the derived features can be compared with independently collected control entities, such as point, line, and planes over the same area. To this end, check points situated on smoothing surfaces are usually adopted, and the quality of the vertical components is the main factor to be evaluated.

As shown in Figure 11.4, in addition to the conventional way of assessing positional discrepancies for point features, angle and distance measures can also be utilized to assess the quality, including both internal and external indicators for line and plane features, if applicable (Chuang, 2012).

Registration of terrestrial LiDAR datasets and strip adjustment of airborne LiDAR point clouds are two major ways to provide quality assessment and quality control for final geospatial products. In the case of terrestrial laser scanning, it is normal to acquire multiple-point clouds from different standpoints for a complete scene of objects (Li and Ibanez-Guzman, 2020; Muhadi et al., 2020). The manipulation of these data should thus be preceded by having them registered relative to the same reference frame. On the other hand, strip adjustment of airborne LiDAR data serves as a means of producing a best-fit surface through an adjustment process that compensates for small misalignments between adjacent datasets (Filin and Vosselman, 2004; Tao and Li, 2007). The discrepancies among strips are typically caused by the varied performance of the geo-referencing components, and thus show systematic patterns which are more visible in overlap areas rich in objects of simpler geometric shapes, as illustrated in Figure 11.5.

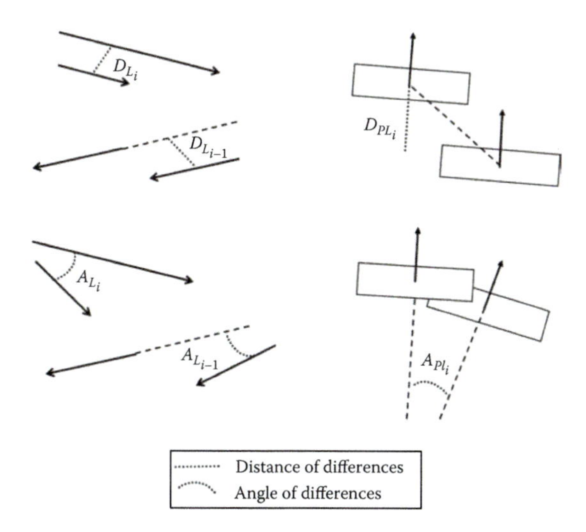

FIGURE 11.4 Illustration of distance and angle differences when using line and plane features.

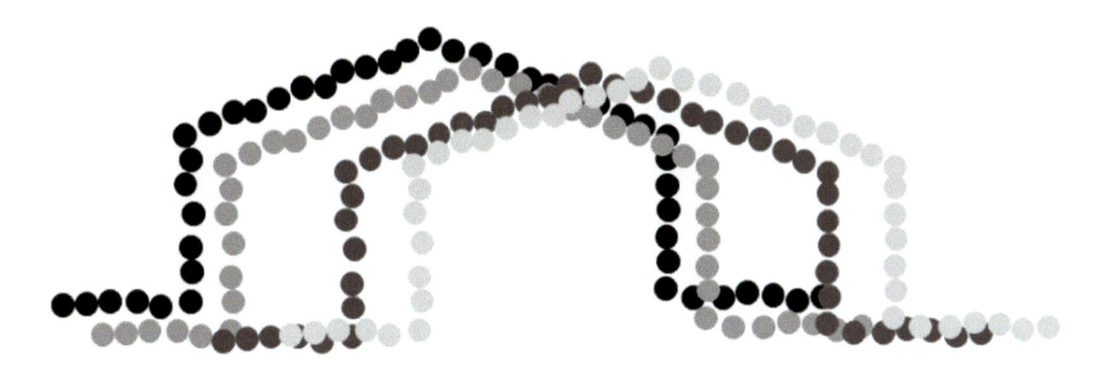

FIGURE 11.5 Illustration of strip discrepancies of airborne LiDAR point clouds.

In summary, the quality assessment of LiDAR systems is restricted to the availability of the raw measurements, which are not usually revealed to the end user. Consequently, quality control is an essential procedure to ensure that the data derived from a LiDAR system is able to meet the requirements of a specific application.

11.3 LiDAR DATA MANAGEMENT

11.3.1 Storage of LiDAR Point Cloud Data

A LiDAR dataset mainly contains the 3D coordinates of observed points, which are commonly called point-cloud data representing the spatial properties of scanned objects (Kovanič et al., 2023; Dong and Chen, 2017; Gargoum and El-Basyouny, 2017; Lohani and Ghosh, 2017; Zhenlong et al., 2018; Cao et al., 2015). The reflected laser intensity of each point is also recorded as the radiometric property. Many LiDAR systems, particularly for airborne LiDAR systems, may record additional information for each point, such as return number, number of returns, scan direction, scan angle, GPS time tag, and so on.

Each manufacturer has developed a particular LiDAR data format for storage, and a standard format is thus needed for data exchange. The simplest way to exchange data is to rewrite it in a text (ASCII) file, although this requires much more storage space than a binary file. In 2003, the American Society of Photogrammetry and Remote Sensing (ASPRS) published a format standard, known LAS 1.0, for LiDAR data exchange (ASPRS, 2013). The LAS format is now used as the standard format of LiDAR data for both users and hardware and software manufacturers. Waveform data can also be included in the two latest versions of this format, announced in 2010 and 2011.

The LAS file format provides several optional ways of recording different point data contents. The optional fields include those for the data related to GPS time tag, color (red, green, blue), waveform package, and near infra-red. Users can choose appropriate formats depends on their specific aims. The LAS 1.4 specification provides up to 11 point formats (ID: 0–10). Table 11.2 shows the recorded fields of point format 10, which contains all optional fields.

In addition to point data, the LAS format also contains different numbers of variable length records (VLRs) for storing additional information about the point cloud. VLRs contain various types of data, including projection information, metadata, waveform packet information and user application data.

Although LAS files store data in binary format and consume less storage space than text files, they are not compact. Isenburg (2013) thus developed a lossless compression scheme for LAS files to reduce the storage space down to less than 25% of the original.

TABLE 11.2
Recording Fields for Point Format 10 of LAS 1.4

Item	Format	Size
X	long	4 bytes
Y	long	4 bytes
Z	long	4 bytes
Intensity	unsigned short	2 bytes
Return Number	4 bits (bit 0–3)	4 bits
Number of Returns (given pulse)	4 bits (bit 4–7)	4 bits
Classification Flags	4 bits (bits 0–3)	4 bits
Scanner Channel	2 bits (bits 4–5)	2 bits
Scan Direction Flag	1 bit (bit 6)	1 bit
Edge of Flight Line	1 bit (bit 7)	1 bit
Classification	unsigned char	1 byte
User Data	unsigned char	1 byte
Scan Angle	short	2 bytes
Point Source ID	unsigned short	2 bytes
GPS Time	double	8 bytes
Red	unsigned short	2 bytes
Green	unsigned short	2 bytes
Blue	unsigned short	2 bytes
NIR	unsigned short	2 bytes
Wave Packet Descriptor Index	unsigned char	1 byte
Byte offset to waveform data	unsigned long long	8 bytes
Waveform packet size in bytes	unsigned long	4 bytes
Return Point Waveform Location	float	4 bytes
X(t)	float	4 bytes
Y(t)	float	4 bytes
Z(t)	float	4 bytes

Although the LAS format was specially designed for recording airborne LiDAR data, it is also perfectly good for recording the data produced by more recent, mobile LiDAR systems. However, ground-based LiDAR data are often stored in specially designed proprietary file formats, in order to take advantage of the particular features of the instruments and to aid in the subsequent data processing that occurs using on the software that came with the hardware. Such software usually allows advanced users to export the required attributes of the point data into a text file in order to carry out data processing using custom-developed software. In practice, ground-based LiDAR data may also be stored in LAS format, and the instrument-related data may be stored as VLRs in the related data file. However, users must know the format of the special VLRs in advance to correctly read and use the recorded data.

For the convenience of data management, airborne LiDAR data are usually stored in one file per strip, and ground-based LiDAR data are stored in one file per scan. However, the way in which mobile LiDAR data is stored is usually decided by the user.

11.3.2 Organization and Generalization of LiDAR Point Cloud

The data of scanned points in a LiDAR data file are originally saved in a sequence based on the scanning time. Retrieving a point cloud of a local area or a cluster of neighboring points can thus be very inefficient, especially when there is a large amount of data. Organization and generalization of point clouds are thus important practical issues for LiDAR data processing (Muhadi et al., 2020; Dong and Chen, 2017; Gargoum and El-Basyouny, 2017).

A scheme of spatial indexing is usually proposed for the organization of LiDAR data. Point clouds in a specified region of interest or a cluster of neighboring points at a location can be retrieved efficiently through the use of spatial indices. In order to visualize a huge LiDAR dataset efficiently, the spatial indexing approach should be extended to allow a dataset to be organized in a hierarchical, multi-resolution fashion. Furthermore, a database management system (DBMS) may be needed to handle the long-term collection of LiDAR data.

11.3.2.1 Spatial Indexing

The purpose of using spatial index of LiDAR point clouds is to establish the relationships among neighboring points, so that it is easier to search for and retrieve of points of interest in a large dataset. LiDAR data processing, such as feature extraction and DEM generation, often requires the derivation of meaningful information from the relationships among neighboring points or closely distributed points in a 3D space (Rivera et al., 2023; Mirzaei et al., 2022; Roriz et al., 2022; Xu et al., 2021). However, searching for specific points and their neighbors in a sequentially stored point dataset is a time-consuming and inefficient task. Some ground-based LiDAR systems store points in 2D array fashion. While this means that points obtained from adjacent laser pulses can be easily accessed from the file, these points may not be located near to each other in 3D space. Beside time concern, memory space of computer is another important issue. In addition, the computer being used to process this data may not have enough memory to load all the point data in a file to search and process it. However, these problems may be overcome if a spatial index for points is used.

Building a spatial index for a point cloud requires the assistance of spatial data structures, such as Triangulated Irregular Networks (TIN), trees and grids (Li and Ibanez-Guzman, 2020; Muhadi et al., 2020; Dong and Chen, 2017). The distribution of airborne LiDAR point clouds is similar to a 2.5D dataset, so TIN (Chen et al., 2006; Pu and Vosselman, 2006) and 2D grids (Chen et al., 2007) are often used to handle the neighborhood of points. While ground-based and mobile LiDAR point clouds have a 3D distribution, they are usually handled using 3D grids (or volume elements, known as voxels) (Bucksch et al., 2009; Gorte and Pfeifer, 2004; Wang and Tseng, 2011). Region quadtrees and octrees, which decompose the space of point distribution into regular subspaces, are commonly used to provide regular spatial indexing for data access and to determine the fixed distance neighbors (FDNs) of points. On the other hand, a point *k*-d tree is often used to determine the *k*-nearest neighbors (*k*-NNs) of points (Rabbani, 2006). In practice, because most grids are empty in a 3D grid for a ground-based or mobile LiDAR point cloud, a region octree structure may be used to record the 3D grids to reduce the storage space (Wang and Tseng, 2011).

11.3.2.2 Hierarchical Representation of LiDAR Data

Before starting the data processing task, it is always helpful for users to realize the contents of the LiDAR point clouds by visually inspecting both the outlines and details of the data. Software for viewing LiDAR data thus need the ability to efficiently display a huge amount points on different scales, with a small scale used for the outlines of the point cloud, and a large one for the details (Roriz et al., 2022; Xu et al., 2021). As with an image pyramid that provides hierarchically multi-resolution sub-images of the original image, a hierarchical representation of LiDAR point clouds can be achieved using a similar approach. At a grid-organized point cloud, a representative point, whose coordinates are the average of the points inside the grid, can be obtained for each grid. These representative points form the first level of generalization of the original point cloud, and the complete generalization hierarchy is then established level by level based on this. Figure 11.6 shows the example of hierarchical representation of a point cloud at different levels.

An additional index file is required to store the resulting generalized hierarchy and spatial index. Based on the viewing scale, representative points of a certain interest level can be loaded from the index file and displayed individually. To view the original point cloud at a large scale, the visible points inside the viewing window can be loaded efficiently through the spatial index stored in the index file. In practice, a 2D hierarchy can be simply saved like an image pyramid. In contrast, the 3D hierarchy organized with an octree structure can be saved as a file in the form of a linear tree (Samet, 1990).

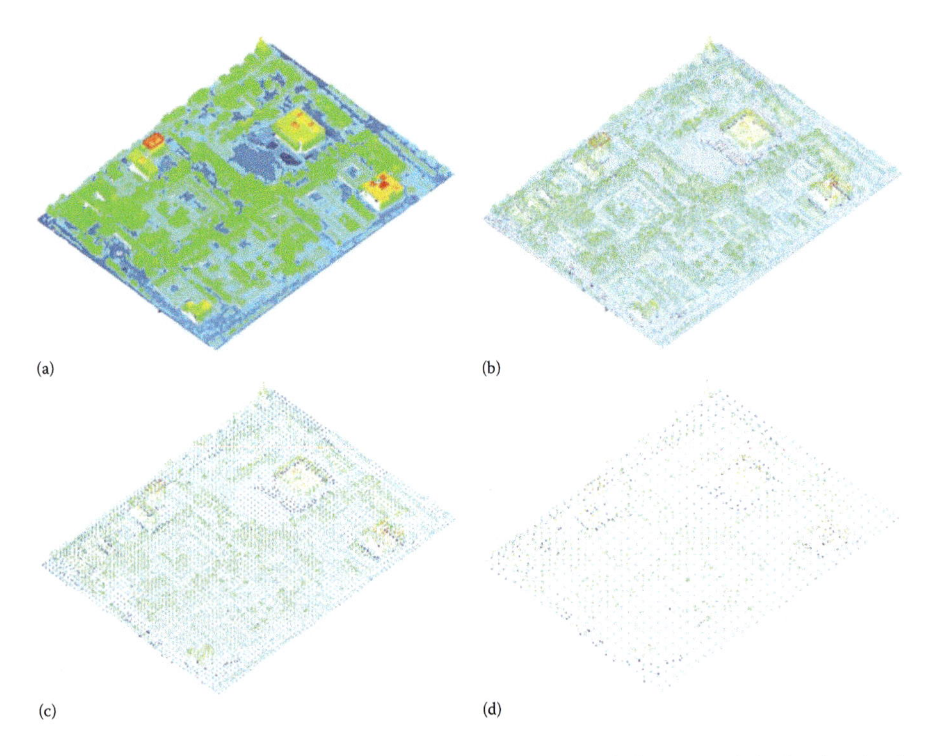

FIGURE 11.6 Results of point cloud generalization at different resolutions: (a) original point cloud, and (b) 4 m, (c) 8 m, (d) 16 m.

Organizing and generalizing a LiDAR point cloud requires extra time, and the results require additional storage space. However, this work only needs to be done one time, and it can then benefit all subsequent data processing and viewing.

11.3.3 LiDAR Database

As the use of LiDAR systems for collecting spatial data getting easier and popular, more and more LiDAR data are collected and accumulated. Thus, the management of the huge amount of LiDAR point clouds becomes an important and inevitable issue (Li and Ibanez-Guzman, 2020; Muhadi et al., 2020; Dong and Chen, 2017). The most intuitive way to manage such huge data is using the well-developed DBMS, which equips with many handy functions such as query, backup, and sorting for manipulating data.

There are two major different ways in using DBMS for point cloud management. The first method partitions point clouds into unified grids and stores the points in each grid separately. The metadata of each grid is also recorded as intermediate for accessing point data. The second method employs the spatial DBMS (SDMBS) to store point clouds. Each point is recorded individually as a spatial element in the SDBMS that makes allowance for optimization of LiDAR processing (Lewis et al., 2012). The generalized hierarchy of point clouds which used for visual inspection purpose can also be store at DBMS.

11.4 LiDAR POINT CLOUD FEATURE EXTRACTION

Being the points sampling from the object surfaces, LiDAR point clouds contain rich spatial information about the scanned targets. Feature extraction of LiDAR point clouds is the recognition of specific geometric shapes or more general smooth surfaces in the point data (Vosselman et al., 2004), and is the primary procedure to the interpretation of the contents of the point clouds. The purpose of LiDAR point cloud feature extraction is to improve the automation of identifying objects of interest and their characteristics from complicated LiDAR datasets. As LiDAR data is composed of discrete point clouds, and there are no relationships among neighboring points, conventional spectral, spatial (texture-based) and photogrammetric feature extraction methods designed for images may not produce satisfactory results if directly applied to the processing and analysis of LiDAR point cloud data. Several algorithms have thus been developed specifically for the feature extraction of LiDAR point clouds and full waveform LiDAR data. In addition, the appropriate method for extracting features from LiDAR data may vary significantly according to the targets of interest and the characteristics of the data. For example, the features required to reconstruct tree canopy models may be very different from those required for building model reconstruction. Similarly, full-waveform LiDAR data consist of features that are not available in conventional LiDAR point clouds.

The features embedded in LiDAR data may be derived from the following properties of LiDAR point clouds (Debnath et al., 2023; Guo et al., 2023; Roriz et al., 2022; Zhenlong et al., 2018; Cao et al., 2015):

- Location (x, y, z)
- Intensity
- Echo (return) number
- Waveform

The features that are derived or extracted may simply be points, lines, plans, surfaces, shapes or other geometric characteristics. On the other hand, the features may also directly represent objects of interest (such as a tree canopy, building model, traffic sign and the like) and physical properties of the objects.

In urban areas, airborne and terrestrial laser scanning (ALS & TLS) are effective data sources for city modeling, especially for the reconstruction of 3D building models. An important issue in building reconstruction from LiDAR data is to filter out ground and occlusion (such as trees) points. Several algorithms have been developed specifically for ground filtering of LiDAR data, and there are thoroughly reviewed in Meng et al. (2010). This filtering is often achieved by employing partitioning algorithms to segment the point clouds, and then collecting the segmentation attributes to classify the segments (Chen et al., 2008; Rabbani et al., 2006; Sampath and Shan, 2007; Vosselman, 2009; Zhang and Whitman, 2005).

In addition to 3D coordinates, LiDAR data also consist of intensity information. Some applications may treat the intensity data as images (after rasterization and resampling) and employ spectral or spatial image analysis algorithms to extract features. For example, a spectral analysis of LiDAR points was used to identify and map volcano lava flows (Mazzarini et al., 2007). Spectral and spatial analyses of LiDAR intensity data were also successfully applied to land-cover classification (Im et al., 2008), extraction of building footprints (Chen et al., 2008; Zhao and Wang, 2014) and other applications.

11.4.1 Spatial Features in LiDAR Data

Because of the blind operating manner of LiDAR systems, the spatial features of the scanned objects are implicitly contained in the point cloud (Muhadi et al., 2020; Dong and Chen, 2017; Gargoum and El-Basyouny, 2017; Lohani and Ghosh, 2017; Zhenlong et al., 2018). Before they can be used in GIS systems, for example, explicit and simple geometrical spatial elements like points, lines, and surfaces, must be extracted from the point clouds for object reconstruction and modelling.

The shapes of 3D objects, especially man-made ones, are composed of simple geometric elements, like points, lines, and surfaces. Being blind remote sensing instruments, the point and line features of objects are difficult to directly and accurately measure by LiDAR systems. In contrast, points distributed on the surfaces of such objects can aid in the extraction of surface features. Point and line features can then be obtained indirectly from the intersection of neighboring surface features.

Methods to extract different spatial features from LiDAR data should thus be designed according to the characteristics of the features of interest. Based on the extraction method used, the spatial features in LiDAR point clouds are classified into three categories, including fitting features, intersection features, and boundary features. Fitting features are obtained by fitting a point set to geometric elements, like surfaces, lines, and points. Intersection features are obtained from the intersection of neighboring surfaces or line features. Boundary features appear at the boundary of surface or line features. All three of these are described in more detail later.

11.4.1.1 Fitting Features

Fitting features are formed if points evenly distribute close to a specific geometric model, such as surfaces or lines. The least-squares estimation which minimizes the squared sum of the normal distance from each point to the selected geometric model is a reasonable method to extract fitting features from a point cloud. Least-squares estimation can be used to determine co-surface or co-linear features. The centroid of a point set, which is the center position of the points, can also be categorized as a particular fitting feature. However, this can be obtained directly by calculating the average coordinates of points, without the need for least-squares estimation. Figure 11.7 shows the basic ideas of fitting surfaces, fitting lines, and gravity centers.

11.4.1.2 Intersection Features

Intersection features are the intersections of existing features, which are usually fitting features. Fitting features should thus be extracted before obtaining intersection features. Intersection features

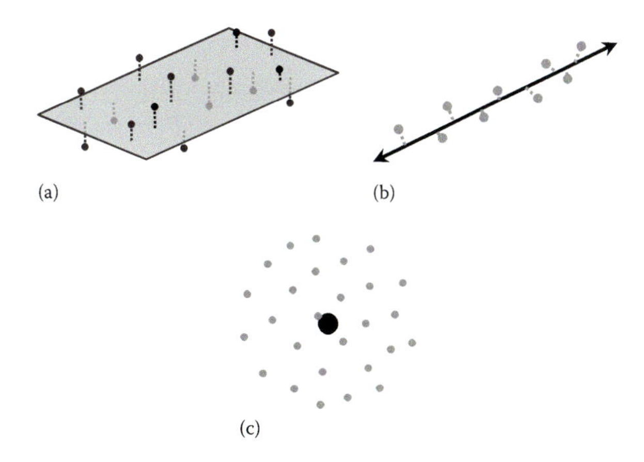

FIGURE 11.7 Basic ideas of (a) fitting plane, (b) fitting line, and (c) gravity center, of points.

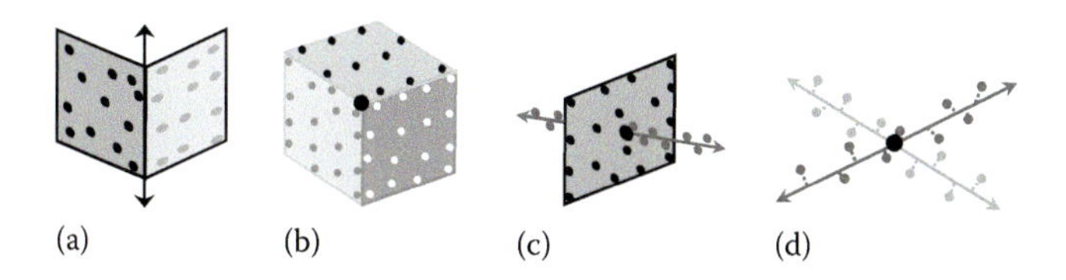

FIGURE 11.8 Basic ideas of intersection of (a) two surfaces, (b) three surfaces, (c) a surface and a line, and (d) two lines.

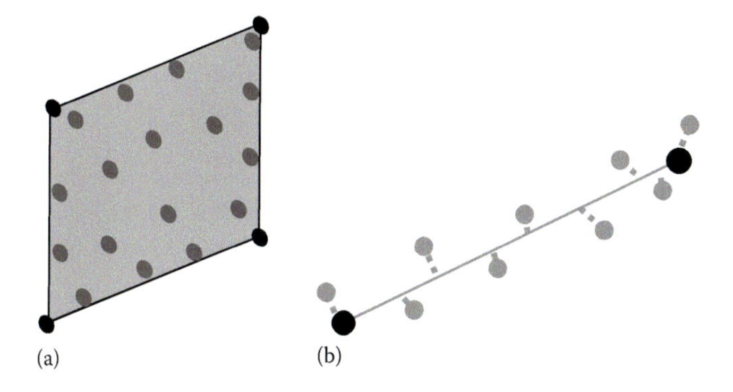

FIGURE 11.9 Basic ideas of boundary features: (a) boundary lines and points of a fitting plane, (b) end points of a fitting line.

include lines and points. Intersection lines can be generated by intersecting two non-parallel surfaces, while intersection points can be obtained by intersecting three non-parallel surfaces, a surface and a line, or two lines. Figure 11.8 shows the basic ideas of several cases of intersection features.

11.4.1.3 Boundary Features

Boundary features appear at the boundary of a point set. If the point set fits to a surface feature, the boundary lines and points form the boundary feature. If the point set fits to a line feature, the two end points of the line are boundary features. Therefore, fitting features must be extracted first, and then the boundary features are determined based on the information contained in the point set, as shown in Figure 11.9.

11.4.2 Methods for the Extraction of Spatial Features

LiDAR point clouds represent the outside appearances of the scanned objects. The shapes of some objects are simple, like pipelines in an industrial installation (Rabbani, 2006), and some are complex, like the terrain of the earth and buildings in a city. If the shape of an object, like cylinder or sphere, can be formulated with a simple equation, it can be extracted directly by determining its geometric parameters using the least squares fitting algorithm (Shakarji, 1998). Complex shapes that cannot be easily formulated are often disassembled into several plane features using segmentation algorithms. The objects are then reconstructed and modeled using the extracted features (Debnath et al., 2023; Guo et al., 2023; Kovanič et al., 2023; Rivera et al., 2023; Dong and Chen, 2017).

In the context of building reconstruction, lines and surfaces are fundamental elements to shape 3D building models. However, extracting line and surface features from LiDAR data is a difficult task, because the point data are usually randomly distributed in the 3D point clouds and lack any topological connections. There are two common approaches to extracting 2D or 3D line features from LiDAR data. The first is to extract 3D line segments directly from the point cloud using scale and rotation invariant point and object features, as demonstrated in Gross and Thoennessen (2006) and Brenner et al. (2013). The other is to intersect identified adjacent 3D planar surfaces to generate line segments (Habib et al., 2005). Since the inherit properties of the distribution of LiDAR points benefit the extraction of surface features, the rest of this section will discuss methods for surface, and especially planar surface, extraction.

11.4.2.1 Determining Feature Parameters Using the Least Squares Fitting Algorithm

The spatial features of simple shapes in LiDAR point clouds can be extracted using the model-based method that simply determines their geometric parameters. Because LiDAR points contain random noises, points are distributed close to the object surfaces with small undulations (Xu et al., 2021; Li and Ibanez-Guzman, 2020; Muhadi et al., 2020; Dong and Chen, 2017; Gargoum and El-Basyouny, 2017). The least squares fitting algorithm is often used to determine the geometric parameters. The object is to minimize the square sums of the normal distance between points and the shape. Linearization and iterated calculation are required if the equation of the shape is nonlinear. The linearization of some shapes can be found in Shakarji (1998).

Among the various shapes of geometric elements, planar surfaces are the simplest case of 3D surfaces. The determination of plane parameters for a point set using the least squares fitting algorithm can be transformed to the principal component analysis (PCA) of points. The solutions are then solved directly using eigensystem analysis, without iteration (Shakarji, 1998; Weingarten et al., 2004).

Two different types of strategies can be applied for curved lines and surfaces. The first is utilizing a semi-automatic, model-driven (or a combination of model- and data-driven) approach to compare point segments with known primitives based on the shapes or characteristic parameters. For example, a cylinder can be described by five parameters while a sphere can be defined by four parameters. However, directly analyzing data in the parametric space may be time consuming and the results may not be reliable. Therefore, processes for reducing the parametric dimensionality or incorporating additional information are often introduced to improve the performance of curved feature extraction. For example, extraction of a cylinder can be separated into two parts: cylinder axis direction and circle plane, while the extraction of a sphere can also utilize the normal vector and other constraints (Vosselman et al., 2004). One thing to note is that this type of feature extraction scheme usually requires a priori knowledge about the objects of interest.

The other strategy is to directly fit the points into parametric surfaces. Figure 11.10 demonstrates an example of constructing curved roof surfaces from TLS and ALS point clouds. Terrestrial laser scanning data acquired from multiple stations (Figure 11.10a) and airborne LiDAR point clouds (Figure 11.10b) were registered first. The roof boundaries were then generated as cubic spline curves using RANSAC and curve fitting algorithms. Finally, the roof surfaces were approximated as ruled

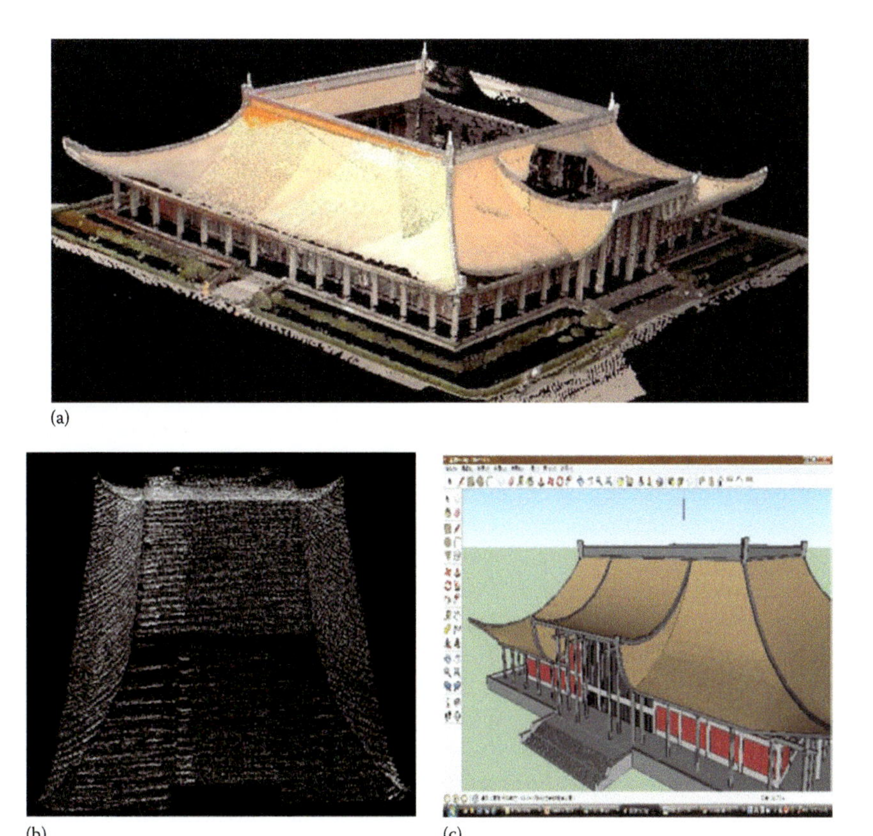

FIGURE 11.10 Parametric surface reconstruction from point clouds: (a) TLS data, (b) ALS point clouds, (c) curved roofs constructed as ruled surfaces.

surfaces from the boundary curves (Figure 11.10c). The advantage of this type of approach is that it is flexible and can deal with complicated shapes and objects, although usually with simplified results.

The features within LiDAR point clouds may be implicit and difficult to extract directly. Therefore, incorporating additional information in the feature extraction process by fusing LiDAR point clouds with other datasets has also been proposed, with successful results reported in different applications. Pu and Vosselman (2009b) combined TLS point clouds with images to reconstruct detailed building facade models. Integration of ground plans or large-scale vector maps and ALS data was developed to reconstruct polyhedral building models (e.g., Vosselman and Dijkman (2001); Chen et al. (2008)). Similar techniques were also applied to generate detailed 3D road models (Chen and Lo, 2009). The idea of the data fusion approach is to take advantage of the explicit information (building layouts, road boundaries and so on) in images and vector datasets to help extract corresponding features from LiDAR point clouds.

11.4.2.2 Extraction of Plane Features Using Segmentation Algorithm

In most cases the appearances of objects are too complex to be formulated by a simple equation. Reconstruction of these objects from LiDAR point clouds requires disassembling the surfaces of an object into several simple geometric elements, and then modeling the object using the extracted results. Many studies use plane features as the primary geometric elements for reconstructing 3D objects from LiDAR point clouds, for the following reasons (Muhadi et al., 2020;

Dong and Chen, 2017; Gargoum and El-Basyouny, 2017; Lohani and Ghosh, 2017; Zhenlong et al., 2018; Cao et al., 2015):

1. The distribution of LiDAR point clouds benefits the extraction of surface features.
2. A general surface is too complex to model using a mathematical function.
3. The shapes of most artificial objects are composed of planar surfaces.
4. Detecting the lines and vertices of structures is not trivial, due to the scanning mechanism of LiDAR systems.
5. The lines and vertices of structures can be obtained from the intersections of neighboring planes.
6. General surfaces can be obtained from the union of neighboring planar patches.

Although determining a plane feature for a point set is easy, as described previously, automatic extraction of all possible plane features from a large amount of points is difficult, and is not a trivial task. Segmentation, a data-driven method, is the most popular approach for the extraction of plane features from LiDAR point clouds (Filin and Pfeifer, 2006; Rabbani et al., 2006; Sithole, 2005; Wang and Tseng, 2010). The strategy of partitioning or segmenting point clouds and extracting linear and planar features has been proved to be an effective approach for identifying building parts, such as roof facets and facades. For example, after segmenting mobile TLS point clouds into planar faces, vertical wall features can be further extracted according to the inclination angles of the segmented planar faces (Rutzinger et al., 2009). Secondary features, such as windows, doors, and curtains, can also be extracted successfully and accurately based on the sizes, relative positions, orientations and other characteristics of the segmented planar surfaces (Pu and Vosselman, 2009a).

Segmentation algorithms deal with the coherence and proximity of points (Melzer, 2007). In other word, points that are distributed closely and have similar geometric properties are grouped together by the segmentation algorithm. Coplanarity is used as the coherence property of points for the extraction of plane features. The segment results are groups of coplanar points, each of which represents a plane feature.

Many successful segmentation algorithms have been developed for extracting spatial features form digital images (Gonzalez and Woods, 1992), and the segmentation algorithms for extracting spatial features from LiDAR point clouds are mainly adopted from these methods. Based on the strategy used to deal with the coherence and proximity of the points, LiDAR point cloud segmentation algorithms can be classified into three categories: clustering, region growing and split-and-merge.

Clustering algorithms perform at the attribute space to group points of similar properties into clusters (Filin, 2004). The normal vector of the plane is the necessary attribute for plane feature extraction. Some algorithms also employ additional attributes to raise the success rate. Because the grouping of points does not consider the distance between points, the points in each of the clusters have to be separated into neighbor point groups according to the proximity criterion. The tensor voting algorithm (Schuster, 2004) and the 3D Hough transform (Vosselman et al., 2004) are two types of clustering algorithm.

Region growing algorithms (Hoover et al., 1996; Rabbani, 2006; Vosselman et al., 2004) start with the selection of a set of coherent and neighboring points as the seed of a point group. The neighboring points of the seed are added to the point group one by one if they satisfy the coherence criterion, until no more neighboring points can be added to the group. Then, another seed is selected and the growing procedure continues until all points are processed.

Split-and-merge algorithms (Wang and Tseng, 2010) include two parts: the split and merge processes. The split process starts with the examination of the coherence of all points of the point cloud. If the points cannot satisfy the coherence criterion, the space is split into eight equal-size subspaces. The same procedure is then performed on the points contained in each subspace, until all points satisfy the coherence criterion. During the merge step the points contained in neighboring subspaces are merged if they satisfy the coherence criteria. The merge procedure then continues until no more points can be merged and the segmentation process is complete. In this method the proximity criterion of points is used during both the split and merge processes.

The profile segmentation algorithm (Sithole, 2005), which is adopted from the scan line segmentation algorithm designed for range image segmentation (Jiang and Bunke, 1994), is a special kind of split-and-merge algorithm. In this method, the space of a point cloud is sliced into connected or cross profiles in different directions. Each profile is treated as a scan line, and the adopted scan line segmentation algorithm is performed on the contained points to obtain collinear point groups. The collinear points at neighboring profiles are then merged to form coplanar point groups.

After the plane features are extracted, line and point features can be obtained by the intersection of neighboring plane features. With some constraints, for example, the included angle, neighboring planes can also be merged to form curved surfaces. However, a general curved surface, like a terrain, is difficult to express using a simple equation. TIN meshes are thus used with most GIS software to represent general curved surfaces.

11.4.2.3 Extraction of Line and Plane Features Using the Hough Transform

The Hough Transform is a classical method of surface extraction. Points belonging to a 2D straight line can be represented as a group of (r, θ) values in the Hough space, where r is the distance between the line and the origin, and θ is the angle of the vector orthogonal to the line and pointing toward the half upper plane. Similarly, points on a 3D planar surface can be described as a collection of (ρ, θ, φ) in a 3D Hough space represented in spherical coordinates (Equation 11.1).

$$(\theta,\phi) \rightarrow \rho = cos\theta cos\phi x + sin\theta cos\phi y + sin\phi z \quad (11.1)$$

11.4.2.4 Extraction of Plane Features Using RANSAC

Another popular approach for plane extraction from LiDAR point clouds is to employ the RANdom SAmple Consensus (RANSAC) algorithm, which was proposed by Fischler and Bolles (1981). RANSAC is a resampling technique to estimate model parameters, and is designed to deal with datasets containing a large portion of outliers. RANSAC starts with the smallest set of the data and proceeds to enlarge this dataset with consistent data points (Fischler and Bolles, 1981). A typical RANSAC-based algorithm for extracting a planar feature from a LiDAR point cloud is outlined in Algorithm 1.

Algorithm 1 RANSAC for plane extraction

1. Randomly select three points.
2. Construct the plane model (solve the parameters of the plane equation).
3. Calculate the distance of a point to the plane, d_i , for all points.
4. Find inliers (points whose d_i is less than a predefined threshold).
5. If the ratio of the inliers to the total number of points is greater than a predefined threshold, reconstruct the plane based on all the identified inliers and terminate the process.
6. Otherwise, repeat steps 1 through 5 until reaching the maximum number of iterations, *N*.

The final plane reconstruction (step 5 in Algorithm 1) is usually based on a least-squares estimation with all identified inliers. The maximum number of iterations, *N*, should be high enough that there is a probability *p*, which is usually set as 0.99, that at least one set of randomly selected samples does not include any outlier. Let *u* be the probability that a selected point is an inlier and *v=1-u* is the probability of an outlier, then

$$1-p=\left(1-u^3\right)^N \quad (11.2)$$

and

$$N = \frac{log(1-p)}{log\left(1-(1-v)^3\right)} \tag{11.3}$$

The disadvantage of the classical RANSAC algorithm is that it may be very time consuming. However, limiting the selection of samples from pre-segmented regions of point clouds will significantly improve the efficiency of RANSAC process. For example, when extracting the roof facets or wall facades of a building, the selection of points should be focused on regions with the same normal orientation.

In conclusion, feature extraction is an essential step of LiDAR processing and analysis. The features of interest vary with different targets and applications. From the feature extraction point of view, LiDAR point clouds provide abundant information, and the methods used for extraction are as important as the data itself. The algorithm chosen depends on the characteristics of the data, the objects of interest, and the aim of the application. There is unlikely to be a single method or piece of software that can adequately address the varied needs of all users, and thus it is necessary to explore the possibility of different algorithms to identify the appropriate methodology for the feature extraction of LiDAR point clouds, as well as data fusion with other geospatial datasets.

11.5 3D CITY MODELING FROM LiDAR DATA

11.5.1 Properties of LiDAR Data in a City Area

The importance of 3D city modeling is increasing due to rapid urbanization and the need for accurate 3D spatial information for urban planning, construction, and management (Kovanič et al., 2023; Rivera et al., 2023; Dong and Chen, 2017; Gargoum and El-Basyouny, 2017). City modeling is mainly based on images and LiDAR point clouds. The uniqueness of the data characteristics indicates different perspectives for city modeling. Image sensors provide spectral information that can be used to derive the well-defined 3D corner and 3D linear features which are implicit in stereo images. LiDAR systems provide abundant 3D shape information for reconstruction of city model. Figure 11.11 compares the images of a building taken by an image sensor and produced by a LiDAR system. The ridge line is over-saturated in the image, while the ridge line can easily be distinguished from shaded LiDAR triangles. The benefits of using LiDAR point clouds for city modeling include: (1) high vertical accuracy for even low texture surfaces; (2) the ability to directly and accurately obtain the 3D shape information; (3) the ability to establish the non-planar surfaces of objects in a city; and (4) it is able to provide different viewpoints (e.g., top-view and front-view) for object interpretation. LiDAR thus has great potential with regard to producing 3D spatial information for a city model.

Airborne LiDAR acquires data from the air to the ground and is used to obtain the 3D points on building rooftop and object surface. On the other hand, terrestrial LiDAR usually acquire the 3D points on building façade and object surface. The scanning distance and beam divergence angle of airborne LiDAR is larger than terrestrial LiDAR and consequently the point density of airborne LiDAR is lower than terrestrial LiDAR. Therefore, airborne LiDAR is more suitable for city-scale object reconstruction while terrestrial LiDAR is suitable for building-scale detailed object reconstruction. For example, the OGC CityGML (OGC, 2012) LOD1 and LOD2 city objects can be reconstructed by airborne LiDAR while the terrestrial LiDAR is usually used for LOD3 and LOD4 city objects.

11.5.2 Object Reconstruction Strategies

Object reconstruction strategies can be classified into three categories, that is, model-driven, data-driven, and hybrid approaches (Brenner, 2005). The model-driven approach is a top-down strategy

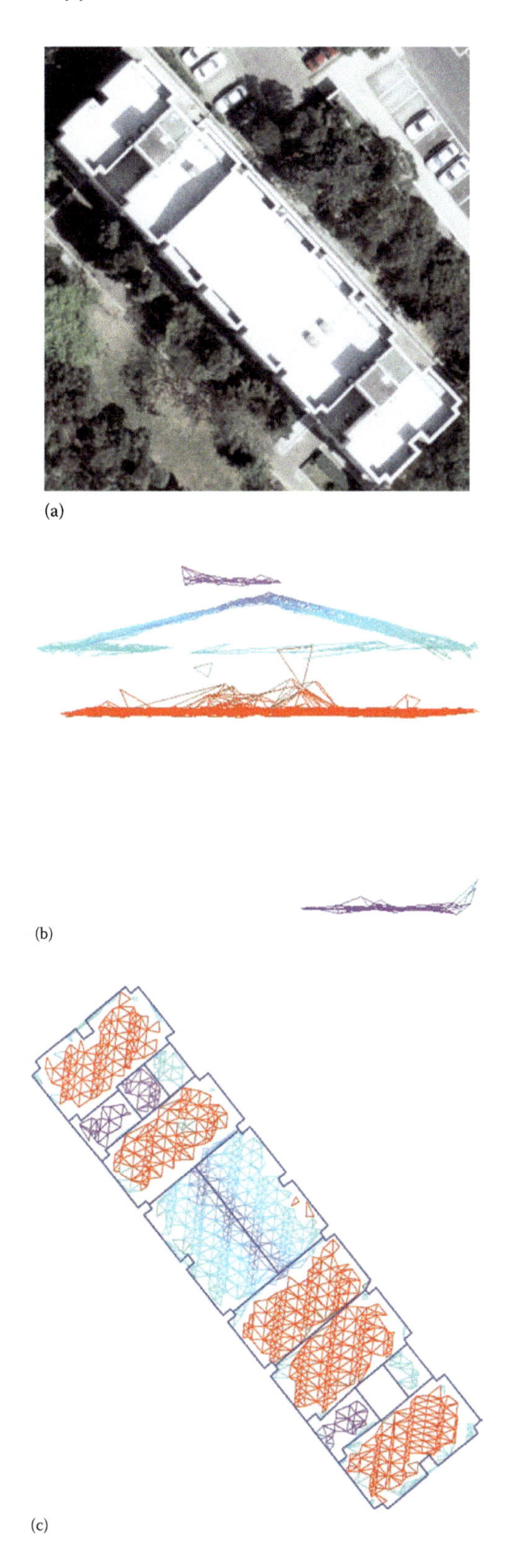

FIGURE 11.11 An example of a building ridge line in an aerial image and airborne LiDAR point clouds: (a) aerial image, (b) LiDAR triangle mesh, and (c) horizontal view of LiDAR triangle mesh.

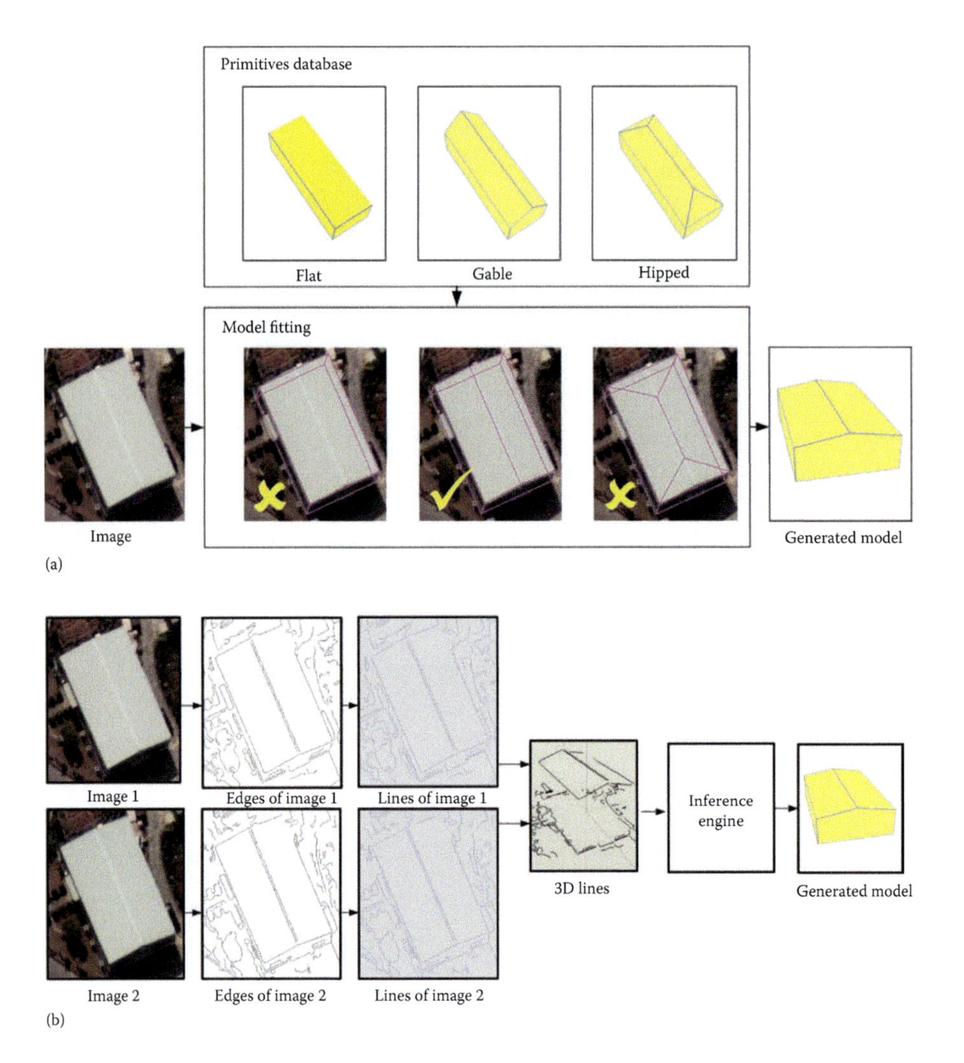

FIGURE 11.12 Illustration of (a) model-driven and (b) data-driven strategies.

that starts with a hypothetical building model which is verified by the consistency of the model with the existing data. This method needs to define a database of building primitives, and thus a parametric model is generally used, as shown in Figure 11.12a. The data-driven approach is a bottom-up strategy in which the building features, such as point, linear and planar features, are extracted at the beginning, then grouped into a building model through a hypothesizing process, as shown in Figure 11.12b. Finally, the hybrid approach integrates the ideas of both model-driven and data-driven methods

11.5.3 Building Extraction

Buildings are the most attractive elements in a 3D city. There are three major steps to establish a building model, and these are detection, reconstruction and attribution (Gruen, 2005). Building detection is composed of various methods to detect regions of interest for subsequent building reconstruction. Building reconstruction is the determination of the 3D geometrical description of buildings located in a given region of interest. Finally, building attribution assigns the depiction of building properties, such as type of building, semantic attribution, textures, and so on. All three steps are described in more detail later.

11.5.3.1 Building Detection

The role of building detection is to extract the location of regions where there are buildings (Guo et al., 2023; Dong and Chen, 2017). Once a building's location is established, the reconstruction process can be focused in a specific region rather than considering the whole dataset. This advantage not only saves computing time, but also reduces ambiguity. The idea of building detection is carried out using different characteristics, such as spectral, texture, shape, roughness and others, to separate buildings and non-buildings. A general procedure to obtain a building region from LiDAR includes the following steps: (1) generate aboveground objects by subtracting the digital terrain model (DTM) from the digital surface model (DSM); (2) calculate the different features for each object. An object's features can be shape, texture, roughness, echo ratio, and so on; and (3) separate building and non-building regions based on the object's features using a classifier.

If LiDAR point clouds are combined with a multispectral image, then building regions can be detected by simultaneously considering spectral and shape information. The detection rate may reach 80% when the automatic approach is adopted, although this figure could be worse if the point cloud density is too low or the area of the building is too small. For example, (Rottensteiner et al., 2005) used LiDAR data with an average point distance of 1.2 m and multispectral images with 0.5 m spatial resolution to perform building detection. For a building larger than 40m^2 the detection rate was between 50~90%, but for a building smaller than 40 m^2 the detection rate was lower than 50%. Point density and building size are two major factors that will influence the detection rate. Increasing point density may thus improve the accuracy of building detection. Figure 11.13 shows an example of building detection using a LiDAR and multispectral image.

11.5.3.2 Building Reconstruction

The objective of building reconstruction is to build up the geometry of a building. The irregular point clouds need to be structuralized for further processing. There are different data structures for LiDAR data in building reconstruction. The first one uses point data to perform the 3D Hough transform to extract the building models from LiDAR point clouds (Hoffman, 2004). The second one applies TIN data to analyze the planar parameters (Chen et al., 2008). The third one groups and linearizes the building models based on grid data.

The general methods used for building reconstruction based on LiDAR data can be classified into two types. In the first type, planar features are extracted and then the extracted planes are used to derive the line features (Sampath and Shan, 2007). The building model is obtained by integrating the planar and linear features. In the second type, linear features are first extracted and then the extracted lines are used to trace the building polygons. The building model is obtained by shaping the top of the building polygons (Hu, 2003). Each approach has less planimetric accuracy when a regular density (e.g., 1~2 pts./m^2) is employed. Figure 11.14 shows examples of building reconstruction using LiDAR contours and the boundary regularization method.

11.5.3.3 Building Attribution

A building model is an object in a geographical information system (GIS). It consists of spatial information and attributes. Building reconstruction is used to shape the spatial information of a building, while building attribution is used to assign the attributes of a building model. Building attributes enable the 3D analysis to produce more valuable results. There are three kinds of attributes. The first can be generated from the data itself, like floor number, area and volume. The second are the semantic attributes, which can be obtained from exterior data, like a topographic map. Semantic attributes include the names of buildings, materials of buildings, and so on. The third attributes are based on the building texture, which make the building model more photo-realistic.

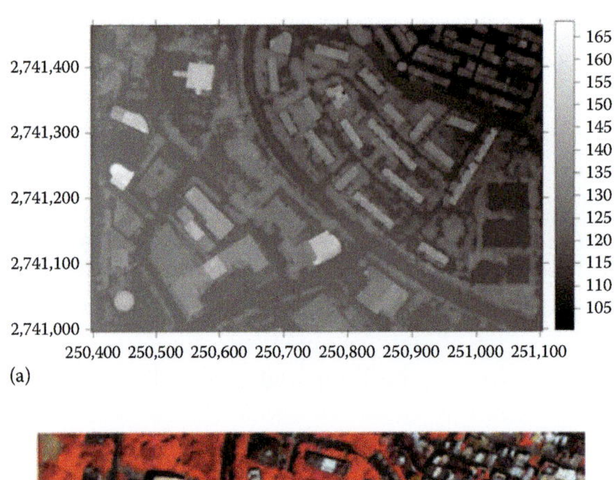

(a)

(b)

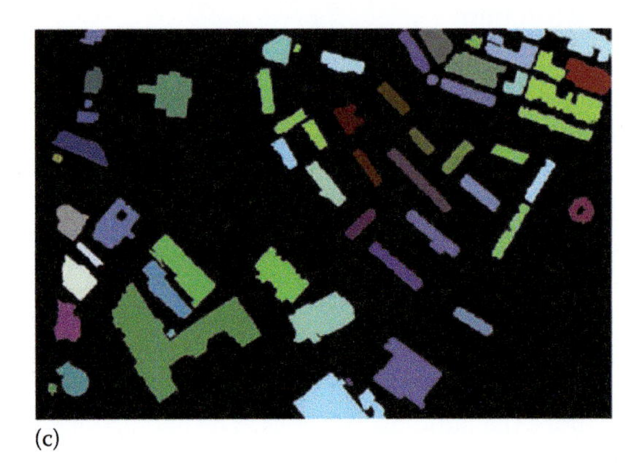

(c)

FIGURE 11.13 An example of building detection: (a) LiDAR digital surface model, (b) multispectral image, and (c) detected building regions.

11.5.4 Road Extraction

A 3D road network is one of the important infrastructures needed for an intelligent transportation system (ITS) and GIS, and these are applied to transportation management, maintenance, planning, analysis, and navigation. The traditional 2D road centerlines and 2D road boundaries are insufficient to represent the 3D reality of actual road systems, especially multi-layered ones. Since LiDAR data may provide a large number of 3D points on the road surface, these can be used to generate 3D road models (Oude Elberink and Vosselman, 2009). Road extraction also includes three major steps, that is, detection, reconstruction and attribution. Road detection is used to detect the 2D road

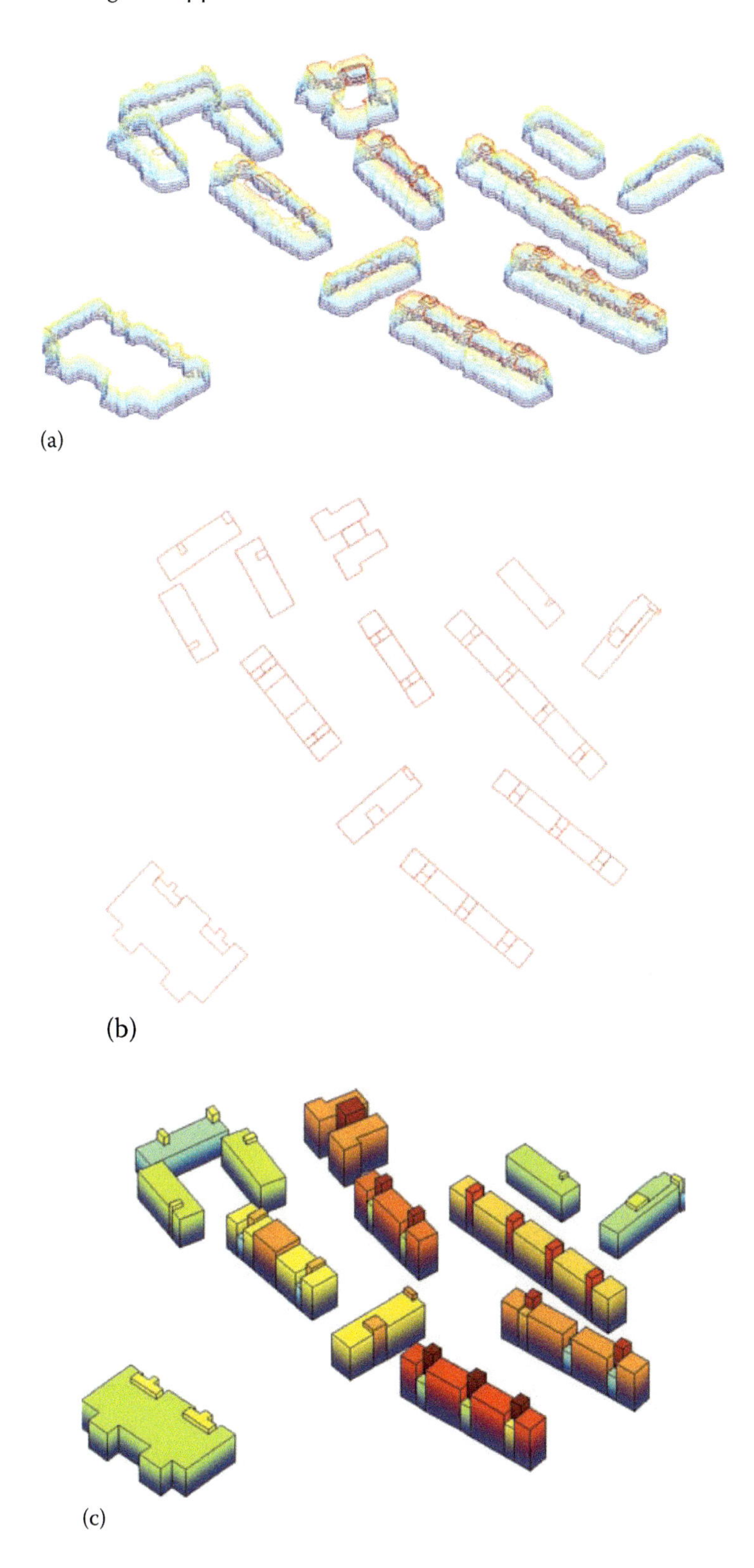

FIGURE 11.14 Examples of building reconstruction: (a) building contours, (b) building boundaries after regularization, and (c) building models.

region, while road reconstruction is to shape the 3D road surface. Road attribution is then used to assign the attributes to the road models.

Road detection from LiDAR data usually analyzes the LiDAR intensity, flatness and continuity. The near infrared of the LiDAR system may have a low return signal in asphalt-covered region, and this is an important attribute that can be used to detect roads. The road design should meet the related regulations, which means the flatness and slope of roads can be estimated to detect road regions. Finally, the continuity of a road network is another constraint that should be considered (Hu, 2003). Note that LiDAR points are irregular points which do not always model the step edge or roadside. The road boundaries extracted from LiDAR points are thus not as accurate as the road boundaries obtained from vector maps. An alternative approach to road detection is adopting the reliable 2D road boundaries from existing topographic maps, with LiDAR points then used for road surface shaping.

In road reconstruction, the major tasks are surface modeling and making connections between road segments (Chiu et al., 2013). Once the road regions are detected, the 3D road surface modeling process will focus on them. A general procedure of 3D road surface modeling includes the following steps: (1) remove the non-road points based on road profile fitting; (2) carry out 3D road surface fitting by using polynomial functions; and (3) check the consistency of 3D road segments. The main challenge in road reconstruction is the shaping of multilayer roads. Because of the occlusion of multilayer roads, an inference engine is needed to restore the missing parts under the upper roads. Figure 11.15 shows an example of multilayer road extraction using LiDAR points (Chen and Lo, 2009).

FIGURE 11.15 An example of road extraction: (a) LiDAR points, (b) road regions, and (c) road models.

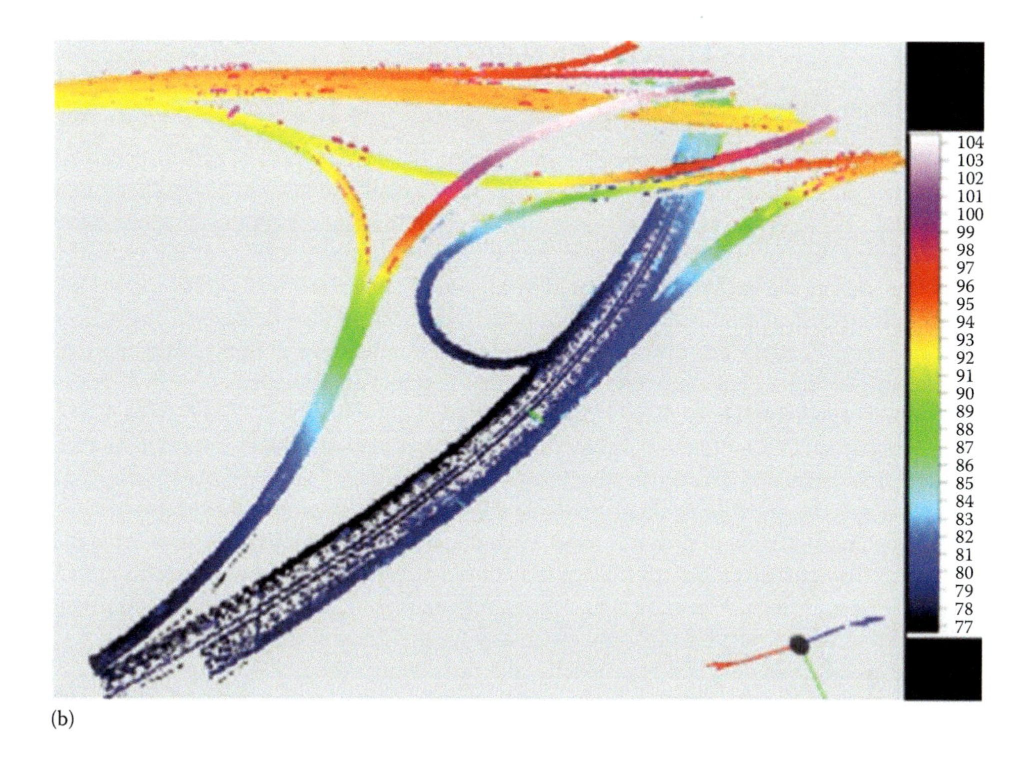

(b)

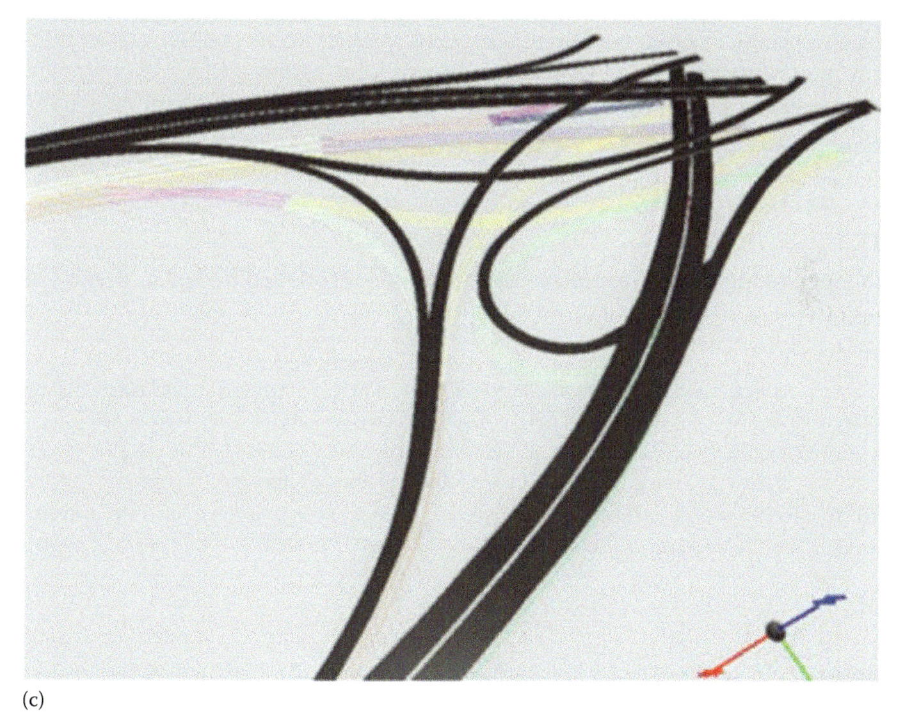

(c)

FIGURE 11.15 (Continued)

11.6 FULL-WAVEFORM AIRBORNE LiDAR

11.6.1 Introduction

Recent advances in laser scanning technology have led to the development of new airborne LiDAR commercial systems, called waveform LiDAR systems (Lohani and Ghosh, 2017; Zhenlong et al., 2018; Cao et al., 2015). A waveform LiDAR system is capable of recording the complete waveform of a return signal (Mallet and Bretar, 2009), which is the digitized intensity of the backscattered signals obtained from the surfaces illuminated by an emitted laser beam. In addition to range measurements, other physical features of the illuminated surfaces can also be derived from the waveform data (Guo et al., 2011). For example the echo width, amplitude, and backscatter cross-section obtained from waveforms have proven to be useful for object classification of urban (Alexander et al., 2010) and vegetation (Heinzel and Koch, 2011; Neuenschwander et al., 2009) areas. Compared with conventional LiDAR systems, a waveform scanner provides additional information about the physical characteristics and geometric structures of the illuminated surfaces. A waveform dataset can be used to generate finer point clouds than the dataset originally provided by the system, so that it improves the interpretation of physical measurements. It is also possible to retrieve some missing points by detecting surface responses known as echoes in waveforms, and the detection process is especially useful with regard to weak and overlapping echoes. This ability is very significant for some applications of airborne LiDAR, and especially for the study of forest areas.

With improvements in data storage capacity and processing speed, waveform airborne LiDAR systems have become increasingly available, and first appeared on the commercial market in 2004 (Hug et al., 2004). Table 11.3 shows the technical specifications of waveform airborne LiDAR systems manufactured by Leica, Riegl, Optech, and Trimble. Such systems store the entire waveform of each received pulse, and thus the sum of reflections from all intercepted surfaces within the laser footprint is preserved. Compared to discrete-return airborne systems, which only provide a single range distance to a target, waveform systems record the entire time history of laser pulses with a

TABLE 11.3
Technical Specifications of Waveform Laser Scanners Produced by Leica, Riegl, Optech and Trimble

Manufacturer	Leica ALS70	Leica ALS60	Riegl LMS Q780i	Riegl LMS Q680i	Optech ALTM Pegasus	Optech ALTM 3100	Trimble Harrier 68i	TrimbleHarrier 56
Beam deflection	Oscillating mirror	Oscillating mirror	Rotating polygon	Rotating polygon	Oscillating mirror	Oscillating mirror	Rotating polygon	Rotating polygon
Flying Height (m)	1600~5000	200~5000	<4700	<1600	300~2500	<2500	30 ~ 1600	<1000
Laser wavelength (nm)	1064	1064	1064	1550	1064	1064	1556	1550
Pulse width (ns)	5 ~ 6	5	3	3	7	8	3	< 4
Pulse Rate (kHz)	250 ~ 500	<50	100 ~ 400	<400	<400	<200	<400	<240
Beam divergence (mrad)	0.22	0.22	0.25	0.5	0.2	0.3 or 0.8	≤0 .5	0.3 or 1.0
Field of view (degrees)	0 ~ 75	75	0 ~ 60	±30	75	±25	45–60	45 or 60
Sampling interval (ns)	1	1	1	1	1	1	1	1

high-resolution sampling interval. Moreover, such waveforms are able to store more details about illuminated targets than discrete systems are. Users can then apply their own pulse detection methods to identify return signals in a more effective manner. The waveform attributes, which represent characteristics of each scanned object, can then be extracted and used as criteria for distinguishing the points into different types.

11.6.2 Waveform Data Analysis

The waveform data record a series of return signals related to each individual laser shot. The data sampling interval is usually 1 nanosecond, which corresponds to a 15 cm interval along the path of the laser beam. The intensity of the backscattered signal is then quantified within a range of specified digital levels (e.g., 8 bits), so that it is finally converted to a digital data stream. In general, it is understood that the higher the sampling rate and the signal-quantified levels are, the more details of the raw analogue signal that are kept.

As the waveform data contain valuable information about the object distribution and reflectance along the laser path, a specially designed process of waveform data analysis is needed to extract the information contained in the waveform data (Debnath et al., 2023; Guo et al., 2023; Kovanič et al., 2023; Rivera et al., 2023; Mirzaei et al., 2022; Roriz et al., 2022; Xu et al., 2021; Li and Ibanez-Guzman, 2020; Muhadi et al., 2020; Dong and Chen, 2017; Gargoum and El-Basyouny, 2017; Lohani and Ghosh, 2017; Zhenlong et al., 2018; Cao et al., 2015). Echo detection and waveform feature extraction are two common processes of waveform data analysis, which will both be introduced in this section. However, before this the shape of the received waveform is first discussed.

11.6.2.1 Shape of Received Waveform

As the shape of a received waveform varies subject to differences in the target structures (e.g., the distribution of illuminated objects and surface reflectance along the laser path), it is necessary to study the types of waveforms caused by some typical target structures. Based on the size of the focused surface geometry within the laser beam (footprint: d, wavelength: λ), Jutzi et al. (2005) categorized illuminated targets into three types, which were macro, meso, and micro structures. Macro structures are considered to have a more extended surface than the footprint d (e.g., building roofs). Meso structures are considered to have a less extended surface than the footprint d, but much greater than wavelength λ. Such structures result in a mixture of different range values, for example in an area with small elevated objects, a slanted plane, large roughness or vegetation. Finally, micro structures are considered to have a surface less extended than the wavelength λ. It is also noted that the shapes of the received waveforms vary depending on different surface structures illuminated, especially for meso structures.

Jutzi and Stilla (2006a) further examined the relationship between surface structures and the shape of waveforms (Figure 11.16). A sloped surface or an area with randomly distributed small objects causes the deformation (widening) of the backscattered waveform with one distinct peak (Figures 11.16b and e). With an increase in the height difference between two elevated surfaces, the two peaks of a waveform are increasingly close together (Figure 11.16d).

Moreover, Wagner et al. (2006) pointed out that when the two scattering clusters are at a distance comparable to or smaller than the range resolution, which is the pulse duration of the LiDAR system, a waveform with a scattering cluster is presented. Jutzi and Stilla (2006a) claimed that the waveform of overlapping responses at the distance of ≤0.85 times pulse width become a single widened peak, and is unlikely to be separated as two returns by peak detection. It is also apparent that when considering the effects from various reflectance properties of multiple returns within the travel path of the laser pulse, the deformation of the received waveforms will be more complex than for those affected by surface geometry alone. As a result, since reflected laser pulses are distorted by surface variations within the footprint, the shapes of received waveforms important information in relation to surface roughness, slope, and reflectivity (Gardner, 1992).

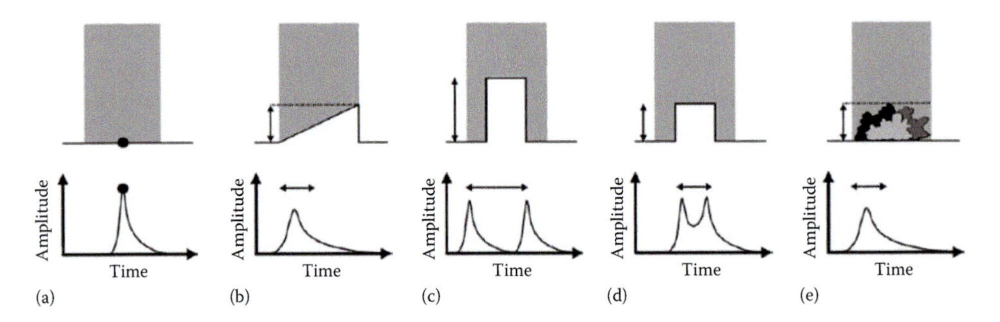

FIGURE 16.16 Surface structure and waveform: (a) plane surface, (b) sloped surface, (c) two significantly different elevated areas, (d) two slightly different elevated areas, and (e) randomly distributed small objects. (Adapted from Jutzi and Stilla, 2006a.)

11.6.2.2 Echo Detection

As full-waveform systems provide an entire time history of laser pulses, users are able to apply their own pulse detection methods to detect successive returns and determine the elapsed time. It even becomes possible to design a combined pulse detection method becomes possible to extract information of interest for specific applications (Wagner et al., 2004). For this purpose, Harding (2009) pointed out that the laser pulse width, detector sensitivity and response time, the system's signal-to-noise performance, detection threshold, and implementation of the ranging electronics all influence the ability to detect discrete returns. Wagner et al. (2004) reported that the characteristics of the effective scattering cross section, object distance and noise level can affect the performance of the detectors. For received pulses with a complex form, the number and arrival timing of the returns are critically dependent on the detection methods employed.

Many echo detection algorithms are available for waveform data. Wagner et al. (2004) compared the results of using methods based on the threshold (TH), center of gravity (CG), maximum (MA), zero crossing of the first deviation (ZC), and constant fraction (CF). It was found that the ZC algorithm has the best discrimination with regard to detecting overlapping echoes among the five algorithms, especially when the noise is at minimum. The CF algorithm achieved worse results than the ZC algorithm, while the TH, CG, and MA algorithms failed at resolving overlapping echoes.

Current state-of-the-art echo detection techniques use small-footprint full-waveform data, for example, the correlation (Levanon and Mozeson, 2004), Deconvolution (Jutzi and Stilla, 2006a), wavelet-based and Gaussian decomposition methods. Since the waveforms can be considered as a convolution between system waveform and apparent cross-section of scatters (Wagner et al., 2006), the surface response (cross-section of scatters) can be recovered by deconvolution algorithms (Jutzi and Stilla, 2006b; Wu et al., 2011). However, prior knowledge such as the original laser pulse shape needs to be known, and sampling of the return signal also needs to be sufficient for such approaches to be applied, and for many of today's LiDAR systems, these prerequisites are not met (Morsdorf et al., 2009). The wavelet-based (WB) echo detector is based on a wavelet transformation algorithm, which measures the similarity between the signal and the wavelet. A wavelet (i.e., a Gaussian wavelet for detecting echoes of LiDAR waveforms) is thus chosen if its shape is similar to that of the echoes being detected. The wavelet can be scaled by a scale factor and shifted along the signal time domain by a translation factor, and so echoes can be located by detecting significant peaks in the wavelet coefficients (the results of wavelet transform). Gaussian decomposition is widely used to model small-footprint waveforms. With this method, it is assumed that pulses are transmitted with a Gaussian-like distribution (i.e., the impulse response is Gaussian) and that the received signal is a sum of individual Gaussian distributions (Hofton et al., 2000; Wagner et al., 2006). Fitting Gaussian functions to waveform data provides each return with a parametric description which can be used to store pulse shape information and decrease the effect of noise. In this way the Gaussian parameters for each return, which are the amplitude, temporal position of Gaussian peak and Gaussian width, can be extracted.

The Gaussian model is a classical curve fitting algorithm (Wagner et al., 2006) as described in Equation (11.4),

$$f_j(x_i) = A_j exp\left(-\frac{(x-\mu)^2}{2W_j^2}\right) \tag{11.4}$$

where A_j and W_j are the pulse amplitude and width of the laser pulse; and μ_j is the center of the echo. Roriz et al. (2022) proposed fitting waveforms into higher order spline curves to better describe the laser waveform patterns. However, high-order spline curve fitting requires more intensive computation, and waveform fitting is usually based on the peak locations of the echo signals. However, since most full-waveform laser scanners provide only a limited number of signal peak locations, certain subtle but useful waveform features may be overlooked. Tsai and Philpot (1998, 2002) proposed derivative analysis to detect the occurrence of local peak positions more accurately, and to generate a more complete waveform of the returned laser pulse.

11.6.2.3 Waveform Feature Extraction

With the development of full-waveform LiDAR, waveform related characteristics have also become valuable features for LiDAR analysis and applications. Once the waveform has been decomposed, related waveform features can be derived from the results. Typical surface features extracted from Gaussian parameters are range, roughness, and reflectance (Jutzi and Stilla, 2005), and thus the estimated parameters provide a direct physical interpretation of the surface targets. Additionally, Wagner et al. (2006) adapted the radar equation order to convert the received power into a backscatter cross-section. As this is a measure of how much energy is scattered backwards towards the sensor, and is very useful for comparing the physical quantities between different surveys, Wagner et al. (2008) suggested that backscatter cross-sections should be also considered as a standard product in laser scanning. The related waveform attributes are as follows.

1. Pulse Width: The pulse width can be described by the standard deviation W_j in Equation (11.4). It can be a quantitative indicator to evaluate the extent of the pulse-broadening. Several studies into small-footprint waveforms have observed that the pulse width of vegetation points is generally larger than that of terrain points (Persson et al., 2005; Wagner et al., 2008). Forest terrain also tends to exhibit larger pulse widths than open terrain (Mücke, 2008), with trees and meadows also generating larger pulse widths than buildings (Stilla and Jutzi, 2008). The distinction of smooth surfaces from plants, bushes, trees, and even short hedges thus seems to be relatively clear (Wagner et al., 2008). This implies that pulse width can be an additional factor used to discriminate between ground returns and vegetation returns, although Stilla and Jutzi (2008) suggested that additional information, such as the 3D geometrical relationships of the returns, is required to classify each return pulse as a specific surface type.
2. Amplitude: the amplitude can be represented by A_j in Equation (11.4). The estimated amplitude for each return by post-processing waveforms provides reflectance information about the illuminated targets. However, without calibration of such data, the information is too noisy to be utilized for classification purposes. It has been observed that there is a large overlap among the amplitude histograms obtained from different land cover classes (Ducic et al., 2006; Mücke, 2008). At low amplitudes, the reliability for the estimate of pulse width can be problematic (Wagner et al., 2008). Since the emitted power is constant, pulse-broadening of received returns also results in a reduction of the amplitude.

3. Backscatter Cross-Section: The backscatter cross-section is one of the unique features available from full-waveform LiDAR data. The backscatter cross-section provides information in relation to the range and scattering properties of the targets (Wagner et al., 2006), and it depends strongly on the number of returns (Wagner et al., 2008). For each individual backscatter cross-section for each return, it has been observed that single returns produce stronger estimates than multiple returns. Grass and gravel can be distinguished well, with a small overlay of cross-section values estimated. Vegetation returns generally produce lower cross-section values than terrain returns, which may be useful to discriminate terrain from vegetation points (Wagner et al., 2008). In addition, the backscatter cross-section varies within forest canopies, which may be useful for tree species classification (Wagner et al., 2008).

The backscatter cross-section feature (σ) of a laser echo can be computed from the reflectance (ρ), range (R), and laser beam divergence angle (β) of the target (Wagner et al., 2006), as described in Equation (11.5).

$$\sigma = \pi \rho R^2 \beta^2 \tag{11.5}$$

Equation (11.5) represents the general form of the apparent cross section of each surface within the laser footprint. For a Gaussian waveform, the equation can be rewritten as

$$\sigma = C_{cal} R^4 A_j W_j \tag{11.6}$$

$$C_{cal} = \frac{\rho \pi \beta^2}{R^2 A_j W_j} \tag{11.7}$$

where C_{cal} is a calibration constant that can be obtained through laboratory experiments (Alexander et al., 2010).

3. Backscatter Coefficient γ: the backscatter coefficient is the backscatter cross-section per unit-illuminated area. Alexander et al. (2010) found that the backscatter coefficient is more useful than the amplitude and backscatter cross-section for discriminating road and grass. Wagner (2010) stated that it is helpful to employ the backscattering coefficient γ when comparing datasets acquired by different sensors and/or different flight campaigns.

11.6.3 Applications

Waveform LiDAR data has been used to explore weak and overlapping returns that were missed by the on-the-fly detection process of LiDAR systems. The point clouds extracted from the waveform data can then better describe the sensed landscape. The canopy height (Hancock et al., 2011; Wang et al., 2013), forest biomass (Clark et al., 2011; Yao et al., 2011), and digital terrain model can thus be improved.

The land cover classification is another important application of waveform LiDAR data. The waveform features introduced in Section 11.1.2.3 have been demonstrated in 3D point cloud classification. Integrations with other information (e.g., geometric and spectral information) provide even more criteria that can aid in achieving advanced applications, such as forestry investigation (Buddenbaum et al., 2013) and building modeling.

Comparing between waveform and multi-return LiDAR point clouds, the main applications are not much difference. For examples, DEM generation, Building reconstruction, and Forest parameter estimation, and so on, can be done by using both kinds of point clouds. However, a major

TABLE 11.4
Strengths and Limitations of Waveform and Multi-Return LiDAR Point Cloud

Waveform Point Cloud	Multi-Return Point Cloud
Number of return is unlimited	Number of return is limit
Detection of error or missing echoes become possible	Echo detection is a black-box operation. Detection of error or missing cannot be evaluated.
Dead zone problem can be eased by a user developed detector.	Multi-return system often suffers from LiDAR dead zone problem (overlapping echoes problem)
More features (waveform features) can be attached to the point cloud.	Only one additional information (intensity) can be attached to each point.

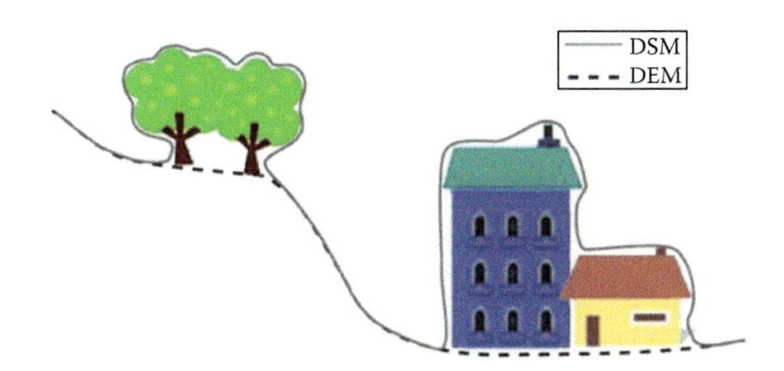

FIGURE 11.17 The general definitions of DEM and DSM.

improvement for waveform point cloud is that the landscape or the target can be descripted more detail. Thus the detection or the reconstruction of the interested targets can be better accomplished. Besides, the waveform feature is another advantage for waveform point cloud applications, especially on the purpose of point cloud classification. Table 11.4 lists the strengths and limitations of waveform LiDAR point cloud and multi-return point cloud.

11.7 DEM AND DSM GENERATION FROM AIRBORNE LiDAR DATA

11.7.1 Introduction

A DEM is a digital model (or the height information) representing the ground surface of a terrain. In general, the ground surface means the earth's surface without any objects like plants, buildings and other man-made structures. However, permanent earthwork structures, such as dams and road embankments, are usually counted as the ground surface. In contrast, a DSM, in general, represents the visual surface of the earth, including all the objects on it. However, in most cases, we do not treat the surface of a water body as part of DSM. Figure 11.17 shows the general definitions of DEM and DSM. A DEM or DSM can be represented as a raster (a grid of squares data structure) or as a vector-based TIN.

Airborne LiDAR data have proven effective for the generation of high-resolution and high-accuracy DEMs and DSMs of various terrestrial environments. In general, a DEM is formed with the ground points extracted from a dataset of LiDAR point clouds, while a DSM is formed with the surface (first-return) points extracted from LiDAR data. The extracted ground and surface points are treated as measured points to form DEMs and DSMs (Mongus and Žalik, 2012).

Accurate DEMs are often used to derive morphometric parameters (e.g., slope, aspect, and surface ruggedness) for the morphological analysis of an area of interest. Morphological analysis is an important tool for characterizing environmental changes in both temporal and spatial scales. For example, the data derived from DEMs allow the identification of morphologically homogeneous zones, for example the identification of different volcanic edifices, structural domains, and evolutionary stages (Norini et al., 2004). DEMs have been applied to reveal the variation in elevation of hills and river beds for landslide and inundation investigations focusing on Chenyulan River (Yang et al., 2011). High-resolution DEMs generated by LiDAR have now become more readily available., and advances in computing technology make it now possible to produce spatially explicit, fully distributed hydraulic and hydrological models and hydro-geomorphological assessments (Biron et al., 2013; Vaze et al., 2010).

11.7.2 Data Processing for DEM and DSM Generation

DEM generation usually employs a filter (or a classifier) to separate ground points from non-ground points. Popular filters, such as those based on TINs (Axelsson, 2000) and morphology (Pingel et al., 2013), are mainly developed by exploiting the detailed geometrical description of the ground topology in LiDAR point clouds. For DSM generation, the surface points can be obtained by choosing all of the first-returning echoes of the LiDAR data. The process of DSM generation is much easier than that of DEM generation.

The overall data processing for DEM and DSM generation using airborne LiDAR data includes preprocessing, boresight calibration, and point cloud generation, strip adjustment, ground and surface point filtering and interpolation to form DEM and DSM. A LiDAR dataset may be contaminated by systematic errors, if the procedures of boresight calibration and strip adjustment are not well performed. The point density of the source data constrains the resolution of the generated DEMs and DSMs (Florinsky, 1998). Accordingly, the resolution of DEM and DSM is constrained mostly by the density of the input terrain data. McCullagh (1988) suggested that the number of grid cells should be similar to the number of terrain data in a covered area.

Since the extracted ground points are treated as measured points to form DEMs, the accuracy of the generated DEMs is subject to the correctness of ground point extraction. Under this circumstance, ground objects, which prevent laser beams from reaching the ground surface, may significantly affect the quality of DEMs. Given the decreased penetration capability of LiDAR systems in dense building or forest areas, the filtering process may mistakenly register non-ground points as ground points, or vice versa. This error depends on the parameter setup of the specific filter employed. To remedy the misclassification of point clouds and obtain a reliable DEM, a manual editing procedure is conducted in the registration of point clouds to the correct class. However, this manual editing procedure requires visual inspection of the thin profiles of point clouds for the whole survey area, and thus is extremely time consuming and costly.

Different levels of DEM quality can therefore be related to the LiDAR DEM generation procedures, which are subject to the processes of calibrating possible systematic errors, checking misclassification errors, and maintaining topographical features. In accordance with the typical demands of users, DEM products are usually categorized into three levels, as follows.

1. Level 1 DEM
   DEMs generated by means of a fast and automatic procedure, without any quality control process, belong to this category. The products are not verified via a careful strategy for the removal of systematic errors, and are thus likely to involve a number of mistakes. As the land cover objects of the observed region become denser and more complicated, so the proportion of the inappropriate filtered points of the DEM could become higher.

2. Level 2 DEM
   DEMs generated by means of a careful double-check procedure with a quality assessment and control process belong to this category. The procedures of LiDAR system calibration and strip adjustment are usually required to reduce the systematic errors. In addition, a procedure of manual editing is also needed to improve the quality of ground point extraction (filtering) and ensure the reliability of final products.
3. Level 3 DEM
   If the level 2 DEMs are generated with consideration of water boundaries and geomorphological features, then they belong to this category. To meet this requirement, the filtered point cloud should be manually inspected and edited to improve the precision of interpolation by adding certain points of water boundaries and geomorphology features (e.g., ridges, valleys, steep slopes, and so on).

11.7.3 DEM Quality Assessment and Control

The process of DEM quality assessment and control (QA/QC) is to ensure the quality of the generated DEMs through a series of verification procedures. The quality of a DEM may be affected by three major factors, including the quality of the original LiDAR point cloud, the quality of the filtering, and the quality of the DEM interpolation (if applicable, including thinning and break line modeling) (Pfeifer and Mandlburger, 2008). These three factors affect the DEM quality at different stages of the generation procedure. The QA/QC must be performed step by step on each stage of the DEM generation workflow. For many practical projects, aerial photos of the survey area are also acquired during the LiDAR point cloud acquisition. An orthophoto of the survey area can then be produced based on these photos and the LiDAR point clouds, which can provide helpful visual clues for manual point cloud classification. The QA/QC procedure should thus take into account the use of aerial photos. An effective the DEM QA/QC procedure may include the following six items:

1. Verification of airborne LiDAR mission plan: to be performed before the scanning operation, including inspection of flight plan and LiDAR system calibration report. The nominal point density should be checked according to the flight plan.
2. Verification of ground control survey: the quantity, distribution and quality of ground control points used for LiDAR strip adjustment.
3. Verification of LiDAR points and strip adjustment: to check the LiDAR strip coverage and overlaps, overall point density achieved, and external and internal errors of the strips.
4. Verification of LiDAR point cloud filtering: including visual inspections and field survey checks.
5. Verification of aerial photography acquisition: to check overall image coverage and quality.
6. Verification of orthophoto: to check the quality of the orthoimages, including the overall image continuity and possible distortion of ground objects.

The quality index of a DEM production depends on the requirements of the project. For general purposes, the quality index provided in the USGS LiDAR Base Specification Version 1.0 (Heidemann, 2012) is an appropriate reference.

At forest area, airborne LiDAR laser may penetrate the gap between leaves and detect the ground points under tree canopy, which is difficult to be obtained from aerial photos. Therefore, DEMs produced from LiDAR data may have higher accuracy than that from photogrammetry. At an earlier study, the height accuracy of DEM derived from airborne LiDAR data is about ±10 cm while that derived from photogrammetry is about ±25 cm (Kraus and Pfeifer, 1998).

11.8 TERRESTRIAL/VEHICLE-BORNE LiDAR DATA PROCESSING

11.8.1 Terrestrial LiDAR Data Processing

Terrestrial laser scanners (TLS) are stationary, ground-based LiDAR systems that scan surrounding objects. The data captured are also point clouds, which are distributed as layers corresponding to the surfaces of scanned objects. A growing number of applications, such as change detection and deformation analysis (Lim et al., 2005; Monserrat and Crosetto, 2008; Santibanez et al., 2012), rapid modeling of industrial infrastructure, architecture, agriculture, construction or maintenance of tunnels and mines, facility management, and urban and regional planning, have been carried out using TLS, and demonstrated that it is a key surveying tool for capturing and modeling highly-detailed geospatial data.

The main advantages of TLS are as follows: (1) the direct measurement of 3D coordinates; (2) the high degree of automation; (3) the easy-to-use hardware; and (4) a massive sampling capability (Monserrat and Crosetto, 2008).

11.8.1.1 Properties of Terrestrial LiDAR Data

The measurement mechanism of a TSL is similar to that of a total station, a surveying instrument used for angle and distance measurement. The difference is that a TSL allows automated and near-simultaneous measurements of hundreds or thousands of nonspecific points in the area surrounding the position where the instrument is set up. In Figure 11.1, a slant range by the laser rangefinder and two orthogonal angles by angular encoders in the horizontal and vertical directions are measured simultaneously. The point clouds are then captured point by point by rotating the reflecting mirrors whose horizontal/vertical angles are gradually increased, with a measuring rate of 1,000 Hz or more. These simultaneous measurements of distance and angle are carried out in a highly automated manner (Petrie and Toth, 2008b).

The TSL can be mounted on a tripod for fixed positions or on a vehicle (or a moving platform) for mobile mapping. The first case also refers to static scanning, in which the exterior orientation of the platform is constant for one scan position, and two-dimensional coverage in the angular domain is performed by rotating components of the device. In the second case, the scanning is performed by a univariate beam deflection unit, and area-wise data acquisition is established by the movement of the scanning platform, that is, dynamic scanning is achieved (Pfeifer and Briese, 2007). Many different kinds of static and dynamic TSLs that are used for topographic mapping and modeling operations are described in (Petrie and Toth, 2008b).

11.8.1.2 Data Processing Strategy

The 3D point cloud captured with a TLS contains detailed geometric information, but further data processing is still needed for certain applications. In general, the processing of TLS LiDAR data can be divided into three procedures: geo-referencing, feature extraction and visualization. The task of geo-referencing is to transform the original point cloud, registered in a local coordinate system, into a ground reference coordinate system. The task of feature extraction converts the point cloud data to be meaningful information for further processing, and the types of derivative features are subject to the involved applications, such as object classification, surface and building reconstruction, and so on. Visualization and interactive operations implemented on a computer are usually needed for manual interpretation and extraction of meaningful object features. The traditional field work of mapping surveys can now be performed on a computer system with the TLS data.

11.8.1.2.1 Registration and Geo-Referencing

The point clouds needed for 3D object modeling may be acquired from different scan stations, although the local point clouds must be transformed into a common coordinate system for further processing. If only the focal object itself is of interest, it is sufficient to determine the relative orientation between scans using registration. If the object also has to be placed in a superior coordinate system, absolute orientation is also needed. If the superior coordinate system is earth fixed, thus then becomes a geo-referencing task (Pfeifer and Briese, 2007).

One of the most popular methods of registration is the ICP (iterative closest point) algorithm developed by Besl and McKay (1992). Some variations and improvements based on ICP methods have been proposed by Chen and Medioni (1992), Zhang (1994), Okatani and Deguchi (2000), and Segal et al. (2009). The ICP is based on the search for pairs of nearest points between two scans, and the rigid body transformation is estimated and applied to the points of one scan. The ICP procedure is iterated until convergence is achieved. Another method for point cloud registration using least squares matching was proposed by Gruen and Akca (2005). The LS3D (least squares 3D surface matching) algorithm estimates the transformation parameters between two or more fully 3D surfaces, using the Generalized Gauss–Markoff model, minimizing the sum of the squares of the Euclidean distances between the surfaces. This formulation makes it possible to match arbitrarily oriented 3D surfaces simultaneously, without using explicit tie points.

The geo-referencing of TLS data is a procedure of coordinate transformation in which the 3D point cloud is transformed to a ground reference coordinate system (GRCS), and the method of geo-referencing used may be direct or indirect (dos Santos et al., 2013). The direct geo-referencing method is based on integrating additional POS (position orientation system) sensors, such as a GNSS and an inertial measurement unit (IMU), so that the platform's position and orientation at each moment of data acquisition can be determined accurately. The direct geo-referencing of the TLS data has been investigated primarily for terrestrial mobile mapping (Haala et al., 2008; Hunter et al., 2006; Talaya et al., 2004). In indirect geo-referencing, a set of pre-surveyed ground control points (GCPs) are required to transform one or multiple scans into a superior coordinate system. In this case, the 3D similarity transformation model is regularly utilized, and the transformation parameters are estimated using least squares adjustment.

11.8.1.2.2 Point Cloud Organization and Segmentation

TSL scans contain detailed geometric information, but still require interpretation of the data to make it useable for mapping purposes. Point cloud organization and segmentation are the early steps in LiDAR data processing, which are not directly linked to an application (Pfeifer and Briese, 2007).

Modern static and kinematic TLS have the ability to acquire point cloud data with large amount of points. A well-organized data structure for TLS point clouds will be helpful for the acceleration of the data storage, processing and visualization. The commonly used data structures include TIN (triangulated irregular network), and grid and octree data structures (Elseberg et al., 2013; Wang and Tseng, 2011).

Segmentation and clustering can also be used to organize discrete points into homogeneous groups (Pfeifer and Briese, 2007), and many algorithms have been proposed for extracting planar surfaces from point clouds using segmentation methods for model reconstruction. Usually one of three distinct methods is employed for segmenting points: region growing (Dold and Brenner, 2004; Hoffman et al., 2002; Pu and Vosselman, 2006), clustering of features (Biosca and Lerma, 2008; Filin, 2002; Filin and Pfeifer, 2006; Hoffman, 2004; Lerma and Biosca, 2005) or model fitting (Bauer et al., 2003; Boulaassal et al., 2007; Bretar and Roux, 2005).

11.8.1.2.3 Visualization

The visualization system is the foundation for several interactive analysis tools for quality control, extraction of survey measurements, and the extraction of isolated point cloud features (Kreylos et al., 2008). Staiger (2003) treated the visualization of point cloud data into six different ways:

1. Point clouds in a 3D projection, with a color or greyscale-coded representation of intensity, are often used as a first visual check of the acquired data.

2. Point clouds can also be combined with derived geometrical elements.
3. "True" ortho-photos are realized by the fusion of digital images (point information) and the registered point clouds (geometry).
4. 3D contour plans
5. 3D models
6. "Virtual flights" through the modeled scene.

11.8.2 Object Extraction from Vehicle-Borne LiDAR Data

Vehicle-borne LiDAR technology enables the real-time capture of high resolution 3D spatial information, which is not possible with static terrestrial LiDAR scanning survey technology. This approach is an important supplement to photogrammetry and remote sensing.

Vehicle-borne and static terrestrial LiDAR data are different in three respects (Boulaassal et al., 2011). First, they are different in terms of the level of accuracy that can be obtained. Because the vehicle-borne LiDAR data require synchronization of the positioning and orientation components, the resulting accuracy is less than that of static LiDAR data. Second, because the system operates on a moving platform, the density and resolution of vehicle-borne LiDAR data are significantly lower than those seen with the stationary method, and so fewer details of objects are obtained. Moreover, the number of points acquired by vehicle-borne LiDAR systems is often higher than that acquired by several successive stationary terrestrial LiDAR stations.

Because vehicle-borne LiDAR systems are operated on roads when collecting data, they record numerous points of various objects on and nearby the road. This point data can then be used to reconstruct and model road surfaces, guardrails, pavements, utility facilities or pole-like features (e.g., power poles, traffic sign poles, and light poles), building façades, bushes, and trees, and this is especially useful for obtaining details of objects in an urban area. The features on the road surface, for example, road markings, can also be extracted from this data.

Since vehicle-borne LiDAR systems move on a road, all the objects around the road that appear in the field of view of the scanner can reflect the laser beam and generate points, although this also means that the system has limited views in some directions. Consequently, the collected data may often not fully cover the target objects of interest, making feature extraction and identification very difficult. Furthermore, when the vehicle is moving at high speed during data collection, the point and scan-line intervals may become rather large and make further data processing very complex. Certain strategies designed to handle vehicle-borne LiDAR data thus need to be adopted to extract, reconstruct, and model the objects along or nearby roads.

First, LiDAR data should be organized in advance using auxiliary data structures, such as scan lines, 2D grids or 3D grids (voxels). The LiDAR data are then classified into point clusters of various objects, like road surfaces, building faces, utility poles, and so on. It is often useful to classify LiDAR data into road surface points and non-road surface points based on the knowledge of the actual scene. For example, the point height on road surfaces should usually be lower than that on non-road surfaces. Additionally, most of the points collected from the road surface will have the same height, and the density of road points is higher than that seen in other places due to the small range from the scanned road surface to the scanner. The height of the road surface varies smoothly, with a very small deviation along the width of the road (Manandhar and Shibasaki, 2001). Based on the road surface points, the road surface can then be modeled by a triangulated network (TIN) or road boundaries.

Non-road surface points may be located on the surfaces of pole-like features (e.g., power poles, traffic sign poles, and light poles), building façades, or trees. Some important knowledge about the objects in the scanning scene should thus be generalized in order to extract and model the objects from discrete and incomplete LiDAR data in a complicated scanning environment. In general, geometrical constraints for objects of interest are often employed. For example, the building façades should be vertical to the ground or at least be planar. To avoid the interference from irrelevant points with regard to the extraction of object points for modeling, the algorithms developed for extracting object points for object model reconstruction have to be robust, and knowledge about the scanner can sometimes be used to achieve this. For example, the scanning mechanism of the scanner can be considered. If the scan line is perpendicular to the ground, the points on the scan lines can be segmented into several vertical lines belonging to the walls. These vertical lines can then be grouped into one plane belonging to the wall.

Manandhar and Shibasaki (2001), Goulette et al. (2006), and Li et al. (2004) discussed the aforementioned concepts with regard to feature extraction, reconstruction, and modeling. More specifically, Manandhar and Shibasaki (2001) and Goulette et al. (2006) organized LiDAR data by scan lines before processing while Li et al. (2004) organized LiDAR data into 2D grids before processing it. In addition to organizing LiDAR data with scan lines or 2D grids, additional information can be attributed into 2D grids for advanced classification. Douillard et al. (2009) developed a cell-wise semantic classification approach for ground cells. This involves the use of color imagery to classify cells into one of two classes: asphalt or grass. Classification of ground cells proceeds according to three main steps: (1) generation of Regions of Interest (ROIs) in the image, (2) feature extraction within each ROI, and (3) feature-based classification.

In addition, organizing LiDAR data using voxels before further data processing is another good strategy. Aijazi et al. (2013) presented a method to classify urban scenes based on a super-voxel segmentation of sparse 3D data obtained from LiDAR sensors. The 3D point cloud is first segmented into voxels, which are then characterized by several attributes transforming them into super-voxels. These are joined together by using a link-chain method rather than the usual region growing algorithm to explore objects. These objects are then classified using geometrical models and local descriptors. Schmitt and Vogtle (2009) converted a raw irregular point cloud into regular voxels, and then extracted planar features extracted by merging adjacent voxels with collinear normal vectors. Wu et al. (2013) presented a new Voxel-based Marked Neighborhood Searching (VMNS) method for efficiently identifying street trees and deriving their morphological parameters from vehicle-borne LiDAR point cloud data. The VMNS method consists of six technical components: voxelization, calculating values of voxels, searching and marking neighborhoods, extracting potential trees, deriving morphological parameters, and eliminating pole-like objects other than trees. The method was validated and evaluated through two case studies, with the results showing that the completeness and correctness of the proposed method for street tree detection are both over 98%. The derived morphological parameters, including tree height, crown diameter, diameter at breast height (DBH), and crown base height (CBH), were also in a good agreement with the field measurements. This method provides an effective tool for extracting various morphological parameters for individual street trees from Vehicle-borne LiDAR point cloud data.

11.9 CONCLUSIONS

LiDAR is a promising approach for the fast and robust acquisition of 3D information from scanned surfaces, and has been widely used in various applications (Debnath et al., 2023; Guo et al., 2023; Kovanič et al., 2023; Rivera et al., 2023; Mirzaei et al., 2022; Roriz et al., 2022; Xu et al., 2021; Li and Ibanez-Guzman, 2020; Muhadi et al., 2020; Dong and Chen, 2017; Gargoum and El-Basyouny, 2017; Lohani and Ghosh, 2017; Zhenlong et al., 2018; Cao et al., 2015). For example, airborne LiDAR systems can obtain nationwide DSM and DEM data, while ground-based and mobile LiDAR systems have been used more often in urban areas for 3D city modeling.

Similar to all other remote sensing methods, careful handling and processing of LiDAR data are required to ensure high quality results. The standard procedures provided in this chapter should be suitable for general applications. However, for more specific applications, for example, very high density point clouds, DEM data for a very dense forest, and so on, it is recommended that consultations with professional service providers are carried out ahead of time, as some technical issues may arise that are not considered here.

REFERENCES

Aijazi, A.K., Checchin, P. and Trassoudaine, L., 2013. Segmentation Based Classification of 3D Urban Point Clouds: A Super-Voxel Based Approach with Evaluation. *Remote Sensing*, 5(4): 1624–1650.

Alexander, C., Tansey, K., Kaduk, J., Holland, D. and Tate, N.J., 2010. Backscatter Coefficient as an Attribute for the Classification of Full-Waveform Airborne Laser Scanning Data in Urban Areas. *ISPRS Journal of Photogrammetry and Remote Sensing*, 65(5): 423–432.

ASPRS, 2013. *LASer (LAS) File Format Exchange Activities*, http://www.asprs.org/Committee-General/LASer-LAS-File-Format-Exchange-Activities.html (last date accessed: November 11, 2013).

Axelsson, P., 2000. DEM Generation from Laser Scanner Data Using Adaptive TIN Models. In *International Archives of Photogrammetry and Remote Sensing*, Leibniz University Hannover, Institute of Photogrammetry and GeoInformation, Hannover, Germany, 33, Part B4, Amsterdam, pp. 110–117.

Baltsavias, E.P., 1999a. Airborne Laser Scanning: Basic Relations and Formulas. *ISPRS Journal of Photogrammetry and Remote Sensing*, 54(2–3): 199–214.

Baltsavias, E.P., 1999b. Airborne Laser Scanning: Existing Systems and Firms and Other Resources. *ISPRS Journal of Photogrammetry and Remote Sensing*, 54(2–3): 164–198.

Bauer, J., Karner, K., Klaus, A., Zach, C. and Schindler, K., 2003. Segmentation of Building Models from Dense 3D Point-Clouds. *Proceedings on 27th Workshop of the Austrian Association on Pattern Recognition*, Institute of Computer Graphics and Vision (7100), Vienna, Austria, pp. 253–258.

Besl, P.J. and McKay, N.D., 1992. A Method for Registration of 3-D Shapes. *IEEE Transactions on Pattern Analysis and Machine Intelligence*, 14(2): 239–256.

Biosca, J.M. and Lerma, J.L., 2008. Unsupervised Robust Planar Segmentation of Terrestrial Laser Scanner Point Clouds Based on Fuzzy Clustering Methods. *ISPRS Journal of Photogrammetry and Remote Sensing*, 63(1): 84–98.

Biron, P.M., Choné, G., Buffin-Bélanger, T., Demers, S. and Olsen, T., 2013. Improvement of Streams Hydro-Geomorphological Assessment Using LiDAR DEMs. *Earth Surface Processes and Landforms*, 38(15): 1808–1821.

Boulaassal, H., Landes, T. and Grussenmeyer, P., 2011. 3D Modelling of Facade Features on Large Sites Acquired by Vehicle Based Laser Scanning. *Archives of Photogrammetry, Cartography and Remote Sensing Edited by Polish Society for Photogrammetry and Remote Sensing*, 22: 215–226.

Boulaassal, H., Landes, T., Grussenmeyer, P. and Tarsha-Kurdi, F., 2007. Automatic Segmentation of Building Facades Using Terrestrial Laser Data. *International Archives of Photogrammetry, Remote Sensing and Spatial Information Sciences*, 36, Part 3.

Brenner, C., 2005. Building Reconstruction from Images and Laser Scanning. *International Journal of Applied Earth Observation and Geoinformation*, 6(3–4): 187–198.

Brenner, M., Wichmann, V. and Rutzinger, M., 2013. Eigenvalue and Graph-Based Object Extraction from Mobile Laser Scanning Point Clouds. In *ISPRS Annals of the Photogrammetry,* Remote Sensing and Spatial Information Sciences, Hong Kong SAR, China, pp. 55–60.

Bretar, F. and Roux, M., 2005. Hybrid Image Segmentation Using LiDAR 3D Planar Primitives. *International Archives of Photogrammetry, Remote Sensing and Spatial Information Sciences*, 16: 72–78.

Bucksch, A., Lindenbergh, R. and Menenti, M., 2009. SkelTre-Fast Skeletonisation for Imperfect Point Cloud Data of Botanic Trees. In *Eurographics Workshop on 3D Object Retrieval.* Eurographics, München, Germany, p. 8.

Buddenbaum, H., Seeling, S. and Hill, J., 2013. Fusion of Full-Waveform Lidar and Imaging Spectroscopy Remote Sensing Data for the Characterization of Forest Stands. *International Journal of Remote Sensing*, 34(13): 4511–4524.

Burman, H., 2002. Laser Strip Adjustment for Data Calibration and Verification. *International Archives of Photogrammetry and Remote Sensing*, 34(Part 3A/B): 67–72.

Cao, V.-H., Chu, K.-X., Le-Khac, N.-A., Kechadi, M.-T., Laefer, D. and Truong-Hong, L., 2015. Toward a New Approach for Massive LiDAR Data Processing. *2015 2nd IEEE International Conference on Spatial Data Mining and Geographical Knowledge Services (ICSDM)*, Fuzhou, China, 2015, pp. 135–140, http://doi.org/10.1109/ICSDM.2015.7298040.

Chen, L.C. and Lo, C.-Y., 2009. 3D Road Modeling Via the Integration of Large-Scale Topomaps and Airborne LIDAR Data. *Journal of the Chinese Institute of Engineers*, 32(6): 811–823.

Chen, L.C., Teo, T.A., Hsieh, C.H. and Rau, J.Y., 2006. Reconstruction of Building Models with Curvilinear Boundaries from Laser Scanner and Aerial Imagery. *Lecture Notes in Computer Science*, 4319: 24–33.

Chen, L.C., Teo, T.A., Kuo, C. and Rau, J., 2008. Shaping Polyhedral Buildings by the Fusion of Vector Maps and Lidar Point Clouds. *Photogrammetric Engineering and Remote Sensing*, 74(5): 1147–1157.

Chen, Q., Gong, P., Baldocchi, D. and Xie, G., 2007. Filtering Airborne Laser Scanning Data with Morphological Methods. *Photogrammetric Engineering and Remote Sensing*, 73(2): 175–185.

Chen, Y. and Medioni, G., 1992. Object Modelling by Registration of Multiple Range Images. *Image and Vision Computing*, 10(3): 145–155.

Chiu, C.M., Teo, T.A. and Chen, C.T., 2013. Three-dimensional Modelling of Multilayer Road Networks Using Road Centerlines and Airborne Lidar Data. *The International Symposium on Mobile Mapping Technology 2013 May 1–3*, Tainan, Taiwan, pp. CD-ROM.

Chuang, T.Y., 2012. *Feature-Based Registration of LiDAR Point Clouds*, National Taiwan University, Taipei, Taiwan.

Clark, M.L., Roberts, D.A., Ewel, J.J. and Clark, D.B., 2011. Estimation of Tropical Rain Forest Aboveground Biomass with Small-Footprint Lidar and Hyperspectral Sensors. *Remote Sensing of Environment*, 115(11): 2931–2942.

Csanyi, N. and Toth, C.K., 2007. Improvement of LIDAR Data Accuracy Using LIDAR Specific Ground Targets. *Photogrammetric Engineering and Remote Sensing*, 73(4): 385–396.

Debnath, Sourabhi, Paul, Manoranjan and Debnath, Tanmoy, 2023. Applications of LiDAR in Agriculture and Future Research Directions. *Journal of Imaging*, 9(3): 57. https://doi.org/10.3390/jimaging9030057

Dold, C. and Brenner, C., 2004. Automatic Matching of Terrestrial Scan Data as a Basis for the Generation of Detailed 3D City Models. *International Archives of Photogrammetry, Remote Sensing and Spatial Information Sciences*, 35(B3): 1091–1096.

Dong, P. and Chen, Q., 2017. *LiDAR Remote Sensing and Applications*. Taylor and Francis Series in Remote Sensing Applications. CRC Press, Taylor and Francis Group, Boca Raton, FL. ISBN: 978-1-4822-4301-7 (Hard Cover), p. 199.

dos Santos, D.R., Dal Poz, A.P. and Khoshelham, K., 2013. Indirect Georeferencing of Terrestrial Laser Scanning Data Using Control Lines. *The Photogrammetric Record*, 28(143): 276–292.

Douillard, B., Brooks, A., Ramos, F. and Durrant-Whyte, H., 2009. Combining Laser and Vision for 3D Urban Classification. *Proceedings of Neural Information Processing Systems Conference (NIPS)*, Vancouver, British Columbia, Canada.

Ducic, V., Hollaus, M., Ullrich, A., Wagner, W. and Melzer, T., 2006. 3D Vegetation Mapping and Classification Using Full-Waveform Laser Scanning. *Workshop on 3D Remote Sensing in Forestry*: 211–217.

Elseberg, J., Borrmann, D. and Nüchter, A., 2013. One Billion Points in the Cloud–an Octree for Efficient Processing of 3D Laser Scans. *ISPRS Journal of Photogrammetry and Remote Sensing*, 76: 76–88.

Filin, S., 2001. *Calibration of Airborne and Spaceborne Laser Altimeters Using Natural Surfaces.*Ph.D. Thesis, The Ohio State University, Columbus, OH, 140 p.

Filin, S., 2002. Surface Clustering from Airborne Laser Scanning Data. *International Archives of Photogrammetry and Remote Sensing*, 34(Part 3A/B): 119–124.

Filin, S., 2003. Recovery of Systematic Biases in Laser Altimeters Using Natural Surfaces. *International Archives of Photogrammetry and Remote Sensing*, 35, WG III/3, Graz, Austria.

Filin, S., 2004. Surface Classification from Airborne Laser Scanning Data. *Computers & Geosciences*, 30: 1033–1041.

Filin, S. and Pfeifer, N., 2006. Segmentation of Airborne Laser Scanning Data Using a Slope Adaptive Neighborhood. *ISPRS Journal of Photogrammetry and Remote Sensing*, 60(2): 71–80.

Filin, S. and Vosselman, G., 2004. Adjustment of Airborne Laser Altimetry Strips. *International Archives of Photogrammetry and Remote Sensing*, 35(B3): 285–289.

Fischler, M.A. and Bolles, R.C., 1981. Random Sample Consensus: A Paradigm for Model Fitting with Applications to Image Analysis and Automated Cartography. *Communications of the ACM*, 24(6): 381–395.

Florinsky, I.V., 1998. Combined Analysis of Digital Terrain Models and Remotely Sensed Data in Landscape Investigations. *Progress in Physical Geography*, 22(1): 33–60.

Fornaciai, A., Bisson, M., Landi, P., Mazzarini, F. and Pareschi, M.T., 2010. A LiDAR Survey of Stromboli Volcano (Italy): Digital Elevation Model-Based Geomorphology and Intensity Analysis. *International Journal of Remote Sensing*, 31(12): 3177–3194.

Gardner, C.S., 1992. Ranging Performance of Satellite Laser Altimeters. *Geoscience and Remote Sensing, IEEE Transactions on*, 30(5): 1061–1072.

Gargoum, S. and El-Basyouny, K. 2017. Automated Extraction of Road Features Using LiDAR Data: A Review of LiDAR Applications in Transportation. *2017 4th International Conference on Transportation Information and Safety (ICTIS)*, Banff, AB, Canada, pp. 563–574, http://doi.org/10.1109/ICTIS.2017.8047822.

Glennie, C., 2007. Rigorous 3D Error Analysis of Kinematic Scanning LIDAR Systems. *Journal of Applied Geodesy Jag*, 1(3): 147–157.

Gonzalez, R.C. and Woods, R.E., 1992. *Digital Image Processing*, Addison-Wisley, Reading, MA, 716 p.

Gorte, B. and Pfeifer, N., 2004. Structuring Laser-Scanned Trees Using 3D Mathematical Morphology. *International Archives of Photogrammetry and Remote Sensing*, 35(B5): 929–933.

Goulette, F., Nashashibi, F., Abuhadrous, I., Ammoun, S. and Laurgeau, C., 2006. An Integrated On-Board Laser Range Sensing System for On-the-Way City and Road Modelling. *Proceedings of the ISPRS Commission I Symposium, "From Sensors to Imagery,"* Paris, France, p. 43.

Gross, H. and Thoennessen, U., 2006. Extraction of Lines from Laser Point Clouds. *Symposium of ISPRS Commission III: Photogrammetric Computer Vision*, 36, pp. 86–91.

Gruen, A., 2005. Towards Photogrammetry 2025. *Photogrammetric Week*.

Gruen, A. and Akca, D., 2005. Least Squares 3D Surface and Curve Matching. *ISPRS Journal of Photogrammetry and Remote Sensing*, 59(3): 151–174.

Guo, L., Chehata, N., Mallet, C. and Boukir, S., 2011. Relevance of Airborne Lidar and Multispectral Image Data for Urban Scene Classification Using Random Forests. *ISPRS Journal of Photogrammetry and Remote Sensing*, 66(1): 56–66.

Guo, Q., Su, Y. and Hu, T., 2023. *LiDAR Principles, Processing, and Applications in Forestry Ecology*. Higher Education Press, Academic Press, Beijing. ISBN 978-0-12-823894-3, p. 496.

Haala, N., Peter, M., Kremer, J. and Hunter, G., 2008. Mobile LiDAR Mapping for 3D Point Cloud Collection in Urban Areas—A Performance Test. *The International Archives of the Photogrammetry, Remote Sensing and Spatial Information Sciences*, 37: 1119–1127.

Habib, A., Al-Durgham, M., Kersting, A. and Quackenbush, P., 2008. Error Budget of Lidar Systems and Quality Control of the Derived Point Cloud. *Proceedings of the XXI ISPRS Congress, Commission I*, 37: 203–209.

Habib, A., Bang, K., Kersting, A.P. and Lee, D.-C., 2009. Error Budget of LiDAR Systems and Quality Control of the Derived Data. *Photogrammetric Engineering and Remote Sensing*, 75(9): 1093–1108.

Habib, A., Ghanma, M., Morgan, M. and Al-Ruzouq, R., 2005. Photogrammetric and Lidar Data Registration Using Linear Features. *Photogrammetric Engineering and Remote Sensing*, 71(6): 699–707.

Habib, A., Kersting, A.P., Bang, K.I. and Lee, D.-C., 2010. Alternative Methodologies for the Internal Quality Control of Parallel LiDAR Strips. *Geoscience and Remote Sensing, IEEE Transactions on*, 48(1): 221–236.

Hancock, S., Disney, M., Muller, J.P., Lewis, P. and Foster, M., 2011. A Threshold Insensitive Method for Locating the Forest Canopy Top with Waveform Lidar. *Remote Sensing of Environment*, 115(12): 3286–3297.

Harding, D., 2009. Pulsed Laser Altimeter Ranging Techniques and Implications for Terrain Mapping. In *Topographic Laser Ranging and Scanning Principles and Processing*, Topographic Laser Ranging and Scanning, CRC Press, pp. 173–194.

Heidemann, H.K., 2012. Lidar Base Specification Version 1.0. *US Geological Survey Techniques and Methods*: 63.

Heinzel, J. and Koch, B., 2011. Exploring Full-Waveform LiDAR Parameters for Tree Species Classification. *International Journal of Applied Earth Observation and Geoinformation*, 13(1): 152–160.

Hoffman, A.D., 2004. Analysis of TIN-Structure Parameter Spaces in Airborne Laser Scanner Data for 3-D Building Model Generation. *International Archives of Photogrammetry, Remote Sensing and Spatial Information Sciences*, 35(Part B3): 302–307.

Hoffman, A.D., Maas, H.G. and Streilein, A., 2002. Knowledge-Based Building Detection Based on Laser Scanner Data and Topographic Map Information. *International Archives of Photogrammetry and Remote Sensing*, 34, Part 3A/B Sept. 9–13, Graz, Austria, pp. 169–174.

Hofton, M.A., Minster, J.B. and Blair, J.B., 2000. Decomposition of Laser Altimeter Waveforms. *IEEE Transactions on Geoscience and Remote Sensing*, 38(4): 1989–1996.

Hoover, A. et al., 1996. An Experimental Comparison of Range Image Segmentation Algorithms. *IEEE Transactions on Pattern Analysis and Machine Intelligence*, 18(7): 673–689.

Hu, Y., 2003. *Automated Extraction of Digital Terrain Models, Roads and Buildings Using Airborne Lidar Data*. Ph.D. Thesis, University of Calgary, Calgary, Alberta, 223 p.

Hug, C., Ullrich, A. and Grimm, A., 2004. Litemapper-5600-a Waveform-Digitizing LiDAR Terrain and Vegetation Mapping System. *International Archives of Photogrammetry, Remote Sensing and Spatial Information Sciences*, 36(Part 8): W2.

Huising, E.J. and Pereira, L.M.G., 1998. Errors and Accuracy Estimates of Laser Data Acquired by Various Laser Scanning Systems for Topographic Application. *ISPRS Journal of Photogrammetry and Remote Sensing*, 53(5): 245–261.

Hunter, G., Cox, C. and Kremer, J., 2006. Development of a Commercial Laser Scanning Mobile Mapping System–StreetMapper. *The International Archives of the Photogrammetry, Remote Sensing and Spatial Information Sciences*, 36.

Im, J., Jensen, J.R. and Hodgson, M.E., 2008. Object-Based Land Cover Classification Using High-Posting-Density LiDAR Data. *GIScience & Remote Sensing*, 45(2): 209–228.

Isenburg, M., 2013. LASzip: Lossless Compression of LiDAR Data. *Photogrammetric Engineering and Remote Sensing*, 79(2): 209–217.

Jiang, X. and Bunke, H., 1994. Fast Segmentation of Range Images into Planar Regions by Scan Line Grouping. *Machine Vision and Applications*, 7(2): 115–122.

Jutzi, B., Neulist, J. and Stilla, U., 2005. High-Resolution Waveform Acquisition and Analysis of Laser Pulses. *Measurement Techniques*, 2(3.1): 2.

Jutzi, B. and Stilla, U., 2005. Measuring and Processing the Waveform of Laser Pulses. *Optical*: 194–203.

Jutzi, B. and Stilla, U., 2006a. Range Determination with Waveform Recording Laser System Using a Wiener Filter. *ISPRS Journal of Photogrammetry and Remote Sensing*, 61(2): 95–107.

Jutzi, B. and Stilla, U., 2006b. Range Determination with Waveform Recording Laser Systems Using a Wiener Filter. *ISPRS Journal of Photogrammetry and Remote Sensing*, 61(2): 95–107.

Kager, H., 2004. Discrepancies Between Overlapping Laser Scanner Strips–Simultaneous Fitting of Aerial Laser Scanner Strips. *International Archives of Photogrammetry, Remote Sensing and Spatial Information Sciences*, 35(B1): 555–560.

Kovanič, Ľudovít, Topitzer, Branislav, Peťovský, Patrik, Blišťan, Peter, Gergeľová, Marcela Bindzárová, and Blišťanová, Monika, 2023. Review of Photogrammetric and Lidar Applications of UAV. *Applied Sciences*, 13(11): 6732. https://doi.org/10.3390/app13116732

Krabill, W. et al., 2000. Airborne Laser II.' Assateague National Seashore Beach. *Photogrammetric Engineering & Remote Sensing*, 66(1): 65–71.

Kraus, K. and Pfeifer, N., 1998. Determination of Terrain Models in Wooded Areas with Airborne Laser Scanner Data. *ISPRS Journal of Photogrammetry and Remote Sensing*, 53(4): 193–203.

Kreylos, O., Bawden, G. and Kellogg, L., 2008. Immersive Visualization and Analysis of LiDAR Data. *Advances in Visual Computing*: 846–855.

Latypov, D. and Zosse, E., 2002. LIDAR Data Quality Control and System Calibration Using Overlapping Flight Lines in Commercial Environment. *ACSM-ASPRS 2002 Annual Conference* April 22–26, Washington, DC.

Lerma, J. and Biosca, J., 2005. Segmentation and Filtering of Laser Scanner Data for Cultural Heritage. *CIPA 2005 XX International Symposium, 26 September – 01 October, 2005, Torino, Italy.*

Levanon, N. and Mozeson, E., 2004. *Radar Signals*. John Wiley & Sons, Hoboken, NJ.

Lewis, P., Mc Elhinney, C.P. and McCarthy, T., 2012. LiDAR Data Management Pipeline; from Spatial Database Population to Web-Application Visualization. *Proceedings of the 3rd International Conference on Computing for Geospatial Research and Applications*, p. 16. https://doi.org/10.1145/2345316.23453

Li, B., Li, Q., Shi, W. and Wu, F., 2004. Feature Extraction and Modeling of Urban Building from Vehicle-Borne Laser Scanning Data. *International Archives of Photogrammetry, Remote Sensing and Spatial Information Sciences*, 35: 934–939.

Li, Y. and Ibanez-Guzman, J., 2020. Lidar for Autonomous Driving: The Principles, Challenges, and Trends for Automotive Lidar and Perception Systems. *IEEE Signal Processing Magazine*, 37(4): 50–61, July. http://doi.org/10.1109/MSP.2020.2973615.

Lichti, D.D. and Licht, M.G., 2006. Experiences with Terrestrial Laser Scanner Modelling and Accuracy Assessment. *The International Archives of the Photogrammetry, Remote Sensing and Spatial Information Sciences*, 36(5): 155–160.

Lim, M. et al., 2005. Combined Digital Photogrammetry and Time-of-Flight Laser Scanning for Monitoring Cliff Evolution. *The Photogrammetric Record*, 20(110): 109–129.

Lohani, B. and Ghosh, S., 2017. Airborne LiDAR Technology: A Review of Data Collection and Processing Systems. *Proceedings of the National Academy of Sciences, India Section A: Physical Sciences*, 87: 567–579. https://doi.org/10.1007/s40010-017-0435-9

Mallet, C. and Bretar, F., 2009. Full-Waveform Topographic Lidar: State-of-the-Art. *ISPRS Journal of Photogrammetry and Remote Sensing*, 64(1): 1–16.

Manandhar, D. and Shibasaki, R., 2001. Feature Extraction from Range Data. *Paper Presented at the 22nd Asian Conference on Remote Sensing*, 5, Copyright (c) 2001 Centre for Remote Imaging, Sensing and Processing (CRISP), National University of Singapore; Singapore Institute of Surveyors and Valuers (SISV); Asian Association on Remote Sensing (AARS), p. 9.

Mazzarini, F. et al., 2007. Lava Flow Identification and Aging by Means of Lidar Intensity: Mount Etna Case. *Journal of Geophysical Research: Solid Earth (1978–2012)*, 112(B2).

McCullagh, M., 1988. Terrain and Surface Modelling Systems: Theory and Practice. *The Photogrammetric Record*, 12(72): 747–779.

Melzer, T., 2007. Non-Parametric Segmentation of ALS Point Clouds Using Mean Shift. *Journal of Applied Geodesy*, 1(3): 159–170.

Meng, X., Currit, N. and Zhao, K., 2010. Ground Filtering Algorithms for Airborne LiDAR Data: A Review of Critical Issues. *Remote Sensing*, 2(3): 833–860.

Mirzaei, K., Arashpour, M., Asadi, E., Masoumi, H., Bai, Y. and Behnood, A., 2022. 3D Point Cloud Data Processing with Machine Learning for Construction and Infrastructure Applications: A Comprehensive Review. *Advanced Engineering Informatics*, 51(2022): 101501. ISSN 1474-0346. https://doi.org/10.1016/j.aei.2021.101501. https://www.sciencedirect.com/science/article/pii/S1474034621002500

Mongus, D. and Žalik, B., 2012. Parameter-Free Ground Filtering of LiDAR Data for Automatic DTM Generation. *ISPRS Journal of Photogrammetry and Remote Sensing*, 67: 1–12.

Monserrat, O. and Crosetto, M., 2008. Deformation Measurement Using Terrestrial Laser Scanning Data and Least Squares 3D Surface Matching. *ISPRS Journal of Photogrammetry and Remote Sensing*, 63(1): 142–154.

Morin, K.W., 2002. *Calibration of Airborne Laser Scanners*. Master of Science Thesis, University of Calgary, Calgary, Alberta, 134 p.

Morsdorf, F., Nichol, C., Malthus, T. and Woodhouse, I.H., 2009. Assessing Forest Structural and Physiological Information Content of Multi-Spectral LiDAR Waveforms by Radiative Transfer Modelling. *Remote Sensing of Environment*, 113(10): 2152–2163.

Mücke, W., 2008. *Analysis of Full-Waveform Airborne Laser Scanning Data for the Improvement of DTM Generation*. M. Sc. Thesis, Institute of Photogrammetry and Remote Sensing, Vienna University of Technology, Vienna, Austria.

Muhadi, Nur Atirah, Abdullah, Ahmad Fikri, Bejo, Siti Khairunniza, Mahadi, Muhammad Razif and Mijic, Ana, 2020. The Use of LiDAR-Derived DEM in Flood Applications: A Review. *Remote Sensing*, 12(14): 2308. https://doi.org/10.3390/rs12142308

Neuenschwander, A.L., Magruder, L.A. and Tyler, M., 2009. Landcover Classification of Small-Footprint, Full-Waveform Lidar Data. *Journal of Applied Remote Sensing*, 3.

Norini, G., Groppelli, G., Capra, L. and De Beni, E., 2004. Morphological Analysis of Nevado de Toluca volcano (Mexico): New Insights into the Structure and Evolution of an Andesitic to Dacitic Stratovolcano. *Geomorphology*, 62(1): 47–61.

OGC, 2012. *OGC City Geography Markup Language (CityGML) Encoding Standard*. Version 2.0. https://www.ogc.org/about-ogc/

Okatani, I.S. and Deguchi, K., 2000. A Method for Fine Registration of Multiple View Range Images Considering the Measurement Error Properties. *Pattern Recognition, 2000. Proceedings. 15th International Conference on*, 1, pp. 280–283. https://doi.org/10.1109/ICPR.2000

Oude Elberink, S.J. and Vosselman, G., 2009. 3D Information Extraction from Laser Point Clouds Covering Complex Road Junctions. *The Photogrammetric Record*, 24(125): 23–36.

Persson, Å., Söderman, U., Töpel, J. and Ahlberg, S., 2005. Visualization and Analysis of Full-Waveform Airborne Laser Scanner Data. *International Archives of Photogrammetry, Remote Sensing and Spatial Information Sciences*, 36(3/W19): 103–108.

Petrie, G. and Toth, C.K., 2008a. Airborne and Spaceborne Laser Profilers and Scanners. In: J. Shan and C.K. Toth (eds.), *Topographic Laser Ranging and Scanning Principles and Processing*, CRC Press, Boca Raton, FL, pp. 29–85.

Petrie, G. and Toth, C.K., 2008b. Terrestrial laser scanners. In: J. Shan and C.K. Toth (eds.), *TOPOGRAPHIC LASER RANGING AND SCANNING Principles and Processing*, CRC, Boca Raton, FL, pp. 87–127.

Pfeifer, N. and Briese, C., 2007. Geometrical Aspects of Airborne Laser Scanning and Terrestrial Laser Scanning. *International Archives of Photogrammetry, Remote Sensing and Spatial Information Sciences*, 36(3/W52): 311–319.

Pfeifer, N. and Mandlburger, G., 2008. LiDAR Data Filtering and DTM Generation. In: J. Shan and C.K. Toth (eds.), *Topographic Laser Ranging and Scanning Principles and Processing*, CRC Press, Boca Raton, FL, pp. 307–333.

Pingel, T.J., Clarke, K.C. and McBride, W.A., 2013. An Improved Simple Morphological Filter for the Terrain Classification of Airborne LIDAR Data. *ISPRS Journal of Photogrammetry and Remote Sensing*, 77: 21–30.

Pu, S. and Vosselman, G., 2006. Automatic Extraction of Building Features from Terrestrial Laser Scanning. *International Archives of Photogrammetry, Remote Sensing and Spatial Information Sciences*, 36(Part 5): 5 pages (on CD-ROM).

Pu, S. and Vosselman, G., 2009a. Building Facade Reconstruction by Fusing Terrestrial Laser Points and Images. *Sensors*, 9(6): 4525–4542.

Pu, S. and Vosselman, G., 2009b. Knowledge Based Reconstruction of Building Models from Terrestrial Laser Scanning Data. *ISPRS Journal of Photogrammetry and Remote Sensing*, 64(6): 575–584.

Rabbani, T., 2006. *Automatic Reconstruction of Industrial Installations Using Point Clouds and Images*.Ph.D. Thesis, Delft University of Technology, 175 p.

Rabbani, T., van den Heuvel, F.A. and Vosselman, G., 2006. Segmentation of Point Clouds Using Smoothness Constraint. *International Archives of Photogrammetry, Remote Sensing and Spatial Information Sciences*, 36(5): 248–253.

Reshetyuk, Y., 2006. Calibration of Terrestrial Laser Scanners Callidus 1.1, Leica HDS 3000 and Leica HDS 2500. *Survey Review*, 38(302): 703–713.

Rivera, G., Porras, R., Florencia, R. and Patricia Sánchez-Solís, J., 2023. LiDAR Applications in Precision Agriculture for Cultivating Crops: A Review of Recent Advances. *Computers and Electronics in Agriculture*, 207(2023): 107737. ISSN 0168-1699. https://doi.org/10.1016/j.compag.2023.107737. https://www.sciencedirect.com/science/article/pii/S0168169923001254

Roriz, R., Cabral, J. and Gomes, T., 2022. Automotive LiDAR Technology: A Survey. *IEEE Transactions on Intelligent Transportation Systems*, 23(7): 6282–6297, July. http://doi.org/10.1109/TITS.2021.3086804

Rottensteiner, F., Trinder, J., Clode, S. and Kubik, K., 2005. Using the Dempster-Shafer Method for the Fusion of LIDAR Data and Multi-spectral Images for Building Detection. *Information Fusion*, 6(4): 283–300.

Rutzinger, M., Elberink, S.O., Pu, S. and Vosselman, G., 2009. Automatic Extraction of Vertical Walls from Mobile and Airborne Laser Scanning Data. *The International Archives of Photogrammetry, Remote Sensing and Spatial Information Sciences*, 38(Part 3): W8.

Samet, H., 1990. *Applications of Spatial Data Structures: Computer Graphics, Image Processing, and GIS*, Addison-Wesley, Reading, MA, 507 p.

Sampath, A. and Shan, J., 2007. Building Boundary Tracing and Regularization from Airborne LiDAR Point Clouds. *Photogrammetric Engineering and Remote Sensing*, 73(7): 805–812.

Santibanez, S.F., dos Santos, D.R. and Faggion, P.L., 2012. Influence of Fitting Models and Point Density Sample in the Detection of Deformations of Structures Using Terrestrial Laser Scanning. *Applied Geomatics*, 4(1): 11–19.

Schenk, T., 2001. *Modeling and Analyzing Systematic Errors in Airborne Laser Scanners.* Technical Notes in Photogrammetry No 19. Department of Civil and Environmental Engineering and Geodetic Science, The Ohio State University, Columbus, OH.

Schmitt, A. and Vogtle, T., 2009. An Advanced Approach for Automatic Extraction of Planar Surfaces and Their Topology from Point Clouds. *Photogrammetrie-Fernerkundung-Geoinformation*, 2009(1): 43–52.

Schuster, H.-F., 2004. Segmentation of LIDAR Data Using the Tensor Voting Framework. *International Archives of Photogrammetry, Remote Sensing and Spatial Information Sciences*, 35: 1073–1078.

Segal, A., Haehnel, D. and Thrun, S., 2009. Generalized-ICP. *Robotics: Science and Systems*, 2: 4.

Shakarji, C.M., 1998. Least-Squares Fitting Algorithms of the NIST Algorithm Testing System. *Journal of Research of National Institute of Standards and Technology*, 103: 633–641.

Sithole, G., 2005. *Segmentation and Classification of Airborne Laser Scanner Data*.Ph.D. Thesis, Delft University of Technology, 203 p.

Skaloud, J. and Lichti, D., 2006. Rigorous Approach to Bore-sight Self-calibration in Airborne Laser Scanning. *ISPRS Journal of Photogrammetry and Remote Sensing*, 61(1): 47–59.
Staiger, R., 2003. Terrestrial Laser Scanning Technology, Systems and Applications. *2nd FIG Regional Conference Marrakech, Morocco*, 1.
Stilla, U. and Jutzi, B., 2008. Waveform Analysis for Small-Footprint Pulsed Laser Systems. In: J. Shan and C.K. Toth (eds.), *Topographic Laser Ranging and Scanning Principles and Processing*, CRC Press, Boca Raton, FL, pp. 215–234.
Talaya, J. et al., 2004. Integration of a Terrestrial Laser Scanner with GPS/IMU Orientation Sensors. *Proceedings of the XXth ISPRS Congress*, 35: 1049–1055.
Tao, C.V. and Li, J., 2007. *Advances in Mobile Mapping Technology: ISPRS Series*, 4, CRC Press, Boca Raton, FL.
Tsai, F. and Philpot, W.D., 1998. Derivative Analysis of Hyperspectral Data. *Remote Sensing of Environment*, 66(1): 41–51.
Tsai, F. and Philpot, W.D., 2002. A Derivative-Aided Hyperspectral Image Analysis System for Land-Cover Classification. *Geoscience and Remote Sensing, IEEE Transactions on*, 40(2): 416–425.
Vaze, J., Teng, J. and Spencer, G., 2010. Impact of DEM Accuracy and Resolution on Topographic Indices. *Environmental Modelling & Software*, 25(10): 1086–1098.
Vosselman, G., 2009. Advanced Point Cloud Processing. *Photogrammetric Week*, 9: 137–146.
Vosselman, G. and Dijkman, S., 2001. 3D Building Model Reconstruction from Point Clouds and Ground Plans. *International Archives of Photogrammetry and Remote Sensing*, 34, 3/W4, Annapolis, Maryland, pp. 37–43.
Vosselman, G., Gorte, B.G.H., Sithole, G. and Rabbani, T., 2004. Recognising Structure in Laser Scanner Point Clouds. *International Archives of Photogrammetry and Remote Sensing*, 36(part 8/W2): 33–38.
Wagner, W., 2010. Radiometric Calibration of Small-Footprint Full-Waveform Airborne Laser Scanner Measurements: Basic Physical Concepts. *ISPRS Journal of Photogrammetry and Remote Sensing*, 65(6): 505–513.
Wagner, W., Hollaus, M., Briese, C. and Ducic, V., 2008. 3D Vegetation Mapping Using Small-Footprint Full-Waveform Airborne Laser Scanners. *International Journal of Remote Sensing*, 29(5): 1433–1452.
Wagner, W., Ullrich, A., Ducic, V., Melzer, T. and Studnicka, N., 2006. Gaussian Decomposition and Calibration of a Novel Small-Foorprint Full-Waveform Digitising Airborne Laser Scanner. *ISPRS Journal of Photogrammetry and Remote Sensing*, 60(2): 100–112.
Wagner, W., Ullrich, A., Melzer, T., Briese, C. and Kraus, K., 2004. From Single-Pulse to Full-Waveform Airborne Laser Scanners: Potential and Practical Challenges. *International Archives of Photogrammetry and Remote Sensing*, 35(B3): 201–206.
Wang, C. et al., 2013. Wavelet Analysis for ICESat/GLAS Waveform Decomposition and Its Application in Average Tree Height Estimation. *IEEE Geoscience and Remote Sensing Letters*, 10(1): 115–119.
Wang, M. and Tseng, Y.-H., 2010. Automatic Segmentation of LiDAR Data into Coplanar Point Clusters Using an Octree-Based Split-and-Merge Algorithm. *Photogrammetric Engineering and Remote Sensing*, 76(4): 407–420.
Wang, M. and Tseng, Y.-H., 2011. Incremental Segmentation of LiDAR Point Clouds with an Octree-Structured Voxel Space. *The Photogrammetric Record*, 26(133): 32–57.
Weingarten, J.W., Gruener, G. and Siegwart, R., 2004. Probabilistic Plane Fitting in 3D and an Application to Robotic Mapping. *Robotics and Automation, 2004. Proceedings. ICRA'04. 2004 IEEE International Conference on*, 1: 927–932.
Wu, B. et al., 2013. A Voxel-Based Method for Automated Identification and Morphological Parameters Estimation of Individual Street Trees from Mobile Laser Scanning Data. *Remote Sensing*, 5(2): 584–611.
Wu, J.Y., van Aardt, J.A.N. and Asner, G.P., 2011. A Comparison of Signal Deconvolution Algorithms Based on Small-Footprint LiDAR Waveform Simulation. *IEEE Transactions on Geoscience and Remote Sensing*, 49(6): 2402–2414.
Xu, Dandan, Wang, Haobin, Xu, Weixin, Luan, Zhaoqing, and Xu, Xia. 2021. LiDAR Applications to Estimate Forest Biomass at Individual Tree Scale: Opportunities, Challenges and Future Perspectives. *Forests*, 12(5): 550. https://doi.org/10.3390/f12050550
Yang, M.-D. et al., 2011. Landslide-Induced Levee Failure by High Concentrated Sediment Flow—A Case of Shan-An Levee at Chenyulan River, Taiwan. *Engineering Geology*, 123(1): 91–99.

Yao, T. et al., 2011. Measuring Forest Structure and Biomass in New England Forest Stands Using Echidna Ground-Based Lidar. *Remote Sensing of Environment*, 115(11): 2965–2974.

Zhang, K. and Whitman, D., 2005. Comparison of Three Algorithms Filtering Airborne Lidar Data. *Photogrammetric Engineering and Remote Sensing*, 71(3): 313–324.

Zhang, Z., 1994. Iterative Point Matching for Registration of Free-Form Curves and Surfaces. *International Journal of Computer Vision*, 13(2): 119–152.

Zhao, T. and Wang, J., 2014. Use of Lidar-Derived NDTI and Intensity for Rule-Based Object-Oriented Extraction of Building Footprints. *International Journal of Remote Sensing*, 35(2): 578–597.

Zhenlong, Li, Hodgson, Michael E. and Li, Wenwen, 2018. A General-Purpose Framework for Parallel Processing of Large-Scale LiDAR Data. *International Journal of Digital Earth*, 11(1): 26–47. http://doi.org/10.1080/17538947.2016.1269842

Zwally, H. et al., 2002. ICESat's Laser Measurements of Polar Ice, Atmosphere, Ocean, and Land. *Journal of Geodynamics*, 34(3): 405–445.

Part II

Change Detection

12 Forest Clear-Cutting Detection in Subtropical Regions with Time Series Remotely Sensed Data

Guiying Li, Mingxing Zhou, Ming Zhang, and Dengsheng Lu

LIST OF ACRONYMS

BFAST	Breaks for additive season and trend
CCDC	Continuous change detection and classification
CFMASK	C function of mask
ETM+	Enhanced thematic mapper plus
ESA	European space agency
FMPI	Forest management planning inventory
GEE	Google Earth engine
GRD	Ground range detected
LaSRC	Land surface reflectance code
LEDAPS	Landsat ecosystem disturbance adaptive processing system
OLI	Operational land imager
SMILE	Statistical machine intelligence and learning engine
TM	Thematic mapper
VCT	Vegetation change tracker

12.1 INTRODUCTION

Remote sensing due to its capability to repeatedly capture land surface features in a large area with digital data format becomes the major tool to detect land cover change, including specific change trajectories such as forest deforestation and urban expansion, as summarized in previous publications (e.g., Singh, 1989; Lu et al., 2004; Bhagat, 2012; Bai et al., 2022; Shafique et al., 2022). Forest clear-cutting as a special land cover change category is a common practice for timber/pulp harvesting or forest regeneration in which all trees are cleared from a selected site at the same time period, irrelevant of tree ages and species. Although clear-cutting has some advantages such as financial efficiency, ease of tree replanting, and improvement of forest health, it has many negative effects on ecosystems, including land degradation, soil erosion, and loss of diversity (Li, 2004; Jokela et al., 2019). One notable and immediate effect of forest clear-cutting is large emissions of carbon dioxide to the atmosphere (Paul-Limoges et al., 2015). Accurately mapping and understanding the spatial distribution of clear-cutting areas is essential for assessing the potential of forest carbon storage, achieving carbon neutrality goal, and promoting "high-quality development with a high-quality ecological environment."

Previous methods for estimating deforestation over large areas include point sampling and bi-temporal change detection (Grinand et al., 2013; Lu et al., 2013). The accuracy of point sampling

DOI: 10.1201/9781003541158-14

approach is closely related to the sampling designs (Steininger et al., 2009), and it is not able to produce a deforestation map of the entire region. Multi-temporal or bi-temporal change detection methods are often used to detect land cover change based on comparison of two images or classification results (Lu et al., 2004, 2014). The change detection accuracy is influenced by the image quality or the accuracy of the classified results (Li et al., 2019). Individual errors will be multiplied if errors on two maps are assumed to be independent (Fuller et al., 2003). A decade years ago, majority of change detection studies used multi-temporal or bi-temporal images for examining land cover change because of limitation in data availability (Lu et al., 2013; Xi et al., 2016). Many change detection approaches as summarized in Singh (1989) and Lu et al. (2004) are suitable for dealing with bi-temporal images, but cannot effectively handle dense time series images.

Selection of a suitable time interval is an important concern for effectively detecting land cover change (Lu et al., 2014). A long time interval such as 5 or 10 years between two dates of images may be not able to detect forest deforestation date and location due to afforestation or regrowth (Lu et al., 2013). This requires collection of time series images to accurately detect the exact years of forest deforestation. Advances in satellite mapping technologies and availability of Landsat archive at no cost provide new opportunities to achieve this goal, and indeed lead to the development of automated algorithms that can track and characterize forest disturbances occurring in forest dynamics (Kennedy et al., 2010; Verbesselt et al., 2012; Zhu and Woodcock, 2014). For examples, Continuous Change Detection and Classification (CCDC) (Zhu et al., 2020; Zhu and Woodcock, 2014), Landsat-based Detection of Trends in Disturbance and Recovery (LandTrendr) (Kennedy et al., 2010), Vegetation Change Tracker (VCT) (Huang et al., 2010), and Breaks For Additive Season and Trend (BFAST) (Verbesselt et al., 2012) are common algorithms for forest disturbance detection at present. These algorithms can detect not only stand-replacing disturbances (e.g., logging, fire) and non-stand-replacing disturbances (drought, insect infestation, wind, etc.), but also can handle unmasked clouds in time series models (Coops et al., 2020; Zhang et al., 2022). LandTrendr and VCT use a single spectral observation per year to construct annual time series datasets, while BFAST and CCDC use every clear observation from all years to drive a phenology model, helping capture sub-annual changes.

Different algorithms have their own advantages and disadvantages, and their performances depend on the research objectives and datasets used. A comparative analysis of the performances among CCDC, VCT, LandTrendr, and MACD (Moving Average Change Detection) in mapping secondary forest in a subtropical region of China found that CCDC produced the highest accuracy in detecting disturbance among four individual algorithms (Zhang et al., 2023). The authors further developed an ensemble model to integrate a couple of algorithms to improve detection accuracy. These methods have been ported to the Google Earth Engine (GEE), which is an integrated cloud computing platform for remote sensing and geographic information processing, facilitating the analysis of huge-volume satellite data in large areas (Hamunyela et al., 2020). However, little work has been done in identifying signal characteristics related to clear-cutting, which might be the most obvious disturbances. Although there were some studies using time series Landsat data to differentiate disturbance causes (Schroeder et al., 2011; Zhang et al., 2022), few studies have focused on the subtropics and tropics of China where clear-cutting is commonly practiced, especially for plantations such as eucalyptus and Chinese fir. In addition to Landsat, another common optical sensor data is Sentinel-2 for land surface disturbance detection (Zhang Y. et al., 2021b). However, the relatively short time series Sentinel-2 data comparing with Landsat make it less application in reality, especially for the studies requiring a long time period of data sources.

Time series data are capable of portraying trends in the ground surface, yet a single sensor is often unable to obtain a comprehensive view of the surface state (Colson et al., 2018). The use of time series multi-sensor data allows detecting disturbances with high spatial and temporal scales (Ienco et al., 2019). Commonly used multi-sensor data include Landsat, MODIS, Sentinel-1, and Sentinel-2 (Long et al., 2021; Yang et al., 2023; Yin et al., 2022). Most studies focused on local and regional scales, and the time range is concentrated around 20 years (Du et al., 2023; Li et al., 2023;

Long et al., 2021). The currently used algorithms can be categorized into two main groups: traditional methods and deep learning methods. Traditional methods include the aforementioned algorithms such as LandTrender, CCDC, and BFAST. These methods generally do not require a large number of samples, but feature engineering (e.g., calculation of a specific vegetation index) is often required before employing specific algorithms, so is the time alignment of the time series multi-sensor data (e.g., filling data to equal time interval) before data input (Long et al., 2021). Many studies have used the LandTrender algorithm for fusion of time series multi-sensor data, especially Landsat and Sentinel-2 data (Yin et al., 2022; Zhang Y. et al., 2021b). Deep learning methods mainly include Convolutional Neural Networks (Grings et al., 2020), Recurrent Neural Networks (Lattari et al., 2022), and Transformer (Du et al., 2023). They have a strong feature extraction capability of capturing temporal features directly from time series images using an end-to-end architecture to detect disturbances with high spatiotemporal accuracy (Gómez et al., 2016; Du et al., 2023). However, deep learning usually requires a large amount of training data, which creates difficulties for their extensive application. For both traditional and deep learning methods, the fusion of time-series multi-sensor data is still at a rudimentary stage, and the use of time series multi-sensor data in forest disturbance detection has not been fully examined. Many studies simply treat different data sources as multiple variables in the model (Li et al., 2023; Long et al., 2021). How to effectively integrate multi-sensor data, including optical and radar data, into a disturbance detection model (traditional or deep learning), needs to be further explored.

Time series remotely sensed data have been extensively employed for land cover or forest disturbance detection (Ienco et al., 2019; Zhang et al., 2017), however, how the time series data can be used for forest clear-cutting detection in subtropical regions have not been fully examined. This research takes two regions as case studies to (1) examine the role of time series Landsat data in mapping forest clearing-cutting distribution in Fujian Province, (2) examine the capability of using time series multi-sensor data to detect forest clear-cutting distribution in Pu'er City of Yunnan Province. Through this research, we can better understand the important role of time series remotely sensed data, including the single sensor and multiple sensor data, to detect forest clear-cutting.

12.2 APPLICATION OF TIME SERIES LANDSAT IMAGES TO FOREST CLEAR-CUTTING DETECTION IN FUJIAN PROVINCE

12.2.1 Study Area and Datasets

12.2.1.1 Study Area—Fujian Province

Fujian Province is located on the southeast coast of mainland China, facing Taiwan across the Taiwan Strait. It borders Zhejiang Province to the north, Guangdong Province to the south, and Jiangxi Province to the west (Figure 12.1). It covers an area of about 124,000 km^2 with mostly mountainous regions. Fujian has a subtropical climate with hot and humid summers and mild winters. The average annual temperature ranges from 17° to 21°C, and average annual precipitation ranges from 1400 mm to 2000 mm. Fujian is abundant in forest resources with a forest coverage rate of 66.8% in 2023, ranking the first in China for 44 consecutive years. The dominant vegetation type is evergreen broadleaf forest, accounting for 83.0% of the total vegetated land, followed by evergreen coniferous forest (14.4%) and shrubland (1.8%) (see Figure 12.1).

12.2.1.2 Datasets—Time Series Landsat Data

The available Landsat Collection 2 surface reflectance data covering Fujian Province (Table 12.1) were all processed using GEE platform. The collection has been preprocessed already, including atmospheric correction, terrain correction, and cloud and cloud-shadow detection. The Thematic Mapper (TM) and Enhanced Thematic Mapper Plus (ETM+) datasets were atmospherically corrected using the LEDAPS (Landsat Ecosystem Disturbance Adaptive Processing System) (Masek et al., 2013), while the OLI (Operational Land Imager) dataset was atmospherically corrected using

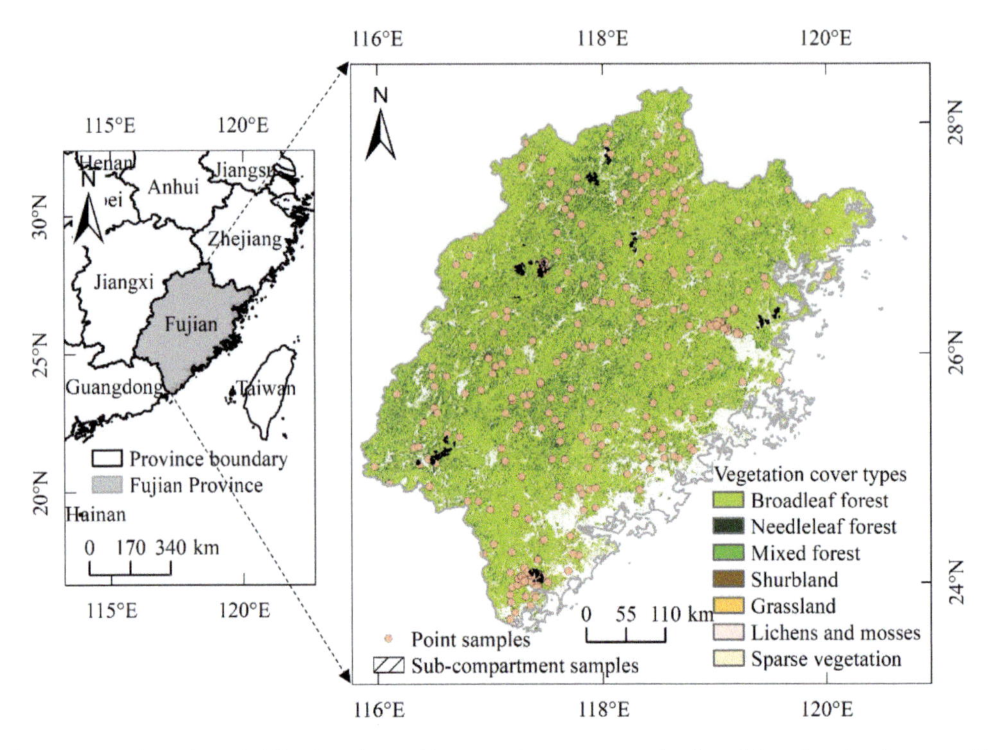

FIGURE 12.1 Location of Fujian Province with vegetation cover map developed from Landsat images which were acquired in 2022 and with spatial distribution of the samples.

TABLE 12.1

A Summary of Landsat Images Used in This Study

	Dataset Availability	Collection ID	No. of Images
Landsat 4 TM	8/22/1982–6/24/1993	LANDSAT/LT04/C02/T1_L2	17
Landsat 5 TM	3/16/1984–5/5/2012	LANDSAT/LT05/C02/T1_L2	5140
Landsat 7 ETM+	5/28/1999–	LANDSAT/LE07/C02/T1_L2	5147
Landsat 8 OLI	3/18/2013–	LANDSAT/LC08/C02/T1_L2	2528
Landsat 9 OLI	10/31/2021–	LANDSAT/LC09/C02/T1_L2	255

the LaSRC (Land Surface Reflectance Code) (Vermote et al., 2016). The cloud and cloud shadow pixels were masked according to the QA_PIXEL band, which was generated by the CFMASK (C Function of Mask) algorithm (Foga et al., 2017). These image collections were filtered for the study area and the time period (from 1 January 1986 to 31 December 2022). The spatial resolution of 30 m and temporal revisit interval of 16 days were adopted. Three visible bands (Blue, Green, and Red), one near-infrared (NIR) band, two short-wave infrared (SWIR1, SWIR2) bands, and the normalized burn index (NBR) were used for each image.

12.2.2 Forest Clear-Cutting Detection Using CCDC

The main components of this study include: (1) data preparations; (2) change detection based on the CCDC algorithm; and (3) use of the Smile random forest for attributing disturbance patches to disturbance event types (Figure 12.2).

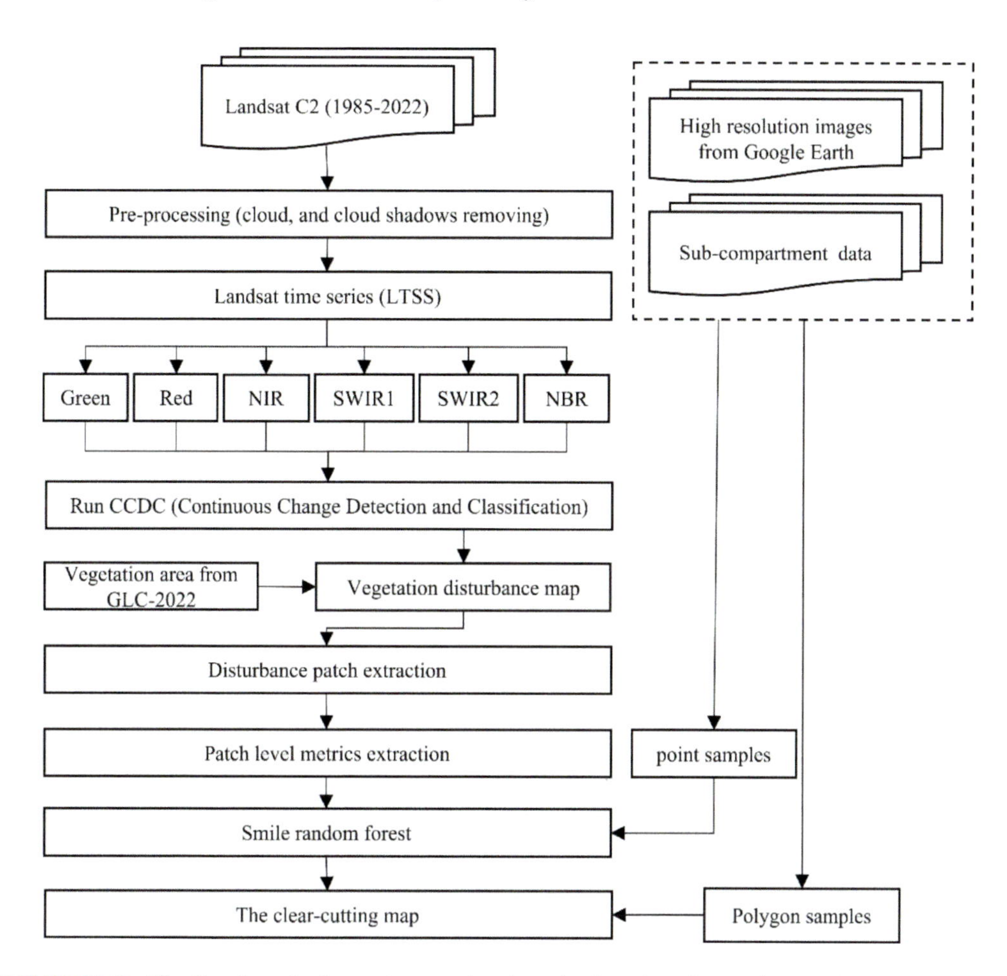

FIGURE 12.2 The flowchart for forest clear-cutting detection based on time series Landsat data.

12.2.2.1 Data Preparation

A total of 341 point samples were interpreted from high resolution images and time series Landsat images for use of training samples (Figure 12.1 and 12.3). The interpretation includes whether or not a change occurred during the study period, the timing of a change, and the type of a disturbance event (clear-cutting event vs. non-stand replacement). A total of 4,280 polygon samples were collected from the Forest Management Planning Inventory (FMPI) (Forest Inventory Level II in China) for validation (Figure 12.1 and 12.3). Four filtering rules were used to ensure the sample's quality: (1) removing polygons with age zero, which typically are non-forested land; (2) removing polygons with an area of less than 11 pixels, given that the minimum mapping unit is defined as one hectare; (3) removing elongated polygons (e.g., some logging roads); and (4) updating the age as of December 31, 2022, based on Landsat spectral trajectories and high-resolution images (Google Earth images slice) due to different survey dates of FMPI.

12.2.2.2 CCDC Algorithm

Vegetation change detection in Fujian Province was implemented using the CCD (Continuous Change Detection) algorithm, which is a change detection component of CCDC (Zhu and Woodcock, 2014). The CCDC application programming interface (https://github.com/parevalo/gee-ccdc-tools) was provided by Arévalo et al. (2020) on the GEE platform. This API (Application Programming

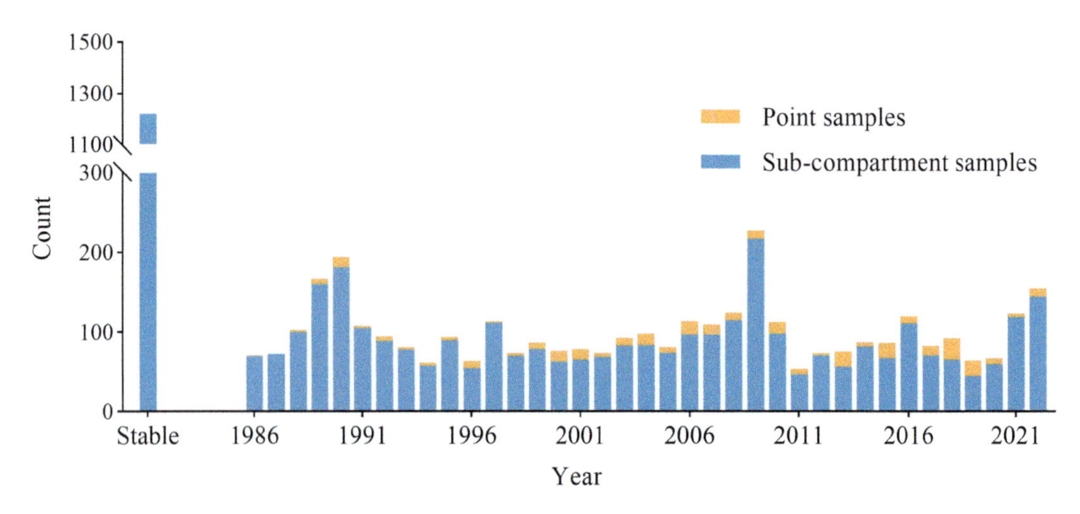

FIGURE 12.3 The number of point and sub-compartment samples each year.

Interface) facilatates the straitforward adoption and implementation of the CCDC algorithm on a large scale.

CCDC is a statistical boundary algorithm (Zhu et al., 2020). First, a specific harmonic model (Equation 12.1) is used to perform an initial fit based on a pre-given number of observations; A "break" is labeled if the predicted reflectance significantly deviates from the observed reflectance for a certain number (four in this study). After a break, another model is initiated again until the next break has been found or the observations have been exhausted. Finally, all observations are split into multiple time-segments, indicating the multiple changes (Xu et al., 2021).

$$\hat{p}(i,t)_{RIRLS} = c_{0,i} + c_{1,i}t + \sum_{n=1}^{3} a_{n,i}\cos(wnt) + b_{n,i}\sin(wnt) \quad (12.1)$$

where n and t are the order and the Julian date; i is the ith Landsat spectral band (Blue, Green, Red, SWIR1, SWIR2, and NBR); w is the base annual frequency ($2\pi/365.25$); $a_{n,i}$ and $b_{n,i}$ are the estimated nth order harmonic coefficients for the ith Landsat band; $c_{0,i}$ and $c_{1,i}$ are the estimated intercept and slope coefficients for the ith Landsat band. The CCD contains two important parameters: chiSquareProbability and minObservations. The higher their values, the fewer breakpoints are detected by the model. After testing the change detection accuracies of different combinations, the values of chiSquareProbability and minObservations were set as 0.99 and 4, respectively. To depict patches, adjacent pixels experiencing disturbance in the same year were assumed to have experienced the same event and were grouped together using an eight-neighbor adjacency rule. Patches with 11 or fewer pixels (e.g., patches smaller than one ha) were removed from the dataset to avoid small, usually unverifiable disturbance events. The resultant disturbance map formed the basis for all further analysis.

Six metrics were calculated to describe the spectral characteristics of clear-cutting agents: disturbance duration, disturbance magnitude, VI (vegetation index) values before and after the disturbance (Pre_VI, Post_VI), and VI slopes before and after the disturbance (Pre_VI_slope, Post_VI_slope). The disturbance duration is the difference between the end time of the preceding segment and the start time of the succeeding segment (C-B) (see Figure 12.4). The disturbance magnitude refers to the difference between the mean of the last five observations before the disturbance and the mean of the first five observations after the disturbance (Arévalo et al., 2020). The pre- and post-disturbance VI values refer to the synthetic values on a particular day prior to the end time of the preceding

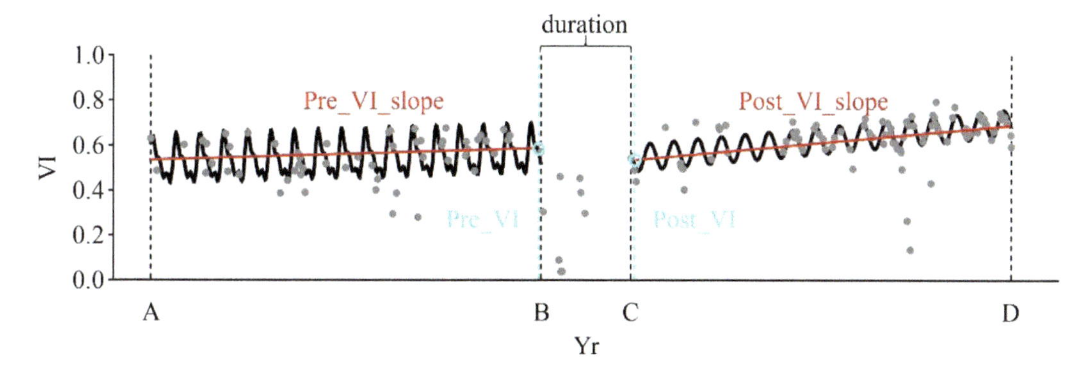

FIGURE 12.4 Conceptual figure explaining how to quantify a disturbance using spectral signal metrics: Pre_VI_slope, Post_VI_slope, Pre_VI, Post_VI, disturbance duration, and disturbance magnitude.

segment and on a particular day after the start time of the succeeding segment, respectively. This particular day was set as the first day of each year. The slopes before and after the disturbance refer to the slopes of the preceding segment and the succeeding segment, respectively. NBR and NIR were selected as VIs to characterize forest clear-cutting. These metrics were calculated at the pixel level (Figure 12.4) and were subsequently averaged at the patch level.

The Random Forest is introduced in GEE as a modified version under the name "smile random forest" (sRF), which was created in the JavaScript language using the framework of statistical machine intelligence and learning engine (SMILE) (Pham et al., 2023). The sRF has a similar structure to the original RF, except for the tree size regularization using the parameters maxNode and nodeSize. The sRF model has three main user-defined parameters: the number of trees (ntree), the number of variables per split (mtry), and the maximum number of leaf nodes in each tree (maxNodes). For mtry and maxNodes, the default values recommended by sRF were used. Compared to other parameters, ntree has the greatest impact on the performance of the sRF. In this study, the ntree was set as 100. Point samples were used to generate a trainer, which was then employed to train all disturbance patches detected by the CCDC to generate the final forest clear-cutting maps for each year.

12.2.2.3 Accuracy Assessment

The clear-cutting detection accuracy was assessed through polygon-wise comparisons in which the detected polygons were overlaid on the reference polygons. The specific steps are shown in Figure 12.5. Only if the total overlapping area of a reference polygon and a detected polygon is more than 50% of the area of the reference polygon and the theme of the reference polygon is clear-cutting can the detected polygon be considered a correct detection; otherwise, it is an incorrect detection. The value of 50% was deemed a reasonable balance that is neither too permissive nor too demanding (Linke et al., 2017). The producer's accuracy (PA), user's accuracy (UA), and overall accuracy (OA) were calculated to evaluate the forest clear-cutting detection results.

The temporal accuracy was assessed based on the spatially correct detection identified early. The majority was used to represent the clear-cutting year for the detected clear-cutting polygon. The OA was calculated from the agreements between the detected years and the true years of clear-cutting polygons. In addition, if the difference between the detected year and the reference year is not greater than 1, it is considered a correct identification, and OA (±1 year) was computed.

12.2.3 Results

The detection accuracy in the spatial and temporal domains is reported in Table 12.2 and Table 12.3, respectively. The OA of 83.0% in the spatial domain indicates a relatively high detection accuracy.

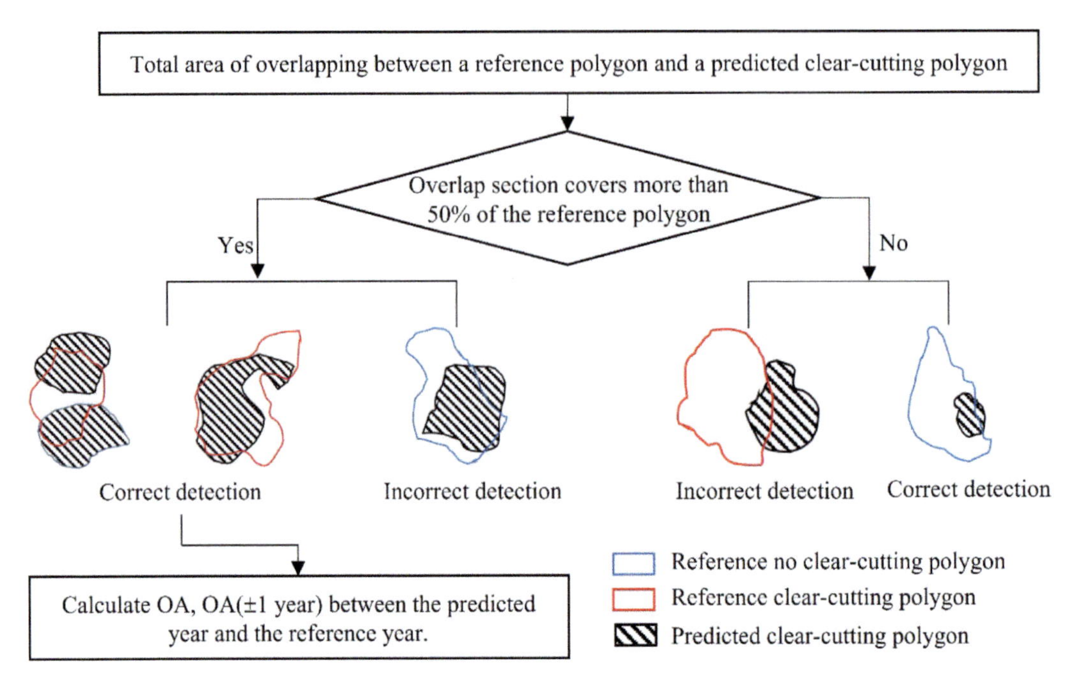

FIGURE 12.5 The strategy to determine forest clear-cutting samples for accuracy assessment.

TABLE 12.2
Accuracy Assessment Results in the Spatial Domain

	Reference Data			
	Stable	**Clear-Cutting**	**UA (%)**	**PA (%)**
Stable	1532	636	70.7	94.5
Clear-cutting	90	2022	95.7	76.1
OA (%)	83.0			

TABLE 12.3
Accuracy Assessment Results in the Temporal Domain

No. of changed samples correctly detectd	No. of changed samples in which changed years correctly identified	No. of changed samples in which changed years (±1 year) correctly identified	OA	OA (±1 year)
2022	1208	1709	59.7%	84.5%

The high UA (95.7%) and the low PA (76.1%) imply low commission and high omission of the clear-cutting detection. The OA of 59.7% (the identified clear-cutting date match the exact date inferred from the ages of the sub-compartment) in the temporal domain indicates a relatively poor overall performance. However, if we make the agreements between the identified cutting years and the reference years looser by one year, the OA (±1 year) rearches 84.5%.

The number of detected clear cuttings for each pixel within the detection period varies from zero to nine. Totally 30% of the vegetation area in Fujian Province experienced the clear-cutting agents during the period of 1986–2022, most of which experienced once, accounting for 78.4% of the total clear-cutting area (Figure 12.6). The clear-cutting areas are mainly located in the western and southern parts of Fujian Province. The detected cutting dates were greatly different across the province. For example, most clear-cutting happened before 2000 in the western part, such as Changting County, while most occurred after 2010 in the south, such as Zhangzhou (Figure 12.6). This distribution was mainly attributed to forestry policies and plantation species. More stricter measures for ecological restoration were implemented around 2000 in Changting County, which is known for its severe soil erosion problem (Cao et al., 2009), prohibiting forest cutting since then. Plantation species in the west part are dominated by Pinus massoniana and Cunninghamia

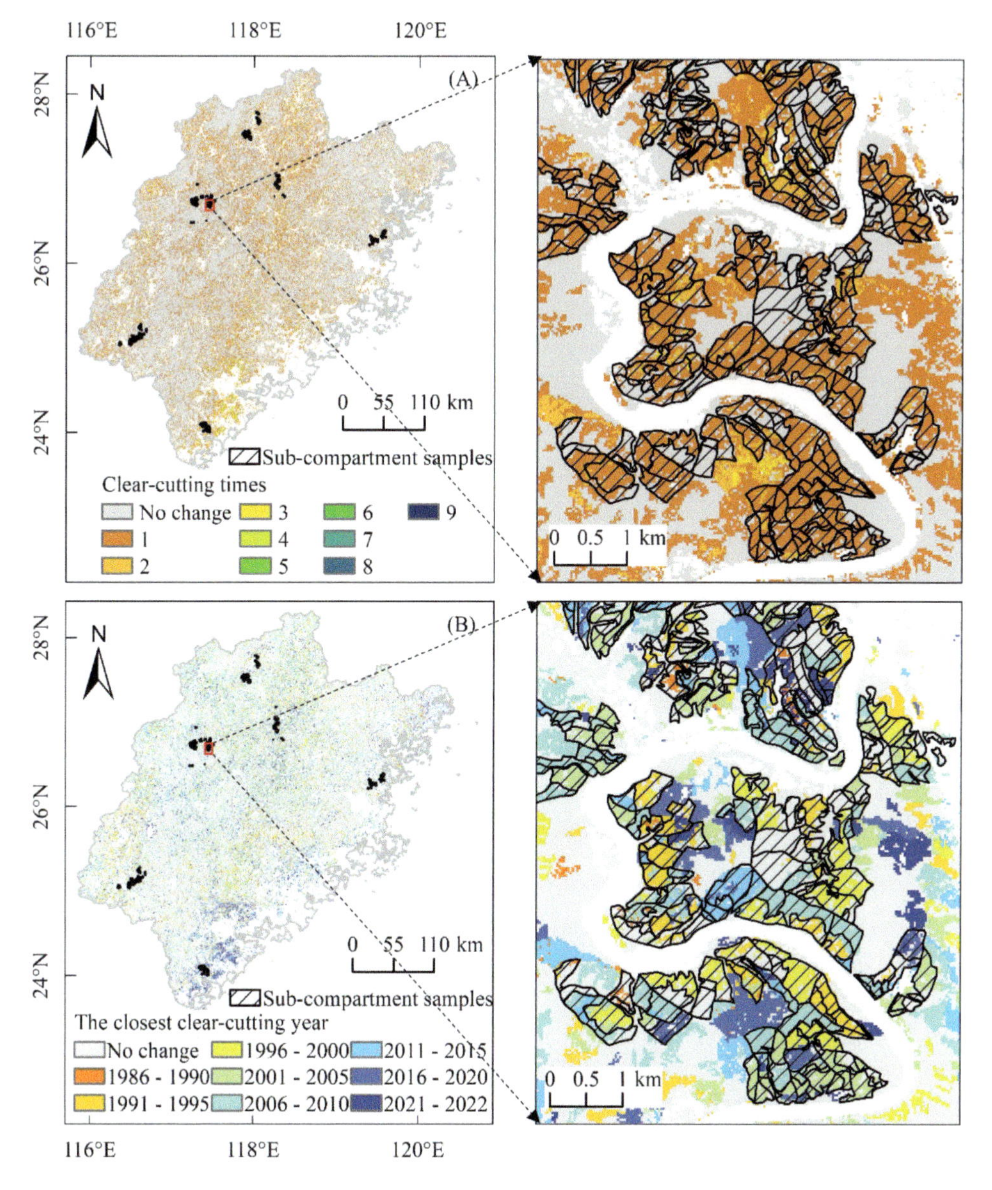

FIGURE 12.6 The distribution of the clear-cutting times and the latest clear-cutting year based on time series Landsat images.

lanceolata having a relatively long rotation (>25 years), whereas plantations in the south part are dominated by Eucalyptus having a short rotation (four to six years) (Lin et al., 2023).

12.2.4 Summary

Forest disturbances in Fujian Province were detected using the CCDC algorithm based on time series Landsat, and the locations and dates of clear-cutting areas were identified using random forest. The accuracy assessment indicates that this method can detect the spatiotemporal distribution of clear-cutting areas with an overall accuracy of 83.0% in the spatial domain and an overall accuracy (±1 year) of 84.5% in the temporal domain. This experiment shows that Earth observation can provide timely, independent, transparent, and consistent detection of clear-cutting across large geographical areas.

12.3 APPLICATION OF TIME SERIES MULTI-SENSOR DATA TO FOREST CLEAR-CUTTING DETECTION IN PU'ER CITY

12.3.1 Study Area and Datasets

12.3.1.1 Study Area—Pu'er City

Simao District of Pu'er City in the southern part of Yunnan Province was selected as a case study (Figure 12.7). The district has a total area of 3,928 km², of which 98.34% are mountainous areas. This region belongs to a south subtropical monsoon climate on the low-latitude plateau. The vertical

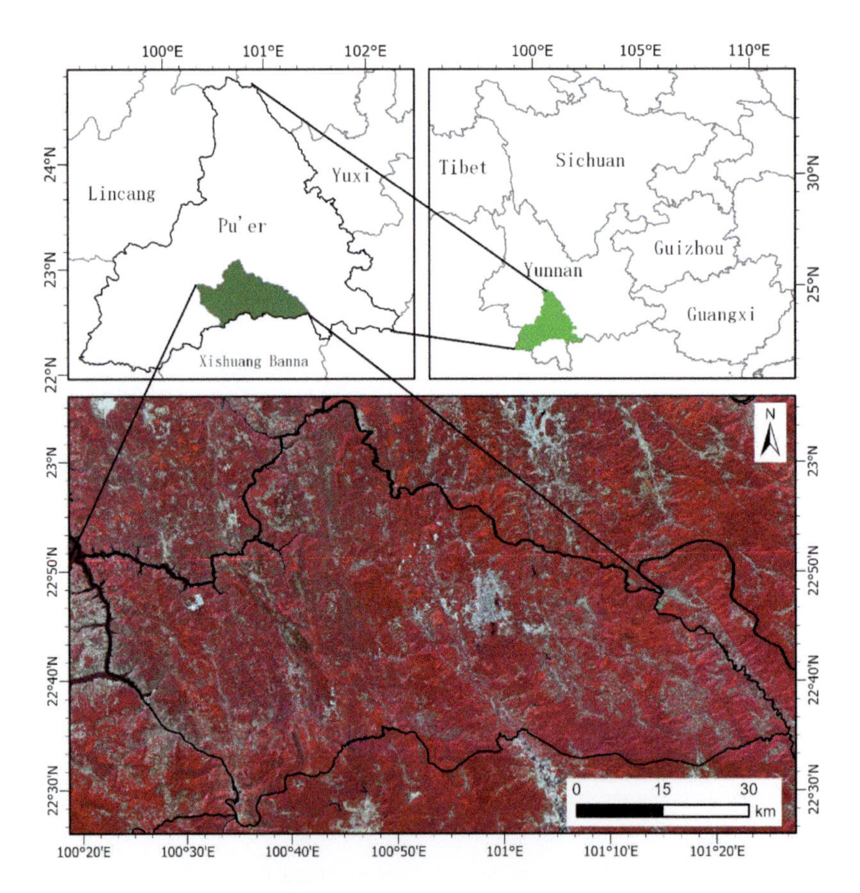

FIGURE 12.7 The location of Simao District in south Yunnan Province.

climate transition from low altitude to high altitude is the northern tropical, south subtropical, central subtropical, and northern subtropical in sequence. The average annual temperature is 18.9° C and average annual rainfall is 1,487 mm. The forest coverage rate of Simao district was 73.36% in 2022, making it the second largest forest district in Yunnan Province. The main forest types include Simao pine, Southwestern birch (*Betula alnoides*), oak, eucalyptus, and mixed coniferous and broadleaf forests.

12.3.1.2 Datasets—Combination of Sentinel-1 SAR and Sentinel-2 Multispectral Data

Field surveys which were conducted in March 2023 and visual interpretations on high spatial resolution images from Google Earth were used to collect clear-cutting samples. A total of 486 polygon samples were collected, covering an area of approximately 19.1 km^2.

Sentinel-1 level 1 GRD (Ground Range Detected) under the ascending orbit and IW mode (Interferometric Wide-Swath) and Sentinel-2 L2A multispectral data covering the study area (Tiles 47QPF and 47 QQF) acquired between 2020 and 2022 were obtained using Google Earth Engine, which were provided by the European Space Agency (ESA) (https://scihub.copernicus.eu/dhus/#/home). For Sentinel-1 radar data, dual-polarization bands (VV+VH) were preprocessed with thermal noise removal and radiometric correction, then resampled to a spatial resolution of 10 m to match the cell size of Sentinel-2 data. At the end, the monthly mean synthesis of VV and VH was generated, forming monthly Sentinel-1 time series data for the years 2020–2022. The Sentinel-2 L2A product contains ten spectral bands with a spatial resolution of 10 m. The product was already atmospherically corrected. In this study, Sentinel-2 images with a cloud cover less than 80% were chosen and de-clouded using the cloud probability band. The 12 scenes of Sentinel-2 images produced by monthly mean synthesis for each year were formed into a dataset of monthly time series of Sentinel-2 data spanning three years (2020–2022). Due to frequent cloudy and rainy weathers in subtropical regions, the number of available Sentinel-2 images varies among years. Table 12.4 summarizes the number of Sentinel-2 images and quality information in the study area.

12.3.2 Forest Clear-Cutting Detection with Deep Learning

12.3.2.1 Data Preparation

To implement deep learning model, which is a pixel-based processing algorithm, the time series remotely sensed data and the collected clear-cutting samples need further processing. First, 70% of 486 collected polygon samples were randomly selected as training samples for deep learning algorithm, and the rest as testing samples for accuracy assessment. The training samples were converted to a raster with a pixel size of 10 m × 10 m in alignment with the cell size of Sentinel-1/2 images; then the raster was overlaid on the time series remotely sensed images to extract time series values of each pixel within clear-cutting patches, forming the pixel-based training samples.

TABLE 12.4
Datasets Used in Simao County

Tile Number	Year	Image Number	Cloudiness < 10%	Cloudiness > 50%
47QPF	2020	50	42%	16%
	2021	58	28%	31%
	2022	43	28%	28%
47QQF	2020	92	40%	24%
	2021	94	28%	31%
	2022	76	30%	34%

12.3.2.2 Change Detection with Deep Learning

The Transformer deep learning algorithm was used as the base model for clear-cutting detection in this study. The Transformer consists of numerous self-attentive layers. With strong sequence modeling capabilities, it can automatically learn the dependencies between input sequences and the dynamics over time and effectively recognize and detect forest disturbances in the time series. Figure 12.8 shows the model architecture of the deep learning approach used in this study.

The Input Embedding module performs primary encoding of the time series data and uniformly maps time series of different lengths into vectors with length d_{model} (d_{model} is a hyperparameter that needs to be set by human beings). Assuming that the input time series is x, the process can be formulated as follows:

$$Input\,Embedding(x) = ReLU(W_1 x + b_1) W_2 + b_2$$

Where W_1, W_2, b_1, and b_2 represent the weight matrices and biases of two fully connected layers, ReLU represent the activation function. The Positional Encoding module models the sequential information of the time series observations in a process that can be expressed as follows:

$$PE(x) = \begin{cases} \sin(w_k x_i), if\ i = 2k \\ \cos(w_k x_i), if\ i = 2k+1 \end{cases}$$

$$where\ w_k = \frac{1}{10000^{\frac{2k}{d_{model}}}}$$

$$i = 0, 1, 2, 3, \ldots, \frac{d_{model}}{2} - 1$$

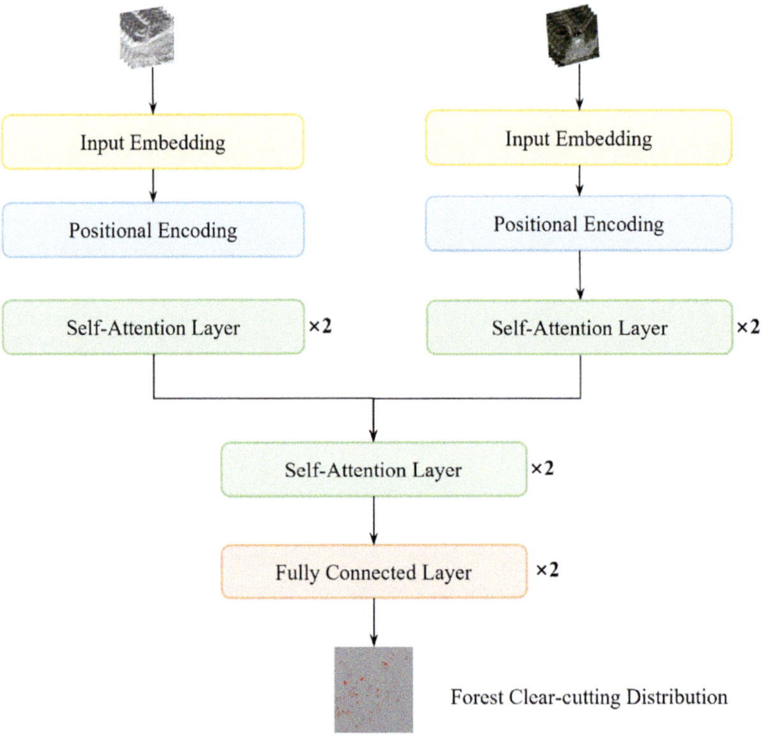

FIGURE 12.8 Architecture of deep learning model used in this research.

The training samples were input to the model, which uses a two-branch structure to encode and extract high-dimensional features from Sentinel-1 and Sentinel-2 time series data independently. To examine the effect of the Sentinel-1 data in clear-cutting detection, the extracted Sentinel-1 features were first used to predict clear-cutting distribution. However, the Sentinel-1 time series could not predict a satisfactory result. Thus, the features from Sentinel-1 and Sentinel-2 time series were fused using two self-attention layers and used to predict a clear-cutting distribution for each year.

12.3.2.3 Accuracy Assessment

The testing polygon samples were overlaid on the pixel-based clear-cutting map. The numbers of clear-cutting and no clear-cutting falling within each of the test polygons were counted and the majority was taken as the detection result of a polygon sample with either clear-cutting or not. The accuracy assessment was conducted based on the agreement between detected results and true values. OA and kappa coefficient were used to evaluate the overall performance, UA and PA were calculated from the error matrix and used to measure the performance of the model for forest clear-cutting class (Congalton and Green, 2019).

12.3.3 Results

The accurancy assessment results (Table 12.5) indicated that OA of 92.8% and a kappa coefficient of 0.856 were obtained using the deep learning method. The UA of clear-cutting (96.3%) is slightly better than the PA (89.0%), suggesting that some of the detected clear-cutting (3.7%) was misclassified as no clear-cutting, but at least 10% of the clear-cutting areas were not recognized. This may be attributed to the lower availability of good-quality Sentinel-2 images in subtropical regions. Sentinel-2 images with high cloud cover cannot recognize spectral differences between clear-cutting and non-clear-cutting, leaving some of the clear-cutting areas out.

The clear-cutting map detected using Transformer deep learning algorithm (Figure 12.9) indicates different spatial patterns of forest clear-cutting areas over time. For example, clear-cutting areas were mainly scattered in the eastern part of Simao District in 2020, while they were concentrated in the northern and central parts in 2021 and in the central part, sporadically distributed in the northern and eastern parts in 2022. In terms of the clear-cutting area, the largest cutting area occurred in 2021, reaching 190 km^2.

12.3.4 Summary

This case study shows the capability of using the Transformer deep learning algorithm for clear-cutting detection based on the combination of Sentiel-1 and Sentinel-2 time series images. The features from Sentinel-1 time series data could not successfully generate clear-cutting results, while the fusion of features extracted from Sentinel-1 and Sentinel-2 effectively detected clear-cutting

TABLE 12.5

Accuracy Assessment of Clear-Cutting Detection Results Using Deep Learning Method

Detected Reference	Clear-Cutting	No Clear-Cutting	UA (%)	PA (%)
Clear-cutting	130	16	96.3	89.0
No clear-cutting	5	140	89.7	96.6
OA: 92.8%				
Kappa: 0.856				

Note: UA and PA represent user's and producer's accuracy, OA represents overall accuracy.

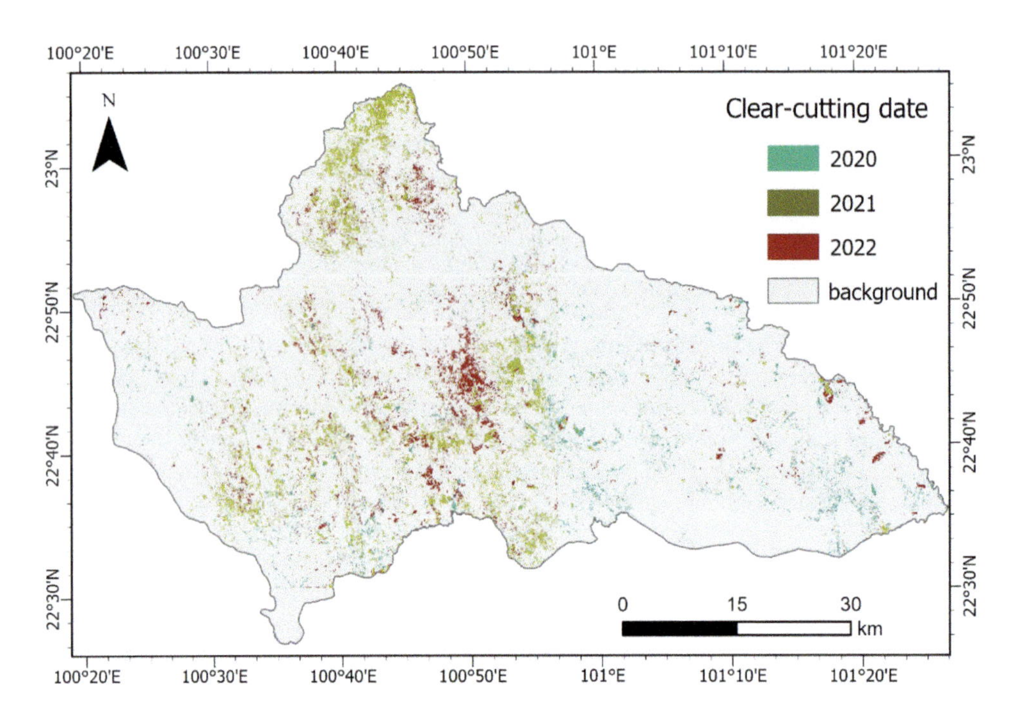

FIGURE 12.9 Forest clear-cutting distribution using deep learning algorithm based on time series Sentinel-1/2 data.

locations with an overall accuracy of 92.8%. However, the lack of sufficient good quality remote sensing data due to the frequent cloudy and rainy weather made it difficult to detect clear-cutting patches completely. This research implies the importance of combining multi-sensor data to form dense time series data for effectively detecting forest clear-cutting areas in the subtropical regions.

12.4 DISCUSSIONS

12.4.1 The Role of Time Series Landsat Data in Forest Clear-Cutting Detection

Before the Landsat data were available at no cost, land cover change detection was mainly based on multitemporal data with a temporal period of several years (e.g., five or even ten years' interval) (Lu et al., 2013). This long time interval may hinder true land cover change; for example, the clear-cutting detection of eucalyptus plantations may be difficult because eucalyptus may experience deforestation and regrowth within its short rotation of five to eight years (Li et al., 2022). To accurately detect forest clear-cutting dates and locations, it is required to use dense time series data, at least annual time interval. In recent decades, Landsat and Sentinel-2 time series data have provided necessary data sources for examining forest disturbances annually (Liu et al., 2017). This research also confirmed that time series Landsat data can be successfully used to detect forest clear-cutting in subtropical regions with an overall accuracy of 83%.

Global forest change detection was originally based on high temporal resolution but coarse spatial resolution datasets such as NOAA/AVHRR and MODIS before time series Landsat data were available in 2008 (Ressl et al., 2009). Since then, time series Landsat TM/ETM/OLI data have become the most commonly used data source because the analysis of long-term time series data can explore the characteristics and patterns of forest cover change that are crucial for an accurate carbon accounting (Goetz et al., 2009). According to the survey, forest patch sizes in subtropical regions of China are often relatively small (100–10,000 m^2) due to topographic variation and natural and anthropogenic disturbances (Shang et al., 2023). At the same time, Forest Harvesting and Management Regulations stipulate that the maximum clear-cutting area should be less than 20 hectares (approximately equal to 222 Landsat pixels or 0.8 MODIS pixels) considering the sustainability of forest resources and the prevention of biodiversity. Landsat images offer an optimal resolution for forest resource management (Townshend et al., 2012; Liu et al., 2017), and the extensively historical records can meet various requirements for resource management and ecological conservation.

12.4.2 The Role of Time Series Multi-Sensor Data in Forest Clear-Cutting Detection

Use of the same sensor data for forest cover change detection has advantages over the use of different sensor data, for example, the same sensor data can minimize the preprocessing in handling geometric errors and radiometric calibrations. However, in tropical and subtropical regions, the cloudy and rainy conditions make it difficult to collect a sufficient number of good-quality optical sensor data, while the use of different sensor data considerably increases new opportunities to collect good-quality images for accurately detecting clear-cutting areas (Lu et al., 2008; Yin et al., 2022). One challenge in using time series multi-sensor data is that existing change detection algorithms are difficult to handle different sensor data for the detection of forest cover change due to their differences in spatial, spectral, and radiometric resolutions, and different lengths of time series. For example, Landsat and Sentinel-2 satellites can acquire more usable images in one year than other satellites such as GaoFeng-1. On the other hand, the acquisition date of satellite observations is not uniformly distributed, and it is almost impossible to acquire good-quality images in seasons with frequent cloudy and rainy weathers. Therefore, how to utilize the temporal information of remote sensing data is the key to further improve the detection accuracy of forest clear-cutting areas. If sufficient training samples are available, deep learning algorithms will be a promising method to detect detailed clear-cutting features. This research confirmed the promise of using both Sentinel-1 SAR and Sentinel-2 multispectral data for successful detection of forest clear-cutting areas in the subtropical region in Yunnan Province. More research is needed on the effective data collection of training samples and the processing of time series multi-sensor data for application of deep learning algorithms to forest disturbance detection.

12.5 PERSPECTIVES

Time series Landsat and Sentinel-2 are currently the dominant data sources for the detection of forest clear-cutting areas, and most algorithms such as BFAST and LandTrendr were proposed based on these datasets. As the user's needs increase for high detection accuracy and effective detection of small patch sizes and exact clear-cutting dates, the spatial resolution of 10–30 m may not meet the requirement; thus different sensor images such as Quickbird, IKONOS, and GaoFeng-1/2/6 with high spatial resolution images (better than 5 m) may be needed to detect the clear-cutting area with small patch sizes, such as less than 1 ha. Current disturbance detection methods such as CCDC, LandTrendr, and VCT do not have the capability to extract this kind of clear-cutting information based on different sensor data. The advance of deep learning algorithms may provide new insights for accurately detecting this kind of changes; however, how to effectively collect the training samples for use in the deep learning algorithm will be critical for successfully conducting this change detection. More research in the future is needed to explore how to effectively use deep learning algorithms to extract detailed clear-cutting information (e.g., date, small parch size) based on multi-sensor data with high spatiotemporal resolutions.

12.6 ACKNOWLEDGMENTS

The research was funded by Intergovernmental International Scientific and Technological Innovation Cooperation Project: China-European Earth Observation Collaboration on Forest Monitoring Technology and Application (grant number 2021YFE0117700-2).

REFERENCES

Arévalo, P., Bullock, E.L., Woodcock, C.E., Olofsson, P., 2020. A suite of tools for continuous land change monitoring in Google earth engine. *Frontiers in Climate*, 2. https://doi.org/10.3389/fclim.2020.576740.

Bai, T., Wang, L., Yin, D., Sun, K., Chen, Y., Li, W., Li, D., 2022. Deep learning for change detection in remote sensing: A review. *Geo-spatial Information Science*. https://doi.org/10.1080/10095020.2022.2085633.

Bhagat, V.S., 2012. Use of remote sensing techniques for robust digital change detection of land: A review. *Recent Patents on Space Technology*, 2, 123–144.

Cao, S., Zhong, B., Yue, H., Zeng, H., Zeng, J., 2009. Development and testing of a sustainable environmental restoration policy on eradicating the poverty trap in China's changting county. *Proceedings of the National Academy of Sciences of the United States of America*, 106, 10712–10716. https://doi.org/10.1073/pnas.0900197106.

Colson, D., Petropoulos, G.P., Ferentinos, K.P., 2018. Exploring the potential of Sentinels-1 & 2 of the Copernicus Mission in support of rapid and cost-effective wildfire assessment. *International Journal of Applied Earth Observation and Geoinformation*, 73, 262–276.

Congalton, R.G., Green, K., 2019. *Assessing the Accuracy of Remotely Sensed Data: Principles and Practices* (third edition). CRC Press, Boca Raton, FL, p. 348. https://doi.org/10.1201/9781420055139

Coops, N.C., Shang, C., Wulder, M.A., White, J.C., Hermosilla, T., 2020. Change in forest condition: Characterizing non-stand replacing disturbances using time series satellite imagery. *Forest Ecology and Management*, 474, 118370. https://doi.org/10.1016/j.foreco.2020.118370.

Du, B., Yuan, Z., Bo, Y., Zhang, Y., 2023. A combined deep learning and prior knowledge constraint approach for large-scale forest disturbance detection using time series remote sensing data. *Remote Sensing*, 15. https://doi.org/10.3390/rs15122963.

Foga, S., Scaramuzza, P.L., Guo, S., Zhu, Z., Dilley, R.D., Beckmann, T., Schmidt, G.L., Dwyer, J.L., Hughes, M.J., Laue, B., 2017. Cloud detection algorithm comparison and validation for operational Landsat data products. *Remote Sensing of Environment*, 194, 379–390. http://doi.org/10.1016/j.rse.2017.03.026.

Fuller, R.M., Smith, G.M., Devereux, B.J., 2003. The characterisation and measurement of land cover change through remote sensing: Problems in operational applications? *International Journal of Applied Earth Observation and Geoinformation*, 4, 243–253. https://doi.org/10.1016/S0303-2434(03)00004-7.

Goetz, S.J., Baccini, A., Laporte, N.T., Johns, T., Walker, W., Kellndorfer, J., Houghton, R.A., Sun, M., 2009. Mapping and monitoring carbon stocks with satellite observations: A comparison of methods. *Carbon Balance Management*, 4, 1–7. https://doi.org/10.1186/1750-0680-4-2.

Gómez, C., White, J.C., Wulder, M.A., 2016. Optical remotely sensed time series data for land cover classification: A review. *ISPRS Journal of Photogrammetry and Remote Sensing*, 116, 55–72. https://doi.org/10.1016/j.isprsjprs.2016.03.008.

Grinand, C., Rakotomalala, F., Gond, V., Vaudry, R., Bernoux, M., Vieilledent, G., 2013. Estimating deforestation in tropical humid and dry forests in Madagascar from 2000 to 2010 using multi-date Landsat satellite images and the random forests classifier. *Remote Sensing of Environment*, 139, 68–80. https://doi.org/10.1016/j.rse.2013.07.008.

Grings, F., Roitberg, E., Barraza, V., 2020. EVI time-series breakpoint detection using convolutional networks for online deforestation monitoring in chaco forest. *IEEE Transaction on Geoscience and Remote Sensing*, 58, 1303–1312. https://doi.org/10.1109/TGRS.2019.2945719.

Hamunyela, E., Rosca, S., Mirt, A., Engle, E., Herold, M., Gieseke, F., Verbesselt, J., 2020. Implementation of BFASTmonitor algorithm on Google Earth engine to support large-area and sub-annual change monitoring using earth observation data. *Remote Sensing*, 12. https://doi.org/10.3390/RS12182953.

Huang, C., Goward, S.N., Masek, J.G., Thomas, N., Zhu, Z., Vogelmann, J.E., 2010. An automated approach for reconstructing recent forest disturbance history using dense Landsat time series stacks. *Remote Sensing of Environment*, 114, 183–198. https://doi.org/10.1016/j.rse.2009.08.017.

Ienco, D., Interdonato, R., Gaetano, R., Minh, D.H.T., 2019. Combining Sentinel-1 and Sentinel-2 Satellite image time series for land cover mapping via a multi-source deep learning architecture. *ISPRS Journal of Photogrammetric and Remote Sensing*, 158, 11–22.

Jokela, J., Siitonen, J., Koivula, M., 2019. Short-term effects of selection, gap, patch and clear cutting on the beetle fauna in boreal spruce-dominated forests. *Forest Ecology and Management*, 446, 29–37. https://doi.org/10.1016/j.foreco.2019.05.027.

Kennedy, R.E., Yang, Z., Cohen, W.B., 2010. Detecting trends in forest disturbance and recovery using yearly Landsat time series: 1. LandTrendr—Temporal segmentation algorithms. *Remote Sensing of Environment*, 114, 2897–2910. https://doi.org/10.1016/j.rse.2010.07.008.

Lattari, F., Rucci, A., Matteucci, M., 2022. A deep learning approach for change points detection in InSAR time series. *IEEE Transactions on Geoscience and Remote Sensing*, 60, 1–16. https://doi.org/10.1109/TGRS.2022.3155969.

Li, D., Lu, D., Wu, Y., Luo, K., 2022. Retrieval of eucalyptus planting history and stand age using random localization segmentation and continuous land-cover classification based on Landsat time-series data. *GIScience & Remote Sensing*, 59(1), 1426–1445. https://doi.org/10.1080/15481603.2022.2118440.

Li, G., Lu, D., Moran, E., Calvi, M.F., Dutra, L.V., Batistella, M., 2019. Examining deforestation and agropasture dynamics along the Brazilian TransAmazon highway using multitemporal Landsat imagery. *GIScience & Remote Sensing*, 56(2), 161–183. https://doi.org/10.1080/15481603.2018.1497438.

Li, M., Zuo, S., Su, Y., Zheng, X., Wang, W., Chen, K., Ren, Y., 2023. An approach integrating multi-source data with LandTrendr algorithm for refining forest recovery detection. *Remote Sensing*, 15. https://doi.org/10.3390/rs15102667.

Li, W., 2004. Degradation and restoration of forest ecosystems in China. *Forest Ecology and Management*, 201, 33–41. https://doi.org/10.1016/j.foreco.2004.06.010.

Lin, X., Shang, R., Chen, J.M., Zhao, G., Zhang, X., Huang, Y., Yu, G., He, N., Xu, L., Jiao, W., 2023. High-resolution forest age mapping based on forest height maps derived from GEDI and ICESat-2 space-borne lidar data. *Agricultural and Forest Meteorology*, 339. https://doi.org/10.1016/j.agrformet.2023.109592.

Linke, J., Fortin, M.J., Courtenay, S., Cormier, R., 2017. High-resolution global maps of 21st-century annual forest loss: Independent accuracy assessment and application in a temperate forest region of Atlantic Canada. *Remote Sensing of Environment*, 188, 164–176. https://doi.org/10.1016/j.rse.2016.10.040.

Liu, S., Wei, X., Li, D., Lu, D., 2017. Examining forest disturbance and recovery in the subtropical forest region of Zhejiang Province using Landsat time-series data. *Remote Sensing*, 9, 479. https://doi.org/10.3390/rs9050479.

Long, X., Li, X., Lin, H., Zhang, M., 2021. Mapping the vegetation distribution and dynamics of a wetland using adaptive-stacking and Google Earth Engine based on multi-source remote sensing data. *International Journal of Applied Earth Observation and Geoinformation*, 102, 102453. https://doi.org/10.1016/j.jag.2021.102453.

Lu, D., Batistella, M., Moran, E., 2008. Integration of Landsat TM and SPOT HRG images for vegetation change detection in the Brazilian Amazon. *Photogrammetric Engineering and Remote Sensing*, 74(4), 421–430.

Lu, D., Li, G., Moran, E., 2014. Current situation and needs of change detection techniques. *International Journal of Image and Data Fusion*, 5(1), 13–38. https://doi.org/10.1080/19479832.2013.868372.

Lu, D., Li, G., Moran, E., Hetrick, S., 2013. Spatiotemporal analysis of land-use and land-cover change in the Brazilian Amazon. *International Journal of Remote Sensing*, 34(16), 5953–5978.

Lu, D., Mausel, P., Brondízio, E., Moran, E., 2004. Change detection techniques. *International Journal of Remote Sensing*, 25(12), 2365–2407.

Masek, J.G., Vermote, E.F., Saleous, N., Wolfe, R., Hall, F.G., Huemmrich, F., Gao, F., Kutler, J., Lim, T.K. 2013. *LEDAPS Calibration, Reflectance, Atmospheric Correction Preprocessing Code*, Version 2. Model Product. ORNL DAAC, Oak Ridge, TN. https://doi.org/10.3334/ORNLDAAC/1146.

Paul-Limoges, E., Black, T.A., Christen, A., Nesic, Z., Jassal, R.S., 2015. Effect of clearcut harvesting on the carbon balance of a Douglas-fir forest. *Agricultural and Forest Meteorology*, 203, 30–42. https://doi.org/10.1016/j.agrformet.2014.12.010.

Pham, H.T., Nguyen, H.Q., Le, K.P., Tran, T.P., Ha, N.T., 2023. Automated mapping of wetland ecosystems: A study using google earth engine and machine learning for lotus mapping in central vietnam. *Water (Switzerland)*, 15. https://doi.org/10.3390/w15050854.

Ressl, R., Lopez, G., Cruz, I., Colditz, R.R., Schmidt, M., Ressl, S., Jiménez, R., 2009. Operational active fire mapping and burnt area identification applicable to Mexican Nature Protection Areas using MODIS and NOAA-AVHRR direct readout data. *Remote Sensing of Environment*, 113, 1113–1126. https://doi.org/10.1016/j.rse.2008.10.016.

Schroeder, T.A., Wulder, M.A., Healey, S.P., Moisen, G.G., 2011. Mapping wildfire and clearcut harvest disturbances in boreal forests with Landsat time series data. *Remote Sensing of Environment*, 115, 1421–1433. https://doi.org/10.1016/j.rse.2011.01.022.

Shafique, A., Cao, G., Khan, Z., Asad, M., Aslam, M., 2022. Deep learning-based change detection in remote sensing images: A review. *Remote Sensing*, 14, 871. https://doi.org/10.3390/rs14040871

Shang, R., Chen, J.M., Xu, M., Lin, X., Li, P., Yu, G., He, N., Xu, L., Gong, P., Liu, L., Liu, H., Jiao, W., 2023. China's current forest age structure will lead to weakened carbon sinks in the near future. *The Innovation*, 100515. https://doi.org/10.1016/j.xinn.2023.100515.

Singh, A., 1989. Digital change detection techniques using remotely-sensed data. *International Journal of Remote Sensing*, 10, 989–1003.

Steininger, M.K., Godoy, F., Harper, G., 2009. Effects of systematic sampling on satellite estimates of deforestation rates. *Environmental Research Letter*, 4. https://doi.org/10.1088/1748-9326/4/3/034015.

Townshend, J.R., Masek, J.G., Huang, C., Vermote, E.F., Gao, F., Channan, S., Sexton, J.O., Feng, M., Narasimhan, R., Kim, D., Song, K., Song, D., Song, X.P., Noojipady, P., Tan, B., Hansen, M.C., Li, M., Wolfe, R.E., 2012. Global characterization and monitoring of forest cover using Landsat data: Opportunities and challenges. *International Journal of Digital Earth*, 5, 373–397. https://doi.org/10.1080/17538947.2012.713190.

Verbesselt, J., Zeileis, A., Herold, M., 2012. Near real-time disturbance detection using satellite image time series. *Remote Sensing of Environment*, 123, 98–108. https://doi.org/10.1016/j.rse.2012.02.022.

Vermote, E., Justice, C., Claverie, M., Franch, B., 2016. Preliminary analysis of the performance of the Landsat 8/OLI land surface reflectance product. *Remote Sensing of Environment*, 185, 46–56.

Xi, Z., Lu, D., Liu, L., Ge, H., 2016. Detection of drought-induced hickory disturbances in western Lin An County, China, using multitemporal Landsat imagery. *Remote Sensing*, 8, 345. https://doi.org/10.3390/rs8040345.

Xu, H., Qi, S., Li, X., Gao, C., Wei, Y., Liu, C., 2021. Monitoring three-decade dynamics of citrus planting in Southeastern China using dense Landsat records. *International Journal of Applied Earth Observation and Geoinformation*, 103, 102518. https://doi.org/10.1016/j.jag.2021.102518.

Yang, B., Wu, L., Ju, Z., Liu, X., Liu, M., Zhang, T., Xu, Y., 2023. Sub-annual scale landtrendr: Sub-annual scale deforestation detection algorithm using multi-source time series data. *IEEE Journal of Selected Topics in Applied Earth Observations and Remote Sensing*, 16, 8563–8576. https://doi.org/10.1109/JSTARS.2023.3312812.

Yin, X., Kou, W., Yun, T., Gu, X., Lai, H., Chen, Y., Wu, Z., Chen, B., 2022. Tropical forest disturbance monitoring based on multi-source time series satellite images and the landtrendr algorithm. *Forests*, 13. https://doi.org/10.3390/f13122038.

Zhang, C., Smith, M., Lv, J., Fang, C., 2017. Applying time series Landsat data for vegetation change analysis in the Florida Everglades Water Conservation Area 2A during 1996–2016. *International Journal of Applied Earth Observation and Geoinformation*, 57, 214–223. https://doi.org/10.1016/j.jag.2017.01.007

Zhang, S., Yu, J., Xu, H., Qi, S., Luo, J., Huang, S., Liao, K., Huang, M. 2023. Mapping the age of subtropical secondary forest using sense Landsat time series data: An ensemble model. *Remote Sensing*, 15, 2067. https://doi.org/10.3390/rs15082067.

Zhang, X., Liu, L., Chen, X., Gao, Y., Xie, S., Mi, J., 2021a. GLC_FCS30: Global land-cover product with fine classification system at 30 m using time-series Landsat imagery. *Earth System Science Data*, 13, 2753–2776. https://doi.org/10.5194/essd-13-2753-2021.

Zhang, Y., Ling, F., Wang, X., Foody, G.M., Boyd, D.S., Li, X., Du, Y., Atkinson, P.M., 2021b. Tracking small-scale tropical forest disturbances: Fusing the Landsat and Sentinel-2 data record. *Remote Sensing of Environment*, 261, 112470. https://doi.org/10.1016/j.rse.2021.112470.

Zhang, Y., Woodcock, C.E., Chen, S., Wang, J.A., Sulla-Menashe, D., Zuo, Z., Olofsson, P., Wang, Y., Friedl, M.A., 2022. Mapping causal agents of disturbance in boreal and arctic ecosystems of North America using time series of Landsat data. *Remote Sensing of Environment*, 272, 112935. https://doi.org/10.1016/j.rse.2022.112935.

Zhu, Z., Woodcock, C.E., 2014. Continuous change detection and classification of land cover using all available Landsat data. *Remote Sensing of Environment*, 144, 152–171. https://doi.org/10.1016/j.rse.2014.01.011.

Zhu, Z., Zhang, J., Yang, Z., Aljaddani, A.H., Cohen, W.B., Qiu, S., Zhou, C., 2020. Continuous monitoring of land disturbance based on Landsat time series. *Remote Sensing of Environment*, 238, 111116. https://doi.org/10.1016/j.rse.2019.03.009.

Part III

Integrating Geographic Information Systems (GIS) and Remote Sensing in Spatial Modeling Framework for Decision Support

13 Geoprocessing, Workflows, and Provenance

Jason A. Tullis, David P. Lanter, Aryabrata Basu, Jackson D. Cothren, Xuan Shi, W. Fredrick Limp, Rachel F. Linck, Sean G. Young, Jason Davis, and Tareefa S. Alsumaiti

LIST OF ACRONYMS

ACSM	American Congress of Surveying and Mapping
AI	Artificial intelligence
APIs	Application programming interfaces
CAST	Center for Advanced Spatial Technologies
CDAT	Climate Data Analysis Tool
CI	Cyberinfrastructure
CNNs	Convolutional neural networks
DBMS	Database management system
DL	Deep learning
ESSW	Earth Systems Science Workbench
FGDC	Federal Geographic Data Committee
GIS	Geographic information systems
GUI	Graphical user interface
HMM	Hidden Markov model
IP	Intellectual property
IoT	Internet of things
LULC	Land use land cover
ML	Machine learning
MODAPS	MODIS adaptive data processing system
NCDCDS	National Committee for Digital Cartographic Data Standards
NISO	National Information Standards Organization
NLP	Natural language processing
NSF	National Science Foundation
OMIDAPS	OMI data processing system
OPM	Open Provenance Model
REST	Representational state transfer
RRR	Repeat, reproduce, and/or replicate
SIPS	Science investigator-led processing system
SOAP	Simple object access protocol
SOI	Service-oriented integration
SOC	Service-oriented computing
SQL	Structured query language
XML	Extensible markup language
XSEDE	eXtreme Science and Engineering Development Environment
UAS	Unmanned aircraft system

DOI: 10.1201/9781003541158-16

USGS	U.S. Geological Survey
URI	Uniform resource identifiers
VCS	Visualization control system
VDC	Virtual data catalog
VDL	Virtual data language
VDS	Virtual data schema
VTK	Visualization toolkit
WPS	Web processing service

13.1 INTRODUCTION

Integrated remote sensing and GIS-assisted problem solving now supports a remarkable array of human activities (e.g., food and agricultural security, public health, critical infrastructure, weather and climate adaptation, forest management, heritage preservation, urban and regional planning, etc.), and is being configured in a great variety of technical means. Given the sheer quantity of innovations reported in journals and books (including *Remote Sensing Handbook*), any one expert may only be keenly aware of a fraction of the detailed remote sensing and related geospatial methods available to address a given problem. Regardless of the remote sensing application under study or review, some reliance (whether implied or reported) is always made upon the *geoprocesses* and *workflows* associated with any artifacts produced. In the context of a geospatial decision support artifact (e.g., a map of predicted crop yield in kg/ha), a record of the specific geoprocesses may be termed geospatial *provenance* (or *lineage*; see Section 13.1.1). *Geoprocessing, Workflows, and Provenance* explores how remote sensing-assisted geoprocessing and related GIS workflows have been or may be combined with digital provenance information to augment scientific reproducibility and replicability (R&R), comparison and discovery, privacy and confidentiality, trust, stakeholder engagement, or to otherwise improve remote sensing-assisted decisions.

Increasingly of interest in computer systems, digital provenance has relatively early geospatial origins that date back to at least the 1980s (e.g., Chrisman 1983), with a definite resurgence around 2009 (e.g., Yue et al. 2010a). The early and expanded geospatial interest and connection to provenance is driven in large part by the questions of methodological innovation and R&R. For remote sensing and GIS integration to best improve the quality of tools across a range of applications and domains, it seems reasonable that, if possible, such innovation must first be machine recognizable. Unfortunately, many geospatial tools lack suitable means to repeat, reproduce, and/or replicate (RRR) their findings, and innovation reported is naturally bracketed by complex questions of accuracy, fitness for ethical use, and a variety of other qualitative and quantitative metrics related to reliability and trust. So, while there is broad conceptual agreement that machine interpretable source and process history records are vital and may even be scientifically transformative in the modern era, questions remain unanswered on how provenance information may simultaneously benefit multiple domains (including the geospatial domain) and stakeholders, and what mechanisms for its digital capture and exchange will most successfully convey those benefits.

There are at least two reasons to believe that even partial success towards machine interpretable geospatial process history records will be transformative. First, correct expert interpretation of the full scope of relevant methods, procedures, algorithms, and expert knowledge is subject to entropy and constitutes an increasingly complex, even daunting companion to 21st century "big [geospatial] data" (Hey et al. 2009). Second, as remote sensing and other geospatial techniques are communicated in the scientific literature, there is an expectation and scholarly requirement that scientific findings previously published are replicable and carefully acknowledged for their relevant achievements and/or limitations. Failure to increasingly harness machine power on these two fronts (but to continue interpretations by experts alone), especially in an age of rapid expansion of artificial intelligence (AI), is probably not a viable long-term option. In a relevant example from computer systems, Buneman (2013) notes that the under-appreciated machine-managed provenance in software version control systems has helped prevent a total disaster in software engineering.

Absent the kinds of methodological analyses enabled in part through careful exchange of provenance information, an increasingly data intensive "geo-cyberinfrastructure" (Di et al. 2013a) renders comprehensive remote sensing-assisted geospatial workflow interpretations, comparisons, and knowledge transfers ever more difficult by experts alone. Furthermore, depending on the geospatial laboratory setting and the capabilities of a given research team, the actual digital methods linked to published materials may overlap significantly with previously reported work, may offer similar results using a more or less computationally efficient means of problem solving, and/or may be idiosyncratic to individual skills and experience. In an integrated geoprocessing, workflow and provenance cycle, expert refinement of remote sensing-assisted decision support knowledge may be augmented by software agents capable of automated exchange and recognition of innovation (Figure 13.1).

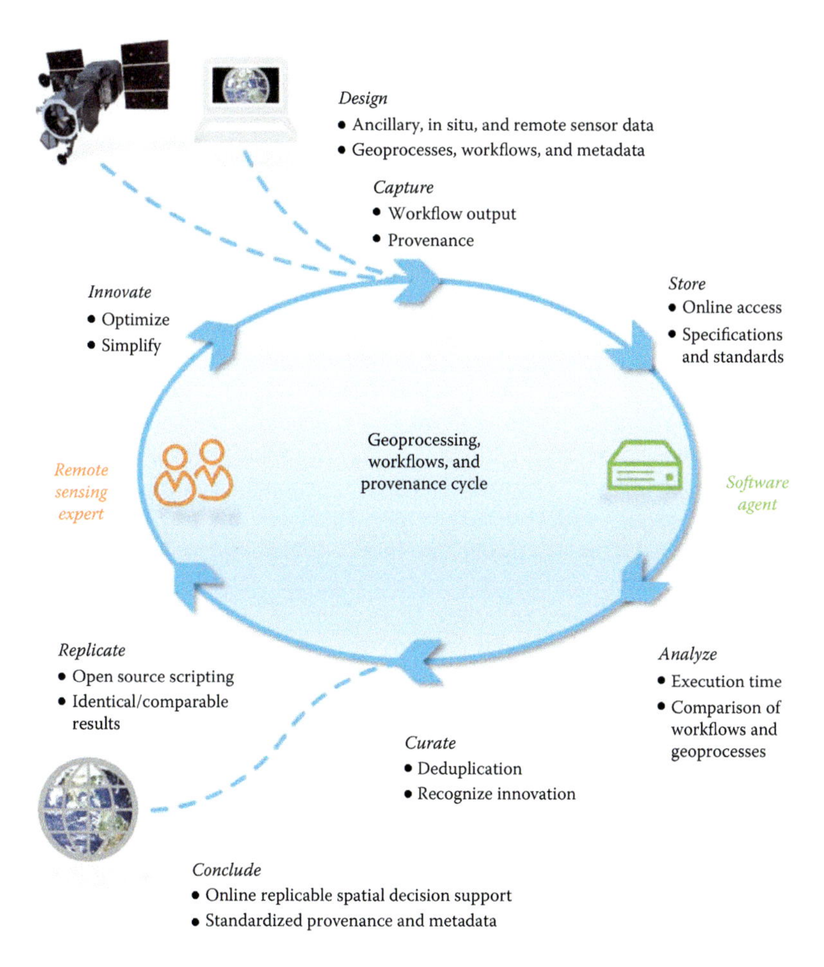

FIGURE 13.1 Integrated geoprocessing, workflows, and provenance may be conceptualized as a positive developmental cycle used to refine remote sensing knowledge before decision support is communicated. Highlighted aspects of this cycle suggest a capacity of remote sensing experts, in conjunction with software agents, to cooperatively capture, store, analyze, curate, RRR (repeat, reproduce, and/or replicate), and innovate remote sensing-assisted decision support methods. (Artist image of WorldView-3 courtesy of DigitalGlobe 2014.)

Over the past 35 years, various prototype forms of geospatial provenance have been implemented in shared workflow environments, including those specialized for high performance capabilities. Despite the potential of these prototypes, single user/workstation geoprocessing and workflow design continues to be a dominant tradition with many active options. There is therefore a discrepancy between futuristic collaborative goals and actual state-of-the-art remote sensing-assisted platforms. There are also variations in how provenance and closely related concepts are defined, whether specifically in a remote sensing or geospatial-related forum, or more broadly in computer systems. It therefore seems reasonable to report progress in terms of what the actual computational environments entail, and which definitions are implied.

13.1.1 Working Definitions

Though commonly understood in a broad remote sensing and geospatial computation parlance, Wade and Sommer (2006) define *geoprocessing* in the context of the many tools available in one software platform (Esri's ArcGIS) with an emphasis on input GIS datasets, operations performed, and associated outputs. More generically its root, *process*, implies an instance of a computer program execution, and this is naturally compatible with a geospatial data processing software context. Of course, identical geospatial computer programs operating on identical input datasets may produce different results as a function of additional configuration parameters. For example, raster-based geoprocessing tools in Esri's ArcGIS platform allow for an "environment setting" called *Snap Raster*. This setting allows the user to specify the spatial grid on which computations are made. In practice, use of this parameter allows pixels in an output raster layer to be exactly aligned with another raster having the same cell size. To a novice the resulting sub-pixel geometric shift may seem inconsequential at the overview scale. However, when geoprocessing tools are chained together into a *workflow* (in the present context, a repeatable sequence of geoprocesses of interest to a person or group), environment settings like *Snap Raster* can affect the logic of a decision support conclusion.

Provenance traces back to 1294 in Old French as a derivative of the Latin "provenire," and while Merriam-Webster (2014) emphasizes provenance as a *concept* (e.g., ownership history of a painting), Oxford University Press (2014) highlights the *record* of such provenance (Moreau 2010). In the art domain where the term is very well established, provenance entails an artifact's complete ownership history, but ideally will also include artistic, social, and political influences upon the work from its creation to the present day. There is an established research process for obtaining an artifact's trusted provenance and the information is highly valued, particularly to authenticate real versus fraudulent works (Yeide et al. 2001; IFAR 2013). As a related term, provenance is now increasingly used in a broad range of fields (e.g., archaeology, computer science, forestry, geology, etc.) with usually overlapping definitions.

Computational definitions of provenance are more numerous than in other domains, largely because of (1) the difference between *concepts* of digital records and *actual* digital records, and (2) variation in software environment such as a database management system (DBMS) versus file-based processing (Moreau 2010). Understanding provenance *within* DBMS queries requires more computationally detailed observations than understanding provenance at a more generalized workflow level (where one step in the workflow may entail multiple database queries). Various traditions further influence how provenance is viewed, for example, whether it is conflated with *metadata* or *trust*, two closely related but distinct concepts (Gil et al. 2010). Given the infrastructural importance of the web in remote sensing-assisted decision support, the W3C Provenance Incubator Group's working definition of provenance (in a web resource context) carries significant weight and is adopted for this chapter of *Remote Sensing Handbook*:

> Provenance of a resource is a record that describes entities and processes involved in producing and delivering or otherwise influencing that resource. Provenance provides a critical foundation for

> assessing authenticity, enabling trust, and allowing reproducibility. Provenance assertions are a form of contextual metadata and can themselves become important records with their own provenance.
>
> *(Gil et al. 2010)*

It should be noted that while *provenance* and *lineage* are here used interchangeably, one can argue there are subtle differences in their meanings. Process history seems to fit more easily with the many definitions attributed to provenance, and lineage implies a kind of genealogy or data pedigree record relative to a remote sensing-assisted decision support artifact. While these semantic differences are not a point of present focus, each word will appear in its historical context (beginning with lineage). Also, a number of surveys have been conducted on provenance including some with a geoprocessing and workflows flavor. For example, Yue and He (2009) provide a review covering various aspects of geospatial provenance. For a broader perspective, Bose and Frew (2005) provide a review covering provenance in geospatial as well as other domains. More recently, Di et al. (2013b) provide an overview of geoscience data provenance.

In late 2017, a 15-person committee from the National Academies of Sciences, Engineering, and Medicine chaired by Harvey Fineberg (Gordon and Betty Moore Foundation) began a series of 12 meetings to study the problem of reproducibility and replicability (R&R) in science and engineering. Prompted by a request of the National Science Foundation (NSF) and under legislative guidance, the committee led an unprecedented Consensus Study Report on *Reproducibility and Replicability in Science*, in which the National Academies (2019) takes an official position that clarifies historically divergent and inconsistent definitions of R&R. "*Reproducibility* is obtaining consistent results using the same input data; computational steps, methods, and code; and conditions of analysis . . . [and] is synonymous with 'computational reproducibility'" (p. 6). "*Replicability* is obtaining consistent results across studies aimed at answering the same scientific question, each of which has obtained its own data. Two studies may be considered to have replicated if they obtain consistent results" (p. 6). The report, which includes a valuable discussion on provenance, had a large impact and research activities surrounding R&R surged after its publication.

Interestingly, after discussions with the National Information Standards Organization (NISO), the Association for Computing Machinery (ACM 2020) altered its definitions of R&R to be in harmony with those clarified by National Academies (2019). In their continuing definition of *repeatability*, a

> measurement can be obtained with stated precision by the same team using the same measurement procedure, the same measuring system, under the same operating conditions, in the same location on multiple trials. For computational experiments, this means that a researcher can reliably repeat her own computation.
>
> *(ACM 2020)*

13.2 HISTORICAL CONTEXT

The earliest work in geospatial lineage was spurred in the United States through the formation of the National Committee for Digital Cartographic Data Standards (NCDCDS) by the American Congress of Surveying and Mapping (ACSM) in 1982 (Bossler et al. 2010). In 1988, chaired by Dr. Harold Moellering from Ohio State University, NCDCDS proposed five fundamental components of a geospatial data quality report, including (1) lineage, (2) positional accuracy, (3) attribute accuracy, (4) logical consistency, and (5) completeness. NCDCDS described lineage in detail, which they presented as the *first* quality component. Less than a third of their description for lineage follows (Moellering et al. 1988, p. 132):

> The lineage section of a quality report shall include a description of the source material from which the data were derived, and the methods of derivation, including all transformations involved in producing the final digital files. The description shall include the dates of the source material and the dates of ancillary information used for update.

As geospatial workflows began to transition from analog to digital environments, it became clear that lineage implied geoprocesses would need to be tracked from their origins, through revisions to the data, and finally to the output (Moore 1983). Chrisman (1983) noted that unfortunately over its lifetime, lineage information in quality records would be subject to entropy or fragmentation as a result of continuous GIS maintenance. He described "reliability diagrams" (for intelligence and other reliability-sensitive applications) embedded with lineage-related geometry and attributes (e.g., polygons identifying specific aerial photographic sources) and recommended they be incorporated in GIS design. While not typically portrayed as lineage or provenance today, this type of lineage-related geodata such as DigitalGlobe image collection footprints accessible in Google Earth, is extremely useful for visualization purposes and may resist digital entropy due to established geodata interoperability.

Beyond the challenges presented by digital records of lineages for multiple geodata versions, Langran and Chrisman's (1988) emphasis on multi-temporal GIS highlighted additional record complexity that would be required. Nyerges' (1987) discussion on geodata exchange implied that quality metadata (including lineage information) could eventually facilitate geoprocessing design (workflows) with the two being mutually dependent. Others including Grady (1988) reasoned that lineage need not only support records of data quality but could in turn be used to record societal mandates (e.g., legislative drivers of geodata development) in the lineage information. While the existence of these additional complexities and potential requirements for geospatial lineage/provenance did not thwart attempts to forge ahead with possible software solutions, they pointed to significant challenges.

13.2.1 Digital Provenance in Remote Sensing and Geospatial Workflows

In recent decades there has been significant attention given to understanding lineage/provenance in computer systems, and a variety of formalisms have been developed to understand their role in scientific workflows (e.g., Buneman and Davidson 2010; Hey et al. 2009; Bose and Frew 2005; Simmhan et al. 2005). Next, we highlight pioneering digital advances with geospatial lineage (circa 1990s) and more recent geo-cyberinfrastructure advances in provenance (circa 2000s to present).

13.2.1.1 Pioneering Work in Geospatial Lineage

As Chrisman (1986) suggested, "evaluation and judgment of fitness of use must be the responsibility of the user, not the producer. To carry out this responsibility, the user must be presented with much more information to permit an informed decision" (p. 352). Moellering et al. (1988) later emphasized producers' obligation to first document and update the lineage of their data in order to trace all the work (whether analog or digital) from original source materials through the intermediate processes to final digital output. It became obvious that both GIS software and international standards would be needed to facilitate the development and maintenance of such records.

An early version of Esri's ARC/INFO Geographic Information System featured a LIBRARIAN module capable of capturing and querying some aspects of geospatial lineage. Using the module's CATALOG command, a database administrator could retrieve information on map production status as well as review time stamps and coordinates of recent map updates (Aronson and Moorehouse 1983). In the mid-1980s, the U.S. Geological Survey (USGS) began development of a GIS-linked automated cartographic workflow system called Mark II with partial lineage capabilities. An important part of Mark II's design was its capacity to track the location (e.g., network address) of datasets and their progress from curated archive toward final map products (Anderson and Callahan 1990; Guptill 1987). While it was envisioned this system would play a key role in fulfilling the National Mapping Program's mission through 2000, the agency focus transitioned by the mid-1990s toward GIS data development including the National Map. The first reported development of a system to specifically and directly address geospatial lineage was David P. Lanter's *Geolineus* project commenced in the late 1980s as part of his doctoral research at University of South Carolina's Department of Geography (Lanter 1989). As the prototype pioneering work in geospatial lineage/provenance, this is reviewed in detail with added explanation.

Lanter invented a method and means to capture, structure, and process geospatial lineage to determine and communicate the meaning and integrity of the contents of a GIS database (Lanter 1993a). His metadata and processing algorithms track and document remotely sensed and other geodata sources and analytic transformations applied to them to derive new datasets. In addition to differentiating between source and derived datasets, Lanter further distinguished intermediate and product derived datasets. More concisely, let

$$Datasets = \{Dataset_i : i = source,\ derived\},$$

$$Dataset_{derived} = \{Dataset_{derived.k} : k = intermediate,\ product\}.$$

Source datasets can be the results of in situ sampling and data collection, remote sensing, or ancillary data (e.g., digitization of maps, or thematic data resulting from digital processing of remotely sensed data). Initially, only source datasets are available for geoprocessing and transformation into a derived dataset (Figure 13.2; $n \geq 1, m = 0$). Later, new datasets can be generated exclusively from derived datasets ($n = 0, m \geq 1$) using spatial analysis transformations such as reclassification, distance measurement (buffering), connectivity, neighborhood characterization, and summary calculations. Alternatively, new datasets can be derived from inputs that include sources, derived, or both ($n + m > 1$) using multi-input transformations such as arithmetic, statistical, and logical overlays, as well as drainage network and viewshed determinations.

Lanter classified datasets into source, intermediate and product types (Figure 13.2), and related them to one another as inputs and outputs of each data processing step of an analytical application. He gave input datasets "parent" links pointing to output datasets they were used to create (Who am I the parent of?) and provided output datasets "child" links connecting them back to their input datasets. (Who am I the child of?) Each parent and child relationship was defined as an ordered pair of input and output datasets. Lanter's parent relationship identified the derived output given a source or derived input dataset, while his child relationship would identify a derived or source dataset when given an output dataset.

Child links connecting output datasets to their inputs enabled automatic deduction of which datasets within an analytic database are sources and which are derived (Lanter 1993b). Derived datasets are connected to their inputs by child links, while sources lack such links. Lanter defined his *child* operator to take a derived dataset, access its child links, and identify inputs used to create it. His *Ancestors* algorithm applied the *child* operator and by a recursive function traced the child links to identify datasets used to create a derived dataset, including any sources in the geoprocessing application. Lanter defined the *parent* operator to take a source or derived dataset as input, and

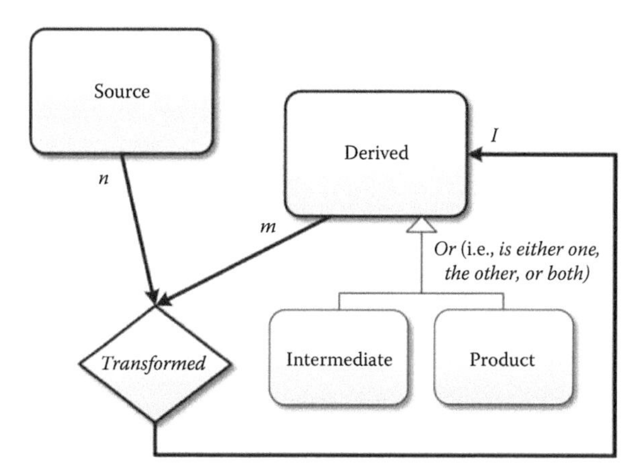

FIGURE 13.2 Relationship among source and derived datasets, where each instance of the latter may be either an intermediate or product dataset, or both.

access and traverse its parent links to identify all the outputs derived from it. His *Descendants* function recursively traced parent links and identified all datasets derived from a source or other derived input dataset used within a geoprocessing application.

Classification of datasets into source, intermediate, and product paved the way to structuring additional lineage metadata attributes. Lanter used the artificial intelligence "frame" data structure to organize knowledge about the metadata properties of source, intermediate and product dataset types. Each source dataset was provided a frame for storing source properties such as its name, feature type(s), date(s), responsible agency, scale, projection, and accuracy attributes. He provided each derived dataset with a frame for storing detailed metadata elements about where it is physically stored, the command applied to derive it, the command's parameters, who derived it, and other aspects of its derivation. Lanter saw products as derived datasets that were provided an additional frame for metadata detailing the analysis goal the dataset was intended to meet, intended audience/users of the dataset, when it was released, etc.

More formally, each $Dataset_i$ (i.e., source, intermediate, or product) was provided an ordered list of metadata properties, A_j, such that $A_j = \{A_{j1}, A_{j2}, I, A_{jk=f(i)}\}$. Specifically,

$$Dataset_{source}\mathrm{A}_{source} = \{Name, Features, Data, Scale, Projection, Agency, AccuracI...\},$$

$$Dataset_{intermediate}\mathrm{A}_{intermediate} = \{Name, Command, Parameters, User, DaIe,...\}, \text{ and}$$

$$Dataset_{product}\mathrm{A}_{product} = \{Goal, Audience, Release\ Date, IntendedIse,...\}.$$

Given $w \in \{source,\ intermediate,\ product\}$, m a metadata property of w, and a_{wm} a value of A_{wm} then a $dataset_w = (Ia_{w2},...,\ a_{wl})$.

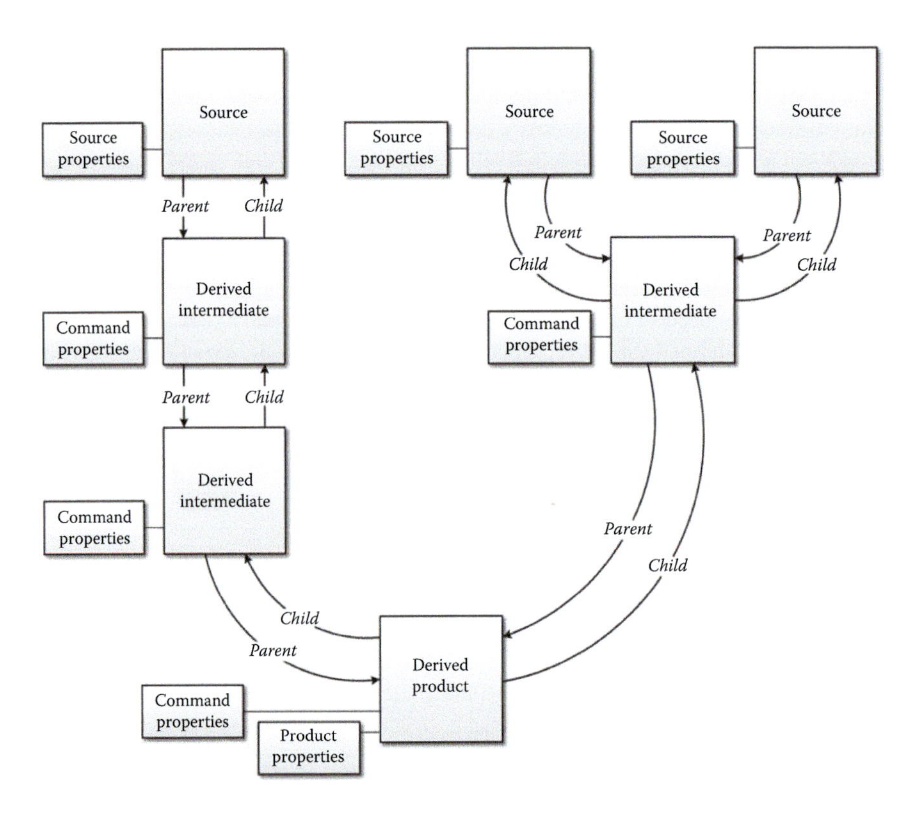

FIGURE 13.3 Lineage represented as structured metadata consisting of parent and child links connecting source, intermediate, and product datasets. While each source possesses a frame containing metadata properties, frames for derived datasets detail the GIS command used in its creation. In addition, derived product datasets possess a frame describing analytic goals, release date, and users.

Lanter's lineage metadata structure represented datasets as nodes coupled with source, command and product properties, and connected them with parent and child links (Figure 13.3). Lanter adapted the *Ancestors* function to respond to lineage queries, and to report on data sources and the sequence of processing (i.e., data lineage) applied to sources and intermediates to derive a target dataset (Lanter 1991). He integrated the *Ancestors* function with a rule-base processor that checked the inputs of each user-entered GIS command, determined their related sources and evaluated their metadata to detect and warn users when they were entering commands that would otherwise combine datasets of incompatible properties such as projections, scales, dates, etc. (Lanter 1989). Lanter subsequently modified the *Descendants* function to automatically generate and run GIS scripts and propagate new source data to update dependent intermediates and products (Lanter 1992a).

GEOLINEUS

Lanter and Essinger designed the "lineage diagram," an icon-based flowchart graphical user interface (GUI), to enable users direct interaction with lineage metadata to understand, modify and maintain their analytical applications and Esri's ARC/INFO's spatial data contents (Essinger and Lanter 1992; Lanter and Essinger 1991), and implemented it in *Geolineus*—(Lanter 1992b)—the first lineage-enabled geospatial workflow system.

Geolineus enabled users of Esri's ARC/INFO and GRID (for image processing) to capture, create, save, exchange, analyze, and reuse lineage metadata to maintain their GIS databases. Geolineus' user interface included a lineage data flow diagram within one panel, coupled with another panel containing its own command line processor in place of the command line processor of ARC/INFO and GRID. As users added source datasets, they were presented with a form to document them, after which they were displayed along the top of the data flow diagram, each with a square icon with a bar at its top. Symbols within the icons would identify if the dataset contained points, lines, polygons, raster grids, and/or value attribute tables. Icons further down the flowchart represent datasets derived

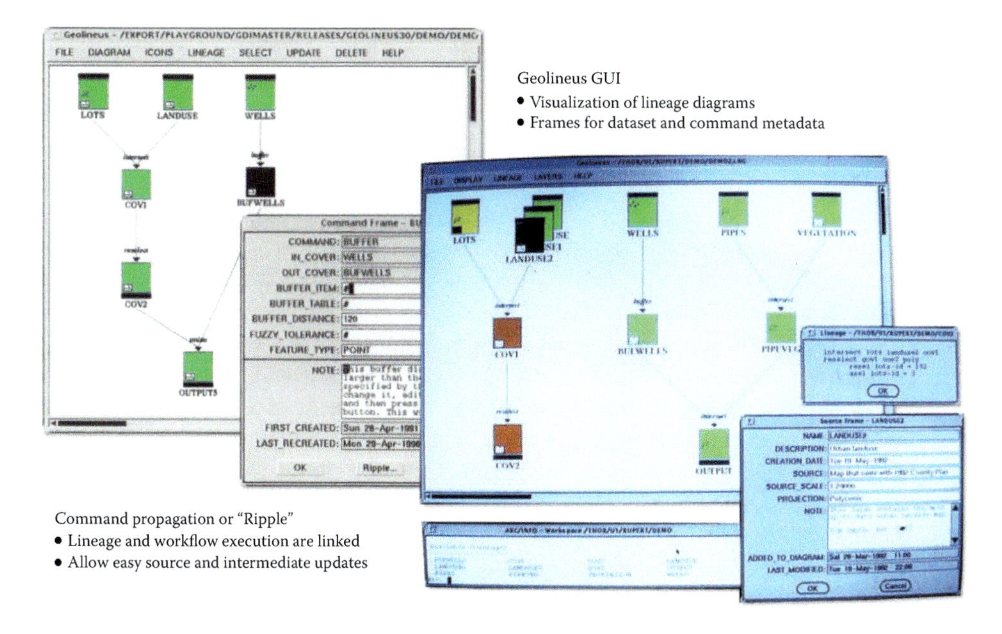

FIGURE 13.4 Examples of Geolineus' interactive lineage diagram GUI. The left screen shot illustrates linkage of a command frame to the BUFWELLS dataset highlighted in black; clicking on the "Ripple" button at the bottom of the command frame propagated changed buffer command parameters throughout the workflow. The right screen shot illustrates the LANDUSE2 dataset's source frame, and commands applied to that source to derive the COV1 and COV2 datasets.

with geospatial analysis operations such as CLASSIFY, BUFFER, and INTERSECT. Geolineus would create icons and arrows connecting them to the flowchart automatically as these commands were used. Icons at the bottom of the flowchart signifying products, that is, derived datasets that represent the final step in the geospatial application, each included a bar along its bottom edge.

Written in Common LISP, Geolineus used multiprocessing capabilities of UNIX to run the geospatial processing software as a background job while providing its own command line window to the user. As the user would enter a command transforming one or more spatial datasets to derive a new one (e.g., classify, union, intersect, etc.), Geolineus would parse, extract the identities of the inputs and output datasets and the command and its parameters, and pass the command off to the geospatial processing software running in the background. Geolineus monitored the processing and feedback messages returned from the geospatial processing software and presented them to the user within its own command line window in real-time to provide the user with the illusion that they were interacting directly with the geospatial software. The system detected whether the processing successfully completed, and if so the input/output relationships and command information would be stored within its meta-database and the data flow diagram dynamically updated with a new icon for the output dataset connected by dotted arrows (labeled with the command) emanating from its data sources. When the final data product was reached, the user could click on its icon, and fill in the displayed product form to document the analytical goal it represented (e.g., wells at risk from nearby leaking pipes), and who should be contacted if it was updated or changed.

Geolineus also monitored each dataset in the diagram to determine if it was edited or replaced. If a source or derived dataset was found to be modified, its icon would turn yellow in the diagram. If the dataset needed its topology rebuilt in response to an edit, the polygon or line feature symbol within the icon would turn red. If a derived dataset was potentially out-of-date because one of the sources it was derived from was edited, its icon would turn orange. Users could click on a source icon to view metadata about what it represented, when it was created, where it came from, and cause a propagation ("ripple") of its data through sequences of commands updating intermediate and product datasets originally derived and created from it. Users could also click on a derived dataset to rerun the commands necessary to pull new, updated and modified source data through the flowchart's processing logic and update the derived geodata (Lanter 1994b). Geolineus enabled users to save, exchange and import lineage metadata in ASCII file format to meet the Federal Geographic Data Committee's (FGDC's) Content Standard for Digital Geospatial Metadata, document exchanged datasets, accompany source datasets, provide logic for use within other instances of the software to reconstitute a derived geospatial database, and serve as reusable analytic application logic templates to snap to replacement source datasets associated with different study areas.

Lanter and Veregin (1992, 1991) modified the lineage metadata to store error measures and demonstrated new algorithms for mathematically modeling how error measures of data sources are transformed and combined through a sequence spatial analysis functions to determine the quality of a derived spatial analytic product dataset. They added properties to the source frame for storing user entered measures of data source error, and properties to the command frame for storing derived errors measures for each derived dataset. Modifications to Geolineus' *Ancestors* and *Descendants* functions enabled them to access error properties of input datasets, select and apply an appropriate error propagation function to derive, store and present the error measure of the derived geospatial dataset as the user typed in their spatial analysis commands. Lanter (1993b) followed this by modifying the lineage metadata and *Ancestors* and *Descendants* functions to use commercial costs of data storage and central processing time to calculate and compare the relative costs of storing versus using lineage metadata to re-derive intermediate and product datasets when needed. The results enabled Geolineus to determine an optimal spatial database configuration and choose which datasets to delete and recreate when needed. Veregin and Lanter (1995) modified the metadata frames and *Ancestors* and *Descendants* functions to demonstrated lineage metadata-based error propagation techniques for identifying the best data source to improve based on cost value per product quality improvement achieved. Geolineus was programmed to systematically vary the error value of each source, iteratively applying mathematical error propagation functions and determining its effect on product quality. Comparing slopes of lines graphing source error versus resulting product enables determination of relative impact each data source has on data product quality.

To help analysts and auditors understand undocumented preexisting analytically derived GIS datasets, Lanter provided Geolineus with capabilities to extract lineage metadata and create a lineage diagram from ARC/INFO log files. As with the history list the UNIX operating system recorded user commands into, the ARC/INFO GIS copied user entered GIS commands into log files which it stored and maintained within operating system file system directories or workspaces. Geolineus' "Create from log" option automatically extracted lineage metadata and created a lineage diagram reflecting the commands contained in the log file of a targeted workspace. While the log files contained the name of the dataset and the file system path indicating where the dataset was stored, they did not include other source metadata (i.e., thematic feature type, date, agency, scale, projection accuracy, etc.) necessary for achieving a clear understanding of contents and qualities of each source. To resolve this, analysts and auditors working with Geolineus clicked on the source icons within the lineage diagram, brought up source frames, and filled in missing source metadata if available.

Lanter (1994a) formulated metadata comparison functions that enabled him to automatically determine if two spatial analytic datasets were equivalent and if two geospatial datasets were similar. These were implemented within Geolineus to identify common and unique geospatial data processing conducted in and among multiple GIS workspaces (Lanter 1994b). His search for datasets common to different lineage metadata representations began with a determination of source equivalence. Source datasets were considered equivalent when their source metadata properties were found to have equivalent values, assuming these properties are sufficient to uniquely identify their contents and qualities. This enabled detection of equivalent and possibly redundant source datasets which are stored in different file system locations but contain equivalent content. Source data equivalence was implemented in Geolineus' "Merge" function, which enabled users to analyze log files of data processing applications run in different workspaces and produce a single unified lineage diagram illustrating their common and unique data sources (Figure 13.5).

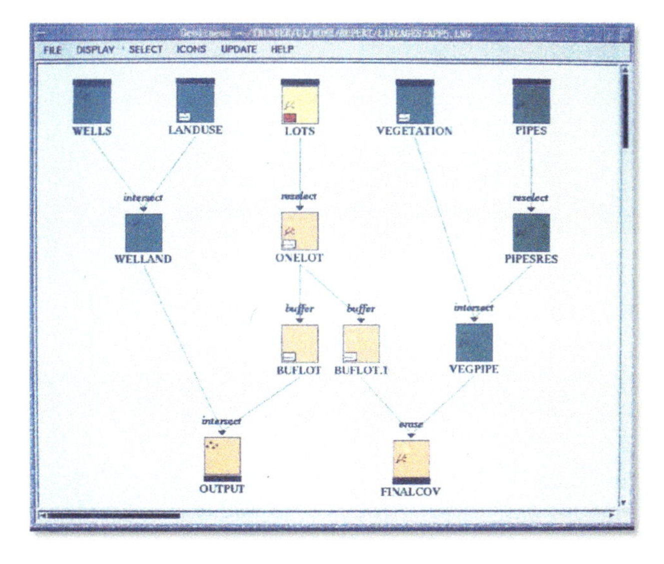

FIGURE 13.5 Geolineus GUI illustrating the results of the "Merge" function unifying two lineage diagrams at their common source LOTS dataset, and "Condense" function, which removed redundant processing and derived datasets unifying them at their common intermediate ONELOT dataset. The red mark on the yellow LOTS dataset indicated an edit and need for polygon topology repair, and the orange color in derived datasets reflected the need for changes to be propagated using the "Ripple" function to update the OUTPUT and FINALCOV products.

In turn, Lanter considered derived datasets equivalent when (1) their input datasets were equivalent, and (2) when transformations applied to compute them from their inputs were found to be equivalent. Derived data equivalence was implemented in Geolineus' "Condense" function. Condense enabled Geolineus' users to: detect the lineage representations of redundant processing and resulting copies of derived data stored under different names or in different file system locations, remove the redundant data, and consolidate the transformational logic applied in their derivation within the unified metadata and lineage diagram.

Lanter and Surbey (1994) put Geolineus' capabilities to work in the first enterprise GIS database and geoprocessing quality audit. They systematically evaluated the geospatial data sources, products, and geoprocessing applied to derive 40 GIS data products, developed within 14 projects, for eight departments of a large south-western electric utility. Lanter and Surbey identified 54 data sources among the 806 raster (GRID) and vector (ARC/INFO) GIS datasets produced for the electric utility's decision makers. They interviewed the department's GIS specialists, filled in as much missing source metadata that could be recalled and confirmed, and noted findings about what was unknown about the source data. In addition to assessing adequacy of source data documentation, Lanter and Surbey analyzed the resulting lineage diagrams they created and measured the complexity of spatial analysis logic employed within the 14 GIS applications projects.

Lanter (1994a) extended his dataset equivalence tests and formulated a set of source and derived data similarity tests in order to detect patterns of data usage and derivations within workflows. He coupled these with a geospatial data taxonomy (e.g., Anderson et al. 1976) and a GIS command language taxonomy (e.g., Giordano et al. 1994) to generalize analytic logic employed within prior applications of GIS and find common data analysis patterns. Lanter presented a suite of lineage-based metadata analysis methods for detecting and communicating commonalities and differences among particularly useful spatial analysis applications, with the intent of improving geographers' basic understandings of spatial analytic reasoning, and to provide a method and means to answer fundamental geographic questions including:

- Are there a finite number of spatial relationships studied within and among different GIS applications areas? If so, what are they?
- Within particular applications areas, are certain spatial relationships stressed more than others? If so, what are they?
- Are common patterns of analytic logic used to build up certain complex spatial relationships? If so, what are they?
- Are certain spatial relationships consistently sought at different spatial, thematic, and temporal scales?

GEO-OPERA

Also incorporating geospatial lineage into its design in the 1990s, Geo-Opera was developed as a prototype geoprocessing support or geospatial workflow management system that would enable interoperability, data recovery, process history records, and data version monitoring in commercial GIS (Alonso and Hagen 1997). Geo-Opera was based on a modular architecture composed of interface, process, and database modules. It used its own process scripting language and was based off the OPERA distributed operating system that allowed for data distribution and process scheduling within a local area network. In Geo-Opera, geodata first had to be registered before being utilized, thus mitigating the common problem (that persists today) of lack of source metadata.

13.2.1.2 Expansion of Limited Provenance in Commercial and Public Geoprocessing

As commercial and public (that is, free and open source) GIS applications rapidly matured and grew in analytical power it became necessary to provide a way for users to build and track workflows involving interactions among many complex and varied geoprocessing operations. At least two

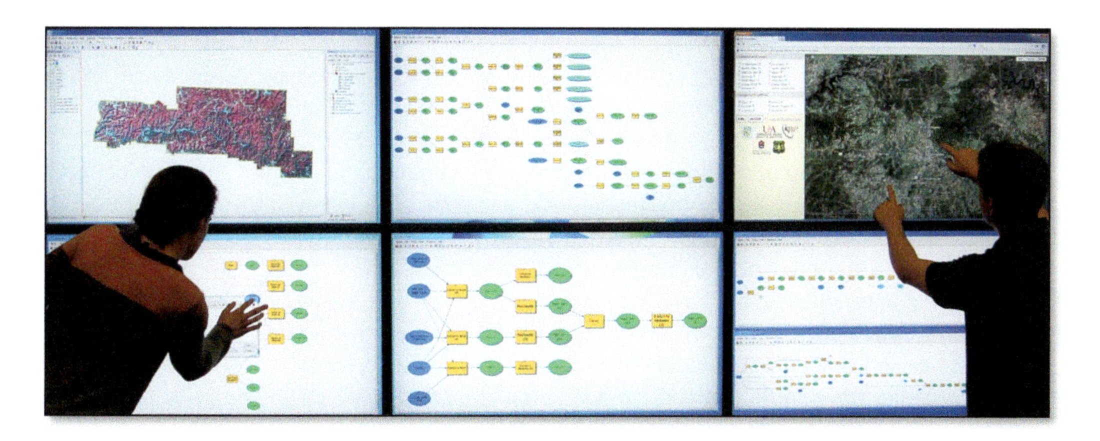

FIGURE 13.6 Geospatial scientists interact with the ASA Hazard Map (Tullis et al. 2012), a remote sensing-assisted silviculture assessment spatial decision support system, and its five downloadable ArcGIS 10 ModelBuilder workflows using a collaborative multi-touch display. Each yellow rectangle represents an ArcGIS tool (e.g., for estimating incoming solar radiation using a LIDAR-derived DEM), and together with inputs, outputs, and other parameters (colored ovals), constitutes a geoprocess. After execution, geoprocesses are marked with shadows that may be cleared only by resetting or changing geoprocess parameters including geoprocessing environment settings. User interaction with shaded geoprocesses effectively provides access to workflow-level provenance information for the most recent execution and facilitates dependent geoprocess updates after any modifications are made.

approaches to create and manage workflows have emerged—graphical block programming and integrated database style querying. The first is essentially a visual interface to programming, while the second approach appeals to users trained in database management. While both enable at least some form of provenance, enterprise database systems can provide record level transaction management which, at least in detail, is beyond the scope of this chapter.

By far the most common approach, due in large part no doubt to its ease of use and graphic nature, is the graphic block programming approach. Commercial GIS and remote sensing applications such as Esri's ArcGIS, Hexagon Geospatial's ERDAS IMAGINE, and PCI's Geomatica expose their complex processing tools in this way (e.g., Figure 13.6). The free and open source GRASS GIS also provides a visual programming environment for both vector and raster operations. Through a recent collaboration with Harvard's Center for Geographic Analysis, the open source visual programming KNIME platform, in development since 2004, has incorporated a geospatial analytics extension to expand within geospatial applications (Singh 2023). All these environments capture and store some degree of provenance including in some cases important environmental settings that can significantly affect geoprocessing results. It is important to note that visual programming interfaces can normally be bypassed by skilled users familiar with application programming interfaces (APIs) or scripting languages integrated with GIS.

A less common approach is incorporated almost exclusively in enterprise databases which have integrated spatial operators and native spatial data objects. With this level of integration, spatial operators become just another type of operation exposed through (often extended) structured query language (SQL) interfaces. At a minimum, the SQL commands used to manipulate spatial data objects are recorded and may be inspected in a variety of graphical environments. Most spatially-enabled databases have operators of particular interest in remote sensing workflows. For example, technologies such as the Oracle Spatial and Graph option for Oracle Database enables image algebra in addition to other remote sensing-oriented capabilities such as LIDAR data processing.

13.2.1.3 Interest in Provenance as a Component in Geo-Cyberinfrastructure

Cyberinfrastructure (CI) is a concept that has been extensively used since Atkins et al. (2003) *Revolutionizing Science and Engineering through Cyberinfrastructure: Report of the National Science Foundation Blue-Ribbon Advisory Panel on Cyberinfrastructure.* As a common infrastructure for scientific data and computing, a variety of components and topics are involved in CI construction, including hardware, software, network, data, and most importantly people. The development of CI can be traced back to the construction of the TeraGrid infrastructure in the 1990s that was replaced by eXtreme Science and Engineering Development Environment (XSEDE) in 2012. By linking supercomputers through high-speed networks, TeraGrid and XSEDE have provided a powerful computing environment and capability to support petascale to exascale scientific computation.

In a broader and general domain, the internet can be regarded as the CI since all computers can be linked together through the network. When varieties of data and databases can be hosted and connected on the internet, data processing and analytics can be conducted through service-oriented computing (SOC). In early 2000, web service technology was proposed to be the solution for software interoperability. In this vision of interoperable software engineering and integration, a service is an API defined in Web Services Description Language, while communication between the service provider and the service requester is based on the Simple Object Access Protocol (SOAP). Meanwhile, Representational State Transfer (REST) services are based on HTTP protocol using its GET/POST methods for mashup online resources (Fielding 2000). Both SOAP- and REST-based services can be deployed for remote procedure calls. Furthermore, with the advancement of telecommunication infrastructure and technology, wireless networking has been providing another approach for data sharing and network computing, while varieties of sensor networks can be connected through wireless networks.

Today, different computing networks can be linked together. Supercomputers on the XSEDE can be accessed through a web portal, while wireless sensor networks can be accessed on the internet. Such a huge but heterogeneous CI increases the difficulty and complexity for geoprocessing, workflows, and provenance research (Wang et al. 2008). In 2007, the NSF released the DataNet program that would support comprehensive data curation research over the CI, and NSF's Data Infrastructure Building Blocks program "will support development and implementation of technologies addressing a subset of elements of the data preservation and access lifecycle, including acquisition; documentation; security and integrity; storage; access, analysis and dissemination; migration; and deaccession," as well as "cybersecurity challenges and solutions in data acquisition, access, analysis, and sharing, such as data privacy, confidentiality, and protection from loss or corruption" (NSF 2014), which are all topics relevant to the themes in provenance.

13.2.2 Geospatial Provenance through Version Control

Linus Torvalds created Git in 2005 to support the development of the Linux kernel; fast and reliable, Git is based on complete snapshots of files over time and can accommodate hundreds of parallel "branches." Geospatial technologies are increasingly being interchanged in a multiuser paradigm through GitHub, GitLab, Bitbucket, Google Code, Code Ocean, etc. (All of these and many more are heavily dependent on Git.) It is important to note that taking advantage of Git-based version control does not require writing code.

13.2.3 Specifications and International Standards for Implementation of Shared Provenance-Aware Remote Sensing Workflows

Since the Moellering et al. (1988) proposal identifying geospatial lineage as the first component in a data quality report, a variety of provenance-related standards have been developed including those at the international level. A widely adopted standard in use is the International Standards Organization's ISO 19115-2, which has been endorsed by the Federal Geographic Data Committee (ISO 2009).

13.2.3.1 Metadata Interchange Standards

In the United States, the FGDC has been coordinating the development of the National Spatial Data Infrastructure by developing policies and standards for sharing geographic data. The Content Standard for Digital Geospatial Metadata defines common geospatial metadata about identification information, spatial reference, status information, metadata reference information, source information, processing history information, distribution information, entity/attribute information, and contact information of the geodata creator.

Partially based on the FGDC's 1994 metadata standards, the ISO Technical Committee (TC) 211 published ISO 19115 Metadata Standard, covering a conceptual framework and implementation approach for geospatial metadata generation. ISO/TC 211 suggests that metadata structure and encoding are implemented based on the Standard Generalized Markup Language that has the same format as the Extensible Markup Language (XML). The XML-based ISO metadata standard has exemplified the advantage in implementation covering a variety of elements in standard definition. ISO 19115 Metadata Standards contain a data provenance component in defining the data quality within the metadata. Unfortunately, while Gil et al. (2010) defined provenance in part as "a form of contextual metadata," their emphasis on the clear distinction between provenance and traditional metadata is not reflected in metadata interchange standards for provenance. For instance, geodata cardinality between a land use land cover (LULC) map and its metadata is one-to-one; in contrast, geodata cardinality between an LULC map and its provenance is potentially one-to-many, thus leading to the extensive recording of duplicate information in a "provenance as metadata" paradigm.

13.2.3.2 Provenance-Specific (Non-Metadata) Interchange Standards

Provenance-specific (non-metadata) standards have been developed at different levels and in a variety of domains. ISO 8000 has a series of standards that address data quality. ISO 8000-110 specifies requirements that can be checked by computer for the exchange, between organizations and systems, of master data that consists of characteristic data. It provides requirements for data quality, independent of syntax. ISO 8000-120 specifies requirements for capture and exchange of data provenance information and supplements the requirements of ISO 8000-110. ISO 8000-120 includes a conceptual data model for data provenance where a given "*provenance_event* records the provenance for exactly one *property_value_assignment*," and every "*property_value_assignment* has its provenance recorded by one or many *provenance_event* objects" (ISO 2016).

In order to trace the changing information and the provenance of data (and by implication geodata) over the web, W3C has recently published a series of documents and recommendations (starting with the term PROV) to guide the provenance interchange on the web. Specifically, the current PROV data model for provenance (PROV-DM; Moreau and Missier 2013) "defines a core data model for provenance for building representations of the entities, people and processes involved in producing a piece of data or thing in the world" (Gil and Miles 2013).

To illustrate PROV-DM in a remote sensing and geoprocessing context, the provenance of a 2001–2006 canopy change layer incorporated in the ASA Hazard Map (Jones et al. 2014; Tullis et al. 2012) can be represented using PROV-DM structures. This may be encoded (Figure 13.7; Table 13.1) as agents (e.g., a specific version of PCI Geomatica as a software agent), entities (e.g., a Landsat image clipped to a forest boundary), activities (e.g., ATCOR2 atmospheric correction based on specific calibration and other parameters), and relationships (e.g., wasInfluencedBy to represent the influence of Wang et al. (2007) on the change detection methodology). It is important to note that PROV-DM is extensible such that subtypes of agents, entities, activities, and relationships can be identified as needed for domain-specific applications (Moreau and Missier 2013). While PROV-DM does not specify structural categories such as sources, intermediate datasets, and products, such information can be derived within the context of a given provenance record.

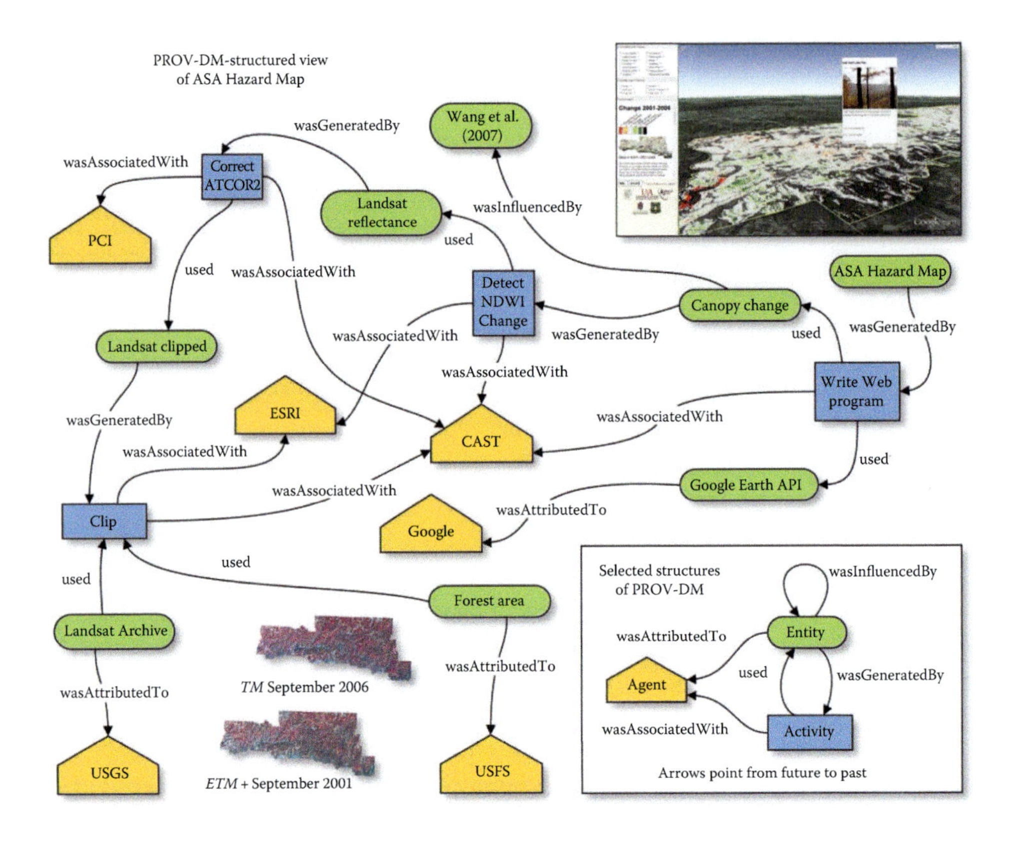

FIGURE 13.7 Selected provenance of the ASA Hazard Map (Jones et al. 2014; Tullis et al. 2012) structured according to W3C's PROV Data Model (PROV-DM; Gil and Miles 2013; Moreau and Missier 2013; Table 13.1). Arrows (relationships) point from future to past, first from the online ASA Hazard Map to its 2001–2006 canopy change layer, then to various agents, entities, and activities involved in the canopy change layer's creation. Some entities (e.g., "Landsat Archive") represent PROV-DM collections of entities (e.g., individual Landsat images available from USGS), and many potential PROV-DM details are not shown.

In the geospatial domain, the efforts of the Open Geospatial Consortium (OGC) initially (late 1990s and early 2000s) focused on the development of specifications that encouraged geospatial data interoperability such as the OGC Simple Features Specification. While not directly related to provenance, this effort has led to common ontologies and semantic structures that are foundational to the integration of geoprocessing, workflows, and provenance. In the 2000s, the OGC's attention shifted to web processing and interoperability of various web services. The OpenGIS Web Processing Service (WPS) specification (Schut 2007) has a *lineage* element in defining the request message to *execute* the spatial operation. In case lineage is defined as *true*, the response message from WPS will contain a copy of input parameter values specified in the service request definition. The OGC also developed the Sensor Web Enablement standard, in which the OpenGIS Sensor Model Language (Botts and Robin 2007) has one specific element that documents the observation lineage to describe how an observation is obtained. Elements of a number of Earth observation process specifications, such as the Catalogue Services Standard 2.0 Extension Package for ebRIM Application Profile: Earth Observation Products (Houbie and Bigagli 2010), the Sensor Observation Service Interface Standard (Bröring et al. 2012), and others, increasingly leverage provenance-related components as key elements. The more recent developments in WaterML and the Open Modeling Interface have increasingly emphasized provenance components.

TABLE 13.1
Characteristics of PROV-DM Structures Including Core Types and Selected Relationships (Moreau and Missier 2013), Each with an Example Provided from the Provenance of the ASA Hazard Map (Jones et al. 2014; Tullis et al. 2012; Figure 13.7)

PROV-DM Structure	Interpretive Highlights	Example from ASA Hazard Map Provenance
Core Types		
agent	Need not be a person but could also represent an organization or even a specific software process	Center for Advanced Spatial Technologies (CAST) *agent* (organization) at University of Arkansas
entity	May be physical, digital, or conceptual	Landsat 5 TM *entity* (satellite image) collected on September 15, 2006, over Ozark National Forest
activity	Involves entities and requires some time to complete	Esri ArcGIS 10 for Desktop "Extract by Mask" *activity* (software tool) used to clip the Landsat 5 TM imagery to the bounds of the study area), together with environment settings (e.g., "Snap Raster")
Selected Relationships		
wasGeneratedBy	Can only represent creation of new entities (that did not already exist)	Clipped Landsat 5 TM image that has been corrected for atmospheric attenuation *was generated by* running the ATCOR2 algorithm
used	Only implies that usage has begun (but not that it is completed)	A GIS model for detecting oak-hickory forest decline or growth *used* a clipped and atmospherically corrected Landsat ETM+ image collected 25 Sep 2001
wasAttributedTo	Links an entity to an agent without any understanding of activities involved	The Google Earth API (used to write a web program to generate the ASA Hazard Map) *was attributed to* Google
wasAssociatedWith	Links an activity to an agent	The ATCOR2 algorithm used to correct Landsat TM and ETM+ imagery for atmospheric attenuation *was associated with* PCI Geomatics through their Geomatica 10 platform
wasInfluencedBy	At a minimum suggests some form of influence between entities, activities, and/or agents; however, highly specific influence may be captured	The 2001–2006 oak-hickory forest canopy change data produced for the ASA Hazard Map *was influenced by* Wang et al. (2007) who used statistical thresholds of change in Landsat-derived normalized difference water index (NDWI) over time to detect oak canopy changes in the Mark Twain National Forest

> The purpose of the Open Modeling Interface (OpenMI) is to enable the runtime exchange of data between process simulation models and also between models and other modeling tools such as databases and analytical and visualization applications. Its creation has been driven by the need to understand how processes interact and to predict the likely outcomes of those interactions under given conditions. A key design aim has been to bring about interoperability between independently developed modeling components, where those components may originate from any discipline or supplier. The ultimate aim is to transform integrated modeling into an operational tool accessible to all and so open up the potential opportunities created by integrated modeling for *innovation* and wealth creation.
>
> *(Vanecek and Moore 2014, p. ix, emphasis added)*

It is likely that future OGC efforts will increasingly focus on provenance. The OGC is a major participant in EarthCube (2014) and in AI provenance (Chester 2023). In 2011, NSF's Cyberinfrastructure and Geosciences Divisions established the EarthCube community to promote geosciences data discovery and interoperability. The OGC plays a major role in this community, which has received NSF-funded research and implementation grants pertaining to provenance records in geoprocessing.

13.3 WHY PROVENANCE IN REMOTE SENSING WORKFLOWS

As Buneman (2013) argues, a "change of attitude" is in order regarding the role for provenance across a range of computer system-supported domains and activities, including (by implication), remote sensing workflows. He makes the comparison between scientific activities where it is considered obvious that such information should be recorded, and other domains where there is little or no awareness of process history or its value. He concludes that "we should worry less about what provenance is and concentrate more on what we can do with it once we have it" (p. 11).

13.3.1 Remote Sensing Questions That Only Provenance Can Answer

For volumes that contain primarily raw or unprocessed geodata (e.g., imagery telemetered directly from a satellite platform), provenance (as used in the present context) may not offer much over traditional metadata. However, when looking at geodata products resulting from complex geoprocessing workflows, there is much valuable information that metadata is ill-equipped to capture and store.

There is sometimes confusion concerning what provenance offers in terms of valuable information to an end user over the far more common and better supported (in terms of software integration) metadata. One way to structure such a discussion is to look at some of the questions data users might ask that can only be reasonably answered using (at least in part) detailed provenance information. For instance, one might ask the following regarding a remote sensing-derived product:

1. What was the processing time necessary to create this product, and what system configuration (hardware, software, firmware, AI setup, etc.) was implemented?
2. In what exact order were processing steps taken, and what precise parameters were used during each intermediate step? Was the process completely automated, or were manual steps (such as onscreen digitization) included in the workflow?
3. What datasets, both source and derived, were used to create this product and how did each contribute to the product?
4. In what ways are geospatial privacy, intellectual property, and export control being addressed in the workflow?
5. How were errors expressed and propagated during the product's creation? Is the result statistically valid?

In addition to the preceding, several questions could be asked trying to identify the source of errors or anomalies in the data. For example, one might wonder at what point in geoprocessing did a specific region get assigned null values and why? Using provenance data, it should be possible to analyze two similar data products and compare their processing history to see how and why they differ (Bose and Frew 2005; Lanter 1994a). The opportunity to better understand and manage the complexity of spatial scale in remote sensing-assisted workflows is a further justification for provenance-enabled geoprocessing (Tullis and Defibaugh y Chávez 2009). Finally, provenance-aware systems could be used to enable and support temporal GIS analyses which require detailed history of a dataset's change over time to properly function (Langran 1988).

The value of provenance tracking, and visualization was demonstrated in a study conducted at the Regional Geospatial Service Center at the University of Texas, El Paso (Del Rio and da Silva 2007). In this study, conducted as part of NSF's GEON Cyberinfrastructure project, Web services

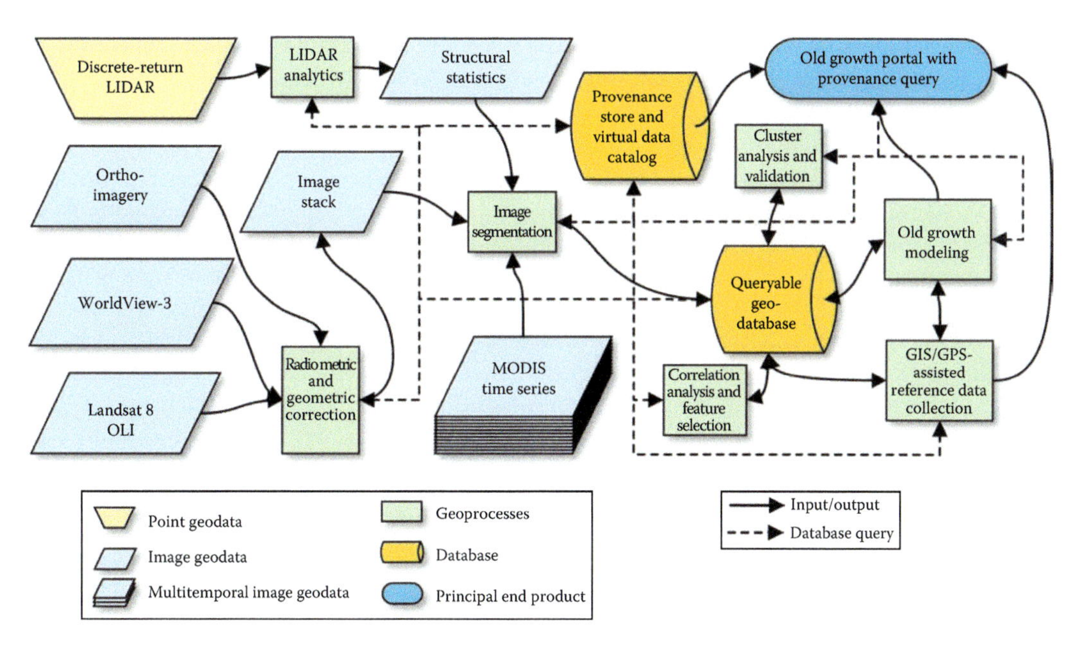

FIGURE 13.8 A provenance-enabled workflow for extracting old growth bald cypress (*Taxodium distichum*) forest quality and biophysical parameters from airborne LIDAR and orbital multispectral imagery.

were built to perform geoprocessing tasks (filtering, gridding and contouring) required to create a contoured gravity map from a raw gravity dataset. Del Rio and da Silva generated multiple contour maps with incorrect parameters (e.g., a grid size parameter larger than important anomalies in the gravity field) and participants in the study were asked to evaluate each contour map with and without provenance information. Without provenance information, subject matter experts were only able to detect errors in 50% of the cases, and to explain cause in only 25% of the cases. Non-subject matter experts fared much worse (11% and 11%). However, when provenance information was provided and visualized using Del Rio and da Silva's (2007) ProbeIt! (a provenance visualization tool), the subject matter experts were able to detect and explain all the errors. A more impressive result, though, is that provenance improved the ability of the non-subject matter experts to detect and explain errors by a factor of seven (78%).

It can be argued that extending a traditional remote sensing workflow to include provenance information offers several immediate advantages. For example, a web portal displaying a detailed map of old growth forest in the Southeastern United States could include provenance query that enables (1) detailed methodological transparency, (2) detailed transparency of the accuracy assessment, and (3) and auto-generated scripts for replication of the detailed workflow even though it includes a complicated blend of commercial and open source software, including cyber-enabled high performance geoprocessing tools (Figure 13.8).

13.3.2 Provenance as a Prerequisite for Remote Sensing Stakeholder Requirements

In the landscape of remote sensing, stakeholders from scientific researchers to policymakers demand reliable data. A cornerstone to meeting this demand is provenance as the comprehensive documentation of data's origins and journey. This section elucidates how provenance underpins stakeholders' needs by ensuring traceability, enhancing reproducibility, affirming data integrity, and fostering trust. We explore methods to capture and utilize provenance, illustrating with case studies where provenance has been pivotal. Ultimately, provenance is not just beneficial but essential for fulfilling the diverse and rigorous requirements of remote sensing stakeholders.

13.3.2.1 Repeatability, Reproducibility, and Replicability

Without provenance in some form, RRR would not be possible. When captured as a record, whether or not standardized, provenance can be communicated or distributed in a way that supports RRR in its central role as a requirement for integrity and utility of scientific inquiry. In a special issue on "Reproducibility and Replicability in Remote Sensing Workflows" published in *Remote Sensing*, Howe and Tullis (2022) point out that concerted efforts within the remote sensing community to address RRR challenges appear to be in their early stages. While this is not surprising given the many interrelated issues of RRR and provenance increasingly identified and recognized across science and engineering (e.g., National Academies 2019), provenance (whether referenced or not) is an integral part of the remote sensing process (Jensen 2016) and is available in various forms to support RRR.

While repeatability seems like a very minimal standard for remote sensing workflows, this can be remarkably challenging over the course of time. For example, software versions (due in part to inventions like Git), change relatively frequently which can lead to workflow runtime compatibility failures or logical changes. Hardware and firmware changes may influence the computational environment such that the same team cannot easily repeat execution of a specific workflow. When provenance is recorded and accessible at sufficient levels of detail, it can be automatically or manually incorporated into the process of repeating a workflow. Similar principles apply for reproducibility (Figure 13.9a) of remote sensing workflows, but this includes more complicated requirements involving independent teams and computational setups.

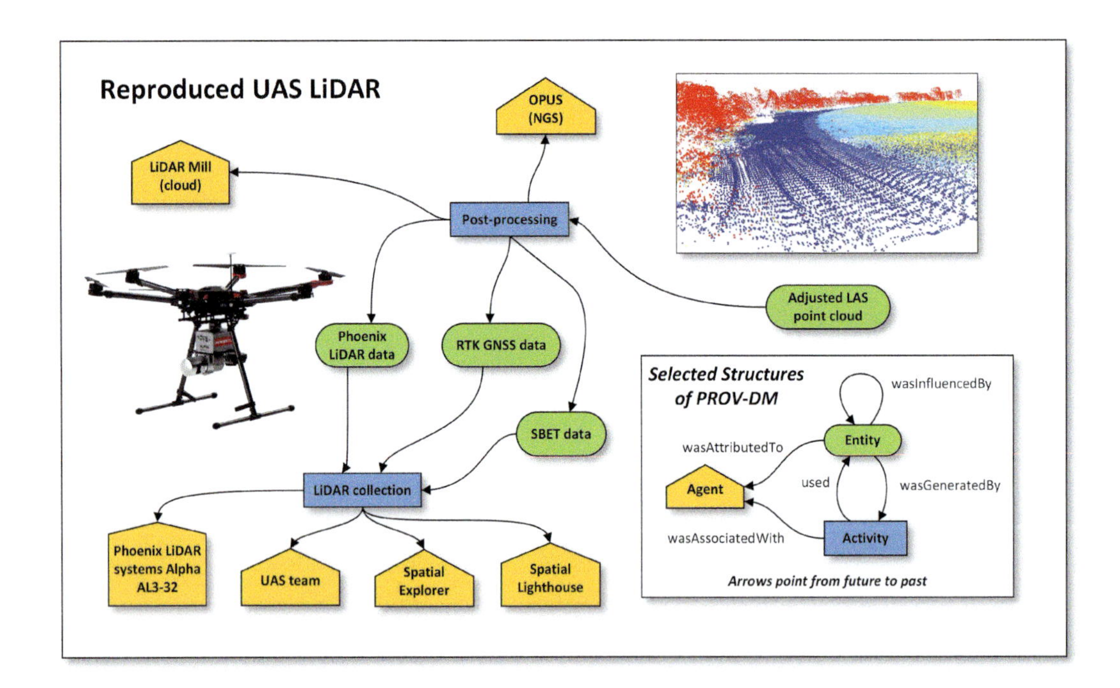

FIGURE 13.9A Provenance information, structured using PROV-DM, that could facilitate reproducibility, and by definition, repeatability. To ensure hardware and software compatibility, a more detailed granularity of this provenance, such as software versions and how to access source data, would be required. (Adapted from Tullis et al. 2019.)

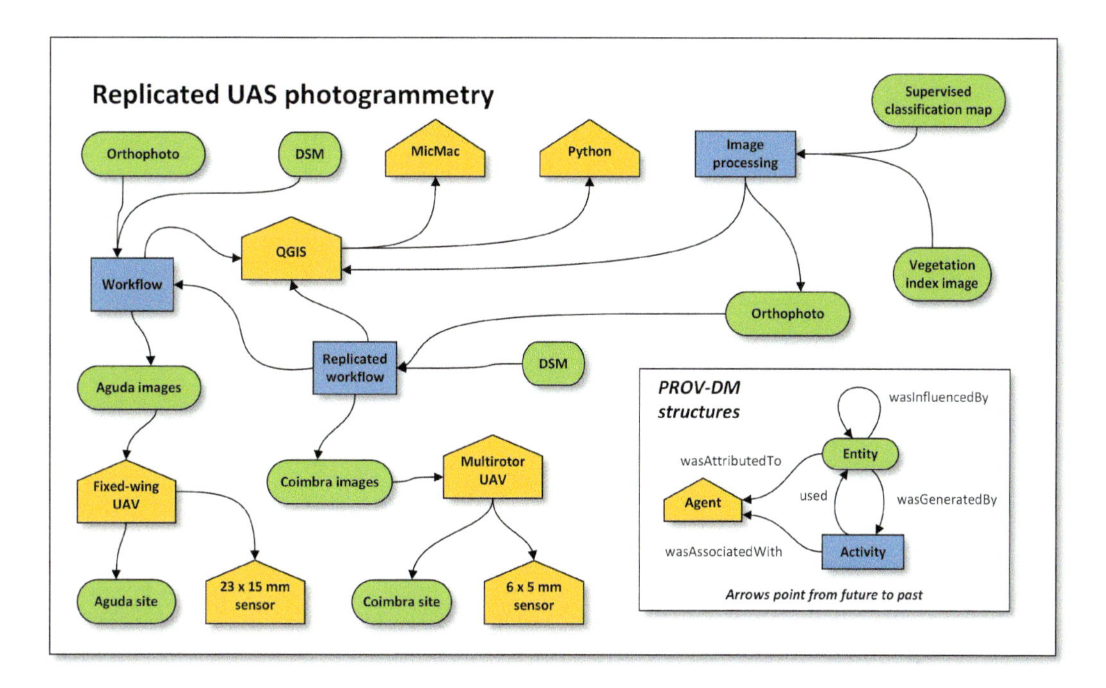

FIGURE 13.9B Provenance information, structured using PROV-DM, that documents a replicated UAS photogrammetry workflow. The original study area is Aguda and the new study site is Coimbra. (Adapted from Howe and Tullis 2022.)

Unlike repeatability and reproducibility, replicability involves making changes to me source data and/or methods. Geography, by definition, introduces a spatially intensive requirement of replicability. For example, simply moving from one study area to another, with all other aspects of the new workflow being held as close to the original as possible, can introduce a series of anticipated and unanticipated challenges associated with replicability (Figure 13.9b). As noted by Howe and Tullis (2022), barriers to RRR in remote sensing workflows include (1) awareness of requirements, (2) access to version control, metadata, and provenance, (3) understanding the geographic nature of the problems in question, and (4) the fact that observations vary geographically.

13.3.2.2 Geospatial Interoperability

The advent of diverse remote sensing technologies and the explosion of geospatial data have underscored the critical need for interoperability within the geospatial domain. Geospatial interoperability refers to the ability of different systems, platforms, and datasets to work together seamlessly, enabling the sharing, processing, and analysis of geospatial data across various applications and among stakeholders. This interoperability is pivotal for enhancing the utility of remote sensing data, facilitating collaborative efforts, and driving innovation in geographic information science.

13.3.2.3 Privacy, Confidentiality, and Intellectual Property

As remote sensing technologies continue to advance, they are increasingly capable of capturing detailed geospatial data across vast areas of the Earth's surface. This capability, while invaluable for environmental monitoring, urban planning, and disaster response, among other applications, also raises significant concerns regarding privacy, confidentiality, and the protection of intellectual property and national exports (Tullis and Kar 2020). The balance between leveraging remote sensing for societal benefits and safeguarding individual and organizational rights is a complex challenge that requires careful consideration and robust frameworks.

Some privacy concerns in remote sensing revolve around the potential for high-resolution imagery to identify individuals, private properties, or sensitive locations without consent. As the resolution of satellite and aerial imagery increases, the risk of infringing on personal privacy increases, necessitating measures to prevent unauthorized surveillance and data misuse.

To address privacy concerns, several strategies can be employed:

- Anonymization of data: Remote sensing data can be processed to anonymize identifiable features, ensuring that individuals or specific properties cannot be distinguished.
- Regulatory compliance: Adherance to legal standards, such as those outlined in data protection regulations, may help ensure that remote sensing practices respect privacy rights.
- Ethical guidelines: Developing and following ethical guidelines for remote sensing operations can help to prevent privacy infringements and foster public trust in these technologies.

In their discussion on location privacy, Tullis and Kar (2020) list 1) cryptography, 2) anonymity, 3) obfuscation, and 4) caching mechanisms as privacy enhancements. They also point to adversarial techniques, including "deep learners that combine contextual information with location information" that may counteract privacy measures. Provenance, they argue, is vital to addressing privacy because it can be used to communicate how this stakeholder interest is being addressed. For example, a particular workflow might make some detailed provenance information available to medical geographers authorized to work with specific kinds of survey data, but not included in final published maps.

Confidentiality issues arise when remote sensing data contains or is combined with sensitive information that could compromise security or competitive advantages. For instance, data revealing the location and characteristics of critical infrastructure or proprietary agricultural practices requires careful handling to prevent unauthorized access and potential exploitation.

The question of intellectual property (IP) in remote sensing is multifaceted, involving the rights to data, images, and derived products. As remote sensing datasets and analytical methods become increasingly valuable, establishing clear IP rights is crucial for promoting innovation while ensuring fair use and distribution of information.

- Licensing frameworks: Implementing licensing agreements that specify the use, distribution, and modification rights of remote sensing data can help manage IP concerns.
- Open data initiatives: Balancing IP rights with open data initiatives can enhance accessibility and applicability of remote sensing data for societal benefits, while still protecting the interests of data providers.

13.3.2.4 Geospatial Artificial Intelligence

The integration of AI with geospatial science, commonly referred to as Geospatial AI or GeoAI, represents a transformative leap in the capability to analyze, interpret, and leverage vast amounts of geospatial data. GeoAI combines the power of AI, including machine learning (ML) and deep learning (DL), with geographic information systems (GIS) and remote sensing technologies, paving the way for unprecedented insights into environmental patterns, urban development, climate change, and more. This section explores the evolution, applications, and future directions of GeoAI.

GeoAI has emerged from the convergence of AI and geospatial sciences, driven by the explosion in geospatial data availability and the advancements in computational power and AI algorithms. Core technologies underpinning GeoAI include:

- Machine learning and deep learning: Algorithms that enable the modeling of complex patterns and predictions based on geospatial data.
- Computer vision: Techniques that allow machines to interpret and analyze imagery and video data captured by satellites, drones, and ground-based sensors.

- Natural language processing (NLP): Enables the extraction of geospatial information from text data, such as social media posts or news articles, providing insights into human activities and sentiments across different regions.

GeoAI has found applications across various domains, demonstrating its versatility and impact:

- Environmental monitoring: AI models analyze remote sensing data to track deforestation, monitor biodiversity, and assess water quality.
- Disaster response and management: GeoAI facilitates real-time analysis of disaster-impacted areas, optimizing rescue operations and assessing damage more efficiently.
- Urban planning: AI-driven analysis of geospatial data supports sustainable urban development, traffic management, and infrastructure planning.
- Agriculture: Precision agriculture benefits from GeoAI through crop monitoring, yield prediction, and optimization of resources.

Despite its potential, the deployment of GeoAI faces challenges, including location privacy, ethical concerns, and the need for robust, interpretable models. The accuracy of GeoAI predictions heavily depends on the quality and quantity of the geospatial data, requiring continuous efforts in data collection and curation. The future of GeoAI lies in addressing these challenges while harnessing opportunities such as:

- Integration with IoT: Combining GeoAI with the Internet of Things (IoT) technologies for real-time environmental monitoring and management.
- Advancements in computational technologies: Leveraging quantum computing and edge computing to enhance the processing capabilities for GeoAI applications.
- Interdisciplinary collaboration: Fostering collaboration across disciplines to innovate and apply GeoAI in addressing global challenges.

Artificial intelligence (AI) is increasingly being used in geospatial workflows and big data processing (Sun et al., 2020). Regular advancements in computer vision and deep learning algorithms are demonstrating increasingly impressive abilities in image recognition, pixel classification, and object detection workflows. These AI advancements have potential to become prominent analysis tools in agricultural systems to process large volumes of field imagery and ultimately better inform or even automate field level production decisions. One example is using remote sensing and AI image processing to assist with site-specific treatment considerations that reduce field inputs and supplement in situ field scouting efforts.

Davis (2022) evaluated a novel UAS image stream and AI processing workflow that developed site-specific treatment maps based on broadleaf weed detections. High spatial resolution (0.4 cm) unmanned aircraft system (UAS) image streams were collected, annotated, and used to train object detection convolutional neural networks (CNNs; RetinaNet, Faster R-CNN, Single Shot Detector, and YOLO v3) through transfer learning. The deep learning models developed were used to make weed detections in field imagery. The maps generated through this processing workflow were evaluated in production field conditions and a 48% reduction in applied pesticides was reported when compared with traditional broadcast applications. Agricultural AI workflows have similarly been evaluated to reduce field inputs in irrigation, nutrient management, plant disease detection, and vegetative health monitoring.

The workflow described in Davis (2022) and commonly in other AI-geoprocessing workflows leveraged a stream of software (workflow software pipeline—UAS navigation, senser, photogrammetry, georeferencing, computer hardware driver, object detection analysis, and field equipment) that are often version specific or version sensitive. While AI tools show much promise in facilitating the processing of large data volumes and facilitating production decisions, the complex network of processing software culminating as a decision-making tool are often viewed as black box technologies (Toda & Okura, 2019). Given the complex architecture of machine learning models and associated workflows, AI is more often contributing to the reproducibility crisis (Sun et al., 2020).

While AI assistance in geoprocessing workflows may add significant ambiguity to data provenance, other AI tools are making documenting and tracking data provenance in these workflows easier. Sun et al. (2020) proposed a scientific framework coined "Geoweaver" that centralizes and records intermediate data throughout a geoprocessing workflow across multiple software, hardware, programing languages, and data sources.

13.3.3 Provenance and Trust in the Remote Sensing Process

When provenance of sources, intermediates, and products is captured and maintained throughout the digital remote sensing process lifecycle, data quality improves, an audit trail is available for reviewers and users, reproducibility is straightforward, attribution is streamlined and the interpretability of geodata products is enhanced (Simmhan et al. 2005). However, any one of these advantages by itself is not as critical as establishing trust in the remote sensing process, which is clearly not immune to being oversold (Jensen 2007). Statistical validity is a key to establishing trust in scientific processes and has been extensively developed in remote sensing workflows (Congalton 2010). It requires documentation of data sources and transformational processing to enable RRR. Simply maintaining a source data inventory and audit trail that demonstrates sound choices in data sources, statistical processing parameters and methods is one way to facilitate RRR so that trust is not erroneously called into question.

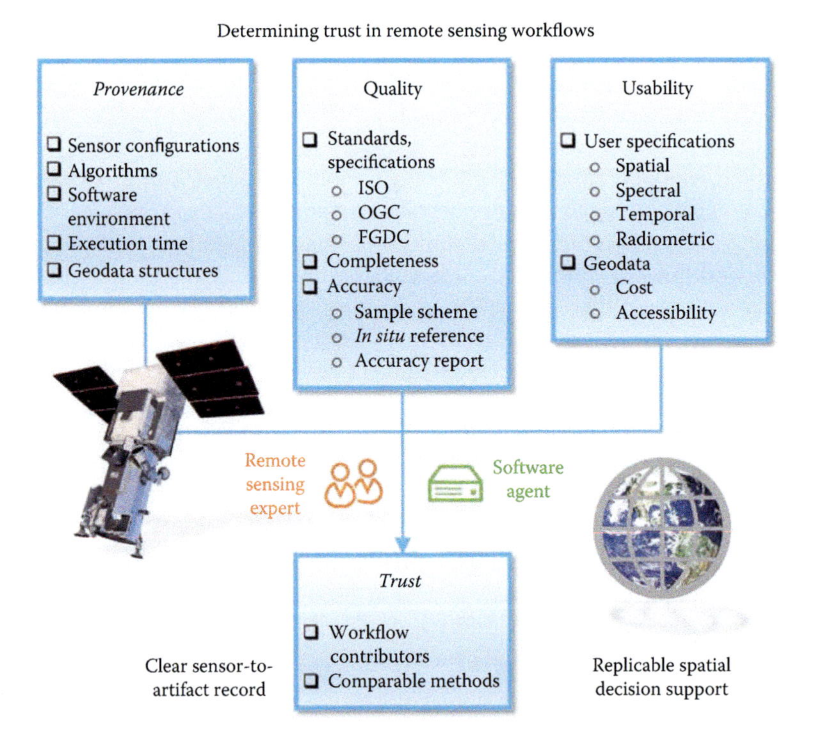

FIGURE 13.10 Provenance, quality, and usability can be used by remote sensing experts to make a subjective decision on workflow trust (Malaverri et al. 2012; Gamble and Goble 2011; Jensen 2005); a sample of their characteristics is shown. As geoprocessing, workflows, and provenance are integrated, software agents can objectively influence replicable spatial decision support. (Artist image of WorldView-3 courtesy of DigitalGlobe, 2014.)

In addition to maintaining source documentation and an audit trail, statistical techniques can be applied to a given workflow to determine trust, and recent work has focused on a measure of trust by using a workflow's provenance. While not demonstrated in a remote sensing context, a non-stationary Hidden Markov Model (HMM) can be used to provide a measure of this trust (Naseri and Ludwig 2013). Provenance can also be useful in helping to determine workflow trust for data on the web. When handling linked data provenance, data authentication may serve as an estimation of data trustworthiness. Uniform Resource Identifiers (URI) and digital signatures can be used as a measure of authentication (Hart and Dolbar 2013).

Traditional indicators of trust may also be used in conjunction with provenance. Recommendation, authority, believability, reputation, and objectivity can all serve as indicators of trust (Gamble and Goble 2011). In addition to provenance, a workflow's quality and usability should also be evaluated when determining trust. As geoprocessing, workflows, and provenance are integrated, software agents can objectively influence replicable spatial decision support (Figure 13.10). Until more quantitative techniques are developed for measuring trust of geographic workflows using provenance, measures of quality and usability used in conjunction with subjective trust indicators, should be examined before making a decision to trust a workflow and its lineage.

13.4 SELECTED RECENT AND PROPOSED PROVENANCE-AWARE SYSTEMS

Many provenance-aware systems have largely been concerned with provenance capture, and this capability is critical for synergistic geoprocessing, workflows, and provenance of interest in remote sensing applications (Figure 13.1). The characteristics of the captured provenance information can greatly influence how it may benefit the remote sensing process. Of particular significance to geospatial applications is the level of provenance detail or granularity. Fine grain provenance is obtained at a data level and can even deal with individual pixels (Woodruff and Stonebraker 1997), whereas coarse grain provenance represents the workflow level (Tan 2007) and can thus be used to facilitate scientific reproducibility. As Henzen et al. (2013) point out, the quality of provenance communication is also very important even when presented in a text format. Several recent and proposed provenance-aware approaches and systems related to remote sensing (Table 13.2) have addressed these and other issues.

13.4.1 General Approaches

The general approach to provenance-aware systems focuses on enabling efficient management and execution of data-driven workflows. By integrating advanced computing resources and supporting flexible scripting environments, these systems facilitate the creation of transparent, reproducible research processes. Emphasizing FAIR (findable, accessible, interoperable, and reusable) principles helps foster collaboration across the scientific community. This approach underlines the importance of user-friendly interfaces for executing and sharing workflows, aiming to streamline research and enhance data integrity.

13.4.1.1 Inversion

Inversion was developed for fine-grain data lineage and provenance in database transaction and transformation (Buneman et al. 2001; Cui et al. 2000; Woodruff and Stonebraker 1997). Database queries or processing functions that generate a view, table, or new data product can be registered in a database system (or provenance store). Registered database transformations can be inverted so as to trace the lineage between the data product and the sources that derive the product. For example, when a view is created or updated, the inversion method can help identify the source tables from which the view is generated. Although inversion can be applied in data provenance for geospatial data, not all functions are invertible. However, a weak or general inversion could be substituted to

TABLE 13.2
Characteristics of Selected Provenance-Aware Systems Reported in Remote Sensing and Other Geodata Applications

Geodata Application	Successes	Limitations or Future Work	References
Earth System Science Workbench and ES3			
Track processing of a laboratory's raw satellite imagery (e.g., AVHRR) into higher level products	(a) Automates geodata provenance capture from running processes (b) Stores provenance in both XML documents and in a searchable online store	Predates recent geodata interoperability standards and specifications	Frew and Bose (2001) Frew (2004) Frew and Slaughter (2008)
MODAPS and OMIDAPS			
Manage MODIS and other NASA satellite imagery and its provenance	(a) Automates version tracking of geodata processing algorithms (b) Reduces geodata storage via on-demand processing based on a virtual archive	Identifies science community's lack of appreciation for provenance information	Tilmes and Fleig (2008)
Karma			
Capture provenance for Japan's AMSR-E passive microwave radiometer on Aqua	(a) Modularizes architecture to facilitate web service interoperability (b) Compatible with Open Provenance Model (OPM) and ISO 19115-2 metadata standards	Requires additional geodata interoperability standards to facilitate geodata (scientific) reproducibility	Simmhan et al. (2008) Plale et al. (2011) Conover et al. (2013)
Data Quality Provenance System			
Assess quality of agricultural mapping based on SPOT satellite imagery	(a) Assigns geodata quality index based on provenance information (b) Compatible with OPM and FGDC metadata standards	Needs to address geodata quality dependencies on provenance granularity	Malaverri et al. (2012)
VisTrails			
Model habitat suitability using WorldView-3 and LIDAR-derived forest structure	(a) Provides Python-based open source provenance and workflow management (b) Allows key focus on provenance in rapidly changing workflows (e.g., during remote sensing process development)	Designed to be domain-generic; VisTrails may have a steep learning curve	Freire et al. (2012) Talbert (2012)
UV-CDAT			
Analyze large scale remote sensing-derived climate data	(a) Built on top of VisTrails with an extensible modularized architecture that supports high performance workflows (b) First end-to-end provenance-enabled tool for large scale climate research	Future work could adapt UV-CDAT successes in climate change for other geodata application areas	Santos et al. (2012)
GeoPWProv			
Visualize and navigate city planning geodata (e.g., LIDAR-derived elevation data) provenance via a map	(a) Allows geodata provenance to be visualized and explored in a map environment (b) Provides for several geospatial levels of provenance query (e.g., via a single polygon versus a larger dataset)	Future work could support geoprocessing replication	Sun et al. (2013)

Geodata Application	Successes	Limitations or Future Work	References
GeoWeaver			
Automate and manage complex AI-assisted workflows involving a large variety of sources and processes in local or distributed environments	(a) Facilitates binding public and private sources into one AI workflow (b) Demonstrates process-oriented, as opposed to data-centric provenance	Future work can further facilitate access to broader community	Sun et al. (2020)
Code Ocean			
Focus development of remote sensing and bioinformatics workflows toward reproducibility and collaboration	(a) Facilitates provisioning and scaling in the cloud (b) Uses containerization to ensure that computational environments can be reproduced as hardware environments change	Its current focus is on bioinformatics community, and there is growth potential in remote sensing roots of the project	Code Ocean (2024)
KnowWhereGraph			
Incorporate remote sensing and other geospatial data into a knowledge graph	(a) Demonstrates a new paradigm in use cases for knowledge graphs (b) Allows provenance to be traced using PROV-O	This represents a paradigm shift in the early phase of growing influence	Janowicz et al. (2022)

approximate the provenance by returning a fraction (or a projection) of the desired provenance. Examples of inversion can be found in some spatial databases including Oracle Database, Microsoft SQL Server, IBM DB2, and PostGIS. One early approach was implemented for vector operations in Intergraph's (now Hexagon's) Geomedia product.

13.4.1.2 Service Chaining

In a vision of SOC and service-oriented integration (SOI), different web services can be found, composed and invoked to accomplish certain tasks. The sequence of service discovery, composition, and execution looks like a chain, while alternatively, the composition processes can be constructed through different approaches, such as service orchestration or choreography that could be applied in enabling business processes and transactions on the web. Service composition and chaining could represent a workflow in which scientific computation can be implemented through the SOC/SOI approach. Capturing the provenance information within service-oriented workflows has been explored in geospatial applications (Del Rio and da Silva 2007; Yue et al. 2010a, 2010b, 2011), though feasible and convincing approaches for provenance in SOC/SOI need further exploration and investigation.

13.4.1.3 Virtual Data Catalog Service

A virtual data catalog (VDC) is a provenance approach to trace the derivation route of data products and virtual data generated in the workflow in order to enable scientists to reproduce and validate workflows and related simulations. The intermediate data may be generated within a workflow but may not exist physically in a database or computer system (e.g., due to storage limitations). For this reason, such data are called virtual data because it is "the representation and manipulation of data that does not exist, being defined only by computational procedures" (Foster et al. 2002, 2003).

Within virtual data systems, such as Chimera, which is a virtual data grid managing the derivation and analysis of data objects, Virtual Data Language (VDL) is developed to define the workflow, while VDC is a service in the virtual data systems. The latter is defined and implemented based on the virtual data schema (VDS). The VDS defines the relevant data objects and relationships, and

VDC can be queried by VDL to construct data derivation procedures from which derived data and output can be recomputed (Foster et al. 2002, 2003; Glavic and Dittrich 2007; Simmhan et al. 2005).

13.4.2 Earth System Science Workbench and ES3

The Earth Systems Science Workbench (ESSW) was an early attempt at automated provenance management and storage. It was a nonintrusive system that made use of Perl scripting techniques and Java to store data provenance as XML documents (Frew and Bose 2001). It contained a registry for provenance, and a server for making the information searchable on the web. ESSW was followed up by the Earth System Science Server (ES3) which allowed for more flexibility in client-side implementation but used essentially the same structure as the ESSW (Frew 2004). ES3, unlike many other systems, automatically captures provenance from running processes. It can also create provenance graphs in XML that can then be visualized using third party tools like yEd (Frew and Slaughter 2008).

13.4.3 MODAPS and OMIDAPS

The MODIS Adaptive Data Processing System (MODAPS) and OMI Data Processing System (OMIDAPS) were designed for use by NASA to manage satellite imagery and provenance from MODIS sensors on the Terra and Aqua satellites, and the OMI sensor on Aura respectively (Tilmes and Fleig 2008). Both systems use a scripting process to track changes in versions of geodata processing algorithms. Using this technique, there is no need to store workflow iterations because enough information is retained so that previous versions of the data can be recreated. Further, these systems periodically are tasked with reprocessing past data using the most up-to-date algorithms to maintain a consistent and improved series of data products. MODAPS in particular makes use of these features to maintain a "virtual archive" with provenance information that persists after a geodata product is deleted, allowing the system to recreate data products on demand rather than keeping extensive archives. Data recreation as implemented in these systems is unique and is something that could be useful in other geospatial provenance systems.

13.4.4 Karma

Plale et al. (2011) make use of the Karma system designed by Simmhan et al. (2008) to collect provenance data for the Advanced Microwave Scanning Radiometer—Earth Observing System (AMSR-E) flown on the Aqua satellite. One of the biggest benefits of Karma is its modular architecture, which simplifies interoperability with Java and other web services. Karma's architecture for this application consists of an application layer, web service layer, core service layer, and a database layer (Plale et al. 2011). The inclusion of Open Provenance Model specifications and XML extends its interoperability further (Moreau et al. 2011). Conover et al. (2013) also made use of Karma to retrofit a legacy system for provenance capture. They chose the NASA Science Investigator-led Processing System (SIPS) for the AMSR-E sensor on the Aqua satellite. Their system uses a two-tiered approach that captures provenance for individual data files as well as collections or series, both automatically and via manual entry using the ISO 19115-2 lineage metadata standard. Query and display are handled with a database-driven web interface called the Provenance Browser.

13.4.5 Data Quality Provenance System

Taking into account a source's trustworthiness and the data's age, Malaverri et al. (2012) created a provenance system that allows a quality index to be assigned. This approach is based on a combination of the Open Provenance Model (OPM) and FGDC geospatial metadata standards. Criteria considered in the quality index include: granularity, accuracy of attribute descriptions, completeness of the data, a logical measure of the data, and spatial positional accuracy. Although measures of trust can be very subjective in nature, in this case requiring a domain experts' input, this approach is somewhat unique in that it attempts to quantify data quality (Malaverri et al. 2012).

13.4.6 VisTrails

VisTrails is a free and open-source scientific workflow and provenance management system (Freire et al. 2012). Written in Python/Qt and designed to be integrated with existing workflow systems, VisTrails has been used in a number of research applications ranging from climate (including UV-CDAT described next) to ecology and biomedical research. Talbert (2012) created software based on VisTrails to capture the details of habitat suitability and species distribution modeling. One of the major advantages of VisTrails is that as an open-source project built in Python it is interoperable, easily customizable, and benefits from a large community of developers contributing code. A key focus of VisTrails is rapidly changing workflows. The information contained in how workflows are developed (and change over time) may provide highly valuable insight into the creative and development aspects of the remote sensing process.

13.4.7 UV-CDAT

Climate Data Analysis Tools (CDATs) are cutting-edge domain-specific tools for the climate research community, but they are ill-equipped to handle very large geodata and provenance information. UV-CDAT is a provenance system for handling large amounts of climate-based data (Santos et al. 2012). UV-CDAT uses a highly extensible modular design and makes use of a Visualization Control System (VCS) and Visualization Toolkit (VTK)/ParaView infrastructure, which allows for high performance parallel streaming data analysis and visualization. Its loosely coupled modular design allows for integration with third party tools such as R and MATLAB® for both analysis and visualization. UV-CDAT is unique in that it is the first end-to-end application for provenance-enabled analysis and visualization for large scale climate research. It has already been distributed and is widely used by scientists throughout the climate change field.

13.4.8 GeoPWProv

GeoPWProv is a provenance system specializing in displaying geospatial provenance as an easily accessible interactive map layer. GeoPWProv has the capability to capture provenance at the feature, dataset, service, or knowledge level (Sun et al. 2013). Comparisons can be made between entities in each level or between various levels. In addition to displaying provenance as a map layer, GeoPWProv supports displaying provenance in a workflow or in the more traditional text-based format. Implementation on the client side through use of a browser and *Open Layers* allows for ease of use. GeoPWProv's display of provenance in different formats and at different levels allows for a customizable user experience when evaluating a workflow.

13.4.9 GeoWeaver

GeoWeaver represents a notable advancement in provenance-aware systems, building on concepts developed through GeoPWProv. It emphasizes the management, sharing, and reuse of complex AI workflows incorporating geospatial data. GeoWeaver's interface allows for the execution of processes on chosen hosts, either locally or remotely, while capturing code and execution logs to facilitate easy sharing and collaboration within the Earth science community. It supports scalability from local computational resources to cloud-based services, ensuring that large datasets can be handled efficiently. The system's architecture is designed to dynamically scale computational resources based on the workflow's requirements, optimizing processing time and resource utilization. It addresses the challenge of hybrid workflows (public and private resources, remote file access, behind-the-scenes data flow, code-machine separation, and process-based (as opposed to data centric) treatment of provenance (Sun et al. 2020).

13.4.10 Code Ocean

A project developed through the Jacobs Technion-Cornell Institute, Code Ocean (2024), is highly focused on reproducibility and transparency. Interestingly, Code Ocean co-founder Simon Adar developed this platform with a background in hyperspectral remote sensing. Code Ocean simplifies workflow sharing using containerization (Dockerfile management) to ensure that researchers can obtain the same computational environments and allows them to alter parameters, sources, study areas, etc. without losing access to specific code libraries. Code Ocean manages its data, code, and metadata using Git version control. It is currently being developed for the bioinformatics community but as noted has remote sensing origins and valuable capacity in the remote sensing and geospatial communities.

13.4.11 KnowWhereGraph

After Google launched its Knowledge Graph in 2012, its search engine capabilities were dramatically augmented to "know" how search terms are related (Singhal 2012). Development of knowledge graph technologies is of particular importance for provenance and can be harmonized with PROV-DM (Moreau and Missier 2013) to enable questions not just about how things are related but also how they were created or influenced. The KnowWhereGraph (Janowicz et al. 2022) is a remarkable project because it takes recent developments in knowledge graphs and incorporates geography. "For instance, [KnowWhereGraph can] model where a fire took place, which events it triggered, and which regions have been affected, for example, by heavy smoke" (Janowicz et al. 2022, p. 32). Behind the scenes, the data in KnowWhereGraph and other knowledge graphs are stored as "triples" in a database management systems. The tables containing these triples can become very large (many billions of records), and the KnowWhereGraph project is developing innovative ways to manage these in the context of GIS and broad usability, and the basic concepts it has developed are gaining momentum. Esri, for example, recently launched its new ArcGIS Knowledge platform (Esri 2024) with an entirely new capacity for tracing provenance.

13.5 CONCLUSIONS AND RESEARCH IMPLICATIONS

Integrated geoprocessing, workflows, and provenance is increasingly understood as a key to high quality, replicable remote sensing-assisted spatial decision support. In early discussions in the 1980s, it soon became clear that provenance (or lineage) in particular is a fundamental element in understanding Earth observation-related and other geodata quality (Moellering et al. 1988). As commercial GIS accelerated during the early 1990s, the Geolineus project (Lanter 1992b) demonstrated how software dedicated to lineage/provenance capture, management, and visualization can enable such gains as replicable geospatial workflows, automated workflow comparison, data quality modeling, data update management, and increased sharing of expert knowledge of geodata creation. Now with increasingly heightened awareness of provenance in computer systems (Yue et al. 2010a; Ikeda and Widom 2009; Bose and Frew 2005; Simmhan et al. 2005), there has been a maturing appreciation of the need to computationally address provenance capture, management and exchange in an increasingly big data scenario.

While definitions of geodata provenance have varied, it is quite arguably distinct from and offers unique benefits over traditional metadata in large part because it encompasses process *history*. Regardless of definitions, the practical application of provenance benefits in remote sensing-assisted decision support workflows cannot be realized without development and demonstration of collaborative software architectures including those in a geo-cyberinfrastructure. Provenance has and will be of increasing interest to and a focus of organizations that create and encourage international specifications and standards (e.g., ISO, W3C, and OGC). As these organizations formulate procedures for the specification of provenance, we will see software developers add this capability to their products in a far more complete implementation than is currently the case. Even before the emerging international standards begin to mature, research is critically needed to demonstrate and fully understand *practical* benefits that user-friendly and integrated geoprocessing, workflows and provenance can offer. A new capacity for provenance is beginning to take root within remote sensing and geospatial geoprocessing environments and workflows, and this is contributing to new avenues of inquiry of vital importance for multiple stakeholder interests. With additional research and development, geospatial provenance has a high potential to benefit quality, trust, and innovation related to remote sensing-assisted spatial decision support.

REFERENCES

Alonso, G., & Hagen, C. (1997). Geo-Opera: Workflow Concepts for Spatial Processes. *Advances in Spatial Databases*, 238–258.

Anderson, J. R., Hardy, E. E., Roach, J. T., & Witmer, R. E. (1976). *A Land Use and Land Cover Classification System for Use with Remote Sensor Data*. U.S. Government Printing Office.

Anderson, K. E., & Callahan, G. M. (1990). The Modernization Program of the U.S. Geological Survey's National Mapping Division. *Cartography and Geographic Information Systems*, *17*(3), 243–248.

Aronson, P., & Morehouse, S. (1983). The ARC/INFO Map Library; A Design for a Digital Geographic Database. *Auto-Carto Six; Proceedings of the Sixth International Symposium on Automated Cartography*, *1*, 372–382.

Association for Computing Machinery. (2020, August 24). *Artifact Review and Badging*. https://www.acm.org/publications/policies/artifact-review-badging

Atkins, D. E., Droegemeier, K. K., Feldman, S. I., Garcia-Molina, H., Klein, M. L., Messerschmitt, D. G., Messina, P., Ostriker, J. P., & Wright, M. H. (2003). *Revolutionizing Science and Engineering Through Cyberinfrastructure: Report of the National Science Foundation Blue-Ribbon Advisory Panel on Cyberinfrastructure*. National Science Foundation. https://arizona.openrepository.com/arizona/handle/10150/106224

Bose, R., & Frew, J. (2005). Lineage Retrieval for Scientific Data Processing: A Survey. *ACM Computing Surveys*, *37*(1), 1–28.

Bossler, J. D., Campbell, J. B., McMaster, R. B., & Rizos, C. (Eds.). (2010). *Manual of Geospatial Science and Technology* (Second Edition). CRC Press.

Botts, M., & Robin, A. (Eds.). (2007). *OpenGIS Sensor Model Language (SensorML) Implementation Specification*. Open Geospatial Consortium. http://portal.opengeospatial.org/files/?artifact_id=21273

Bröring, A., Stasch, C., & Echterhoff, J. (Eds.). (2012). *OGC Sensor Observation Service Interface Standard*. Open Geospatial Consortium.

Buneman, P. (2013). The Providence of Provenance. In G. Gottlob, G. Grasso, D. Olteanu, & C. Schallhart (Eds.), *Big Data* (Vol. 7968, pp. 7–12). Springer. http://link.springer.com/10.1007/978-3-642-39467-6_3

Buneman, P., & Davidson, S. B. (2010). *Data Provenance – The Foundation of Data Quality*. Carnegie Mellon University Software Engineering Institute. https://www.semanticscholar.org/paper/Data-provenance-%E2%80%93-the-foundation-of-data-quality-Buneman-Davidson/9ec4275fed43df7145dec34cba9743a9186dc972

Buneman, P., Khanna, S., & Tan, W.-C. (2001). Why and Where: A Characterization of Data Provenance. *Database Theory – ICDT*, *2001*, 316–330.

Chester, S. (2023, September 26). OGC Adopts Training Data Markup Language for Artificial Intelligence Conceptual Model as Official Standard. *Open Geospatial Consortium*. https://www.ogc.org/press-release/ogc-adopts-training-data-markup-language-for-artificial-intelligence-conceptual-model-as-official-standard/

Chrisman, N. R. (1983). The Role of Quality Information in the Long-Term Functioning of a Geographic Information System. *Automated Cartography*, *6*, 302–312.

Chrisman, N. R. (1986). Obtaining Information on Quality of Digital Data. In M. Blakemore (Ed.), *AutoCarto Proceedings of the International Symposium on Computer-Assisted Cartography* (Vol. 1, pp. 350–358). Cartography and Geographic Information Society.

Code Ocean. (2024). *Code Ocean.* https://codeocean.com

Congalton, R. (2010). Remote Sensing: An Overview. *GIScience & Remote Sensing, 47*(4), 443–459.

Conover, H., Ramachandran, R., Beaumont, B., Kulkarni, A., McEniry, M., Regner, K., & Graves, S. (2013). Introducing Provenance Capture into a Legacy Data System. *IEEE Transactions on Geoscience and Remote Sensing, 51*(11), 5098–5014.

Cui, Y., Widom, J., & Wiener, J. L. (2000). Tracing the Lineage of View Data in a Warehousing Environment. *ACM Transactions on Database Systems, 25*(2), 179–227.

Davis, J. A. (2022). *Precision Weed Management Based on UAS Image Streams, Machine Learning, and PWM Sprayers* [Ph.D., University of Arkansas]. https://www.proquest.com/pqdtglobal/docview/2778692747/abstract/2A23DA7925984CFDPQ/1

Del Rio, N., & da Silva, P. P. (2007). Probe-it! Visualization Support for Provenance. In *Advances in Visual Computing* (pp. 732–741). Springer. http://link.springer.com/chapter/10.1007/978-3-540-76856-2_72

Di, L., Shao, Y., & Kang, L. (2013a). Implementation of Geospatial Data Provenance in a Web Service Workflow Environment With ISO 19115 and ISO 1911-2 Lineage Model. *IEEE Transactions on Geoscience and Remote Sensing, 51*(11), 5082–5089.

Di, L., Yue, P., Ramapriyan, H. K., & King, R. L. (2013b). Geoscience Data Provenance: An Overview. *IEEE Transactions on Geoscience and Remote Sensing, 51*(11), 5065–5072. https://doi.org/10.1109/TGRS.2013.2242478

DigitalGlobe. (2014). *DigitalGlobe.* http://www.digitalglobe.com/

EarthCube. (2014). *EarthCube: Transforming Geosciences Research.* http://earthcube.org/

Esri. (2024). *ArcGIS Knowledge.* https://www.esri.com/en-us/arcgis/products/arcgis-knowledge/overview

Essinger, R., & Lanter, D. P. (1992). User-centered Software Design in GIS: Designing an Icon-based Flowchart that Reveals the Structure of ARC/INFO Data Graphically. *Proceedings of the Twelfth Annual ESRI User Conference.* Twelfth Annual Esri User Conference, Palm Springs, CA.

Fielding, R. T. (2000). *Architectural Styles and the Design of Network-based Software Architectures.* University of California, Irvine. http://www.ics.uci.edu/~fielding/pubs/dissertation/top.htm

Foster, I., Vöckler, J., Wilde, M., & Zhao, Y. (2002). Chimera: A Virtual Data System for Representing, Querying, and Automating Data Derivation. *Proceedings of the 14th International Conference on Scientific and Statistical Database Management*, 37–46. http://ieeexplore.ieee.org/xpls/abs_all.jsp?arnumber=1029704

Foster, I., Vöckler, J., Wilde, M., & Zhao, Y. (2003). The Virtual Data Grid: A New Model and Architecture for Data-Intensive Collaboration. *Proceedings of the First Biennial Conference on Innovative Data Systems Research (CIDR), 3*, 12.

Freire, J., Koop, D., Santos, E., Scheidegger, C., Silva, C., & Vo, H. T. (2012). VisTrails. In A. Brown & G. Wilson (Eds.), *The Architecture of Open Source Applications: Elegance, Evolution, and a Few Fearless Hacks: Vol. Volume I.* aosabook.org. http://aosabook.org/en/vistrails.html

Frew, J. (2004). *Earth System Science Server (ES3): Local Infrastructure for Earth Science Product Management.* NASA's Earth Science Technology Conference 2004, Palo Alto, CA. http://esto.gsfc.nasa.gov/conferences/estc2004/papers/a4p3.pdf

Frew, J., & Bose, R. (2001). Earth System Science Workbench: A Data Management Infrastructure for Earth Science Products. *Proceedings of the International Conference on Scientific and Statistical Database Management*, 180–189. https://doi.org/10.1109/SSDM.2001.938550

Frew, J., & Slaughter, P. (2008). ES3: A Demonstration of Transparent Provenance for Scientific Computation. In J. Freire, D. Koop & L. Moreau (Eds.), *Provenance and Annotation of Data and Processes* (pp. 200–207). Springer. http://link.springer.com/chapter/10.1007/978-3-540-89965-5_21

Gamble, M., & Goble, C. (2011). Quality, Trust, and Utility of Scientific Data on the Web: Towards a Joint Model. *Proceedings of the 3rd International Web Science Conference, 15.* http://dl.acm.org/citation.cfm?id=2527048

Gil, Y., Cheney, J., Groth, P., Hartig, O., Miles, S., Moreau, L., & Pinheiro da Silva, P. (Eds.). (2010). *Provenance XG Final Report.* W3C. http://www.w3.org/2005/Incubator/prov/XGR-prov-20101214/

Gil, Y., & Miles, S. (Eds.). (2013). *PROV Model Primer.* W3C. http://www.w3.org/TR/prov-primer/

Giordano, A., Veregin, H., Borak, E., & Lanter, D. P. (1994). A Conceptual Model of GIS-Based Spatial Analysis. *Cartographica: The International Journal for Geographic Information and Geovisualization, 31*(4), 44–57. https://doi.org/10.3138/H150-2445-3448-4382

Glavic, B., & Dittrich, K. R. (2007). Data Provenance: A Categorization of Existing Approaches. *Proceedings of the 12th GI Conference on Database Systems in Business, Technology, and Web (BTW)*, *7*, 227–241.

Grady, R. K. (1988). The Lineage of Data in Land and Geographic Information Systems. *Proceedings of GIS/LIS '88 American Congress on Surveying and Mapping: Data Lineage in Land and Geographic Information Systems*, *2*, 722–730.

Guptill, S. C. (1987). Techniques for Managing Digital Cartographic Data. *Proceedings of the 13th International Cartographic Conference*, *4*(16), 221–226.

Hart, G., & Dolbar, C. (2013). *Linked Data: A Geographic Perspective* (1st ed.). CRC Press.

Henzen, C., Mas, S., & Bernard, L. (2013). Provenance Information in Geodata Infrastructures. In *Geographic Information Science at the Heart of Europe* (Vol. III, pp. 133–151). Springer International Publishing.

Hey, T., Tansley, S., & Tolle, K. (Eds.). (2009). *The Fourth Paradigm: Data-Intensive Scientific Discovery*.Microsoft Research.

Houbie, F., & Bigagli, L. (2010). *OGC Catalogue Services Standard 2.0 Extension Package for ebRIM Application Profile: Earth Observation Products*. http://portal.opengeospatial.org/files/?artifact_id=35528

Howe, C., & Tullis, J. A. (2022). Context for Reproducibility and Replicability in Geospatial Unmanned Aircraft Systems. *Remote Sensing*, *14*(17), 4304. https://doi.org/10.3390/rs14174304

IFAR. (2013). *Provenance Guide*. International Foundation for Art Research. http://www.ifar.org/provenance_guide.php

Ikeda, R., & Widom, J. (2009). *Data Lineage: A Survey* [Technical Report]. Stanford University InfoLab. http://ilpubs.stanford.edu:8090/918/

ISO. (2009). *ISO 19115- 2:2009 Geographic Information—Metadata—Part 2: Extensions for Imagery and Gridded Data*. International Organization for Standardization.

ISO. (2016). ISO 8000-120:2016 *Data Quality—Part 120: Master Data: Exchange of Characteristic Data: Provenance*. International Organization for Standardization.

Janowicz, K., Hitzler, P., Li, W., Rehberger, D., Schildhauer, M., Zhu, R., Shimizu, C., Fisher, C. K., Cai, L., Mai, G., Zalewski, J., Zhou, L., Stephen, S., Gonzalez, S., Mecum, B., Lopez-Carr, A., Schroeder, A., Smith, D., Wright, D., . . . Currier, K. (2022). *Know, Know Where, Know Where Graph: A Densely Connected, Cross-Domain Knowledge Graph and Geo-Enrichment Service Stack for Applications in Environmental Intelligence*. https://onlinelibrary.wiley.com/doi/10.1002/aaai.12043

Jensen, J. R. (2005). *Introductory Digital Image Processing: A Remote Sensing Perspective* (3rd ed.). Prentice Hall.

Jensen, J. R. (2007). *Remote Sensing of the Environment: An Earth Resource Perspective* (2nd ed.). Prentice Hall.

Jensen, J. R. (2016). *Introductory Digital Image Processing: A Remote Sensing Perspective* (4th ed.). Pearson Education.

Jones, J. S., Tullis, J. A., Haavik, L. J., Guldin, J. M., & Stephen, F. M. (2014). Monitoring Oak-Hickory Forest Change During an Unprecedented Red Oak Borer Outbreak in the Ozark Mountains: 1990 to 2006. *Journal of Applied Remote Sensing*, *8*(1), 1–13.

Langran, G. (1988). Temporal GIS Design Tradeoffs. *Proceedings of GIS/LIS*, *88*, 890–899.

Langran, G., & Chrisman, N. R. (1988). A Framework for Temporal Geographic Information. *Cartographica*, *25*(3), 1–14.

Lanter, D. P. (1989). *Techniques and Method of Spatial Database Lineage Tracing*.University of South Carolina.

Lanter, D. P. (1991). Design of a Lineage-Based Meta-Data Base for GIS. *Cartography and Geographic Information Systems*, *18*(4), 255–261.

Lanter, D. P. (1992a). *GEOLINEUS: Data Management and Flowcharting for ARC/INFO* (92–2). National Center for Geographic Information & Analysis. http://www.ncgia.ucsb.edu/Publications/tech-reports/91/91-6.pdf

Lanter, D. P. (1992b). Propagating Updates by Identifying Data Dependencies in Spatial Analytic Applications. *Proceedings of the Twelfth Annual ESRI User Conference*. Twelfth Annual Esri User Conference, Palm Springs, CA.

Lanter, D. P. (1993a). A Lineage Meta-Database Approach Toward Spatial Analytic Database Optimization. *Cartography and Geographic Information Systems*, *20*(2), 112–121.

Lanter, D. P. (1993b). *Method and Means for Lineage Tracing of a Spatial Information Processing and Database System* (United States Patent and Trademark Office Patent 5193185).

Lanter, D. P. (1994a). A Lineage Metadata Approach to Removing Redundancy and Propagating Updates in a GIS Database. *Cartography and Geographic Information Systems*, *21*(2), 91–98.

Lanter, D. P. (1994b). Comparison of Spatial Analytic Applications of GIS. In W. K. Michener, J. W. Brunt & S. G. Stafford (Eds.), *Environmental Information Management and Analysis: Ecosystem to Global Scales* (pp. 413–425). CRC Press.

Lanter, D. P., & Essinger, R. (1991). *User-Centered Graphical User Interface Design for GIS* (91–6). National Center for Geographic Information & Analysis. http://www.ncgia.ucsb.edu/Publications/tech-reports/91/91-6.pdf

Lanter, D. P., & Surbey, C. (1994). Metadata Analysis of GIS Data Processing: A Case Study. In T. C. Waugh & R. G. Healey (Eds.), *Advances in GIS Research: Proceedings of the 6th International Symposium on Spatial Data Handling* (pp. 314–324). Taylor & Francis Ltd.

Lanter, D. P., & Veregin, H. (1991). A Lineage Information Program for Exploring Error Propagation in GIS Applications. *Proceedings of the 15th Conference of the International Cartographic Association*, 468–472. https://icaci.org/icc-archive/

Lanter, D. P., & Veregin, H. (1992). A Research Paradigm for Propagating Error in Layer-based GIS. *Photogrammetic Engineering and Remote Sensing*, *58*(6), 825–833.

Malaverri, J. E. G., Bauzer Medeiros, C., & Camargo Lamparelli, R. (2012). A Provenance Approach to Assess Quality of Geospatial Data. *27th Symposium on Applied Computing*. SAC 2012, Riva del Garda (Trento), Italy.

Merriam-Webster. (2014). *Merriam-Webster Online*. http://www.merriam-webster.com/

Moellering, H., Fritz, L., Franklin, D., Marx, R. W., Dobson, J. E., Edson, D., Dangermond, J., Davis, J., Hagan, P., Boyle, A. R., Nyerges, T., Merchant, D., & Calkins, H. (1988). The Proposed Standard for Digital Cartographic Data. *The American Cartographer*, *15*(1), 9–140.

Moore, H. (1983). The Impact of Computer Technology in the Mapping Environment. *Proceedings of the Sixth International Symposium on Automated Cartography*, *1*, 60–68.

Moreau, L. (2010). The Foundations for Provenance on the Web. *Foundations and Trends in Web Science*, *2*(2–3), 99–241. https://doi.org/10.1561/1800000010

Moreau, L., Clifford, B., Freire, J., Futrelle, J., Gil, Y., Groth, P., Kwasnikowska, N., Miles, S., Missier, P., Myers, J., Plale, B., Simmhan, Y., Stephan, E., & den Bussche, J. V. (2011). The Open Provenance Model Core Specification (v1.1). *Future Generation Computer Systems*, *27*(6), 743–756. https://doi.org/10.1016/j.future.2010.07.005

Moreau, L., & Missier, P. (Eds.). (2013). *PROV-DM: The PROV Data Model*. http://www.w3.org/TR/2013/REC-prov-dm-20130430/#section-example-two

Naseri, M., & Ludwig, S. A. (2013). Evaluating Workflow Trust Using Hidden Markov Modeling and Provenance Data. In Q. Liu, Q. Bai, S. Giugni, D. Williamson, & J. Taylor (Eds.), *Data Provenance and Data Management in eScience* (Vol. 426, pp. 35–58). Springer-Verlag.

National Academies of Sciences, Engineering, and Medicine. (2019). *Reproducibility and Replicability in Science*. National Academies Press. https://www.nap.edu/catalog/25303

NSF. (2014). *Data Infrastructure Building Blocks (DIBBs)*. http://www.nsf.gov/pubs/2014/nsf14530/nsf14530.htm

Nyerges, T. (1987). GIS Research Needs Identified During a Cartographic Standards Process: Spatial Data Exchange. *International Geographic Information Systems Symposium: The Research Agenda*, *1*, 319–330.

Oxford University Press. (2014). *Oxford English Dictionary*. http://www.oed.com/

Plale, B., Cao, B., Herath, C., & Sun, Y. (2011). Data Provenance for Preservation of Digital Geoscience Data. In A. K. Sinha, D. Arctur, I. Jackson & L. C. Gundersen (Eds.), *Societal Challenges and Geoinformatics* (pp. 125–137). Geological Society of America.

Santos, E., Koop, D., Maxwell, T., Doutriaux, C., Ellqvist, T., Potter, G., Freire, J., Williams, D., & Silva, C. T. (2012). Designing a Provenance-Based Climate Data Analysis Application. In P. Groth & J. Frew (Eds.), *Provenance and Annotation of Data and Processes* (pp. 214–219). Springer-Verlag.

Schut, P. (Ed.). (2007). *OpenGIS Web Processing Service*. Open Geospatial Consortium.

Simmhan, Y. L., Plale, B., & Gannon, D. (2005). A Survey of Data Provenance in e-Science. *SIGMOD Record*, *34*(3), 31–36.

Simmhan, Y. L., Plale, B., & Gannon, D. (2008). Karma2: Provenance Management for Data-Driven Workflows. *International Journal of Web Services Research*, *5*(2), 1–22.

Singh, Gurmeet. (2023, February 1). *Harvard & KNIME: Geospatial Analytics for All*. https://www.knime.com/blog/harvard-geospatial-analytics-for-all

Singhal, A. (2012, May 16). *Introducing the Knowledge Graph: Things, Not Strings*. Google. https://blog.google/products/search/introducing-knowledge-graph-things-not/

Sun, Z., Di, L., Burgess, A., Tullis, J. A., & Magill, A. B. (2020). Geoweaver: Advanced Cyberinfrastructure for Managing Hybrid Geoscientific AI Workflows. *ISPRS International Journal of Geo-Information*, *9*(2), 119. https://doi.org/10.3390/ijgi9020119

Sun, Z., Yue, P., Hu, L., Gong, J., Zhang, L., & Lu, X. (2013). GeoPWProv: Interleaving Map and Faceted Metadata for Provenance Visualization and Navigation. *IEEE Transactions on Geoscience and Remote Sensing*, *51*(11), 5131–5136.

Talbert, C. (2012). *Software for Assisted Habitat Modeling Package for VisTrails (SAHM: VisTrails) v.1* [Computer Software]. USGS Fort Collins Science Center. https://www.fort.usgs.gov/products/23403

Tan, W.-C. (2007). Provenance in Databases: Past, Current, and Future. *Bulletin of the IEEE Computer Society Technical Committee on Data Engineering*, *30*(4), 3–12.

Tilmes, C., & Fleig, A. J. (2008). Provenance Tracking in an Earth Science Data Processing System. In J. Freire & D. Koop (Eds.), *Provenance and Annotation of Data and Processes* (pp. 221–228). Springer-Verlag. http://ebiquity.umbc.edu/_file_directory_/papers/445.pdf

Toda, Y., & Okura, F. (2019). How Convolutional Neural Networks Diagnose Plant Disease. *Plant Phenomics*, *2019*, 9237136. https://doi.org/10.34133/2019/9237136

Tullis, J. A., Corcoran, K., Ham, R., Kar, B., & Williamson, M. (2019). Multiuser Concepts and Workflow Replicability in sUAS Applications. In J. B. Sharma (Ed.), *Applications of Small Unmanned Aircraft Systems* (pp. 35–56). CRC Press. https://www.crcpress.com/Applications-of-Small-Unmanned-Aircraft-Systems-Best-Practices-and-Case/Sharma/p/book/9780367199241

Tullis, J. A., & Defibaugh y Chávez, J. M. (2009). Scale Management and Remote Sensor Synergy in Forest Monitoring. *Geography Compass*, *3*(1), 154–170.

Tullis, J. A., & Kar, B. (2020). Where Is the Provenance? Ethical Replicability and Reproducibility in GIScience and Its Critical Applications. *Annals of the American Association of Geographers*, 1–11. https://doi.org/10.1080/24694452.2020.1806029

Tullis, J. A., Stephen, F. M., Guldin, J. M., Jones, J. S., Wilson, J., Smith, P. D., Sexton, T., Cullpepper, B., Gorham, B., Riggins, J. J., & Defibaugh y Chávez, J. M. (2012, August 10). *Applied Silvicultural Assessment (ASA) Hazard Map*. University of Arkansas Forest Entomology's Applied Silvicultural Assessment. http://asa.cast.uark.edu/hazmap/

Vanecek, S., & Moore, R. (2014). *OGC Open Modelling Interface Interface Standard, Version 2.0*. https://portal.opengeospatial.org/files/?artifact_id=59022

Veregin, H., & Lanter, D. P. (1995). Data-quality Enhancement Techniques in Layer-based Geographic Information Systems. *Computers Environment and Urban Systems*, *19*(1), 23–36. https://doi.org/10.1016/0198-9715(94)00032-8

Wade, T., & Sommer, S. (2006). *A to Z GIS: An Illustrated Dictionary of Geographic Information Systems*. Esri Press. http://esripress.esri.com/display/index.cfm?fuseaction=display&websiteID=102&moduleID=0

Wang, C., Lu, Z., & Haithcoat, T. L. (2007). Using Landsat Images to Detect Oak Decline in the Mark Twain National Forest, Ozark Highlands. *Forest Ecology and Management*, *240*, 70–78.

Wang, S., Padmanabhan, A., Myers, J. D., Tang, W., & Liu, Y. (2008, November 5). *Towards Provenance-Aware Geographic Information Systems*. 16th ACM SIGSPATIAL International Conference on Advances in Geographic Information Systems, Irvine, CA. http://acmgis08.cs.umn.edu/papers.html#posterpapers

Woodruff, A., & Stonebraker, M. (1997). Supporting Fine-Grained Lineage in a Database Visualization Environment. In W. A. Gray & P.-Å. Larson (Eds.), *Proceedings of the Thirteenth International Conference on Data Engineering* (pp. 91–102). IEEE.

Yeide, N. H., Akinsha, K., & Walsh, A. L. (2001). *The AAM Guide to Provenance Research*. American Association of Museums.

Yue, P., Gong, J., & Di, L. (2010a). Augmenting Geospatial Data Provenance Through Metadata Tracking in Geospatial Service Chaining. *Computers & Geosciences*, *36*, 270–281.

Yue, P., Gong, J., Di, L., He, L., & Wei, Y. (2010b, November 7). Semantic Provenance Registration and Discovery using Geospatial Catalogue Service. *Proceedings of the Second International Workshop on the Role of Semantic Web in Provenance Management (SWPM 2010)*. Second International Workshop on the role of Semantic Web in Provenance Management, Shanghai, China.

Yue, P., & He, L. (2009, August 12). Geospatial Data Provenance in Cyberinfrastructure. *Proceedings of the 17th International Conference on Geoinformatics*. The 17th International Conference on Geoinformatics, Fairfax, VA.

Yue, P., Wei, Y., Di, L., He, L., Gong, J., & Zhang, L. (2011). Sharing Geospatial Provenance in a Service-oriented Environment. *Computers, Environment and Urban Systems*, *35*(4), 333–343. https://doi.org/10.1016/j.compenvurbsys.2011.02.006

14 Toward Democratization of Geographic Information

GIS, Remote Sensing, and GNSS Applications in Everyday Life

Gaurav Sinha, Barry J. Kronenfeld, and Jeffrey C. Brunskill

LIST OF ACRONYMS

AGI	Association for geographic information
ALIC	Australian Land Information Council
CGIS	Canada Geographic Information System
DIME	Dual independent map encoding
Geo-ICTs	Geospatial-information and communication technologies
GIS	Geographic information system
GIScience	Geographic information science
GNSS	Global navigation satellite system
GRASS	Graphical resources analysis support system
LBS	Location based services
LiDAR	Light detection and ranging
LKKCAP	Local knowledge and climate change adaptation project
MODIS	Moderate resolution imaging spectroradiometer
NCGIA	National Center for Geographic Information and Analysis
NGOs	Non-governmental organizations
OSM	OpenStreetMap
PPGIS	Public participation GISs
PGIS	Participatory GISs
SDI	Spatial data infrastructures
TIGER	Topologically integrated geographic encoding and referencing
UAV	Unmanned aerial vehicle
WMS	Web mapping service

14.1 RETHINKING GEOGRAPHIC INFORMATION AND TECHNOLOGIES IN THE 21ST CENTURY

There are very few academic domains that remain unaffected by the reach of geo-information technologies. Geography and other closely affiliated environmental science disciplines rely heavily on such technologies for their empirical work, but environmental engineering, social science, humanities, health science, and the business communities have also found geo-technologies and the geographic perspective crucial to many of their disciplinary pursuits. There was a time, until very recently, when a statement such as this would suffice to describe the geo-information revolution. Yet, today, the growing influence of geographic/geospatial theory, information and technology goes

DOI: 10.1201/9781003541158-17

far beyond the academe and the workplace—it has become the story of our everyday lives. There is no better example of this than the life story of a poor Indian boy who lost his family in 1986, ending up on a wrong train that took him all the way across India to the bustling city of Calcutta, where he had to beg for a living, but from where ultimately, he was transported far away to Australia through adoption by a loving couple, the Brierleys. Saroo Brierley ended up in much more comfortable settings than he would have probably been entitled in the impoverished settings of his earlier life, but he never gave up the urge to find his lost family. He had some memories of his first neighborhood, and that it was within a 14 hour train ride to Calcutta, but there was no way to use that information gainfully—until Google Earth was released and satellite imagery and maps made it possible to visually search places anywhere remotely using a computer. Saroo drew a circle to locate all places within 14 hours train ride of Calcutta (~1200 km), and while it seemed like finding the proverbial needled in a haystack, his persistence paid off when recognizes landmarks in in Khandwa, Madhya Pradesh, several hundred miles from Kolkata (formerly Calcutta, when Saroo got lost). Saroo eventually went to India, and tracked down his biological family, 25 years after he was separated—with mostly Google Earth to thank. The story of his life is now a bestseller book *A Long Way Home* (Brierley, 2014) and a great ode to the power of personal spatial memories and geospatial technologies—especially high resolution satellite imagery based mapping services that anybody can use.

Google Earth's success and popularity is symbolic of many different transitions that signify today geographic information creation and consumption is no longer under the tight control of institutions, experts, and the powerful few. Their grips have been loosened incrementally and sometimes disruptively by certain developments that finally have given individuals, novices, and the traditionally disenfranchised a strong say in how places, events, and perceptions are mapped and communicated. We now live in a world where the average person is increasingly aware of the everyday benefits and drawbacks for geo-information and associated technologies (Bittner, 2017; Young et al., 2021). This includes powerful tools that support navigation and exploration, communication, social interaction, public participation, security, and the like. Our increasing reliance on mobile devices (e.g., phones, tablets, and notebooks) is an important catalyst for many of these developments. Such developments also give rise to increased concerns over the quality and quantity of the spatially-oriented data we collect, and maintaining our privacy as personal communication devices, web services, electronic cards and surveillance cameras (in developed countries) create "digital shadows" (Klinkenberg, 2007) of our everyday lives. In a short time, we have gone from representing places and features to populating our geospatial databases with our doppelgangers, whose controls, ironically, lie in hands other than our own, and which will continue to "roam" within large databases, long after we cease to exist ourselves.

The start of this geo-information revolution goes back to the 1960s, when three distinct clusters of major geo-information technologies started to form the expansive geo-information landscape of today: remote sensing, geographic information system (GIS) and global navigation satellite system (GNSS) technologies (Merschdorf and Blaschke, 2018; Sudmanns, 2023). At their core, both remote sensing and GNSSs are geospatial data collection systems, while GISs have multiple functional aspects. To define them briefly in the context of this chapter:

- Remote Sensing is the science of collecting, processing and interpreting information of the Earth from aircraft or spacecraft equipped with instruments for sensing signals emitted or reflected by the surface (or atmosphere) of the Earth.
- A GIS is an epistemology institutionalized through *software* and *social practices* for processing, managing, analyzing, modeling, visualizing and communicating about geospatial datasets of various types.
- A GNSS is a suite of satellites in orbit around the Earth transmitting location and timing data that may be used by receivers worldwide to determine location on or near the Earth.

These technologies have been constantly evolving, and lately through synergistic combinations, they have revolutionized every aspect of geographic information creation, analysis and visualization in all the major sectors of the modern economy: scientific, commercial, educational, and governmental. Their collective contribution to the world economy has grown significantly over time to several tens of billions of dollars (Karagiannopoulou et al., 2022). Jerome Dobson, a pioneer who helped establish the field of GIScience, envisions remote sensing, GIS, GNSS, and related technologies, collectively as a "macroscope" for viewing large-scale phenomena in finer detail (similar to the microscope which magnifies truly small-scale phenomena, or the telescope which magnifies distant, apparently small phenomena) (Dobson, 2011). Over the last decade, the development of newer forms of geospatial-information and communication technologies (Geo-ICTs) and Web 2.0 services have played a key role in widening the "scope" of the geospatial macroscope by making the collection, processing, and sharing of geographically referenced information a part of our everyday lives. We afford ourselves the functionalities of these technologies not just in professional contexts, but also for managing our personal information. In doing so we uncover another important context—the fascinating relationship between technology and society. As the German philosopher Martin Heidegger observed "the essence of technology is by no means anything technical" (Heidegger, 1977: 4, as quoted in Crampton, 2010: 6); technology, science and society do not evolve independent of one another. As such, questions of society have become increasingly important as geotechnologies have become more personalized (Bozsik et al., 2022; Denwood et al., 2022).

This chapter complements other chapters of this book with a socio-technical discussion of how remote sensing, GIS, and GNSS are co-dependent and will continue to strengthen the foundations of our newly geo-enabled societies (Bozsik et al., 2022; Flaherty et al., 2022). We avoid the deep technical trenches to keep the reader focused on the "macroscopic" trends, since technical discussions of these technologies abound elsewhere. The material presented here is intended to help readers recognize these technologies at work in ordinary life situations—not just in scientific and professional settings (which is already covered extensively in other chapters of this book). With this in mind, the remainder of this chapter is organized as follows. In Section 14.2, we review the historical development of remote sensing, GIS and GNSS technologies as a context for discussing the democratization of the geospatial domain in the 21st century. In Section 14.3, we explore several application domains to showcase the diverse ways in which these technologies are being integrated into our everyday lives. In Section 14.4, we provide a critical analysis of the technical challenges in trying to make geo-information and related technologies easily accessible to the public, as well as the political, organizational, social and ethical issues that must be considered in an increasingly geospatially enabled society. We conclude with a brief summary of this chapter in Section 14.5.

14.2 TOWARD DEMOCRATIZATION OF GEO-INFORMATION TECHNOLOGIES

For centuries, paper maps were the state-of-the-art technology for representing, analyzing and communicating geographic information (Haworth, 2018; Nelson et al., 2022). While they still play a significant role in many spheres of our personal and professional lives today, paper maps are being replaced by services on mobile phones, virtual globes, Google Glass and other Geo-ICTs. The history and development of spatial representations from pre-historic maps to modern spatially-enabled mobile devices is characterized by Goodchild et al. (2007) according to three stages of development, namely *historic, enlightened*, and *contemporary*. During the *historic* phase, the longest of the three phases, map production was uncoordinated, incomplete, and undertaken by both public and private entities for various human needs. The *enlightened* phase began toward the beginning of the 20th century with the advent of state sponsored mapping initiatives and evolved into the *contemporary* phase starting in the last decade of the 20th century. This last phase is ongoing and characterized by increasing emphasis on cooperation and data sharing, as well a movement towards greater access to the modes of constructing and using geographic information. In this chapter, we extend Goodchild et al.'s (2007) original model with a *transition* phase to clearly distinguish the intervening period between the *enlightened* and *contemporary* phases. This period started with intellectual debates about how and if GIS can matter to societal development, recognition of the importance of geographic information and technologies by governments, development of internet technologies and emergence of the open source movement. This short period was quite critical in setting the stage for the defining developments of the ongoing contemporary phase, especially those related to the democratization of geographic information.

In this section, we present a representative, but not exhaustive, chronology of developments during the *enlightened, transition* and *contemporary* to provide a basic background for reasoning about the events that have fostered the public's awareness and understanding of geo-information. We do not discuss the *historical* phase hereafter, since our concern is mostly with later geo-information technologies and events which have primarily shaped the 21st century democratization of geographic information.

14.2.1 Enlightened Phase (20th Century): State Sponsored Geo-Information Technologies

The *enlightened* phase started in earnest with the massive increase in surveying and mapping needs during the two World Wars, and with the rise of the scientific method in the early to middle twentieth century. Large scale investments in geo-information technologies by governments (as detailed in Table 14.1) led to the development of national mapping agencies (both military and civil) with large budgets. The agencies were typically charged with the task of producing high-accuracy maps of the physical and social state of nations at several scales. This fostered the development of several new technologies including aerial photogrammetry, remote sensing with satellites (e.g., TIROS, LandSat), automated cartography, GIS, and the first GNSS—the U.S. NAVSTAR GPS (Nelson et al., 2022). The technological developments and emerging benefits to society and science were unquestionable. They were, however, not equally distributed among nations as most of the developments occurred in the United States, followed by the UK and other European countries, and Australia. The U.S. led the world in topographic mapping, remote sensing, GIS and GNSS. In addition to government-sponsored initiatives, significant contributions were made by academics, and by

TABLE 14.1
Important Geo-Information Technology Developments during the Enlightened Phase

Year	Enlightened Phase Events (1960–1989)
1960	First non-military satellite, TIROS-1, launched by US for space based meteorological observations.
1964	Canada Geographic Information System (CGIS) project launched.
1967	U.S. Census Bureau develops Dual Independent Map Encoding (DIME) vector topological data model.
1969	ESRI, the leading GIS software and services company today, is founded.
1972	Landsat 1, the first civilian Earth-monitoring satellite, is launched.
1978	Map Overlay and Statistical System (MOSS), the first full-fledged interactive vector and raster GIS, deployed at many US Federal agencies.
1982	ARC/INFO released by ESRI, to become most popular commercial GIS.
1983	U.S. announces GPS will be available for civilian sector when operational.
1985	GPS becomes operational.
1985	Graphical Resources Analysis Support System (GRASS) software, a raster focused, open-source GIS, is released by U.S. Army Corps.
1986	The Australian Land Information Council (ALIC) is established.
1987	"Handling Geographic Information" report (Chorley Report) released by U.K.
1987	National Center for Geographic Information and Analysis (NCGIA) established.
1989	U.S. Census Bureau releases Topologically Integrated Geographic Encoding and Referencing (TIGER) products into the public domain.

the commercial sector with the creation of popular cartographic, GIS and remote sensing software. The development of commercial software, in particular, facilitated collaborations between government and private entities and led to the creation of a special "class" of geospatial professionals. At the time, the average person, however, was still dependent on various types of maps produced by the professionals to understand their own spaces. It is for this reason that this era, particularly from the late 1980s onward, has been criticized for creating a hegemonic system that perpetuated the grip of powerful agencies and trained professionals on geographic information. This critique marks the beginning of the *transition* phase, during which the importance of reconceiving mapping technologies as products of and for society became evident.

14.2.2 Transition Phase (20th Century): Governance and Scholarship *for* Society

During the transition phase (Table 14.2), some social theorists began to argue that the seductive power of technologies had blinded many users from the more complex socio-political realities of technological development (Chrisman, 1987; Taylor, 1990; Panek and Netek, 2019). The epistemological interpretation of GISs (and related technological developments) was considered quite intellectually impoverished since the positivist, and absolute space focused view of conventional GIS, was in denial of the social roots and impacts of GIS and the social constructivist nature of geographic information (Warf and Sui, 2010). This led to the so-called GIS Wars, instigated by social theorists' criticism that GIS was imperialistic and used to subjugate local and indigenous perspectives of space, place, and culture in favor of the majority's perspectives. GIS and its practitioners were accused of helping strengthen hegemonic narratives serving the powerful elite, who spared no thought for marginalization of those on the less fortunate side of the economic, social, and digital

TABLE 14.2
Important Geo-Information Technology Related Developments during the Transition Phase That Paved the Way for Contemporary Processes of Democratization of Geo-Information

Year	Transition Phase Events (1989–1998)
1989	Association for Geographic Information (AGI) lobbying/advisory group established in UK.
1990	U.S. Federal Geographic Data Committee (FGDC) is established.
1990	"GIS Wars" erupt between social theorists and GIS theorists/practitioners.
1991	Miniature GPS receiver technology becomes widely available.
1992	The term *Geographic Information Science* is coined (Goodchild 1992).
1992	U.S. Remote Sensing Act of 1992 opens skies to commercial satellites.
1993	The NCGIA "Geographic Information and Society" meeting is held at Friday Harbor, Maine
1993	European Umbrella Organisation for Geographic Information (EUROGI) is established.
1994	U.S. President Bill Clinton signs Executive Order 12906 "Coordinating Geographic Data Acquisition and Access: The National Spatial Data Infrastructure (NSDI)."
1994	OpenGIS Consortium (now: Open Geospatial Consortium) is formed.
1995	The U.S. NAVSTAR GPS attains full operational capability.
1995	The Predator unmanned aerial vehicle (UAV) becomes operational.
1996	MapQuest and MultiMap, the earliest interactive web-mapping services based on government collected, public domain data, are released in U.S. and U.K.
1996	NCGIA Initiative 19: GIS & Society Workshop is organized.
1996	Public Participation GIS (PPGIS) is defined at an NCGIA workshop at the University of Maine.
1996	U.S. Federal Communications Commission requires wireless carriers to determine and transmit the location of callers who dial 9-1-1.

divide (Pickles, 1994; Sheppard, 1995; Harley, 2002). Others criticized specifically the military origins and "historical complicity of remote sensing, GIS and GNSS in military, colonial, racist and discriminatory practices" (Crampton, 2010: 7) and sought to expose the role of such technologies as weapons of intelligence and war (Smith, 1992; Monmonier, 2002). These social critiques were ultimately channelized and a set of critical research issues were identified at the Friday Harbor NCGIA meeting in 1996, (Harris and Weiner, 1996). The results from this and some other similar engagements provided the foundation for a broad GIS and Society research agenda that has thrived since the late 1990s.

A particularly long-lasting, tangible outcome of the GIS and society debates was the rise of grassroots mapping projects led by communities and facilitated by GIS and mapping experts. These projects, which may take on many forms, have been called participatory or community mapping projects, community integrated GISs, public participation GISs (PPGIS), or participatory GISs (PGIS), depending on the goals, participation, mapping process and technologies used (Craig et al., 2002; Sieber, 2006; Elwood, 2011; Brown, 2012; Bittner, 2017). These methods of geographic information creation and/or use are context and issue-driven, rather than technology-driven (Dunn, 2007). Over the last 20 years, hundreds of such mapping efforts have been undertaken worldwide. The projects, ranging from basic sketch-mapping to sophisticated online map-based surveys, have been undertaken to help communities broaden their spatial, environmental, social and political perspectives, address land ownership disputes, conserve and manage natural resources, document and protect cultural heritage, revitalize neighborhoods, and challenge existing narratives of their community. The projects are fertile grounds for the convergence of cartography, GISs, GNSSs, and remote sensing. Topographic maps, aerial photos, satellite imagery, GNSS collected waypoints and tracks, and GIS-based demographic and environmental data may all contribute to a group's mapping and decision making. This is evidenced by the PPgis.net electronic forum and the IAPAD website,[1] which are world-wide resources for those seeking to develop PGIS/PPGIS/participatory mapping projects, especially in developing countries.

The development of the GIS and society research agenda coincided with the development of a new multidisciplinary field called Geographic Information Science (GIScience), a field that explores the conceptual foundations, design, and the application contexts of geospatial technologies (Goodchild, 1992). While the definition and scope of the field has continued to evolve (Blaschke and Merschdorf,2014; Goodchild, 2012; Panek and Netek, 2019; Boland et al., 2022; Bozsik et al., 2022), the one thing that remains constant is the idea that GIScience research proceeds in at least three dimensions: computer, individual and society (Longley et al., 2010). The issues of interest to GIScience specialists are diverse, pertaining not just to the computational contexts of geographic data use, but also to the cognitive, behavioral, social, legal and ethical factors that govern the creation, dissemination and consumption of geographic information. Arguably, society still remains the weakest dimension of GIScience, but compared to the time of origination of the GIS & Society debates, there is much better understanding today of the two-way relationship between geo-information technologies and society (Warf and Sui, 2010). This is supported by the success of the PPGIS/PGIS movement, and the more recent establishment of Critical GIS as an important area of research within GIScience (Schuurman, 1999; O'Sullivan, 2006; Wilson and Poore, 2009).

During the transition phase, several key technological developments occurred to lay the groundwork for broader access and so-called democratization of Geo-ICTs in the current contemporary phase. These included the replacement of older, more cumbersome GPS devices with miniature receivers, operational completion of the GPS satellite system, development of unmanned aerial vehicles (UAVs), and, of course, the popularization of the World Wide Web. Development of open source GIS, which started with GRASS in 1985, progressed enough to justify establishment of a consortium of open source GIS software systems. Meanwhile, the commercial sector began to see the potential market in interactive web mapping services such as MapQuest and MultiMap.

As a final point of interest, the *transition* phase should also be recognized for the establishment of government-run agencies, national spatial data infrastructures (SDIs) and related organizations (e.g., U.S. FGDC), which were designed to coordinate the collection and sharing of geographic data between government agencies and with the public (Haworth, 2018; Nelson et al., 2022; Sudmanns, 2023). Today, SDIs have been implemented in more than 100 countries (Masser, 2011). SDIs are still largely one-way vehicles for sharing government data, and still have to evolve to become two-way vehicles, allowing the public to also contribute to the SDI (Budhathoki et al., 2008). Despite this limitation, the existence of these first-generation SDIs made many *contemporary* phase developments possible. During the *transition* phase, several private/public groups such as Association for Geographic Information (AGI) and EUROGI were also established to further the interests of geographic information communities.

14.2.3 Contemporary Phase (21st Century): The Vision of Democratization

The *contemporary* phase began during the first decade of the 21st century (see Table 14.3 for a list of important events). Unlike the *enlightened* phase, in which geo-information technologies were managed by large institutions in a fundamentally top-down hierarchy, the *contemporary* phase is defined by a much more democratized approach to geo-information (Karagiannopoulou et al., 2022; Nelson et al., 2022; Sudmanns, 2023). Contemporary uses of GISs and related technologies have become more context-sensitive, issue-driven, with some (but not nearly enough) recognition to issues of power, commodification, and surveillance (Warf and Sui, 2010). This democratized framework is only partly the result of the government SDIs, academic critiques and research agendas, and grassroots participatory mapping initiatives—a major push came from the rise of several private sector companies that invested heavily in online and mobile mapping technologies. By the start of the new century, government agencies across the world were searching for new ways to provide geospatial services in a time of reduced budgets, Geo-ICTs that dropped the entry barrier for geographic information collection and sharing, and an intellectual shift from mapping static places to representing dynamic activities of people across places over time (Goodchild et al., 2007). In the *contemporary* phase, governments still play a major role, but increasingly as facilitators (e.g., maintaining national spatial data infrastructures). They are also more inclusive, involving citizens in topographic and resource mapping initiatives, environmental monitoring, disaster preparation, and health and emergency services. The development of inclusive, bottom-up (individual and collaborative) processes for creating and sharing geographic is made possible by integrating the functionality of traditional geospatial technologies with newer Geo-ICTs (Goodchild, 2007; Sui et al., 2013). The development of technologies like OpenStreetMap, Microsoft Virtual Earth, Google Earth, and Google Glass are prime examples.

At the end of the 20th century, a seminal event occurred in 1998, when U.S. Vice President Al Gore presented an inspiring vision called Digital Earth as "a multi-resolution, three-dimensional representation of the planet, into which we can embed vast quantities of geo-referenced data" (Gore, 1998). Gore proposed the Digital Earth as a comprehensive information system composed of many distributed components, together providing access to all historical and current information about the entire planet (including the activities of its inhabitants), and supported by modeling tools for predicting future conditions. Gore envisioned the resource as a publicly available one-stop virtual environment in which anybody—child or adult—could explore information about the Earth effortlessly (e.g., on a "magic carpet" that could fly through space and also back in time). In many ways, Gore's presentation of his vision can be seen as the start of the *contemporary* phase. It encouraged existing initiatives and also stimulated new initiatives in the government and the private sector where. For example, the U.S. National Aeronautics and Space Administration (NASA) initiated the "Digital Earth Initiative" in 1998 and, among other things, created the current Web Mapping Service (WMS) standard, crucial for seamless sharing of geographic data and services on the web (Grossner et al., 2008). NASA's *World Wind* was an early virtual globe inspired by the Digital Earth vision. Today Google's *Google Earth*, Microsoft's *Bing*), ESRI's *ArcGIS Explorer*, and SkyGlobe's

TABLE 14.3
Important Geo-Information Technology Related Developments That Have Proven to Be Central to the Ongoing Democratization of Geographic Information in the Contemporary Phase

Year	Contemporary Phase Events (1998–Present)
1998	U.S. Vice President Gore presents the Digital Earth vision.
1998	Microsoft launches TerraServer in partnership with United States Geological Survey (USGS).
1998	NASA starts the Digital Earth Initiative and creates the Web-Mapping System (WMS) data sharing standard
1999	The first U.S. commercial satellite, IKONOS, is successfully launched with a very high panchromatic spatial resolution of about 0.82 meters.
2000	Selective Availability, which degraded NAVSTAR GPS signals for civilians, is switched off.
2001	Keyhole's Earthviewer and GeoFusion's GeoPlayer virtual globes are released.
2001	The Infrastructure for Spatial Information in Europe (INSPIRE) is launched by the European Union.
2001	Wikipedia, a crowdsourced encyclopedia, is launched.
2002	USGS launches *The National Map* web-service.
2002	Friendster, a social networking website, goes online.
2003	U.S. E-Government data access initiative "Geospatial One-Stop" goes online.
2003	URISA (Urban and Regional Information Systems Association) approves Code of Ethics for geospatial industry professionals.
2004	The National Imagery and Mapping Agency (NIMA) is reoriented as National Geospatial-intelligence Agency (NGA).
2004	OpenStreetMap launched as first crowdsourced, public domain street mapping database.
2004	Facebook and Flickr are launched.
2005	The Google Earth, Google Maps, and Microsoft Virtual Earth geobrowsers are released.
2005	Global Earth Observation System of Systems (GEOSS) ten-year implementation plan adopted by intergovernmental Group on Earth Observations (GEO).
2006	Twitter is launched.
2006	The term *neogeography* is proposed to describe creation and use of geospatial information by non-experts.
2007	Concepts of citizens as censors and *volunteered geographic information* (VGI) are promoted and gain a foothold (Goodchild, 2007).
2007	The National Science Foundation (NSF) promotes a cyberinfrastructure to foster collaborative research and data sharing.
2007	Apple releases the first iPhone and Google releases the Android mobile operating system.
2008	Ushahidi VGI platform launched to track violence during Kenya post-election crisis.
2010	Google announces its autonomous car project.
2013	Google Glass is launched for testing purposes.
2014	The U.S. government relaxes restrictions on satellite imagery to allow image resolutions below 50 centimeters for commercial purposes.

TerraExplorer are the popular virtual globe platforms.[2] While NASA does not maintain special units to support this vision, the vision is now promoted by the International Society for Digital Earth, spearheaded by China.[3] The Digital Earth vision today has evolved significantly since Gore's speech, and today it is understood more pragmatically as a globally distributed set of technological services and practices serving as a collective geographic knowledge organization and information retrieval system (Craglia et al., 2008; Grossner et al., 2008). In this regard, the virtual globes of today can be imagined as small private digital earths that offer access to the global Digital Earth system.

The popularity of web-mapping services and virtual globes established *geobrowsing*, that is, the search for geographic information via a map interface (Lemmens, 2011) as a near ubiquitous phenomenon, with new users, services, and regions still being added regularly. Geobrowsing is an intrinsically geospatial endeavor that is so intuitive that even young school kids can engage in it—because of intelligent obfuscation of certain peculiarities of geographic data (e.g., scale and projection). The value of geobrowsing lies in the how it brings mapping down to the level of the everyday user, and encourages people to engage with geographic information. For example, geobrowsers can be used to explore places, assist with everyday navigation and wayfinding tasks, visualize data, and create new maps and datasets. They highlight the manner in which our everyday activities are tied to space. With the growing popularity of geobrowsing applications on mobile devices (Google Maps was the most frequently downloaded smartphone application in 2014[4]), it seems reasonable to suggest that there is a strong desire in people from all walks in life and professions to directly engage and manipulate geographic information, rather than passively absorb it from static maps. Clearly, in the 21st century, we have moved on from merely *looking* at maps to *participating and interacting* with geobrowsers.

Fueled partly by the success of the virtual globes, web-mapping services (i.e., geobrowsers), and easy to use GNSS units, the *contemporary* phase has witnessed an explosion of user-generated geographic information—a phenomenon called "neogeography." Neogeography refers to the practice of using mapping techniques and tools for and by non-expert individuals or communities (Turner, 2006). Its applications are generally not analytical or formal but mostly descriptive and visual. The rise of neogeography lies in a synergy that has developed between various technologies (e.g., Geo-ICTs, GNSS-enabled mobile devices, and Web 2.0) and social data collection practices (e.g., crowdsourcing and volunteered geographic information) that allow everyday users to create mashups of map services and individualized content (Goodchild, 2007, 2009; Sui et al., 2013; Wilson and Graham, 2013; Bittner, 2017; Haworth, 2018; Merschdorf and Blaschke, 2018; Panek and Netek, 2019). Evidence of this synergy can be seen in the popularity of crowdsourcing repositories like OpenStreetMap (OSM)[5] and Wikimapia.[6] OSM is a free-to-use global map database of geographic features built by individual volunteers relying on a mix of resources including satellite imagery, GNSS tracks, and knowledge of place names. Wikimapia is a comparable database that actually contains many more user-generated entries for place names than any official list of place-names. These databases are important repositories of socially generated geographic information, and, in some countries, may be the only digital maps available to the public for economic or political reasons (see Figure 14.1 for an example). Further, even if official or commercial maps are available, crowdsourced maps offer diversity by recording users' perspectives of space. As Monmonier (1996) states, multiple maps can be made for the same place, with the same data, for the same situation. The diverse nature and versatility of these datasets have been invaluable in efforts to support humanitarian relief following natural disasters, and to empower communities through grassroots mapping (as will be discussed in the next section).

Over the last 15 years, many of the barriers to democratization of geographic information have disappeared, or at least become less relevant. This is evidenced, to a certain degree, by the Digital Earth initiative and by neogeography efforts to design collaborative, bottom-up processes for creating and sharing geographic information. Yet, despite this progress, efforts to democratize geographic information still have a long way to go—neogeography has limited connection to the academic domain of geography or GIScience (Goodchild, 2009) and its claims of democratization and making geographic information available to anyone, anywhere, anytime (Turner, 2006) have been shown to be premature and shallow (Haklay, 2013). Still, the developments that define the contemporary phase suggest that the traditional technologies of remote sensing, GISs, and GNSSs *can* be used to empower people and help them view the world from a perspective other than a traditional top-down, institutional frameworks of the recent past.

FIGURE 14.1 As the high resolution satellite image indicates, the slum in Kibera is a heavily populated part of Nairobi. While OSM Map captures substantial detail of the structures and roads collected from community mapping efforts, Google Maps and Bing Maps do not map most structures in Kibera since they are deemed illegal by local authorities. (Screenshot from www.tools.geofabrik.de/mc.)

14.3 DISCOVERING REMOTE SENSING, GIS, AND GNSS APPLICATIONS IN OUR DAILY LIVES

In this 21st century contemporary age, the functionalities offered by remote sensing, GISs, and GNSSs are finally becoming increasingly integrated into our daily lives. This integration is largely the result of new Geo-ICTs, as well as an increasing public awareness of the value of *location* as a search parameter when seeking information about the world. Over three decades worth of research exists on the conceptual and technical aspects of efforts to integrate these technologies (Gao, 2002; Mesev, 2007; Merchant and Narumalani, 2009; Bittner, 2017). GNSSs are used to georeference aerial photos and satellite images and are a core component of high-resolution light detection and ranging (LIDAR) remote sensing of the Earth. Both remotely sensed data and GNSS data can be imported and overlaid with other data layers in a GIS environment. In a GIS these layers can be used to create and analyze a variety of built (e.g., buildings, roads) and environmental (e.g., elevation, hydrography, land use/land cover) datasets. While it is more common to incorporate remotely sensed and GNSS derived data in a GIS, GIS-derived datasets are also used extensively in the rectification, classification, mapping and dissemination of remotely sensed imagery. The integration of these technologies is also evident in these way data are often seamlessly distributed to professional users. For example, the U.S. Federal government's Geospatial One-Stop project provides a single portal to geographic information. One of its services, The National Map[7] (see Figure 14.2) web service, supports both interactive online visualization and free downloading of a wide array of historical and current datasets including satellite imagery, elevation, land cover, transportation, hydrography, boundaries, and geographic names for the U.S.

In addition to the more standard methods for integrating remote sensing, GISs and GNSSs, outlined earlier, a variety of new increasingly democratized methods have evolved during the *contemporary* phase of geospatial development. For example, PPGIS projects have taken advantage of

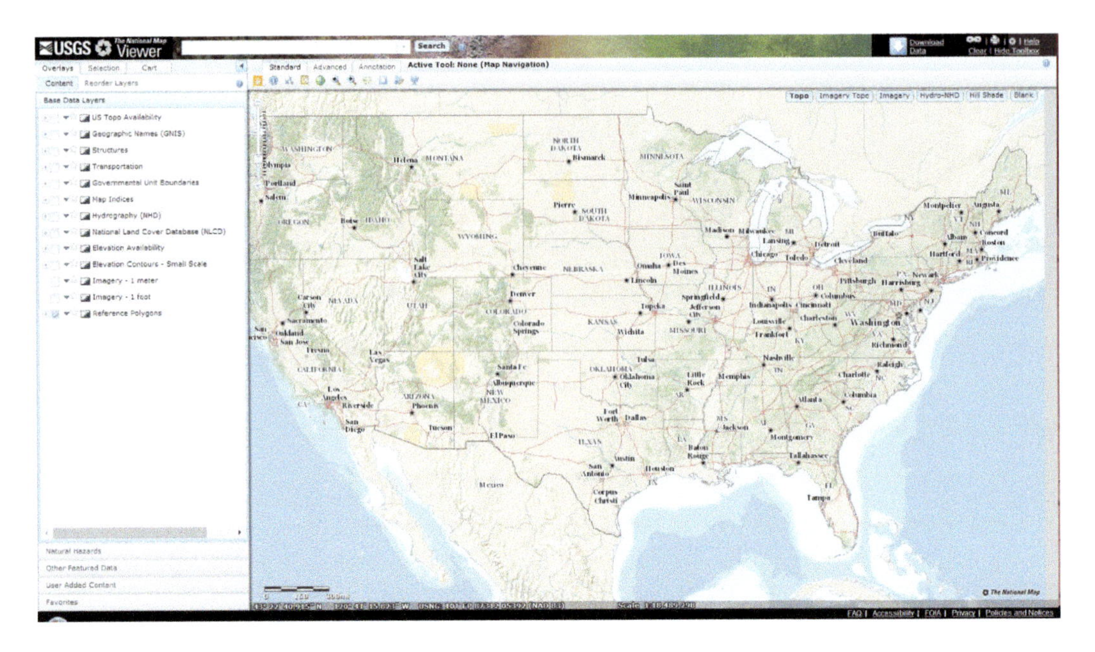

FIGURE 14.2 The USGS National Map provides free topographic mapping and data downloading services for the entire U.S. (Screenshot from www.viewer.nationalmap.gov.)

low-cost, high-accuracy GNSSs to map local communities and develop representations of features that are not evident on existing maps or from satellite imagery alone. In communities where current and/or high-resolution imagery are not available, amateur aerial photographs have been developed with make-shift aerial cameras consisting of GNSS-enabled digital cameras attached to helium filled balloons. The aerial cameras have been used to map small neighborhoods at cost of less than 35 U.S. cents per hectare (Seang et al., 2008). Perhaps, the most obvious example for owners of mobile Geo-ICTs arises in the context of geobrowsing in which high-resolution satellite images and aerial photos, made available by a service provider (e.g., Google or Microsoft), are used as basemaps for displaying thematic data layers. These integrations are often referred to as *map mash-ups*, or simply *mashups*. Similar combinations of imagery and data layers presented in three-dimensional virtual globe software (e.g., Google Earth, NASA World Wind) are also used to create virtual fly-through animations. In addition to government run geospatial data clearinghouses, public/private clearinghouses like OpenStreetMap (OSM) have also been developed that allow everyday users to contribute and use geospatial data in a community driven setting. The ability to both edit and download the entire OSM database locally is a tremendous benefit that no other map service provides.

The integration of geo-information data and functionalities provide the backbone for many of the services and tools that are developed by government agencies, non-profit agencies, and businesses to help people make everyday decisions (Yang et al., 2010; Lemmens, 2011). Such services include, but are not limited to: emergency response, disaster response and planning, natural resources management, environmental and public health services, weather forecasting, floodplain mapping, precision farming, agricultural services, community planning, property and utility mapping, crime monitoring and analysis, operations research, traffic monitoring, real estate services, and even K–12 education. To explore this topic in greater detail, we present four broad contexts in which Geo-ICTs are interwoven in people's everyday lives: location based services (LBSs), disaster relief and emergency management, participatory mapping, and participatory sensing of our everyday physical and social environments. In presenting these examples, particular attention is paid to the manner in which these applications facilitate democratization of geographic information.

(a) (b) (c)

FIGURE 14.3 (a) A car retrofitted by Google for collecting 360° panoramic street view photos. (b) A car retrofitted by Google to be tested as part of its self-driving car fleet. (c) Google's prototype of a fully automated self-driving car with numerous sensors but no manual traditional controls (steering wheel, and accelerator and brake pedals).

14.3.1 Location-Based Services (LBSs)

The term location based service, or LBS, generally refers to software applications that use location as a basis for providing information or performing a service for a user (Zipf and Jöst, 2011; Bittner, 2017). The start of LBSs can be traced to a 1996 U.S. Federal Communications Commission mandate that set a minimum accuracy for determining the location of an emergency E911 call by a wireless device. Although this policy was oriented towards emergency response, resulting improvements in location accuracy have fostered innovation in the marketplace by catalyzing the development of mobile mapping programs and LBSs. As a particularly interesting and futuristic example, consider the recent initiative by Google and others to develop automated cars (see Figure 14.3). In 2010, Google announced that it was working on a project to develop self-driving cars (Thrun, 2010). Not surprisingly, the project leader was also the co-inventor of Google's Street View mapping technology. Other automotive companies and universities are also developing prototypes and testing them on roads. Google is ahead of most though, since it has already conducted more than half a million miles of pilot testing. Driverless cars will be legal on the road from January 2015 U.K. and already so in the U.S. states of California, Florida and Nevada, with legislation pending in many others (Weiner and Smith, 2014). Autonomous cars are stellar examples of Geo-ICT technology and services integration. Such cars now depend on LIDAR and/or 360° cameras and computer vision technology, and currently data from driving patterns of cars driven by human beings, to create real-time high-resolution three-dimensional models of a car's surroundings. It is estimated that they may need to sense about 1GB of data *every second* to work optimally today! The GNSS-derived location and LIDAR model also need to be combined with information from a GIS database of static infrastructure (e.g., telephone poles, crosswalks, and traffic lights) to identify all types of static and moving objects (e.g., pedestrians, cyclists) to plot a safe path through space (Fisher, 2013). Similar to cars, lightweight unmanned aerial vehicles (UAVs) or drones may be found at the disposal of any citizen in the future for collecting remote sensing data about any area of interest for business, environmental, recreational and other purposes. Apart from such futuristic vehicles, smartphones and the recently launched Google Glass are also great platforms for LBS integration.

LBSs today represent such an important category of business that it is being fiercely competed over by information technology, telecommunication, mobile phone, and increasingly other types of companies. Today, LBSs are designed to help people make decisions in the contexts of an incredibly wide variety of personal and professional matters (Raper et al., 2007) including, but not limited to:

- *Navigation*: turn-by-turn navigation; public transit; traffic and road condition updates; roadside emergency assistance; user-centered route selection; fuel consumption; autonomous car; vehicle to vehicle communication;

- *Retail/Business services*: retail advertising; store locators; shopping aisle information services; retail store/mall maps; real estate services; credit/bank card fraud prevention;
- *Recreation*: online and real world games; sky gazing; geo-social networking; mobile place guides; location-based augmented reality; travel/tourism services;
- *Societal services*: location-based warnings and alerts; seeking/providing emergency help; disabled people mobility; toll collection; weather services; environmental services; agricultural services; participatory community planning;
- *Mapping*: crowdsourcing points of interest; monitoring environmental conditions; infrastructure maintenance; participatory/citizen sensing; mapping and monitoring personal health; documenting geocoded events; mapping disease outbreaks; finding or tracking people/animals/objects; mashup services.

With such a diverse list of activities, it should be obvious that the commercial sector is agog with LBS innovations and improvements. LBSs almost always combine data and functionalities from GIS, mobile cartography, and GNSS (Merschdorf and Blaschke, 2018; Young et al., 2021; Nelson et al., 2022). For example, Google Maps provides several types of map services ranging from locating and describing landmarks, routing directions for different modes of transport (driving, walking, biking and public transport), traffic patterns, traffic alerts. In Figure 14.4, a photomontage derived from several screenshots depicts shows examples of these LBSs as they would look on a single map view. Bing Maps, MapQuest and other mapping sites also provide similar services. Apart from such map or imagery based LBSs, more specialized map based LBSs include weather advisory services, field surveying by professionals or by untrained volunteers, and disaster relief work after earthquakes, fires, floods, and storms. LBSs to post real time alerts about environmental conditions, manage drone delivery services, and support augmented reality Geo-ICTs such as Google Glass should be quite popular soon. The reach of LBSs into both mundane and critical decision-making processes will also continue to get deeper, with technological developments such as three-dimensional mobile cartography, ubiquitous positioning (Mannings, 2008) and ubiquitous

FIGURE 14.4 Examples of various map based services available through Google Maps (www.maps.google.com). Note this image is a composite of several computer screenshots, since these services cannot be simultaneously viewed on the same map view (as of August, 2014).

computing/Internet of Things (Weiser, 1991) making mobility patterns of people, not just their locations, the focus of LBSs. The implications of such deep LBS integration are obviously both exciting and scary at the same time. There are many issues and challenges that will come to the forefront regarding service reliability, information overload, energy use, human-machine interaction, personal privacy, provision of social, environmental and health services, perception and experience of places, social networking, and cognitive and behavioral modifications

14.3.2 Disaster Relief and Emergency Management

On November 8, 2013, Super Typhoon Haiyan swept through the Eastern Visayas region of the Philippines, killing 6,000 people, displacing 1.9 million and impacting over nine million (REACH et al., 2014). Even before the typhoon hit, a Humanitarian OpenStreetMap Team (HOT) was created in anticipation to map infrastructure and damage in the affected region. Harnessing the power of global volunteers to quickly evaluate post-disaster satellite imagery, the HOT performed damage assessment and facilitated prioritization of scarce resources. Maps created from the effort were visible to the public within days of the disaster (Buchanan et al., 2013).

The Haiyan HOT was just the latest in a long list of geospatial crowdsourcing efforts for disaster relief. This growth can be traced back at least to Hurricane Katrina in 2005, when a computer programmer in Austin, Texas, created a website that allowed users to post descriptions of local conditions onto a map (Singel, 2005). In 2007, a local radio station contributed to coverage of the San Diego wildfires in California by maintaining a dynamic web map of user-reported conditions (Zook et al., 2010). And, in 2009, OSM was first used for humanitarian purposes in the Philippines in response to Tropical Storm Ondoy. While crisis mapping[8] as a concept was beginning to take hold in 2009 it came to the fore with the formation of HOT following the January 12, 2010, earthquake in Haiti (Soden and Palen, 2014).

In the days that followed the Haiti earthquake, relief workers posted information on locations of trapped survivors, damaged buildings and triage centers; a global network of volunteers translated information from Haitian Kreyol, geotagged it and placed it into online mapping services (created based on remote sensing imagery) for other relief workers to use. The effort was made possible by the maturation of communities such as OSM, Ushahidi,[9] Geocommons[10] and CrisisCamp Haiti (Pool, 2010), the cooperation of private companies (e.g., Google, Microsoft, and GeoEye), government agencies, non-governmental organizations, and GEO-CAN—a unique voluntary network of more than 600 experts from 23 countries. These entities relied heavily on information produced from the convergence of GIS software and data, GNSS field data, several sources of aerial and satellite imagery, and expert created information products (Duda and Jones, 2011; van Aardt et al., 2011). Without any centralized command, these disparate actors and resources converged within hours, and the collective, organic systems that were created in the short time easily surpassed what could ever be achieved by traditional top-down, centralized government efforts.

In a different context, Goodchild and Glennon (2010) discuss the case of VGI being useful in making quick real-time decisions during four wild fires near Santa Barbara between 2007 and 2009. The volunteers used GNSS-enabled cameras and phones to report information as text, photos and videos over the internet. For one wild fire, some citizens were able to access and interpret comparatively fine temporal and spatial resolution imagery from the Moderate Resolution Imaging Spectroradiometer (MODIS) satellite sensor, and, using services such as Google Maps, created several maps that were often more up-to-date than official maps. For another fire, volunteer map sites were set up to readily synthesize the volunteer and official information as it became available. This real time life-saving participatory sensing of an endangered environment bears testimony to the power of democratization of geo-information and placing remote sensing, GISs, GNSSs and other Geo-ICTs in people's homes.

Until only a few years ago, it was the norm that government and trans-governmental agencies (e.g., the United Nations) would lead disaster relief, with non-governmental organizations (NGOs)

playing a supporting role. Emrich et al. (2011) provide an excellent review of the history of use of remote sensing and GISs in preparing or responding to emergencies. Remote sensing and GISs have been used in all phases of emergency management for multiple decades now. While the early focus was on understanding the physical processes, at the beginning of the 21st century, geo-information technologies were being used in all phases of emergency management—though still primarily to support the top-down hazard research, analysis, and disaster response command structure (Emrich et al., 2011; Zook et al., 2010). It is only since 2010, due to technological advances, supporting services, and changing mindsets, that virtually every major disaster has stimulated VGI/crowdsourcing mapping. These have included the Queensland and Australian Floods in 2010–2011, the earthquake in Christchurch, New Zealand in 2011, the earthquake and tsunami in northern Japan in 2011, the 2013 Super Typhoon Haiyan in the Philippines, the 2013 tornadoes in Illinois, USA, and even the search for the missing Malaysian Airlines aircraft MH-370 in 2014, which enlisted millions of volunteers scanning imagery for evidence of the missing aircraft. The trend is likely to as adoption of mobile devices continues in all parts of the world.

While the "crowd" and volunteer cannot be a replacement for domain experts, these successful relief efforts still clearly demonstrate two things. The first is that crowdsourcing, aided by appropriate online and field Geo-ICTs, can often provide critical geospatial information much faster than traditional methods. The second is that crowdsourcing is highly suited to the large and sudden demand created by natural and man-made disasters. Local volunteer groups and government agencies can also use crowdsourcing to gather crucial information that may be needed by both officials and communities to plan ahead of time for disasters in areas especially prone to storms, floods, landslides, fires, and disease outbreaks. On the other hand, this transition from traditional command-and-control structures to distributed crowdsourcing efforts is not without its own problems and raises numerous technological, organizational and ethical issues that will need to be addressed in coming years. Some of these issues are touched upon in Section 14. 4.

14.3.3 Community Building

The story of crowdsourcing in Haiti after the earthquake could easily have dissipated as the disaster started slipping away from the media spotlight. Fortunately, the HOT that formed initially to respond to the earthquake, dug its heels and worked for about a year and half after the earthquake to help people claim ownership of their broken and impoverished communities (Soden and Palen, 2014). The Haitian HOT is recognized today as remarkable for several reasons: it established HOT as a serious techno-humanitarian group, changed the perception of OSM and its contributors, became a quintessential example for champions of neogeography, and blurred the boundary between crowdsourcing, VGI, and PGIS through this long-term collaborative mapping initiative. Above all, HOT inspired Haitians to reclaim legal, cultural, and emotional ownership of the land, codify their local knowledge through geospatial data and maps, thus becoming more resilient against future threats. Undoubtedly, the project will be long cited as the quintessential example of how combining geo-information technologies can produce extremely valuable geo-information for the masses—by the masses.

The mapping success of the Haitian HOT can be traced back to another iconic OSM led PGIS/collaborative mapping effort—the Map Kibera project.[11] Kibera, an informal settlement of about 170,000 people in Nairobi, Kenya, has captured worldwide attention in recent years as supposedly the largest slum in Africa. Despite such population, and a tremendous presence of NGOs, it is still invisible on most maps of the Nairobi, appearing only as a large blank space or an uninhabited forest. Similar to most informal settlements, Kibera developed on occupied public lands, and its illegality prevents it from being officially mapped or becoming eligible to receive government services. Its residents had little awareness of its geography or their collective socio-political capital until November 2009 when OSM contributors Mikel Maron and Erica Hagen started the Map Kibera project, which helped Kiberans create a community information system to empower themselves

(Hagen, 2011). The two project leaders slowly incentivized and built a mapping team of youth by teaching them GNSS mapping, videography, and journalism skills— rather than paying them as NGOs had become accustomed to doing. The achievement of OSM's Map Kibera project can be easily shown by simply comparing satellite imagery for the area with map views from Google Maps, Bing Maps™ and OSM. As can be seen in Figure 14.1, Google and Bing map the area with only a few main streets. The emptiness in these officially sanctioned maps stands in stark contrast to the dense settlement visible from the satellite imagery and OSM's map, which is being used by local residents to conceptualize their neighborhoods, demand services from the government, and fight many social problems. Based on the success of the Map Kibera project, mapping efforts have now spread to other slums in Nairobi, and elsewhere in Africa and Asia.

It should be kept in mind that OSM facility and infrastructure mapping projects are only one kind of approach to PGIS. Another approach is to pair trained geospatial professionals with local communities to help them record their spatial histories and explore and communicate about community problems through maps. Currently, one of the authors of this chapter (Sinha) is part of an international collaborative research project called the Local Knowledge and Climate Change Adaptation Project (LKKCAP),[12] under the auspices of which, several community mapping projects were organized in Mwanga district, Tanzania. LKCCAP was funded by the U.S. National Science Foundation to explore the relationship between local knowledge, institutions, and climate change adaptation practices embedded in rural livelihoods of Tanzania. The community mapping projects were designed to let village residents explore social, political, and environmental dimensions of livelihood adaptation to climate change. The project involved sketch mapping (Figure 14.5) and GPS field mapping, with collected data being digitized into GIS databases and represented on poster-sized maps to help the community hold discussions with district officials and aid agencies.

FIGURE 14.5 A participatory sketch mapping session in Kirya village located in the Mwanga District of Tanzania for helping villagers map water infrastructure and ecological resources.

We have focused on OSM and rural mapping projects to highlight that geo-information technologies are currently being used by non-professionals, even in the poorest of communities. PGIS/PPGIS and collaborative mapping are, obviously, equally relevant and popular in more prospering communities as well. Community concerns about neighborhood planning, natural resource conservation, climate change, flood mapping, environmental pollution, water scarcity, economic revitalization, infrastructure development, social discrimination, substance abuse, indigenous culture preservation and many similar issues become the starting point for the PPGIS/PGIS projects. Special online collaborative spatial decision support systems have been developed for many studies (Jankowski, 2011). Various social networking platforms and online community planning forums, such as PlaceSpeak,[13] Mind Mixer,[14] and MetroQuest[15] may be used in combination with OSM, Google Earth, Google Map Maker™, and other Geo-ICTs to engage communities, planners and governments in more developed countries (and also some urban communities in developing countries). As Geo-ICTs, such as augmented reality, indoor positioning and mapping, and mobile ground remote sensing devices become more common in developed societies, communities will be able to generate three-dimensional virtual and augmented reality models of their neighborhoods—resulting in seamless boundaries between virtual and real world experiences of places.

14.3.4 Participatory Sensing

Participatory sensing, a mapping process in which citizens and community groups engage in sensing and documenting their daily activities, has become more common in recent years. The scope of participatory sensing ranges from individuals making personal observations, to the combination of data from hundreds, or even thousands, of individuals that reveals patterns across an entire city (Goldman et al., 2009; Panek and Netek, 2019; Young et al., 2021). Participatory sensing leads to incredibly rich models of environments and spaces that people define through their activities (Sagl et al., 2014). Participatory sensing is a good example of citizen science, since the projects are generally designed for a specific purpose, often by researchers interested in learning about a phenomenon. It is this characteristic that distinguishes it from agenda free VGI or politically motivated PPGIS/PGIS projects. The most common type of collaborative sensing and mapping involves the use of citizens' mobile devices, which act as nodes of a sensor network to collectively measure any number of interesting geocoded environmental parameters such as traffic characteristics, air quality, noise levels, luminosity, temperature, humidity, radiation (Burke et al., 2006). Participatory sensing projects such as these have been fostered by advances in mobile phone technology, increased access to remote sensing imagery and relevant spatial datasets, LBSs, and the increasing social appeal of crowdsourcing.

When participatory-sensed data are integrated via mashups, or displayed on virtual globes, they can be used to assess environmental variables at multiple spatial scales, or in relation to other geographic datasets (e.g., air quality data). For example, as part of the NoiseTube[16] project, citizens in Paris used mobile phones to measure their personal exposure to noise in their everyday environment and created collective noise maps by sharing their geo-localized and annotated measurements with the community (Maisonneuve et al., 2010). Figure 14.6 maps the results from one such NoiseTube project using Google Earth satellite imagery as a basemap, with a pie-chart also providing the breakdown for the variously tagged sources of noise pollution by the participants. A different implementation was seen for the Common Scents project of the SENSEable City Laboratory at the Massachusetts Institute of Technology, USA, which equipped Copenhagen citizens' bikes with location aware environmental sensors to develop a mobile sensor network. The project's goal was to gather real-time, fine grained, air quality data to allow citizens and local officials to assess environmental conditions (Resch et al., 2009). Another project, called the Personal Environmental Impact Report (PEIR) used location records from mobile phones to calculate estimates of an individual's impact and exposure to different environments (e.g., carbon monoxide emissions from cars) (Mun et al., 2009). A similar approach based on mobile phone mobility analysis also helped identify and

FIGURE 14.6 Map of noise pollution made from participatory sending data collection by the NoiseTube project in Paris, France. (Image reproduced with permission from the NoiseTube project, www.noisetube.net.)

characterize variations in human activity in urban environments (Sagl el al., 2014). In all such projects, citizens are given the ability to pool their data to reveal impacts and exposures in numbers that could be useful to both the individual, as well as city planners.

Participatory sensing also includes projects for monitoring of environments, mapping live traffic, and collecting data and responding to environmental hazards real time, particularly those that need to be monitored real-time (e.g., storms, forest fires, floods, volcanic eruptions) and might require evacuation or help from emergency responders within a short period of time after a critical event. The informal network of storm chasers/spotters, operational meteorologists, and television media outlets is a classic example of volunteers and experts collaborating monitor weather threats and to help people make decisions in their daily life (Palmer and Kraushaar, 2013). Another project, undertaken by more than 100 volunteers trained by the New Jersey statewide environmental agency, involved the collection of ground truth verification data for a database of 13,000 ephemeral vernal pools (breeding grounds for many endangered or threatened species) that had been derived from satellite imagery. A web-based mapping system was set up to collect the verification data, which must verified by the state biologists to meet legal requirements. Although the project is far from complete due to lack of budget and inaccessibility of many of the pools, the projects of this nature highlight the potential of using citizens to collect important cultural and scientific data. Thus, Environmental Collaborative Monitoring Networks (ECMN) have been proposed (Gouveia and Fonseca, 2008) as a framework for combining traditional environmental monitoring networks with the bottom-up, participatory, and open source movement.

Personal mobile health monitoring has also become an extremely popular activity/sport for many people in urban centers. This includes the use of personal fitness devices for tracking physiological parameters such as heart rate, perspiration, steps or distance traveled, and velocity. Many devices are also geo-enabled, thereby allowing users to map their physiological parameters in relation to the places they visit during the day. The benefits of geo-enabled personal health monitoring range from casual monitoring of personal fitness to developing early warning systems for high-risk segments of the population. The current trend clearly seems to suggest that such information, contextualized with other forms of static and real time geo-information about the local environment, will provide an important basis for personalized or governmental/institutional environmental health decision making in the future.

14.4 DEMOCRATIZATION OF GEO-INFORMATION: CIRCUMSCRIBING ISSUES AND CHALLENGES

From the start of the geo-information revolution in the 1960s remote sensing, GIS and GNSS technologies have evolved to play an increasingly significant role in our everyday lives. The resoundingly clear message in the previous discussions of LBSs, disaster relief, community mapping, and participatory sensing is that people now have incredibly diverse opportunities to collect and use geographic information to improve their lives. These changes are remarkable and have undoubtedly saved countless lives and empowered large new segments of the global population. At the same time, it is important to recognize that technological revolutions have a tendency to outpace critical theories and institutional structures designed to assure that their use furthers the social and cultural goals of society at large (Chrisman, 1987). The democratization of Geo-ICTs is also affecting the way data are analyzed and translated into actionable information. In this section, we examine some impediments to democratization, and also remind readers that uncritical submission to the easy appeal of trendy Geo-ICTs is regressive and problematic (Panek and Netek, 2019).

14.4.1 Quality of Information and Services

As the task of geospatial data collection has been transferred from geospatial specialists to the general public, many concerns related to quality (accuracy, completeness, geographic diversity, lack of metadata) of the data and metadata have naturally arisen (Haworth, 2018; Merschdorf and Blaschke, 2018; Flaherty et al., 2022; Karagiannopoulou et al., 2022). The lack of obligatory commitment of "free" volunteers or the crowd is a further problem since interest in projects can easily dampen or die, leaving many data collection and mapping efforts incomplete. To objectively assess quality of such volunteered and crowdsourced data, a couple of studies explicitly compared OSM data to authoritative Ordinance Survey data in England (Haklay, 2010) and German ATKIS data for Munich (Fan et al., 2014). The errors were not large (less than 20 m offset for roads in England, and 4 m offset and simplification of building footprints in Munich). One way to judge the quality of VGI quality is the number of peers who have reviewed and edited the content, a principle known as Linus's law (Elwood et al., 2011). Large numbers of peers can yield even higher quality data than obtained the traditional authoritative sources. This is often evident during emergencies, when fast decisions must be made real-time, so that the risks associated with volunteered information are often outweighed by the benefits, as was found in the case of the wildfires near Santa Barbara between 2007 and 2009 (Goodchild and Glennon, 2010). Similarly, McDougall (2012) examined three disasters in Australia, New Zealand and Japan, respectively, and concluded that crowd sourced mapping provided a unique bottom-up perspective unattainable through the conventional emergency service "command and control" structure or implementation. Moreover, post-disaster mapping is not just useful for guiding immediate development efforts, but also creates much needed information about places with important long-term use-value. For example, as shown in Figure 14.7, Typhoon Haiyan resulted in a dramatic increase in volunteer mapping activity in affected areas, which now have a much higher density of mapped buildings than other places with comparable population densities. Otherwise, only Manila, the political and cultural capital has a high density of mapped buildings.

Yet, volunteered and crowdsourced datasets do have a lot of quality issues, which, if ignored, lead to faulty decisions. For example, ground truthing of building damage assessment done by

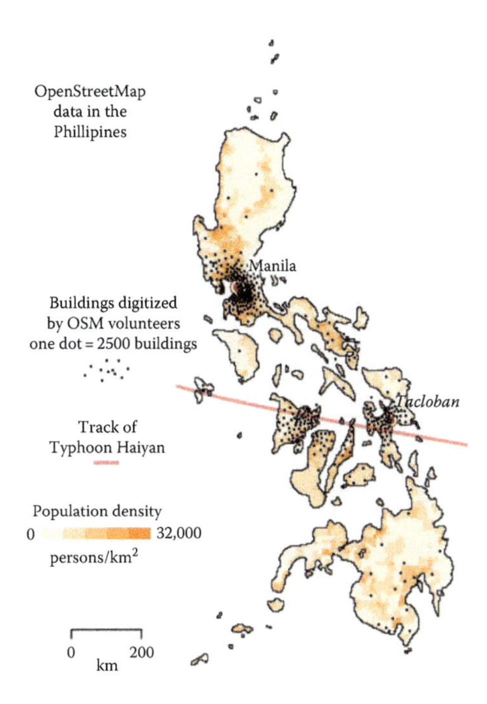

FIGURE 14.7 Map of buildings digitized by OpenStreetMap volunteers in the Philippines until end of July 2014. Typhoon Haiyan resulted in a dramatic increase in volunteer activity in the affected areas, whereas buildings in other areas, except around Manila, remain sparsely mapped.

volunteers looking at satellite imagery after Typhoon Haiyan in Philippines found overall classification accuracy to be merely 36%, primarily due to overestimation of building damage (REACH et al., 2014). For the 2011 Christchurch earthquake, a similar analysis found large errors of omission (56% in downtown, 86% in residential areas) but virtually no errors of commission (3% in downtown, 0% in residential areas) (Foulser-Piggot et al., 2013). Reasons for such errors are related to low spatial resolution, lack of pre-disaster reference imagery, lack of expertise and visual/spatial interpretation skills, imagery georectification errors, and overly simplistic instructions for volunteers. Figure 14.8 shows examples of building digitization errors by volunteers for Typhoon Haiyan HOT.

It would be naïve to expect untrained volunteers with no background in geospatial technology and science to realize or rectify such errors. Champions of neogeography must understand that a lot of expertise is needed to engage in deep geographic thinking and geospatial analysis, quite contrary to the scope of neogeography (Goodchild, 2009). Moreover, the need to process data fast and reliably along with lack of incentives for deep theoretical analysis means that many technologies are being deployed in non-professional settings in ways that lead to lowering or relaxation of quality and consistency. This is occurring not only at the level of individual projects, but also within the context of institutional and professional norms. A prime example of this is the *de facto* establishment of Web Mercator as the standard coordinate system for web mapping in all geobrowsers. Established for reasons related to server efficiency, Web Mercator is inappropriate for many (perhaps most) cartographic and quantitative applications (Battersby et al., 2014). Another example is the normalized practice of relying on users to uncover database errors, enabling a faster product release cycle. The botched launch of Apple's iOS6 Maps in 2012, for which Apple CEO Tim Cook had to offer the first open apology in the history of the company,[17] demonstrates that this practice can also disenfranchise users who expect high quality spatial data.

FIGURE 14.8 Screenshot from OSM showing substantial offsets for some buildings digitized by volunteers for Typhoon Haiyan HOT. While the buildings on the right align closely with the satellite image, the three buildings on the left are considerably offset, because volunteers likely had access to different images at different times. (Screenshot created July 12, 2014.)

In many cases, the appeal of spatial data is leading to misappropriation of technologies that are not designed for geospatial accuracy. For example, location inference methods used by social networking services, for which location is just a secondary parameter, often rely on coarse methods of positioning, and using such inaccurate locations for applications where precise positions are needed creates unanticipated problems. Worse, many LBS are not designed properly for efficient decision making, especially for real-time decisions while on the move (Raubal, 2011). One has only to think of the many limitations of automated navigation systems that people routinely experience and have to address based on their own contextual geographic knowledge. Such problems may be ignored in many contexts, but in emergency response situations inaccurate locations can put lives in danger. Resolutions of such errors are as much about technology improvement, as exploring how human cognitive systems function and differ across use contexts and for individuals based on gender, culture, native language, age and many other specific factors.

14.4.2 Privacy and Confidentiality

The constant demand for geospatial data and increasing use of linked social-spatial data raises concerns about the ability to protect confidentiality and privacy (NRC, 2007). Geosurveillance as a concept has been around from the early 1990s when GIS was first attacked for being an extension of the state for monitoring and surveillance (Smith, 1992; Pickles, 1991). The concept has now evolved to include a much wider spectrum of Geo-ICTs, and the surveillance is not just being done by governments, but also powerful corporates, and that too with our implicit acquiescence. The exaflood of big data (Sui et al., 2013) that results from massive data collection efforts is simultaneously a source of exciting new possibilities and a threat to our privacy, and thus, our freedoms. Data mining makes it possible to derive "deep information" on individuals' behaviors in virtual and physical space (Sui and Goodchild, 2011). High resolution imagery today allows us to explore our world in fascinating detail and simultaneously empowers us to counter secretive government strategies (Perkins and Dodge, 2009).

Yet, much of this big data is owned by either corporations or government intelligence agencies (Sui and Goodchild, 2011)—with a nexus between the two also possible. Online search is currently only possible through commercial search engine platforms, with no practical open-source or non-profit alternative. Companies benefit from free user generated content and by monitoring and mining user behavior through advertising and sale of such data to third parties. Should we not be quite concerned that the online information we receive is filtered through commercial search engines and that the digital shadows (Klinkenberg, 2007) we create are not in our control? As the benefits derived from joining social media networks and other Geo-ICTs have increased, so has the willingness of individuals to share geospatial data about themselves and their family and friends, without realizing the implications of their choices. When such information is combined across space, time and the multiple thematic dimensions of our lives, the analysis reveals hidden and surprising patterns about our individual and collective behaviors. Why is that the data miners are not obligated to share such information they find with the very people that unsuspectingly helped generate such information?

It is also important to distinguish the concept of data protection from that of individual privacy, the latter being much more nebulous and culturally dependent. In this new age of the geoweb and Geo-ICTs, the very notion of privacy needs is being reconceptualized to reflect the experienced impacts of the use of such technologies (Elwood and Leszczynski, 2011; Nouwt, 2008). A significant problem in this era of instant multimedia communication and proliferation of social networks is that individuals do not realize they may be revealing unwelcome, unauthorized, or even incriminating information about others inadvertently. Moreover, impacts might be several years away or other future action and in the context of other information. Crowdsourcing is another development that presents similar problems of lack of awareness of participants. Despite the bottom-up and openness of crowdsourced mapping efforts, it is not clear that participants realize that they may be collecting unauthorized information, or and others are not citizen sensors, but the ultimate beneficiaries of all such projects. Similarly, in the context of PGIS, community trust needs to be respected and nurtured and it should be transparent what kind of data are being collected and shared, about whom, and if required permissions have been acquired legitimately. This has not always been the case, due to breaches of confidentiality or insensitive protocols for data collection and sharing.

In response to the deluge of technologies and information, an interesting new approach to privacy is that of sousveillance (inverse surveillance), first proposed in Mann et al. (2002) as the philosophy and practice of wearing computing devices visibly to "surveil the surveillers reflectively"—as a form of protest against state and company surveillance measurements. Pervasive sousveillance would mean we are all being constantly monitored by everybody around us, which in itself raises many moral, legal and ethical problems. Using phones today or Google Glass in the future, and many participatory sensing activities, can be interpreted as forms of sousveillance. In contrast, proponents of the "post-privacy" movement believe that since technology makes it impossible to control ownership of information, society should practically abandon the notion of privacy in favor of a completely transparent society, where everybody knows what's going on most of the time (Brin, 1999; Jarvis, 2011). Social networking giant Facebook's founder and CEO Mark Zuckerberg's has even made the claim that people have accepted that privacy is no longer a "social norm" (The Economist, 2013). Perhaps, it is under such assumptions that Facebook (and researchers employed at Cornell University) and intentionally manipulated Facebook news feeds to investigate their influence on users' emotional states (Meyer, 2014). The study drew considerable indignation for Facebook, but the company is hardly unique in its designs to profit from manipulating people's behaviors—any marketing or lobbying campaign with biased information can be accused of such infringements.

If the manipulation of users for experimental purposes strikes many as unethical, the very premise of location-based services (LBS) may need to be reexamined, since it involves manipulation of content based on user location. For example, while Google's automated car initiative has the potential to offer great benefits to society by preventing accidents, reducing fuel usage and carbon emissions, minimizing traffic congestion, and saving commuting times, the company is not motivated by altruistic reasons, but inspired by the company's long-term goals to collect and eventually monetize information about its customers. Clearly, Google needs and wants to be at the forefront regarding the design and sale of automated driving technology and information systems, the collection and sale of data on driving habits and navigation patterns, and the ability to provide LBSs to drivers. How Google will monetize such information remains to be seen, but one thing is certain—it will create many controversies. Google's Street View technology raised tremendous privacy concerns in several contexts and the company has been perennially engaged in legal battles and paid fines for alleged invasion of privacy in several countries.[18] Its latest technology, Google Glass is already generating similar concerns, and polarizing early adopters from the skeptics.

No matter what the technology, the fundamental ethical question for all LBSs is quite simple: how can ethical guidelines be constructed to allow for "legitimate" manipulation of content while excluding more subversive purposes? There is rarely a context in which an LBS technology cannot be subverted to collect information against the will of the user, and the more accurate it gets the more harm that can be inflicted. Several years ago, Dobson and Fisher (2003) coined the term "geoslavery" to alert us to this kind misuse of geo-information technologies to spy on people—by government, corporations, and individuals aided by powerful technologies. Several questions arise, therefore, in the quest of democratization and freedom from geoslavery. How do we prevent ourselves from becoming geoslaves and fight the commoditization of our lives? What kind of in-built controls and preventative measures can be provided to users? Can we through our own choices send strong messages about what we can accept and what must change? What legislations need to be passed to rein in companies?

Originally, the concept of geosurveillance was associated only with governments, whose data collection powers far outreach that of corporations. While government may offer recourse against corporations, there may be none against the government itself, especially in non-democratic countries. Even in USA, one of the better functioning democracies, there is currently great concern about military and intelligence uses of the technology, against citizens and foreign nationals. One only has to follow the Edward Snowden leaks of the National Security Agency's documents to realize how much power and technical ability is at the disposal of the intelligence community.[19] Research of the U.S. intelligence economy is enough to justify all concerns. The U.S. government is the largest consumer of geospatial data and Geo-ICTs in the world, with most of that budget allocated for intelligence gathering—and, not surprisingly, satellite imagery is the largest component of intelligence budget (Crampton et al., 2014). There also exists a massive "contracting nexus" of more than 50,000 contractor companies, universities and nonprofits, who have received a staggering *3.75 trillion* dollars between 2000 and 2012 from the Department of Defense and intelligence agencies (Crampton et al., 2014). These figures make one thing clear—geo-information technologies and data are being developed and used in the name of national security in diverse ways by diverse entities, while the public is largely unaware of what the data may or may not be used for and who may have access to it.

14.4.3 Empowerment and Equity

Aided by digital mapping technologies and social networking practices of Web 2.0, the neogeography movement has been critical in advancing the democratization of geographic information and broadened the ontological scope of the formal systems for representing and analyzing geographic information. Neogeographical knowledge can be interpreted as arising from contextual personal and communal interpretive interactions of space and place, and, thus intrinsically about the culture of everyday life (Warf and Sui, 2010). Despite all its successes, neogeography is no panacea to the digital divide between the technological elite and the much larger group of untrained and uncritical laborers who are not empowered by the use of the technology (Dodge and Kitchin, 2013; Haklay, 2013). The "free" labor from neogeography is opportunistic capital for many entrepreneurial companies who benefit directly from such user generated content through advertising and sale of user generated information (including users' online behavior), with the users receiving no share of such profits. However, without these laborers, we still would be limited to the sparse and barren maps of Kibera, when the reality, as shown in Figure 14.1, is quite different. This reveals the true meaning of democratization, which is the empowerment of people who would otherwise be underrepresented. Through individual and coordinated organic acts of mapping, neogeographers have put on the map the stories of people who live far on the unprivileged side of the digital divide—those who cannot even read or have access to the basic amenities; those displaced due to war; or persecuted by their governments. Motivated by curiosity, purpose, and humanitarian instinct, neogeographers have challenged academic researchers to engage in a *place-centric* GIS grounded in qualitative human discourse, rather than stay limited to the conventional spatial perspective (Elwood et al., 2013).

Despite neogeographers having been a prime force behind many recent developments, their efforts may fail to serve the original purpose, or contribute toward purposes they did not sanction explicitly. Corporations such as Google and Microsoft have retained copyright on volunteer-collected spatial data using their platforms. This has impeded data sharing during humanitarian operations such as the 2011 Haiti earthquake response (Zook et al., 2010). Disasters and other non-urgent crowdsourcing efforts may even be seen as opportunities for data collection for private gain. Digital Globe manages the TomNod[20] online platform to crowdsource identification of interesting objects and places (often in the context of a disaster) from high resolution satellite imagery. Volunteers are often recruited for such campaigns through "feel-good" messages, without providing concrete information regarding how the data will be used, who owns the data or whether it will be made available to the volunteers themselves. At the very least, more transparency is needed to clarify how their data are being used. Google similarly owns all maps and data created by users. Similar to the concerns associated with the dangers to individuals arising from data mining of their personal information from variety of sources, VGI and crowdsourced information can be easily used without permission for enhancing profit margins or, worse, for unethical or illegal purposes—by corporations, governments, and individuals.

Linus' law also implies, unfortunately, that volunteered and crowdsourced information about accessible, popular, or populated places and readily observable phenomena will be more accurate than information about remote places and not so readily observable phenomena. A recent study (Graham et al., 2014) analyzed the geography of Wikipedia articles, only to discover that there is disproportionate amount of information about North America, much of Europe, and heavily populated parts of Asia, and not much about the rest of the world. A large part of the geographic variation was statistically explained by three variables: population, availability of broadband internet, and number of edits originating from a country. Despite Wikipedia being one of the ten most popular websites in the world, its content reflects existing and possibly also creates new geographies of the digital divide. Other sources of biases in people's perception of places and how they act on such perceptions include lack of cultural diversity, lack of geographic education, limited exposure to places, and disproportionate media attention. For example, Typhoon Haiyan related mapping efforts exhibited a clear "media effect" since OSM volunteers disproportionately mapped damage in Tacloban City, which was the focus of media coverage (REACH et al., 2014), but ignored other affected places. Since OSM maps were used to guide relief efforts, this may have had consequences as to how much aid was received by different locales. This geographic unevenness in OSM efforts can be observed more generally at the national scale as well. As shown in Figure 14.7, buildings in Manila, being the well-known capital of Philippines and in areas (especially Tacloban) affected by Typhoon Haiyan have been digitized at a disproportionately higher rate compared to other areas with comparable population density, but lacking special appeal.

Yet another form of mapping bias creeps in due to the profit motive and involvement of corporates, which engage in public service often for self-serving public relations purposes. Are recent campaigns sponsored by DigitalGlobe through their TomNod platform to locate the missing airplane MH370, map damage after the 2013 Illinois tornado outbreak, and identify invasive species in Hawaii examples of corporate good deeds or media attention-grabbing? Though both perspectives contain some truth, the bigger picture is that as corporations begin to see public relations benefits of sponsoring VGI efforts for "good causes," they might co-opt (perhaps unwittingly) the power to decide which causes are worth sponsoring. There is little doubt that the charitable activities of large corporations such as Google and DigitalGlobe, as well as non-profit organizations such as OpenStreetMap, have contributed immensely to recent disaster relief efforts and other causes. While these initiatives should continue to be supported, guidelines should also be developed from within the geospatial community to ensure that volunteers derive tangible benefits from participating in VGI projects, and that societal values such as equity, political freedom, information sharing, and so on, be supported in such efforts.

14.5 CONCLUSION

The goal of this chapter has been to diligently cover the brief history of democratization of geographic information and through some selective examples highlight the immense impact it has on our ability to not just function better, but completely change the course of our lives. There is no better example of this than Saroo Brierley, whose discovery of his family using Google Earth serves as a shining example of Digital Earth technologies as life-altering solutions, and, less radically, as technologies for creating vicarious place experiences, and expanding our geographic understanding. Saroo found his family because of these developments but think of a future where there exist advanced geo-information retrieval programs through which Saroo could express and search automatically places resembling mental map of his neighborhood. Saroo would not have to spend months browsing imagery to find his family, in such a future. Indeed, the democratization of geo-information is not just about people's involvement, and accessibility to information and tools, but must also depend on innovations in computational reasoning, human-computer interfaces (Google Earth's success was primarily due to its ease-of-use factor), and growth of public information infrastructures. While universities and government settings are better suited for more scientific and socially responsible research, commercial entities have played an important role through technological innovations and expanding the reach of Geo-ICTs to the common citizen.

To come back to Heidegger's quote from earlier in the chapter, technological problems are not technical—and technical strategies cannot alone help us resolve the conflicting demands for data access, data quality, and confidentiality (NRC, 2007). Clearly, technology does not operate in a vacuum and cannot be viewed as devoid of social, political and economic contexts. However, if there is to be true democratization of geographic information and many of its supporting technologies, citizens must step up to nurture this fledgling democracy. Neogeography, VGI, crowdsourcing, citizen science, Web 2.0 and many other related neologisms capture many such citizen led empowering forms of technological progress. Unfortunately, these new practices have also created new problems of data quality, devaluation of established knowledge or capital production systems, exploitation of volunteers, and biasing discourses in favor of popular and easily accessible information.

As much as we need technologies to promote our development as individuals and societies, we must also never lose sight of how they are being used by those who control them—be it individuals, communities, corporations, institutions, or governments. As mentioned earlier, information retrieval is practically hostage to commercial search engines and our online behavior is being harvested for profit purposes. We do benefit from such services through more targeted advertising and more relevant information retrieval, but the downsides are often hidden from us. Even the relationships of large corporations with colleges and universities needs to be monitored to limit unethical influences, even if subtly, on teaching, research and administrative decisions of grants, scholarships, endowments, hiring, and free/discounted software made available by companies. Governmental and private geo-surveillance may be unavoidable in practice, simply because of geo-political complexities and ubiquity of monitoring devices, but that is no excuse for different branches of the government not exercising valid checks on each other, and other violators. Furthermore, if knowledge privileged professionals and intellectuals become accessible to the populace outside their exalted circles, many insidious impacts of Geo-ICTs can be avoided. Ultimately, it falls on all individuals and communities to be vigilant about developments that violate or can help protect our basic freedoms.

At the same time, critical analyses that merely pander to insecurities and fan the politics of hope and fear need to be countered intelligently and collectively. As Monmonier (2002) and Klinkenberg (2007) have reasoned, the reasonable approach is to stay critical, but only to negate the "evils" of technologies—and, to ensure that all strata of society are involved in the path to socially positive technologies. There has undoubtedly been considerable progress in geospatial technology, its use, and its discourse in just a few decades, which became the inspiration and provided the content for this chapter on democratization of geographic information. What remains to be seen now is whether the democratization process flourishes in practice and transforms our lives, or if it remains a mere *cause célèbre* in the academic community.

NOTES

1 www.iapad.org
2 See http://en.wikipedia.org/wiki/Virtual_globe for a complete list of virtual globes.
3 http://www.digitalearth-isde.org
4 www.businessinsider.comgoogle-smartphone-app-popularity-2013–9#infographic (accessed July 17, 2014).
5 www.openstreetmap.org
6 www.wikimapia.org
7 www.NationalMap.gov
8 www.crisismappers.net
9 www.ushahidi.com
10 www.geocommons.com
11 www.mapkibera.org
12 www.tzclimadapt.ohio.edu
13 www.placespeak.com
14 www.mindmixer.com
15 www.metroquest.com
16 www.noisetube.net
17 www.apple.com/letter-from-tim-cook-on-maps
18 www.en.wikipedia.org/wiki/Google_Street_View_privacy_concerns
19 www.theguardian.com/world/the-nsa-files
20 www.tomnod.com

REFERENCES

Battersby, S. E., Finn, M. P., Usery, E. L., & Yamamoto, K. H. (2014). Implications of web mercator and its use in online mapping. *Cartographica: The International Journal for Geographic Information and Geovisualization*, *49*(2), 85–101.

Bittner, C. (2017). Diversity in volunteered geographic information: comparing OpenStreetMap and Wikimapia in Jerusalem. *GeoJournal*, *82*, 887–906.

Blaschke, T., & Merschdorf, H. (2014). Geographic information science as a multidisciplinary and multiparadigmatic field. *Cartography and Geographic Information Science*, *41*(3), 196–213.

Bozsik, S., Cheng, X., Kuncham, M., & Mitchell, E. (2022). Democratizing housing affordability data: Open data and data journalism in Charlottesville, VA," *2022 Systems and Information Engineering Design Symposium (SIEDS)*. Charlottesville, VA, pp. 178–183.

Brierley, S. (2014). *A Long Way Home: A Memoir*. Putnam Adult.

Brin, D. (1999). *The Transparent Society: Will Technology Force Us to Choose Between Privacy and Freedom*? Basic Books.

Brown, G. (2012). Public participation GIS (PPGIS) for regional and environmental planning: Reflections on a decade of empirical research. *URISA Journal*, *24*(2), 7–18.

Buchanan, L., Fairfield, H., Parlapiano, A., Peçanha, S., Wallace, T., Watkins, D., & Yourish, K. (2013). *Mapping the Destruction of Typhoon Haiyan*. Retrieved July 14, 2014, from http://www.nytimes.com/interactive/2013/11/11/world/asia/typhoon-haiyan-map.html

Budhathoki, N., Bruce, B., & Nedović-Budić, Z. (2008). Reconceptualizing the role of the user of spatial data infrastructure. *GeoJournal*, *72*(3), 149–160.

Burke, J. A., Estrin, D., Hansen, M., Parker, A., Ramanathan, N., Reddy, S., & Srivastava, M. B. (2006). Participatory sensing. *Center for Embedded Network Sensing*. Retrieved from http://escholarship.org/uc/item/19h777qd

Chrisman, N. (1987). Design of geographic information systems based on social and cultural goals. *Photogrammetric Engineering and Remote Sensing*, *53*(10), 1367–1370.

Craig, W., Harris, T., and Weiner, D. (2002). Introduction. In W. Craig, T. Harris, and D. Weiner (Eds.), *Community Participation and Geographic Information Systems* (pp. 1–6). Taylor & Francis.

Craglia, M., Goodchild, M. F., Annoni, A., Camara, G., Gould, M., Kuhn, W., . . . Parsons, E. (2008). Next-generation digital earth: A position paper from the vespucci initiative for the advancement of geographic information science. *International Journal of Spatial Data Infrastructures Research*, *3*, 146–167.

Crampton, J. W. (2010). *Mapping: A Critical Introduction to Cartography and GIS: A Critical Introduction to GIS and Cartography*. Wiley-Blackwell.

Crampton, J. W., Roberts, S. M., & Poorthuis, A. (2014). The new political economy of geographical intelligence. *Annals of the Association of American Geographers*, *104*(1), 196–214.

Denwood, T., Huck, J. J., & Lindley, S. (2022). Participatory mapping: A systematic review and open science framework for future research. *Annals of the American Association of Geographers*, *112*(8), 2324–2343.

Dobson, J. (2011). Through the macroscope: Geography's view of the world. *ArcNews*, Winter.

Dobson, J., & Fisher, P. (2003). Geoslavery. *IEEE Technology and Society Magazine*, Spring, 47–52.

Dodge, M., & Kitchin, R. (2013). Crowdsourced cartography: Mapping experience and knowledge. *Environment and Planning A*, *45*(1), 19—36.

Duda, K. A., & Jones, B. K. (2011). USGS remote sensing coordination for the 2010 Haiti Earthquake. *Photogrammetric Engineering and Remote Sensing*, *77*(9), 899–907.

Dunn, C. (2007). Participatory GIS – A people's GIS? *Progress in Human Geography*, *31*(5), 616–637.

The Economist. (2013, November 16). The people's panopticon. *The Economist*. Retrieved from http://www.economist.com/news/briefing/21589863-it-getting-ever-easier-record-anything-or-everything-you-see-opens

Emrich et al. (2011). GIS and emergency management. In *The SAGE Handbook of GIS and Society* (pp. 321–343). SAGE Publications Ltd.

Elwood, S. (2011). Participatory approaches in GIS and society research: Foundations, practices, and future directions. In *The SAGE Handbook of GIS and Society* (pp. 381–399). SAGE Publications Ltd.

Elwood, S., Goodchild, M. F., & Sui, D. Z. (2011). Researching volunteered geographic information: Spatial data, geographic research, and new social practice. *Annals of the Association of American Geographers*, *102*(3), 571–590.

Elwood, S., Goodchild, M. F., & Sui, D. Z. (2013). Prospects for VGI research and the emerging fourth paradigm. In D. Sui, S. Elwood and M. Goodchild (Eds.), *Crowdsourcing Geographic Knowledge* (pp. 361–375). Springer.

Elwood, S., & Leszczynski, A. (2011). Privacy, reconsidered: New representations, data practices, and the geoweb. *Geoforum*, *42*(1), 6–15.

Fan, H., Zipf, A., Fu, Q., & Neis, P. (2014). Quality assessment for building footprints data on OpenStreetMap. *International Journal of Geographical Information Science*, *28*(4), 700–719.

Fisher, A. (2013, September 18). Inside Google's quest to popularize self-driving cars. *Popular Science*. Retrieved from http://www.popsci.com/cars/article/2013-09/google-self-driving-car

Flaherty, E., Sturm, T., & Farries, E. (2022). The conspiracy of Covid-19 and 5G: Spatial analysis fallacies in the age of data democratization. *Social Science & Medicine*, *293*(2022), 114546. ISSN 0277-9536. https://doi.org/10.1016/j.socscimed.2021.114546.

Foulser-Piggot, R., Spence, R., & Brown, D. (2013). *The Use of Remote Sensing for Building Damage Assessment Following 22nd February 2011 Christchurch Earthquake: The GEOCAN Study and Its Validation*. Cambridge Architectural Research, Ltd. Retrieved from http://www.willisresearchnetwork.com/assets/templates/wrn/files/GEOCAN%20Christchurch%20Report.pdf

Gao, J. (2002). Integration of GPS with remote sensing and GIS: Reality and prospect. *Photogrammetric Engineering & Remote Sensing*, *68*(5), 447–453.

Goldman, J., Shilton, K., Burke, J., Estrin, D., Hansen, M., Ramanathan, N., . . . West, R. (2009). *Participatory Sensing: A Citizen-Powered Approach to Illuminating the Patterns That Shape Our World*. Woodrow Wilson International Center for Scholars. Retrieved from http://wilsoncenter.org/topics/docs/participatory_sensing.pdf

Goodchild, M. F. (1992). Geographical information science. *International Journal of Geographic Information Systems*, *6*(1), 31–45.

Goodchild, M. F. (2007). Citizens as sensors: The world of volunteered geography. *GeoJournal*, *69*(4), 211–221.

Goodchild, M. F. (2009). NeoGeography and the nature of geographic expertise. *Journal of Location Based Services*, *3*(2), 82–96.

Goodchild, M. F. (2012). GIScience in the 21st century. In W. Shi, M. F. Goodchild, B. Lees and Y. Leung (Eds.), *Advances in Geo-Spatial Information Science* (pp. 3–10). CRC Press.

Goodchild, M. F., Fu, P., & Rich, P. (2007). Sharing geographic information: An assessment of the geospatial one-stop. *Annals of the Association of American Geographers*, *97*(2), 250–266.

Goodchild, M. F., & Glennon, J. A. (2010). Crowdsourcing geographic information for disaster response: A research frontier. *International Journal of Digital Earth*, *3*(3), 231–241.

Gore, A. (1998). *The Digital Earth: Understanding our planet in the 21st Century*. California Science Center, Los Angeles, CA. Retrieved from http://www.isde5.org/al_gore_speech.htm

Gouveia, C., & Fonseca, A. (2008). New approaches to environmental monitoring: The use of ICT to explore volunteered geographic information. *GeoJournal*, *72*(3–4), 185–197.

Graham, M., Benie H., Straumann, R.K., & Medhat, A. (2014). Uneven geographies of user-generated information: Patterns of increasing informational poverty. *Annals of the Association of American* Geographers, *104*(4), 746–764.

Grossner, K. E., Goodchild, M. F., & Clarke, K. C. (2008). Defining a digital earth system. *Transactions in GIS*, *12*(1), 145–160.

Hagen, E. (2011). Mapping change: Community information empowerment in Kibera (Innovations Case Narrative: Map Kibera). *Innovations: Technology, Governance, Globalization*, *6*(1), 69–94.

Haklay, M. (2010). How good is volunteered geographical information? A comparative study of OpenStreetMap and Ordnance Survey datasets. *Environment and Planning B: Planning and Design*, *37*(4), 682–703.

Haklay, M. (2013). Neogeography and the delusion of democratisation. *Environment and Planning A*, *45*(1), 55–69.

Harley, J. B. (2002). *The New Nature of Maps: Essays in the History of Cartography.*Baltimore, MD: Johns Hopkins University Press.

Harris, T. M., & Weiner, D. (1996). *GIS and Society: The Social Implications of How People, Space, and Environment Are Represented in GIS. Scientific Report for the Initiative 19 Specialist Meeting* (Scientific Report for the Initiative 19 Specialist Meeting No. #96–7). Koinonia Retreat Center, South Haven, MN: National Center for Geographic Information and Analysis (NCGIA).

Haworth, B. T. (2018). Implications of volunteered geographic information for disaster management and GIScience: A more complex world of volunteered geography. *Annals of the American Association of Geographers*, *108*(1), 226–240.

Heidegger, M. (1977). *The Question Concerning Technology, and Other Essays* (W. Lovitt, Trans.). Harper Torchbooks.

Jankowski, P. (2011). Designing public participation geographic information systems. In T. Nyerges, H. Couclelis & R. B. McMaster (Eds.), *The SAGE Handbook of GIS and Society*. SAGE Publications Ltd.

Jarvis, J. (2011). *Public Parts: How Sharing in the Digital Age Improves the Way We Work and Live*. Simon & Schuster.

Karagiannopoulou, A., Tsertou, A., Tsimiklis, G., & Amditis, A. (2022). Data fusion in earth observation and the role of citizen as a sensor: A scoping review of applications, methods and future trends. *Remote Sensing*, *14*(2022), 1263.

Klinkenberg, B. (2007). Geospatial technologies and the geographies of hope and fear. *Annals of the Association of American Geographers*, *97*(2), 350–360.

Lemmens, M. (2011). *Geo-Information: Technologies, Applications and the Environment.*Springer.

Longley, P. A., Goodchild, M., Maguire, D. J., & Rhind, D. W. (2010). *Geographic Information Systems and Science* (3rd ed.). Wiley.

Maisonneuve, N., Stevens, M., & Ochab, B. (2010). Participatory noise pollution monitoring using mobile phones. *Information Polity*, *15*(1), 51–71.

Mann, S., Nolan, J., & Wellman, B. (2002). Sousveillance: Inventing and using wearable computing devices for data collection in surveillance environments. *Surveillance & Society*, *1*(3), 331–355.

Mannings, R. (2008). *Ubiquitous Positioning.*Artech House.

Masser, I. (2011). Emerging frameworks in the information age: The spatial data infrastructure (SDI) phenomenon. In *The SAGE Handbook of GIS and Society* (pp. 271–286). SAGE Publications Ltd.

McDougall (2012). An assessment of the contribution of volunteered geographic information during recent natural disasters. In *Spatially Enabling Government, Industry and Citizens: Research and Development Perspectives* (pp. 201–214). GSDI Association Press.

Merchant, J. W., & Narumalani, S. (2009). Integrating remote sensing and geographic information systems. In T. A. Warner, M. D. Nellis, & G. M. Foody (Eds.), *The SAGE Handbook of Remote Sensing*. SAGE Publications Ltd.

Merschdorf, H., & Blaschke, T. (2018). Revisiting the role of place in geographic information science. *ISPRS International Journal of Geo-Information*, *7*(2018), 364.

Mesev, V. (2007). *Integration of GIS and Remote Sensing*. Wiley.

Meyer, R. (2014, June 28). Everything we know about Facebook's secret mood manipulation experiment. *The Atlantic*. Retrieved July 13, 2014, from http://www.theatlantic.com/technology/archive/2014/06/everything-we-know-about-facebooks-secret-mood-manipulation-experiment/373648/

Monmonier, M. (1996). *How to Lie with Maps* (2nd ed.). University Of Chicago Press.

Monmonier, M. (2002). *Spying with Maps: Surveillance Technologies and the Future of Privacy.*University of Chicago Press.

Mun, M., Reddy, S., Shilton, K., Yau, N., Burke, J., Estrin, D., . . . Boda, P. (2009). PEIR, the personal environmental impact report, as a platform for participatory sensing systems research. In *Proceedings of the 7th International Conference on Mobile Systems, Applications, and Services* (pp. 55–68). ACM.

Narumalani, S., & Merchant, J. W. (2009). Integrating remote sensing and geographic information systems. In T. A. Warner, M. D. Nellis & G. M. Foody (Eds.), *The SAGE Handbook of Remote Sensing*. SAGE Publications Ltd.

National Research Council (NRC) (2007). *Putting People on the Map: Protecting Confidentiality with Linked Social-Spatial Data*. The National Academies Press.

Nelson, T. A., Goodchild, M. F., & Wright, D. J. (2022). Accelerating ethics, empathy, and equity in geographic information science. *PNAS*, *119*(19), e2119967119.

Nouwt, S. (2008). Reasonable expectations of geo-privacy? *SCRIPTed*, *5*(2), 375–403.

O'Sullivan, D. (2006). Geographical information science: Critical GIS. *Progress in Human Geography*, *30*(6), 783–791.

Palmer, M. H., & Kraushaar, S. (2013). Volunteered geographic information, actor-network theory, and severe-storm reports. In D. Sui, S. Elwood & M. Goodchild (Eds.), *Crowdsourcing Geographic Knowledge* (pp. 287–306). Springer.

Panek, J., & Netek, R. (2019). Collaborative mapping and digital participation: A tool for local empowerment in developing countries. *Information*, *10*(2019), 255.

Perkins, C., & Dodge, M. (2009). Satellite imagery and the spectacle of secret spaces. *Geoforum*, *40*(4), 546–560.

Pickles, J. (1991). Geography, GIS, and the surveillant society. *Papers and Proceedings of Applied Geography Conferences*, *4*, 80–91.

Pickles, J. (1994). *Ground Truth: The Social Implications of Geographic Information Systems.*The Guilford Press.

Pool, B. (2010). Crisis Camp Haiti: Techno-types volunteer their computer skills to aid quake victims. *Los Angeles Times*. Retrieved from http://articles.latimes.com/2010/jan/16/world/la-fg-haiti-crisiscamp17-2010jan17

Raper, J., Gartner, G., Karimi, H., & Rizos, C. (2007). Applications of location–based services: A selected review. *Journal of Location Based Services*, *1*(2), 89–111.

Raubal, M. (2011). Cogito ergo mobilis sum. In *The SAGE Handbook of GIS and Society* (pp. 159–173). SAGE Publications Ltd.

REACH, American Red Cross, & USAID. (2014). *Groundtruthing OpenStreetMap Building Damage Assessment, Haiyan Typoon, The Philippines, Final Assessment Report*. Retrieved July 13, 2014, from http://www.reach-initiative.org.

Resch, B., Mittlboeck, M., Lipson, S., Welsh, M., Bers, J., Britter, R., & Ratti, C. (2009). Urban sensing revisited – common scents: Towards standardised geo-sensor networks for public health monitoring in the city. In *Proceedings of the 11th International Conference on Computers in Urban Planning and Urban Management – CUPUM2009*. Hong Kong, 16–18 June 2009.

Sagl, G., Delmelle, E., & Delmelle, E. (2014). Mapping collective human activity in an urban environment based on mobile phone data. *Cartography and Geographic Information Science*, *41*(3), 272–285.

Schuurman, N. (1999). Critical GIS: Theorizing an emerging discipline. *Cartographica: The International Journal for Geographic Information and Geovisualization*, *36*(4), 1–108.

Seang, T. P., Mund, J.-P., & Symann, R. (2008). Low cost amateur aerial pictures with balloon and digital camera. *MethodFinder*. Retrieved July 14, 2014 from http://methodfinder.net.

Sheppard, E. (1995). GIS and society: Toward a research agenda. *Cartography and Geographic Information Systems*, *22*(1), 5–16.

Sieber, R. (2006). Public participation geographic information systems: A literature review and framework. *Annals of the Association of American Geographers*, *96*(3), 491–507.

Singel, R. (2005). *A Disaster Map "Wiki" Is Born. WIRED.* Magazine. Retrieved July 14, 2014, from http://archive.wired.com/software/coolapps/news/2005/09/68743

Smith, N. (1992). History and philosophy of geography: real wars, theory wars. *Progress in Human Geography*, *16*(2), 257–271.

Soden, R., & Palen, L. (2014). From crowdsourced mapping to community mapping: The post-earthquake work of OpenStreetMap Haiti. In R. Chiara, L. Ciolfi, D. Martin, & B. Coneien (Eds.), COOP 2014: *Proceedings of the 11th International Conference on the Design of Cooperative Systems* (pp. 311–326), 27–30 May 2014, Nice, France.

Sudmanns, M. et al. (2023). Think global, cube local: An Earth observation data cube's contribution to the digital Earth vision. *Big Earth Data*, *7*(3), 831–859. http://doi.org/10.1080/20964471.2022.2099236

Sui, D., & Goodchild, M. (2011). The convergence of GIS and social media: Challenges for GIScience. *International Journal of Geographical Information Science*, *25*(11), 1737–1748.

Sui, D., Goodchild, M., & Elwood, S. (2013). Volunteered geographic information, the exaflood, and the growing digital divide. In D. Sui, S. Elwood & M. Goodchild (Eds.), *Crowdsourcing Geographic Knowledge* (pp. 1–12). Springer.

Taylor, P. (1990). Editorial comment: GKS. *Political Geography Quarterly*, *9*, 211–212.

Thrun, S. (2010). What we're driving at. *Google Official Blog*. Commercial. Retrieved from http://googleblog.blogspot.com/2010/10/what-were-driving-at.html

Turner, A. (2006). *Introduction to Neogeography*. O'Reilly Media.

Van Aardt, J. A. N., McKeown, D., Faulring, J., Raqueño, N., Casterline, M., Renschler, C., . . . Gill, S. (2011). Geospatial disaster response during the Haiti Earthquake: A case study spanning airborne deployment, data collection, transfer, processing, and dissemination. *Photogrammetric Engineering & Remote Sensing*, *77*(9), 943–952.

Warf, B., & Sui, D. (2010). From GIS to neogeography: Ontological implications and theories of truth. *Annals of GIS*, *16*(4), 197–209.

Weiner, G., & Smith, B. W. (2014). *Automated Driving: Legislative and Regulatory Action. The Center for Internet and Society*. Retrieved July 15, 2014, from http://cyberlaw.stanford.edu

Weiser, M. (1991). The computer for the 21st century. *The Scientific American*, *256*(3), 94–104.

Wilson, M. W., & Graham, M. (2013). Guest editorial. *Environment and Planning A*, *45*(1), 3–9.

Wilson, M. W., & Poore, B. S. (2009). Theory, practice, and history in critical GIS: Reports on an AAG panel session. *Cartographica: The International Journal for Geographic Information and Geovisualization*, *44*(1), 5–16.

Yang, C., Wong, D., Miao, Q., & Yang, R. (Eds.). (2010). *Advanced Geoinformation Science*. CRC Press.

Young, J. C., Lynch, R., Boakye-Achampong, S. et al. (2021). Volunteer geographic information in the Global South: Barriers to local implementation of mapping projects across Africa. *GeoJournal*, *86*, 2227–2243.

Zipf, A., & Jöst, M. M. (2011). Location-based services. In W. Kresse & D. M. Danko (Eds.), *Springer Handbook of Geographic Information*. Springer.

Zook, M., Graham, M., Shelton, T., & Gorman, S. (2010). Volunteered geographic information and crowdsourcing disaster relief: A case study of the Haitian Earthquake. *World Medical & Health Policy*, *2*(2), 7–33.

15 Frontiers of GIScience
Evolution, State of the Art, and Future Pathways

May Yuan

LIST OF ACRONYMS

ABM	Agent based modeling
ACM	Association for Computing Machinery
CA	Cellular automata
COSIT	Conference on spatial information theory
FGDC	Federal Geographic Data Committee
GIS	Geographic information systems
GIScience	Geographic information science
NCGIA	National Center for Geographic Information and Analysis
NSDI	National Spatial Data Infrastructure
NSF	National Science Foundation
SfM	Structure from motion
UAV	Unpiloted aviation vehicles
UCGIS	University Consortium for Geographic Information Science
VGI	Volunteered geographic information

15.1 INTRODUCTION

Geographic Information Science (GIScience) is the science that underlies Geographic Information Systems (GIS) technology. Roger Tomlinson introduced GIS in his report on computer mapping and analysis to the National Land Inventory in the Canada Department of Agriculture (Tomlinson, 1962). Yet, when GIS is broadly defined as a system that deals with geographic information, it can be traced far back into the time when humans started recording and sharing knowledge about the environment. Before computer-based GIS technology, oral traditions and maps were the primary means to encode and communicate geographic information. Nowadays, GIS technologies are diverse and thriving in mapping, spatial analysis and modeling, location-based services, cyber geographical applications, and crowd-sourcing and citizen sensing for timely and large-scale ground observations. GIS technologies are now important research tools for environmental sciences, biological and agricultural sciences, public health, urban planning, and economic, political, and social studies. GIScience to subserve the conceptual, theoretical, and computational foundations for these technologies.

The advent of GIS technologies traces back to the world's oldest map: a mural, dated to ~6600 BC, found at the Neolithic site of Catalhöyük (Schmitt et al., 2014). The early adoption of maps is of no surprise. Maps are intuitive and effective tools to give directions, express spatial arrangements of features, and plan spatial activities. Humans made maps long before they invented writing. Likewise, the world's first GIS, Tomlinson's Canadian GIS, was motivated by using computers to automate map analysis and production. To date, map libraries continue the important role of curating and providing access to atlases, aerial photographs and spatial data in digital forms, while

DOI: 10.1201/9781003541158-18

massive and diverse geographic information is also widely available from government agencies, businesses, organizations, and communities.

While mapping is essential, a GIS also consists of tools to process geospatial data, manage geospatial databases, integrate data through geo-referencing and compatible attribute definitions, analyze embedded spatial patterns, model geographic phenomena and processes, and render data and findings in multiple ways. The technology was initially developed out of application needs, and its conceptual and computational frameworks were fragmented across solutions. GIScience research contributes to developing fundamental frameworks for GIS technologies and takes the technological challenges to improve understanding geographic information, producing and communicating geographic knowledge, and supporting spatial predictions and decision making.

The recent convergence of GIScience and data science has extended spatial statistics with large-scale geospatial computing with an increasing reliance on cyberinfrastructure. Spatial analysis and models traditionally build upon descriptive and inferential statistics with assumptions about the distribution and variance of a given population. Advances in GIScience, leveraging big data and data science, equip data analytics that embed semantic assumptions in analytical procedures to elicit insightful patterns from data of high velocity, volume, variance, variety, and voracity, Moreover, GIScience's relation to remote sensing science is akin to its relation to statistics. Traditionally, GIScience differs from remote sensing in its emphasis on mapping, geographic data modeling, and spatial modeling. The rise of machine learning and GeoAI has blurred the RS-GIS divide by the formation of data process pipelines that streamline data acquisition, knowledge production, and user interactions. Deep learning methods are now common in image analytics with a continuing workflow that ingests images, extracts features, and makes spatial predictions. The continuation of RS-GIS pipelines is essential to realize the vision of the digital twin of the Earth and improve the model predictability of the Earth systems, including human systems.

This chapter aims to highlight the past, present, and future of GIScience research. As a field of interdisciplinary and multidisciplinary research, GIScience enjoys outstanding advances in both breadth and depth as evidenced by the multitude of names associated with the discipline, such as Geospatial Science, Spatial Science, Spatial Information Science, Geoinformatics, Geomatics, and Spatial Data Science (Table 15.1). Consequently, it is challenging to capture the full scope of research development in the field. What follows reflects the author's perspectives on the evolution, state of the art, and future pathways of GIScience. Since the chapter is focused on GIScience, the discussions here emphasize the key intellectual development of spatial concepts, theories, and computational approaches. GIS applications are not GIScience research and, therefore, are beyond the scope of this chapter. The next section elaborates on the evolution of GIS technologies to GIScience.

TABLE 15.1

Term	Primary Communities	Special Emphases
Spatial science (or geospatial science to stress the space of interest at a geographic scale)	Geography, statistics, and other domains applying mapping and spatial methods in their fields, such as spatial social science	Mapping, spatial statistics, spatial analysis, spatial modeling
Spatial information science	Computer science and management information science	Spatial computing, spatial database management, and visualization and communication of spatial information
Geoinformatics	Geoscience, computer science, geostatistics	Information computing and management for geoscience data, most commonly with earth science data
Geomatics	Geodetics, survey engineering, geophysics, earth science.	Geodetics, Geoid, datums, land surveys, mapping.

From the early emphases on the transitions from technological advances in mapping, spatial database building, and inventory and planning applications to scientific inquiries into the nature of geographic information, spatial computing, and geographical understanding. Section 15.3 highlights the active GIScience research directions in cognition, representation, integration, and computation. The chapter concludes with promising pathways for future GIScience development.

15.2 EVOLUTION

Computer-based GIS technology revolutionized the processes of recording and disseminating geographic information and invoked new possibilities to represent, analyze and compute geography. Since its conception, the term GIS was often referenced exclusively to computer-based GIS. Coppock and Rhind (1991) characterized the early development of computerized GIS into four general phases from 1960 to 1990:

1. **A phase of pioneers** from the early 1960s to 1975. Key leaders included Howard Fisher of the Harvard Laboratory for Computer Graphics, Roger Tomlinson of the Canadian Geographic Information System, and David Bickmore at the Experimental Cartographical Unit in the United Kingdom.
2. **A phase of national drivers** from 1973 to early 1980s. Key agencies included Canada's Department of Agriculture, the U.S. Bureau of the Census, and the Ordnance Survey in Great Britain. In the United States, GIS technology attracted great interest from many federal agencies such as the Department of Defense, Central Intelligence Agency, U.S. Forest Service, Fish and Wildlife Service, and Department of Housing and Urban Development, as well as state and local governments including California, Maryland, Minnesota, New York, and others.
3. **A phase of commercial dominance** from early to late 1980s, most noticeably the Environmental Systems and Research Institute (ESRI, now Esri) and Integraph. The companies not only developed GIS software packages but also designed and implemented GIS projects for government agencies. These GIS packages were adopted in college courses, and to date they remain the primary tools for learning GIS and doing GIS projects. In 1988, the United States National Science Foundation (NSF) awarded a grant to establish the National Center for Geographic Information and Analysis (NCGIA) with the University of California at Santa Barbara, State University of New York at Buffalo, and University of Maine. The NSF grant provided $10 million for eight years of NCGIA leadership that transformed GIS to GIScience and resulted in lasting impacts in education and research in the U.S. and around the world.
4. **A phase of user dominance** since early 1990s with the rise of desktop GIS that emphasized ease of use and promoted wide adoption of GIS technology beyond research universities, large government organizations and big companies. In 1994, U.S. Executive Order 12906 established the Federal Geographic Data Committee (FGDC) as the executive branch leadership to develop the National Spatial Data Infrastructure (NSDI) marked the first multi-agency nation-wide efforts to coordinate GIS data management and access. The expanded availability of free GIS data stimulated many geospatial research and business opportunities and popularized GIS technology in a wide range of domain applications.

In a short period of 30 years (1960s to 1990s), GIS started with a few visionaries who sought ways to use computers for mapping and analyzing geographic data and then grew to a generation of researchers and professionals that brought GIS into mainstream college curricula, government functions, and business operations. With this growth, research efforts went beyond mapping and spatial data handling. Researchers ventured into the unique complexity of geographic information and ensuing challenges in acquiring and using spatial data to understand geographic processes and make spatial predictions. *The International Journal of Geographical Information Systems* (*IJGIS*) was launched in 1987 and was recognized as the primary academic journal in the field (Caron et al., 2008). Goodchild published a landmark paper in *IJGIS*, entitled Geographic Information Science (Goodchild, 1992). The paper highlighted scientific problems unique to geographical data and established the topical content for GIScience.

Since then, many organizations and journals adopted the term GIScience over GIS. Efforts of the academic community, with most participants from Geography, established the University Consortium for Geographic Information Science (UCGIS) in 1994 and, through community efforts, defined GIScience as *the development and use of theories, methods, technology, and data for understanding geographic processes, relationships, and patterns. The transformation of geographic data into useful information is central to geographic information science* (UCGIS, 2002; Mark, 2003). It is important to note that GIScience research is not about using GIS technologies to solve scientific problems. This is similar to statistics and mathematics; applications of statistical or mathematical methods to solve a biological problem contribute to biological science, not to statistics or mathematics.

The early development of GIScience can be attributed to the NCGIA's leadership in a series of initiatives as well as the UCGIS community efforts to identify and articulate research challenges. In 1997, the *International Journal of Geographical Information Systems* was renamed *International Journal of Geographical Information Science* marking its second decade of publication (Fisher, 2006). However, the tendency to use GIScience as a synonym for GIS was quite common in early 2000 (Mark, 2003) and remains rather persistent today. Many programs offer GIScience courses with the same instructional materials for GIS, and many do not differentiate GIScience research from research using GIS. Nevertheless, leading journals (such as *IJGIS* and *Geoinformatics*) and conferences in GIScience (such as GIScience and ACM-SIGSPATIAL) emphases papers with contributions to conceptual, theoretical, and computational innovations.

Foundational work in cartography, spatial statistics, and spatial modeling has significantly contributed to the development of GIS, and these continue to be important subjects in GIScience research today. Computer cartography made notable progress in line generalization (Douglas and Peucker, 1973), map generalization (Buttenfield and McMaster, 1991), cartographic label placement (Marks and Shieber, 1991), and interactive digital atlases production (MacEachren, 1998). Landmark spatial studies led to new methods that account for local variations and local processes, such as Map Algebra (Tomlin, 1994), Local Indicator of Spatial Autocorrelation, aka LISA (Anselin, 1995), Geographically Weighted Regression, aka GWR (Brunsdon et al., 1998), and Geo-Algebra (Takeyama and Couclelis, 1997). Furthermore, spatial modeling advanced new approaches to simulate hydrological processes (Olivera and Maidment, 1999) and urban systems (Couclelis, 1997; Batty, 2007) by leveraging dynamic methods from other fields, such as distributed modeling, cellular automata (CA), and agent based modeling (ABM). The popularity of Web 2.0 and location-aware devices has led to ubiquitous data on the whereabouts of human activities and movements. With an abundance of movement data, GIS analysis has extended traditional emphases on spatial patterns and relationships among points and polygons to polylines (e.g., trajectories or flows). New research emerges to the call for a new integrated science of movement (Demšar et al., 2021).

Moreover, arguments were made that the foundations of GIScience should be tied closely to Information Science (Mark, 2003). Information Science studies the means and processes of information transmission among humans and/or computers. Syntactic form, semantic content, and contextual relevance are key elements in determining the value and optimal means of information flows from transmitters to receivers (Worboys, 2003). However, any judgment about value and optimality of the key elements must rely on a common understanding of the domain between transmitters and receivers. Geographic ontologies became an important subject in GIScience research (Agarwal, 2005), and research on spatial ontologies and representation along with other issues related to the nature of geographic information, was prominent in NCGIA research initiatives and UCGIS research challenges. Fundamental GIScience research has been promoted through the Conference on Spatial Information Theory (COSIT) starting in 1993 and the International Conference on

GIScience (GIScience) which began in 2000. Since, the two conferences have been held in alternate years and locations between Europe and North America. In addition, the Auto-Carto International Symposium on Automatic Cartography,[1] International Cartographical Conference, and International Symposium on Spatial Data Handling all have a long history as primary academic venues for cartographers and geographers working in GIScience. Computer Scientists interested in spatial databases and information started the annual Association for Computing Machinery (ACM) Workshop on Advances in Geographic Information Systems in 1993. They successfully expanded the annual workshop to the annual ACM-GIS International Symposium in 1998 and furthermore established the ACM Special Interest Group on Spatial Information (SIGSPATIAL) as the catalyst for research on spatially-related information among computer scientists (Samet et al., 2008).

These pioneer efforts established a strong foundation for GIScience. Research has migrated from GIS enabling computerization of geographic data processing and mapping to GIScience inquiries into the essence of geographic information and epistemology. Goodchild (2014) highlighted research and institutional accomplishments in the 20 years of progress since the introduction of GIScience in 1994. On measurements, research foci shifted from spatial errors in the 1980s to spatial uncertainty in the 1990s. On representation, research advanced from vector/raster in the 1980s and objects/fields in the 1990s, to complex object-fields and field-objects in 2000s. GIScience research on analysis progressed from spatial autocorrelation in the 1970s to spatial heterogeneity in the 1990s. These efforts also built a strong foundation that supports GIScience research and development into the mainstream of information technology. Recent advances popularize Spatial Data Science (Anselin, 2016), digital twin (Tomko and Winter, 2019; Deng et al., 2021) and GeoAI (Janowicz et al., 2020) in general and the surge of geospatial data empowers research attention to spatial dynamics and drives the development of city science (Batty, 2013) and movement science (Demšar et al., 2021). The GIScience evolution is summarized in Figure 15.1.

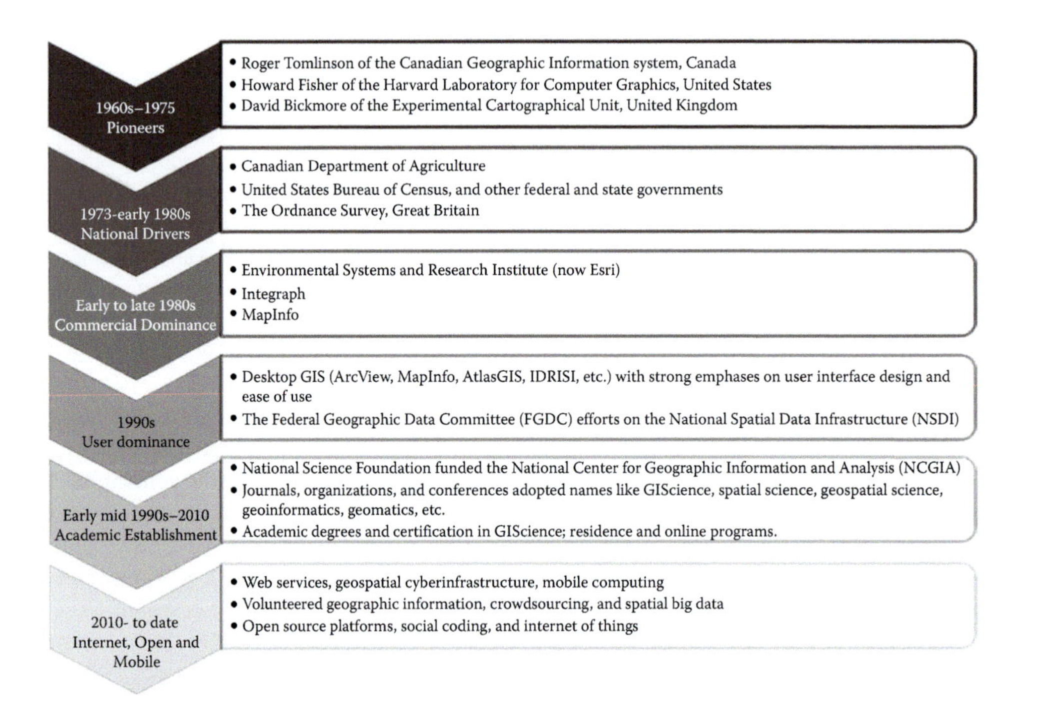

FIGURE 15.1 A summary of the evolution of GI systems and GIScience.

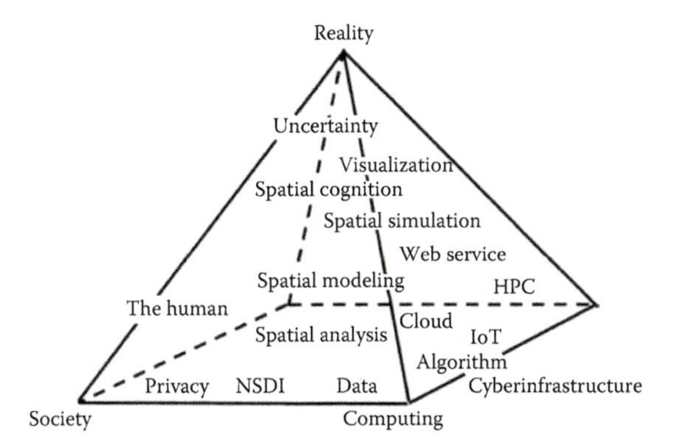

FIGURE 15.2 A conceptual framework for research themes in GIScience, whereas HPC stands for high performance computing, and IoT is the Internet of Things. (Modified from Goodchild, M. F. Journal of Spatial Information Science, (1), 3, 2014.)

Over the years, GIScience research frontiers were articulated through 21 NCGIA research initiatives UCGIS research priorities and Computing Community Consortium Spatial Computing Visioning. Goodchild (2014) listed some of the topics that resulted from discussions in GIScience communities in several venues and summarized in a conceptual framework for GIScience that connects the dimensions of human, society, and computer. Expanded upon his conceptual framework, Figure 15.2 incorporates major developments in cyberinfrastructure and computing that have transformed the interplays among the human, society, and computers as well as how we perceive and understand human, physical, biological, and many other dimensions of reality. There is no shortage of research challenges in GIScience. This becomes evident with a quick search on Google Scholar which results in more than 5,000 publications on the subject. Some fundamental topics remain outstanding and are likely to persist at the core of GIScience research, such as geographic ontologies, space-time representation, spatial algorithms, spatial cognition, geovisualization, and spatial decision support.

15.3 STATE OF THE ART

GIScience continues to evolve with an increasing attention to what is local rather than global, individual rather than aggregated, collaborative rather than authoritative, culturally aware rather universal, open rather than exclusive, mobile rather than desktop, and data-driven rather than model-driven. Moreover, models and methods are being developed to represent and visualize multidimensional and multimedia data, and more models and methods fashion learning than hypothesis testing. Leveraged by the internet, new GIS platforms are being realized on the World Wide Web, with cyberinfrastructure, and in the Cloud. All these developments have profound influences on what is summarized here as 3 A's: *Abstraction, Algorithms*, and *Assimilation* throughout GIScience epistemology.

15.3.1 ABSTRACTION

Abstraction takes place at multiple levels in GIScience research. It is concerned with how we conceptualize geographic worlds and spatial problems and subsequently how we represent, compute, and communicate all the relevant concepts and findings. As spatial data are from different sources, integration can be challenging at each level of abstraction. To-date the most popular abstraction used in GIScience remains the so-called data layers (Figure 15.3). While the data-layer abstraction is intuitive, GIScience research examines issues in cognition, ontology, and statistics (e.g., sampling)

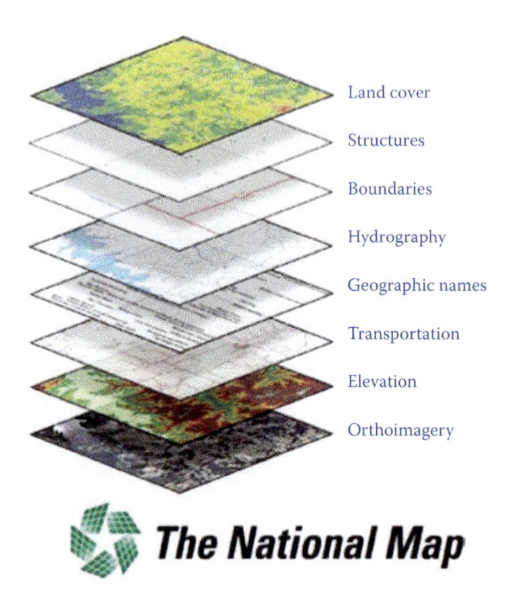

FIGURE 15.3 An example of geographic abstractions: the National Map from the United States Geological Survey.

for better abstraction of reality. The rise of Machine Learning pushes for learning representations from data without pre-defined geographic conceptualization or abstraction (Mai et al., 2022).

Across all the levels of abstraction, cognitive research helps understand how people learn and organize geographic knowledge. Such cognitive understanding can improve GIS usability and communication. Montello (2009) summarized five main areas of cognitive research in GIScience since 1992: human factors of GIS, geovisualization (including spatialization), navigation systems, cognitive geo-ontologies, spatial thinking and memory, and cognitive aspects of geographic education. Much of the cognitive research confirms the complexity of geographic information and knowledge due to indeterminacy, vagueness, and the interdependency of individuals and geographic context. As a result, geographic categorization and reasoning may vary from person to person or place to place. For example, cognitive geo-ontologies recognize that people may see things differently, and their conceptualizations may vary due to environmental, cultural or linguistic differences (Mark et al., 2011; Wellen and Sieber, 2013; Turk and Stea, 2014). Such differences have profound implications for information sharing and integration, spatial data infrastructure building, spatial decision support, and many other issues that deal with the usefulness of GIS technologies and intrinsic technological biases. Spatial cognition research now commonly adopts eye-tracking and EEG technologies to collect empirical evidence on human attention to spatial information displays or spatial features on screen or in physical or virtual environments (Çöltekin et al., 2010). The empirical evidence not only gives insights into our spatial cognitive models but also improves user interface design for spatial knowledge communication and user performance on spatial tasks, such as environmental perception and wayfinding.

Information sharing and integration was the initial motivator of ontological research through the rise of the Semantic Web that extends the World Wide Web for people to share and reuse data across application boundaries. Ontological approaches are now commonly used to define specifications of geographic abstractions in a problem domain (Jung et al., 2013; Ujang and Rahman, 2013), achieve semantic consistency for data integration and complex query support (Wiegand, 2012), and assure interoperability across systems and over web services (Shi and Nellis, 2013). Different frameworks have been proposed for geo-ontologies. Frank (2001) argued for a tiered ontology to assure consistency constraints based on how different kinds of things are conceptualized and from where they are abstracted. Tier 0 ontologies are for *human-independent reality* where natural laws prevail regardless of human observers. Tier 1 ontologies are for *observations of physical world* with measurements and statistics. Tier 2 ontologies are for *objects with properties* that can be used to identify individuals and determine categorical memberships with necessary and sufficient properties. Tier 3 ontologies are for *social reality* that is subject to social, cultural, and linguistic contexts. Finally, Tier 4 ontologies are for *subjective knowledge*, which may be incomplete or partial, used by individuals or institutions for reasoning or decision making. Couclelis (2010) articulated the need for geographic information constructs as the core of ontologies in GIScience. Her framework centers on an ontological hierarchy to connect intentionality and relevant information. There are seven levels of semantic resolution in the hierarchy. In the order of low to high levels, the semantic levels of resolution include *existence, observables, similarities, simple objects, composite objects, function*, and *purpose*. She introduced the idea of *semantic contraction* to generalize semantic richness from higher, more complex levels to a lower level of simpler semantics, and *object of discourse* to represent entities as composites of geographic information constructs at the higher levels of the hierarchy. This ontological research expanded our understanding of semantic granularity (Fonseca et al., 2002) and spatial tasks (Wiegand and García, 2007) and laid the foundation for building theories of geographic information.

In addition to ontologies, abstraction also accounts for means by which geographic information can be effectively acquired, analyzed, and communicated. Traditionally, geography is abstracted in forms of data from field surveys, maps, imagery, tables, graphs, and text. Advances have opened new means to acquire geographic information with new kinds of geographic abstraction. For example, data from dynamic geosensor networks (Llaves and Kuhn, 2014), tweets (Tsou et al., 2013), geotagged photos (Samet et al., 2013), and location-related information from various social media (Croitoru et al., 2013; Jiang and Miao, 2014) offer real-time or near real-time environmental and social abstractions which enable detection of events and activities as they unfold. Online reviews from Yelp or TripAdvisor become data about human experiences at locations. As the geographic world captured by these data is transitory and ephemeral, so is the ensuing geographic abstraction. Volunteered geographic information (VGI), crowdsourced geographic information (Goodchild, 2007; Goodchild and Glennon, 2010), and ambient geospatial information (Stefanidis et al., 2013) commonly condense information entries to point locations. Consequently, geographic abstraction is generally reduced to individuals and collections of points. Spatial synthesis would be more appropriate than analysis to decipher these data. Location-aware devices, such as smartphones or mobile computers, become a major apparatus for these user-contributed geographic data. Furthermore, location-aware devices sequentially provide location points about the users, enabling the collection of users' movement trajectories. Traditionally, most GIS analysis functions apply to points or polygons. Traditionally, GIS line functions are mostly for network analysis. Spatial trajectories present a recent addition to GIS abstractions for modeling lifelines, accessibility, and environmental exposure (Kwan, 2009) and give rise to the uncertain geographic context program (Kwan, 2012). Similarly, advances in spatiotemporal flows drive new approaches to identify and examine collective movements, such as traffic or migration (Gu et al., 2023).

Novel geospatial data sources continue to grow. Besides geosensor and social media data, multimedia data incorporate "near sensing" images, video, audio, virtual reality, and augmented reality to represent geography (Camara and Raper, 1999; Qiao and Yuan, 2021). Street View images provide local perspectives on the ground, useful for identifying geospatial features and functional zones in urban landuse (Fang et al., 2021). Videos may be interviews, documentary films, surveillance recordings, or animation of temporal information. Audios may be oral stories, narration by a native speaker, testimonies, songs, or animal sounds. Besides extracting features, actions, or events, video and audio greatly enrich the geographic context for spatial abstractions. Cube sat, Ballon sat, Unpiloted Aviation Vehicles (UAV), and Structure from Motion (SfM) photos provide geospatial data that are focused, timely, highly intensive, and multi-dimensional. Virtual reality and augmented reality, usually with 3D visualization, supplement spatial abstraction with videos, audios, photographs, digital documents, and labels in a dynamic context-aware immersive environment. The granularity of geographic abstraction becomes finer or coarser depending on the user's location and view. Virtual geographic environment (Lin et al., 2013) leverages virtual reality and multidimensional GIS to provide a digital platform for geographic experiments through collaborative visualization and simulation. Collaboration requires shared geographic abstraction of both declarative knowledge and procedural knowledge as the basis for communication and integration, which in turn rests on cognitive and ontological compatibility.

15.3.2 Algorithms

Algorithms are step-by-step instructions for calculations. Here, algorithms are broadly defined as approaches to data processing, analysis, modeling, and simulation. As geographic abstraction shifts emphases to semantics, the development of spatial algorithms also attempts to reveal local meanings and individual behaviors in space and time. New spatial filtering (Griffith and Chun, 2019) and geographically weighted regression (Zhou et al., 2023) attend to spatial heterogeneity and local effects in large, heterogeneous data. Similarly, machine learning algorithms consider one data item at a time to gradually adjust model parameters for local curve fitting through gradient descent. A wide range of deep learning algorithms have been applied to geospatial classification or prediction in geospatial feature identification, traffic flows, zonal functions, hazard risks, human dynamics, public health etc. (*c.f.*, Gao, 2021).

The rise of Critical GIS (O'Sullivan, 2006; Schuurman, 2006) reflects the needs to engage social critiques in GIS-based geographic knowledge production in terms of basic concepts, representation, participation, and social implications. Volunteered geographic information (VGI) and web map services partially address the needs by empowering ordinary citizens to create geographic data and participate in geographic knowledge production. Many critical GIS researchers also echo the criticisms of positivist's biases in GIS and advocate for qualitative GIS (Cope and Elwood, 2009) to address the needs to incorporate contextual details and interpretations of the described situation and processes. Broadening GIS methodology and the programming environmental allow qualitative methods that are commonly used by sociologists and humanities scholars, like coding, triangulating source materials, and content analysis in recursive and iterative forms to produce knowledge, such as Geo-Narrative (Kwan and Ding, 2008).

VGI is only one of the many unstructured crowdsourced data sources available from the We (Yuen et al., 2011). For geospatial data, crowdsourced systems usually provide web map services or web feature services that support map mash-ups by which geospatial data from remote servers can be visually overlapped in a browser on a client site. As discussed in the abstraction section, the Semantic Web transforms webpage content to data as Web 2.0. Various social media facilitate crowdsourcing and provide ambient geographic information that can be exploited to recognize social pulses (Croitoru et al., 2013) or validate environmental conditions (See et al., 2013). Crowdsourced data are either directly requested by a project web service, such as "Did you feel it?" web portal by the United States Geological Surveys (USGS) Earthquake Hazards Program[2] or harvested from social media feeds via application programming interfaces (API), such as OpenStreetMap API.[3] Heipke (2010) provided a good introduction on crowdsourcing geospatial data with highlights of successful projects, the basic technologies and comments on data quality.

Since VGI and crowdsourced data lack statistical sampling schemes and are collected from various sites, researchers need to develop customized algorithms for web scraping, data preprocessing, mapping and analysis. Of great challenge is the fact that these data violate most, if not all, sampling assumptions based on which conventional statistical methods are founded. Location information associated with VGI, crowdsourced data, and data from web crawling may be explicitly tagged through GPS readings as latitude, longitude or other x, y coordinate pairs. Alternatively, location may be implicitly noted in forms of place names or addresses. Addresses can geocoded against street network databases. For place names, toponym resolution and gazetteer matching will be necessary for georeferencing (Adelfio and Samet, 2013). More generally, conceptual and computational frameworks are being developed to transform text to a rich geospatial data source (Vasardani et al., 2013; Yuan et al., 2014). While several studies showed that VGI and crowdsourced data are timely and at times more representative of geographic reality than authoritative data (Goodchild and Glennon, 2010), most VGI and geospatial crowdsourced projects remain primitive and do not go beyond visualization, animation, and frequency graphing (Batty et al., 2010). Because crowdsourced data collection does not follow any statistical sampling methods and are often subject to demographic or geographic biases, they cannot be applied to established statistical models. Sentiment analysis of postings and messages is often based on keywords without cross-references to the content.

Besides VGI and crowdsourced data, GIScience researchers are active in cyberinfrastructure research and cloud computing. CyberGIS integrates GIS and spatial analysis and modeling into cyberinfrastructure that provides high performance (terra grid) computing and large-scale data repositories (Wang et al., 2013), particularly "Roger," the first supercomputer designated for GIScience research and education. Cyberinfrastructure transforms GIS from an isolated platform to a cyber-network of supercomputers, virtual organizations, and massive shared data and computational resources. The fundamental differences in computing platforms require new algorithms for data processing, management, analysis, and modeling, and much has been implemented as middleware. While also taking the advantage of internet information technologies, cloud computing leverages four types of services: Infrastructure as a Service (IaaS), Platform as a Service (PaaS), Software as a Service (SaaS), and Data as a Service (DaaS), with open-source resources Hadoop and MapReduce to offer elastic, distributed, and on demand computing facilities.

The ideas of spatial cloud computing encompass not only utilization of existing cloud services for intensive spatial computing but also the development of data and tool services for geospatial applications that are made accessible over the web (Yang et al., 2011). Cloud computing provides elastic, advanced resources for experimenting with ideas, application development, and app distribution. Location-aware or spatially enabled apps are widely available to map property values, routes, crime incidents, restaurants, and gas stations, for example. Cloud computing opens GIS workflows to tightly connect to resources on the web and transforms GIS into a service that frees users from desktop computers to anywhere and any platform with internet access. CloudGIS emerges as a promising platform for large-scale geospatial computing and opens many opportunities for mobile GIS, geosensors, and spatial big data (Bhat et al., 2011; Shekhar et al., 2012; Fan et al., 2013).

Even before the introduction of big data, many spatial data are big and grow exponentially, especially imagery data and sensor data. The popularity of location-aware devices and geosensors have motivated algorithm development for trajectory analysis, geometrically (Li et al., 2011) or semantically (Yan et al., 2013) among other methods for movement modeling (Long and Nelson, 2012). Intensive observation updates of location-aware and sensor data further challenge hardware and software capacity. CudaGIS is an example of GIS design with GPUs to provide parallel data processing capabilities (Zhang and You, 2012). GPU algorithms are being developed to enable rapid urban simulation (Ma et al., 2008), Lidar data processing (Sugumaran et al., 2011), viewshed analysis and other data intensive computation (Steinbach and Hemmerling, 2012). Some of the parallel and GPU algorithms have been implemented in open source GRASS GIS for fast spatial computing and rendering (Osterman, 2012). With on-demand IaaS, GPU-based cloud computing has shown to be effective for intelligent transportation management (Wang and Shen, 2011.

15.3.3 Assimilation

Assimilation is defined in the Oxford English Dictionary as the action of becoming conformed to or conformity with. In this paper, assimilation is broadened to processes that bring individual's contributions to a commons for a greater good or to reveal a bigger picture. With the definition, assimilation efforts in GIScience have flourished through Open-Source GIS, Social Coding, Open GIS, and Spatial Turns.

Open-source GIS, like GRASS, Quantum GIS, and PostGIS, were developed by individuals through community efforts (Neteler and Mitasova, 2008) and have gained significant momentum since 2005. To date, there are more than 350 free and open source GIS software packages available.[4] Along with the free software are free data and documents to serve as a foundation for building a learning society in which source code, algorithms, and models can be tested and continuously improved upon. Steiniger and Bocher (2009) reviewed 10 free and open-source GIS software packages and argued for the use of open source practices and software in research for transparency, testability, and adaptability to other projects. Many diverse open source GIS communities thrived in 2012 (Steiniger and Hunter, 2013). Assimilation of individual's contributions for tool development and code improvement in an open source environment collectively results in richer and better GIS resources for all. Social coding follows a similar idea of collaboration, but instead of working towards a package, social coding can be any project or program codes initiated by individuals rather than a community. Perhaps, the most popular social coding site is Github[5] where people can freely copy and modify codes to assimilate into other projects. There are many geospatial projects on Github, such as CartoDB, GeoNode, Spatial4J, OSGeo, and geopython. It is noteworthy that Esri is also active at Github with a suite of open source projects.

Another large-scale collaborative assimilation is the R project, an open source environment for statistical computing and graphics built upon the R language developed by Ihaka and Gentleman (1996). R users can study the source code to understand the underlying statistical procedures and assimilate their new modules with existing R methods, which facilitates advances in methodological research and opportunities to submit proposed models from publications for testing and reuse. Over the last 15 years, R has gained strong volunteer support in building various extensions, including packages for spatial statistics, for example, SpatStat (Baddeley and Turner, 2005), GeoXp (Laurent et al., 2009), and spacetime (Pebesma, 2012). The call for integration of GIS and spatial data analysis (Goodchild et al., 1992) was originally intended to add more spatial analysis capabilities to a GIS. Instead, much success has been realized by assimilating spatial data and methods into R statistics.[6] Currently, R consists of a large suite of spatial modules covering raster analysis, interpolation and geostatistics, spatiotemporal simulation models, spatial autocorrelation, spatial econometrics, spatial structure models, spatial Bayesian models, spatiotemporal cluster analysis, and various mapping and graphing tools (Bivand et al., 2013). In addition, efforts are being made to apply R directly to GRASS GIS database files (Bivand, 2000) or port R scripts to Quantum GIS (Solymosi et al., 2010). Similarly, spatial analysis functions are being assimilated into the Python programming environment, most notably PySAL module (Rey and Anselin, 2010), and many spatial functions has been refactored to support parallelization (Rey et al., 2013). Free spatial data analysis packages such as GeoDa, although not open source nor extendable, offer a graphic user interface and tight coupling of GIS and exploratory spatial analysis tools (Anselin et al., 2006).

The assimilation of GIScience into other disciplines led to exciting new approaches, such as spatial ecology, spatial epidemiology, spatial history, spatial humanities, and spatial social sciences. In addition to spatial analysis and modeling, a suite of geospatial online data processing, information services and computational methods popularizes web mapping and applications. Figure 15.4 illustrates an example of web applications for spatial ecology research that assimilates species, ecological, and environmental data in the Gulf of Mexico (Simons et al., 2013). Location-awareness is now common in research and development in computing and information science (Hazas et al., 2004). Programming libraries are being developed to improve the integration of GIS and remote sensing (Karssenberg et al., 2007; Bunting et al., 2014). Besides mapping and visualization, these spatial turns not only provide new analytical innovations and leveraged space as a problem framing and reasoning framework but also invoked new perspectives to improve understanding in natural sciences (Rosenberg and Anderson, 2011), social sciences (Raubal et al., 2014), and humanities (Bodenhamer, 2013). It is important to make clear that these assimilating efforts are developing new spatially integrated thinking and methodologies, not just applying exiting GIS technologies in domain sciences.

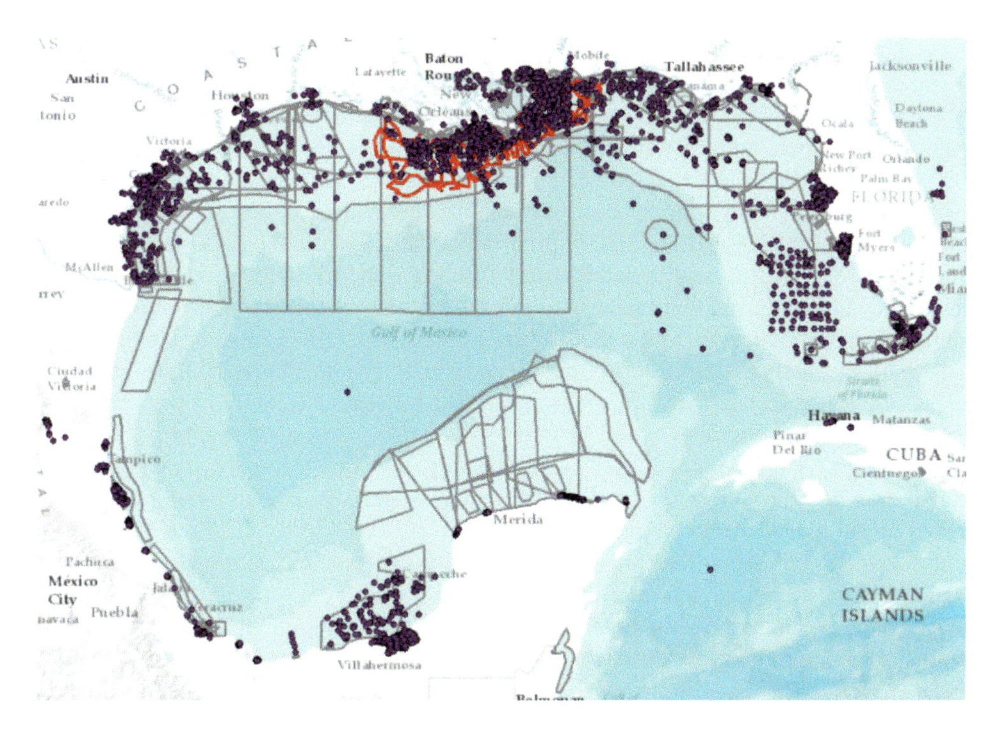

FIGURE 15.4 An example of spatial ecology research in the Gulf of Mexico. Species, ecology, and environmental data for 4,092 sites were extracted from 449 references to support meta-analysis of fish habitat and trophic dynamics.

15.4 FUTURE PATHWAYS AND CONCLUDING REMARKS

As spatial abstraction, algorithms, and assimilation continue evolving, GIScience thrives for multi-perspective, distributed, and collaborative research across people, platforms, and domain sciences. CyberGIS and Cloud GIS foster high performance and ubiquitous spatial computing. Both technologies not only accelerate spatial data processing but transform the ways of doing GIScience and developing GIS applications. Wright (2012) sketched *a post-GISystems world where GIS is subsumed into a broader framework known simply as "the web," divorced from the desktop* in a new paradigm (p. 2197). The future of GIScience will manifest itself in the grand scheme of computational, environmental and social sciences. While time and themes are common axes along which disciplines build knowledge, GIScience distinguishes itself with the emphasis of using space as the first-order principle to acquire, organize, and compute information as well as discover and share knowledge. The distinction was already apparent in early GIS development and initial discussions on GIScience (Goodchild, 1992; Mark, 2003). It will be even more prominent in the big data era when data from location-aware devices continue to grow exponentially in volume and complexity, and spatial contextualization and integration are becoming more effective to sensing making and prediction.

The emphasis of space (e.g., spatiality, location, and situation) will continue to be the focus in pathways for future GIScience development in a world where we have access to needed information everywhere, any time, that is, *an IEWAT world*, enabled through online-offline integration, the Internet of Things, cloud-mobile computing, collaborative information seeking and knowledge building, and integrative cyber-physical-social systems. Clearly, these are also hot topics in the broader computational, environmental, and social sciences. In other words, the future pathways of GIScience are intimately intertwined with those of computational, environmental, and social sciences, and furthermore, GIScience should contribute substantially through understanding of space and the use of space to achieve the vision of an IEWAT world.

Recent developments in GIScience have built strong foundations in all the three areas of spatial abstraction, algorithms, and assimilation as discussed in Section 3. The pathways forward for an IEWAT world would extend the three areas into a multiverse of a truly diverse, distributed, and collaborative nature. Every location, every person, and everything is becoming a data producer. Data are from everywhere and anytime with different ontological notions. Algorithms are being developed, coded, modified, and forked by many over the web. Information is being analyzed and synthesized dynamically and continuously to reflect real-time and near-real time situations in the environment and our society. Online and offline computational platforms are being transitioned seamlessly to maximize the efficiency of mobile computing anywhere and anytime. Fully integrative cyber-physical-social systems inform us of the past, present, and future of what things/people are, where they reside, how they work, how they may evolve, where we should go, and what we should do.

To date, a GIS is no longer confined to a computer system or as a software package. GIS is immersed in the greater web computing environment and heading to an IEWAT world of truly ubiquitous spatial computing. Ontological and cognitive understandings of geospatial categorization and reasoning are essential to properly conceptualize geospatial problems and realize geospatial abstractions to connect reality and GIS databases. Spatial programming, web programming, and statistical programming are essential skills to analyze data and develop geospatial solutions. Spatial analysis, spatial data mining, mapping, geovisualization, visual analytics, data science, and GeoAI remain critical to geospatial data exploration, information understanding and knowledge discovery. Moreover, the pathway that will revolutionize GIScience is heading in the direction in which the common mode of GIScience practices is not confined to conventional research groups but involves scientists, practitioners, and citizens in a collaborative social cloud environment. It will be an IEWAT world of the people, by the people, and for the people.

Acknowledgment: Revisions were made through work supported by (while serving at) the National Science Foundation. Any opinion, findings, and conclusions or recommendations expressed in this material are those of the author(s) and do not necessarily reflect the views of the National Science Foundation.

NOTES

1 Starting 2024, the AutoCarto International Symposium becomes the International Symposium of Cartography and Geographic Information Science (CaGIS).
2 http://earthquake.usgs.gov/earthquakes/dyfi
3 http://wiki.openstreetmap.org/wiki/API_v0.6
4 http://freegis.org
5 http://github.com
6 One may use R as a GIS (http://pakillo.github.io/R-GIS-tutorial/; accessed March 22, 2014).

REFERENCES

Adelfio, M. D. and H. Samet (2013). GeoWhiz: Toponym resolution using common categories. *Proceedings of the 21st ACM SIGSPATIAL International Conference on Advances in Geographic Information Systems*. ACM.

Agarwal, P. (2005). Ontological considerations in GIScience. *International Journal of Geographical Information Science* 19(5): 501–536.

Anselin, L. (1995). Local indicators of spatial association—LISA. *Geographical Analysis* 27(2): 93–115.

Anselin, L. (2016). Spatial data science. In *International Encyclopedia of Geography: People, the Earth, Environment and Technology*. Taylor & Francis, pp. 1–6.

Anselin, L., I. Syabri and Y. Kho (2006). GeoDa: An introduction to spatial data analysis. *Geographical Analysis* 38(1): 5–22.

Baddeley, A. and R. Turner (2005). Spatstat: An R package for analyzing spatial point patterns. *Journal of Statistical Software* 12(6): 1–42.
Batty, M. (2007). *Cities and Complexity: Understanding Cities with Cellular Automata, Agent-Based Models, and Fractals*. The MIT Press.
Batty, M. (2012). Building a science of cities. *Cities* 29: S9–S16.
Batty, M. (2013). *The New Science of Cities*. MIT Press.
Batty, M., A. Hudson-Smith, R. Milton and A. Crooks (2010). Map mashups, Web 2.0 and the GIS revolution. *Annals of GIS* 16(1): 1–13.
Bhat, M. A., R. M. Shah and B. Ahmad (2011). Cloud computing: A solution to geographical information systems (GIS). *International Journal on Computer Science & Engineering* 3(2): 594–600.
Bivand, R. S. (2000). Using the R statistical data analysis language on GRASS 5.0 GIS database files. *Computers & Geosciences* 26(9): 1043–1052.
Bivand, R. S., E. Pebesma and V. Gómez-Rubio (2013). *Applied Spatial Data Analysis with R*. Springer.
Bodenhamer, D. J. (2013). *Beyond GIS: Geospatial Technologies and the Future of History. History and GIS*. Springer, pp. 1–13.
Brunsdon, C., S. Fotheringham and M. Charlton (1998). Geographically weighted regression. *Journal of the Royal Statistical Society: Series D (The Statistician)* 47(3): 431–443.
Bunting, P., D. Clewley, R. M. Lucas and S. Gillingham (2014). The remote sensing and GIS software library (RSGISLib). *Computers & Geosciences* 62: 216–226.
Buttenfield, B. P. and R. B. McMaster (1991). *Map Generalization: Making Rules for Knowledge Representation*. Longman Scientific & Technical.
Camara, A. S. and J. Raper (1999). *Spatial Multimedia and Virtual Reality*. CRC Press.
Caron, C., S. Roche, D. Goyer and A. Jaton (2008). GIScience journals ranking and evaluation: An international delphi study. *Transactions in GIS* 12(3): 293–321.
Çöltekin, A., S. I. Fabrikant and M. Lacayo (2010). Exploring the efficiency of users' visual analytics strategies based on sequence analysis of eye movement recordings. *International Journal of Geographical Information Science* 24(10): 1559–1575.
Cope, M. and S. Elwood (2009). *Qualitative GIS: A Mixed Methods Approach*. Sage.
Coppock, J. T. and D. W. Rhind (1991). The history of GIS. *Geographical Information Systems: Principles and Applications* 1(1): 21–43.
Couclelis, H. (1997). From cellular automata to urban models: New principles for model development and implementation. *Environment and Planning B* 24: 165–174.
Couclelis, H. (2010). Ontologies of geographic information. *International Journal of Geographical Information Science* 24(12): 1785–1809.
Croitoru, A., A. Crooks, J. Radzikowski and A. Stefanidis (2013). Geosocial gauge: A system prototype for knowledge discovery from social media. *International Journal of Geographical Information Science* 27(12): 2483–2508.
Demšar, U., J. A. Long, F. Benitez-Paez, V. Brum Bastos, S. Marion, G. Martin, S. Sekulić, K. Smolak, B. Zein and K. Siła-Nowicka (2021). Establishing the integrated science of movement: Bringing together concepts and methods from animal and human movement analysis. *International Journal of Geographical Information Science* 35(7): 1273–1308.
Deng, Tianhu, Keren Zhang and Zuo-Jun Max Shen (2021). A systematic review of a digital twin city: A new pattern of urban governance toward smart cities. *Journal of Management Science and Engineering* 6(2): 125–134.
Douglas, D. H. and T. K. Peucker (1973). Algorithms for the reduction of the number of points required to represent a digitized line or its caricature. *Cartographica: The International Journal for Geographic Information and Geovisualization* 10(2): 112–122.
Fan, X., S. Wu, Y. Ren and F. Deng (2013). An approach to providing cloud GIS services based on scalable cluster. *Geoinformatics (GEOINFORMATICS), 2013 21st International Conference on*. IEEE.
Fang, F., Y. Yu, S. Li, Z. Zuo, Y. Liu, B. Wan and Z. Luo (2021). Synthesizing location semantics from street view images to improve urban land-use classification. *International Journal of Geographical Information Science* 35(9): 1802–1825.
Fisher, P. F. (2006). *Introduction Twenty Years of IJGIS*. Classics from IJGIS, CRC Press, pp. 1–6.
Fonseca, F., M. Egenhofer, C. Davis and G. Câmara (2002). Semantic granularity in ontology-driven geographic information systems. *Annals of Mathematics and Artificial Intelligence* 36(1–2): 121–151.

Frank, A. U. (2001). Tiers of ontology and consistency constraints in geographical information systems. *International Journal of Geographical Information Science* 15(7): 667–678.
Gao, S. (2021). *Geospatial Artificial Intelligence (GeoAI)*. Oxford University Press.
Goodchild, M. F. (1992). Geographical information science. *International Journal of Geographical Information Systems* 6(1): 31–45.
Goodchild, M. F. (2007). Citizens as sensors: The world of volunteered geography. *GeoJournal* 69(4): 211–221.
Goodchild, M. F. (2014). Twenty years of progress: GIScience in 2010. *Journal of Spatial Information Science* 1: 3–20.
Goodchild, M. F. and J. A. Glennon (2010). Crowdsourcing geographic information for disaster response: A research frontier. *International Journal of Digital Earth* 3(3): 231–241.
Goodchild, M. F., R. Haining and S. Wise (1992). Integrating GIS and spatial data analysis: Problems and possibilities. *International Journal of Geographical Information Systems* 6(5): 407–423.
Griffith, D. A. and Y. Chun (2019). Implementing Moran eigenvector spatial filtering for massively large georeferenced datasets. *International Journal of Geographical Information Science* 33(9): 1703–1717.
Gu, Y., M. J. Kraak, Y. Engelhardt and F. B. Mocnik (2023). A classification scheme for static origin–destination data visualizations. *International Journal of Geographical Information Science*: 1–28.
Hazas, M., J. Scott and J. Krumm (2004). Location-aware computing comes of age. *Computer* 37(2): 95–97.
Heipke, C. (2010). Crowdsourcing geospatial data. *ISPRS Journal of Photogrammetry and Remote Sensing* 65(6): 550–557.
Ihaka, R. and R. Gentleman (1996). R: A language for data analysis and graphics. *Journal of Computational and Graphical Statistics* 5(3): 299–314.
Janowicz, K., S. Gao, G. McKenzie, Y. Hu and B. Bhaduri (2020). GeoAI: Spatially explicit artificial intelligence techniques for geographic knowledge discovery and beyond. *International Journal of Geographical Information Science* 34(4): 625–636.
Jiang, B. and Y. Miao (2014). The evolution of natural cities from the perspective of location-based social media. *arXiv preprint arXiv:1401.6756*.
Jung, C.-T., C.-H. Sun and M. Yuan (2013). An ontology-enabled framework for a geospatial problem-solving environment. *Computers, Environment and Urban Systems* 38: 45–57.
Karssenberg, D., K. de Jong and J. Van Der Kwast (2007). Modelling landscape dynamics with Python. *International Journal of Geographical Information Science* 21(5): 483–495.
Kwan, M. P. (2009). From place-based to people-based exposure measures. *Social Science & Medicine* 69(9): 1311–1313.
Kwan, M. P. (2012). The uncertain geographic context problem. *Annals of the Association of American Geographers* 102(5): 958–968.
Kwan, M.-P. and G. Ding (2008). Geo-narrative: Extending geographic information systems for narrative analysis in qualitative and mixed-method research. *The Professional Geographer* 60(4): 443–465.
Laurent, T., A. Ruiz-Gazen and C. Thomas-Agnan (2009). GeoXp: An R package for exploratory spatial data analysis. *TSE Working Paper 9*.
Li, Z., J. Han, M. Ji, L.-A. Tang, Y. Yu, B. Ding, J.-G. Lee and R. Kays (2011). MoveMine: Mining moving object data for discovery of animal movement patterns. *ACM Transactions on Intelligent Systems and Technology* 2(4): 1–32.
Lin, H., M. Chen and G. Lu (2013). Virtual geographic environment: A workspace for computer-aided geographic experiments. *Annals of the Association of American Geographers* 103(3): 465–482.
Llaves, A. and W. Kuhn (2014). An event abstraction layer for the integration of geosensor data. *International Journal of Geographical Information Science* (ahead-of-print): 1–22.
Long, J. A. and T. A. Nelson (2012). A review of quantitative methods for movement data. *International Journal of Geographical Information Science*: 1–27.
Ma, C., Y. Qi, Y. Chen, Y. Han and G. Chen (2008). VR-GIS: An integrated platform of VR navigation and GIS analysis for city/region simulation. *Proceedings of The 7th ACM SIGGRAPH International Conference on Virtual-Reality Continuum and Its Applications in Industry*. ACM.
MacEachren, A. M. (1998). Cartography, GIS and the world wide web. *Progress in Human Geography* 22: 575–585.
Mai, G., K. Janowicz, Y. Hu, S. Gao, B. Yan, R. Zhu, L. Cai and N. Lao (2022). A review of location encoding for GeoAI: Methods and applications. *International Journal of Geographical Information Science* 36(4): 639–673.

Mark, D. M. (2003). Geographic information science: Defining the field. *Foundations of Geographic Information Science*: 3–18.
Mark, D. M., A. G. Turk, N. Burenhult and D. Stea (2011). *Landscape in Language: Transdisciplinary Perspectives*. John Benjamins Publishing.
Marks, J. and S. M. Shieber (1991). *The Computational Complexity of Cartographic Label Placement*. Citeseer.
Montello, D. R. (2009). Cognitive research in GIScience: Recent achievements and future prospects. *Geography Compass* 3(5): 1824–1840.
Neteler, M. and H. Mitasova (2008). *Open Source Software and GIS*. Springer.
Olivera, F. and D. Maidment (1999). Geographic Information Systems (GIS)-based spatially distributed model for runoff routing. *Water Resources Research* 35(4): 1155–1164.
Osterman, A. (2012). Implementation of the r. cuda. los module in the open source GRASS GIS by using parallel computation on the nvidia cuda graphic cards. *Elektrotehniski Vestnik* 79(1–2): 19–24.
O Sullivan, D. (2006). Geographical information science: Critical GIS. *Progress in Human Geography* 30(6): 783.
Pebesma, E. (2012). Spacetime: Spatio-temporal data in r. *Journal of Statistical Software* 51(7): 1–30.
Qiao, Z. and X. Yuan (2021). Urban land-use analysis using proximate sensing imagery: A survey. *International Journal of Geographical Information Science* 35(11): 2129–2148.
Raubal, M., G. Jacquez, J. Wilson and W. Kuhn (2014). Synthesizing population, health, and place. *Journal of Spatial Information Science* 7: 103–108.
Rey, S. J. and L. Anselin (2010). PySAL: A Python library of spatial analytical methods. *Handbook of Applied Spatial Analysis*. Springer, pp. 175–193.
Rey, S. J., L. Anselin, R. Pahle, X. Kang and P. Stephens (2013). Parallel optimal choropleth map classification in PySAL. *International Journal of Geographical Information Science* 27(5): 1023–1039.
Rosenberg, M. S. and C. D. Anderson (2011). PASSaGE: Pattern analysis, spatial statistics and geographic exegesis. Version 2. *Methods in Ecology and Evolution* 2(3): 229–232.
Samet, H., M. D. Adelfio, B. C. Fruin, M. D. Lieberman and J. Sankaranarayanan (2013). PhotoStand: A map query interface for a database of news photos. *Proceedings of the VLDB Endowment* 6(12): 1350–1353.
Samet, H., W. G. Aref, C.-T. Lu and M. Schneider (2008). Proposal to ACM for the establishment of SIGSPATIAL, ACM-SIGSPATIAL.
Schmitt, A. K., M. Danišík, E. Aydar, E. Şen, İ. Ulusoy and O. M. Lovera (2014). Identifying the volcanic eruption depicted in a neolithic painting at Çatalhöyük, Central Anatolia, Turkey. *PLoS ONE* 9(1): e84711.
Schuurman, N. (2006). Formalization matters: Critical GIS and ontology research. *Annals of the Association of American Geographers* 96(4): 726–739.
See, L., A. Comber, C. Salk, S. Fritz, M. van der Velde, C. Perger, C. Schill, I. McCallum, F. Kraxner and M. Obersteiner (2013). Comparing the quality of crowdsourced data contributed by expert and non-experts. *PLoS ONE* 8(7): e69958.
Shekhar, S., V. Gunturi, M. R. Evans and K. Yang (2012). Spatial big-data challenges intersecting mobility and cloud computing. *Proceedings of the Eleventh ACM International Workshop on Data Engineering for Wireless and Mobile Access*. ACM.
Shi, X. and M. D. Nellis (2013). Semantic web and service computation in GIScience applications: A perspective and prospective. *Geocarto International* (ahead-of-print): 1–18.
Simons, J. D., M. Yuan, C. Carollo, M. Vega-Cendejas, T. Shirley, M. L. Palomares, P. Roopnarine, L. Gerardo Abarca Arenas, A. Ibanez and J. Holmes (2013). Building a fisheries trophic interaction database for management and modeling research in the Gulf of Mexico large marine ecosystem. *Bulletin of Marine Science* 89(1): 135–160.
Solymosi, N., S. E. Wagner, Á. Maróti-Agóts and A. Allepuz (2010). Maps2WinBUGS: A QGIS plugin to facilitate data processing for Bayesian spatial modeling. *Ecography* 33(6): 1093–1096.
Stefanidis, A., A. Crooks and J. Radzikowski (2013). Harvesting ambient geospatial information from social media feeds. *GeoJournal* 78(2): 319–338.
Steinbach, M. and R. Hemmerling (2012). Accelerating batch processing of spatial raster analysis using GPU. *Computers & Geosciences* 45: 212–220.
Steiniger, S. and E. Bocher (2009). An overview on current free and open source desktop GIS developments. *International Journal of Geographical Information Science* 23(10): 1345–1370.
Steiniger, S. and A. J. Hunter (2013). The 2012 free and open source GIS software map–A guide to facilitate research, development, and adoption. *Computers, Environment and Urban Systems* 39: 136–150.

Sugumaran, R., D. Oryspayev and P. Gray (2011). GPU-based cloud performance for LiDAR data processing. *Proceedings of the 2nd International Conference on Computing for Geospatial Research & Applications*. ACM.

Takeyama, M. and H. Couclelis (1997). Map dynamics: Integrating cellular automata and GIS through Geo-Algebra. *International Journal of Geographical Information Science* 11(1): 73–91.

Tomko, M. and S. Winter (2019). Beyond digital twins–A commentary. *Environment and Planning B: Urban Analytics and City Science* 46(2): 395–399.

Tomlin, C. D. (1994). Map algebra: One perspective. *Landscape and Urban Planning* 30(1): 3–12.

Tomlinson, R. (1962). An introduction to the use of electronic computers in the storage, compilation and assessment of natural and economic data for the evaluation of marginal lands. *The National Land Capability Inventory Seminar. Ontario, Canada, The Agricultural Rehabilitation and Development Administration of the Canada Department of Agriculture*. Ottawa, p. 11.

Tsou, M.-H., J.-A. Yang, D. Lusher, S. Han, B. Spitzberg, J. M. Gawron, D. Gupta and L. An (2013). Mapping social activities and concepts with social media (Twitter) and web search engines (Yahoo and Bing): A case study in 2012 US Presidential Election. *Cartography and Geographic Information Science* 40(4): 337–348.

Turk, A. and D. Stea (2014). David Mark's contribution to ethnophysiography research. *International Journal of Geographical Information Science* (ahead-of-print): 1–18.

UCGIS. (2002). *UCGIS Bylaws*. Retrieved March 6, 2014, from http://ucgis.org/basic-page/laws

Ujang, U. and A. A. Rahman (2013). Temporal three-dimensional ontology for geographical information science (GIS): A review. *Journal of Geographic Information System* 5(3).

Vasardani, M., S. Timpf, S. Winter and M. Tomko (2013). From descriptions to depictions: A conceptual framework. In *Spatial Information Theory*. Springer, pp. 299–319.

Wang, K. and Z. Shen (2011). Artificial societies and GPU-based cloud computing for intelligent transportation management. *IEEE Intelligent Systems* 26(4): 22–28.

Wang, S., L. Anselin, B. Bhaduri, C. Crosby, M. F. Goodchild, Y. Liu and T. L. Nyerges (2013). CyberGIS software: A synthetic review and integration roadmap. *International Journal of Geographical Information Science* 27(11): 2122–2145.

Wellen, C. C. and R. Sieber (2013). Toward an inclusive semantic interoperability: The case of Cree hydrographic features. *International Journal of Geographical Information Science* 27(1): 168–191.

Wiegand, N. (2012). Ontology for the engineering of geospatial systems. In *Geographic Information Science*. Springer, pp. 270–283.

Wiegand, N. and C. García (2007). A task-based ontology approach to automate geospatial data retrieval. *Transactions in GIS* 11(3): 355–376.

Worboys, M. F. (2003). Communicating geographic information in context. In *Foundations of Geographic Information Science*. Routledge, Taylor & Francis Group, pp. 33–45.

Wright, D. J. (2012). Theory and application in a post-GISystems world. *International Journal of Geographical Information Science* 26(12): 2197–2209.

Yan, Z., D. Chakraborty, C. Parent, S. Spaccapietra and K. Aberer (2013). Semantic trajectories: Mobility data computation and annotation. *ACM Transactions on Intelligent Systems and Technology (TIST)* 4(3): 49.

Yang, C., M. Goodchild, Q. Huang, D. Nebert, R. Raskin, Y. Xu, M. Bambacus and D. Fay (2011). Spatial cloud computing: How can the geospatial sciences use and help shape cloud computing? *International Journal of Digital Earth* 4(4): 305–329.

Yuan, M., J. McIntosh and G. De Lozier (2014). GIS as a narrative generation platform. In *Spatial Narrative and Deep Mapping*. Eds. J. C. D. Bodenhamer and T. Harris. Indiana University Purdue University Indianapolis Press.

Yuen, M.-C., I. King and K.-S. Leung (2011). A survey of crowdsourcing systems. *Privacy, Security, Risk and Trust (Passat)–2011 IEEE Third International Conference on Social Computing* (Socialcom). IEEE.

Zhang, J. and S. You (2012). CudaGIS: Report on the design and realization of a massive data parallel GIS on GPUs. *Proceedings of the Third ACM SIGSPATIAL International Workshop on GeoStreaming*. ACM.

Zhou, X., R. Assunção, H. Shao, C. C. Huang, M. Janikas and H. Asefaw (2023). Gradient-based optimization for multi-scale geographically weighted regression. *International Journal of Geographical Information Science*: 1–28.

16 Object-Based Regionalization for Policy-Oriented Partitioning of Space

Stefan Lang, Stefan Kienberger, Michael Hagenlocher, and Lena Pernkopf

LIST OF ACRONYMS

AVHHR	Advanced very high resolution radiometer
CCCI	Cumulative climate change impact
DEM	Digital elevation model
DFO	Dartmouth Flood Observatory
EAC	East African Community
ESP	Estimation of scale parameter
IQR	Interquartile range
LV	Local variance
NIR	Near-infrared
OBA	Object-based analysis
OBIA	Object-based image analysis
RBD	River basin districts
RBC	Recognition-by-components
SD	Spectral distance
UNEP	United Nations Environment Programme
VBD	Vector-borne disease
VGI	Volunteered geographic information
VHI	Vegetation health index

16.1 MAPPING MULTI-DIMENSIONAL PHENOMENA

Multi-dimensional spatial phenomena can be operationalized and mapped by integrating multiple sets of indicators over space. Spatial representation of such complex phenomena, in order to be useful to decision makers in different domains, needs to be efficient and robust. Drawing from the experiences in object-based image analysis (OBIA) and the spatial integration of multidimensional image data, regionalization techniques can be used to map, analyze, and represent these phenomena in a spatial explicit way.

16.1.1 Ambitious Policy Targets Require a Systemic View

The complexity of global challenges faced today entails a holistic and systemic viewpoint, in order to better understand their cause-effect relationships and mutual dependencies. Assessing and monitoring the status of sustainable development policies in a comprehensive and holistic sense relies on integrative, synthesizing techniques (Mirchooli et al., 2023; Sur et al., 2022; Fiksel, 2006; Stahl

DOI: 10.1201/9781003541158-19

et al., 2011). The global warming debate, for example, entails a systemic view on the interrelationships among the environmental sphere and the social sphere (including cultural, institutional, and economic aspects) in integrated assessment models (Parson, 1994; Weyant et al., 1996). These spheres are complex themselves with several sub-spheres to be considered: for example in the ecological sphere one may investigate the behavior of ecological sensitivity to changing climatic conditions; in the socio-cultural sphere one may study social vulnerability to natural hazards. The conceptual operationalization aims at breaking down the complexity of these systems into graspable, and ultimately measurable, compartments. For example, the concept of vulnerability has been operationalized by Kienberger et al. (2009) in several stages of domains and sub-domains, until the level of measurable indicators is reached (see Figure 16.1).

When reaching "down to the bottom," we approach the problem by separate aspects, that is, analytically, likely neglecting the full picture for the detail. Measuring and monitoring every contributing aspect separately is a fundamental requirement; still, more important, but also more challenging, is the strive to capture their cumulative effect (Hagenlocher et al., 2014b). Thus, target-oriented strategic commitments increasingly adopt an integrated assessment approach (Mirchooli et al., 2023; Sur et al., 2022; Ghorbanzadeh et al., 2021; Al Rahhal et al., 2018; Lang, 2018; Bianchini et al., 2017; Yu et al., 2017; Risbey et al., 1996; Jakeman and Letcher, 2003). On implementation level, policies and directives on national, regional, contintental (EU, ECOWAS, ASEAN, etc.), and global level, aim towards holistic concepts and integrated assessments, greatly influenced by systems thinking (Capra and Luisi, 2014). At the same time they strive for pragmatic ways to communicate complexity in a "simplified" manner. Comprehensive yet integrative: these premise assumes that through systemic behavior new qualities emerge, which to capture is both a challenge and a chance. Vulnerability, resilience, mitigation, etc. can be considered as systemic properties that require integrative, synthesizing mapping techniques (Capra and Luisi, 2014). In other words, spatial analysis

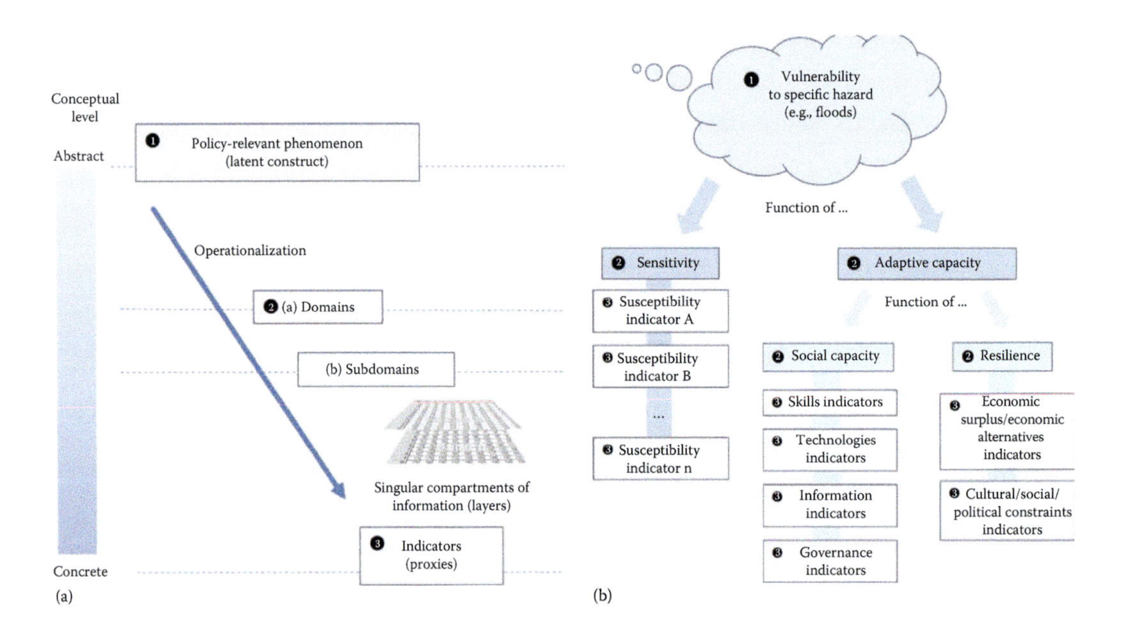

FIGURE 16.1 Operationalization—breaking down multi-dimensional phenomena to measurable indicators: (a) From abstract, conceptual level (1) to measurable indicators. (b) Example of operationalizing "vulnerability to a specific hazard" (e.g., a flood, an earthquake) and its conceptualization via domains, subdomains and sets of indicators.

tools that are capable to assimilate compartments of information and overcome the classical, analytical concept of single geospatial data layers are demanded (Lang, 2018; Bianchini et al., 2017; Yu et al., 2017; Lang et al., 2014b). A suitable integrative geographic technique for this is regionalization, also known as spatial classification (Wise et al., 2001), that is also the methodological core element of the geon approach presented in this chapter.

16.1.2 Approaching Complex Phenomena in a Spatial Focus

Abstract or complex phenomena that cannot be observed directly are termed *latent* variables, or factors (Byrne, 1998). Byrne concludes if "latent variables are not observed directly, [. . .] they cannot be measured directly" (p. 4). A way out of this dilemma is to "operationally define the latent variable of interest in terms of behavior believed to represent it" (Byrne, 1998)(p. 4). The underlying construct is constituted by the direct measurement of the observed (or manifest) variable. Manifest variables are presumed to represent the underlying construct and serve as indicators of it (ibid.). When using such indicators, a conceptual bridge is built between the measurable and the latent part of the phenomenon, a fact that requires further investigation.

In the given context, complex phenomena are for example "vulnerability to hazards," "adaptability to climate change," "landscape sensitivity to human impact," "quality of life," among others. All these cannot be directly measured with a specific measuring device and hardly reduced to a single indicator. A remedy is to approach such complex properties by aggregating manifest variables or indicators into an index or composite indicator, a procedure that works on multidimensional attribute spaces (Marti et al., 2020; Lang, 2018; Bianchini et al., 2017; Yu et al., 2017; Nardo et al., 2005). However, such procedures neglect the true spatial distribution of such properties when using predefinded geographies (such as administrative boundaries) that obscure or bias the actual spatial distribution.

Implementing policies that address such complex phenomena on multiple geographical scales adopts a synthetic view on multi-domain and multi-source geospatial datasets. The reasonable granularity of recent global datasets enables a multi-purpose, effective usage for the public sector in fulfilling public tasks (Wise et al., 2001), which are more and more interdependend. An appropriate unitization of space shall trace the topic-related functional characteristics of the phenomenon addressed, both in terms of scale and thematic discontinuities. In other words, boundaries should appear where they reflect a significant change in behavior of this phenomenon, not anywhere else. An example is the European Water Framework Directive (2000/60/EC), where so-called river basin districts (RBD) transgress national boundaries and represent catchments rather than existing administrative units. Generalizing from this example, we argue that the spatial variability of the actual phenomenon being addressed requires a more flexible spatial unitization within the geographical policy scope.

16.1.3 Geons—Terminology and Conceptual Background

For providing such policy-related spatial units, a methodological framework has been developed (Li et al., 2019; Lang, 2018; Bianchini et al., 2017; Yu et al., 2017; Lang et al., 2014b). Exposed to different policy domains such as regional development planning (Tiede et al., 2010), disaster risk reduction and vulnerability mapping (Kienberger et al., 2009), sensitivity units for strategic environmental assessment (Pernkopf and Lang, 2011), and hotspot analysis for regional climate change adaptation (Hagenlocher et al., 2014b). The framework builds on a workflow to regionalize continuous geospatial data, resulting in a set of geons. The latter are units that synthetically aggregate domain-specific information with a uniform response regarding the complex phenomenon under concern. A geon-set represents the spatial explicit distribution of this aggregated information. Table 16.1 and the Case Studies section contain examples to illustrate the concept. The term "geon" was initially introduced in cognitive psychology by Irving Biederman (1987) in his theory of recognition-by-components (RBC). It is based on the concept of volumetric primitives, defined as geons (geometric ions), and the hypothesis that cognitive objects can be decomposed into basic shapes or components. Geons or gaeons (Peuquet, 2002) in Biederman's view are basic volumetric bodies such as cubes, spheres, cylinders and wedges. Conformity with the original geon concept is discussed by Lang et al. (2008) in regard to (1) the role of generalization for the definition and strength of a geon (though: scale dependent); (2) the significance of the spatial organization of the elements, which leads to emergent properties and specific qualities; (3) the possibility of recovering objects in the presence of occlusions (i.e., data errors, measure failures, lack of data, mismatch of data due to bad referencing). The concept presented here is related, but not identical to Biederman's idea. The term geon has been proposed by Lang (2008) to widen the scope of the original concept and adapt it to the domain of GIScience. Lang et al. (2014b) proposed the following redefinition:

A *geon* (derived from Greek *gē* (Γῆ) = land, Earth and the suffix *-on* = something being) is a type of region, semi-automatically delineated with expert knowledge incorporated, scaled and of uniform response to a phenomenon under space-related policy concern. The aim of generating geons is to map policy-relevant spatial phenomena in an adaptive and expert-validated manner, commensurate to the respective scale of intervention.

The geon approach aims at to delineate regions, but does not relate to administrative regions with predefined, normative boundaries. Instead of the commonly used term "analytical regions" (Duque and Suriñach, 2006), we prefer to use the term "synthetic [*sensu* Sui (2011)] regions" to emphasize the integrative character of geons. Thus geon stands for a systems-theory driven, scale-dependent "Earth-object." Policy relevant units are based on advanced geodata integration, expert knowledge and user validation. Lang et al. (2014b) distinguish between two types of geons, composite and integrated geons. We focus on the latter in the present chapter (see Section 16.2). "Geonalytics" are built upon a comprehensive pool of techniques, tools and methods for (1) geon generation (i.e., transformation of continuous geoinformation into discrete objects by algorithms for interpolation, segmentation, and regionalization); (2) analyzing spatial arrangements and characterize form and spatial organization, investigate spatial emergent properties, and issues of scale; (3) monitoring of modifications and changes and evaluation of the status of geons.

The key issue is spatial regionalization of a complex spatial reality. This reality is represented exhaustively and in a range of interdependent, inherent scales. The strategy of the approach is *per se* geographically motivated. Geographic location is key to an integrative assessment of complex and multidimensional phenomena that have a spatial component. With the maturity of geospatial technology the integrative power of space has been boosted over recent years, also across disciplines, on conceptual, technological and methodological level (Lang et al., 2014a). Integrated spatial analysis methods support the shift from a rather mechanistic and analytical view, to a more systemic one. This includes a change in perspective from a focus on objects to relationships, from (quantitative) measuring to (qualitative) mapping (Capra and Luisi, 2014). Undoubtedly, the ubiquity in spatial data repositories (both public and private, cf. "big data") and a cross-cultural technology adoption (e.g., through smartphones) has recently prepared the ground for "spatializing" (i.e., adding coordinates to) all kinds of societal or physical phenomena. But a gap remains between the factual capacity of spatial analysis and decision support tools and their actual usage. Commonly, socioeconomic indicators are integrated on pre-defined administrative or regular grid cells (Nardo et al., 2005), which underexploits the full potential of geographical synthesis (Lang et al., 2014a). While this is often due to conventional or pragmatic reasons, we do have the means to represent complex phenomena in a more appropriate way.

Just like multi-resolution image segmentation is used to represent tangible real-world objects in multidimensional feature spaces (Chapter 10), latent phenomena can be represented by regionalizing multiple sets of spatialized indicators. "Spatialized" in this context means that indicators are represented as continuous fields in geographic space. The attribute of "continuous" is scale depending though, as it implies a sampling of the indicator using any kind of tessellation into small units, say regular (grid cells, hexagons, etc.) or irregular (TINs, enumeration units, etc). These again may source from data collected on finer level (irregular sensor measurement grids, socioeconomic data collected on household level, etc. and then interpolated over the area). Often the term "spatial" is used for reasons of simplification, but strictly speaking "spatial indicators" measure actual spatial properties (size, distribution, proximity, etc.) of spatial units (Bock et al., 2005) and not just the variability of a phenomenon over space.

Looking at the spatio-temporal variability of the phenomena under concern, we need spatialized indicators presumed to represent the single dimensions of such phenomena in a spatially explicit way. The number of such indicators varies depending on the phenomenon studied (Table 16.1).

TABLE 16.1

Operationalization of Multi-dimensional (Latent, Complex) Phenomena in a Certain Policy Scope and Geographical Scale Using a Set of Spatialized Indicators

Multi-dimensional Phenomenon (Policy Scope)	Geographic Scale (Modeling Scale)	Domains	Dimensionality of Indicator Space	# of Geons (Geon Size)
Vulnerability units/ Climate change context	Salzach river catchment (Austria) [1 km²]	Sensitivity, adaptive capacity (socioeconomic dimension)	52	1462 [36/4.0/3.8]
Vulnerability units/ Disaster risk reduction context	Salzach river catchment [1 km²]	Exposure, susceptibility, lack of resilience (Social dimension)	16	181 [147/32.6/28.6]
		Economic dimension	6	300 [118/19.7/18.5]
		Environmental dimension	21	314 [376/18.8/38.5]
		Physical dimension	22	248 [147/23.8/22.1]
Vulnerability units/ Disaster risk reduction context	District of Búzi, Mozambqiue [1 km²]	Susceptibility, adaptive capacity (Social dimension)	11	307 [108/23.4/18.7]
		Economic dimension	6	225 [127/32.0/23.9]
		Environmental dimension	5	391 [120/18.4/33.9[
		Physical dimension	4	213 [801/33.8/105.9]
Sensitivity units/ Strategic impact assessment context	Marchfeld region, Austria [250m]	Ability of the ecosystem to resist to external disturbances	9	151 [46.3/5.8/6.5]
		Ecological significance of the affected areas	6	121 [55.3/7.2/9.5]
		Societal significance	6	139 [40.8/6.3/6.8]
Climate change hot spots/climate change adaptation context	Sahel and western Africa (sub-continental) [~16 km²]	n/a	4 (temperature, precipitation, drought occurrences and major flood events)	2,283 [645/63.5/42.1]

Note: Dimensionality of indicator space = number of indicators, modeling scale = minimum resolution. The delineated geons are characterized by number (#) as well as size in km² (max/mean/std-dev).

Ecosystem integrity may be grossly approximated by a vegetation index (e.g., the Normalised Difference Vegetation Index, NDVI) relating the red (R) and near-infrared (NIR) bands of a remotely sensed imagery for indicating tree species diversity and tree stand vitality within a deciduous forest. For a comprehensive assessment of a forest ecosystem's integrity, more parameters are required, like groundwater regime, timber usage and other management regimes, climatic conditions, and so on. Recalling the example of societal vulnerability to flood hazards, multi-dimensional problems require conceptual, expert-based models to operationalize them (Kienberger et al., 2013a). For pragmatic and performance reasons, the number of indicators should be handable: a commensurate number of indicators to best represent the phenomenon might be the methodologically soundest way (Moldan and Dahl, 2007).

We try to approximate the phenomenon under concern by a set of proxies to decompose its multi-dimensionality into compartments exhibiting spatial variation and autocorrelation that we are able to map. The actual integration of indicators controls the level of *spatial composite indicators* (Hagenlocher et al., 2014b) or *meta-indicators* (Lang et al., 2008). Lang (2008) uses the term "conditioned information" to underline that this process entails the creation of new geographies as a flexible, yet statistically robust and (user) validated unitization of space.

16.2 DOMAIN-SPECIFIC REGIONALIZATION

Integrated geons (for a definition, see Section 16.2.2) are delineated in a semi-automated way incorporating expert knowledge and adhering to statistical robustness and scale optimization. Thus we are able to map units of uniform response to the phenomenon under concern, commensurate to the scale of policy intervention measures and stable in their aggregation.

16.2.1 Principles of Regionalization

Within spatial science the term "regionalization" implies both a top-down (i.e., disaggregating) and a bottom-up (i.e., aggregating) notion. Disaggregating a larger whole into smaller regions is often associated with a political hierarchy of administrative units, for which the European Statistical Office uses the term normative regions (Sur et al., 2022; Ghorbanzadeh et al., 2021; Csatáriné et al., 2020; Marti et al., 2020; Eurostat, 2006). Scientific regionalization usually follows a more bottom-up strategy applying routines to group neighboring sub-units, that is, small geographical units, pixels, raster cells, etc., into larger ones. While some regionalization routines do imply a (technical) top-down component (e.g., the "split-and-merge" algorithm), most regionalization methods are implemented in a bottom up fashion, performing any kind of spatially constrained aggregation method (Duque and Suriñach, 2006). Such aggregation is done in a way that the resulting analytical regions are "conveniently related to the phenomena under examination" (ibid., p. 2).

The main objectives of regionalization (Berry, 1967) are summarized by Lang (2018); Lang et al. (2014b):

1. To organize, visualize, and synthesize the information contained in multivariate spatial data (Long et al., 2010);
2. To reduce data dimensionality (Ng and Han, 2002) while minimizing information loss (Nardo et al., 2005);
3. To control the effects of outliers or inaccuracies in the data and to facilitate the visualization and interpretation of information by maps;
4. To limit the sensitivity due to data fidelity by aggregating the original units (e.g., pixels) into larger zones (Blaschke and Strobl, 2001; Wise et al., 2001).

Regionalization is based on the principle of spatial auto-correlation, assuming that neighboring areas tend to have similar properties or uniform behavior (Tobler, 1970). Region-building assumes

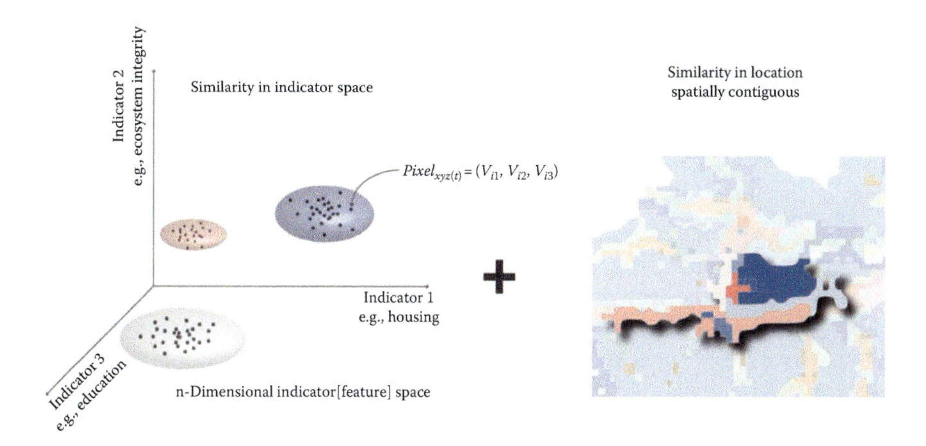

FIGURE 16.2 Regionalization—similarity in attribute and real space. Next to similarity in feature space, the classification of spatial data is controlled by location. Spatially constraint classification is called regionalization.

that such uniform behavior exists as long as transitions occur leading to a different behavior along a certain gradient or boundary. While, due to the principle of spatial auto-correlation, the internal structure of continuous spatial data enables an empirical construction of regions, there is no *a priori* fixed set of regions to be built. As with classification in general, the aggregation of data is to some degree arbitrary (Johnston, 1968), and the areal units to be built by spatial aggregation can be done at different scales and in different (though equally plausible) ways (Wise et al., 2001). Like for statistical data in general, the problem of aggregation effects applies in particular to spatial studies (modifiable areal unit problem, MAUP). As Openshaw (1984) points out there are no standards or international conventions for spatial aggregation and it is subject to the "whims and fancies [. . .] of whoever is doing the aggregation" (p. 3). We will return to this problem and see how we (partly) cope with it.

16.2.2 Integrated Geons

16.2.2.1 From Multispectral Image Segmentation to Multidimensional Regionalization

Experiences in analyzing high-fidelity, multispectral imagery using geographic object-based image analysis, (GE-)OBIA (Ghorbanzadeh et al., 2021; Csatáriné et al., 2020; Al Rahhal et al., 2018; Lang, 2018; Hay and Castilla, 2008; Blaschke, 2010) can be transferred to address complex spatial phenomena. The transfer of OBIA techniques for the analysis of non-image data has been discussed for univariate spatial variables such as a digital elevation model (DEM) by Drăgut and Eisank (2011). Kienberger et al. (2009) used object-based analysis (OBA) to regionalize an n-dimensional indicator space for assessing socioeconomic vulnerability to flood hazards. Non-image data in this context are gridded geospatial data provided by regular sampling of a continuous phenomenon, the interpolation of point samples over space, or based on spatially disaggregated indicators. Principles and conceptual findings from social sciences with technical achievements from OBIA are bridged to generate so-called integrated geons (Lang et al., 2014b). Next to the challenge of incorporating expert knowledge and adhering to statistical robustness as well as scale optimization, an issue remains in assigning nominal categories (labels) to the generated units and characterize their evolution over time.

Just like image segmentation is used to represent tangible real-world objects, complex phenomena can be represented by regionalizing multiple indicators (Lang, 2018; Lang et al., 2014a). The indicators may be mapped as singular layers in a GIS environment and evaluated separately. A common strategy is to approach such complex properties by aggregating variables/indicators towards an index or composite indicator, a procedure that works on multidimensional attribute spaces (Nardo et al., 2005). However, such procedures often neglect the true spatial distribution of such properties when using *a priori* geographies (such as district boundaries).

Our strategy is to extend multivariate clustering from feature space to real space, thereby reaching a flexible partioning of space, composing (new) geographies. Just like multispectral imagery can be clustered into homogenous objects by applying segmentation techniques, we represent complex spatial phenomena by regionalizing multi-dimensional attribute/indicator spaces. This is different to overlaying or intersecting geospatial layers when treating the spatial dimension as a constraint, but flexible in terms of scale. Regionalization techniques that utilize rule-bases with flexible strategies for geodata integration under consideration of and expert knowledge follow the methodological framework of OBIA.

By applying segmentation routines we go one step further toward spatial explicitness (see Figure 16.3). "Classical" aggregation that uses a raster overlay with a local operator performed on regular grid cells or given reference units does not change the level of spatial explicitness, while regionalization does. Here, spatial explicitness is not meant in terms of more spatial detail, but spatial synthesis. Regionalization does not increase the spatial detail, but introduces additional spatial characteristics in the aggregation process, and thereby "spatializes" it.

16.2.2.2 How to Delineate Geons

An initial workflow to delineate geons was developed by Kienberger et al. (2009) in assessing place-based vulnerability in a climate change context. Ever since the approach has been refined and transferred to various application domains (see Case Studies), including, for example, hot spot analysis of cumulative climate change impacts (Marti et al., 2020; Sudmanns et al., 2020; Al Rahhal et al., 2018; Lang, 2018; Bianchini et al., 2017; Yu et al., 2017; Hagenlocher et al., 2014b), landscape sensitivity in the context of environmental impact assessment for infrastructure projects (Pernkopf and Lang, 2011) or vulnerability to vector-borne diseases (Kienberger et al., 2014). Currently available methods to model integrated geons build on four major stages (see Table 16.2). These stages serve as a framework and can be adapted accordingly.

In a **first stage** the identification and definition of associated concepts of the phenomena under investigation is carried out. Subsequently this is followed in a **second stage** by the identification of

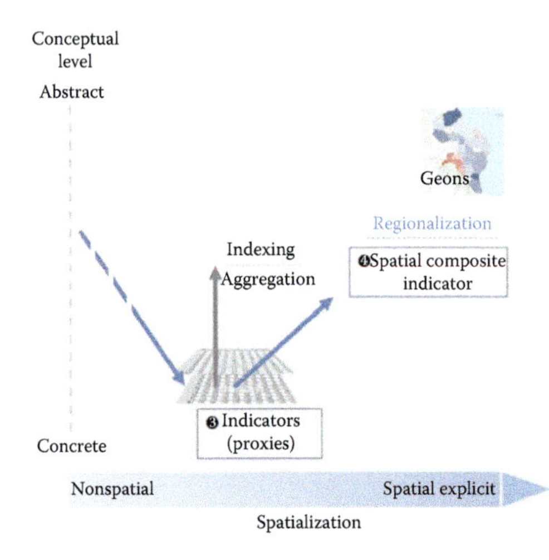

FIGURE 16.3 Regionalization versus indexing. Note that – conceptually – this is a continuation of Figure 16.1A. Indexing, for example, aggregating on a per-cell basis does not generate new spatial information. Instead, regionalization provides additional spatial-explicit information by its generation of (new) spatial units.

TABLE 16.2
Stages of the Geon Workflow and Relevant Aspects to Be Considered

#	Stage	#.n	Key steps involved	Issues to be considered
1	**Definition of conceptual framework**	1.1	Conceptual operationalization	• Soundness • Practicability • Expert opinion • Scientific communities/schools, etc.
		1.2	Domains and dimensions	• Comprehensiveness
		1.3	Identification of indicators	• Literature/expert knowledge • Selection criteria: salience, credibility, legitimacy • Data availability
2	**Data pre-processing**	2.1	Data acquisition	• Scale: global, national, etc • Data availibilty: public domain, authority mandate, commercial • Resolution: admin units, grid
		2.2	Pre-processing	• Resampling to continuous grids • Data transformation • Data imputation and outlier treatment • Normalization
		2.3	Multivariate analysis	• Multicollinearity analysis • PCA
		2.4	Local sensitivity analysis	• Evaluating the influence of indicator choice, normalization method, aggregating method and weighting
3	**Regionalization**	3.1	Indicator weighting	• Equal weights • Expert weighting • Statistical weights
		3.2	Unit delineation	• (multi-)scale assessment and segmentation • Composite index (based on vector magnitude)
4	**Visualization/ geonalytics**	4.1	Mapping	• Legend and intervals
		4.2	Explorative analysis	• Shape and size variation, diversity
		4.3	Monitoring	• For example, using WebGIS solutions

Source: Kienberger et al., 2009, Kienberger/Hagenlocher et al., 2014b.

indicators and corresponding data collection and pre-processing. In the **third** and **core stage**, the multidimensional indicator framework is integrated through regionalization, by different weightings of the indicators, including sensitivity analysis. In a **final** and **fourth stage** the results are appropriately communicated and visualized to the intended user group.

In the following, these four steps are described in detail: latent, multi-dimensional phenomena are concept-driven and need to be defined accordingly. The first stage has a strong link to different theories originating from various disciplines such as sociology, ecology, or geography—to name a few. Such concepts and frameworks may also change over time and within different schools of thought. Therefore it is important to define an "appropriate" framework depending on the context of the study. Associated to the selection of appropriate theories and useful frameworks, a first set of possible indicators emerges. Irrespective of the availability of data one should consider how the different dimensions of the chosen theoretical framework could be characterized and measured. It is likewise important to have an in-depth understanding of the underlying framework that comprises causal relationships and logical dependencies between the single domains and indicators relevant for operationalizing the phenomena.

Once a preliminary indicator set is identified, the second step aims to establish a quantitative, valid and representative indicator framework. This includes the collection and quality assurance of different datasets to populate the indicators, as well as different statistical pre-processing routines. Such pre-processing routines are well known in the composite indicator community (see, e.g., Nardo et al. 2005), and can be applied to derive a statistically sound indicator framework. This includes the transformation of the indicators to continuous grids, the transformation from absolute into relative measures, the identification and treatment of outliers, missing data and multi-collinearities in the data, as well as the application of normalization techniques to render the different indicator values comparable (e.g., min-max normalization, *z*-score standardization, etc.). Alternatively, value function approaches (Beinat, 1997) can be applied, where values are normalized on expert or empirically defined relationships.

Once a final indicator framework is set up, the grid-based indicator data are regionalized to derive homogenous regions of the investigated phenomena. Geons are delineated using capable routines such as the multi-resolution segmentation algorithm of Baatz and Schäpe (2000) as implemented in the eCognition software environment (Trimble Geospatial). This is a region-based, globally optimised approach that merges image segments according to the degree of fitting (Baatz and Schäpe, 2000). It allows for controlling two complementary criteria of similarity of neighboring segments: similarity in "color" and "form." The latter refers to the influence of geometric compactness in the regionalization process. A scale-factor, based on the internal variance of the delineated units, allows for a user-driven control of scaled representations. Here we shortly reflect the algorithm as discussed earlier by Kienberger et al. (2009): The difference between adjacent objects (ibid.) is expressed by the spectral distance (SD) of two pixels or objects p1, p2 in a feature space:

$$SD=\sqrt{\sum_{d=1}^{n} (p_1-p_2)^2} \tag{16.1}$$

or noted as vector difference for a three-dimensional feature space as:

$$SD = (\overrightarrow{v1} - \overrightarrow{v2});$$

$$with\ \overrightarrow{v1} = \begin{pmatrix} d_{11} \\ d_{12} \\ d_{13} \end{pmatrix} \text{ and } \overrightarrow{v2} = \begin{pmatrix} d_{21} \\ d_{22} \\ d_{23} \end{pmatrix} \tag{16.2}$$

To optimize the degree of homogeneity between two neighboring pixels or objects, the specific heterogeneity is minimized at every merge. The current degree of fitting (h_{diff}) is characterized by the change in heterogeneity (Equation 16.3):

$$h_{diff} = h_{\min} - \frac{| SD_l + SD_2}{2} \tag{16.3}$$

By weighting heterogeneity with object size the requirement of producing objects of similar size can be accomplished. Form homogeneity is considered by relating actual boundary length (perimeter) to the perimeter of the most compact form of the same size. This ideal form is a circle, the deviation of which can be expressed by the shape index:

$$SHP = \frac{p}{2\sqrt{\pi * s}} \tag{16.4}$$

where p equals the perimeter and s equals the size of an object.

Identifying an appropriate scale factor has been subject to much debate, as a common approach is to use "trial-and-error." To overcome this subjectivity in object generation, a routine suggests initial

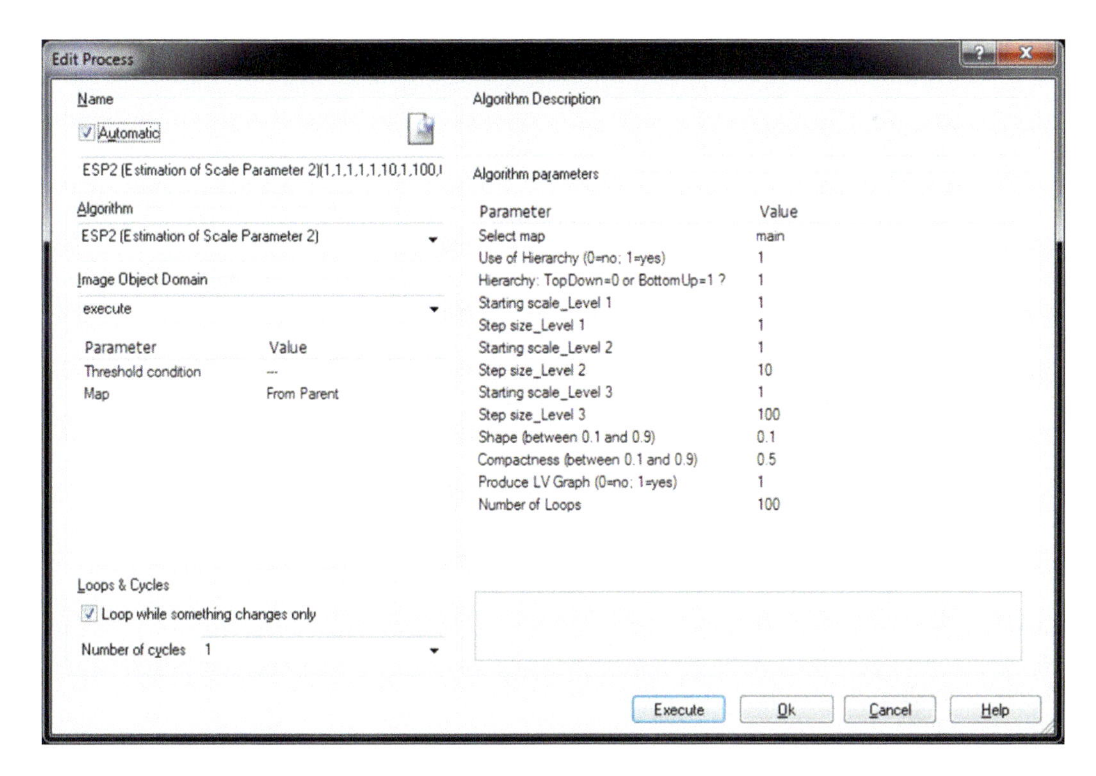

FIGURE 16.4 Graphical user interface of the ESP II tool.

parameters for the multi-resolution segmentation (Drăgut et al., 2014). The so-called Estimation of Scale Parameter (ESP) tool is implemented as a generic tool for eCognition software (Figure 16.4).

The tool utilizes local variance (LV) graphs (Woodcock and Strahler, 1987) to detect scale domains in geospatial data (Dragut et al., 2010). It segments continuous datasets iteratively applying a multi-resolution segmentation algorithm, from smaller to coarser scale in a small, constant increment. The mean LV value is computed for each image level that is created as the ratio between the sum of the LVs for each layer (LV_1–LV_n) and the number of layers (n) used in the image segmentation. When a scale level records an LV value that is equal to or lower than the previous value, the iteration ends, and the objects segmented in the previous level are retained (Drăgut et al., 2014).

According to different indicator weights, the algorithm also considers their relative contribution to the resulting composite value. To assign a quantitative value to each of the identified units or regions, a (weighted) vector magnitude is calculated, which measures the vector distance of each unit within its n-dimensional indicator (feature) space (see Equation 16.1). Alternative aggregation approaches, such as weighted sum or geometric mean, are also available, while the vector magnitude is more compelling in this context. Finally, the values of the resulting spatial composite indicator are normalized to a defined classification scheme (e.g., from zero to one) to ease its interpretation. Additionally for each unit the contribution of the underlying indicators to each integrated geon through different metrics (mean values, standard deviation, etc.) can be assessed.

So far we discussed how geons are generated by integrating a set of indicators and dimensions. All of the modeling stages described earlier introduce uncertainty in the results since there is a range of plausible alternatives at each stage (Hagenlocher et al., 2014a), looking at the impact of the set of indicators, the experts weighting, and statistical features such as the normalization method and the regionalization parameters. For illustrative purposes, we discuss the impact of the final choice of indicators and indicator weights on the modeling outputs in Section 16.3.

We recommend a combination of data-driven (statistical) and normative (expert-based) approaches to generate concept-related *fiat* objects (Chapter 9) of this kind and to deliver a consolidated output of domain-relevance. This especially applies to the choice of indicators and the setting of weights (Decancq and Lugo, 2012). In all cases, the modeling of integrated geons should be complemented with robustness tests to determine to what degree the results are driven by these influences. Different approaches to uncertainty and sensitivity analysis have been proposed by the composite indicators community (Nardo et al., 2005; Saisana et al., 2005; Saltelli et al., 2008). Whereas uncertainty analysis quantifies the overall uncertainty in the output as a result of the uncertainties in the model input, sensitivity analysis evaluates how changes in each individual input parameter affect the final output and how the variation in the output can be apportioned, qualitatively or quantitatively, to different sources of variation in the assumptions. A combination of both practices, uncertainty and sensitivity analysis, provides insights into the robustness of the modeling results. Testing the robustness of geons confronts us with the challenge of changing geometries in combination with changing index values.

The generated units, though considered plausible, do not directly correspond with observable real-world objects. The validity of the spatial extent and the pre-categorical nature of such units can be indirectly assessed by collecting evidence data on the prevalence of the phenomenon's impact. Examples are damage scenarios or loss estimation scenarios in the case of vulnerability to natural hazards or, in the case of disease vulnerability, to assess the spatial distribution of the disease combined with patient data on socioeconomic background and demographic setting.

In a final stage integrated geons are visualized. A grey scale or continuous color scheme is applied for the composite indicator value, in order to be compared to the other regions. To allow an interactive exploration of the different regions, and its underlying indicators, integrated geons are best visualized through interactive, web-based mapping tools, for example, Kienberger et al. (2013b). Instead of an abstract continuous value, geons may be characterized more qualitatively. Riedler and Lang (2018) categorize geons according to the dominating indicators or a specific combination of indicators, while Lang (2018) demonstrates biophysical parameters and qualitative assessments are integrated in the characterization of geons.

16.2.2.3 Monitoring Geons

Once geons are generated in agreement to the policy realm, they can be used to monitor the spatio-temporal dynamics of the considered (latent) phenomenon over time (see Section 16.2.6).

Understanding the status quo of a complex spatial phenomenon includes its observation over time, treating geons as space-time entities (Mirchooli et al., 2023; Li et al., 2019; Gudiyangada et al., 2019; Yu et al., 2017; Peuquet, 2002). While a particular intervention may be drawn from the situation "as is," the comparison to previous stages deepens its specific meaning. The impact of such intervention requires a repetitive analysis of altering conditions. A multi-temporal geon set allows for a re-assessment of intervention measures (Lang et al., 2008). Working in a spatial-explicit object-based environment considers the shapes of single geons on individual level, as well as their spatial arrangement and distribution on collective level. Spatio-temporal dynamics help understand the underlying trend. As for any change analysis, noise needs to be told from real changes. Dealing with aggregated units rather than single cells, helps overcome the problem of data fidelity. Originating from OBIA, the concept called object-fate analysis, OFA (Chapter 9) (Schöpfer et al., 2008; Hernando et al., 2012), can be used to characterize the behavior of geons over time, and their possible transitions to other stages. When monitoring the state of geons over time we may encounter a certain threshold beyond which we would consider it "favorable", where no policy intervention is required. Countermeasures have to be taken above (or below) a certain threshold, which we may call the fair-state range (see Figure 16.5).

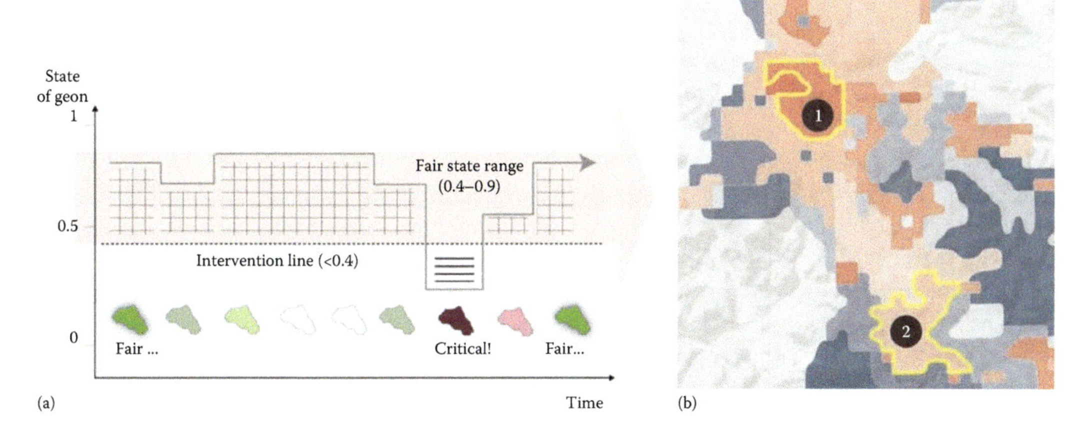

FIGURE 16.5 (a) Monitoring integrated geons supports a scenario of "threshold-based intervention planning." Fair state range: "natural" dynamic of a geon state through time. Green: geon in fair/good state, no action required, red: below fair-state line, action to be considered, dark red/dashed: intervention required. (b) Two geons (representing "vulnerability to floods") in different states below fair state. The dark red tone (1) would indicate required measures. See Figure 16.7 for larger extent and additional context. (From Lang et al., 2014a.)

16.2.2.4 MAUP and Scale Dependency

Until today, many assessments of complex spatial phenomena are carried out using administrative boundaries or continuous grids as the final reporting unit. As policies and interventions are often spatially targeted at the administrative level, results reported on administrative units match the scale of current policy interventions. Policy and decision makers on the ground have a long tradition in using them for analysis, benchmarking, reporting and planning purposes. However, as pointed out by Openshaw (1984), the results and their interpretation have to be treated with caution as they are "biased" by the fact that these units are modifiable, artificial areal units. This holds also true for continuous grids, or pixels, which—like administrative units—do not have an intrinsic geographical meaning (Yu et al., 2017; Hagenlocher et al., 2014a).

The MAUP concept has been developed for addressing the drawbacks of mapping any measurement, model outcome, or statistical value on existing geographies (units, e.g., census or other administrative units). In relation to that, ecological fallacy problematizes the fact that all individuals (or sub-units) in a given spatial unit are treated collectively, how close (spatially or property-wise) they may be to the reported phenomenon (Openshaw, 1984). MAUP can also be linked to the observation that smaller units (e.g., pixels in an image, cells in a raster representation, etc.) are aggregated to larger units quite arbitrarily. While there is an infinite number of possible combinations or groupings, regionalization techniques aim to algorithmically or heuristically group neighboring units according to any kind of similarity criterion. Even though merging neighboring pixels assigned to the same category (class) may depend on the classification routine and/or a given size constraint applied, there is an inherent reason why these units are generated and they bear any kind of homogenous internal behavior, in particular when compared to neighboring units (basic principle of delineation or demarcation). In the case of integrated geons, the mapping of a spatial composite indicator is embedded in a conceptual framework; its methodological realization, provided the basic principles of multivariate data analysis are obeyed, does provide inter-subjectively relevant units. Boundaries are generated where the latent phenomenon under concern changes its behavior. In contrast to administrative units and grids geons provide a representation of the real spatial distribution of the respective phenomenon (Hagenlocher et al., 2014a). Although geons do not necessarily match the scale of current policy interventions they provide a powerful means for planning as their internal heterogeneity is minimized (Lang, 2018; Lang et al., 2014b).

Integrated geons can be delineated at various scale levels, ranging from local to national, continental or global. It is challenging to find a commensurate scale for finding the respective gradient to represent the phenomenon best (Marceau, 1999). Next to the nature of the phenomenon, this is also influenced by the availability of data which are appropriate to be integrated in a regionalization approach, the level of spatial detail that is required, and the scope of the analysis (Hagenlocher et al., 2014b). The three different kinds of scale mentioned by Wu and Li (2006) are likewise important when modeling integrated level geons: the intrinsic scale, the modeling/observational scale and the policy scale. As we postulate that integrated level geons are policy relevant, we have to consider the relevant scale level at which policies occur. For instance if a district government is responsible for disaster risk reduction activities, decision makers require relevant information on (sub-)district level in a meaningful and disaggregated manner. However, this has to be in line with the intrinsic scale level of the certain phenomena investigated (e.g., where flood occurs; vulnerability at a local level is characterized different than one at the global one). The intrinsic scale level may be difficult to identify in an objective manner as it can also rely on our perception, however certain assumptions need to be made. This scale level needs to bridge towards the observational and modeling scale. It is essential that the observational scale (e.g., resolution of raw data) and the final modeling scale match, and are valid to establish the bridge between the intrinsic and final policy scale level. These considerations need to be taken into account, when identifying indicators (do they reflect the intrinsic sale at the district level appropriately?), the associated input data and modeling domain (e.g., resolution of raster-based data), and finally the policy scale level, for which the results will provide decision makers with relevant information. Therefore, consistency among the different kinds of scales needs to be maintained in order to provide valid results, while developers of spatial composite indicators should be aware of the implications of spatial scales for analyzing and monitoring latent phenomena (Hagenlocher et al., 2014b)

16.2.2.5 How to Validate Geons

User validation is meant in the sense of (geographic) object validity (Sur et al., 2022; Li et al., 2019; Lang, 2018; Lang et al., 2010) that comprises both policy relevance and expert-proven functional relevance. From a data model point of view the generated geons are areal (i.e., polygonal) objects. The vector data model suggests crispness of the generated boundaries and soundness of the assigned label. If units are not *a priori*, can they be objective and inter-subjectively acceptable? Johnston, cited in Hancock (1993), stated in 1968 that "all approaches to classification are actually subjective," so the question arises: are such units valid beyond a narrow thematic domain? When using integrated geons, we aggregate values of the underlying indicators and their weighted contribution in a multi-dimensional variable space. At the same time we deal with spatial constraints that check the similarity of neighboring cells. After that we visualize the resulting composite values: depending on the number of intervals (decile, quantile, etc.), the geon-scape will appear differently, while the delineated units as such remain. Particularly, when aiming at deriving hot spots of these complex phenomena, thresholds need to be defined, which indicate whether or not an object will be marked as such a hot spot. Defining these thresholds adds a certain level of subjectivity to the final results, as *a priori* thresholds do not exist and must be defined by either making use of expert knowledge or a needs-driven approach (e.g., specific number of hot spots in an area). A challenge is to assign nominal categories ("labels") to the generated units, which are at first characterized by value ranges of the computed meta-indicator. As long as conceptual links are missing, we can hardly shift from ordinal scaled ranks (low, medium, high) to nominal categories; just as we transform continuous elevation data into landform classes (Drăgut and Eisank, 2011).

We argue that the farther we move from *bona fide* objects, the trickier—and more domain subjective—validation gets. Lang et al. (2014b) characterise integrated geons as concept-related *fiat* entities, which cannot be assessed by established accuracy assessment methods. The term "object validity" (Lang et al., 2010) reflects the limited power of binary assessments in judging thematic accuracy of a given object label. Object validity should ensure a purpose-oriented judgment whether the product meets the users demand.

16.2.3 In Depth: Systemic Areal Units

Systems thinking has been widely referred to in spatial science literature as dealing with scaled representations of continuous data, for example, in multi-scale image analysis (Hay et al., 2001) and in addressing spatial patterns over several scales, such as the "scaling ladder" concept (Wu, 1999) in landscape ecology. Originating in a non-spatial or at least not spatial explicit context, systems theory (Bertalanffy, 1969) deals with the hierarchical organizations of concrete systems, which on each level exhibit systemic, that is, emergent properties (in other words: a "level" is constituted where emergent properties become obvious). Koestler (1967) coined the term holon to underline the nested behavior of systemic units. The geon concept as a holistic regionalization approach intends to address the level of the generated units by abstracting from specific thematic topics or application. The common denominator is to generate units that are not *a priori* (administrative or normative) but "demanded", i.e., representing inherently the spatial dimension of a given policy framework. Geons show uniform response regarding this framework and are expert validated in terms of relevance and usability. They bear integrated spatial information and representatives of systemic areal units.

Terminologically, "geo-on" suggests synonymy with "geo-atom" (Goodchild et al., 2007) or even more radical, the "Elementary_geoParticle" (Voudouris, 2010). But geons do not claim "atomicity" (Masolo and Vieu, 1999) as something undividable. In fact, a geon is not the smallest possible unit in space and/or time as topons or chronons (Couclelis, 2010), but the smallest valid one in a certain context ("as small as necessary, as big as possible") (Lang et al., 2014b). In generating an aggregate that has its own meaning and stability, the composition of a geon will start with, but go beyond, a purpose-driven exercise. In the composite indicator community, it is widely acknowledged that integrating single indicators mirror systemic effects (Nardo et al., 2005). In other words, the variance of a systemic whole is much lower than the sum of the variances of its components, which leads to emergent properties. However, raster cells (with composite values) that can represent an Elementary_geoParticle would not qualify as a geon, due to its technically fixed spatial definition. A geon is a spatial object with systemic stability features such as minimized inner variance and gradients towards the outside through vector encoding. Regionalization implies some loss of information (Hancock, 1993) and suppresss redundant detail that acts as noise in the context under concern.

In congruence with systems thinking (Laszlo, 1972), we argue that spatial systemic units also bear a dual character in terms of self-assertive tendencies (whole-ness) and self-integrative ones (part-ness) (Lang et al., 2004). This nested behavior can be applied to systemic areal units as a geographical correspondence to holons (Wu, 1999). Geons that are composed from spatial elementary units show such systemic properties in certain functional qualities, which are best to be treated as a conceptual whole. Geons are a meaningful product when applying spatial classification of an integrated set of spatialized indicators (Kienberger et al., 2009).

16.3 CASE STUDIES

Here we present several examples how the geon approach was utilized in different application domains including disaster risk reduction, public health, climate change adaptation and environmental impact assessment.

The case studies section (Mirchooli et al., 2023; Sur et al., 2022; Ghorbanzadeh et al., 2021; Csatáriné et al., 2020; Marti et al., 2020; Sudmanns et al., 2020; Li et al., 2019; Gudiyangada et al., 2019; Al Rahhal et al., 2018; Lang, 2018; Bianchini et al., 2017; Yu et al., 2017) will use practical examples to illustrate the following summary statements:

- **Conceptual frameworks** build the basis for operationalizing the multi-dimensional phenomenon addressed; they guide the assessment and foster the reproducibility of results as well as the transferability of the approach to different scales or regions.
- Several plausible methods exist for constructing **spatial composite indicators** (e.g., normalization, weighting, aggregation); important to make the approach transparent and to conduct a sensitivity analysis; for geons this is a field under development.
- In some stages, **expert judgment** has to be made (selection of indicators, weighting of indicators, regionalization parameter etc.), which might affect the result.
- **Transferability** across geographic/policy-scales (local to regional) as well as across domains (disaster risk reduction, climate change adaptation, public health, strategic planning) is a key requirement, see Figure 16.6.
- The geon approach works with multi-domain, multi-source datasets and independent from the number of indicators; however, indicators and indicator weights must be chosen carefully to best reflect the **idiosyncrasies** of the phenomenon that is addressed.
- **Validation** of latent phenomena is tricky (how to validate the "immeasurable"?), because we hardly find *a-priori* in-field observations to calibrate the model.

16.3.1 Socioeconomic Vulnerability to Hazards

16.3.1.1 Multi-Dimensional Phenomenon Addressed

Currently much emphasis is put on conceptualizing and mapping vulnerabilities in the context of *disaster risk reduction* and *climate change adaptation*. Here we illustrate an example of how to

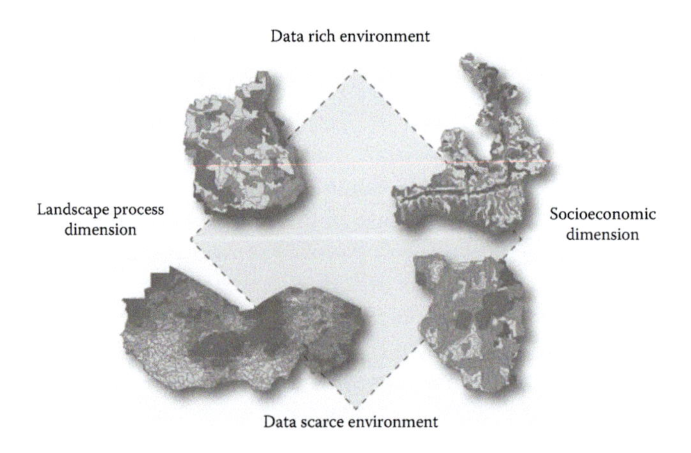

FIGURE 16.6 Geon-scapes—exemplary illustration considering different policy domains and scales, dimensions and data conditions.

perform an integrated, holistic assessment of social, economic, and environmental vulnerability to flood hazards. The study has been performed in the Salzach River Basin (Austria) by and discussed in greater detail by (Sur et al., 2022; Ghorbanzadeh et al., 2021; Al Rahhal et al., 2018; Lang, 2018; Kienberger et al., 2014) in the context of flood hazards at the catchment scale level for the social, economic and environmental domains. The Salzach catchment is characterized through its alpine upstream area and highly dynamic river valleys with the city of Salzburg in the downstream area. Recent severe floods have occurred in 2002, 2005, and 2013 affecting large parts of the river catchment. The majority of the population lives in the city of Salzburg (approx. 150,000 inhabitants) and its urban fringe and connected urban centres where the surplus of population settles. The area is highly attractive for tourism and dynamic in regard to its economic development, putting pressure on adequate spatial planning policies in an area with a topographically limited amount of suitable land.

To achieve the objective to protect people against floods, Austria has been implementing flood hazard mappings, improvement of river regulations, and technical flood protection measures along with sophisticated early warning systems. Despite of these measures, floods are not prevented per se. They allow for enough time to prepare, warn, and evacuate people and minimize economic damage and loss of human lives based on specific flood protection activities.

16.3.1.2 Indicators and Datasets

The study aimed to model homogenous vulnerability regions for the social economic and environmental dimension. As such, homogenous regions in terms of their degree as well as their inherent characteristic of vulnerability should facilitate the improved identification of place-specific intervention measures for disaster risk reduction.

An expert-based approach was selected due the fact that vulnerability is not directly measurable in its complex dimension and social construction. In order to model the spatial distribution of a complex phenomenon, established methodologies such as multicriteria decision analysis, Delphi exercises, and regionalization approaches to derive homogenous vulnerability regions were integrated. At the initial stage of the workflow, the concept of vulnerability is addressed, which in this case builds on the adapted MOVE risk and vulnerability framework (Birkmann et al., 2013). An essential step is the identification of indicators and available datasets, the weighting of the different indicators, and its aggregation to sub-indices. As the final list of appropriate indicators identified from expert choice, literature and community perspective, five key experts were asked to weight the indicators through budget allocation methods. Data availability has been an advantage in the Salzach catchment, as important key datasets are available in the spatial data infrastructure established at the government of Salzburg, as well as the availability of detailed raster-based census data for Austria, which specifically helped to apply the regionalization approach. The data used range from infrastructure data to different socioeconomic parameters such as the size of companies, means of subsistence, age, and workforce in economy sectors, origin and education level of the population. They originated from the census survey in 2001 and are not only provided on the basis of different administrative units but also for a standardized grid with different available spatial resolutions. To allow comparability, data were normalized through linear min–max normalization within an 8-bit scale range (values with 0–255). The modeling scale level was based on the standardized grid with a regular cell size of 1000 m. In the next step, the modeling of homogenous vulnerability regions for each dimension (as integrated geons) was implemented through the application of multiresolution segmentation in combination with a smoothing algorithm (polynomial approximation). Therefore, regions of vulnerability have been delineated that share a commonality in regard to their underlying indicator values as well as a spatial constraint. A vulnerability index was calculated through a weighted vector magnitude in the multidimensional indicator space. Final index values were then normalized within a scale range of 0–1, and visualized by applying a centile classification with exponential kernel algorithm.

16.3.1.3 Results

The results allow the identification of spatially-explicit regions with different levels of vulnerability ("hot spots") independent from administrative boundaries. Furthermore, the results provide decision makers with place-specific options for targeting disaster risk reduction interventions that aim to reduce vulnerability and ultimately the risk of impacts from floods. The most vulnerable regions in the three dimensions are located along the Salzach river and its tributaries. However, the most vulnerable region in all the dimensions is the city of Salzburg and its surroundings. These results are due to the density of the built up area, a big concentration of historic buildings and wide-spread infrastructure. Therefore, the highest vulnerability degree in regard to the social dimension is concentrated in the city of Salzburg as one of the largest settlements located along the Salzach river. The economic dimension has its hot spot also in this urban region due to the presence of employment sources, and the city being an important node in the transport network in the country. The environmental dimension shows the highest degree of vulnerability in the stretch of the Salzach river from the city to the north, associated to river fragmentation.

16.3.2 Social Vulnerability to Malaria

16.3.2.1 Multi-Dimensional Phenomenon Addressed

Despite the global decline in malaria cases over the past decades, malaria remains the most prevalent vector-borne disease (VBD). Caused by the bite of an infected Anopheles mosquito, malaria resulted in approximately 207 million infections and has causes approximately 627,000 malaria-attributable deaths in 2012 (WHO, 2013), primarily among African children. Thus, in addition to modeling transmission potentials, it is of utmost importance for the planning of (preventive) interventions to assess prevailing levels of malaria vulnerability in a spatially explicit manner (Li et al., 2019; Gudiyangada et al., 2019; Yu et al., 2017; Hagenlocher et al., 2014b). Based on an adapted

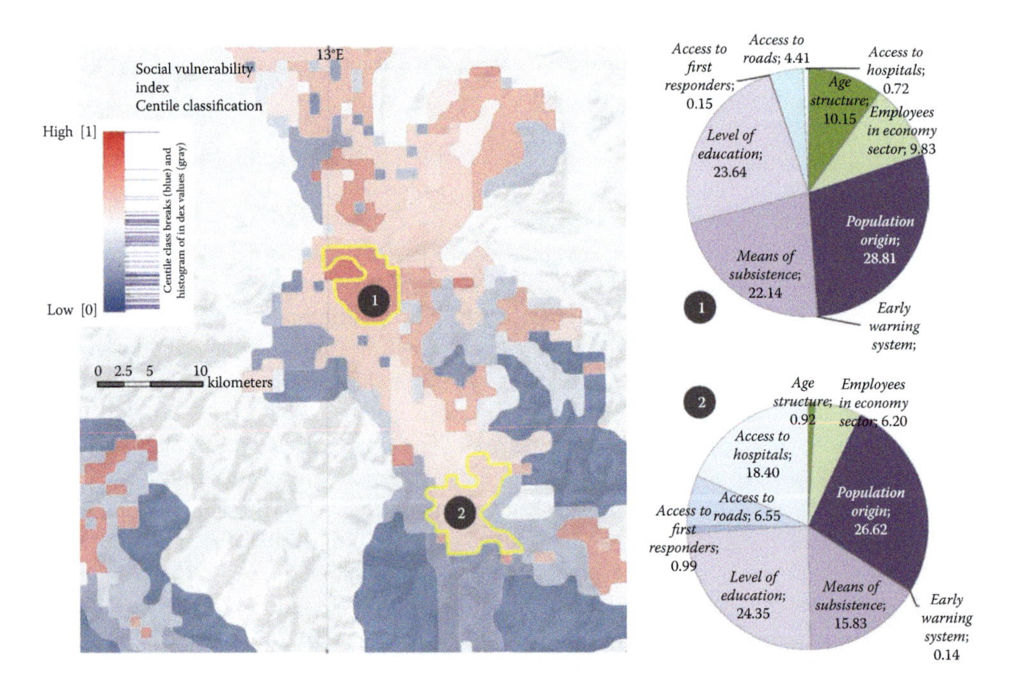

FIGURE 16.7 Geonscape showing social vulnerability to flood hazards in the Salzach river catchment. Geons score from low to high in range between 0 and 1. See legend. Single geons can be explored in terms of the contributing indicators and their respective contribution to the overall score (see pie charts 1 and 2).

version of the MOVE risk and vulnerability framework, relative levels of social vulnerability to malaria were modeled for the East African Community (EAC) region using the geon approach (Kienberger et al., 2014).

16.3.2.2 Indicators and Datasets

Based on the outcomes of a systematic review of literature, the consultation of domain experts, and data availability, (1) a preliminary set of 15 biological and disease-related (e.g., immunity, age, pregnancy, etc.), (2) socioeconomic (e.g., socioeconomic status, poverty, nutritional status, education, etc.), as well as (3) accessibility-related indicators (e.g., access to health services, etc.) were identified. Data for these indicators was acquired from multiple sources, including remote sensing data (e.g., Land Use/Land Cover information, etc.), survey data (e.g., Demographic and Health Survey data), and volunteered geographic information (VGI) as provided by OpenStreetMap. After statistical pre-processing, as described in Section 16.2, one of the 15 indicators was omitted to reduce existing multicollinearities in the data.

16.3.2.3 Results

Figure 16.8 shows the spatial distribution of social vulnerability to malaria for the EAC region. In the map areas of high vulnerability are displayed in red, while areas of low vulnerability are displayed in blue, indicating high levels of malaria vulnerability in the highland areas where immunity of the population is currently low.

A sensitivity analysis was carried out to assess the impact of (1) the individual indicators on the vulnerability index and (2) the choice of the weighting scheme (expert weights vs. equal weights) on the size and shape of the geons as well as on the final vulnerability index for each geon. The influence of the single indicators on the vulnerability index was evaluated by discarding one indicator at a time, while keeping all other settings, including the geometry of the geons, constant. Figure 16.9 shows the influence of the vulnerability indicators on the vulnerability index. The vulnerability index revealed a high overall robustness in regards to the final choice of vulnerability indicators,

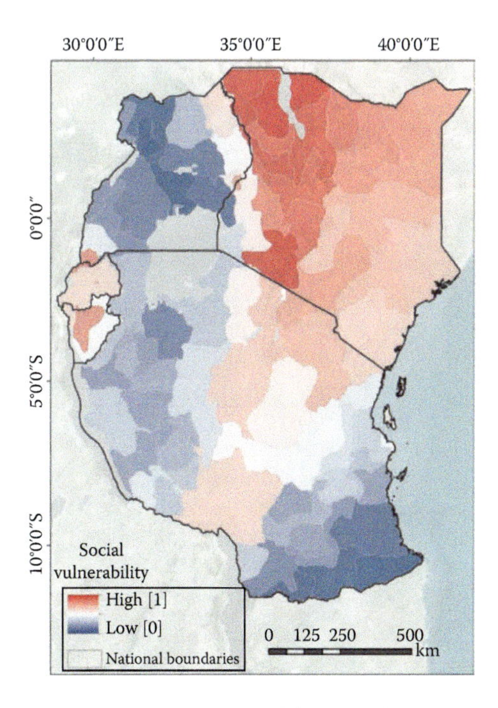

FIGURE 16.8 Social vulnerability to malaria in East Africa. Pie-charts show the contribution of the single vulnerability indicators for each geon (Source: Kienberger & Hagenlocher, 2014.)

with the exception of the Immunity indicator which has a marked impact on the composite vulnerability index (Kienberger & Hagenlocher, 2014).

Figure 16.10 compares the modeling outputs based on expert weights (Figure 16.10, panel 1) and equal weights (Figure 16.10, panel 2), revealing several interesting findings regarding the geometry of the delineated integrated geons and the vulnerability index. First, the relevance of

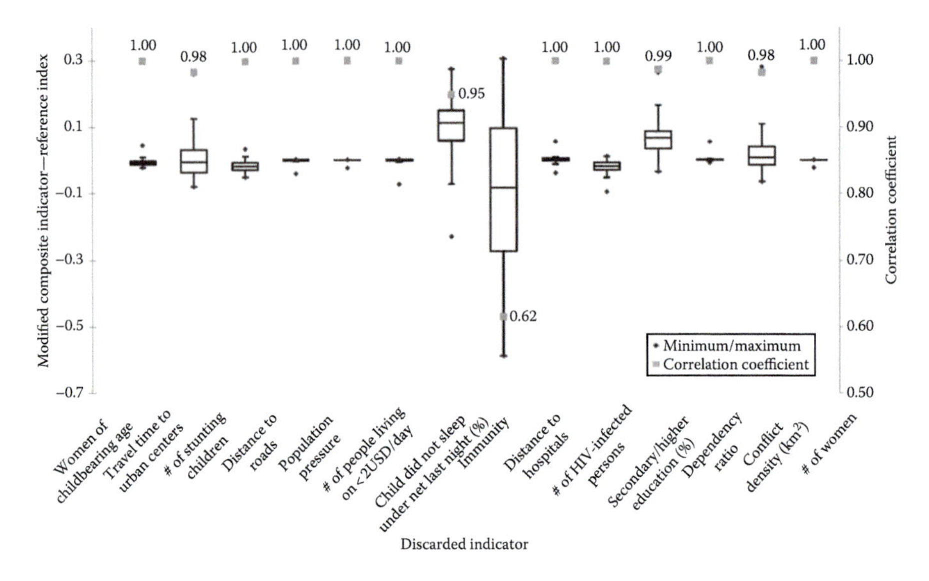

FIGURE 16.9 Influence of the single vulnerability indicators on the vulnerability index. The higher the interquartile range (IQR), the higher the impact of the respective indicator on the final index. (Source: Kienberger & Hagenlocher, 2014.)

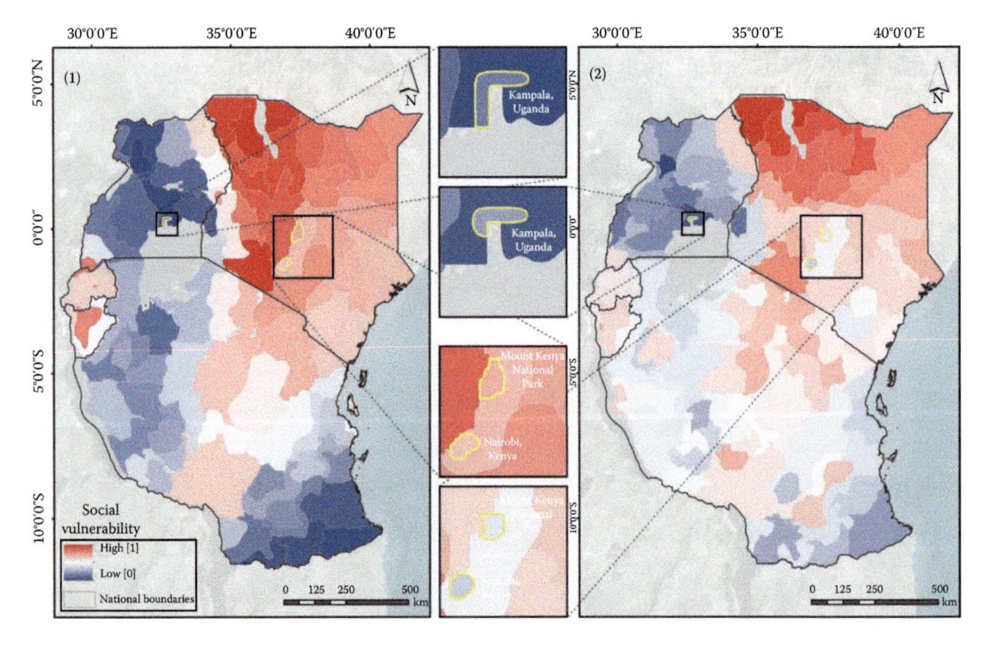

FIGURE 16.10 Evaluating the impact of the choice of the weighting scheme (panel 1: expert weights; panel 2: equal weights) on the modeling outputs.

the Immunity indicator, which was ranked one of the most important indicators by the expert, is clearly visible when comparing both maps. The expert-based map clearly shows that vulnerability is high in highland areas where immunity to malaria is generally low. As the impact of the Immunity indicator is lowered when assigning equal weights to all indicators, the marked impact of immunity on the vulnerability index, and thus the clear distinction between high levels of vulnerability in the highlands and low levels in the lowlands becomes less distinct when using equal weights. Second, it becomes obvious that the geometry of geons (size and shape) is clearly influenced by the weights assigned to the individual indicators. At the same time both approaches clearly demarcate urban centers which differ from their vicinity in terms of both socioeconomic and demographic characteristics, such as Kampala (Uganda) or Nairobi (Kenya).

16.3.3 Landscape Sensitivity to Road Construction

16.3.3.1 Multi-Dimensional Phenomenon Addressed

Major infrastructure plans such as the construction of a new road require an integrative assessment of their likely impacts on the environment. In addition to environmental factors, also social and economic should be considered to minimize potentially negative effects (Sur et al., 2022; Ghorbanzadeh et al., 2021; Csatáriné et al., 2020; Al Rahhal et al., 2018; Lang, 2018). The degree of sensitivity of a particular landscape determines how it is able to accommodate developments. Landscape sensitivity varies both spatially and temporally, and can only be gauged through use of indicators. Pernkopf and Lang (2011) have presented a spatial composite indicator that combines the different aspects of sensitivity in an integrative approach instead of taking them separately, which is standard practice in environmental impact assessment. The approach was tested in the Marchfeld region, an area of 1,000 km² in northeastern Austria, where a highway connection was planned between two major cities (Figure 16.11).

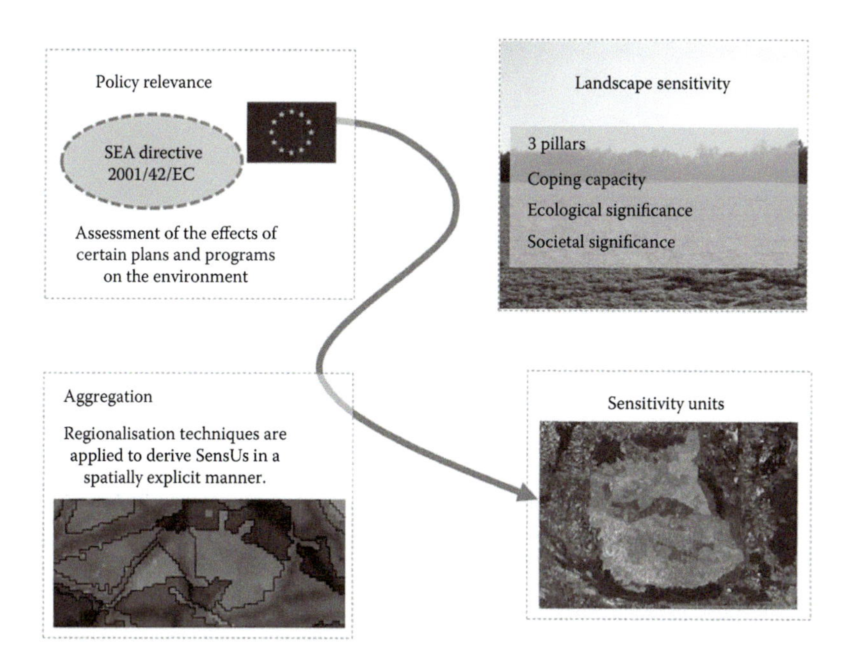

FIGURE 16.11 Using regionalization techniques to model landscape sensitivity in an integrated way.

16.3.3.2 Indicators and Datasets

Based on a review of environmental reports for different development plans, a set of 21 indicators was selected to model the sensitivity of the landscape. Biophysiscal, as well as socioeconomic factors were considered in three domains of sensitivity: (1) ability to resist to external disturbances, (2) ecological significance, and (3) societal significance. Each of them consists of 5–10 indicators (e.g., land use, degree of landscape fragmentation, conservation status, and population density), which are weighted by experts according to their relative importance.

16.3.3.3 Results

Figure 16.12 (panel 1) shows the spatial variability of landscape sensitivity to the road infrastructure project in the Marchfeld region. The resulting sensitivity units are discrete spatial units with similar characteristics in terms of their sensitivity to the impact. The results provide aggregated information to decision makers while preserving the detail provided by single indicators. Some areas, even within protected landscapes, are more sensitive than others, as indicated by geon boundaries that are independent from predefined administrative units. Thus, the approach indicates particular landscape areas which would be more suitable or less sensitive for the construction of a road and to avoid an undesirable development. The results shall not substitute an in-depth analysis of the impact effects, but support a more integrated and transparent decision making in environmental impact assessment.

The influence of single indicators on the sensitivity index is shown in Figure 16.12 (panel 2) using the example of agricultural land use. The intensity of land use is considered as one indicator of the societal significance of the landscape. Discarding this indicator has an effect on the resulting geon geometry as well as on the final sensitivity index values. In most areas, modeling outputs are relatively robust to changes in the indicator set because agricultural land use is only one of 21 weighted indicators. Areas with an intensive land use (wine growing regions and irrigated areas)

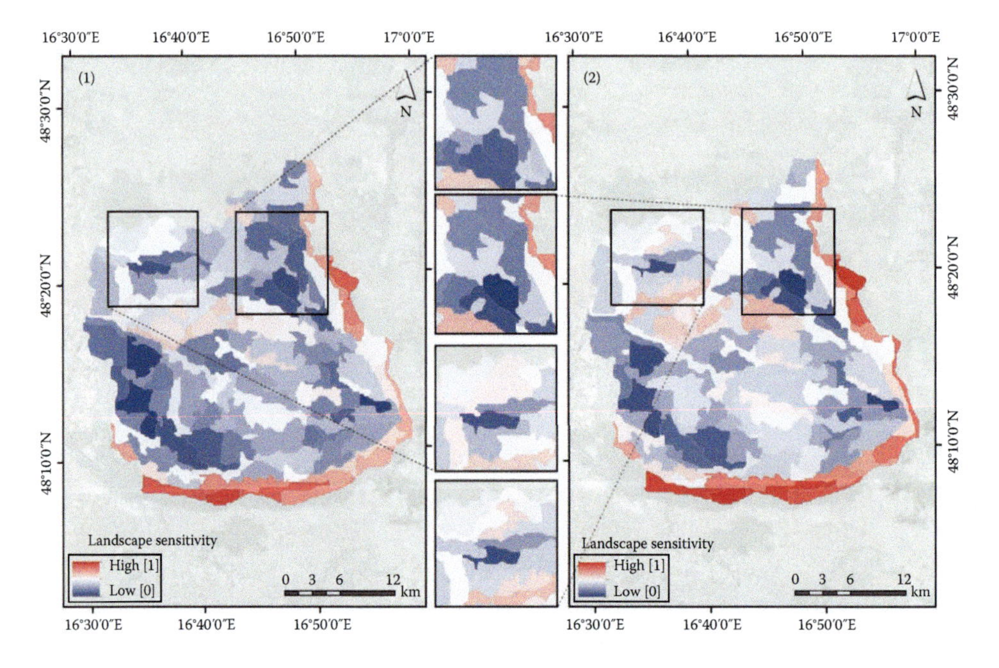

FIGURE 16.12 Evaluating the impact of individual indicators on the modeling outputs (panel 1: use of full indicators set; panel 2: agricultural land use indicator was discarded) on the modeling outputs.

show changes in the size and shape of the sensitivity units as well as different index values, which are typically lower where the socioeconomic value of agriculture is less considered in the landscape sensitivity to a road infrastructure project. In some areas of extensive agricultural land use and other, more important, sensitivity aspects, the index increases as well.

16.3.4 Climate Change Susceptibility (Cumulative CC Impact)

16.3.4.1 Multi-Dimensional Phenomenon Addressed

The recently published IPCC WGII contribution to the Fifth Assessment Report (AR5) highlights that the risk of climate-related impacts results from the interaction of climate-related hazards with the vulnerability of exposed human and natural systems (Field et al., 2014). Based on this concept, an integrative, spatial explicit assessment of cumulative climate change impact (CCCI) was carried out using the Sahel and Western Africa as study region. The focus was placed on identifying, mapping and evaluating "hotspots" of climate change impact to provide conditioned information for targeted climate change adaptation measures (Hagenlocher, 2013; Hagenlocher et al., 2014b).

16.3.4.2 Indicators and Datasets

To provide information on the hazards component of the IPCC risk framework, a set of four climate-related datasets was identified by domain experts from the United Nations Environment Programme (UNEP), including long-term average seasonal (1) precipitation and (2) temperature trends, as well as frequency of extreme events, such as (3) drought occurrences, and (4) major flood events over the past decades. The analysis was based on time-series of freely available continuous datasets, including remote sensing data. Data on the two essential climate variables, monthly mean temperature and precipitation, was acquired from the Climatic Research Unit (CRU) using their time series (TS 3.0) datasets. These comprise 1,224 monthly grids of observed climate for the period from 1901 to 2006, as reported by more than 4,000 weather stations around the globe. Data on drought frequency was derived from the NOAA NESDIS-STAR vegetation health index (VHI) dataset, which is based on measurements of the Advanced Very High Resolution Radiometer (AVHHR) on-board the NOAA satellite, while data on flood frequency was acquired from the Dartmouth Flood Observatory (DFO). They provide an active archive of large flood events which is updated based on remote sensing information (e.g., MODIS, Landsat, etc.) as well as various other sources (news, governmental reports, etc.). In line with the IPCC definition of climate, which was defined as a period of 30 years, the observation period was set to the past 24 to 36 years (depending on data availability).

16.3.4.3 Results

Following a statistical data pre-processing routine as described in Section 16.2, the four singular climate-related indicators were mapped and visualized. In addition to mapping singular trends in the four indicators, 1,233 units showing homogeneous regions of CCCI were delineated using the geon concept. Based on the statistical analysis of the resulting geon-scape a set of 19 climate change hotspots were identified. These represent areas most affected by changes in climate conditions (precipitation, temperature) and related extreme events (drought and flooding) over the past decades (Hagenlocher et al., 2014b). Next to the location and size of the hotspots, the proportional contribution of each of the four climate-related indicators was analyzed and mapped by means of a pie-chart for each of the hotspots by decomposing each hotspot into its underlying indicators (Figure 16.13).

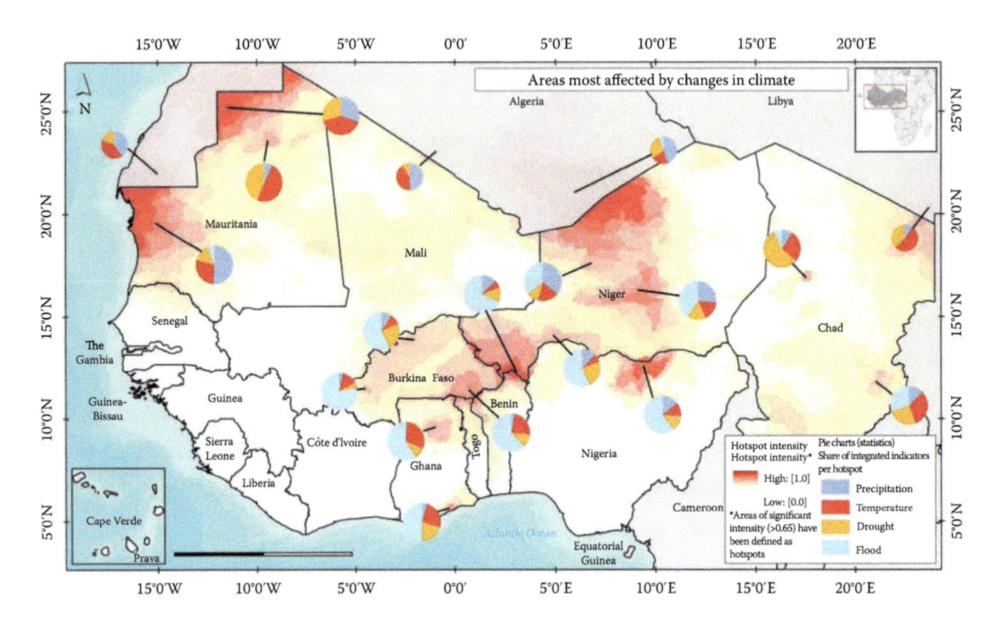

FIGURE 16.13 CCCI in the Sahel and western Africa based on the aggregation of a set of four climate-/hazard-related indicators (temperature, precipitation, drought, and flooding). Hotspots are displayed in red color. (Source: Hagenlocher et al., 2014.)

16.4 CONCLUSION

In this chapter we have presented the geon concept as a conceptual approach and methodological toolset to operationalize spatial multi-dimensional ("latent") phenomena (Mirchooli et al., 2023; Sur et al., 2022; Ghorbanzadeh et al., 2021; Csatáriné et al., 2020; Marti et al., 2020; Sudmanns et al., 2020; Li et al., 2019; Gudiyangada et al., 2019; Al Rahhal et al., 2018; Lang, 2018; Bianchini et al., 2017; Yu et al., 2017). Geons were demonstrated as an appropriate means to present geographical information more effectively and thus support efficient policy implementation or intervention planning. We showed how geons are constructed through integrating dedicated geospatial datasets applying regionalization techniques. We discussed aspects that influence the robustness of geon delineation, also in relation to MAUP. The generic and widely applicable concept reduces complexity by applying a systemic perspective while maintaining the level of integration at the specific level of intervention. By this we treat complex phenomena, such as vulnerability, sensitivity or climate change impact as a spatial phenomenon emerging from the specific arrangement of units generated by an expert-based partitioning of space.

REFERENCES

Al Rahhal, Mohamad M., Yakoub Bazi, Taghreed Abdullah, Mohamed L. Mekhalfi, Haikel AlHichri and Mansour Zuair. 2018. Learning a multi-branch neural network from multiple sources for knowledge adaptation in remote sensing imagery. *Remote Sensing* 10(12): 1890. https://doi.org/10.3390/rs10121890

Baatz, M. and A. Schäpe. 2000. *Multiresolution segmentation: An optimization approach for high quality multi-scale image segmentation.* Salzburg, Wichmann Verlag.

Beinat, E. 1997. *Value functions for environmental management.* Dordrecht, Kluwer Academic Publishers.

Berry, B. J. L. 1967. Grouping and regionalizing: An approach to the problem using multivariate analysis. In *Quantitative geography.* Eds. W. L. Garrison and D. F. Marble. Northwestern Studies in Geography: 219–251.

Bertalanffy, L. V. 1969. *General system theory – foundations, developments, applications.* New York, George Braziller.
Bianchini, S., F. Raspini, A. Ciampalini, D. Lagomarsino, M. Bianchi, F. Bellotti and N. Casagli. 2017. Mapping landslide phenomena in landlocked developing countries by means of satellite remote sensing data: The case of Dilijan (Armenia) area. *Geomatics, Natural Hazards and Risk* 8(2): 225–241. http://doi.org/10.1080/19475705.2016.1189459
Biederman, I. 1987. Recognition-by-components: A theory of human image understanding. *Psychological Review* 94(2): 115–147.
Birkmann, J., O. D. Cardona, M. L. Carreño, A. H. Barbat, M. Pelling, S. Schneiderbauer, S. Kienberger, M. Keiler, D. Alexander, P. Zeil, and T. Welle. 2013. Framing vulnerability, risk and societal responses: The MOVE framework. *Natural Hazards* 67(2): 193–211.
7Blaschke, T. 2010. Object based image analysis for remote sensing. *ISPRS International Journal of Photogrammetry and Remote Sensing* 65(1): 2–16.
Blaschke, T. and J. Strobl. 2001. What's wrong with pixels? Some recent developments interfacing remote sensing and GIS. *Zeitschrift für Geoinformationssysteme* 14(6): 12–17.
Bock, M., et al. 2005. Spatial indicators for nature conservation from European to local scale. *Ecological Indicators* 5: 322–338.
Byrne, B. M. 1998. *Structural Equation Modeling with LISREL, PRELIS, and SIMPLIS.* Mahwah, NJ, Lawrence Erlbaum Associates, Inc.
Capra, F. and P. L. Luisi. 2014. *The systems view of life. A unifying vision.* New York, Cambridge University Press.
Couclelis, H. 2010. Ontologies of geographic information. *International Journal of Geographical Information Science* 24(12): 1785–1809.
Csatáriné Szabó, Zsuzsanna, Tomáš Mikita, Gábor Négyesi, Orsolya Gyöngyi Varga, Péter Burai, László Takács-Szilágyi and Szilárd Szabó. 2020. Uncertainty and overfitting in fluvial landform classification using laser scanned data and machine learning: A comparison of pixel and object-based approaches. *Remote Sensing* 12(21): 3652. https://doi.org/10.3390/rs12213652
Decancq, K. and M. A. Lugo. 2012. Weights in multidimensional indices of wellbeing: An overview. *Econometric Reviews* 32(1): 7–34.
Drăgut, L., O. Csillik, C. Eisank and D. Tiede. 2014. Automated parameterisation for multi-scale image segmentation on multiple layers. *ISPRS Journal of Photogrammetry* 88: 119–127.
Drăgut, L. and C. Eisank. 2011. Object representations at multiple scales from digital elevation models. *Geomorphology* 129: 183–189.
Dragut, L., D. Tiede and S. R. Levick. 2010. ESP – a tool to estimate scale parameter for multiresolution image segmentation of remotely sensed data. *International Journal of Geographical Information Science* 24(6): 859–871.
Duque, J. C. and R. R. Suriñach. 2006. Supervised regionalization methods – a survey. *San Diego* 8: 31.
Eurostat. 2006. *Regions in the European union – nomenclature of territorial units for statistics NUTS 2010/EU-27.* Luxembourg, Publications Office of the European Union.
Field, C. B., et al. 2014. *Part A: Global and sectoral aspects. Contribution of working group II to the fifth assessment report of the intergovernmental panel on climate change.* Cambridge, Cambridge University Press.
Fiksel, J. 2006. Sustainability and resilience: Toward a systems approach. *Sustainability: Science, Practice, & Policy* 2(2): 14–21.
Ghorbanzadeh, O., S. R. Meena, H. Shahabi Sorman Abadi, S. Tavakkoli Piralilou, L. Zhiyong and T. Blaschke. 2021. Landslide mapping using two main deep-learning convolution neural network streams combined by the dempster–shafer model. *IEEE Journal of Selected Topics in Applied Earth Observations and Remote Sensing* 14: 452–463. http://doi.org/10.1109/JSTARS.2020.3043836
Goodchild, M. J., Y. M. and T. J. Cova. 2007. Towards a general theory of geographic representation in GIS. *International Journal of Geographical Information Science* 21(3): 239–260.
Gudiyangada Nachappa, Thimmaiah, Sepideh Tavakkoli Piralilou, Omid Ghorbanzadeh, Hejar Shahabi and Thomas Blaschke. 2019. Landslide susceptibility mapping for Austria using geons and optimization with the dempster-shafer theory. *Applied Sciences* 9(24): 5393. https://doi.org/10.3390/app9245393
Hagenlocher, M. 2013. Identifying and evaluating hotspots of climate change in the Sahel and Western Africa. In *From social vulnerability to resilience: Measuring progress toward Disaster Risk Reduction.* Eds. S. L. Cutter and C. Corendea. Bonn, Publication Series of UNU-EHS: 93–107

Hagenlocher, M., S. Kienberger, S. Lang and T. Blaschke. 2014a. Implications of spatial scales and reporting units for the spatial modelling of vulnerability to vector-borne diseases. In*GI_Forum 2014. Geospatial innovation for society*. Eds. R. Vogler, A. Car, J. Strobl and G. Griesebner. Berlin, Wichmann Verlag: 197–206.

Hagenlocher, M., S. Lang, D. Hölbling, D. Tiede, and S. Kienberger. 2014b. Modeling hotspots of climate change in the Sahel using object-based regionalization of multi-dimensional gridded datasets. *IEEE Journal of Selected Topics in Applied Earth Observations and Remote Sensing* 7: 229–234.

Hancock, J. R. 1993. *Multivariate regionalization: An approach using interactive statistical visualization.* AutoCarto XI: 218–227.

Hay, G. J. and G. Castilla. 2008. Geographic object-based image analysis (GEOBIA): A new name for a new discipline. In *Object-based image analysis: Spatial concepts for knowledge-driven remote sensing applications*. Eds. T. Blaschke, S. Lang and G. J. Hay. Berlin, Springer.

Hay, G. J., D. J. Marceau, P. Dubé and A. Buchard. 2001. A multiscale framework for landscape analysis: Object-specific analysis and upscaling. *Landscape Ecology* 16(6): 471–490.

Hernando, A., D. Tiede, F. Albrecht and S. Lang. 2012. Spatial and thematic assessment of object-based forest stand delineation using an OFA-matrix. *The International Journal of Applied Earth Observation and Geoinformation* 19: 214–225.

Jakeman, A. J. and R. A. Letcher. 2003. Integrated assessment and modelling: Features, principles and examples for catchment management. *Environmental Modelling & Software* 18(6): 491–501.

Johnston, R. J. 1968. Choice in classification – the subjectivity of objective methods. *Annals of the Association of American Geographers* 58: 575–589.

Kienberger, S., T. Blaschke and R. Z. Zaidi. 2013a. A framework for spatio-temporal scales and concepts from different disciplines: the 'vulnerability cube'. *Natural Hazards* 68: 1343–1369.

Kienberger, S., D. Contreras and P. Zeil. 2014. Spatial and holistic assessment of social, economic, and environmental vulnerability to floods — Lessons from the Salzach River Basin, Austria. In *Assessment of vulnerability to natural hazards: A European perspective*. Eds. J. Birkmann, D. Alexander and S. Kienberger. Elsevier: 53–74.

Kienberger, S., M. Hagenlocher, E. Delmelle and I. Casas. 2013b. A WebGIS tool for visualizing and exploring socioeconomic vulnerability to dengue fever in Cali, Colombia. *Geospatial Health* 8(1): 313–316.

Kienberger, S., S. Lang and P. Zeil. 2009. Spatial vulnerability units—expert-based spatial modeling of socio-economic vulnerability in the Salzach catchment, Austria. *Natural Hazards and Earth System Sciences* 9: 767–778.

Koestler, A. 1967. *The ghost in the machine.*London, Hutchinson.

Lang, S. 2008. Object-based image analysis for remote sensing applications: modeling reality—dealing with complexity. In *Object-based image analysis – spatial concepts for knowledge-driven remote sensing applications*. Eds. T. Blaschke, S. Lang and G. J. Hay. Berlin, Springer: 3–28.

Lang, S. 2018. Urban green valuation integrating biophysical and qualitative aspects. *European Journal of Remote Sensing* 51(1): 116–131. http://doi.org/10.1080/22797254.2017.1409083

Lang, S., F. Albrecht, S. Kienberger and D. Tiede. 2010. Object validity for operational tasks in a policy context. *Journal for Spatial Science* 55(1): 9–22.

Lang, S., C. Burnett and T. Blaschke. 2004. Multi-scale object-based image analysis: A key to the hierarchical organisation of landscapes. *Ekologia* 23(Supplement): 1–9.

Lang, S., S. Kienberger, L. Pernkopf and M. Hagenlocher. 2014a. Object-based multi-indicator representation of complex spatial phenomena. *South-Eastern European Journal of Earth Observation and Geomatic* 3(2s): 625–628.

Lang, S., S. Kienberger, D. Tiede, M. Hagenlocher and L. Pernkopf. 2014b. Geons—domain-specific regionalization of space. *Cartography and Geographic Information Science* 41(3): 214–226.

Lang, S., P. Zeil, S. Kienberger and D. Tiede. 2008. Geons—policy-relevant geo-objects for monitoring high-level indicators. *GI Forum. A. Car and J. Strobl. Salzburg*. Geospatial Crossroads @ GI_Forum' 08: 180–186.

Laszlo, E. 1972. *The systems view of the world.* New York, George Braziller.

Li, Z., B. Xu and J. Shan. 2019. Geometric object based building reconstruction from satellite imagery derived point clouds. *The International Archives of the Photogrammetry, Remote Sensing and Spatial Information Sciences* XLII-2/W13: 73–78. https://doi.org/10.5194/isprs-archives-XLII-2-W13-73-2019, 2019.

Long, J., T. Nelson and M. Wulder. 2010. Regionalization of landscape pattern indices using multivariate cluster analysis. *Environmental Management* 46: 134–142.

Marceau, D. J. 1999. The scale issue in the social and natural sciences. *Canadian Journal of Remote Sensing* 25: 347–356.

Marti, Renaud, Zhichao Li, Thibault Catry, Emmanuel Roux, Morgan Mangeas, Pascal Handschumacher, Jean Gaudart, Annelise Tran, Laurent Demagistri, Jean-François Faure et al. 2020. A mapping review on urban landscape factors of dengue retrieved from earth observation data, GIS techniques, and survey questionnaires. *Remote Sensing* 12(6): 932. https://doi.org/10.3390/rs12060932

Masolo, C. and L. Vieu. 1999. *Atomicity vs. infinite divisibility of space*. London, Springer: 235–250.

Mirchooli, F., Z. Dabiri, J. Strobl, A. Khaledi Darvishan and S. Hamidreza Sadeghi. 2023. Spatial and temporal dynamics of rangeland ecosystem services across the shazand watershed, iran. *Rangeland Ecology & Management* 90(2023): 45–55. ISSN 1550-7424. https://doi.org/10.1016/j.rama.2023.05.005. https://www.sciencedirect.com/science/article/pii/S1550742423000635

Moldan, B. and A. L. Dahl. 2007. Challenges to sustainability indicators. In *Sustainability indicators – a scientific assessment*. Eds. T. Hák, B. Moldan and A. L. Dahl. Washington, DC, Island Press: 1–26.

Nardo, M., M. Saisana, A. Saltelli, S. Tarantola, A. Hoffmann and E. Giovannini. 2005. *Handbook on constructing composite indicators – methodology and user guide*. OECD Working Papers. O. f. E. C.-o. a. Development. Handbook on Constructing Composite Indicators: Methodology and User Guide|OECD Statistics Working Papers|OECD iLibrary (oecd-ilibrary.org).

Ng, R. T. and J. Han. 2002. CLARANS: A method for clustering objects for spatial data mining. *IEEE Transactions on Knowledge and Data Engineering* 14: 1003–1016.

Openshaw, S. 1984. *The modifiable areal unit problem.*Norwich, Geo Books.

Parson, E. A. 1994. Searching for integrated assessment: A preliminary investigation of methods and projects in the integrated assessment of global climatic change. *3rd meeting of the CIESIN Harvard commission on global environmental change information policy*. Washington, DC.

Pernkopf, L. and S. Lang. 2011. Spatial meta-indicators: Assessing landscape sensitivity in the context of SEA. RegioResources, Dresden.

Peuquet, D. J. 2002. *Representations of space and time.*New York, The Guilford Press.

Riedler, B. and S. Lang. 2018. A spatially explicit patch model of habitat quality, integrating spatio-structural indicators. *Ecological Indicators* 94: 128–141.

Risbey, J., M. Kandlikar and A. Patwardhan. 1996. Assessing integrated assessments. *Climatic Change* 34: 369–395.

Saisana, M., A. Saltelli and S. Tarantola. 2005. Uncertainty and sensitivity analysis techniques as tools for the quality assessment of composite indicators. *Journal of the Royal Statistical Society* 168(2): 307–323.

Saltelli, A., et al. 2008. *Global sensitivity analysis. The primer*. John Wiley and Sons, Chichester.

Schöpfer, E., S. Lang and F. Albrecht. 2008. Object-fate analysis – spatial relationships for the assessment of object transition and correspondence. In *Object-based image analysis – spatial concepts for knowledge-driven remote sensing applications*. Eds. T. Blaschke, S. Lang and G. J. Hay. Berlin, Springer: 785–801.

Stahl, C., A. Cimorelli, C. Mazzarella and B. Jenkins. 2011. Toward sustainability: A case study demonstrating transdisciplinary learning through the selection and use of indicators in a decision-making process. *Integrated Environmental Assessment and Management* 7(3): 483–498.

Sudmanns, M., D. Tiede, H. Augustin and S. Lang. 2020. Assessing global Sentinel-2 coverage dynamics and data availability for operational Earth observation (EO) applications using the EO-Compass. *International Journal of Digital Earth* 13(7): 768–784. http://doi.org/10.1080/17538947.2019.1572799

Sur, Ujjwal, Prafull Singh, Sansar Raj Meena and Trilok Nath Singh. 2022. Predicting landslides susceptible zones in the lesser Himalayas by ensemble of per pixel and object-based models. *Remote Sensing* 14(8): 1953. https://doi.org/10.3390/rs14081953

Tiede, D., S. Lang, F. Albrecht and D. Hölbling. 2010. Object-based class modeling for cadastre constrained delineation of geo-objects. *Photogrammetric Engineering and Remote Sensing*: 193–202.

Tobler, W. 1970. A computer movie simulating urban growth in the Detroit region. *Economic Geography* 46(2): 234–240.

Voudouris, V. 2010. Towards a unifying formalisation of geographic representation: the object–field model with uncertainty and semantics. *International Journal of Geographical Information Science* 24(12): 1811–1828.

Weyant, J. P., et al. 1996. Integrated assessment of climate change: An overview and comparison of approaches and results. *Climate change 1995: Economic and social dimensions. Contribution of working group III to the second assessment report of the intergovernmental panel on climate change*. Eds. P. Bruce, H. Lee and E. F. Haites. Cambridge, Cambridge University Press.

WHO. 2013. *World Malaria report 2013*. World malaria report 2013 (who.int).

Wise, S., R. Haining and J. Ma. 2001. Providing spatial statistical data analysis functionality for the GIS user: The SAGE project. *International Journal of Geographical Information Science* 3(1): 239–254.

Woodcock, C. E. and A. H. Strahler. 1987. The factor of scale in remote sensing. *Remote Sensing of Environment* 21(3): 311–332.

Wu, J. 1999. Hierarchy and scaling: Extrapolating information along a scaling ladder. *Canadian Journal of Remote Sensing* 25(4): 367–380.

Wu, J. and H. Li. 2006. Concepts of scale and scaling. In *Scaling and uncertainty analysis in ecology: Methods and applications*. Eds. J. Wu, K. B. Jones, H. Li and O. L. Loucks. Berlin, Springer.

Yu, Guirui, Chen Zhi, Zhang Leiming, Peng Changhui, Chen Jingming, Piao Shilong, Zhang Yangjian, Niu Shuli, Wang Qiufeng, Luo Yiqi, Ciais Philippe, and Baldocchi D. Dennis 2017. Recognizing the scientific mission of flux tower observation networks—lay the solid scientific data foundation for solving ecological issues related to global change. *Journal of Resources and Ecology* 8(2): 115–120, 1 March. https://doi.org/10.5814/j.issn.1674-764x.2017.02.001

Part IV

Summary and Synthesis of Volume II

17 Summary Chapter, Volume II, Remote Sensing Handbook (Second Edition)

Image Processing, Change Detection, GIS, and Spatial Data Analysis

Prasad S. Thenkabail

LIST OF ACRONYMS

AGB	Aboveground biomass
AI	Artificial intelligence
ALS	Airborne laser scanning
ANN	Artificial neural networks
CART	Classification and regression trees
CNN	Convolution neural network
CT	Classification trees
DGGS	Discrete global grid systems
DL	Deep learning
GEDI	Global ecosystem dynamics investigation
GEOBIA	Geographic object-based image analysis
GIS	Geographic information system
GLAS	Geoscience Laser Altimeter System
HNB	Hyperspectral narrowband
ICESat	Ice, Cloud, and Land Elevation Satellite
ISODATA	Iterative self-organizing data analysis technique
kNN	k-Nearest neighbor
LiDAR	Light detection and ranging
LULC	Land use/land cover
MBB	Multispectral broadband
MESMA	multiple endmember spectral mixture analysis
ML	Machine learning
MLR	Multinomial logistic regression
MP	Morphological profile
MRFs	Markov random fields
NDVI	Normalized difference vegetation index
OBAA	Object-based accuracy assessment
OBCD	Object-based change detection

DOI: 10.1201/9781003541158-21

OBIA	Object based image analysis
OCNN	Object-based CNN
OFA	Object fate analysis
RF	Random forest
RHSeg	Recursive hierarchical segmentation
ROSIS	Reflected optics spectrographic imaging system
SMT	Spectral matching technique
SVM	Support vector machines
USGS	United States Geological Survey
VGI	Volunteer geographic information
VHRI	Very high spatial-resolution imagery
WS4GEE	Web services for Google Earth engine

This chapter provides a summary of each of the 16 chapters in Volume II of the six-volume *Remote Sensing Handbook* (Second Edition). The topics covered in the chapters of Volume II include (Figure 17.0): (1) digital image processing fundamentals and advance; (2) digital image classifications for applications such as urban, land use, and land cover; (3) hyperspectral image processing methods and approaches; (4) thermal infrared image processing principles and practices; (5) image segmentation; (6) object oriented image analysis (OBIA) including geospatial data integration techniques in OBIA; (7) image segmentation in specific applications like land use and land cover; (8) LiDAR digital image processing; (9) change detection; and (10) integrating geographic information systems (GIS) with remote sensing in geoprocessing workflows, democratization of GIS data and tools, frontiers of GIScience, and GIS and remote sensing policies. One or more chapters cover each of the preceding broad topics. For example, 11 chapters discuss image classification methods (Comber et al., 2012) and approaches, the most essential topic extensively for beginners as well as advanced professionals. In a nutshell, these chapters provide a comprehensive overview of these critical topics, capture the advances over the last 60+ years, and provide a vision for further development in the years ahead. By reading this summary chapter, a reader can have a quick understanding of what is in each of the chapters of Volume II, see how the chapters interconnect and intermingle, and get an overview of the importance of various chapters in developing complete and comprehensive knowledge of the remote sensing Data Characterization, Classification, and Accuracies.

Chapter 17: Summary Chapter for
Remote Sensing Handbook (Second Edition, Six Volumes): Volume II

Volume II: Image Processing, Change Detection, GIS and Spatial Data Analysis

Chapter 1: Digital Image Processing Methods and Techniques

Chapter 2 and 3: Image Classification for land use\land cover, Croplands, Urban Studies

Chapters 4 and 5: Hyperspectral Remote Sensing including Digital Image Processing of Hyperspectral Data

Chapters 6 : Thermal Remote Sensing including Digital Image Processing of Thermal Data

Chapters 7 to 10: Object Based Image Analysis (OBIA), Image Segmentation, Land Categorization

Chapters 11: LiDAR data Processing and Applications

Chapter 12: Change Detection

Chapter 13 and 14 Geoprocessing, Workflows, Provenance and Democratization of Geospatial Data

Chapter 15 and 16: GIScience and Regionalization of Geospatial Data for Policy

FIGURE 17.0. Overview of the chapters in Volume II of the *Remote Sensing Handbook* (Second Edition).

17.1 DIGITAL IMAGE PROCESSING: METHODS AND TECHNIQUES

Chapter 1 by Dr. Sunil Narumalani and Dr. Paul Merani provides a step-by-step approach to understanding the fundamental methods and techniques involved in digital image processing. These steps are summarized in Figure 17.1 and are broadly grouped into eight image-processing categories: (1) quality assessment, (2) pre-processing, (3) enhancement, (4) registration and re-projection, (5) mosaicking and megafile (Latif et al., 2023; Cravero et al., 2022; Thenkabail et al., 2009a, 2009b) creation, (6) analysis, (7) classification, and (8) accuracy assessment. Fundamental methods and approaches (see Burger and Burge, 2022; Richards, 2022; Petrou and Kamata, 2021; Sabins and Ellis, 2020; Jensen, 1996; Lillesand et al., 2008; Richards and Xiuping, 2006) have remained firm over the years and are essential for anyone wanting to master the remote sensing data processing methods. In recent years numerous advances have been made in digital image processing (Richards, 2022; Burger and Burge, 2022; Nagamalai et al., 2011) that include data fusion (Jia et al., 2020), data integration, object-based image analysis (OBIA) or geographic OBIA (GEOBIA) (Chen et al., 2018a), support vector machines (SVMs), decision tress, spectral un-mixing, artificial neural networks, phenological matrices, Fourier transforms, and spectral matching techniques (SMTs). In data composition, multi-date, multispectral data are time composited to long time periods into single mega file data cubes (Latif et al., 2023; Cravero et al., 2022; Thenkabail et al., 2009a, 2009b), akin to hyperspectral data cubes (Oppelt et al., 2012). This has facilitated applying hyperspectral data analysis techniques like SMTs for multi-date multispectral data composited over long time periods. Such data enables the analysis of time-series phenological matrices. All aspects of image analysis (Figure 17.1, Lillesand et al., 2008) have become efficient as a result of speed in computing power as well as smarter and better algorithms.

Advances in modern remote sensing are many. First, remote sensing has become ubiquitous and acquiring data repeatedly for the planet Earth in a wide array of spatial, spectral, radiometric, and temporal resolutions. Second, in addition to traditional satellites like the Landsat-series, SPOT-series, IRS-series, we now have data from diverse acquisitions from sensors, such as hyperspatial (e.g., Planetscope Doves and Super Doves), hyperspectral (e.g., DESIS, PRISMA, EnMAP), and Radar (e.g., NISAR). Third, long-term archives of data from various platforms, such as NOAA AVHRR GIMMS-series (1979–present) (Pinzon et al., 2023), Terra/Aqua MODIS (2000-present) (Román et al., 2024), Landsat-series (1972–present) (Crawford et al., 2023; USGS, 2024) and others like the SPOT-series (1984–present) (Hafner et al., 2022; Courtois and Traizet, 1986) and IRS-series (1988–present) (Murthy et al., 2022) are only increasing. These mega-datasets provide a great opportunity to study our planet over 60+ years with consistent, objective, high-quality data for a wide range of applications. Nevertheless, this data proliferation over time and space and acquisition across the electromagnetic spectrum means we are now dealing with datasets that no traditional computer systems can store, process, and analyze. As a result, a whole new dimension of remote

FIGURE 17.1 Standard digital image processing steps.

sensing data storage, processing, and analysis is required. More recent developments in addressing the preceding challenges are discussed in various other chapters of this volume. Readers should also refer to Burger and Burge (2022), Richards (2022), and Sabins and Ellis (2020) for a very comprehensive coverage of the various aspects of the subject. With very high spatial (hyperspatial) resolution imagery, incorporating texture information has become critical. This topic is specifically dealt with in Petrou and Kamata (2021). With the proliferation of hyperspectral data, a modern understanding of hyperspectral data analysis is critical (Khan et al., 2018). The advent of machine learning, deep learning (DL; Bai et al., 2023), and cloud computing requires a new perspective of image analysis (Thenkabail et al., 2021; Khan et al., 2018; Yuan et al., 2017; Li et al., 2019; Maxwell et al., 2018).

17.2 IMAGE CLASSIFICATION METHODS IN LAND COVER AND LAND USE AND CROPLAND STUDIES

Expanding upon Chapter 1, Chapters 2 and 3 provide image processing methods and approaches for specific applications. Chapter 2 by Dr. Mutlu Ozdogan discussed classification methods for land cover and land use (e.g., Figure 17.2), vegetation, croplands, and other land themes. The classification methods are broadly classified into two categories:

1. Parametric
2. Non-parametric

Parametric classification algorithms assume a known (often normal) statistical distribution of data. Parametric classification methods include minimum distance to mean, Gaussian maximum likelihood, Mahalanobis distance, and parallelepiped (Van der Sluijs et al., 2023).

Non-parametric methods are not limited by any statistical distribution assumption of data and are based on class statistics that include mean vectors and covariance matrices. This is specifically suitable in conditions where there is great heterogeneity in landscape and variances between classes are high. Non-parametric classification methods include classification and regression trees (CART), artificial neural networks (ANN), k-nearest neighbor (kNN), and SVMs.

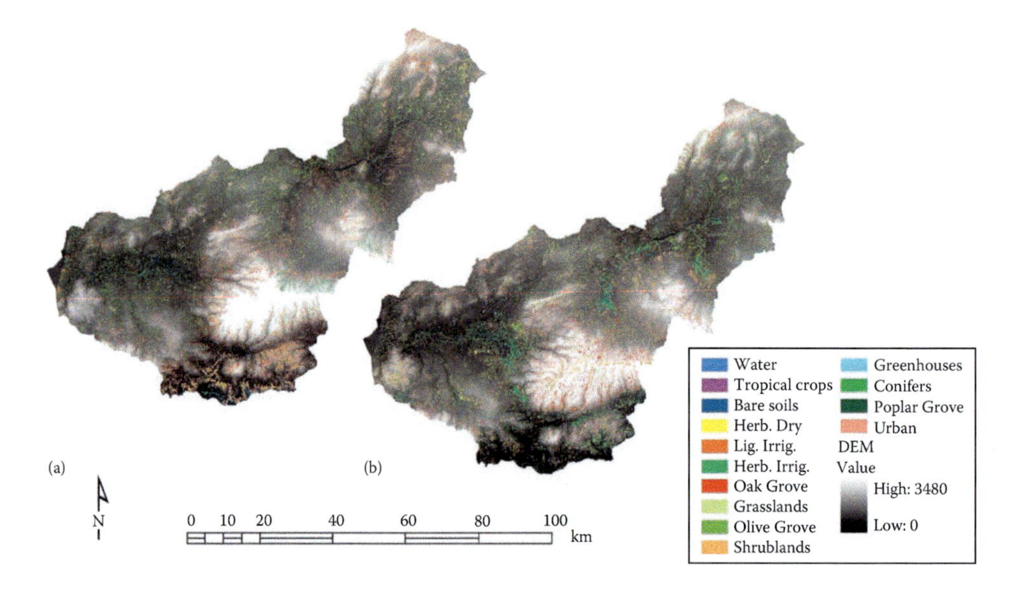

FIGURE 17.2 Land-cover change maps between classification trees (CT) and random forest (RF) for Granada Province: (a) RF and (b) classification tree. (Source: Rodriguez-Galiano et al., 2012.)

These classification algorithms are applied to satellite data through either supervised or unsupervised logic (Macarringue et al., 2022; Pandey et al., 2021; Abdi, 2020; Keshtkar et al., 2017). In a supervised approach, classes are trained through available knowledge (e.g., ground data, accurate secondary maps, and expert knowledge). Widely used supervised methods are Gaussian maximum likelihood, minimum distance to mean decision tree algorithms, Random Forest (RF), SVM, ANN, and nearest neighbor classifier are also supervised classifications but fall into the non-parametric category (Macarringue et al., 2022). Gaussian maximum likelihood, k-means nearest neighbor, and SVMs, on the other hand, are widely used supervised parametric methods. ISOCLASS k-mean clustering (Thenkabail et al., 2021; Thenkabail et al., 2009a, 2009b) is the widely used unsupervised classifier.

Unsupervised classification algorithms assign classes based on the spectral statistics (e.g., mean, standard deviation, matrices such as covariance, and correlations) of a pixel. Classes are then identified based on available knowledge (e.g., ground data, reference data), which will then turn the information classes generated by the algorithm into land cover classes. Unsupervised classification is also known as clustering and includes iterative self-organizing data analysis technique (ISODATA) clustering (Ulug and Karslıoglu, 2022), k-means clustering, and Narenda-Goldberg clustering (Ulug and Karslioglu, 2022). An unsupervised classification approach is recommended when a priori knowledge of the landscape (e.g., ground data) is unavailable, limited or inadequate. K-means is an unsupervised nonparametric classification algorithm. Sometimes, unsupervised classification is used as a first step, before applying supervised classification. This is known as a hybrid classification.

These image classification methods can be performed based on: (a) Per-pixel, (b) sub-pixel, (c) per-field, and (d) per-object. These methods may provide soft classification (e.g., mixed classes such as vegetation that may include croplands, rangelands, forests), meaning that the class boundaries can inter-mingle to some extent, or hard classifications (e.g., croplands, soils, rangelands) when the class boundaries are distinct without any inter-mingling.

Chapter 2 also discusses other classification approaches that are becoming more popular. For example, object-oriented classifications such as RHSEG (Xiong et al., 2017) or a combination of object-oriented segmentation followed by pixel-based classification as outlined in Thenkabail et al. (2021) and Xiong et al. (2017).

17.3 URBAN IMAGE CLASSIFICATION METHODS AND APPROACHES

Urban Land Use/Land Cover (LULC) information is of great importance for planning and developmental purposes of any city or town, but very difficult to automatically extract detailed information from remote sensing imagery (Wang et al., 2022; Cai et al., 2019; Hersperger et al., 2018; Wei et al., 2014; Xu et al., 2013). It requires a combination of imagery of various resolutions such as Landsat 30 m, Sentinel 10 m, and sub-meter to 5-m very high-resolution imagery such as IKONOS, Quickbird, Geoeye, and PlansetScope's Doves and super doves. However, the need for images of multiple resolutions depends on the sizes of targeted features and the level of detail in the background. A combination of these images helps extract different types of urban features. For example, Landsat 60 m thermal imagery helps map urban heat islands or night lights; Landsat multispectral 30 m data will help establish urban land use patterns; and sub-meter to 5 m imagery from IKONOS, Quickbird, Planetscope's Doves or super doves, and LiDAR data will help in mapping detailed street maps, individual buildings, and individual tree species. Advanced artificial intelligence (AI), machine learning (ML) methods, such as convolution neural network (CNN), SVM, and RFs have made it possible to classify and map urban land dynamics which also help perform operations with speed and accuracy (Chaturvedi and de Vries, 2021; Hartling et al., 2019).

Urban LULC classification methods and approaches are continually evolving because of advances in remote sensing data, methods (Dardel et al., 2014) and approaches. In Chapter 3 by Dr.

Soe W. Myint et al., discussed four distinct methods of image classification for urban mapping that looks back several decade long incremental developments to current period. These four methods, discussed in Chapter 10, are summarized as follows:

1. Per-pixel spectrally based methods (hard classifications, e.g., Figure 17.3a) are used in all types of classification (e.g., land use/land cover, urban or forests categories) using algorithms such as maximum likelihood, ISODATA, and minimum distance-to-mean. In per pixel classifiersre, each pixel falls into a class or a group and forms minimum mapping unit. Per-pixel classifiers have significant limitations in detecting and accurately mapping individual urban features, but they can produce simple and convenient thematic maps with reasonable accuracies (often in the realm of 70%) in producing few urban classes depending on the complexity and scale of the area studied.
2. Sub-pixel methods (soft classifications, e.g., Figure 17.3b) quantify fractions of multiple classes within a pixel. Due to the complexity of the landscape, the presence of multiple classes within a pixel is common. Methods used to resolve sub-pixel composition of pixels include linear methods (e.g., linear spectral mixture analysis, regression trees, and regression analysis) and nonlinear methods (e.g., artificial neural network). The success of linear spectral mixture analysis will depend on the purity of collecting end member (preferably through exact ground knowledge). Typically, the number of end members doesn't exceed the number of bands. However, multiple endmember spectral mixture analysis (MESMA) can overcome this. Nevertheless, it is complex to identify all the possible end members and the many unidentified end members lead to uncertainties.
3. With the advent of very high resolution (sub-meter to 5 m) imagery around the year 2000, object-based image analysis (OBIA) or geospatial object-based image analysis (GEOBIA) (Du et al., 2019; Blaschke T., 2010) became a standard approach to discern numerous urban features like individual buildings, pools, roads, sidewalks, urban trees, and golf courses. This involves, the use of spectral and textural characteristics of the image data to first segment the image into distinct groups and then apply classification algorithms separately on individual segments to identify detailed urban classes accurately. Yet, computing challenges of many segments that need to be carried out based on various scale parameters in object-oriented algorithms remains challenging. Nevertheless, numerous studies have shown an increase of anywhere between 10 to 20% increase in accuracies (reaching the realm of 90% accuracy) using OBIA or GEOBIA have been reported when compared with pixel-based approaches (e.g., Figure 17.3).
4. Some geospatial methods such as spatial co-occurrence matrix, spatial autocorrelation, fractal, and lacunarity approaches capture the local texture similarities into distinct group of pixels, before a classification algorithm is applied to improve classification accuracy. This can be done by local moving windows (e.g., 2 × 2, 3 × 3, 5 × 5). In contrast to the preceding advanced geospatial approaches, a new spatial frequency-based algorithm called wavelets characterizes spatial features in different directions at multiple scale. A window-based approach employing wavelet theory and dyadic decomposition procedures to measure spatial arrangements of features at multiple scales has been determined to be more effective than the preceding advanced geospatial approaches in directly identifying urban classes. While this approach outperformed other spatial approaches, it is limited by a finite decomposition procedure that requires a large window size to perform a classification. An overcomplete wavelet analysis using an infinite-scale decomposition procedure is more powerful in measuring the spatial complexity of geographic features and identify detailed urban classes more effectively. Advanced methods such as the object-based CNN (OCNN) can perform better than other landuse classification algorithms, especially for complex urban land use (Zhang et al., 2018).

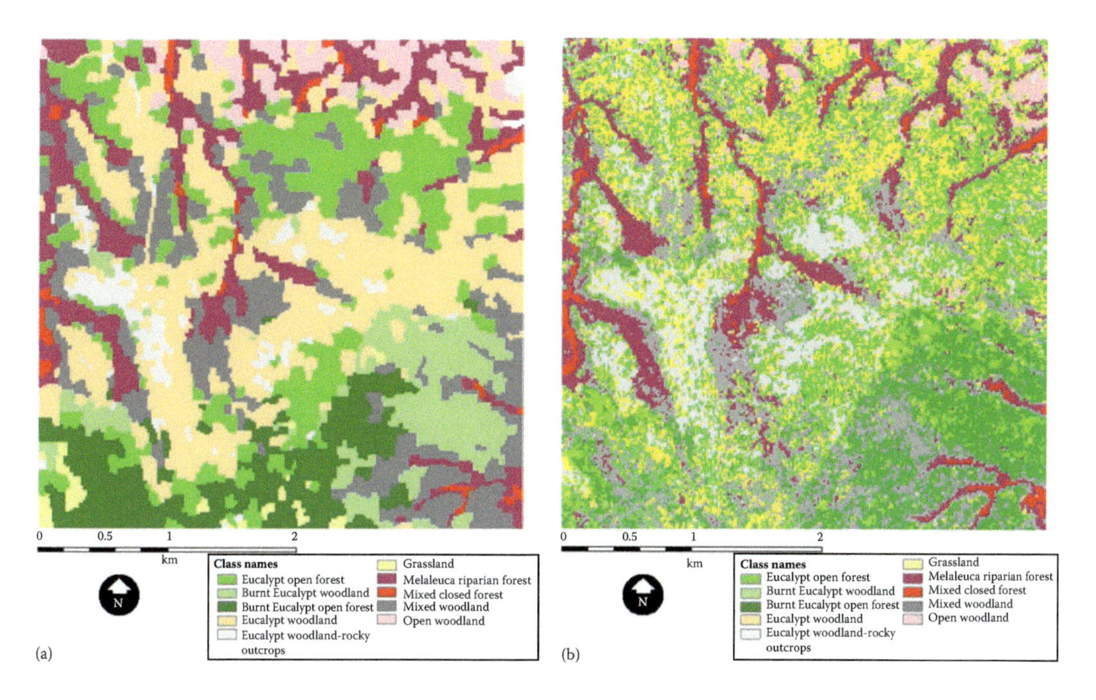

FIGURE 17.3 Resultant images of the object-based (a) and pixel-based (b) classifications. (Source: Whiteside et al., 2011.)

17.4 HYPERSPECTRAL IMAGE PROCESSING METHODS AND APPROACHES

Hyperspectral data are acquired in hundreds or even thousands of hyperspectral narrowbands (HNBs) in a near continuous form along the electromagnetic spectrum. This offers spectral signatures of an object rather than few data points along the spectrum as in multispectral broadband (MBB) data. Nevertheless, Hyperspectral data bring their own challenges of massive data volumes, data redundancy, and complexity of data understanding and analysis. Thereby, effective methods of hyperspectral data processing and analysis to extract best possible information pertaining to multitude of applications. Even though many of these methods of image processing are like those in previous chapters, the hyperspectral data involving hundreds of narrowbands is distinct. As a result of such distinctive data, the approached to image analysis differ. For example, even before classification, one can look at data redundancy and reduce data volumes by removing bands in atmospheric windows. Also, the reference training, testing, and validation data will be increase due to number of bands involved. For more details on these see Thenkabail et al. (2021b).

In Chapter 4, Dr. Jun Li and Dr. Antonio Plaza explore some current trends in hyperspectral image processing. The classification approaches discussed in the chapter include:

1. Supervised classification consisting of discriminant analysis techniques. In these discriminant classifiers discriminant functions such as nearest neighbor, decision trees, linear functions, and nonlinear functions are applied.
2. Kernel methods for supervised classification including the well-known SVM classifier.
3. Spectral-spatial classification approaches, which provide significantly improved classification results compared to spectral-based classifiers. Spatial features are extracted through advanced morphological techniques such as morphological profiles (MPs), Markov random fields (MRFs), and others as described in the chapter.
4. Probabilistic classification methods such as the multinomial logistic regression (MLR).

The main problem with the supervised pixel-based classifiers such as the RF and SVMs is addressing the huge data dimensionality of hyperspectral data in comparison with the limited availability of training samples. Specifically, supervised approaches require sample data from pure locations both for classification and validation. The number of samples required increases with data dimensionality and, as a result, the challenge of using a large number of training samples to address the high dimensionality of hyperspectral data is costly and often prohibitive. This issue, known as Hughes' phenomenon (Ma et al., 2013), is hard to address and is often the bottleneck in the effective use of supervised classification approaches. So, often, the uncertainty in classification results from the supervised hyperspectral image analysis is high relative to unsupervised approaches (Xiong et al., 2017).

The concept of semi-supervised classification, as discussed by Dr. Jun Li and Dr. Antonio Plaza is to not only use the labeled samples for training, but also unlabeled data. Semi-supervised learning is performed, for example, through machine learning SVMs (SVMs), among other strategies discussed in the chapter. However, the unlabeled data may not be confident enough and hence can also lead to uncertainties.

The aforementioned difficulties related to the Hughes' phenomenon in supervised classification methods can be effectively overcome through kernel methods such as the SVM (Liu et al., 2018). The kernel method have fewer samples but more homogeneous samples that have far lesser uncertainties. In classifications, the quality of samples is superior to the quantity. Authors use a well-known and widely used hyperspectral dataset, collected by the reflected optics spectrographic imaging system (ROSIS) over the University of Pavia, Italy (Mueller et al., 2022) to provide a series of results comparing the most commonly used SVM classifier with: (1) several discriminant classifiers, (2) composite SVM obtained using summation kernel, (3) combination of the morphological EMP for feature extraction followed by SVM for classification (EMP/SVM), (4) pixel-wise SVM classifier with the

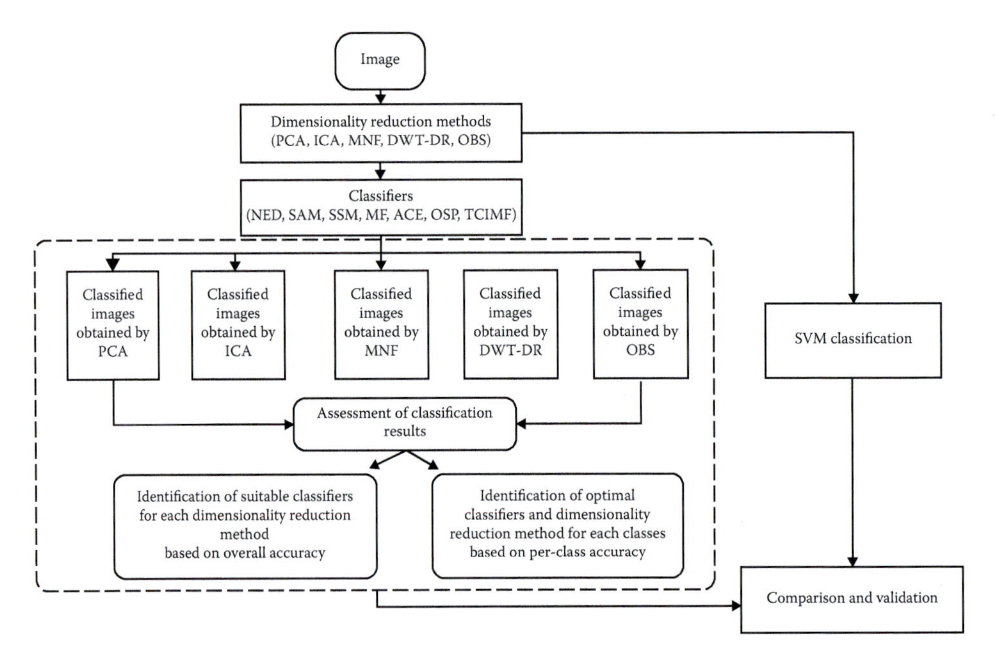

FIGURE 17.4 Flowchart showing hyperspectral image data reduction and classification methods. Dimensionality reduction methods include: PCA = principal component analysis, independent component analysis (ICA), minimum noise fraction (MNF), and discrete wavelet transform based dimensionality reduction (DWT-DR), optimal band selection (OBS). Multiple classifier system includes: normalized Euclidian distance (NED), spectral angle mapper (SAM), spectral similarity measure (SSM), matched filter (MF), adaptive coherence estimation (ACE), orthogonal subspace projection (OSP), and target constraint minimum filter (TCIMF). (Source: Damodaran and Nidamanuri, 2014.)

morphological watershed, and (5) SVM classifier combined with the segmentation result provided by the unsupervised recursive hierarchical segmentation (RHSeg) algorithm (Du et al., 2022; Tilton et al., 2012, 2017), and (6) a subspace-based multinomial logistic regression classifier followed by spatial post processing using a multilevel logistic prior (MLRsubMLL) (Tilton et al., 2017). For this purpose, Tilton et al. (2017) use 9 land use classes and show how SVM performs relative to other methods. In general, the performance of classification increases progressively as we move from step (1) to step (6). The chapter does not discuss hyperspectral image analysis using unsupervised classification, even though object-oriented classification RHSEG (Tilton et al., 2017) is discussed briefly.

Some of the well-known dimensionality reduction and classification methods for hyperspectral data are summarized in Figure 17.4 by Damodaran and Nidamanuri (2014). They concluded that the choice of dimensionality reduction method and classifier significantly influences the classification results obtained in hyperspectral image analysis. Deep learning (DL) (Yuan et al., 2020) has been recognized as a powerful feature-extraction tool to effectively address nonlinear problems and widely used in hyperspectral image processing tasks (Li et al., 2019; Liu et al., 2018). Convolutional neural networks (CNNs) architecture considerably outperforms other state-of-the-art methods (Gao et al., 2018; Yu et al., 2017). A problem with DL classifiers is they require a large number of training data that are extremely difficult to obtain and resource intensive, especially when such data is needed year after year. To overcome this Jia et al. (2021) recommend DL method with few labeled samples.

17.5 ADVANCES IN DIGITAL IMAGE PROCESSING WITH NEW AND OLD GENERATION OF HYPERSPECTRAL DATA

Great advances are taking place in remote sensing with the advent of new generation of hyperspectral sensors. These include data from, already in orbit sensors such as: (1) Germany's Deutsches Zentrum fur Luftund Raumfahrt (DLR's) Earth Sensing Imaging Spectrometer (DESIS) sensor onboard the International Space Station (ISS), (2) Italian Space Agency's (ASI's) PRISMA (Hyperspectral Precursor of the Application Mission), and (3) Germany's DLR's Environmental Mapping and Analysis Program (EnMAP; Uhl et al., 2013, 2014). Further, Planet Labs PBC (Public Benefit Corporation) recently announced the launch of two hyperspectral sensors called Tanager in 2023 (Joshua et al., 2023). NASA is planning for the hyperspectral sensor Surface Biology and Geology (SBG) to be launched in the coming years (Cawse-Nicholson et al., 2023). Further, we already have over 83,000 hyperspectral images of the world acquired from NASA's Earth Observing-1 (EO-1) Hyperion that are freely available to anyone from the U.S. Geological Survey's data archives (USGS, 2019). These suites of sensors acquire data in 200 plus hyperspectral narrowbands (HNBs) in 2.55 to 12 nm bandwidth, either in 400–1,000 or 400–2,500 nm spectral range with SBG also acquiring data in the thermal range. In addition, Landsat-NEXT is planning a constellation of three satellites each carrying 26 bands in the 400–12,000 nm wavelength range (Crawford et al., 2023; USGS, 2024). HNBs provide data as "spectral signatures" in stark contrast to "a few data points along the spectrum" provided by multispectral broadbands (MBBs) such as the Landsat satellite series.

Thereby, there is a need to explore the advances that can be made using new generation of hyperspectral sensors. Further, we need to explore advances in methods and techniques that help decipher information from hyperspectral data more effectively. In this regard Aneece and Thenkabail (2022) selected PRISMA and DESIS images during the 2020 growing season in California's Central Valley to study seven major crops. PRISMA and DESIS images were highly correlated (R2 of 0.9–0.95). Out of the 235 DESIS bands (400–1,000 nm) and 238 PRISMA bands (400–2,500 nm), 26 (11%) and 45 (19%) bands, respectively, were optimal to study agricultural crops. Hyperspectral data poses major challenges for supervised classification methods due to the high dimensionality of the data and the limited availability of training samples (Pooletti et al., 2019). Delogu et al. (2023) compared three supervised classification techniques for agricultural crop recognition (Tilling et al.,

2007) using PRISMA data: RF, artificial neural network (ANN), and CNN and find out that CNN provided better results overall. Machine learning (ML) and DL (Bai et al., 2023), especially when used on the cloud, have revolutionized image analysis, and proved to be a powerful tool for processing high-dimensional remotely sensed data, adapting their behavior to the special characteristics of hyperspectral data (Pooletti et al., 2019). Aneece and Thenkabail (2021) ran the supervised classification algorithms: RF, SVM, and Naive Bayes (NB), and the unsupervised clustering algorithm WekaXMeans (WXM) were run using selected optimal Hyperion and DESIS HS narrowbands (HNBs) and found RF and SVM returned the highest overall producer's, and user's accuracies, with the performances of NB and WXM being substantially lower.

Recently, collaborative representation (CR) methods and approaches have been popular in hyperspectral data analysis. In Chapter 5, Du et al. explore novel collaborative representation (CR) for hyperspectral image classification and detection. CR are type of ML methods, but are popular due to simple implementation, efficient calculation, and excellent performance. Jia et al. (2019) adopt CR approach by evaluating it on four popular hyperspectral image datasets, and found results showing advantages of CR, particularly for a hyperspectral image with high spatial resolution. Chapter 5 explore how CR methods and approaches are used on multiple hyperspectral data in data reduction, data classification, and applications. Conceptual framework of the CR is provided in Figure 17.5. It has three-parts: the generator based on a joint spatial–spectral hard attention module, the discriminator based on convolutional LSTM, and the classification of CA-GAN based on collaborative and competitive learning (Feng et al., 2020). As shown in Figure 17.5 (Feng et al., 2020): (1) in the first part, the noise and the class labels are used as the input of the generator; (2) in the next part, the discriminator is constructed to capture joint spatial–spectral features by merging a convolutional long short-term memory (ConvLSTM) layer after the convolutional layer, and (3) in the final part, the collaborative learning mechanism is constructed based on the generator and discriminator with symmetrical structure. The classification performance of the discriminator is improved through CS (Feng et al., 2020). As more and more advanced hyperspectral data becomes a norm, and as these data develop a time-series history, advance methods such as CR are very much required to take the full benefit of hyperspectral data, to overcome data redundancies, and to address other computational challenges (Khan et al., 2018).

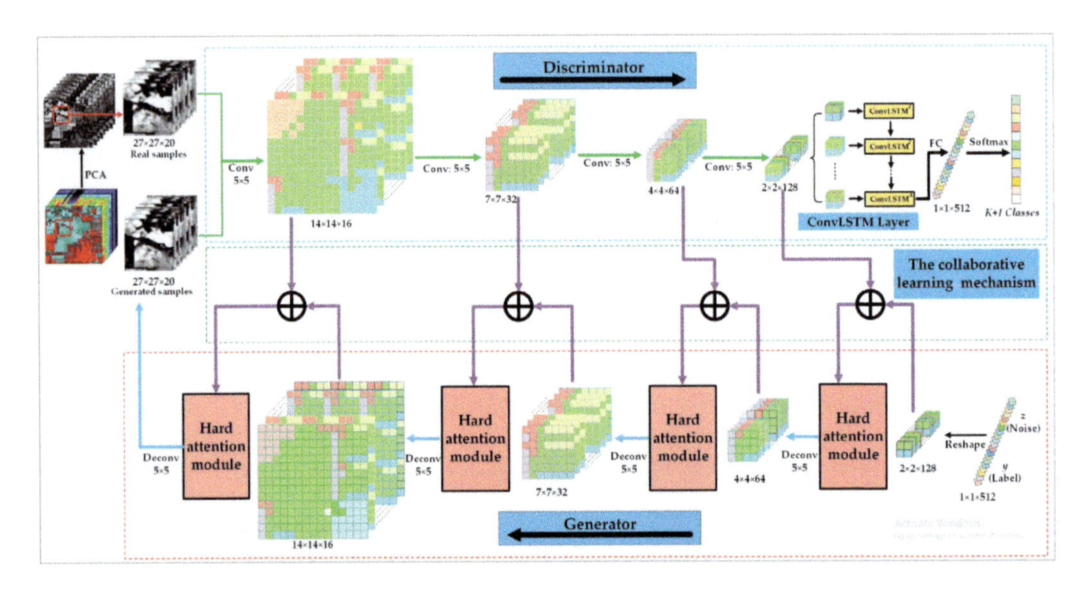

FIGURE 17.5 The framework of convolutional Generative Adversarial Network (GAN) based on collaborative learning and attention mechanism (CA-GAN). (Feng et al., 2020.)

17.6 THERMAL INFRARED REMOTE SENSING: PRINCIPLES AND THEORETICAL BACKGROUND

Different authors define the wavelength range of thermal infrared, TIR, domain differently, but most sensor measurements are undertaken in the 8–14 μm wavelength range. In general, the thermal infrared is the field of remote sensing, which utilizes emitted radiation in the wavelength domain, where our planet has its emission maximum. A large example of TIR data is it independence of an illumination source: TIR sensitive sensors can collect TIR data during the day as well as during the night. The principles and theoretical background of thermal data and thermal data analyses are presented in Chapter 6 by Kuenzer et al. This chapter not only provides a comprehensive overview of the theoretical background including all the laws applicable for the TIR domain, but also elucidates numerous application examples.

TIR data usage is exemplarily demonstrated in Chapter 6 for the analyses the general derivation of land and sea surface temperature, LST, the derivation of urban heat islands, and the detection of hot spots, including thermal anomalies resulting from forest fires, coal fires, or gas flaring activities. Also examples of sea surface temperature products, SST, and thermal water pollution are presented. The authors furthermore explain the valuable principle of diurnal temperature changes (ΔT), which makes use of the fact that different surface types exhibit a different thermal behavior over the day. Some heat up faster than others in the morning, some cool down slower than others, and the extent of the diurnal temperature change (ΔT) contains valuable clues about the physical characteristics of a surface. For example, water bodies warm up very slowly during the day and cool of slowly during the night—overall, their diurnal temperature change (ΔT) of water is very low and usually lies within the range of only few degrees Celsius. ΔT is much more accentuated for—for example—a dark asphalt surface, which heats up very fast during a sunny day, and also cools of relatively fast after dawn. Here, diurnal temperature change is much more accentuated and can reach several tenths of degrees Celsius. This principle can be exploited when diurnal data is available—for example, daytime and nighttime data of the Landsat thermal band, or diurnal data of the MODIS sensor thermal bands, and can support the TIR based mapping of different surface types. Furthermore, statements on the moisture content of the surface can be derived, as—for example—a wet soil will exhibit a lower diurnal temperature change (ΔT) than a dry soil.

The theoretical framework laid out in Chapter 6 on thermal data understanding and its characteristics allows reader to get a better understanding on where and how to use thermal data. Indeed, modern remote sensing integrates remote sensing data from various portions of the electromagnetic spectrum and thermal data is increasingly used for a wide array of applications such as in the study of irrigated and rainfed croplands (Figure 17.6). For example, thermal remote sensing in agriculture includes irrigation scheduling, drought monitoring, crop disease detection, and mapping of soil properties, residues and tillage, field tiles, and crop maturity and yield (Kullberg et al., 2017; Khanal et al., 2017). Multi-/hyperspectral TIR together with those from VNIR/SWIR and SIF sensors within a multi-sensor approach can provide profound insights to actual plant (water) status and the rationale of physiological and biochemical changes, thus helping in identification of plant water stress or drought is of utmost importance to guarantee global water and food supply (Gerhards et al., 2019). Gerhards et al. (2019) evaluate thermal remote sensing indices to estimate crop evapotranspiration coefficients. Airborne TIR data can support the monitoring of irrigated areas (Figure 17.6), agricultural monitoring of crop water stress, can support with the detection of technical accidents and leakages (e.g., pipe bursts, nuclear accidents etc.), and is also frequently employed in medical imaging.

It must be noted that one major limitation of spaceborne thermal data is its relatively low spatial resolution. Whereas optical and multispectral data in the visible and near infrared domains can be collected with resolution of better than one meter pixel size, spaceborne thermal sensor data is collected at between 60 m and 1 km spatial resolution. Whereas Landsat 7 ETM+ has a thermal band acquiring the data at 60 m, the novel Landsat 8 DCM only offers 100 m spatial resolution in the

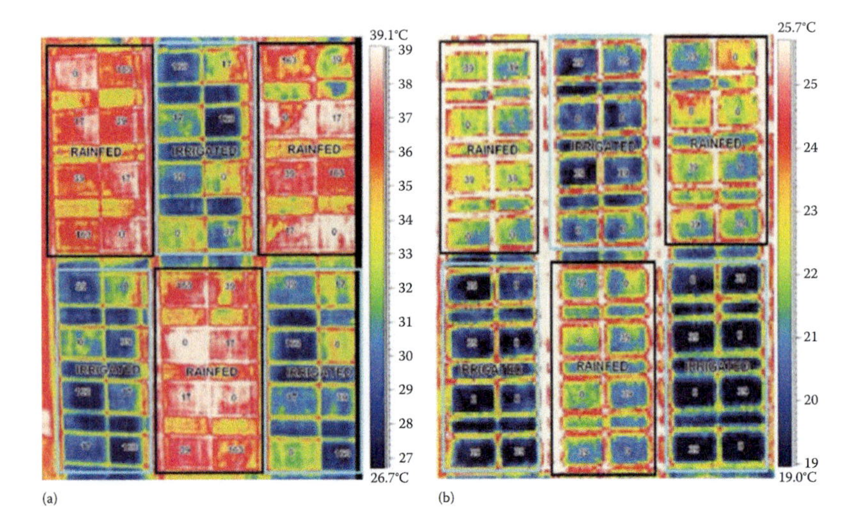

FIGURE 17.6 Thermal images gathered using an airborne ThermaCAM P40 (FLIR systems AB, Danderyd, Sweden) infrared digital imager measuring emitted energy in the range of 7.5–13 µm for an experiment site located near Horsham, Australia (36°44′S, 142°06′ E; elevation 133 m). The study areas comprised 48 plots (18 m × 12 m) planted to wheat arranged in a randomized block design with three replications for October 2004 (a) and October 2005 (b). Rainfed or irrigated blocks are indicated. Numbers denote kg/ha N applied to each plot. Lighter shades correspond to higher temperatures.

thermal bands, and sensors such as AVHRR or MODIS only allow for the monitoring and mapping at 1km resolution. In the TIR domain band widths and spatial resolution usually need to be wider and lower to ensure that a proper amount of incoming energy is collected. Furthermore, higher resolution TIR sensors are very costly to build. However, as demonstrated in Chapter 6 by Kuenzer et al. the number of nations launching satellites including TIR sensors is growing rapidly. We can therefore be confident that remote sensing in the thermal infrared domain will continue to strive.

17.7 REMOTE SENSING IMAGE SEGMENTATION: METHODS, APPROACHES, AND ADVANCES

Object-oriented classification evolved over last two decades and complement/supplement pixel-based spectral classification of imagery. When object-oriented classification is used, in addition to, pixel-based spectral classification, accuracies and precision of the products would be improve significantly, especially for high spatial resolution images (Thenkabail et al., 2021). Geographic object-based image analysis (GEOBIA) is devoted to developing automated methods to partition remote sensing (RS) imagery into meaningful image objects, and assessing their characteristics through spatial, spectral, texture and temporal features, thus generating new geographic information in a GIS-ready format (Gu et al., 2017). Semantic image segmentation is a fundamental task in computer vision (CV) that assigns a label to each pixel, aka pixel-level classification (Yuan et al., 2021). Thorsten and Kuenzer (2020) discuss well performing DL architectures on remote sensing datasets as well as reflect on advances made in CV and narrow the gap between the reviewed, theoretical concepts from CV and practical application in EO.

In Chapter 7, Hossain and Chen begin with background on evolution of image segmentation with proliferation of sub-meter to 5 m very high spatial-resolution imagery (VHRI) from sensors such as IKONOS series of satellites, WorldView-series of satellites, and Planet Labs Doves and Super Doves. The Geographic Object-Based Image Analysis (GEOBIA or OBIA) became key image analysis technique, especially to handle VHRI. Segmentation methods that they discuss in Chapter 7

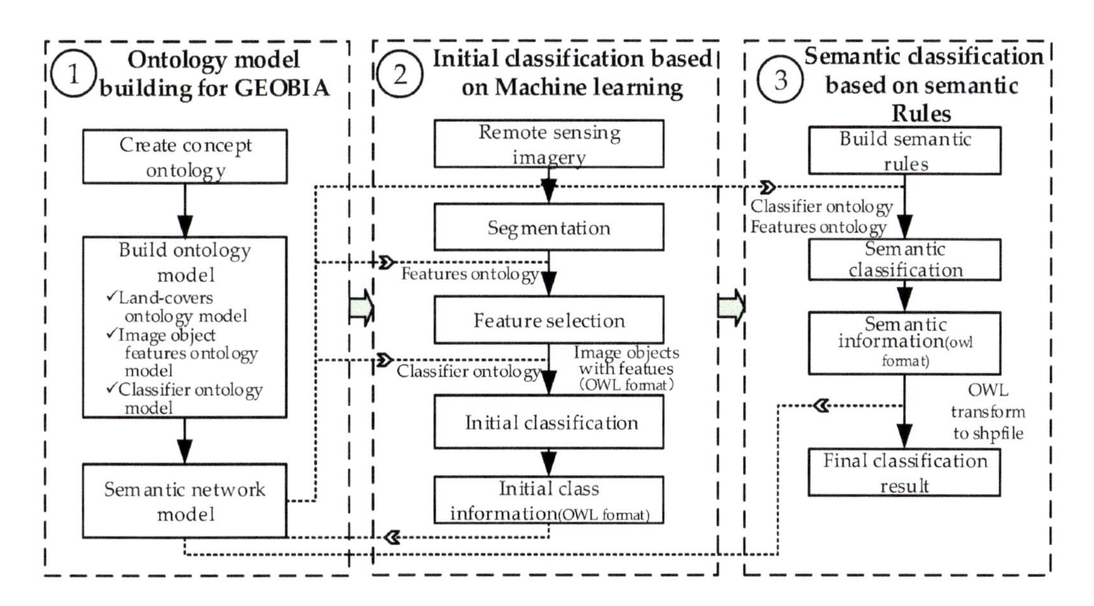

FIGURE 17.7 Overview of the methodology followed in this study. (Source: Gu et al., 2017.)

include: (1) spectrally based methods, (2) spatially based methods, (3) region-based methods, (4) graph-based methods, (5) hybrid methods, and (6) deep-learning (DL) methods. They also discuss, extensively, the DL algorithms along with their strengths and challenges in various applications. Segmentation of imagery types such as LiDAR and hyperspectral are also presented and discussed. They highlight, in conclusions, that integrating semantic segmentation methodologies, mainly through AI and DL, is a current focal point in the remote sensing community.

Gu et al., 2017 provide a lucid three-step workflow of the object-based semantic classification (Figure 17.7): (1) ontology modeling, (2) initial classification based on a data-driven machine learning method, and (3) semantic classification based on knowledge-driven semantic rules. The selection of optimal segmentation parameters' values generates a qualitative segmentation output and has a direct impact on feature extraction and subsequent overall classification accuracy (Kotaridis and Lazaridou, 2021).

17.8 OBJECT-BASED IMAGE ANALYSIS (OBIA): EVOLUTION AND STATE OF THE ART

Object based image analysis (OBIA) or geospatial OBIA (GEOBIA)—the latter is used when emphasizing scales and applications in remote sensing and Geographic Information Science—evolved with the advent of very high spatial resolution satellite imagery (sub-meter to 5 m) around the year 2000. However, the underlying concepts of OBIA have been around far longer as enumerated in Chapter 8 by Dr. Thomas Blaschke et al. OBIA involves segmenting an image into distinct objects and then performing the classification on distinct objects separately by utilizing the spatial, spectral, radiometric, and temporal characteristics of image data and eventually auxiliary data. OBIA addresses object properties such as shape, size, pattern, tone, texture, shadows, and association (see Blaschke et al., 2014; Blaschke, 2010). An overview of GEOBIA approach to classification is shown in Figure 17.8 (Ke et al., 2010) in classifying forest species using very high spatial resolution Quickbird multispectral imagery and LiDAR data. Quickbird imagery is used for spectral-based segmentation and LiDAR data used for LiDAR based segmentations (Figure 17.8).

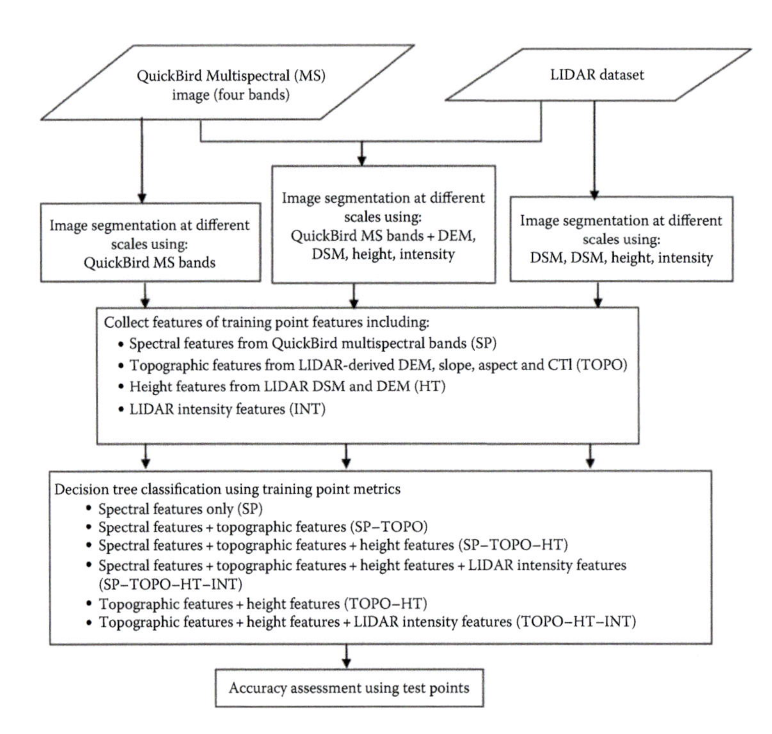

FIGURE 17.8 An object based image analysis (OBIA) protocol for forest species classification through synergistic use of Quickbird multispectral imagery and LiDAR data. (Source: Ke et al., 2010.)

Chapter 8 discusses the evolution of OBIA, concepts of image segmentation and fusion, classification and synthesis, and various applications of GEOBIA. As Chapter 8 shows us, there is substantial increase in accuracies using OBIA as opposed to traditional per-pixel classification results. For example, GEOBIA is now increasingly used for various integration tasks including new technologies such as mobile applications (e.g., locating where you make a call), geo-intelligence, and volunteer geographic information (VGI). Emerging trends of OBIA or GEOBIA were found in multiple subfields of GEOBIA, including data sources, image segmentation, object-based feature extraction, and geo-object-based modeling frameworks (Chen et al., 2018b). Recent research on the convergence of GEOBIA with deep CNNs, is a new form of GEOBIA (Kucharczyk et al., 2020). Recent developments also include use of GEOBIA along with machine learning approaches such as SVM, RF, and CART on cloud computing platforms.

17.9 GEOSPATIAL DATA INTEGRATION IN OBIA AND IMPLICATIONS ON ACCURACY AND VALIDITY

OBIA is now extensively supporting geographic information needs. The type of information derived from OBIA is wide ranging and could include such features as forest species, buildings, road-networks, and farm field boundaries. Availability of very high spatial resolution imagery (sub-meter to 5 m) from various satellite sensors (e.g., IKONOS, Quickbird, Geoeye), LiDAR, and Uncrewed Aerial Vehicles has made deriving objects over large areas feasible. However, remote sensing data is acquired in a number of platforms, in number of resolutions (spatial, spectral, radiometric, and temporal). It is handled by different user's in different way (e.g., different atmospheric correction models (Vermote and El Saleous, 2002); with and without atmospheric correction). Methods and approaches in analyzing the data are often different.

In Chapter 9, Dr. Stefan Lang and Dr. Dirk Tiede discuss strategies and approaches of geospatial data integration in OBIA. The chapter introduces various concepts of object validation that include multi-stage validation, object fate analysis (OFA), object-based accuracy assessment (OBAA), object-based change detection (OBCD), and object linking. Object identification means able to delineate objects accurately through such means as color, texture, shape, size, scale, or form. Recent advances in OBIA have resulted in increasing accuracies in the objects delineated such as the building features and their areas (e.g., Figure 17.9). Beyond that OBIA enables to assess the way how objects are delineated, in terms of appropriate scale, complexity, shape and the fitness

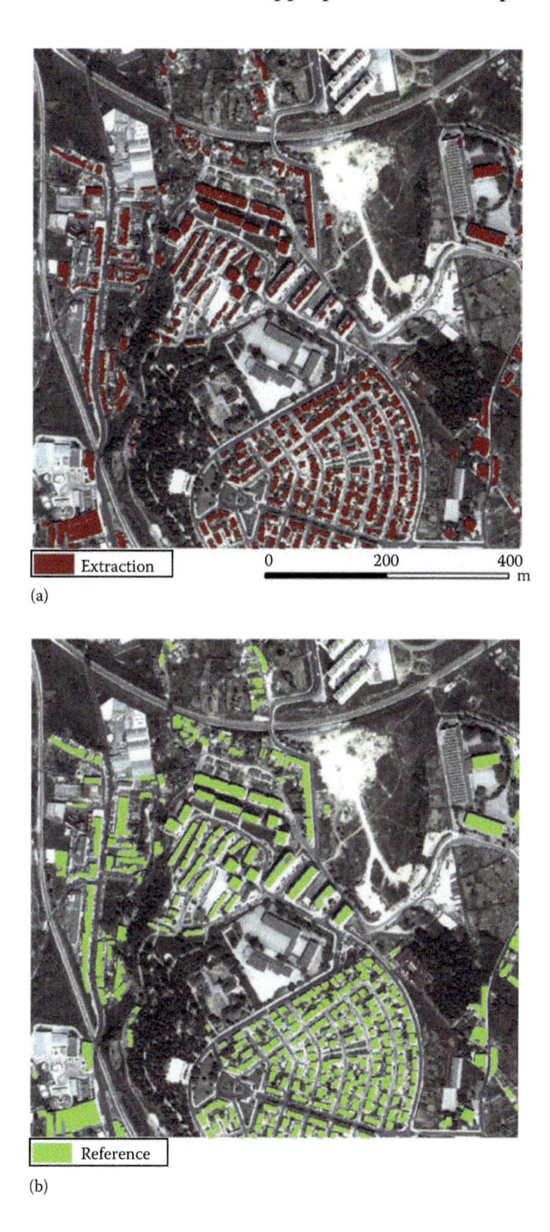

FIGURE 17.9 Assessing accuracy of GEOBIA extracted building features and their areas (top image) relative to reference data (bottom image) for Lisbon, Portugal. Data used for GEOBIA include Quickbird pan sharpened 0.6 m, Quickbird multispectral 2.4 m, Quickbird image derived normalized difference vegetation index (NDVI) 2.4 m, normalized digital surface model 1 m, and reference map of bridges. GEOBIA extracted 317 of the 330 features found in reference data. (Source: Freire et al., 2014.)

to existing geospatial datasets. Hossain and Chen (2019; also see Chapter 7) proposed several algorithms for OBIA/GEOBIA and found none to be ideal. They determined that (Hossain and Chen (2019) (1) edge-based algorithms are easy to implement, but they are missing the contextual information, (2) region-based method generates better results compared to the edge-based method, however, finding appropriate seeds and other parameters is the real challenge in that case, (3) to resolve the seeding problems, superpixels algorithms are introduced in remote sensing image segmentation, and (4) hybrid method, Nevertheless, OBIA or GEOBIA requires significant advances and improvements in its methodological developments concerning sampling design, reference design and accuracy analysis of the per-polygon approaches (Ye et al., 2018). They reviewed 209 journal articles and found (Ye et al., 2018): (1) a third of the articles did not report whether or how the probabilistic sampling was performed; (2) a third of the articles did not present information concerning sample size, and (3) half of the articles used a per-pixel approach, thus ignored a main feature of the OBIA map.

17.10 IMAGE SEGMENTATION ALGORITHMS FOR LAND CATEGORIZATION

Early remote sensing, prior to 2000, heavily depended on land categorization and classification algorithms that are purely spectral based. These methods include simple thresholding (e.g., using NDVI or band reflectivity; Tucker, 1979), single or multi-band classifications using supervised or unsupervised approaches as k-means algorithms, maximum likelihood, and minimum distance to mean, and so on. But, over the years, remote sensing scientists have been looking to improve our understanding of the land categorization with improved accuracies and reduced uncertainties. One of the powerful approaches has been to include image segmentation in the analysis. The pixel-based classification results inevitably include "salt and pepper" noise and disjointed land fragments in practice. Object-based analysis can improve salt and pepper effects and increase classification accuracies over pixel-based image classification (Xiong et al., 2017; Im et al., 2008; Stow et al., 2008). Image segmentation gathers several similar neighbor pixels together as objects, and categorizes or labels objects, which would be further labeled as croplands or non-croplands in the integration step with pixel-based classification (Xiong et al., 2017). Image segmentation procedures have many implementation (Im et al., 2008; Stow et al., 2008) with very high memory and CPU requirements. GEE provides APIs related to image segmentation such as region grow (Espindola et al., 2006).

The Chapter 10 by Dr. Tilton et al. provides a comprehensive overview of various segmentation algorithms and illustrates their strengths and limitations. The chapter also summarizes in a Table various commercial and non-commercial segmentation software package available and discusses on where and how they have been used. Chapter 10 shows that segmentation algorithms involve using spectral, textural, spatial, and other secondary (e.g., elevation, slope) information of the landscape obtained from remote sensing to "segment" the land into unique units or segments (e.g., uplands from lowlands, or agricultural lands from natural vegetation) and then classify them separately. This reduces the complexities within each segment, and also enables shape-based as well as contextual modeling of segments that are not possible using pixel-based approaches. Furthermore, when classification algorithms are applied separately on each segment, we have more meaningful and accurate classes. In addition, we can also get more unique classes. For example, lowland vegetation is often distinct from upland vegetation. When we classify the image that included both upland and lowland segments, we will be often not able to distinguish upland and lowland vegetation with great degree of accuracy to spectral limitations of data. But, when we look at them separately, we will be able to categorize upland vegetation and lowland vegetation with greater degree of certainty. There are many other segmentation algorithms evolving, for instance region-based Image Segmentation Algorithm based on k-means clustering (RISA) (Wang et al., 2010; e.g., Figure 17.10). Chapter 10 concludes that the existing and evolving wide range of image segmentation approaches provide a rich menu for the users to choose the method which is the best adapted

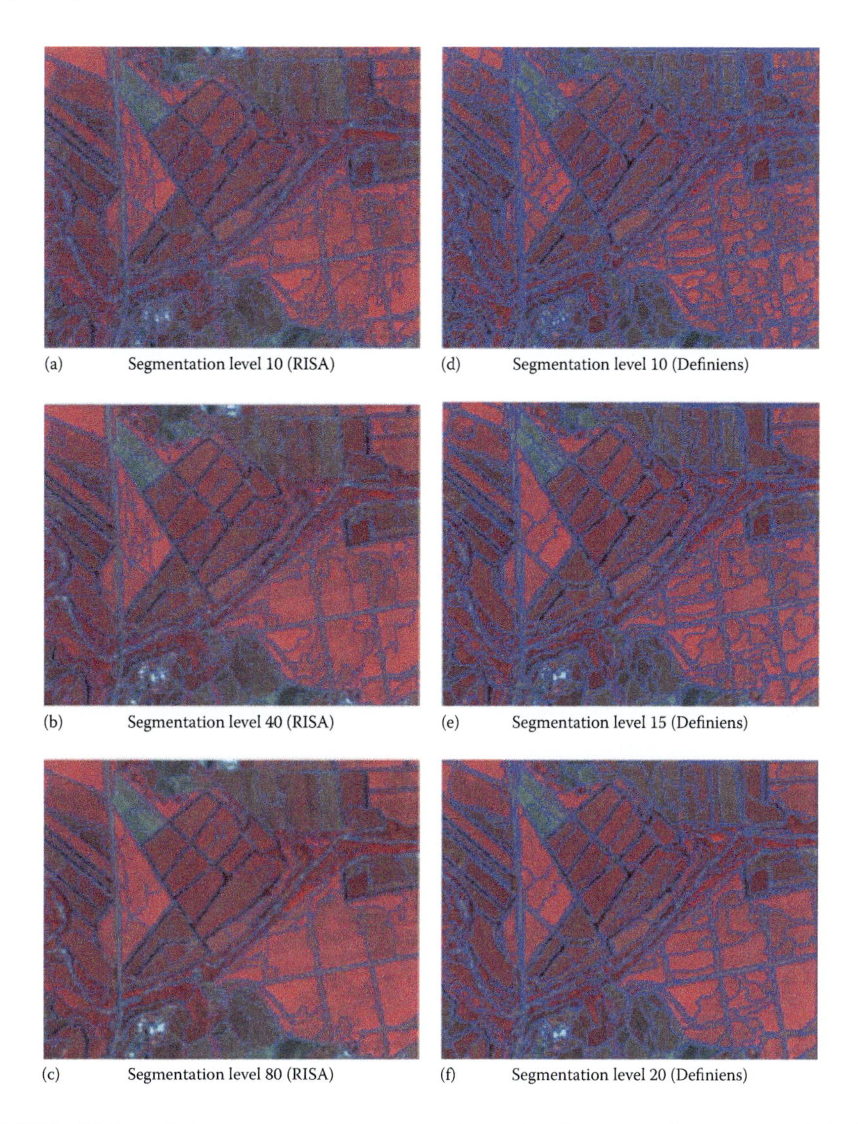

FIGURE 17.10 Multi-scale image segmentation comparison using the proposed algorithm and the Definiens algorithm (Wang et al., 2010). The left three images (a, b, and c) were segmented using Region-based Image Segmentation Algorithm based on k-means clustering (RISA) with the default parameter values and segmentation scales of 10, 40, and 80, respectively. The right three images (d, e, and f) were segmented using the Definiens software with the color criterion of 0.9, shape criterion of 0.1 (compactness 0.2 and smoothness 0.8), and segmentation scales of 10, 15, and 20, respectively. (Source: Wang et al., 2010.)

for each particular application. Many recent fusion based approaches involving the pixel-based classifications and object-based classification have evolved in handling massively big remote sensing datasets on the cloud platforms using machine learning, DL, and AI (Thenkabail et al., 2021; Tilton et al., 2017; Xiong et al., 2017). For example, Xiong et al. (2017) demonstrated that RHSeg when combined with the supervised pixel-based classification algorithms such as the RF and the SVM provided significantly improved classification accuracies in establishing cropland extent of continental Africa using Landsat-30 m data. Massey et al. (2018) developed fusion based cropland classification involving pixel-based RF classification and an object-based segmentation algorithm called RHSeg to show improved cropland extent and areas. Thenkabail et al. (2021) demonstrated significantly improved cropland extent mapping at global-scale using time-series Landsat data.

17.11 LiDAR DATA PROCESSING AND APPLICATIONS

Light Detection and Ranging (LiDAR) data are collected as "point clouds" (Figure 17.11a) with each point having three dimensions (3D) coordinates: horizontal (*x*, *y*), and vertical (*z*). LiDAR data are acquired with a laser scanning system in very high point rate. For each point measurement, a laser pulse is transmitted, and the reflected energy is caught. The return signal can be fully recorded as a waveform or interpreted to be a single or multiple returns to form point clouds. The features embedded in LiDAR "point clouds" include: location (*x,y,z*), intensity, echo, and waveform. The data can be used to model, map, and study various Earth surface features such as trees, buildings, and terrain in 3D. There are four broad categories of LiDAR systems: (1) terrestrial or ground based laser scanning with footprint of 1–10 cm and point accuracy of 1–5 cm; (2) airborne laser scanning (ALS) with footprint of 5–50 cm and point accuracy of 10–100 cm; (3) mobile laser scanning with footprint of 1–10 cm and point accuracy of 1–30 cm; and (4) spaceborne LiDAR with footprint of 30–100 m and vertical accuracy of 25 cm to 10 m (e.g., GLAS, DESDYnI, ICESatII). Waveform LiDAR is recent advance in ALS technology. The ability to capture 3D structure of plant and objects is a major advantage of LiDAR data, unlike other remote sensing. In 3D modeling it has big advantages compared to photogrammetry or land surveying due to its relatively fast acquisitions and automated processing. However, LiDAR data often have small coverage of data collection, large processing volume of data, and affected by cloud and haze.

Chapter 11 by Dr. Yi-Hsing Tseng focuses on LiDAR data processing and applications. The LiDAR data processing involves noise filtering, processing 3D (*x,y,z*) points, independent variables such as laser use time, scanner position, and scanner orientation. The LiDAR data are often delivered in ASPRS LAS format. Data quality is required to meet accuracies of ASPRS Large Scale Mapping Standard and/or the National Spatial Data Accuracy standard. The chapter extensively discusses LiDAR data:

1. Quality assessment and control (e.g., system components, error budgets, quality assessment and control)
2. Management (e.g., storage of LiDAR point cloud, spatial indexing, hierarchical representation)
3. Point cloud feature extraction (e.g., spatial feature in LiDAR data, methods of extraction of spatial features that include extraction of line and plane features)
4. Full-waveform airborne processing (e.g., shape of received waveform, echo detection, feature extraction)

The Chapter 11 enumerates various applications of LiDAR data such as in:

1. Forestry studies: structure, diameter at breast height, crown size, basal area, tree height, aboveground biomass (AGB, e.g., Kronseder et al., 2012; Figure 17.11a). LiDAR ability to penetrate vegetation and look through it and capture information is a huge benefit that is infeasible using other remote sensing techniques.
2. DEM, DTM, DSM, TIN generation.
3. City modeling (e.g., building detection/construction, road extraction);
4. Structures and objects (e.g., dams, bridges);
5. Agriculture (e.g., farm topography, plant height);

6. Hydrology and geomorphology (e.g., stream network, slopes derived off DEM);
7. Land cover classification.
8. Flood mapping and assessment.
9. Disasters (e.g., earthquakes)

Recent studies have demonstrated that LiDAR data when used by integrating with other remote sensing data help map unique products by discerning features that is otherwise infeasible and also help achieve higher accuracies. For example, the Geoscience Laser Altimeter System (GLAS) onboard the Ice, Cloud, and Land Elevation Satellite (ICESat), provides worldwide LiDAR data equipped with three laser sensors, L1–L3, each of which has a 1064-nm laser channel for surface altimetry and dense cloud heights. This is used to map global AGB by using GLAS LiDAR data, optical imagery and forest inventory data (Hu et al., 2016). The Global Ecosystem Dynamics Investigation (GEDI) lidar instrument onboard the International Space Station collecting unique data on vegetation structure was used along with global Landsat analysis-ready data to extrapolate GEDI footprint-level forest canopy height measurements, creating a 30 m spatial resolution global forest canopy height map for the year 2019 (Potapov et al. 2021). Lang et al. (2023) provide global canopy height using GEDI LiDAR waveform data using ensemble of deep convolution neural networks (Figure 17.11b).

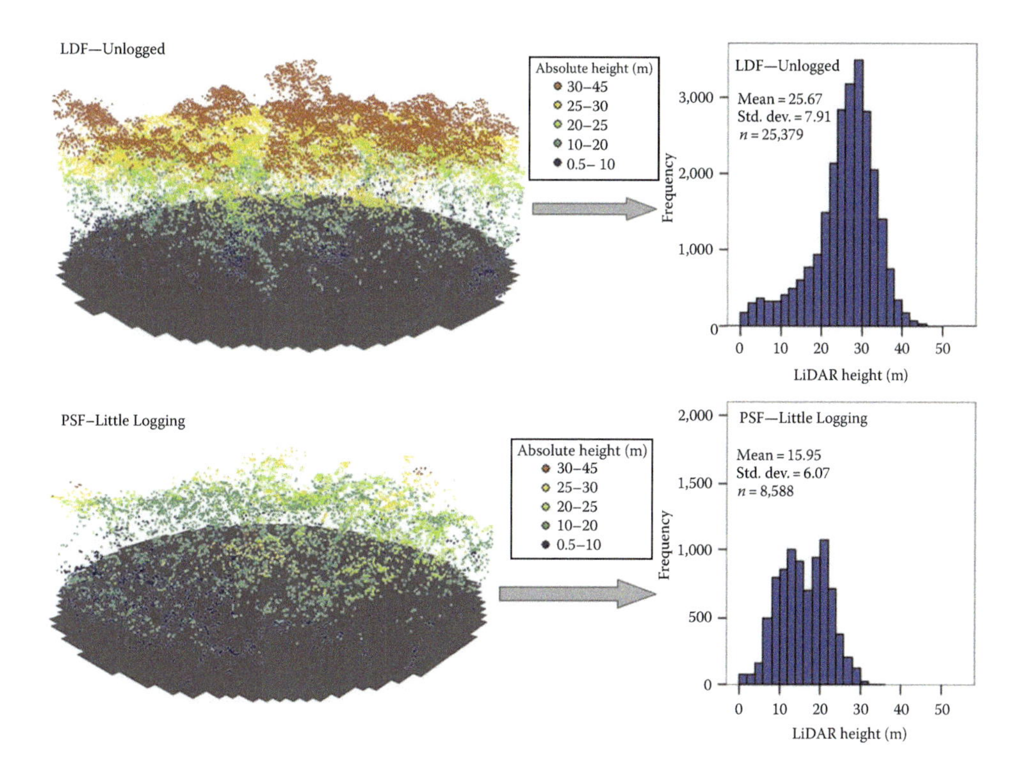

FIGURE 17.11A Distribution of LiDAR point heights within 1 ha plots: comparison of lowland dipterocarp forest and mixed peat swamp forest (LDF: lowland dipterocarp forest; PSF: peat swamp forest). Mean tree height, its standard deviation (std. dev.) and the total number of points higher than 0.5 m (n) is given in the histograms. (Source: Kronseder et al., 2012.)

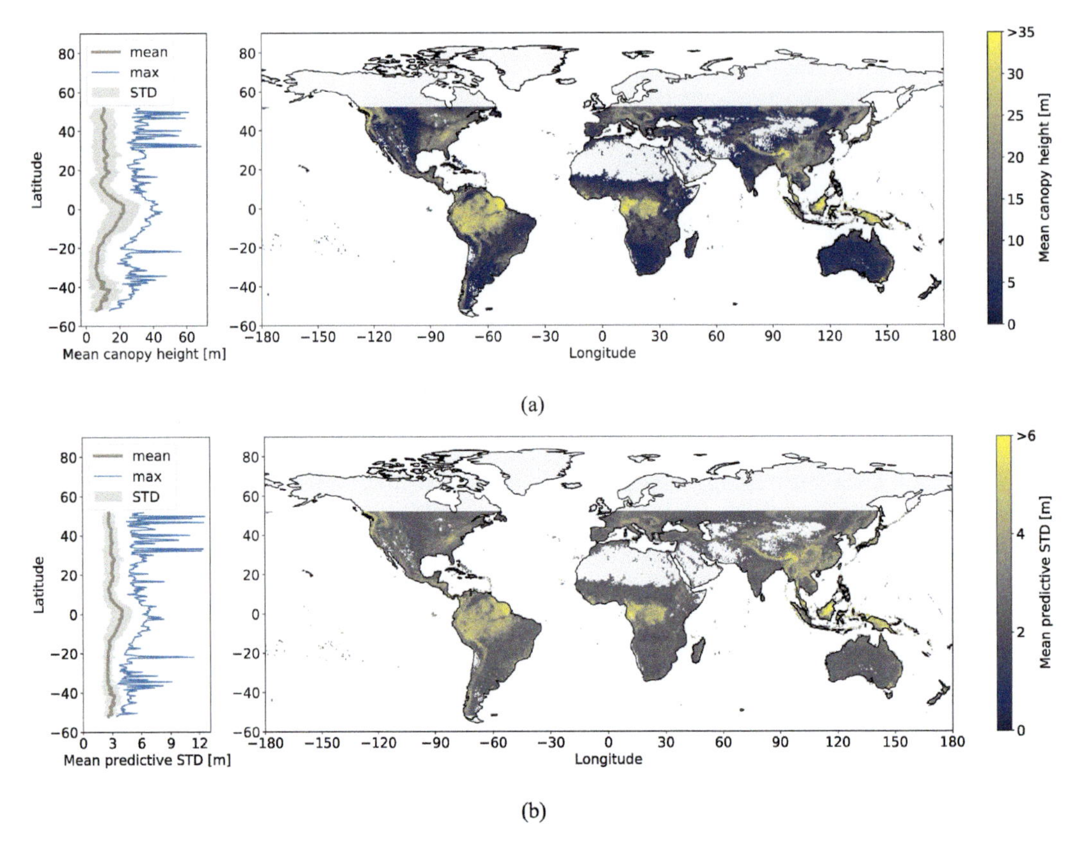

FIGURE 17.11B Global canopy height predictions. (a) Mean canopy top height at 0.5 degree resolution (≈55.5 km on the equator), created based on 346 × 10^6 waveform predictions from the first four months of the GEDI mission (April–July 2019). (b) Mean predictive uncertainty (standard deviation, denoted earlier as STD) at the same raster resolution. On the left, the latitudinal distribution of mean and max heights, integrated around the globe (Source: Lang et al., 2022.)

17.12 CHANGE DETECTION

Remote sensing by its inherent characteristics such as repeat coverage of any locations, and consistency of data offers the best data type for change detection anywhere on planet Earth. So, remote sensing, particularly spaceborne remote sensing, is widely used for change detection of a wide range of factors such as the land cover (physical footprint), land use (anthropogenic), forests, agriculture, glaciers, and biomass. Change detection may be achieved utilizing approaches such as the conventional visualization technique, an object-based change detection technique, a pixel-based change detection approach, and a hybrid change detection technique (Selvaraj and Natarajan, 2022). The first principle for change detection is accurate calibration, harmonization and normalization of multidate, multisensory images (Goward et al., 2012; Chander et al., 2009; Chavez, 1988, 1989; Elvidge et al., 1995).

Daniela Anjos, Dengsheng Lu and others in Chapter 12 provide an outline of change detection methods, reason out why remote sensing is central, identify various approaches for change detection, and present an example of land cover/land use change detection in the Brazilian Amazon. There are four broad approaches to change detection studies:

1. Conversion: detailed land cover change trajectories
2. Binary change: change versus no-change
3. Specific changes: deforestation, urbanization
4. Continuous change: forest disturbance

Chapter 12 presents a complex case of performing land cover/land use change detection using multi-sensor data. They use for one year radar remote sensing data from RADARSAT and for another date use optical remote sensing data from EO-1 ALI. This is not an ideal situation to get best results since using characteristically different data sources for change detection leads to higher degree of uncertainties. Nevertheless, using distinctly different remote sensing data is necessary, especially in areas of high cloud cover such as Amazon or Congo rainforests. The ability of Radar to penetrate the clouds becomes useful. So, in this regard, the case study presented for change detection in Chapter 12 is important. Further, Chapter 12 also shows us the utility of using the direct classification approach of SVMs as a preferred advanced method for change detection. Whereas simple and direct change detection approaches such as the difference imaging (e.g., NDVI differences; Vrieling et al., 2014) can be used, methods such as SVMs allow us to use the power of multiband data and from multiple sensors leading to relatively greater accuracies. Many other methods such as RF classifiers (e.g., Figure 17.12; Haas and Ban, 2014) can be used in change detection studies. With

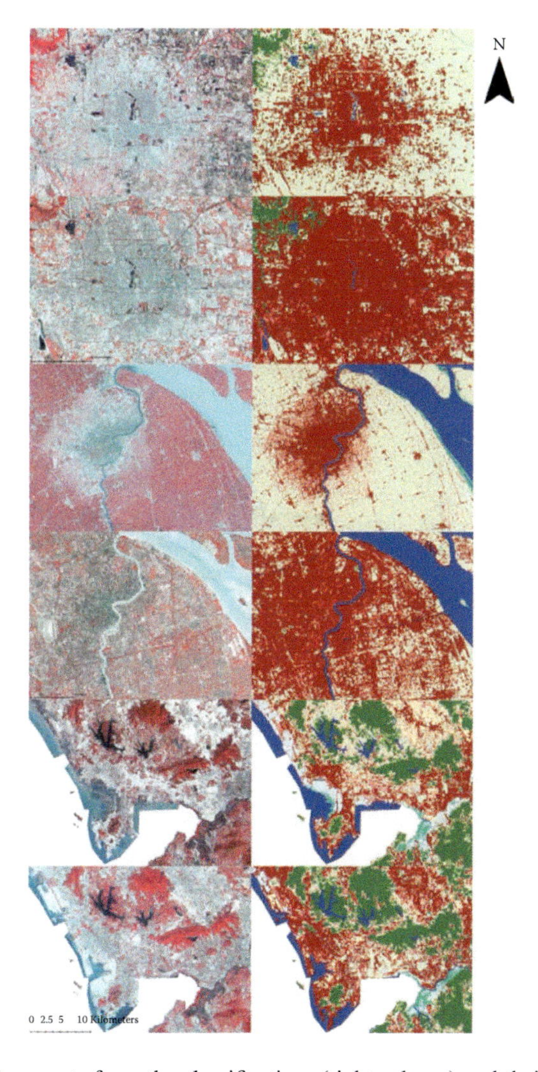

FIGURE 17.12 Detailed excerpts from the classifications (right column) and their respective areas in false color composite (FCC) images in the left. The six rows show the following areas in descending order: Beijing 1990, Beijing 2010, Shanghai 1990, Shanghai 2010, Shenzhen 1990, and Shenzhen 2010. Images used are Landsat TM and HJ-1A/B. Images were studied using random forest classifier, tassel cap transformations, urban indices, and landscape matrices. (Source: Haas and Ban, 2014.)

increasing spatial and spectral resolution of remote sensing data, newer methods and approaches of change detection are required. Ning et al. (2024) proposed and implemented a novel multi-stage progressive change detection network specifically designed for urbanization change detection, which integrates invariant detection, knowledge distillation, and a coarse-to-fine change detection structure utilizing very high spatial-resolution imagery (VHRI) of sub-meter to 2 m resolution. Deep learning has been introduced for automatic change detection and has achieved great success (Khelifi and Mignotte, 2020). Bai et al. (2022) found that DL change detection's (DLCD's) advantages over conventional change detection can be attributed to three factors: (1) improved information representation; (2) improved change detection methods; and (3) performance enhancements. Currently, AI technology has become a research focus in developing new change detection methods (Shi et al., 2020). With availability of remote sensing data from multiple satellites on near-real time and on a continuous basis over time and space, a paradigm shift away from change detection, typically using two points in time, to monitoring, or an attempt to track change continuously in time is proposed by Woodcock et al. (2020).

17.13 GEOPROCESSING, WORKFLOWS, AND PROVENANCE

Today remote sensing data goes through complex chains of events from acquisition, curation, preprocessing, applying analytical methods and approaches to derive and synthesize information, leading to post-processing and dissemination to users (e.g., Figure 17.13). Different laboratories or individuals may apply different models, workflows, inferences and processing chains toward achieving the same goal. For example, a forest change map may be produced using at-sensor surface reflectance from similar but distinct sensor systems (e.g., Landsat 5 TM versus Landsat 8 OLI). Another forest change map may be produced using a combination of optical, radar, and LiDAR data. Yet another study may also use the same combination of input data to produce a forest change map

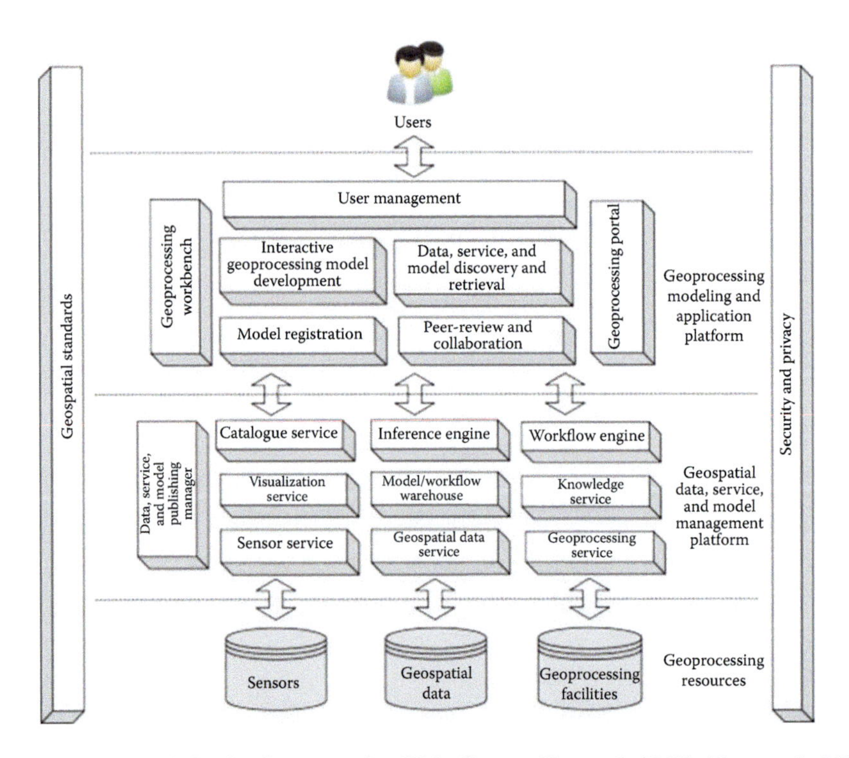

FIGURE 17.13 Framework for the Geoprocessing Web. (Source: Yue et al., 2010b; Zhao et al., 2012.)

but using a different analytical change detection algorithm. Unless a user has a full understanding on how the maps are produced, it is difficult to put real value to them or to even replicate another scientist's remote sensing workflow. This is where careful geospatial provenance (or lineage) becomes very valuable and effective. Provenance in this case entails keeping an exact historical and machine-readable record of the geospatial processes whereby a remote sensing product originated and was developed. Wu et al. (2023) describe a three-level conceptual view of provenance: (1) the automatic capture of provenance in the semantic execution engine; (2) the query, and (3) inference of provenance.

In Chapter 13, Dr. Jason A. Tullis et al. provide a historical context of geospatial provenance with a focus on David P. Lanter's (Lanter and Veregin, 1992) Geolineus project in the early 1990s, show us how remote sensing product quality, reproducibility, and trust depend on provenance, and provide a number of examples of recent provenance-aware remote sensing geoprocessing systems. They discuss current international standards, specifications, and recommendations pertaining to implementation of shared remote sensing provenance information. These include, for example, International Standards Organization's ISO 19115-2 metadata standard, endorsed by the Federal Geographic Data Committee or FGDC (Brodeur et al., 2019), and World Wide Web Consortium's PROV Data Model (Moreau et al., 2015) and accompanying recommendations of keen interest in the provenance research community. They identify questions pertaining to remote sensing data products that only provenance can help answer such as the exact processing steps applied to develop a remote sensing product, processing order, parameters, execution time, and how errors were expressed and propagated. Finally, they discuss provenance-aware systems applied to remote sensing such as MODIS Adaptive Data Processing System and OMI Data Processing System designed for use by NASA to manage satellite imagery and provenance from MODIS sensors on the Terra and Aqua satellites, and the OMI sensor on Aura respectively (Tilmes and Fleig, 2008). In recent advances, Linked Data, including the common data model, standardized data access mechanism, and link-based data discovery, allow effective sharing and discovery of geospatial resources in Spatial Data Infrastructures (Yue et al., 2016). Liang et al. (2023) proposed geospatial web services for Google Earth Engine (WS4GEE) which facilitates the wrapping of GEE datasets, functions, and models as GEE-enabled open geospatial consortium web map services, web feature services, web coverage services, and web processing services. WS4GEE can support direct access to GEE through OGC interfaces thus allowing geoprocessing workflows to use GEE resources (Liang et al., 2023).

17.14 TOWARD DEMOCRATIZATION OF GEOGRAPHIC INFORMATION

Chapter 14 by Dr. Gaurav Sinha et al. discusses the democratization of geo-information and geospatial information communication technologies (Geo-ICTs). Early use of geographic information technologies, much like computing, was an elitist intellectual activity where general public had hardly any participation. The information was processed, analyzed, and disseminated by very few highly specialized experts knowledgeable in remote sensing, GIS, GPS/GNSS, and spatial modeling. This top-down approach was criticized as a "hegemonic system perpetuating the grip of powerful agencies and trained professionals on geographic information" as viewed by several thinkers (Levidow et al., 2023). This was also a period where GIS, remote sensing, GNSS (Jia et al., 2021; Stosius et al., 2011; Dow et al., 2007) were mainly tools of well-funded national and international specialized institutes, and military establishments.

Chapter 14 discusses how geo-information has become ubiquitous, linking web mapping, wireless delivery, crowdsourcing (Heipke, 2010), and other digital technologies through expert collaboration and user participation into a seamless platform for powerful delivery of data and knowledge. One of the earliest influences of this democratization and merger of technologies was U.S. Vice President, Al Gore's "Digital Earth" vision (Annoni et al., 2023). As a result, access to digital geo-information through virtual globes is now integral to billions of people's daily lives. In addition to the accessibility of Geo-ICTs, neogeography (or new geography), participatory GIS, crowdsourcing,

VGI, and social networking have further integrated geo-information into people's lives. For example, there exist several crowdsourcing initiatives, especially for managing disasters like earthquakes (Mori and Takahashi, 2012; USGS, 2012), floods, and tsunamis (Suppasri et al., 2012; Fujima, 2011). One such example are flood databases, such as the International Disaster Database (EM-DAT), ReliefWeb (launched by the United Nations Office for the Coordination of Humanitarian Affairs), the International Flood Network and the Global Active Archive of Large Flood Events (created by the Dartmouth Flood Observatory) (Wan et al., 2014; Figure 17.14). Another global initiative called Geo-Wiki (Fritz et al., 2012) helps improve global land cover assessment and mapping (Belward and Skøien, 2014). Readers can expand on these ideas after reading Chapter 14. Long back, Prof. Duane Marble, one of the pioneers of geographic information science predicted that "all information will be geospatial," meaning that the information will be tied to precise geographic location (Du et al., 2022; Marble, 1990). Today, an enormous amount of information is captured location specific, and delivered with location specificity. However, this ubiquitous nature of geo-information, and the ease with which it can now be captured, stored, shared and sold, has also raised many questions about individual rights and freedom. Degbelo (2022) highlight the increasing need for techniques to make geovisualizations FAIR—Findable, Accessible, Interoperable, and Reusable. For example, development of user-friendly tools for democratizing the knowledge and awareness on possible impacts that might arise from natural hazards in the built environments, with special focus on historical areas are enumerated by Villani et al. (2023). At the same time, there also arise issues related to the quality of information generated and processed by non-professionals. For example, when data is crowdsourced,

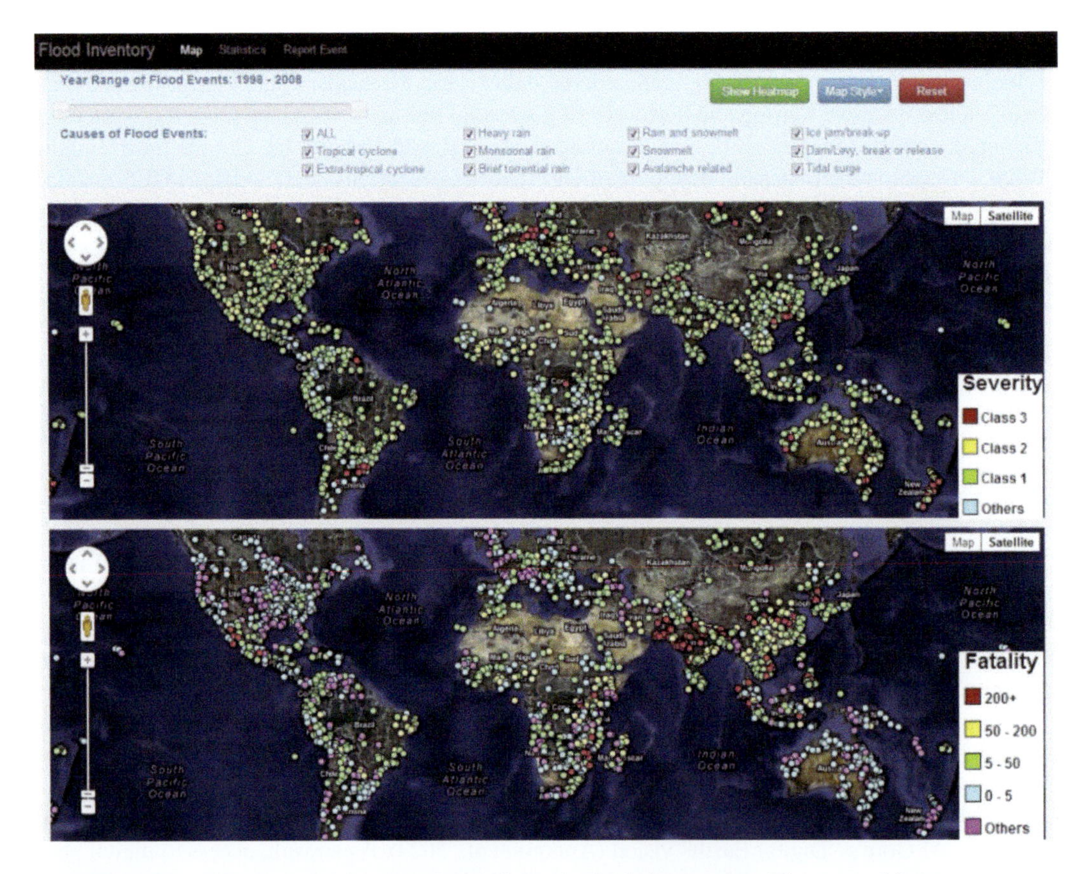

FIGURE 17.14 The map visualization of global flood cyberinfrastructure. The top and bottom maps are color coded by severity and fatalities respectively. (Source: Wan et al., 2014.)

experts as well as non-experts share data. Thus, quality of such data can vary widely depending on the source, which could even be detrimental to the progress of precise geo-information that is reliable and trusted.

17.15 GISCIENCE

GIScience is the science of geographic information and analysis that substrates the development of GIS and a host of related technologies (cartography, geodesy, surveying, photogrammetry, GPS, remote sensing, spatial modeling) for solving problems (e.g., detecting a landfill site; Figure17.15) related to a wide spectrum of themes such agriculture, water, forestry, geography, geology, geophysics, oceanography, ecology, environmental science, and social sciences. GIScience is contributing spatial thinking and spatial computing that propels many disciplines to "take a spatial turn," such as spatial ecology, spatial epidemiology, spatial social sciences, and spatial humanities, etc. For example, adopting a GIScience approach, and in particular, making use of location-based intelligence tools, help improve the shortcomings in data reporting as evidenced by more accurate reporting of how COVID-19 will have a long-term impact on global health (Rosenkrantz et al., 2021). Similarly, GIScience is a powerful tool for cancer research, bringing additional context to cancer data analysis and potentially informing decision making and policy, ultimately aimed at reducing the burden of cancer (Sahar et al., 2019). There are innumerable such examples of the power and the impact of GIScience in multitude of application of great importance to sustainable development and preservation of the planet Earth.

Soon after the advent of computers, early pioneers started the development of GIS technology in 1960s and 1970s. The maturing of GIS technology has led to a wide adoption of mapping and spatial analysis in domain applications with extensive data from multiple sources (e.g., GIS, GPS, remote sensing) and sophisticated algorithms (e.g., image processing algorithms, map algebra, spatial econometrics, etc.) for spatial decision support. GIScience grounds GIS technologies with theories of spatial representation, spatial relationships as well as a range of computational methods that transform the underlying conceptualization and problem-solving approaches. Today, GIScience research expands into data uncertainty, space-time analytics, and social implications on web mapping, crowdsourcing, and cloud computing as well as mobile delivery. In Chapter 15 Dr. May Yuan presents an outline and history of GIScience and highlights its core epistemology: Abstraction, Algorithms, and Assimilation. Also see Chapter 13 (geoprocessing), Chapter 14 (democratization of GIScience), and remote sensing (Chapters 1–11) and how GIScience is intricately linked to these techniques, especially going forward as discussed by Dr. Yuan in this chapter. Discrete global grid systems (DGGS), which are broadly location-coding systems, are gaining traction to be a building block for the next generation of GIS in support of the digital earth vision (Hojati et al., 2022). As enumerated by (Hojati et al., 2022), DGGS has many advantages over traditional raster and vector data models (Purss et al., 2019) and is actively developed by academics (Amiri et al., 2019; Alderson et al., 2018), industry (Uber, 2020), and others like the open geospatial consortium (Gibb et al., 2022). Integration of AI, machine learning, and DL with GIS (GeoAI) are emerging and provide new opportunities for handling and analyzing large spatial datasets, improving the accuracy and efficiency of spatial analysis tasks, and developing new tools and frameworks for integrating GIS with other disciplines (Choi, 2023).

17.16 OBJECT-BASED REGIONALIZATION FOR POLICY-ORIENTED PARTITIONING OF SPACE

One of the biggest challenges facing policy makers and other using remote sensing and other geospatial data derived information (Evangelidis et al., 2014) is to ensure its integrity, robustness, and reliability. Inherently remote sensing data is consistent and robust (with proper calibration and

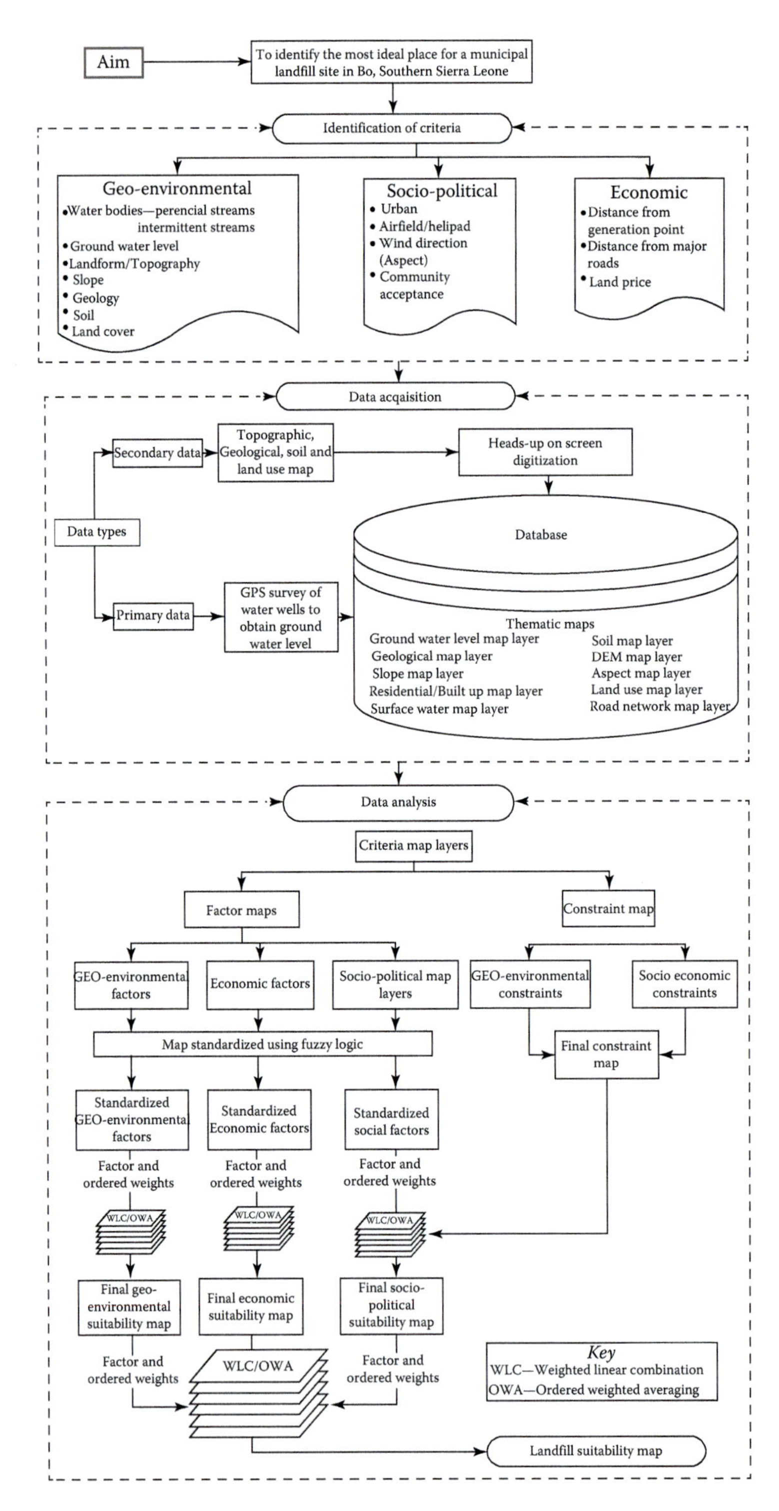

FIGURE 17.15 Spatial model for the identification of a potential landfill site by integrating spatial data from various sources (e.g., remote sensing, GIS, GPS). (Source: Gbanie et al., 2013.)

normalization; Chander et al., 2009). But methods and approaches in analyzing the data are often different. Further, in order to derive new information or enhance existing geospatial information, remote sensing data is used as a data source to be integrated with existing geospatial data in different ways. Resultant products are often dissimilar, depending on the algorithms used, or the weights assigned to each of the information source (e.g., Figure 17.16; Branger et al., 2013). These issues are discussed in Chapter 16 by Dr. Stefan Lang et al., introducing the geon approach as a strategy to integrate multiple sets of geospatial data. Object-based Geons represents meaningful units, and thus, are likely more suitable than per-pixel approaches for planning and mitigation (Nachappa et al., 2020; Nachappa and Meena, 2020).

Chapter 16 begins with definitions of latent, complex, and multidimensional phenomena and the related concept of object-based geons and sets the platform for its spatial representation that is robust and useful to decision makers in different domains. The authors discuss the principles of regionalization, how object-based geons are used in domain-specific regionalization, and object-based image analysis in image segmentation and regionalizations. They take four case studies (socioeconomic vulnerability of hazards, social vulnerability to Malaria, landscape sensitivity, and

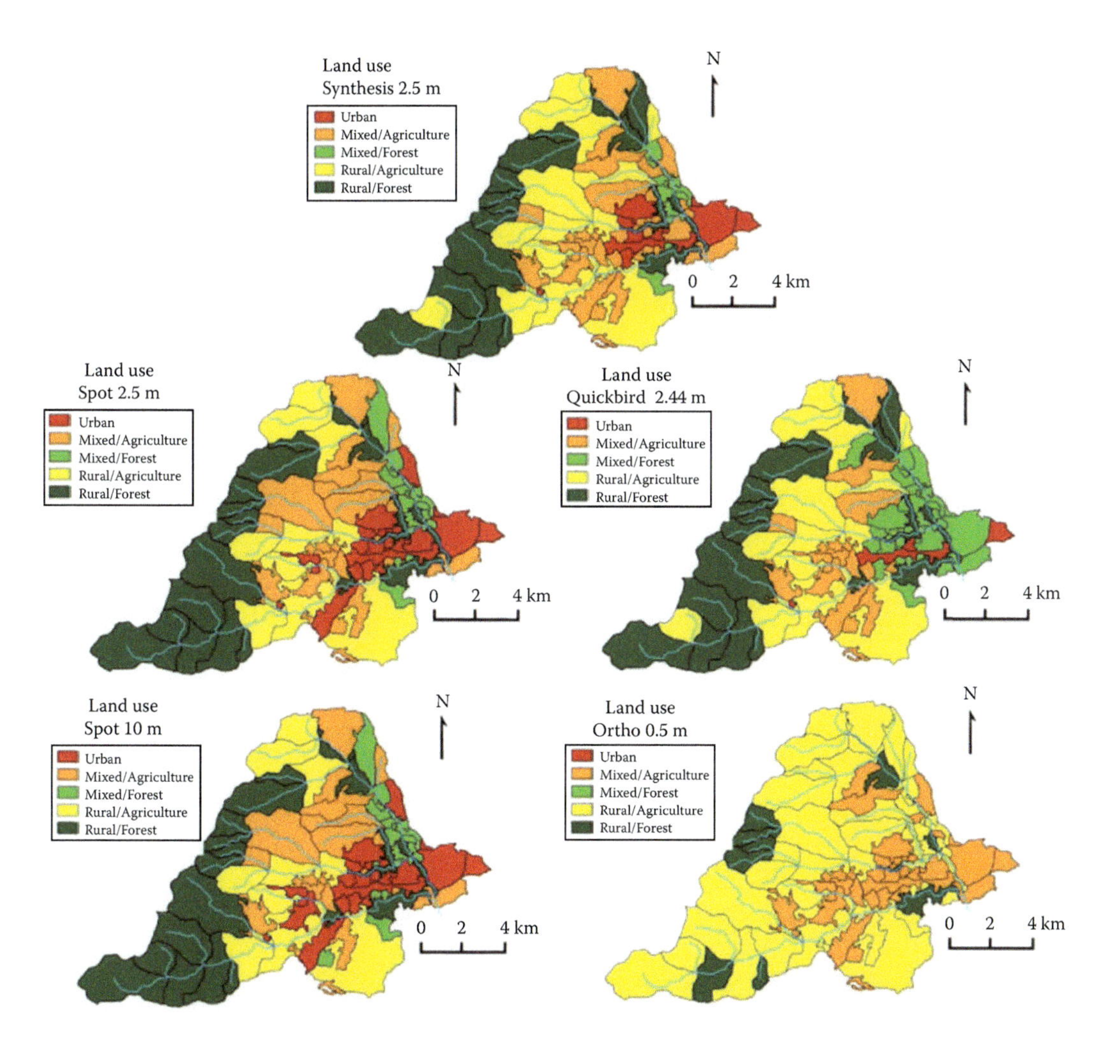

FIGURE 17.16 Model hydrological response units (HRUs) and reaches and land use classification for Synthesis 2008, Spot 2008 (2.5 m and 10 m), Quickbird 2008 and Ortho 2008 land use maps. (Source: Branger et al., 2013.)

climate change susceptibility) to show readers how latent phenomenon is addressed, what indicators and datasets are used, and the results. The object-based Geon methods can achieve equivalent or better mapping results with less computation cost compared with the pixel-based method, making it more suitable for large-scale mapping (Yan et al., 2022). They suggest that object-based geon methods will become more advantageous when the segmentation results of urban regions can be improved. Combined per-pixel and object-based geon approaches yielded better results than the per-pixel approaches alone in flood mapping (Nachappa and Meena, 2020) and in cropland mapping (Xiong et al., 2017; Thenkabail et al., 2021).

17.17 SYNTHESIS OF CHAPTER 17 OF VOLUME II

Volume II of *Remote Sensing Handbook* is focused on image processing, change detection and integration of remote sensing and GIS. The chapters provide introductory as well as advanced image understanding and digital image processing approaches, methods, and techniques. Currently, images are acquired in various spatial, spectral, temporal, and radiometric resolutions (Markham and Barker, 1987) from multitude of government and privately owned satellites. Further, archival of imagery of the planet Earth such as from Landsat (1972–present) is also increasing. In addition, sources of ancillary data that can be used in conjunction with imagery are also increasing by leaps and bounds. All of this calls for rapid advances in image processing. The digital image processing methods and techniques vary depending on these image characteristics and are addressed in different chapters of this volume. For example, when dealing with hyperspectral data, preprocessing steps should reduce data redundancy and remove wavebands in the atmospheric windows. Similarly, image segmentation approaches prior to using pixel-based classifications are found useful in increasing classification accuracies. Further, the image processing methods and techniques differ based on the applications such as land use/land cover, evapotranspiration, agriculture, forestry, and urban studies. For example, actual evapotranspiration and urban heat island studies require thermal emissivity data, the crop biophysical and biochemical quantification requires computation of specific vegetation indices (Price, 1987) and forest height, and biomass are best established using LiDAR data along with optical data. Overall, the new trends and advances in remote sensing data analysis are: (1) Data fusion involving integration of data from multiple sensors, (2) a paradigm shift away from change detection to track changes continuously or near-continuously (e.g., daily) in time (Woodcock et al., 2020), (3) Increasing utilization of image segmentation methods (e.g., OBIA, GEOBIA) in conjunction with pixel-based classifications, (4) Integration of optical, thermal, radar, and LiDAR data to improved estimates of quantities such as biomass, carbon, and crop water use or actual evapotranspiration, (5) Machine learning, DL, AI and cloud computing involving planet-scale petabyte volume big-data processing and analytics, (6) crowdsourcing of reference data for training machine learning algorithms and for testing, and validating products, (7) Automated classification approaches where mature algorithms have ability to past-cast, now-cast, and future-cast, (8) Development of hyperspectral vegetation indices (Thenkabail et al., 2002) for study of specific biophysical, biochemical, plant health, plant structure, and plant moisture quantities, (9) Ability of very high-spatial resolution data (VHRI) to accurately and automatically extract features such as road-networks, field boundaries, and tree crowns, and 10. Web-based analysis integrating remote sensing and GIS for spatial modeling and decision making.

17.18 ACKNOWLEDGMENTS

I would like to thank the lead authors and co-authors of each of the chapters for providing their insights and edits of my chapter summaries. I would like to express my gratitude to the United States Geological Survey (USGS) for the opportunity to work for it and contribute to remote sensing science. Over the years, a great deal of my remote sensing science research was performed through various NASA grants. Very grateful for it. Thanks to my USGS colleagues Dr. Pardhasaradhi

Teluguntla, Dr. Itiya Aneece, Mr. Adam Oliphant, and Mr. Daniel Foley for their support at various times of this book project. Thanks also to Dr. Murali Krishna Gumma of the International Crops Research Institute for the Semi-Arid Tropics (ICRISAT). Use of trade, firm, or product names is for descriptive purposes only and does not imply endorsement by the U.S. government.

REFERENCES

Abdi, A.M. 2020. Land cover and land use classification performance of machine learning algorithms in a boreal landscape using Sentinel-2 data. *GIScience & Remote Sensing*, 57(1), 1–20. http://doi.org/10.1080/15481603.2019.1650447

Alderson, T., Mahdavi-Amiri, A., and Samavati, F. 2018. Offsetting spherical curves in vector and raster form. *The Visual Computer*, 34(6), 973–984.

Amiri, A.M., Alderson, T., and Samavati, F. 2019. Geospatial data organization methods with emphasis on aperture-3 hexagonal discrete global grid systems cartographica. *The International Journal for Geographic Information and Geovisualization*, 54(1), 30–50.

Aneece, I., and Thenkabail, P.S. 2021. Classifying crop types using two generations of hyperspectral sensors (Hyperion and DESIS) with machine learning on the cloud. *Remote Sensing*, 13(22), 4704. https://doi.org/10.3390/rs13224704

Aneece, I., and Thenkabail, P.S. 2022. New generation hyperspectral sensors DESIS and PRISMA provide improved crop classifications. *Photogrammetric Engineering and Remote Sensing*, 88(11), 715–729.

Annoni, A., Nativi, S., Çöltekin, A., Desha, C., Eremchenko, E., Gevaert, C.M., . . . Tumampos, S. 2023. Digital earth: Yesterday, today, and tomorrow. *International Journal of Digital Earth*, 16(1), 1022–1072. https://doi.org/10.1080/17538947.2023.2187467

Bai, T., Wang, L., Yin, D., Sun, K., Chen, Y., Li, W., and Li, D. 2022. Deep learning for change detection in remote sensing: A review. *Geo-Spatial Information Science*, 26(3), 262–288. https://doi.org/10.1080/10095020.2022.2085633

Bai, T., Wang, L., Yin, D., Sun, K., Chen, Y., Li, W., and Li, D. 2023. Deep learning for change detection in remote sensing: A review. *Geo-spatial Information Science*, 26(3), 262–288. http://doi.org/10.1080/10095020.2022.2085633

Belward, A.S., and Skøien, J.O. 2014. Who launched what, when and why; trends in global land-cover observation capacity from civilian earth observation satellites. *ISPRS Journal of Photogrammetry and Remote Sensing*, Available online 28 April 2014. ISSN 0924-2716. http://doi.org/10.1016/j.isprsjprs.2014.03.009.

Blaschke, T. 2010. Object-based image analysis for remote sensing. *ISPRS International Journal of Photogrammetry and Remote Sensing*, 65(1), 2–16.

Blaschke, T., Hay, G.J., Kelly, M., Lang, S., Hofmann, P., Addink, E., Feitosa, R., van der Meer, F., van der Werff, H., Van Coillie, F., and Tiede, D. 2014. Geographic object-based image analysis: A new paradigm in remote sensing and geographic information science. *ISPRS International Journal of Photogrammetry and Remote Sensing*, 87(1), 180–191.

Branger, F., Kermadi, S., Jacqueminet, C., Michel, K., Labbas, M., Krause, P., Kralisch, S., and Braud, I. 2013. Assessment of the influence of land use data on the water balance components of a peri-urban catchment using a distributed modelling approach. *Journal of Hydrology*, 505, 312–325. ISSN 0022-1694. http://doi.org/10.1016/j.jhydrol.2013.09.055.

Brodeur, J., Coetzee, S., Danko, D., Garcia, S., and Hjelmager, J. 2019. Geographic information metadata—An outlook from the international standardization perspective. *ISPRS International Journal of Geo-Information*, 8, 280. https://doi.org/10.3390/ijgi8060280

Burger, W., and Burge, M.J. 2022. Digital image processing. *Texts in Computer Science* (3rd ed.). Springer Nature Switzerland AG, p. 944. ISSN 1868-0941, ISBN 978-3-031-05744-1 (eBook). https://doi.org/10.1007/978-3-031-05744-1_1.

Cai, G., Ren, H., Yang, L., Zhang, N., Du, M., and Wu, C. 2019. Detailed urban land use land cover classification at the metropolitan scale using a three-layer classification scheme. *Sensors*, 19(14), 3120. https://doi.org/10.3390/s19143120

Cawse-Nicholson, K., Raiho, A.M., Thompson, D.R., Hulley, G.C., Miller, C.E., Miner, K.R., Poulter, B., Schimel, D., Schneider, F.D., Townsend, P.A., and Zareh, S.K. 2023. Surface biology and geology imaging spectrometer: A case study to optimize the mission design using intrinsic dimensionality. *Remote Sensing of Environment*, 290(2023), 113534. ISSN 0034-4257. https://doi.org/10.1016/j.rse.2023.113534. https://www.sciencedirect.com/science/article/pii/S0034425723000858

Chander, G., Markham, B.L., and Helder, D.L. 2009. Summary of current radiometric calibration coefficients for Landsat MSS, TM, ETM+, and EO-1 ALI sensors. *Remote Sensing of Environment*, 113, 893–903.

Chaturvedi, V., and de Vries, W.T. 2021. Machine learning algorithms for urban land use planning: A review. *Urban Science*, 5(3), 68. https://doi.org/10.3390/urbansci5030068

Chavez, P.S. 1988. An improved dark-object subtraction technique for atmospheric scattering correction of multispectral data. *Remote Sensing of Environment*, 24, 459–479.

Chavez, P.S. 1989. Radiometric calibration of Landsat thematic mapper multispectral images. *Photogrammetric Engineering and Remote Sensing*, 55, 1285–1294.

Chen, G., Weng, Q., Hay, G.J., and He, Y. 2018a. Geographic object-based image analysis (GEOBIA): Emerging trends and future opportunities. *GIScience & Remote Sensing*, 55(2), 159–182. http://doi.org/10.1080/15481603.2018.1426092

Chen, Y., Zhou, Y., Ge, Y., An, R., and Chen, Y. 2018b. Enhancing land cover mapping through integration of pixel-based and object-based classifications from remotely sensed imagery. *Remote Sensing*, 10, 77. https://doi.org/10.3390/rs10010077

Choi, Y. 2023. GeoAI: Integration of artificial intelligence, machine learning, and deep learning with GIS. *Applied Sciences*, 13(6), 3895. https://doi.org/10.3390/app13063895

Comber, A., Fisher, P., Brunsdon, C., and Khmag, A. 2012. Spatial analysis of remote sensing image classification accuracy. *Remote Sensing of Environment*, 127, 237–246. ISSN 0034-4257. http://doi.org/10.1016/j.rse.2012.09.005

Courtois, M., & Traizet, M. 1986. The SPOT satellites: From SPOT 1 to SPOT 4. *Geocarto International*, 1(3), 4–14. https://doi.org/10.1080/10106048609354050

Cravero, A., Pardo, S., Galeas, P., López Fenner, J., and Caniupán, M. 2022. Data type and data sources for agricultural big data and machine learning. *Sustainability*, 14, 16131. https://doi.org/10.3390/su142316131

Crawford, C.J., Roy, D.P., Arab, S., Barnes, C., Vermote, E., Hulley, G., Gerace, A., Choate, M., Engebretson, C., Micijevic, E., Schmidt, G., Anderson, C., Anderson, M., Bouchard, M., Cook, B., Dittmeier, R., Howard, D., Jenkerson, C., Kim, M., Kleyians, T., Maiersperger, T., Mueller, C., Neigh, C., Owen, L., Page, B., Pahlevan, N., Rengarajan, R., Roger, J.C., Sayler, K., Scaramuzza, P., Skakun, S., Yan, L., Zhang, H.K., Zhu, Z., and Zahn, S. 2023. The 50-year Landsat collection 2 archive. *Science of Remote Sensing*, 8, 100103. ISSN 2666-0172. https://doi.org/10.1016/j.srs.2023.100103. https://www.sciencedirect.com/science/article/pii/S2666017223000287

Damodaran, B.B., and Nidamanuri, R.R. 2014. Assessment of the impact of dimensionality reduction methods on information classes and classifiers for hyperspectral image classification by multiple classifier system. *Advances in Space Research*, 53(12), 1720–1734. ISSN 0273-1177. http://doi.org/10.1016/j.asr.2013.11.027

Dardel, C., Kergoat, L., Hiernaux, P., Mougin, E., Grippa, M., and Tucker, C.J. 2014. Re-greening Sahel: 30 years of remote sensing data and field observations (Mali, Niger). *Remote Sensing of Environment*, 140, 350–364. ISSN 0034-4257. http://doi.org/10.1016/j.rse.2013.09.011

Degbelo, A. 2022. FAIR geovisualizations: definitions, challenges, and the road ahead. *International Journal of Geographical Information Science*, 36(6), 1059–1099. https://doi.org/10.1080/13658816.2021.1983579

Delogu, G., Caputi, E., Perretta, M., Ripa, M.N., and Boccia, L. 2023. Using PRISMA hyperspectral data for land cover classification with artificial intelligence support. *Sustainability*, 15(18), 13786. https://doi.org/10.3390/su151813786

Dow, J.M., Neilan, R.E., Weber, R., and Gendt, G. 2007. Galileo and the IGS: Taking advantage of multiple GNSS constellations. *Advances in Space Research*, 39(10), 1545–1551. ISSN 0273-1177. http://doi.org/10.1016/j.asr.2007.04.064

Du, J., Wang, S., Ye, X., Sinton, D.S., and Kemp, K. 2022. GIS-KG: Building a large-scale hierarchical knowledge graph for geographic information science. *International Journal of Geographical Information Science*, 36(5), 873–897. https://doi.org/10.1080/13658816.2021.2005795

Du, S., Shu, M., and Wang, Q. 2019. Modelling relational contexts in GEOBIA framework for improving urban land-cover mapping. *GIScience & Remote Sensing*, 56(2), 184–209. https://doi.org/10.1080/15481603.2018.1502399

Elvidge, C.D., Yuan, D., Weerackoon, R.D., and Lunetta, R.S. 1995. Relative radiometric normalization of landsat multispectral scanner (MSS) data using an automatic scattergram controlled regression. *Photogrammetric Engineering and Remote Sensing*, 61, 1255–1260.

Espindola, G., Câmara, G., Reis, I., Bins, L., and Monteiro, A. 2006. Parameter selection for region-growing image segmentation algorithms using spatial autocorrelation. *International Journal of Remote Sensing*, 27, 3035–3040.

Evangelidis, K., Ntouros, K., Makridis, N., and Papatheodorou, C. 2014. Geospatial services in the cloud. *Computers & Geosciences*, 63, 116–122. ISSN 0098-3004. http://doi.org/10.1016/j.cageo.2013.10.007

Feng, J., Feng, X., Chen, J., Cao, X., Zhang, X., Jiao, L., and Yu, T. 2020. Generative adversarial networks based on collaborative learning and attention mechanism for hyperspectral image classification. *Remote Sensing*, 12(7), 1149. https://doi.org/10.3390/rs12071149

Freire, S., Santos, T., Navarro, A., Soares, F., Silva, J.D., Afonso, N., Fonseca, A., and Tenedório, J. 2014. Introducing mapping standards in the quality assessment of buildings extracted from very high resolution satellite imagery. *ISPRS Journal of Photogrammetry and Remote Sensing*, 90, 1–9. ISSN 0924-2716. http://doi.org/10.1016/j.isprsjprs.2013.12.009

Fritz, S., McCallum, I., Schill, C., Perger, C., See, L., Schepaschenko, D., van der Velde, M., Kraxner, F., and Obersteiner, M. 2012. Geo-Wiki: An online platform for improving global land cover. *Environmental Modelling & Software*, 31, 110–123. ISSN 1364-8152. http://doi.org/10.1016/j.envsoft.2011.11.015

Fujima, K. 2011. *Tsunami Measurement Data Compiled by IUGG Tsunami Commission*. http://www.nda.ac.jp/cc/users/fujima/TMD/index.html [accessed 02.12.11].

Gao, Q., Lim, S., and Jia, X. 2018. Hyperspectral image classification using convolutional neural networks and multiple feature learning. *Remote Sensing*, 10(2), 299. https://doi.org/10.3390/rs10020299

Gbanie, S.P., Tengbe, P.B., Momoh, J.S., Medo, J., and Kabba, V.T.S. 2013. Modelling landfill location using Geographic Information Systems (GIS) and Multi-Criteria Decision Analysis (MCDA): Case study Bo, Southern Sierra Leone. *Applied Geography*, 36, 3–12. ISSN 0143-6228. http://doi.org/10.1016/j.apgeog.2012.06.013

Gerhards, M., Schlerf, M., Mallick, K., and Udelhoven, T. 2019. Challenges and future perspectives of multi-/hyperspectral thermal infrared remote sensing for crop water-stress detection: A review. *Remote Sensing*, 11(10), 1240. https://doi.org/10.3390/rs11101240

Gibb, R.G., Purss, M.B.J., Sabeur, Z., Strobl, P., and Qu, T. 2022. Global reference grids for big earth data. *Big Earth Data*, 6(3), 251–255. https://doi.org/10.1080/20964471.2022.2113037

Goward, S.N., Ghander, G., Pagnutti, M., Marx, A., Ryan, R., Thomas, N., and Tetrault, R. 2012. Complementarity of resourceSat-1 AwiFS and Landsat TM/ETM+ sensors. *Remote Sensing of Environment*, 123, 41–560. ISSN 0034-4257. http://doi.org/10.1016/j.rse.2012.03.002

Gu, H., Li, H., Yan, L., Liu, Z., Blaschke, T., and Soergel, U. 2017. An object-based semantic classification method for high resolution remote sensing imagery using ontology. *Remote Sensing*, 9(4), 329. https://doi.org/10.3390/rs9040329

Haas, J., and Ban, Y. 2014. Urban growth and environmental impacts in Jing-Jin-Ji, the Yangtze, River Delta and the Pearl River Delta. *International Journal of Applied Earth Observation and Geoinformation*, 30, 42–55. ISSN 0303-2434. http://doi.org/10.1016/j.jag.2013.12.012.

Hafner, E.D., Barton, P., Daudt, R.C., Wegner, J.D., Schindler, K., and Bühler, Y. 2022. Automated avalanche mapping from SPOT 6/7 satellite imagery with deep learning: Results, evaluation, potential and limitations. *The Cryosphere*, 16, 3517–3530. https://doi.org/10.5194/tc-16-3517-2022

Hartling, S., Sagan, V., Sidike, P., Maimaitijiang, M., and Carron, J. 2019. Urban tree species classification using a worldview-2/3 and LiDAR data fusion approach and deep learning. *Sensors*, 19(6), 1284. https://doi.org/10.3390/s19061284

Heipke, C. 2010. Crowdsourcing geospatial data. *ISPRS Journal of Photogrammetry and Remote Sensing*, 65(6), 550–557. ISSN 0924-2716. http://doi.org/10.1016/j.isprsjprs.2010.06.005

Hersperger, A.M., Oliveira, E., Pagliarin, S., Palka, G., Verburg, P., Bolliger, J., and Grădinaru, S. 2018. Urban land-use change: The role of strategic spatial planning. *Global Environmental Change*, 51, 32–42. ISSN 0959-3780. https://doi.org/10.1016/j.gloenvcha.2018.05.001. https://www.sciencedirect.com/science/article/pii/S0959378017306829

Hojati, M., Robertson, C., Roberts, S., and Chaudhuri, C. 2022. GIScience research challenges for realizing discrete global grid systems as a Digital Earth. *Big Earth Data*, 6(3), 358–379. http://doi.org/10.1080/20964471.2021.2012912

Hossain, M.D., and Chen, D. 2019. Segmentation for object-based image analysis (OBIA): A review of algorithms and challenges from remote sensing perspective. *ISPRS Journal of Photogrammetry and Remote Sensing*, 150, 115–134. ISSN 0924-2716. https://doi.org/10.1016/j.isprsjprs.2019.02.009. https://www.sciencedirect.com/science/article/pii/S0924271619300425

Hu, T., Su, Y., Xue, B., Liu, J., Zhao, X., Fang, J., and Guo, Q. 2016. Mapping global forest aboveground biomass with spaceborne LiDAR, optical imagery, and forest inventory data. *Remote Sensing*, 8(7), 565. https://doi.org/10.3390/rs8070565

Im, J., Jensen, J.R., and Tullis, J.A. 2008. Object-based change detection using correlation image analysis and image segmentation. *International Journal of Remote Sensing*, 29, 399–423.

Jensen, J.R. 1996. *Introductory Digital Image Processing: A Remote Sensing Perspective*, 3rd ed. Prentice Hall, p. 318. ISBN: 0-13-145361-0.

Jia, J., Deng, X., Zhu, J., Xu, M., Zhou, J., and Jia, X. 2019. Collaborative representation-based multiscale superpixel fusion for hyperspectral image classification. *IEEE Transactions on Geoscience and Remote Sensing*, 57(10), 7770–7784. https://doi.org/10.1109/TGRS.2019.2916329

Jia, S., Deng, X., Zhu, J., Xu, M., Zhou, J., and Jia, X. 2020. Collaborative representation-based multiscale superpixel fusion for hyperspectral image classification. *IEEE Transactions on Geoscience and Remote Sensing*, 57(10), 7770–7784. https://doi.org/10.1109/TGRS.2019.2916329

Jia, S., Jiang, S., Lin, Z., Li, N., Xu, M., and Yu, S. 2021. A survey: Deep learning for hyperspectral image classification with few labeled samples. *Neurocomputing*, 448, 179–204. ISSN 0925-2312. https://doi.org/10.1016/j.neucom.2021.03.035

Joshua, M., Salvaggio, K., Keremedjiev, M., Roth, K., and Foughty, E. 2023. Planet's upcoming VIS-SWIR hyperspectral satellites. *Optica Sensing Congress 2023* (AIS, FTS, HISE, Sensors, ES), Technical Digest Series (Optica Publishing Group, 2023), paper HM3C.5.

Ke, Y., Quackenbush, L.J., and Im, J. 2010. Synergistic use of QuickBird multispectral imagery and LIDAR data for object-based forest species classification. *Remote Sensing of Environment*, 114(6), 1141–1154. ISSN 0034-4257. http://doi.org/10.1016/j.rse.2010.01.002

Keshtkar, H., Voigt, W., and Alizadeh, E. 2017. Land-cover classification and analysis of change using machine-learning classifiers and multi-temporal remote sensing imagery. *Arabian Journal of* Geosciences, 10, 154. https://doi.org/10.1007/s12517-017-2899-y

Khan, M.J., Khan, H.S., Yousaf, A., Khurshid, K., and Abbas, A. 2018. Modern trends in hyperspectral image analysis: A review. *IEEE Access*, 6, 14118–14129. http://doi.org/10.1109/ACCESS.2018.2812999

Khanal, S., Fulton, J., and Shearer, S. 2017. An overview of current and potential applications of thermal remote sensing in precision agriculture. *Computers and Electronics in Agriculture*, 139(2017), 22–32. ISSN 0168-1699. https://doi.org/10.1016/j.compag.2017.05.001. https://www.sciencedirect.com/science/article/pii/S0168169916310225

Khelifi, K., and Mignotte, M. 2020. Deep learning for change detection in remote sensing images: Comprehensive review and meta-analysis. *IEEE Access*, 8, 126385–126400. http://doi.org/10.1109/ACCESS.2020.3008036

Kotaridis, I., and Lazaridou, M. 2021. Remote sensing image segmentation advances: A meta-analysis. *ISPRS Journal of Photogrammetry and Remote Sensing*, 173, 309–322. ISSN 0924-2716. https://doi.org/10.1016/j.isprsjprs.2021.01.020. https://www.sciencedirect.com/science/article/pii/S0924271621000265

Kronseder, K., Ballhorn, U., Böhm, V., and Siegert, F. 2012. Above ground biomass estimation across forest types at different degradation levels in Central Kalimantan using LiDAR data. *International Journal of Applied Earth Observation and Geoinformation*, 18, 37–48. ISSN 0303-2434. http://doi.org/10.1016/j.jag.2012.01.010

Kucharczyk, M., Hay, G.H., Ghaffarian, S., and Hugenholtz, C.H. 2020. Geographic object-based image analysis: A primer and future directions. *Remote Sensing*, 12(12), 2012. https://doi.org/10.3390/rs12122012

Kullberg, E.G., DeJonge, K.C., and Chávez, J.L. 2017. Evaluation of thermal remote sensing indices to estimate crop evapotranspiration coefficients. *Agricultural Water Management*, 179, 64–73. ISSN 0378-3774. https://doi.org/10.1016/j.agwat.2016.07.007. https://www.sciencedirect.com/science/article/pii/S0378377416302530

Lang, N., Kalischek, N., Armston, J., Schindler, K., Dubayah, R., and Wegner, J.D. 2022. Global canopy height regression and uncertainty estimation from GEDI LIDAR waveforms with deep ensembles. *Remote Sensing of Environment*, 268, 112760. ISSN 0034–4257. https://doi.org/10.1016/j.rse.2021.112760. https://www.sciencedirect.com/science/article/pii/S0034425721004806

Lang, N., Jetz, W., Schindler, K. et al. 2023. A high-resolution canopy height model of the Earth. *Nature Ecology and Evolution*, 7, 1778–1789. https://doi.org/10.1038/s41559-023-02206-6

Lanter, D.P., and Veregin, H. 1992. A research paradigm for error propagation in layer-based GIS. *Photogrammetric Engineering and Remote Sensing*, 58(6), 825–833.

Latif, R.M.A., He, J., and Umer, M. 2023. Mapping cropland extent in Pakistan using machine learning algorithms on Google earth engine cloud computing framework. *ISPRS International Journal of Geo-Information*, 12, 81. https://doi.org/10.3390/ijgi12020081

Levidow, L. 2023. 6: Green new deal agendas: System change versus continuity. In *Beyond Climate Fixes*. Bristol University Press. Retrieved Jul 26, 2024, from https://doi.org/10.51952/9781529222418.ch006

Li, S., Song, W., Fang, L., Chen, Y., Ghamisi, P., and Benediktsson, J.A. 2019. Deep learning for hyperspectral image classification: An overview. *IEEE Transactions on Geoscience and Remote Sensing*, 57(9), 6690–6709. http://doi.org/10.1109/TGRS.2019.2907932

Liang, J., Jin, F., Zhang, X., and Wu, H. 2023. WS4GEE: Enhancing geospatial web services and geoprocessing workflows by integrating the Google Earth Engine. *Environmental Modelling & Software*, 161, 105636. ISSN 1364-8152. https://doi.org/10.1016/j.envsoft.2023.105636. https://www.sciencedirect.com/science/article/pii/S1364815223000221

Lillesand, T.M., Kiefer, R.W., and Chipman, J.W. 2008. *Remote Sensing and Image Interpretation*, 6th ed., p. 768. ISBN: 978-0-470-46555-4.

Liu, B., Yu, X., Zhang, P., Yu, A., Fu, Q., and Wei, X. 2018. Supervised deep feature extraction for hyperspectral image classification. *IEEE Transactions on Geoscience and Remote Sensing*, 56(4), 1909–1921. http://doi.org/10.1109/TGRS.2017.276967

Ma, W., Gong, C., Hu, Y., Meng, P., and Xu, F. 2013. The Hughes phenomenon in hyperspectral classification based on the ground spectrum of grasslands in the region around Qinghai Lake. *Proc. SPIE 8910, International Symposium on Photoelectronic Detection and Imaging 2013: Imaging Spectrometer Technologies and Applications*, 89101G. https://doi.org/10.1117/12.2034457

Macarringue, L., Bolfe, É., and Pereira, P. 2022. Developments in land use and land cover classification techniques in remote sensing: A review. *Journal of Geographic Information System*, 14, 1–28. http://doi.org/10.4236/jgis.2022.141001

Marble, D. 1990. Geographic information systems: An overview. In *Introductory Readings in Geographic Information Systems*, 1st ed. CRC Press, p. 11. eBook ISBN9780429220715

Markham, B.L., and Barker, J.L. 1987. Radiometric properties of U.S. processed Landsat MSS data. *Remote Sensing of the Environment*, 22, 39–71.

Massey, R., Sankey, T.T., Yadav, K., Congalton, R.G., and Tilton, J.C. 2018. Integrating cloud-based workflows in continental-scale cropland extent classification. *Remote Sensing of Environment*, 219, 162–179. ISSN 0034-4257. https://doi.org/10.1016/j.rse.2018.10.013. https://www.sciencedirect.com/science/article/pii/S0034425718304619

Maxwell, A.E., Warner, T.A., and Fang, F. 2018. Implementation of machine-learning classification in remote sensing: An applied review. *International Journal of Remote Sensing*, 39(9), 2784–2817. http://doi.org/10.1080/01431161.2018.1433343

Moreau, L., Groth, P., Cheney, J., Lebo, T., and Miles, S. 2015. The rationale of PROV. *Journal of Web Semantics*, 35, Part 4, 235–257, ISSN 1570-8268. https://doi.org/10.1016/j.websem.2015.04.001. https://www.sciencedirect.com/science/article/pii/S1570826815000177

Mori, N., and Takahashi, T. 2012. The 2011 Tohoku Earthquake Tsunami Joint Survey Group Nationwide post event survey and analysis of the 2011. *Tohoku Earthquake Tsunami Coastal Engineering Journal*, 54(2012), 1250001.

Mueller, A.A., Hausold, A., and Strobl, P. 2022. HySens-DAIS/ROSIS imaging spectrometers at DLR. *Proc. SPIE 4545, Remote Sensing for Environmental Monitoring, GIS Applications, and Geology* (23 January 2002). https://doi.org/10.1117/12.453677

Murthy, M.V.R., Usha, T., and Kankara, R.S. 2022. Three decades of Indian remote sensing in coastal research. *Journal of the Indian Society of Remote Sensing*, 50, 599–612. https://doi.org/10.1007/s12524-021-01342-5

Nachappa, T.G., Kienberger, S., Meena, S.R., Hölbling, D., and Blaschke, T. 2020. Comparison and validation of per-pixel and object-based approaches for landslide susceptibility mapping, Geomatics. *Natural Hazards and Risk*, 11(1), 572–600. http://doi.org/10.1080/19475705.2020.1736190

Nachappa, T.G., and Meena, S.R. 2020. A novel per pixel and object-based ensemble approach for flood susceptibility mapping, Geomatics. *Natural Hazards and Risk*, 11(1), 2147–2175. http://doi.org/10.1080/19475705.2020.1833990

Nagamalai, D., Renaulat, E., and Dhanuskodi, M. 2011. *Advances in Digital Image Processing and Information Technology: First International Conference on Digital Image Processing and Pattern Recognition, in Computer and Information Science*, 2011 ed. Paperback. Springer, p. 478. ISBN-10: 3642240542.

Ning, X., Zhang, H., Zhang, R., and Huang, X. 2024. Multi-stage progressive change detection on high resolution remote sensing imagery. *ISPRS Journal of Photogrammetry and Remote Sensing*, 207, 231–244. ISSN 0924-2716. https://doi.org/10.1016/j.isprsjprs.2023.11.023. https://www.sciencedirect.com/science/article/pii/S0924271623003404

Oppelt, N., Schulze, F., Bartsch, I., Doernhoefer, K., and Eisenhardt, I. 2012. Hyperspectral classification approaches for intertidal macroalgae habitat mapping: A case study in Helgoland. *Optical Engineering*, 51, 111703. http://doi.org/10/1117/1.OE.51.11.111703.

Pandey, P.C., Koutsias, N., Petropoulos, G.P., Srivastava, P.K., and Ben Dor, E. 2021. Land use/land cover in view of earth observation: data sources, input dimensions, and classifiers—a review of the state of the art. *Geocarto International*, 36(9), 957–988. http://doi.org/10.1080/10106049.2019.1629647

Petrou, M.M., and Kamata, S. 2021. *Image Processing: Dealing with Texture*, 2nd ed. Wiley, p. 816. ISBN: 978-1-119-61859-1.

Pinzon, J.E., Pak, E.W., Tucker, C.J., Bhatt, U.S., Frost, G.V., and Macander, M.J. 2023. *Global Vegetation Greenness (NDVI) from AVHRR GIMMS-3G+, 1981–2022*. ORNL DAAC, Oak Ridge, TN. https://doi.org/10.3334/ORNLDAAC/2187

Pooletti, M.E., Haut, J.M., Plaza, J., and Plaza, J. 2019. Deep learning classifiers for hyperspectral imaging: A review. *ISPRS Journal of Photogrammetry and Remote Sensing*, 158, 279–317. ISSN 0924-2716. https://doi.org/10.1016/j.isprsjprs.2019.09.006. https://www.sciencedirect.com/science/article/pii/S0924271619302187

Potapov, P., Li, X., Hernandez-Serna, A., Tyukavina, A., Hansen, M.C., Kommareddy, A., Pickens, A., Turubanova, S., Tang, H., Silva, C.E., Armston, J., Dubayah, R., Blair, J.B., and Hofton, M. 2021. Mapping global forest canopy height through integration of GEDI and Landsat data. *Remote Sensing of Environment*, 253, 112165. ISSN 0034-4257. https://doi.org/10.1016/j.rse.2020.112165. https://www.sciencedirect.com/science/article/pii/S0034425720305381

Price, J.C. 1987. Calibration of satellite radiometers and the comparison of vegetation Indices. *Remote Sensing of the Environment*, 21, 15–27.

Purss, M.B.J., Peterson, P.R., Strobl, P., Dow, C., Sabeur, Z.A., Gibb, R.G., and Ben, J. 2019. Datacubes: A discrete global grid systems perspective. *Cartographica: The International Journal for Geographic Information and Geovisualization*, 54(1), 63–71.

Richards, J.A. 2022. *Remote Sensing Digital Image Analysis*, 6th ed. Springer, p. 567. eBook ISBN: 978-3-030-82327-6. https://doi.org/10.1007/978-3-030-82327-6.

Richards, J.A., and Xiuping, J. 2006. *Remote Sensing Digital Image Analysis*, 4th ed. Springer-Verlag. XXV, p. 439.

Rodriguez-Galiano, V.F., Ghimire, B., Rogan, J., Chica-Olmo, M., and Rigol-Sanchez, J.P. 2012. An assessment of the effectiveness of a random forest classifier for land-cover classification. *ISPRS Journal of Photogrammetry and Remote Sensing*, 67, 93–104. ISSN 0924-2716. http://doi.org/10.1016/j.isprsjprs.2011.11.002

Român, M.O., Justice, C., Paynter, I., Boucher, P.B., Devadiga, S., Endsley, A., Erb, A., Friedl, M., Gao, H., Giglio, L., Gray, J.M., Hall, D., Hulley, G., Kimball, J., Knyazikhin, Y., Lyapustin, A., Myneni, R.B., Noojipady, P., Pu, J., Riggs, G., Sarkar, S., Schaaf, C., Shah, D., Tran, K.H., Vermote, E., Wang, D., Wang, Z., Wu, A., Ye, Y., Shen, Y., Zhang, S., Zhang, S., Zhang, X., Zhao, M., Davidson, C., and Wolfe, R. 2024. Continuity between NASA MODIS collection 6.1 and VIIRS collection 2 land products. *Remote Sensing of Environment*, 302, 113963. ISSN 0034-4257. https://doi.org/10.1016/j.rse.2023.113963. https://www.sciencedirect.com/science/article/pii/S0034425723005151

Rosenkrantz, L., Schuurman, N., Bell, N., and Amram, O. 2021. The need for GIScience in mapping COVID-19. *Health & Place*, 67, 102389. ISSN 1353-8292. https://doi.org/10.1016/j.healthplace.2020.102389. https://www.sciencedirect.com/science/article/pii/S1353829220308431

Sabins, F.F. Jr., and Ellis, J.M. 2020. *Remote Sensing: Principles, Interpretations, and Applications*, 4th ed. Waveland Press Inc., p. 523. ISBN 1-4786-3710-2.

Sahar, L., Foster, S.L., Sherman, R.L., Henry, K.A., Goldberg, D.W., Stinchcomb, D.G., and Bauer, J.E. 2019. GIScience and cancer: State of the art and trends for cancer surveillance and epidemiology. *Cancer*, 125(15), 2544–2560.

Selvaraj, R., and Nagarajan, S. 2022. Chapter 6—Change detection techniques for a remote sensing application: An overview. In Yu-Dong Zhang and Arun Kumar Sangaiah (eds.), *Cognitive Data Science in Sustainable Computing, Cognitive Systems and Signal Processing in Image Processing*. Academic Press, pp. 129–143. ISBN 9780128244104. https://doi.org/10.1016/B978-0-12-824410-4.00015-5. https://www.sciencedirect.com/science/article/pii/B9780128244104000155

Shi, W., Zhang, M., Zhang, R., Chen, S., and Zhan, Z. 2020. Change detection based on artificial intelligence: State-of-the-art and challenges. *Remote Sensing*, 12(10), 1688. https://doi.org/10.3390/rs12101688

Stosius, S., Beyerle, G., Hoechner, A., Wickert, J., and Lauterjung, J. 2011. The impact on tsunami detection from space using GNSS-reflectometry when combining GPS with GLONASS and Galileo. *Advances in Space Research*, 47(5), 843–853. ISSN 0273-1177. http://doi.org/10.1016/j.asr.2010.09.022

Stow, D., Hamada, Y., Coulter, L., and Anguelova, Z. 2008. Monitoring shrubland habitat changes through object-based change identification with airborne multispectral imagery. *Remote Sensing of Environment*, 112, 1051–1061.

Suppasri, A., Futami, T., Tabuchi, S., and Imamura, F. 2012. Mapping of historical tsunamis in the Indian and Southwest Pacific Oceans. *International Journal of Disaster Risk Reduction*, 1, 62–71. ISSN 2212-4209. http://doi.org/10.1016/j.ijdrr.2012.05.003.

Thenkabail, P.S., Aneece, I., Teluguntla, P., and Oliphant, A. 2021b. Hyperspectral narrowband data propel gigantic leap in the earth remote sensing. *Photogrammetric Engineering and Remote Sensing*, 87(7), 461–467. http://doi.org/10.14358/PERS.87.7.461. http://www.asprs.org/a/publications/pers/2021journals/07-21_July_Flipping_Public.pdf. IP-127022.

Thenkabail, P.S., Biradar C.M., Noojipady, P., Dheeravath, V., Li, Y.J., Velpuri, M., Gumma, M., Reddy, G.P.O., Turral, H., Cai, X.L., Vithanage, J., Schull, M., and Dutta, R. 2009b. Global irrigated area map (GIAM), derived from remote sensing, for the end of the last millennium. *International Journal of Remote Sensing*, 30(14), 3679–3733.

Thenkabail, P.S., Lyon, G.J., Turral, H., and Biradar, C.M. 2009a. *Remote Sensing of Global Croplands for Food Security*. CRC Press and Taylor and Francis Group, p. 556 (48 pages in color). Published in June, 2009.

Thenkabail, P.S., Smith, R.B., and De-Pauw, E. 2002. Evaluation of narrowband and broadband vegetation indices for determining optimal hyperspectral wavebands for agricultural crop characterization. *Photogrammetric Engineering and Remote Sensing*, 68(6), 607–621.

Thenkabail, P.S., Teluguntla, P.G., Xiong, J., Oliphant, A., Congalton, R.G., Ozdogan, M., Gumma, M.K., Tilton, J.C., Giri, C., Milesi, C., Phalke, A., Massey, R., Yadav, K., Sankey, T., Zhong, Y., Aneece, I., and Foley, D., 2021, Global cropland-extent product at 30-m resolution (GCEP30) derived from Landsat satellite time-series data for the year 2015 using multiple machine-learning algorithms on Google Earth Engine cloud: U.S. *Geological Survey Professional Paper*, 1868, 63 p. https://doi.org/10.3133/pp1868

Thorsten, H., and Kuenzer, C. 2020. Object detection and image segmentation with deep learning on earth observation data: A review-Part I: Evolution and recent trends. *Remote Sensing*, 12(10), 1667. https://doi.org/10.3390/rs12101667

Tilling, A.K., O'Leary, G.J., Ferwerda, J.G., Jones, S.D., Fitzgerald, G.J., Rodriguez, D., and Belford, R. 2007. Remote sensing of nitrogen and water stress in wheat. *Field Crops Research*, 104(1–3), 77–85. ISSN 0378-4290. http://dx.doi.org/10.1016/j.fcr.2007.03.023

Tilmes, C., & Fleig, A.J. (2008). Provenance tracking in an earth science data processing system. In J. Freire and D. Koop (eds.), *Provenance and Annotation of Data and Processes*. Springer-Verlag, pp. 221–228. http://ebiquity.umbc.edu/_file_directory_/papers/445.pdf

Tilton, J.C., Tarabalka, Y., Montesano, P.M., and Gofman, E. 2012. Best merge region-growing segmentation with integrated nonadjacent region object aggregation. *IEEE Transactions on Geoscience and Remote Sensing*, 50, 4454–4467.

Tilton, J.C., Xiong, J., and Massey, R. 2017. *Approaches to Incorporating RHSeg Image Segmentation into Cropland Extent Mapping*. Workshop on Global Food Security-Support Analysis Data @ 30m (GFSAD30) held @ Dallas Peck Auditorium, USGS HQ, 12201 Sunrise Valley Drive, Reston, VA. https://www.usgs.gov/media/files/rhseg-image-segmentation-gfsad-workshop-reston-va

Tucker, C.J. 1979. Red and photographic infrared linear combinations for monitoring vegetation. *Remote Sensing of the Environment*, 8, 127–150.

Uber. (2020). *Uber H3-js*. Author. uber.github.io/h3

Uhl, F., Oppelt, N., and Bartsch, I. 2013. Mapping marine macroalgae in case 2 waters using CHRIS PROBA. *Proc. of ESA Living Planet Symposium*, September 9–13, Edinburgh (UK), ESA Special Proceedings SP-722 (CD-ROM).

Uhl, F., Oppelt, N., Bartsch, I., Geisler, T., Heege, T., and Nehring, F. 2014. *KelpMap—Development of an EnMAP Approach to Monitor Sublitoral Marine Macrophytes.* (KelpMap—Entwicklung eines EnMAP Verfahrens zur Bestimmung von sublitoralen marinen Makrophyten). Final report of research project FKZ: 50EE1020, funded by the German Federal Ministry of Economy and Technology (BMWi): 36p.

Ulug, R., and Karslıoglu, M.O. 2022. A new data-adaptive network design methodology based on the k-means clustering and modified ISODATA algorithm for regional gravity field modeling via spherical radial basis functions. *Journal of Geodesy*, 96, 91. https://doi.org/10.1007/s00190-022-01681-2

US Geological Survey (USGS). 2012. *Earth's Tectonic Plates—USGS.* http://earthquake.usgs.gov/regional/nca/ . . . /kml/Earths_Tectonic_Plates.kmz [accessed 25.05.12].

USGS. 2019. *USGS EROS Archive – Earth Observing One (EO-1)—Hyperion.* Earth Observing One (EO-1)—Hyperion Digital Object Identifier (DOI) number: /10.5066/P9JXHMO2

USGS. 2024. *U.S. Geological Survey, 2024, Landsat Next* (ver. 1.1, March 25, 2024): U.S. Geological Survey Fact Sheet 2024–3005, 2 p. https://doi.org/10.3133/fs20243005. ISSN: 2327-6932 (online) ISSN: 2327-6916 (print)

Van der Sluijs, J., Peddle, D.R., and Hall, R.J. (2023). Characterizing tree species in Northern Boreal forests using multiple-endmember spectral mixture analysis and multi-temporal satellite imagery. *Canadian Journal of Remote Sensing*, 49(1). https://doi.org/10.1080/07038992.2023.2216312

Vermote, E.F., El Saleous, N.Z., and Justice, C.O. 2002. Atmospheric correction of MODIS data in the visible to middle infrared: first results. *Remote Sensing of Environment*, 83(1–2), 97–111.

Villani, Maria Luisa, Giovinazzi, Sonia, and Costanzo, Antonio. 2023. Co-creating GIS-based dashboards to democratize knowledge on urban resilience strategies: Experience with camerino municipality. *ISPRS International Journal of Geo-Information*, 12(2), 65. https://doi.org/10.3390/ijgi12020065

Vrieling, A., Meroni, M., Shee, A., Mude, A.G., Woodard, J., (Kees) de Bie, C.A.J.M., and Rembold, F. 2014. Historical extension of operational NDVI products for livestock insurance in Kenya. *International Journal of Applied Earth Observation and Geoinformation*, 28, 238–251. ISSN 0303-2434. http://dx.doi.org/10.1016/j.jag.2013.12.010

Wan, Z., Hong, Y., Khan, S., Gourley, J., Flamig, Z., Kirschbaum, D., and Tang, G. 2014. A cloud-based global flood disaster community cyber-infrastructure: Development and demonstration. *Environmental Modelling & Software*, 58, 86–94. ISSN 1364-8152. http://doi.org/10.1016/j.envsoft.2014.04.007

Wang, J., Bretz, M., Ali Akber Dewan, M., and Delavar, M.A. 2022. Machine learning in modelling land-use and land cover-change (LULCC): Current status, challenges and prospects. *Science of the Total Environment*, 822, 153559. ISSN 0048-9697. https://doi.org/10.1016/j.scitotenv.2022.153559. https://www.sciencedirect.com/science/article/pii/S0048969722006519

Wang, Z., Jensen, J.R., and Im, J. 2010. An automatic region-based image segmentation algorithm for remote sensing applications. *Environmental Modelling & Software*, 25(10), 1149–1165. ISSN 1364-8152. http://doi.org/10.1016/j.envsoft.2010.03.019

Wei, Y., Liu, H., Song, W., Yu, B., and Xiu, C. 2014. Normalization of time series DMSP-OLS nighttime light images for urban growth analysis with Pseudo Invariant Features. *Landscape and Urban Planning*, 128, 1–13. ISSN 0169-2046. http://dx.doi.org/10.1016/j.landurbplan.2014.04.015

Whiteside, T.G., Boggs, G.S., and Maier, S.W. 2011. Comparing object-based and pixel-based classifications for mapping savannas. *International Journal of Applied Earth Observation and Geoinformation*, 13(6), 884–893. ISSN 0303-2434. http://dx.doi.org/10.1016/j.jag.2011.06.008

Woodcock, C.E., Loveland, T.R., Herold, M., and Bauer, M.E. 2020. Transitioning from change detection to monitoring with remote sensing: A paradigm shift. *Remote Sensing of Environment*, 238, 111558. ISSN 0034-4257. https://doi.org/10.1016/j.rse.2019.111558. https://www.sciencedirect.com/science/article/pii/S0034425719305784

Wu, Zhaoyan, Li, Hao, and Yue, Peng. 2023. Provenance in GIServices: A semantic web approach. *ISPRS International Journal of Geo-Information*, 12(3), 118. https://doi.org/10.3390/ijgi12030118

Xiong, Jun, Thenkabail, Prasad S., Tilton, James C., Gumma, Murali K., Teluguntla, Pardhasaradhi, Oliphant, Adam, Congalton, Russell G., Yadav, Kamini, and Gorelick, Noel. 2017. Nominal 30-m cropland extent map of continental Africa by integrating pixel-based and object-based algorithms using sentinel-2 and landsat-8 data on Google earth engine. *Remote Sensing*, 9(10), 1065. https://doi.org/10.3390/rs9101065

Xu, H., Huang, S., and Zhang, T. 2013. Built-up land mapping capabilities of the ASTER and Landsat ETM+ sensors in coastal areas of southeastern China. *Advances in Space Research*, 52(8), 1437–1449. ISSN 0273-1177. http://dx.doi.org/10.1016/j.asr.2013.07.026

Yan, Ziyun, Ma, Lei, He, Weiqiang, Zhou, Liang, Lu, Heng, Liu, Gang, and Huang, Guoan. 2022. Comparing object-based and pixel-based methods for local climate zones mapping with multi-source data. *Remote Sensing*, 14(15), 3744. https://doi.org/10.3390/rs14153744

Ye, S., Pontius, R.G., and Rakshit, R. 2018. A review of accuracy assessment for object-based image analysis: From per-pixel to per-polygon approaches. *ISPRS Journal of Photogrammetry and Remote Sensing*, 141, 137–147. ISSN 0924-2716. https://doi.org/10.1016/j.isprsjprs.2018.04.002. https://www.sciencedirect.com/science/article/pii/S0924271618300947

Yu, S., Jia, S., and Xu, C. 2017. Convolutional neural networks for hyperspectral image classification. *Neurocomputing*, 219, 88–98. ISSN 0925-2312. https://doi.org/10.1016/j.neucom.2016.09.010. https://www.sciencedirect.com/science/article/pii/S0925231216310104

Yuan, K., Meng, G., Cheng, D., Bai, J., Xiang, S., and Pan, C. 2017. Efficient cloud detection in remote sensing images using edge-aware segmentation network and easy-to-hard training strategy. *2017 IEEE International Conference on Image Processing (ICIP)*, Beijing, China, 2017, pp. 61–65. https://doi.org/10.1109/ICIP.2017.8296243

Yuan, Q., Shen, H., Li, T., Li, Z., Li, S., Jiang, Y., Xu, H., Tan, W., Yang, Q., Wang, J., Gao, J., and Zhang, L. 2020. Deep learning in environmental remote sensing: Achievements and challenges. *Remote Sensing of Environment*, 241, 111716. ISSN 0034-4257. https://doi.org/10.1016/j.rse.2020.111716. https://www.sciencedirect.com/science/article/pii/S0034425720300857

Yuan, X., Shi, J., and Gu, L. 2021. A review of deep learning methods for semantic segmentation of remote sensing imagery. *Expert Systems with Applications*, 169, 114417. ISSN 0957-4174. https://doi.org/10.1016/j.eswa.2020.114417. https://www.sciencedirect.com/science/article/pii/S0957417420310836

Yue, P., Gong, J., Di, L., Yuan, J., Sun, L., Sun, Z., and Wang, Q. 2010. GeoPW: Laying blocks for geospatial processing Web. *Transactions in GIS*, 14(6), 755–772.

Yue, P., Guo, X., Zhang, M., Jiang, L., and Zhai, X. 2016. Linked data and SDI: The case on Web geoprocessing workflows. *ISPRS Journal of Photogrammetry and Remote Sensing*, 114, 245–257. ISSN 0924-2716. https://doi.org/10.1016/j.isprsjprs.2015.11.009. https://www.sciencedirect.com/science/article/pii/S0924271615002610

Zhang, C., Sargent, I., Pan, X., Li, H., Gardiner, A., Hare, J., and Atkinson, P.M. 2018. An object-based convolutional neural network (OCNN) for urban land use classification. *Remote Sensing of Environment*, 216, 57–70. ISSN 0034-4257. https://doi.org/10.1016/j.rse.2018.06.034. https://www.sciencedirect.com/science/article/pii/S0034425718303122

Zhao, P., Foerster, T., and Yue, P. 2012. The geoprocessing Web. *Computers & Geosciences*, 47, 3–12. ISSN 0098-3004. http://doi.org/10.1016/j.cageo.2012.04.021

Index

Note: Page numbers in *italics* indicate a figure and page numbers in **bold** indicate a table on the corresponding page.

E

F

G

H

I

J

K

L

M

N

P

Q

R

S

T

U

V

W

Z